AF510446

Human T-Lymphotropic Viruses: A Pathological Approach

Human T-Lymphotropic Viruses: A Pathological Approach

Editor: Marie Galvin

FOSTER ACADEMICS

www.fosteracademics.com

www.fosteracademics.com

FOSTER ACADEMICS

Cataloging-in-Publication Data

Human t-lymphotropic viruses : a pathological approach / edited by Marie Galvin.
p. cm.
Includes bibliographical references and index.
ISBN 978-1-64646-621-4
1. HTLV (Viruses). 2. HTLV (Viruses)--Diagnosis. 3. Retroviruses. 4. Virus diseases--Pathophysiology. I. Galvin, Marie.
QR414.5 .H86 2023
576.648 4--dc23

© Foster Academics, 2023

Foster Academics,
118-35 Queens Blvd., Suite 400,
Forest Hills, NY 11375, USA

ISBN 978-1-64646-621-4 (Hardback)

This book contains information obtained from authentic and highly regarded sources. Copyright for all individual chapters remain with the respective authors as indicated. All chapters are published with permission under the Creative Commons Attribution License or equivalent. A wide variety of references are listed. Permission and sources are indicated; for detailed attributions, please refer to the permissions page and list of contributors. Reasonable efforts have been made to publish reliable data and information, but the authors, editors and publisher cannot assume any responsibility for the validity of all materials or the consequences of their use.

Trademark Notice: Registered trademark of products or corporate names are used only for explanation and identification without intent to infringe.

Contents

Preface

The human T-lymphotropic virus, also known as HTLV, refers to a group of human retroviruses which are recognized for causing a type of cancer known as adult T-cell leukemia/lymphoma, and a demyelinating disease called HTLV-1 linked myelopathy/tropical spastic paraparesis. The HTLVs are a part of larger group of primate T-lymphotropic viruses. The members of this family which infect human beings are known as HTLVs and those who infect old world monkeys are known as Simian T-lymphotropic viruses (STLVs). There are four types of HTLVs that have been found, namely, HTLV-1, HTLV-2, HTLV-3, and HTLV-4. HTLV-1 is clinically significant and it is transmitted sexually, through blood transfusion, sharing needles in drug use and breast feeding. The virus illustrates low antigenic variability and natural immunity is possible in humans. This book includes some of the vital pieces of work being conducted across the world, on various topics related to human T-lymphotropic viruses. It will help the readers in keeping pace with the rapid changes in this area of study.

Various studies have approached the subject by analyzing it with a single perspective, but the present book provides diverse methodologies and techniques to address this field. This book contains theories and applications needed for understanding the subject from different perspectives. The aim is to keep the readers informed about the progresses in the field; therefore, the contributions were carefully examined to compile novel researches by specialists from across the globe.

Indeed, the job of the editor is the most crucial and challenging in compiling all chapters into a single book. In the end, I would extend my sincere thanks to the chapter authors for their profound work. I am also thankful for the support provided by my family and colleagues during the compilation of this book.

Editor

1

Comprehensive Antiretroviral Restriction Factor Profiling Reveals the Evolutionary Imprint of the *ex Vivo* and *in Vivo* IFN-β Response in HTLV-1-Associated Neuroinflammation

Fabio E. Leal[1,2†], Soraya Maria Menezes[3†], Emanuela A. S. Costa[4], Phillip M. Brailey[1], Lucio Gama[5], Aluisio C. Segurado[4], Esper G. Kallas[4], Douglas F. Nixon[1], Tim Dierckx[3], Ricardo Khouri[3,6], Jurgen Vercauteren[3], Bernardo Galvão-Castro[7], Rui Andre Saraiva Raposo[1] and Johan Van Weyenbergh[3*]*

[1] Oncovirology Program, Instituto Nacional de Câncer (INCA), Rio de Janeiro, Brazil, [2] Microbiology Immunology and Tropical Medicine, George Washington University, Washington, DC, United States, [3] Department of Microbiology and Immunology, Rega Institute for Medical Research, KU Leuven, Leuven, Belgium, [4] Departamento de Moléstias Infecciosas e Parasitárias, Faculdade de Medicina, Universidade de São Paulo, São Paulo, Brazil, [5] Department of Molecular and Comparative Pathobiology, Johns Hopkins University School of Medicine, Baltimore, MD, United States, [6] Fundação Oswaldo Cruz, Instituto Gonçalo Moniz (IGM), Salvador-Bahia, Brazil, [7] Escola Bahiana de Medicina e Saúde Pública, Salvador-Bahia, Brazil

***Correspondence:**
Fabio E. Leal
fabio.leal@inca.gov.br;
fabioit@me.com
Johan Van Weyenbergh
johan.vanweyenbergh@kuleuven.be;
j.vw@live.be

† Joint first authors.

HTLV-1-Associated Myelopathy (HAM/TSP) is a progressive neuroinflammatory disorder for which no disease-modifying treatment exists. Modest clinical benefit from type I interferons (IFN-α/β) in HAM/TSP contrasts with its recently identified IFN-inducible gene signature. In addition, IFN-α treatment *in vivo* decreases proviral load and immune activation in HAM/TSP, whereas IFN-β therapy decreases tax mRNA and lymphoproliferation. We hypothesize this "IFN paradox" in HAM/TSP might be explained by both cell type- and gene-specific effects of type I IFN in HTLV-1-associated pathogenesis. Therefore, we analyzed *ex vivo* transcriptomes of $CD4^+$ T cells, PBMCs and whole blood in healthy controls, HTLV-1-infected individuals, and HAM/TSP patients. First, we used a targeted approach, simultaneously quantifying HTLV-1 mRNA (HBZ, Tax), proviral load and 42 host genes with known antiretroviral (anti-HIV) activity in purified $CD4^+$ T cells. This revealed two major clusters ("antiviral/protective" vs. "proviral/deleterious"), as evidenced by significant negative (TRIM5/TRIM22/BST2) vs. positive correlation (ISG15/PAF1/CDKN1A) with HTLV-1 viral markers and clinical status. Surprisingly, we found a significant inversion of antiretroviral activity of host restriction factors, as evidenced by opposite correlation to *in vivo* HIV-1 vs. HTLV-1 RNA levels. The anti-HTLV-1 effect of antiviral cluster genes was significantly correlated to their adaptive chimp/human evolution score, for both Tax mRNA and PVL. Six genes of the proposed antiviral cluster underwent lentivirus-driven purifying selection during primate evolution (TRIM5/TRIM22/BST2/APOBEC3F-G-H), underscoring the cross-retroviral evolutionary imprint. Secondly, we examined the genome-wide type I IFN response in HAM/TSP patients, following short-term *ex vivo* culture of PBMCs with either IFN-α or IFN-β. Microarray analysis evidenced 12 antiretroviral genes (including TRIM5α/TRIM22/BST2) were significantly up-regulated by IFN-β, but not IFN-α, in HAM/TSP. This was paralleled by a significant decrease in lymphoproliferation by IFN-β, but not IFN-α treatment. Finally, using published *ex vivo* whole blood transcriptomic data of independent cohorts, we validated the significant positive correlation between TRIM5, TRIM22, and BST2 in HTLV-1-infected individuals and HAM/TSP patients, which was independent of the HAM/TSP disease signature. In conclusion, our results provide *ex vivo* mechanistic evidence for the observed immunovirological effect of *in vivo* IFN-β treatment in HAM/TSP, reconcile an apparent IFN paradox in HTLV-1 research and identify biomarkers/targets for a precision medicine approach.

Keywords: HTLV-1, HIV, retrovirus, evolution, interferon, neuroinflammation, multiple sclerosis, transcriptomics

BACKGROUND

Human T-cell Lymphotropic Virus-1 (HTLV-1), recently renamed Human T-cell Leukemia Virus-1 due to its strong oncogenic potential (Gallo et al., 2017; Tagaya and Gallo, 2017), is also the causative agent of the debilitating neuroinflammatory disorder, HTLV-1-Associated Myelopathy/Tropical Spastic Paraparesis (HAM/TSP) (Osame et al., 1986). HAM/TSP is associated with high HTLV-1 proviral load (PVL) and transcriptional levels of retroviral regulatory genes, *Tax* and *HBZ* (Saito et al., 2009). The pathogenesis of HAM/TSP is complex (Bangham et al., 2015), and a proportion of HTLV-1 asymptomatic carriers (AC) may present a PVL and inflammatory profile similar to HAM/TSP patients but do not develop clinical symptoms. It is unknown why 2–3% of HTLV-1 infected patients develop HAM/TSP after years of latent infection, but an interferon-inducible gene signature has been identified in HAM/TSP and is absent in AC (Tattermusch et al., 2012). However, type I interferon-based clinical trials reported modest clinical benefit, as well as antiviral (decreased PVL, *Tax* mRNA levels) and immunomodulatory effects (decreased T-cell spontaneous proliferation and activation), suggesting IFN-α and IFN-β may hamper disease progression (Izumo et al., 1996; Oh et al., 2005). To reconcile this apparent contradiction, we hypothesize specific IFN-stimulated genes, e.g. antiviral effector genes, can exert deleterious vs. protective roles in HAM/TSP, as we recently suggested for B-cell expression of CD80 and CD86 (Menezes et al., 2014).

In contrast to HTLV-1, several anti(retro)viral effector genes have been classified as "restriction factors" (RFs) for HIV-1 infection, some of which are modulated by type I IFNs (Foster et al., 2017). We have previously compiled and validated a real-time PCR-array of 42 well-characterized RFs with suppressive activity against HIV-1 (Neil et al., 2008; Abdel-Mohsen et al., 2013; Raposo et al., 2013a,b, 2014), including APOBECs, TRIMs, and BST2/Tetherin.. Enhanced HIV-1 infection upon siRNA-mediated silencing in neuroblasts illustrates the relevance of TRIM5α and TRIM22 in neuroinfection (Singh et al., 2014).

To assess the role of antiretroviral genes in HAM/TSP pathogenesis, we determined transcriptional levels of 42 RFs and regulatory HTLV-1 genes *Tax* and *HBZ* in $CD4^+$ T cells from HTLV-1 patients. We identified a strong negative association between expression of HTLV-1 *Tax* and a cluster of RFs, including TRIM5α/TRIM22/BST2. Genome-wide transcriptomic analysis of HAM/TSP patients showed a significant and distinct up-regulation of RFs after *ex vivo* exposure to IFN-β, but not IFN-α. These results provide mechanistic evidence for the immunovirological impact of IFN-β therapy observed *in vivo* in HAM/TSP patients and pave the way to an evidence-based precision medicine approach to this neuroinflammatory disorder.

PATIENTS AND METHODS

All participants signed a written informed consent in accordance with the Declaration of Helsinki, approved by the University Institutional Review Boards of USP (#0855/08) and CPqGM-FIOCRUZ (#022/03) before inclusion in the study. *Ex vivo* analysis of antiretroviral restriction factors was conducted in a cohort of 18 individuals (7 HAM/TSP, 6 HTLV-1-infected asymptomatic carriers (AC) and 5 age- and gender-matched HTLV-1-negative healthy controls (HC)), enrolled at the HTLV-1 Outpatient Clinic at the University of Sao Paulo (USP), Brazil. The second cohort (lymphoproliferation/microarray analysis) consisted of 10 HAM/TSP patients, followed at the Bahiana School of Medicine and Public Health HTLV reference center in Salvador-Bahia, Brazil. Validation of transcriptomic findings was performed in a third, published cohort (London, UK). Demographic and clinical data of all cohorts are detailed in **Table 1**. Clinical status was determined based on WHO criteria for HTLV-1 associated diseases (Osame, 1990). None of the HAM/TSP patients had received prior IFN-based therapy. Blood samples were obtained and processed with Ficoll-Paque PLUS (Amersham Pharmacia Biotech, Uppsala, Sweden) gradient centrifugation and peripheral-blood mononuclear cells (PBMC) were isolated and either cryopreserved in 10 % DMSO in FBS (RF analysis) or used for short-term *ex vivo* culture (48 h for microarray analysis, 96 h for lymphoproliferation), in the absence

TABLE 1 | Demographic and clinical data of 3 HTLV-1 cohorts.

Cohorts	HC	AC	HAM/TSP	Cell type
Cohort 1 (Sao Paulo, BR)				CD4+ T cells
Number of samples	5	6	7	
Age (median, IQR)	54(38–61)	62(54–66)	51(45–62)	
Gender	1M/4F	1M/5F	2M/5F	
PVL (median, ± SD)	N/A	7.6 ± 151.5	181.0 ± 134.5	
Tax mRNA (median, ± SD)	N/A	0.02 ± 0.02	0.68 ± 0.77	
HBZ mRNA (median, ± SD)	N/A	$10^{-5} \pm 5.4\ 10^{-4}$	$2.7\ 10^{-4} \pm 8.1\ 10^{-4}$	
Cohort 2 (Bahia, BR)				PBMC
Number of samples	11	5	10	
Age (median, IQR)			52 ± 6.2	
Gender	4M/7F	2M/3F	4M/6F	
Cohort 3 (published; London, UK)				Whole blood
Number of samples	8	17	10	
Age (median, IQR)	49(46–71)	47(38–62)	54(49–62)	63(44–74)
Gender	4M/4F	0M/17F	4M/6F	
PVL (median, IQR)	N/A	0.6(0.4–0.9)	3.9 (3.4–6.7)	10.1(6.7–17.8)
Total number of individuals	24	23	27	

or presence of 1,000 IU/ml clinical grade IFN-α (gift of Blausiegel Ltda., Sao Paulo-Brazil) or IFN-β (gift of Dr. Brassat, Toulouse-France), as previously described (Moens et al., 2012a; Dierckx et al., 2017).

Proviral Load and mRNA Assessment

Total DNA and RNA were extracted from enriched $CD4^{+}$ T cells using a commercial kit (Qiagen GmbH, Hilden Germany) and cDNA was generated using Superscript VILO cDNA synthesis kit (Invitrogen) following the manufacturer's instructions. HTLV-1 proviral load absolute quantification was performed as previously described (Dehee et al., 2002), normalized to human albumin gene. Samples were assayed in duplicate. HTLV-1 proviral load was calculated as follows: copy number of HTLV-1 per 1,000 $CD4^{+}$ T cells = (copy number of HTLV-1)/(copy number of albumin) × 2 × 1,000 cells.

Transcription levels of *Tax* and *HBZ* were measured as previously described (Saito et al., 2009), using housekeeping gene Ubiquitin C (Life Technologies assay ID# Hs00824723_m1) to calculate *tax* and *HBZ* $2^{-\Delta Ct}$ relative expression. Relative quantification of 42 RFs using mRNA from $CD4^{+}$ T cells from 13 HTLV-1-infected patients and 5 uninfected healthy controls was performed using custom-made TaqMan® Low Density Arrays previously described (Abdel-Mohsen et al., 2013)(Applied Biosystems, Foster City, CA). RF gene descriptions are given in **Table 2.**

Microarray Analysis

Total RNA extraction (RNeasy kit, QIAGEN, Venlo, the Netherlands) and Whole Genome microarray analysis (HuGene 1.0 ST array, Affymetrix, Santa Clara, CA) were performed according to the manufacturers' specifications. Data were analyzed using the Bioconductor limma package. Microarray data used in this study are available at Gene Expression Omnibus under GEO accession number GSE82160 (Brazilian HTLV-1 cohort), GSE29333 (UK HTLV-1 cohort, Tattermusch et al.) and GSE18233 (Swiss HIV cohort, Rotger et al., 2010).

Enrichment and Evolutionary Analysis

Genome-wide enrichment analysis was performed using a modified Fisher's test, considering the total number of annotated genes for which transcripts were detectable by microarray (n = 22,370), followed by stringent FDR (Benjamini–Hochberg) correction. Data and full methodology on human candidate retroviral restriction factors and chimp/human cross-species adaptive evolution and purifying selection during primate evolution are detailed in (Osame, 1990; Singh et al., 2014). In brief, measurements of cross-species adaptive evolution used the McDonald-Kreitman value to compare human sequence to chimpanzee in a set of 15,052 protein-coding genes (Singh et al., 2014) and purifying selection during primate evolution was measured using the proportion of nonsynonymous (K_A) over synonymous (K_S) substitutions per site in five primates (human, chimpanzee, orangutan, rhesus, and common marmoset) in a set of 140 candidate genes selected for their relationship to HIV pathogenesis (Osame, 1990).

Statistical Analysis

Statistical analysis was performed using GraphPad Prism software (GraphPad Software, version 6 and 7, San Diego, CA). Non-parametric statistical tests (Mann–Whitney, Wilcoxon tests, and Spearman correlation) were used for patient data, with Bonferroni correction for multiple comparisons between RFs, PVL and *tax/HBZ* mRNA levels. Transcriptome-wide correlation of mRNA expression levels with either set point viral load (HIV-1) or TRIM5 (HTLV-1) mRNA expression was calculated using Spearman's correlation, followed by FDR correction for multiple testing.

RESULTS

Antiretroviral RFs Separate in Two Major Clusters, Positively and Negatively Associated With HTLV-1 Virological and Clinical Status

Based on a previously defined subset of host genes with significant anti-HIV activity (Abdel-Mohsen et al., 2013), we

TABLE 2 | List of 42 antiretroviral genes measured in CD4+ T cells.

Target Information	NCBI Gene Description
APOBEC3A-H	Apolipoprotein B mRNA editing enzyme, catalytic polipeptide-like 3
BST2/theterin	Bone Marrow Stromal cell antigen 2
SLFN11	Schlafen family member 11
TRIM5α, TRIM11, TRIM21, TRIM22, TRIM26, TRIM28, TRIM32	Tripartite motif family
CPSF6	Cleavage and Polyadenylation Specific Factor 6
SAMHD1	SAM domain and HD domain 1
PML	Promyelocytic Leukemia protein
RNF114	Ring Finger Protein 114
Trex1	Three Prime Repair Exonuclease 1
RPRD2	Regulation of nuclear Pre-mRNA Domain containing 2
CHFR	Checkpoint with forkhead and ring finger domains, E3 ubiquitin protein ligase
ISG15	ISG15 ubiquitin-like modifier
EIF2AK2/PKR	Eukaryotic translation initiation factor 2-alpha kinase 2
CTR9/PAF1/RTF1	CTR9, PAF1, RTF1/RNA polymerase II complex component
IFITM family (3 members)	Interferon induced transmembrane protein
RSAD2 (viperin)	Radical S-adenosyl methionine domain containing 2
MOV10	Moloney leukemia virus 10, homolog
HERC5	HECT domain and RLD 5
MX2	MX dynamin like GTPase 2
CDKN1A	Cyclin-dependent kinase inhibitor 1A
BRD4	Bromodomain containing 4
RNASEL	Ribonuclease L (2′,5′-oligoisoadenylate synthetase-dependent)
CH25H	Cholesterol 25-Hydroxylase
LGALS3BP	Lectin, Galactoside binding Soluble 3 Binding Protein
CNP	2′,3′-cyclic nucleotide 3′ phosphodiesterase
IFI16	Interferon gamma Inducible protein 16

used a custom-made array to quantify the transcriptional levels of 42 RFs (listed in **Table 2**) in peripheral CD4+ T cells from HTLV-1 infected patients or healthy controls. We compared the gene expression profiles of 13 HTLV-1-infected patients (6 AC, 7 HAM/TSP) and 5 uninfected subjects using cluster analysis. As evident from **Figure 1A**, two major RF subsets appeared as separate clusters. First, a minor cluster of RFs (CDKN1A/ISG15 and PAF1) positively correlates with biomarkers of HTLV-1 disease (PVL, mRNA levels of *Tax* and *HBZ,* as well as clinical status, box **Figure 1A**). Second, the largest cluster of RFs was found to correlate negatively with HTLV-1 biomarkers (**Figure 1A**). Within this large cluster, negative correlations between the expression levels of Tax and TRIM5α ($r = -0.86$; $p = 0.0084$), TRIM22 ($r = -0.81$; $p = 0.0336$), BST2 ($r = -0.85$; $p = 0.0126$), and RNASEL ($r = -0.83$; $p = 0.021$) were statistically significant after stringent Bonferroni correction (**Figure 1B**). Conversely, cyclin-dependent kinase inhibitor 1A (CDKN1A, also known as *p21*), a regulator of cell-cycle progression, was positively associated with HTLV-1 proviral load ($r = 0.84$; $p = 0.042$) (**Figure 1C**). Surprisingly, these *ex vivo* correlations recapitulate the *in vivo* findings of IFN-β treatment in an immunovirological trial in HAM/TSP, in which IFN-β was found to significantly decrease both *tax* mRNA levels and lymphoproliferation (Oh et al., 2005), but not PVL. In contrast, *in vivo* IFN-α treatment of HAM/TSP patients was found to decrease PVL and immune activation (Izumo et al., 1996). Therefore, we investigated if IFN-α/β might differentially affect lymphoproliferation and genome-wide expression profiles in HAM/TSP.

Ex Vivo Lymphoproliferation Is Significantly Down-Regulated by IFN-β, but Not IFN-α in HAM/TSP

We found that lymphoproliferation was significantly inhibited by IFN-β ($p < 0.01$), but not by IFN-α (**Figure 2**) This differential effect was not due to a defect in IFN-α bioactivity, since both IFNs displayed equal antiviral activity in a standardized Vesicular Stomatitis Virus (VSV) bioassay (Dierckx et al., 2017). This effect was specific for HAM/TSP, since no significant IFN-induced antiproliferative effect was observed in both healthy controls (**Figure 2**) and HTLV-1-infected carriers (data not shown).

Genome-Wide Analysis Identifies Selective Up-Regulation of RFs by IFN-β, but Not IFN-α, in HAM/TSP

Since RFs can be IFN-regulated, we used microarray analysis to assess whether the proviral and antiviral clusters might be selectively up- or down-regulated by either IFN-α or IFN-β, the two IFN subtypes previously used in HAM/TSP trials (Izumo

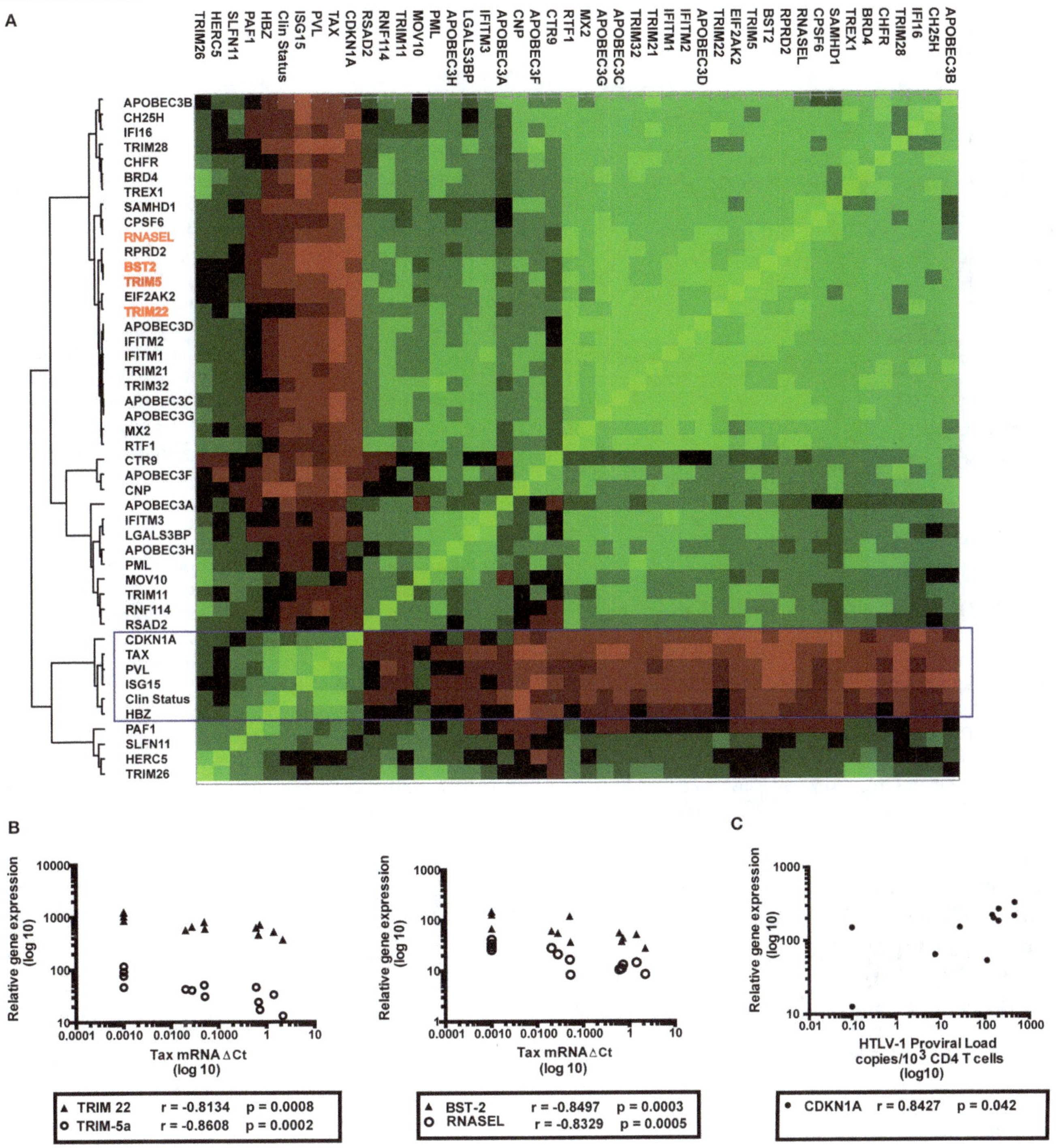

FIGURE 1 | Association between host restriction factors (RFs) expression, HTLV-1 virological parameters and clinical status of HTLV-1 infected individuals. **(A)** Analysis of mRNA levels of 42 RFs, Tax/HBZ and HTLV-1 proviral load (PVL) in $CD4^{+}$ T cells from 6 asymptomatic carriers (AC), 7 HAM/TSP patients and 5 healthy donors (HD). Expression of host RFs was quantified by TLDA, results were clustered according to Spearman's correlation and shown as heatmap (red = negative, green = positive). The boxed cluster indicates a minor subset of "proviral" RFs (CDKN1A, ISG15) clustering with HTLV-1 PVL, *Tax* mRNA, *HBZ* mRNA, and clinical status (HD/AC/HAM), whereas the majority of RFs forms a large "antiviral" cluster. A subset of antiviral RFs (TRIM5α, TRIM22, BST2, and RNASEL) with statistically significant negative correlation to tax mRNA are highlighted in red. **(B)** Correlation between TRIM5α, TRIM22, BST2, RNASEL, and tax mRNA **(C)** Correlation between CDKN1A and tax mRNA in HTLV-1-infected individuals (AC $n = 6$, HAM $n = 7$).

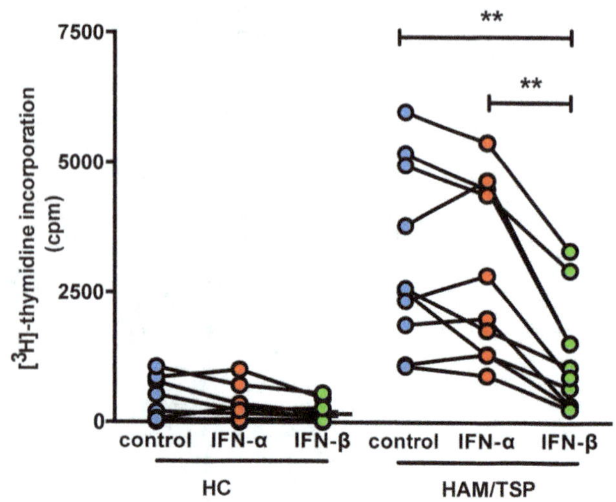

FIGURE 2 | *Ex vivo* lymphoproliferation is significantly down-regulated by IFN-β, but not IFN-α in HAM/TSP. Spontaneous lymphoproliferation was measured by [^{3}H]-thymidine incorporation, as previously described Moens et al. (2012a). The effect of IFN-α and IFN-β (1000 IU/ml) upon *ex vivo* lymphoproliferation was determined in healthy controls (HC, $n = 11$) and HAM/TSP patients ($n = 10$). $^{**}p < 0.01$.

et al., 1996; Oh et al., 2005). As shown in **Figure 3A**, genome-wide analysis identified 12/42 RFs (including TRIM5α, TRIM22, and BST2) were significantly up-regulated by IFN-β ($n = 5$, enrichment $p < 0.0001$), vs. 0/42 by IFN-α treatment ($n = 6$, $p > 0.05$) during short-term (48 h) *ex vivo* culture of PBMC from HAM/TSP patients. In a striking parallel to our results in CD4$^+$ T cells, where TRIM5 displayed the strongest negative correlation to Tax mRNA, TRIM5 was also identified as the top IFN-β-induced gene among the significant genes ($p = 0.0009$).

Genome-Wide Analysis Identifies a Strongly Co-regulated TRIM5α/TRIM22/BST2 Subset in HTLV-1 Infection, Independent of the HAM/TSP Disease Signature

To validate our findings, we analyzed published *ex vivo* whole blood transcriptomes in an additional HTLV-1-infected cohort (Tattermusch et al.; detailed in **Table 1**) (Osame et al., 1986). As shown in **Figure 3B**, significant positive correlations were confirmed between TRIM5α/TRIM22 ($p = 0.005$, $r = 0.50$) and TRIM5α/BST2 ($p < 0.001$, $r = 0.67$) in both HTLV-1-infected AC ($n = 17$) and HAM/TSP patients ($n = 10$). Of note, TRIM5α mRNA expression was not significantly correlated to IFITM1 mRNA levels (data not shown), which is the only RF (among 42 studied herein) that is present in the HAM/TSP disease signature.

RFs Correlate Oppositely to *in Vivo* Retroviral RNA Levels in Untreated HTLV-1 vs. HIV-1 Infection

Surprisingly, the "proviral/deleterious" gene cluster identified by its strong positive correlation to HTLV-1 RNA levels and PVL contains the RFs previously demonstrated to exert a robust protective anti-HIV-1 effect *in vivo* and *in vitro*, namely *CDKN1A, SLFN11, PAF1,* and *ISG15* (Telenti, 2005; Ortiz et al., 2009; McLaren et al., 2015; Nozuma et al., 2017). Therefore, we performed a pairwise analysis of correlation to *in vivo* RNA levels in both retroviral infections, using our HTLV-1 data (**Figure 1A**) and data from the Swiss HIV cohort (Rotger et al., 2010) by calculating transcriptome-wide correlation of viral load set point to each separate gene, followed by FDR correction. As shown in **Figures 4A,B**, a striking opposite effect can be observed for the "proviral" vs. "antiviral" clusters. HTLV-1 "antiviral" cluster genes show significantly increased (higher R-values) correlation to HIV-1 *in vivo* RNA levels ($p < 0.0001$, Wilcoxon signed rank test, **Figure 4A**). The reverse phenomenon can be observed for the HTLV-1 "proviral" cluster genes, showing a tendency for decreased (lower R-values) correlation to HIV-1 *in vivo* RNA levels ($p = 0.06$, Wilcoxon signed rank test, **Figure 4B**).

Lentivirus-Driven Primate Evolution Has Shaped Antiretroviral Activity in Untreated HTLV-1 Infection

Lentivirus infections, to which HIV and its ancestral SIV belong, have had a pronounced effect upon primate evolution (Telenti, 2005), whereas no similar evidence exists for deltaretroviruses, such as HTLV or PTLV. As shown in **Figure 5**, the antiviral RF cluster we identified for HTLV-1 is strongly selected during recent primate evolution, as measured by both cross-species adaptive evolution between chimpanzees and humans (**Figures 5A,B**) and purifying selection throughout primate evolution (**Figure 5C**). Among the 42 anti-HIV RFs selected in this study, ten (TRIM5, APOBEC3A-B-C-F-G, IFI16, HERC5, EIF2AK2, and MX2) were found to display chimp/human adaptive evolution, as described by McLaren et al. (2015). The anti-HTLV-1 effect (measured as R^2 to represent effect size) of these 10 genes was significantly correlated to their adaptive evolution score, for both Tax mRNA ($r = 0.86$, $p = 0.0013$, **Figure 5A**) and PVL ($r = 0.78$, $p = 0.0073$, **Figure 5B**). In agreement with our results in **Figures 1A,B**, the effect size was greater for Tax mRNA (**Figure 5A**—range 0.1–0.74) than for PVL (**Figure 5B**—range 0.0–0.23), underscoring the putative role of these RFs in viral replication, rather than clonal expansion. Only six genes of the proposed "antiviral" cluster (TRIM5, TRIM22, and BST2, as well as APOBEC3F-G-H) underwent lentivirus-driven purifying selection during primate evolution (Ortiz et al., 2009), measured as K_A/K_S. In addition, these six HTLV-1-correlated genes underwent a significantly stronger selection during primate evolution, as compared to all other lentivirus-selected primate genes ($n = 134$, $p < 0.0001$, Mann–Whitney test, **Figure 5C**).

DISCUSSION

The striking negative association between mRNA levels of *Tax* and RFs TRIM5α, TRIM22, BST2, and RNASEL, despite the small number of participants in the study, suggests that these RFs may represent important mechanisms of defense against

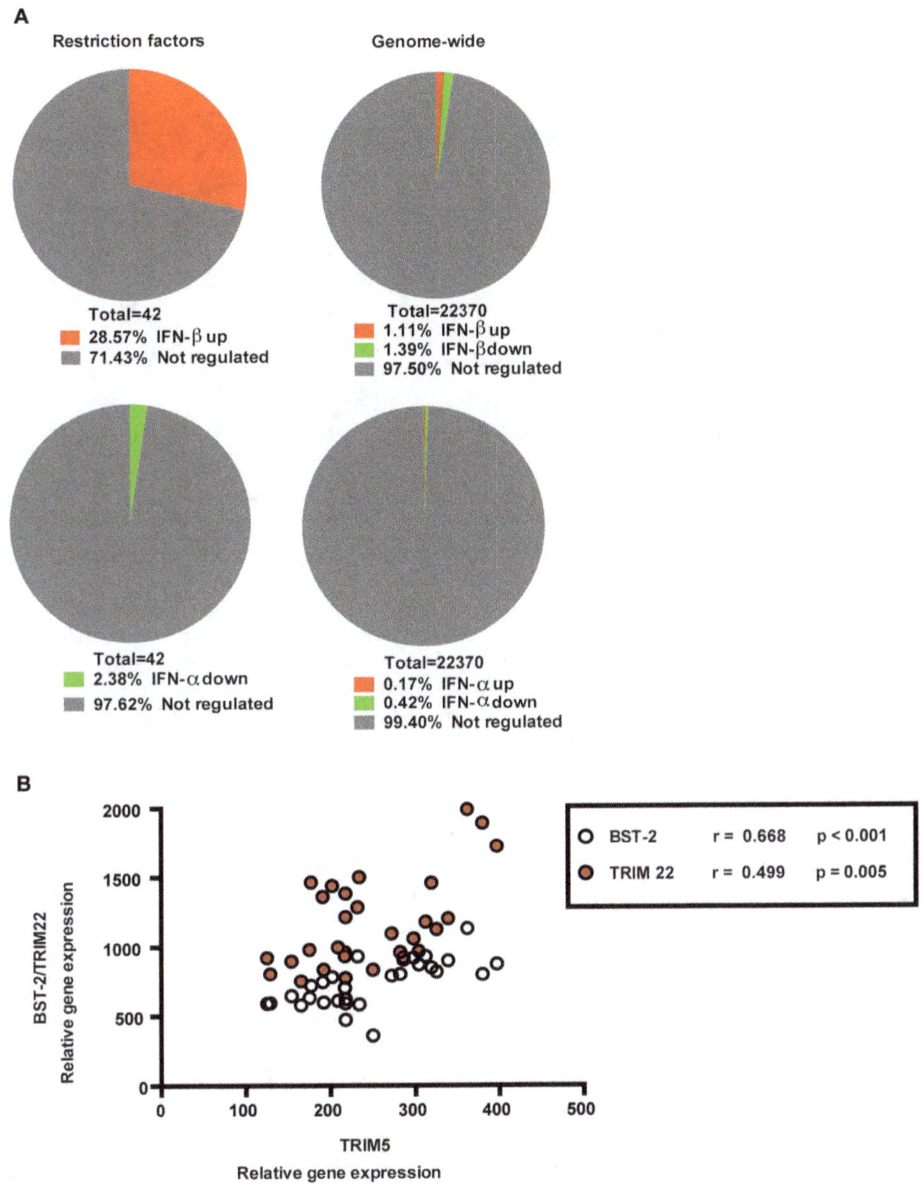

FIGURE 3 | Genome-wide analysis of PBMCs from HAM/TSP patients of RFs after *ex vivo* treatment with type I interferons. **(A)** Microarray analysis to determine genes significantly up- or down-regulated after 48 h *ex vivo* treatment of PBMCs from HAM/TSP patients with 1000 IU/ml IFN-β ($n = 5$), upper right corner pie-chart or IFN-α ($n = 6$), lower right corner pie-chart. RFs were significantly enriched among IFN-β-regulated, (upper left corner, genome-wide enrichment $p < 0.0001$), but not IFN-α-regulated (lower left corner, $p > 0.05$) genes. A total of 22370 genes were analyzed. **(B)** Reanalysis of published *ex vivo* whole blood transcriptomes from an independent HTLV-1 UK cohort (Tattermusch et al., 2012). Correlation between TRIM5α and TRIM22/BST2 in both HTLV-1-infected AC ($n = 17$) and HAM/TSP patients ($n = 10$).

HTLV-1 infection and/or neuroinflammation. Our hypothesized antiviral activity of TRIM5 against HTLV-1 is supported by the recent findings of Nozuma et al. demonstrating a significant association between TRIM5 R136Q polymorphism and lower PVL, specific to HAM/TSP patients (Nozuma et al., 2017). Strikingly, polymorphisms in TRIM5, TRIM22, and BST2, but not APOBECs or TREX1 were significantly associated to another neuroinflammatory disorder, multiple sclerosis (Nexo et al., 2013). Conversely, the apparent proviral effect of PAF-1, ISG15, and CDKN1A are in agreement with the recently demonstrated IFN-inducible HAM/TSP disease signature (Tattermusch et al., 2012). Moreover, the robustness of the antiviral and proviral clusters we identified in this study indicates they represent plausible candidate biomarkers for future HAM/TSP clinical trials. In agreement with Saito et al., who demonstrated a positive correlation between HBZ mRNA levels and HAM/TSP disease severity (Saito et al., 2009), our cluster analysis shows the strongest association between HBZ and clinical status

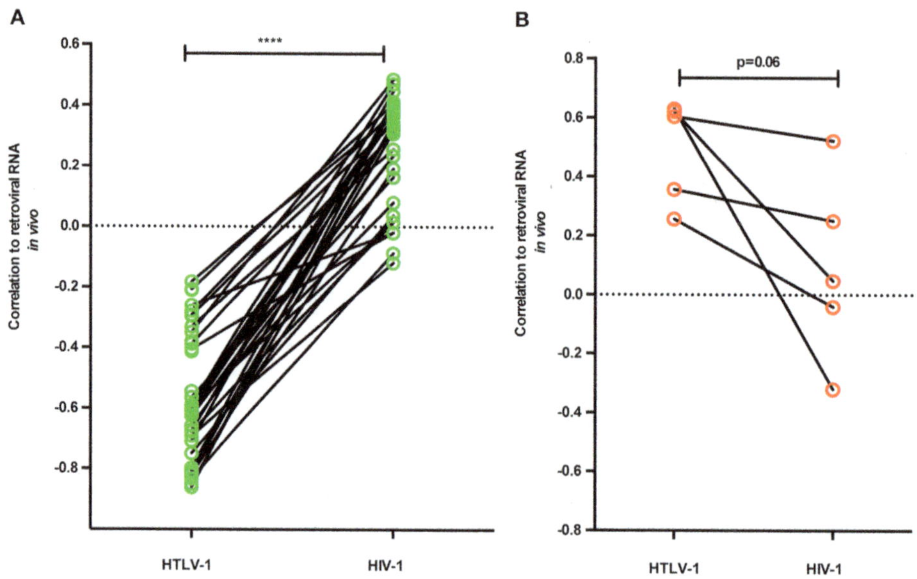

FIGURE 4 | Restriction factors correlate oppositely to *in vivo* retroviral RNA levels in untreated HTLV-1 vs. HIV-1 infection. A pairwise analysis was performed for RFs in antiviral and proviral clusters, by correlating to *in vivo* RNA levels in both retroviral infections, using HTLV-1 data from this study (**Figure 1A**) and data from the Swiss HIV cohort (Rotger et al., 2010). Out of 42 RFs, 40 were matched to unique transcripts in HIV-1 microarray data, followed by calculating transcriptome-wide correlation of viral load set point to each separate gene, using stringent FDR (Benjamini-Hochberg) correction. **(A)** HTLV-1 "antiviral" cluster genes show significantly increased (positive) correlation to HIV-1 *in vivo* RNA levels (****$p < 0.0001$, Wilcoxon signed rank test, $n = 35$). **(B)** HTLV-1 "proviral" cluster genes show a tendency for decreased (positive) correlation to HIV-1 *in vivo* RNA levels ($p = 0.06$, Wilcoxon signed rank test, $n = 5$).

(**Figure 1A**). However, no RFs were significantly correlated to HBZ mRNA levels after correction for multiple testing, in contrast to tax mRNA levels (**Figure 1B**). This might be due to post-transcriptional regulation of HBZ mRNA vs. protein levels, as Baratella et al. recently demonstrated cytoplasmatic HBZ protein as a biomarker able to discriminate between AC and HAM/TSP (Baratella et al., 2017). Due to their significant association with all currently used surrogate markers (PVL, *Tax*, and *HBZ* mRNAs), as well as clinical status, antiviral and proviral clusters are likely to reflect positive vs. negative therapeutic outcomes, respectively. Therefore, our results help explain the "IFN paradox" in HAM/TSP, by reconciling apparently conflicting data in the literature, namely the existence of an interferon-inducible disease signature and the demonstrated beneficial therapeutic effects of type I interferons, both IFN-α and β in HAM/TSP (Izumo et al., 1996; Oh et al., 2005).

In addition, our study also reveals striking differences between these two widely used IFN subtypes. First, a significant antiproliferative effect of β and not α was observed in HAM/TSP patients, corroborating our recent findings in ATL (Dierckx et al., 2017) and revealing a surprising similarity between these two highly distinct HTLV-1-associated pathologies. On the other hand, our study also reveals another parallel between neuroinflammatory disorders HAM/TSP and multiple sclerosis, namely their unique sensitivity to IFN-β over IFN-α. Of note, our results provide mechanistic evidence for the previously described immunovirological and clinical impact observed *in vivo* with IFN-β therapy in HAM/TSP patients, in an open-label trial (Oh et al., 2005), as well as two case reports with remarkable clinical response (Costa et al., 2012; Viana et al., 2014). This superior antiproliferative effect of IFN-β in HAM/TSP, as compared to IFN-α, parallels our findings in the other major HTLV-1-associated disease, Adult T-cell Leukemia (Dierckx et al., 2017).

Surprisingly, the "proviral/deleterious" gene cluster, identified by its strong positive correlation to HTLV-1 RNA levels and PVL, contains the RFs previously demonstrated to exert the strongest protective anti-HIV-1 effect *in vivo* and *in vitro*. Thus, p21 (*CDKN1A*), schlafen 11 (*SLFN11*), and PAF1 were strongly associated with reduced CD4+ T cell-associated HIV RNA during antiretroviral treatment (Abdel-Mohsen et al., 2015). *SLFN11* was also identified as overexpressed in HIV-positive "elite controllers" (Abdel-Mohsen et al., 2013), who maintain undetectable viral load, even in the absence of antiretroviral treatment. PAF1 was identified by a genome-wide screen for HIV RFs (Liu et al., 2011) and restricts HIV-1, HIV-2, and SIV *in vitro*. Finally, CDKN1A was demonstrated as a robust antiviral effect against HIV-1 replication *in vitro*, by different groups (Leng et al., 2014; Farberov et al., 2015).

With up to eight percent of the human genome of retroviral origin, the importance of retroviruses for the evolution of susceptible host organisms cannot be overestimated (Ortiz et al., 2009; Lascano et al., 2015). Over millions of years, Lentivirus infections (retroviruses including HIV and ancestral SIV) have had a pronounced effect upon primate evolution (Ortiz et al., 2009). Yet, no evidence exists for deltaretroviruses (including HTLV and PTLV). Using two different quantitative approaches, i.e., measuring chimp/human cross-species adaptive evolution (**Figures 5A,B**) and purifying selection throughout primate

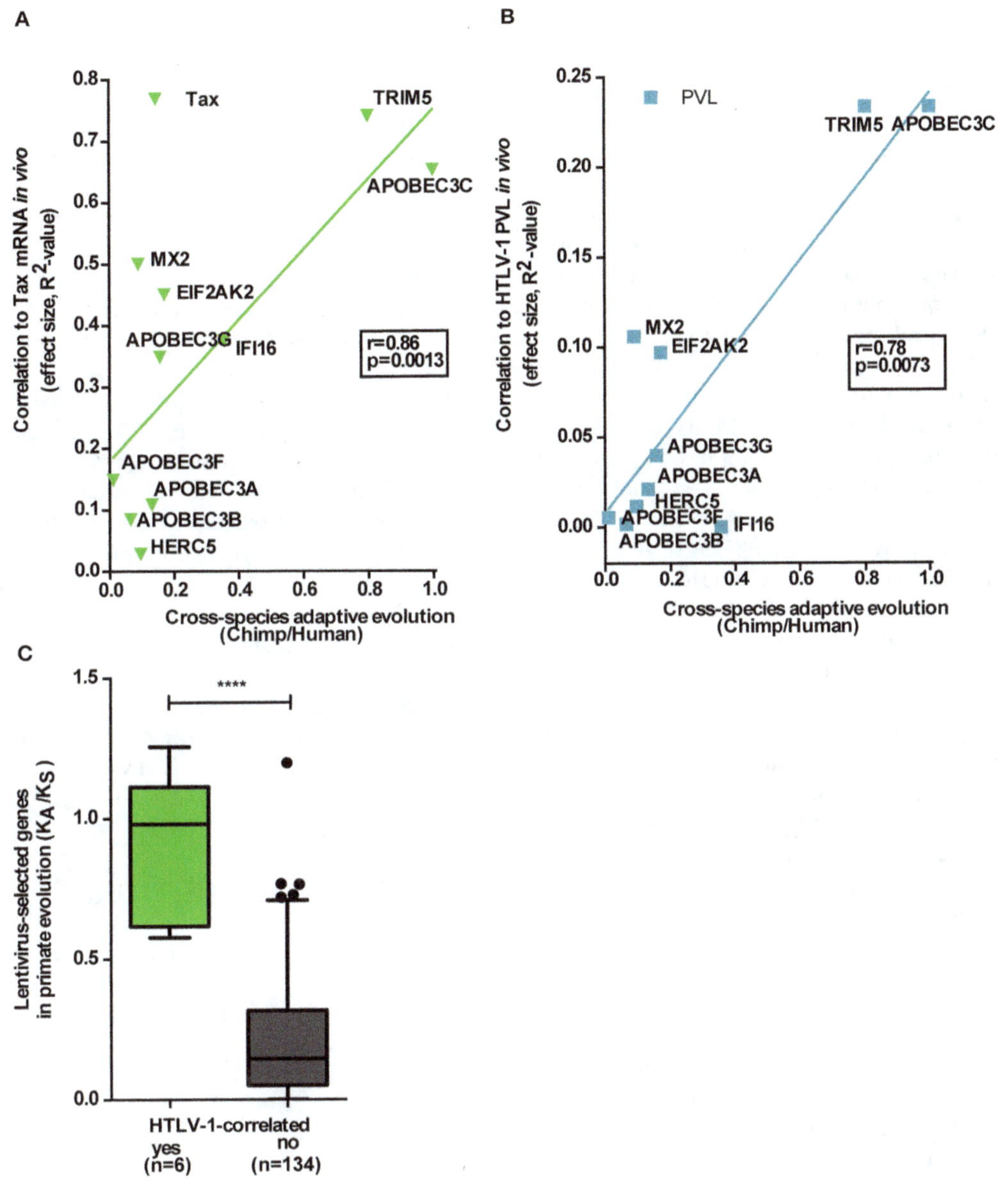

FIGURE 5 | Lentivirus-driven primate evolution has shaped antiretroviral activity in untreated HTLV-1 infection. The antiviral RF cluster is strongly selected during recent primate evolution, as measured by both cross-species adaptive evolution between chimpanzees and humans and purifying selection throughout primate evolution. **(A,B)** Among the 42 anti-HIV RFs selected in this study, 10 were found to display chimpanzee/human cross-species adaptive evolution (McLaren et al., 2015). The anti-HTLV-1 effect (measured as R^2) of these 10 genes was significantly correlated to their adaptive evolution score, for both Tax mRNA ($r = 0.86$, $p = 0.0013$) and PVL ($r = 0.78$, $p = 0.0073$). **(C)** Six genes of the proposed antiviral cluster underwent lentivirus-driven purifying selection (measured as K_A/K_S, Ortiz et al., 2009) during primate evolution, which was significantly increased vs. all other lentivirus-selected primate genes ($n = 134$, $^{****}p < 0.0001$, Mann–Whitney test).

evolution (**Figure 5C**), we found that the antiviral RF cluster we identified for HTLV-1 was under strong purifying selection during recent primate evolution. Our preselection of a small number of candidate RFs, by virtue of their proven antiviral effect against HIV-1, is an obvious limitation of the first part of our study, focusing on purified CD4+ cells. However, we confirm the unique TRIM5/TRIM22/BST2 antiviral cluster by two independent unbiased methods, i.e., PBMC transcriptomics and genome-wide evolutionary analysis of protein-coding genes. By comparing the only two pathogenic human retroviruses, we found striking differences in *in vivo* correlations of "candidate" RFs to retroviral RNA levels, as an indirect measure of their possible anti-HIV and anti-HTLV-1 activity. Although some antiviral effector molecules such as cGAS exhibit "panviral" activity (Schoggins et al., 2014), our results argue against "panretroviral" restriction factors in the human setting. Very few host genes, such as CIITA and PKR (also known as EIF2AK2), have been formally demonstrated to restrict both HIV-1 and HTLV-1 replication *in vitro* (Schoggins et al., 2011; Tosi et al., 2011; Cachat et al., 2013; Kinpara et al., 2013;

Forlani et al., 2016). Interestingly, CIITA interacts and cooperates with TRIM22 in restricting replication of HIV-1 (Forlani et al., 2017). However, CIITA is unique among RFs because it is transcriptionally upregulated by IFN-gamma but downregulated by IFN-β (Lu et al., 1995). In contrast, IFN-α strongly upregulates PKR/EIF2AK2 in HTLV-1-infected cells but does not decrease Tax or Hbz mRNA (Moens et al., 2012b; Cachat et al., 2013; Kinpara et al., 2013). Thus, CIITA and PKR are unlikely candidates for the IFN-β-mediated antiviral and antiproliferative effect we observed *ex vivo* and *in vitro* in HAM/TSP patients. Unfortunately, high-throughput *in vitro* analysis of RFs, such as performed by Schoggins et al. for HIV-1 and several other viruses, are lacking for HTLV-1 (Schoggins et al., 2011). Based on our previous demonstration of a 1000-fold difference in HIV-1 vs. HTLV-1 antiviral effect in co-infected MT-4 cells (Moens et al., 2012b) we anticipate strong differences might be observed for *in vitro* HTLV-1 vs. HIV-1 restriction pathways. As a final layer of complexity, HIV-1 also displays striking discrepancies between *in vivo* and *ex vivo* RF antiviral activity (Rotger et al., 2010; Schoggins et al., 2014). Therefore, we believe the term "restriction factors" should be used with caution in human retrovirology. "Antiviral effector genes" might be a more precise term, if defined within a precise context, i.e., depending on cell type, virus and upstream regulators such as IFN subtypes. Regarding cell type-specificity, TRIM5α and TRIM22 expression at both mRNA and protein level has been shown to predominate in CD4$^+$ T cells, as compared to other mononuclear cell types (Singh et al., 2014), which explains the strong agreement between our RF analysis in CD4$^+$ T cells and IFN response in PBMCs, as well as whole blood analysis from UK cohort participants (Tattermusch et al., 2012).

Mechanistically, both the antiviral vs. "proviral" effects of TRIM5, CDKN1A, and ISG15 might also be explained by broader cellular processes, outside the narrow definition of RFs. TRIM5 is an intracellular protein that exerts its protective effect by disrupting the retroviral capsid as it transports viral nucleic acid into the nucleus. In addition, innate immune signaling might contribute to TRIM5-mediated restriction. Lascano et al. have shown that activation of innate immune signaling is conserved among primate and carnivore TRIM5 orthologs and that such activity is required for TRIM5-mediated restriction activity (Lascano et al., 2015). CDKN1A upregulation by Tax might shorten G1 phase by promoting formation of stable kinase complexes, contributing to cell-cycle progression (Kehn et al., 2004) and proliferation of infected clones. Regarding ISG15, our group has recently demonstrated its anti-inflammatory extracellular cytokine-like activity, through monocyte-specific induction of IL-10 (Dos Santos et al., 2018). However, recent data reveal an unexpected IL-10-induced proliferative switch in HAM/TSP-derived HTLV-1-infected T-cell lines (Sawada et al., 2017). Therefore, ISG15 inducing IL-10 and hence triggering lymphoproliferation might represent, at least in part, a molecular mechanism for the putative "proviral" role of ISG15 in HAM/TSP. Integrating the data from Tattermusch et al. (2012), Dos Santos et al. (2018), Sawada et al. (2017) and this study, we hypothesize that an "IFN-beta like/TRIM5" gene signature vs. a "non-IFN-beta like/ISG15" gene signature might be predictive of HAM/TSP disease progression, as well as of therapeutic outcome with immunomodulatory and/or antiproliferative drugs (IFNs, glucocorticoids and others).

In conclusion, our integrated *ex vivo* approach reveals that antiretroviral genes in HTLV-1 infection and HAM/TSP cluster in two distinct "proviral/antiviral" classes, of which the TRIM5α/TRIM22/BST2 antiviral subset, selected during recent primate evolution, is selectively up-regulated by IFN-β signaling in HAM/TSP. Our results thus provide *ex vivo* mechanistic evidence for the observed *in vivo* immunovirological effect of IFN-β treatment in HAM/TSP and identify biomarkers as well as possible therapeutic targets for a precision medicine approach. Finally, a protective antiviral and IFN-inducible TRIM5α/TRIM22/BST2 gene cluster, independent of the HAM/TSP IFN-inducible disease signature, reconciles the apparent IFN paradox in HTLV-1 research and confirms type I IFN as a two-edged sword in human health and disease.

AUTHOR CONTRIBUTIONS

FL, RR, and JVW: conceived and designed the experiments; SM, FL, EC, PB, and JVW: performed the experiments; FL, SM, LG, RR, RK, JV, TD, and JVW: analyzed the data; EK, DN, AS, BG-C, and LG: contributed reagents, materials, and analysis tools; FL, SM, JVW, and DN: wrote the paper.

FUNDING

This work was supported by CNPq (Pronex, PVE), KU Leuven (Vaast Leysen Leerstoel voor Infectieziekten in Ontwikkelingslanden) and FWO (grant G0D6817N). This work was partially supported by the National Institutes of Health grant to the District of Columbia Center for AIDS Research, (P30AI087714 and P30AI117970), Fundação de Amparo a Pesquisa do Estado de São Paulo (04/15856-9/EGK and 2010/05845-0/EGK/DFN), CNPq/CAPES 056/2012 (DFN).

REFERENCES

Abdel-Mohsen, M., Raposo, R. A., Deng, X., Li, M., Liegler, T., Sinclair, E., et al. (2013). Expression profile of host restriction factors in HIV-1 elite controllers. *Retrovirology* 10:106. doi: 10.1186/1742-4690-10-106

Abdel-Mohsen, M., Wang, C., Strain, M. C., Lada, S. M., Deng, X., Cockerham, L. R., et al. (2015). Select host restriction factors are associated with HIV persistence during antiretroviral therapy. *AIDS* 29, 411–420. doi: 10.1097/QAD.0000000000000572

Bangham, C. R., Araujo, A., Yamano, Y., and Taylor, G. P. (2015). HTLV-1-associated myelopathy/tropical spastic paraparesis. *Nat. Rev. Dis. Primers* 1:15012. doi: 10.1038/nrdp.2015.12

Baratella, M., Forlani, G., Raval, G. U., Tedeschi, A., Gout, O., Gessain, A., et al. (2017). Cytoplasmic localization of HTLV-1 HBZ protein: a biomarker of

HTLV-1-associated myelopathy/tropical spastic paraparesis (HAM/TSP). *PLoS Negl. Trop. Dis.* 11:e0005285. doi: 10.1371/journal.pntd.0005285

Cachat, A., Chevalier, S. A., Alais, S., Ko, N. L., Ratner, L., Journo, C., et al. (2013). Alpha interferon restricts human T-lymphotropic virus type 1 and 2 de novo infection through PKR activation. *J. Virol.* 87, 13386–13396. doi: 10.1128/JVI.02758-13

Costa, D. T., Sundberg, M., Passos, L., Muniz, A. L., and Santos, S. (2012). Interferon Beta-1a improves urinary symptoms, reduces proviral load, and modifies the immune response in a patient with HAM/TSP. *Case Rep. Neurol. Med.* 2012:958786. doi: 10.1155/2012/958786

Dehee, A., Cesaire, R., Desire, N., Lezin, A., Bourdonne, O., Bera, O., et al. (2002). Quantitation of HTLV-I proviral load by a TaqMan real-time PCR assay. *J. Virol. Methods* 102, 37–51. doi: 10.1016/S0166-0934(01)00445-1

Dierckx, T., Khouri, R., Menezes, S. M., Decanine, D., Farre, L., Bittencourt, A., et al. (2017). IFN-β induces greater antiproliferative and proapoptotic effects and increased p53 signaling compared with IFN-α in PBMCs of Adult T-cell Leukemia/Lymphoma patients. *Blood Cancer J.* 7:e519. doi: 10.1038/bcj.2016.126

Dos Santos, P. F., Van Weyenbergh, J., Delgobo, M., Oliveira Patricio, D., Ferguson, B. J., Guabiraba, R., et al. (2018). ISG15-induced IL-10 Is a novel anti-inflammatory myeloid axis disrupted during active tuberculosis. *J. Immunol.* 200, 1434–1442. doi: 10.4049/jimmunol.1701120

Farberov, L., Herzig, E., Modai, S., Isakov, O., Hizi, A., and Shomron, N. (2015). MicroRNA-mediated regulation of p21 and TASK1 cellular restriction factors enhances HIV-1 infection. *J. Cell Sci.* 128, 1607–1616. doi: 10.1242/jcs.167817

Forlani, G., Tosi, G., Turrini, F., Poli, G., Vicenzi, E., and Accolla, R. S. (2017). Tripartite motif-containing protein 22 interacts with class II transactivator and orchestrates its recruitment in nuclear bodies containing TRIM19/PML and cyclin T1. *Front. Immunol.* 8:564. doi: 10.3389/fimmu.2017.00564

Forlani, G., Turrini, F., Ghezzi, S., Tedeschi, A., Poli, G., Accolla, R. S., et al. (2016). The MHC-II transactivator CIITA inhibits Tat function and HIV-1 replication in human myeloid cells. *J. Transl. Med.* 14:94. doi: 10.1186/s12967-016-0853-5

Foster, T. L., Pickering, S., and Neil, S. J. D. (2017). Inhibiting the ins and outs of HIV replication: cell-intrinsic antiretroviral restrictions at the plasma membrane. *Front. Immunol.* 8:1853. doi: 10.3389/fimmu.2017.01853

Gallo, R. C., Willems, L., and Tagaya, Y. (2017). Time to Go back to the original name. *Front. Microbiol.* 8:1800. doi: 10.3389/fmicb.2017.01800

Izumo, S., Goto, I., Itoyama, Y., Okajima, T., Watanabe, S., Kuroda, Y., et al. (1996). Interferon-alpha is effective in HTLV-I-associated myelopathy: a multicenter, randomized, double-blind, controlled trial. *Neurology* 46,1016–1021. doi: 10.1212/WNL.46.4.1016

Kehn, K., Deng, L., de la Fuente, C., Strouss, K., Wu, K., Maddukuri, A., et al. (2004). The role of cyclin D2 and p21/waf1 in human T-cell leukemia virus type 1 infected cells. *Retrovirology* 1:6. doi: 10.1186/1742-4690-1-6

Kinpara, S., Kijiyama, M., Takamori, A., Hasegawa, A., Sasada, A., Masuda, T., et al. (2013). Interferon-α (IFN-α) suppresses HTLV-1 gene expression and cell cycling, while IFN-α combined with zidovudine induces p53 signaling and apoptosis in HTLV-1-infected cells. *Retrovirology* 10:52. doi: 10.1186/1742-4690-10-52

Lascano, J., Uchil, P. D., Mothes, W., and Luban, J. (2015). TRIM5 retroviral restriction activity correlates with the ability to induce innate immune signaling. *J. Virol.* 90, 308–316. doi: 10.1128/JVI.02496-15

Leng, J., Ho, H. P., Buzon, M. J., Pereyra, F., Walker, B. D., Yu, X. G., et al. (2014). A cell-intrinsic inhibitor of HIV-1 reverse transcription in CD4(+) T cells from elite controllers. *Cell Host Microbe* 15, 717–728. doi: 10.1016/j.chom.2014.05.011

Liu, L., Oliveira, N. M., Cheney, K. M., Pade, C., Dreja, H., Bergin, A. M., et al. (2011). A whole genome screen for HIV restriction factors. *Retrovirology* 8:94. doi: 10.1186/1742-4690-8-94

Lu, H. T., Riley, J. L., Babcock, G. T., Huston, M., Stark, G. R., Boss, J. M., et al. (1995). Interferon (IFN) beta acts downstream of IFN-gamma-induced class II transactivator messenger RNA accumulation to block major histocompatibility complex class II gene expression and requires the 48-kD DNA-binding protein, ISGF3-gamma. *J. Exp. Med.* 182, 1517–1525. doi: 10.1084/jem.182.5.1517

McLaren, P. J., Gawanbacht, A., Pyndiah, N., Krapp, C., Hotter, D., Kluge, S. F., et al. (2015). Identification of potential HIV restriction factors by combining evolutionary genomic signatures with functional analyses. *Retrovirology* 12:41. doi: 10.1186/s12977-015-0165-5

Menezes, S. M., Decanine, D., Brassat, D., Khouri, R., Schnitman, S. V., Kruschewsky, R., et al. (2014). CD80+ and CD86+ B cells as biomarkers and possible therapeutic targets in HTLV-1 associated myelopathy/tropical spastic paraparesis and multiple sclerosis. *J. Neuroinflammation* 11:18. doi: 10.1186/1742-2094-11-18

Moens, B., Decanine, D., Menezes, S. M., Khouri, R., Silva-Santos, G., Lopez, G., et al. (2012a). Ascorbic acid has superior *ex vivo* antiproliferative, cell death-inducing and immunomodulatory effects over IFN-α in HTLV-1-associated myelopathy. *PLoS Negl. Trop. Dis* 6:e1729. doi: 10.1371/journal.pntd.0001729

Moens, B., Pannecouque, C., Lopez, G., Talledo, M., Gotuzzo, E., Khouri, R., et al. (2012b). Simultaneous RNA quantification of human and retroviral genomes reveals intact interferon signaling in HTLV-1-infected CD4+ T cell lines. *Virol. J.* 9:171. doi: 10.1186/1743-422X-9-171

Neil, S. J., Zang, T., and Bieniasz, P. D. (2008). Tetherin inhibits retrovirus release and is antagonized by HIV-1 Vpu. *Nature* 451, 425–430. doi: 10.1038/nature06553

Nexo, B. A., Hansen, B., Nissen, K. K., Gundestrup, L., Terkelsen, T., Villesen, P., et al. (2013). Restriction genes for retroviruses influence the risk of multiple sclerosis. *PLoS ONE* 8:e74063. doi: 10.1371/journal.pone.0074063

Nozuma, S., Matsuura, E., Kodama, D., Tashiro, Y., Matsuzaki, T., Kubota, R., et al. (2017). Effects of host restriction factors and the HTLV-1 subtype on susceptibility to HTLV-1-associated myelopathy/tropical spastic paraparesis. *Retrovirology* 14:26. doi: 10.1186/s12977-017-0350-9

Oh, U., Yamano, Y., Mora, C. A., Ohayon, J., Bagnato, F., Butman, J. A., et al. (2005). Interferon-beta1a therapy in human T-lymphotropic virus type I-associated neurologic disease. *Ann. Neurol.* 57, 526–534. doi: 10.1002/ana.20429

Ortiz, M., Guex, N., Patin, E., Martin, O., Xenarios, I., Ciuffi, A., et al. (2009). Evolutionary trajectories of primate genes involved in HIV pathogenesis. *Mol. Biol. Evol.* 26, 2865–2875. doi: 10.1093/molbev/msp197

Osame, M. (1990). "Review of WHO Kagoshima Meeting and diagnostic guidelines for HAM/TSP," in *Human Retrovirology: HTLV*, ed W. Blattner (New York, NY: Raven Press), 191–177.

Osame, M., Usuku, K., Izumo, S., Ijichi, N., Amitani, H., Igata, A., et al. (1986). HTLV-I associated myelopathy, a new clinical entity. *Lancet* 1, 1031–1032. doi: 10.1016/S0140-6736(86)91298-5

Raposo, R. A., Abdel-Mohsen, M., Bilska, M., Montefiori, D. C., Nixon, D. F., and Pillai, S. K. (2013a). Effects of cellular activation on anti-HIV-1 restriction factor expression profile in primary cells. *J. Virol.* 87, 11924–11929. doi: 10.1128/JVI.02128-13

Raposo, R. A., Abdel-Mohsen, M., Deng, X., Hecht, F. M., Pilcher, C. D., Pillai, S. K., et al. (2014). Dynamic regulation of host restriction factor expression over the course of HIV-1 infection *in vivo*. *J. Virol.* 88, 11624–11629. doi: 10.1128/JVI.01771-14

Raposo, R. A., Abdel-Mohsen, M., Holditch, S. J., Kuebler, P. J., Cheng, R. G., Eriksson, E. M., et al. (2013b). Increased expression of intrinsic antiviral genes in HLA-B*57-positive individuals. *J. Leukoc. Biol.* 94, 1051–1059. doi: 10.1189/jlb.0313150

Rotger, M., Dang, K. K., Fellay, J., Heinzen, E. L., Feng, S., Descombes, P., et al. (2010). Genome-wide mRNA expression correlates of viral control in CD4+ T-cells from HIV-1-infected individuals. *PLoS Pathog.* 6:e1000781. doi: 10.1371/journal.ppat.1000781

Saito, M., Matsuzaki, T., Satou, Y., Yasunaga, J., Saito, K., Arimura, K., et al. (2009). *In vivo* expression of the HBZ gene of HTLV-1 correlates with proviral load, inflammatory markers and disease severity in HTLV-1 associated myelopathy/tropical spastic paraparesis (HAM/TSP). *Retrovirology* 6:19. doi: 10.1186/1742-4690-6-19

Sawada, L., Nagano, Y., Hasegawa, A., Kanai, H., Nogami, K., Ito, S., et al. (2017). IL-10-mediated signals act as a switch for lymphoproliferation in Human T-cell leukemia virus type-1 infection by activating the STAT3 and IRF4 pathways. *PLoS Pathog.* 13:e1006597. doi: 10.1371/journal.ppat.1006597

Schoggins, J. W., MacDuff, D. A., Imanaka, N., Gainey, M. D., Shrestha, B., Eitson, J. L., et al. (2014). Pan-viral specificity of IFN-induced genes reveals new roles for cGAS in innate immunity. *Nature* 505, 691–695. doi: 10.1038/nature 12862

Schoggins, J. W., Wilson, S. J., Panis, M., Murphy, M. Y., Jones, C. T., Bieniasz, P., et al. (2011). A diverse range of gene products are effectors of the type I interferon antiviral response. *Nature* 472, 481–485. doi: 10.1038/nature09907

Singh, R., Patel, V., Mureithi, M. W., Naranbhai, V., Ramsuran, D., Tulsi, S., et al. (2014). TRIM5alpha and TRIM22 are differentially regulated according to HIV-1 infection phase and compartment. *J. Virol.* 88, 4291–4303. doi: 10.1128/JVI.03603-13

Tagaya, Y., and Gallo, R. C. (2017). The exceptional oncogenicity of HTLV-1. *Front. Microbiol.* 8:1425. doi: 10.3389/fmicb.2017.01425

Tattermusch, S., Skinner, J. A., Chaussabel, D., Banchereau, J., Berry, M. P., McNab, F. W., et al. (2012). Systems biology approaches reveal a specific interferon-inducible signature in HTLV-1 associated myelopathy. *PLoS Pathog.* 8:e1002480. doi: 10.1371/journal.ppat.1002480

Telenti, A. (2005). Adaptation, co-evolution, and human susceptibility to HIV-1 infection. *Infect. Genet. Evol.* 5, 327–334. doi: 10.1016/j.meegid.2004.11.001

Tosi, G., Forlani, G., Andresen, V., Turci, M., Bertazzoni, U., Franchini, G., et al. (2011). Major histocompatibility complex class II transactivator CIITA is a viral restriction factor that targets human T-cell lymphotropic virus type 1 Tax-1 function and inhibits viral replication. *J. Virol.* 85, 10719–10729. doi: 10.1128/JVI.00813-11

Viana, G. M., Silva, M. A., Souza, V. L., Lopes, N. B., Silva, D. L., and Nascimento Mdo, D. (2014). Interferon beta-1a treatment in HTLV-1-associated myelopathy/tropical spastic paraparesis: a case report. *Rev. Inst. Med. Trop. Sao Paulo* 56, 443–445. doi: 10.1590/S0036-46652014000500013

2

How does HTLV-1 Undergo Oncogene-Dependent Replication Despite a Strong Immune Response?

*Hélène Gazon[1,2], Pradeep Chauhan[1,2], Malik Hamaidia[1,2], Clotilde Hoyos[1,2], Lin Li[1,2], Roghaiyeh Safari[1,2] and Luc Willems[1,2]**

[1] *National Fund for Scientific Research, Molecular and Cellular Epigenetics, Interdisciplinary Cluster for Applied Genoproteomics, Liège, Belgium,* [2] *Molecular Biology, TERRA, Gembloux Agro-Bio Tech, Gembloux, Belgium*

***Correspondence:**
Luc Willems
luc.willems@Uliege.be

In 1987, Mitsuaki Yoshida proposed the following model (Yoshida and Seiki, 1987): "... T-cells activated through the endogenous p40x would express viral antigens including the envelope glycoproteins which are exposed on the cell surface. These glycoproteins are targets of host immune surveillance, as is evidenced by the cytotoxic effects of anti-envelope antibodies or patient sera. Eventually all cells expressing the viral antigens, that is, all cells driven by the p40x would be rejected by the host. Only those cells that did not express the viral antigens would survive. Later, these antigen-negative infected cells would begin again to express viral antigens, including p40x, thus entering into the second cycle of cell propagation. These cycles would be repeated in so-called healthy virus carriers for 20 or 30 years or longer...." Three decades later, accumulated experimental facts particularly on intermittent viral transcription and regulation by the host immune response appear to prove that Yoshida was right. This Hypothesis and Theory summarizes the evidences that support this paradigm.

Keywords: HTLV, BLV, microRNA, cytotoxic T cells, long non-coding RNA, Leukemia

INTRODUCTION

At least 20 million people worldwide are infected with human T-cell leukemia virus type 1 (HTLV-1) (Gessain and Cassar, 2012; Bangham, 2017; Watanabe, 2017). This retrovirus is prevalent in southwestern Japan, sub-Saharan Africa, the Caribbean islands, South America, the Middle East and Austro-Melanesia. Transmission occurs principally from mother to child via milk or between sexual partners through contaminated blood. Infected individuals are at risk of developing a rapidly progressive malignancy, adult T-cell leukemia/lymphoma (ATLL), and a debilitating neurologic condition, HTLV-1 associated myelopathy/tropical spastic paraparesis (HAM/TSP) (Willems et al., 2017).

Although frequently neglected in the field, bovine leukemia virus (BLV) is a useful model to address specific questions that cannot be answered in the HTLV-1 system (Rodríguez et al., 2011; Polat et al., 2017). Both viruses are indeed closely related δ-retroviruses that induce hematological diseases. In the bovine species, the most prevalent clinical manifestation in about one-third of infected animals is persistent lymphocytosis, a benign accumulation of infected B-lymphocytes (Gutierrez et al., 2014). In a minority of cases (about 5–10%), BLV infection can progress to fatal leukemia/lymphoma whose most dramatic consequence is spleen hypertrophy and disruption consecutive to tumor formation. BLV typically persists in less than 1% of peripheral blood cells, leading to an asymptomatic infection in the majority of infected animals. BLV is transmitted horizontally by direct contact, iatrogenic procedures or insect bites upon transfer of infected cells from milk, blood, and body fluids from heavily infected dams (Barez et al., 2015).

THE PARADOX: HTLV-1 REPLICATION IS DRIVEN BY ONCOGENIC PROTEINS THAT EXPOSE THE INFECTED CELL TO THE HOST IMMUNE RESPONSE

According to currently most accepted model, two viral proteins (Tax and HBZ, HTLV-1 bZIP) are hypothesized to have the highest impact on viral replication and cell transformation (Matsuoka and Jeang, 2007; Carpentier et al., 2015). The modes of action of Tax and HBZ are remarkably pleiotropic and involve a variety of cell signaling pathways (CREB, NF-κB, AKT, and TNF) (Twizere et al., 2003; Boxus et al., 2008). Tax inhibits tumor suppressors (p53, Bcl11B, and TP53INP1) and activates cyclin-dependent kinases (CDKs), both of these mechanisms leading to accelerated cell proliferation. Experimental evidence further shows that Tax drives tumor formation in transgenic mouse models, supporting its oncogenic potential. Through an interaction with the helicase complex (mcm2-7), Tax accelerates S phase progression by initiating additional replication origins. By promoting unscheduled cell growth, Tax also induces genomic instability and generates somatic alterations. Another viral protein, HBZ also favors cell proliferation by inhibiting apoptosis/senescence and modulating the cell cycle (Ma et al., 2016). In fact, the HBZ protein counteracts a series of Tax-activated viral and cellular pathways (such as NF-κB, Akt, and CREB). Transgenic expression of HBZ in CD4 + T cells induced T-cell lymphomas and systemic inflammation in mice, resembling diseases observed in HTLV-1 infected individuals (Satou et al., 2011).

A major issue of expression of viral proteins is initiation of the host immune response. Indeed, Tax induces a strong immune response that would be harmful to infected cells (Nagai et al., 2001; Kannagi et al., 2005; Rowan et al., 2014). In comparison, HBZ triggers a less efficient immunity that is consistent with low expression low expression throughout HTLV-1 infection (MacNamara et al., 2010). How does HTLV-1 persist despite a strong immune response against viral oncogenes that promote infected cell replication? The next paragraph lists the experimental evidence pertaining to this paradox.

EXPERIMENTAL EVIDENCE AND INTERPRETATIONS

Evidence 1: Infected Cells Proliferate Faster to Undergo Clonal Expansion

The BLV model has been instrumental to quantify the dynamics of cell turnover *in vivo* (Florins et al., 2007). Experiments based on intravenous injection of bromodeoxyuridine (BrdU) and carboxyfluorescein diacetate succinimidyl ester (CFSE) demonstrate that B-lymphocytes are proliferating significantly faster in BLV-infected sheep than in healthy controls (Debacq et al., 2002, 2006). Excess of proliferation is compensated by an increase in cell death, thereby maintaining homeostasis. Increased cell proliferation is also reported in HTLV-induced HAM/TSP using a similar strategy based on incorporation of deuterated glucose (Asquith et al., 2007). *In vivo*, BLV and HTLV infection is thus characterized by an increased cell turnover, likely driven by viral oncogenes such as Tax and HBZ.

Evidence 2: 5′ LTR Directed Transcription Is Extremely Low *in Situ*

The amount of viral RNA transcribed from the 5′ LTR promoter is extremely low *in vivo*. In primary tumor cells, only very sensitive techniques such as RT-PCR can identify viral RNA transcribed from the 5′ LTR promoter (Lagarias and Radke, 1989; Rovnak and Casey, 1999; Shimizu-Kohno et al., 2011; Demontis et al., 2015). It was initially concluded that the provirus is silent. However, *in situ* experiments showed that rare cells expressed large amounts of viral transcripts (Lagarias and Radke, 1989).

Evidence 3: Sense Transcription Can Be Activated by Various Stimuli

The main regulatory element of the 5′ LTR promoter that is activated by Tax is a 21 bp enhancer that interacts with CREB/ATF transcription factors (Suzuki et al., 1993; Adam et al., 1996). This complex activates sense transcription when cells are isolated *ex vivo* from HTLV-1 carriers or BLV-infected sheep. This reactivation can further be increased by various stimuli such as polyclonal activators, HDAC inhibitors or oxygen deprivation (Kerkhofs et al., 1996; Achachi et al., 2005; Tajima and Aida, 2005; Lezin et al., 2007; Olindo et al., 2011; Kulkarni et al., 2017).

Evidence 4: Immunity against Most Viral Antigens Is Extremely Efficient While HBZ Is Poorly Immunogenic

Persistent infection by BLV and HTLV-1 is characterized by a permanent and vigorous immunity against viral antigens (Kannagi et al., 2005; Burbelo et al., 2008; Bangham et al., 2009; Kattan et al., 2009). This immune response efficiently controls viral replication *in vivo* as demonstrated in the BLV model (Florins et al., 2006, 2009, 2011; Gillet et al., 2013). Among viral proteins, Tax is the immunodominant HTLV-1 antigen in the T-cell response (Kannagi et al., 1991; Goon et al., 2004). In contrast, the HBZ protein is very poorly immunogenic and expressed at very low levels in infected cells (Suemori et al., 2009; Enose-Akahata et al., 2013; Rowan et al., 2014; Raval et al., 2015; Baratella et al., 2017). Humoral immunity against HBZ protein is indeed particularly weak (Raval et al., 2015; Shiohama et al., 2016). However, cytotoxic T cells specific to HBZ but not to the immunodominant Tax are the most effective in the control of HTLV-1 (MacNamara et al., 2010). These results thus focus attention on the extremely low expression and low immunogenicity of HBZ.

Evidence 5: Antisense Transcripts and microRNAs Are Abundantly and Permanently Expressed in Tumors

In contrast to 5′ LTR directed transcription, antisense RNA synthesis initiating at the 3′ LTR is consistently identified in primary ATL cells and BLV tumors (Usui et al., 2008; Saito et al., 2009; Durkin et al., 2016). In contrast to HTLV-1, BLV also abundantly transcribes a cluster of microRNAs from internal pol III promoters (Kincaid et al., 2012; Rosewick et al., 2013; Van Driessche et al., 2016). These microRNAs are required for efficient viral replication and induction of pathogenesis (Gillet et al., 2016).

Evidence 6: HBZ and AS RNAs Are Mainly Localized in the Nucleus Suggesting a Role in Epigenetics

The HBZ RNA is mainly localized in the nucleus, consistent with a low rate of translation (Hivin et al., 2005; Rende et al., 2011; Shimizu-Kohno et al., 2011; Li et al., 2012; Raval et al., 2015). Although the function of the HBZ protein has been clearly evidenced, the dominant nuclear localization of the HBZ RNA thus suggests other regulatory roles such as for example epigenetic modulation of gene expression. In BLV, the antisense transcripts are not predicted to be translated but are rather primarily retained in the nucleus, hinting at a lncRNA-like role (Durkin et al., 2016).

Evidence 7: Untranslated HBZ RNA Has a Function

The oncogenic role of the HBZ RNA was revealed by an untranslatable HBZ RNA (i.e., devoid of initiation codon) able to induce the proliferation of a human IL-2-dependent T-cell line (Kit225) (Satou et al., 2006). Microarray expression analysis reveals that HBZ RNA and protein differentially modulate the transcription of host genes. HBZ RNA activates the transcription of survivin and cell-cycle related genes (Mitobe et al., 2015). Thus, the HBZ gene has bimodal functions in two different molecular forms as a polypeptide and a ribonucleic acid. Notwithstanding important activities as protein, the scarcity of the HBZ polypeptide in tumor cells contrasts with the expression of the HBZ RNA. Therefore, main functions of HBZ are exerted by its RNA form. Although the level of expression of HBZ protein is unquestionably very low, the protein must nevertheless play an essential role in the life-cycle of the virus because the coding sequence has been conserved during evolution in presence of a protective host immune response.

Evidence 8: Alternate Transcription of Sense and Antisense Transcription

Single cell analysis shows that the Tax RNA is expressed in bursts and is exported from the nucleus, whereas the majority of hbz RNA is retained (Billman et al., 2017). Time-lapse imaging of destabilized enhanced green fluorescent protein indicates that Tax expression is transient, fluctuates between on/off states and is detected only in HBZ-negative cells (Jasunaga and Matsuoka, 2017; Mahgoub et al., 2017). Viral persistence is thus characterized by successive cycles of sense/antisense transcription. Evidence from single-molecular RNA-FISH nevertheless indicates a more complex relationship between expression of the two strands that are not transcribed in strict alternation (Billman et al., 2017).

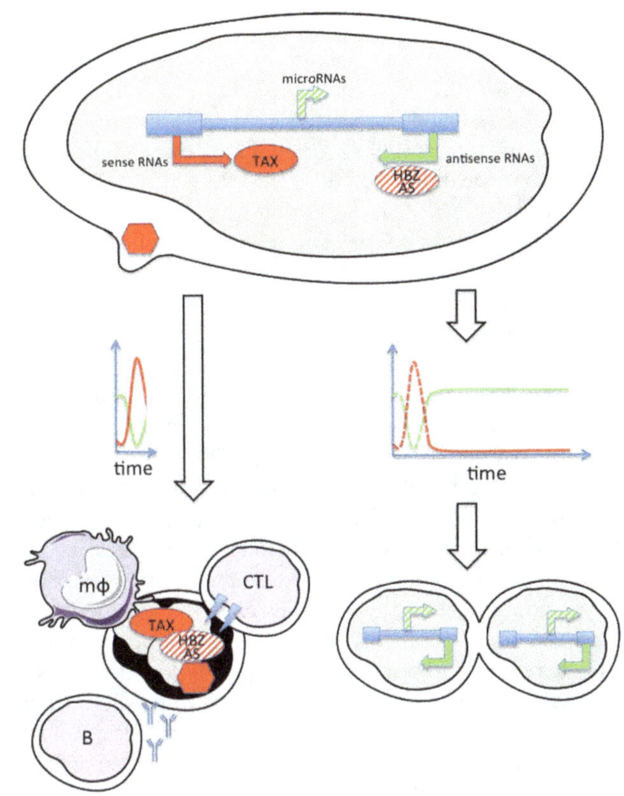

FIGURE 1 | Hypothetical model combining temporal regulation of viral expression allowing escape from host immunity. Sense transcription (red arrow) directed by the 5′ LTR promoter allows translation of the TAX protein (red ellipse) and synthesis of the viral particle (red hexagon). The antisense strand is transcribed in a HBZ (HTLV-1) and (AS) BLV RNA (green arrow) and protein (ellipse hatched in red). BLV, but not HTLV, also encodes RNA polymerase III driven microRNAs (hatched green arrow). Burst of sense transcription (red line on the time graph at the left) transiently exposes infected cells to the immune response: cytotoxic T cells (CTL), antibodies from B-lymphocytes (B) and macrophages (mϕ). Silencing of sense transcription, for example by initiating antisense RNA synthesis (graph on the right) allows further cell proliferation.

A MODEL FOR VIRAL PERSISTENCE UNDER IMMUNE CONTROL

These experimental evidences are consistent with the model presented in **Figure 1** and support Yoshida's paradigm.

BLV/HTLV proviruses are transcribed in both orientations. Sense transcription from the 5′ LTR (red arrow) generates genomic and spliced subgenomic RNAs (e.g., Tax). Translation of these RNAs into oncogenic proteins (Tax) triggers cell proliferation, mitosis, and clonal expansion of the infected cell. Synthesis of structural (gag, env) and enzymatic (protease, reverse transcriptase, and integrase) proteins are required for assembly of the viral particle (red hexagon) that will further colonize new target cells. The provirus is also transcribed in the antisense orientation from the 3′ LTR (green arrow) yielding HBZ and AS RNAs for HTLV and BLV, respectively. HBZ is very poorly translated while the coding potential of AS is ambiguous (hatched in red). Furthermore, the immunogenicity of HBZ is weak compared to all other viral antigens. BLV, but not HTLV, also encodes RNA polymerase III driven microRNAs (hatched green arrow). RNA synthesis from the 5′ LTR is mostly silent but burst of sense transcription (red line on the time graph) transiently exposes infected cells to the immune response (e.g., cytotoxic, humoral, and innate). The only option that allows survival is to silence sense transcription, for example by initiating antisense RNA synthesis (graph on the right). This simplified model is nevertheless incomplete because the infected cell is also exposed to the CTL response to HBZ providing that the protein is expressed. Experimental evidence indicates that the HBZ RNA is mostly nuclear and is therefore not translated. Single cell kinetics of RNA and protein expression would answer to this still unsolved question. Another issue is the role of the intrinsic immunity operating within virus-infected cells, i.e., restriction factors (RFs) inhibiting Tax function or reverse transcription of viral genome (Tosi et al., 2011; Bai and Nicot, 2015; Forlani et al., 2016).

CONCLUSION

Yoshida predicted that the Tax oncogene should be expressed in cycles to allow cell survival. Recent reports describing bursts between sense and antisense transcriptions are consistent with this hypothesis. Regular switches between 5′ and 3′ transcription indeed allows transient Tax expression and fast silencing of viral expression. This mechanism would allow Tax-driven cell proliferation and synthesis of viral particles in presence of the host immunity. This model thus illustrates the dynamic equilibrium between a virus attempting to proliferate under a tight control exerted by the immune response.

AUTHOR CONTRIBUTIONS

All authors have contributed equally to this work. LW drafted the manuscript. All authors corrected, edited, and approved the paper.

FUNDING

This work was supported by the "Fonds National de la Recherche Scientifique" (FNRS), the Télévie, the Interuniversity Attraction Poles (IAP) Program "Virus-host interplay at the early phases of infection" BELVIR initiated by the Belgian Science Policy Office, the Belgian Foundation against Cancer (FBC), the "Centre Anticancéreux près ULg" (CAC), the "Subside Fédéral de Soutien à la Recherche and AgricultureIsLife" projects of Gembloux Agrobiotech (GxABT), the "ULg Fonds Spéciaux pour la Recherche" and the "Plan Cancer" of the "Service Public Fédéral." HG and RS (Télévie), CH (FNRS research fellow), and LW (Research Director) are members of the FNRS. PC was a Marie Curie COFUND postdoctoral fellow supported by European Union. LL and MH previously received support from AgricultureIsLife. MH and PC are presently fellows of the FBC. Nathalie Renotte and Jean-Rock Jacques provided skillful technical help.

ACKNOWLEDGMENTS

The authors thank the GIGA technology platforms.

REFERENCES

Achachi, A., Florins, A., Gillet, N., Debacq, C., Urbain, P., Foutsop, G. M., et al. (2005). Valproate activates bovine leukemia virus gene expression, triggers apoptosis, and induces leukemia/lymphoma regression in vivo. *Proc. Natl. Acad. Sci. U.S.A.* 102, 10309–10314. doi: 10.1073/pnas.0504248102

Adam, E., Kerkhofs, P., Mammerickx, M., Burny, A., Kettmann, R., and Willems, L. (1996). The CREB, ATF-1, and ATF-2 transcription factors from bovine leukemia virus-infected B lymphocytes activate viral expression. *J. Virol.* 70, 1990–1999.

Asquith, B., Zhang, Y., Mosley, A. J., de Lara, C. M., Wallace, D. L., Worth, A., et al. (2007). In vivo T lymphocyte dynamics in humans and the impact of human T-lymphotropic virus 1 infection. *Proc. Natl. Acad. Sci. U.S.A.* 104, 8035–8040. doi: 10.1073/pnas.0608832104

Bai, X. T., and Nicot, C. (2015). MiR-28-3p is a cellular restriction factor that inhibits human T cell leukemia virus, type 1 (HTLV-1) replication and virus infection. *J. Biol. Chem.* 290, 5381–5390. doi: 10.1074/jbc.M114.626325

Bangham, C. R. M. (2017). Human T cell leukemia virus type 1: persistence and pathogenesis. *Annu. Rev. Immunol.* 15, 25–53. doi: 10.1146/annurev-immunol-042617-053222

Bangham, C. R. M., Meekings, K., Toulza, F., Nejmeddine, M., Majorovits, E., Asquith, B., et al. (2009). The immune control of HTLV-1 infection: selection forces and dynamics. *Front. Biosci.* 14:2889–2903. doi: 10.2741/3420

Baratella, M., Forlani, G., Raval, G. U., Tedeschi, A., Gout, O., Gessain, A., et al. (2017). Cytoplasmic localization of HTLV-1 HBZ protein: a biomarker of HTLV-1-associated myelopathy/tropical spastic paraparesis (HAM/TSP). *PLOS Negl. Trop. Dis.* 11:e0005285. doi: 10.1371/journal.pntd.0005285

Barez, P. Y., de Brogniez, A., Carpentier, A., Gazon, H., Gillet, N., Gutiérrez, G., et al. (2015). Recent advances in BLV research. *Viruses* 7, 6080–6088. doi: 10.3390/v7112929

Billman, M. R., Rueda, D., and Bangham, C. R. M. (2017). Single-cell heterogeneity and cell-cycle-related viral gene bursts in the human leukaemia virus HTLV-1. *Wellcome Open Res.* 2:87. doi: 10.12688/wellcomeopenres.12469.1

Boxus, M., Twizere, J.-C., Legros, S., Dewulf, J.-F., Kettmann, R., and Willems, L. (2008). The HTLV-1 Tax interactome. *Retrovirology* 5:76. doi: 10.1186/1742-4690-5-76

Burbelo, P. D., Meoli, E., Leahy, H. P., Graham, J., Yao, K., Oh, U., et al. (2008). Anti-HTLV antibody profiling reveals an antibody signature for HTLV-I-associated myelopathy/tropical spastic paraparesis (HAM/TSP). *Retrovirology* 5:96. doi: 10.1186/1742-4690-5-96

Carpentier, A., Gazon, H., de Brogniez, A., Willems, L., Carpentier, A., Barez, P.-Y., et al. (2015). Modes of human T cell leukemia virus type 1 transmission, replication and persistence. *Viruses* 7, 3603–3624. doi: 10.3390/v7072793

Debacq, C., Asquith, B., Kerkhofs, P., Portetelle, D., Burny, A., Kettmann, R., et al. (2002). Increased cell proliferation, but not reduced cell death, induces lymphocytosis in bovine leukemia virus-infected sheep. *Proc. Natl. Acad. Sci. U.S.A.* 99, 10048–10053. doi: 10.1073/pnas.142100999

Debacq, C., Gillet, N., Asquith, B., Sanchez-Alcaraz, M. T., Florins, A., Boxus, M., et al. (2006). Peripheral blood B-cell death compensates for excessive proliferation in lymphoid tissues and maintains homeostasis in bovine leukemia virus-infected sheep. *J. Virol.* 80, 9710–9719. doi: 10.1128/JVI.01022-06

Demontis, M. A., Sadiq, M. T., Golz, S., and Taylor, G. P. (2015). HTLV-1 viral RNA is detected rarely in plasma of HTLV-1 infected subjects. *J. Med. Virol.* 87, 2130–2134. doi: 10.1002/jmv.24264

Durkin, K., Rosewick, N., Artesi, M., Hahaut, V., Griebel, P., Arsic, N., et al. (2016). Characterization of novel Bovine Leukemia Virus (BLV) antisense transcripts by deep sequencing reveals constitutive expression in tumors and transcriptional interaction with viral microRNAs. *Retrovirology* 13:33. doi: 10.1186/s12977-016-0267-8

Enose-Akahata, Y., Abrams, A., Massoud, R., Bialuk, I., Johnson, K. R., Green, P. L., et al. (2013). Humoral immune response to HTLV-1 basic leucine zipper factor (HBZ) in HTLV-1-infected individuals. *Retrovirology* 10:19. doi: 10.1186/1742-4690-10-19

Florins, A., de Brogniez, A., Elemans, M., Bouzar, A.-B., Francois, C., Reichert, M., et al. (2011). Viral expression directs the fate of B cells in bovine leukemia virus-infected sheep. *J. Virol.* 86, 621–624. doi: 10.1128/JVI.05718-11

Florins, A., Gillet, N., Asquith, B., Boxus, M., Burteau, C., Twizere, J., et al. (2007). Cell dynamics and immune response to BLV infection: a unifying model. *Front. Biosci.* 12, 1520–1531. doi: 10.2741/2165

Florins, A., Gillet, N., Asquith, B., Debacq, C., Jean, G., Schwartz-Cornil, I., et al. (2006). Spleen-dependent turnover of CD11b peripheral blood B lymphocytes in bovine leukemia virus-infected sheep. *J. Virol.* 80, 11998–12008. doi: 10.1128/JVI.01447-06

Florins, A., Reichert, M., Asquith, B., Bouzar, A. B., Jean, G., François, C., et al. (2009). Earlier onset of δ-retrovirus-induced leukemia after splenectomy. *PLOS ONE* 4:e6943. doi: 10.1371/journal.pone.0006943

Forlani, G., Abdallah, R., Accolla, R. S., and Tosi, G. (2016). The MHC class II transactivator CIITA inhibits the persistent activation of NF-kB by human T cell lymphotropic virus type-1 Tax-1 oncoprotein. *J. Virol.* 90, 3708–3721. doi: 10.1128/JVI.03000-15

Gessain, A., and Cassar, O. (2012). Epidemiological aspects and world distribution of HTLV-1 infection. *Front. Microbiol.* 3:388. doi: 10.3389/fmicb.2012.00388

Gillet, N. A., Gutiérrez, G., Rodriguez, S. M., de Brogniez, A., Renotte, N., Alvarez, I., et al. (2013). Massive depletion of bovine leukemia virus proviral clones located in genomic transcriptionally active sites during primary infection. *PLOS Pathog.* 9:e1003687. doi: 10.1371/journal.ppat.1003687

Gillet, N. A., Hamaidia, M., de Brogniez, A., Gutiérrez, G., Renotte, N., Reichert, M., et al. (2016). Bovine leukemia virus small noncoding RNAs are functional elements that regulate replication and contribute to oncogenesis in vivo. *PLOS Pathog.* 12:e1005588. doi: 10.1371/journal.ppat.1005588

Goon, P. K. C., Biancardi, A., Fast, N., Igakura, T., Hanon, E., Mosley, A. J., et al. (2004). Human T cell lymphotropic virus (HTLV) type-1-specific CD8 + T cells: frequency and immunodominance hierarchy. *J. Infect. Dis.* 189, 2294–2298. doi: 10.1086/420832

Gutierrez, G., Rodríguez, S. M., De Brogniez, A., Gillet, N., Golime, R., Burny, A., et al. (2014). Vaccination against δ-retroviruses: the bovine leukemia virus paradigm. *Viruses* 6, 2416–2427. doi: 10.3390/v6062416

Hivin, P., Frédéric, M., Arpin-André, C., Basbous, J., Gay, B., Thébault, S., et al. (2005). Nuclear localization of HTLV-I bZIP factor (HBZ) is mediated by three distinct motifs. *J. Cell Sci.* 118, 1355–1362. doi: 10.1242/jcs.01727

Jasunaga, J., and Matsuoka, M. (2017). Oncogenic spiral by infectious pathogens: the cooperation of multiple factors in cancer development. *Cancer Sci.* doi: 10.1111/cas.13443 [Epub ahead of print].

Kannagi, M., Harada, S., Maruyama, I., Inoko, H., Igarashi, H., Kuwashima, G., et al. (1991). Predominant recognition of human T cell leukemia virus type I (HTLV-I) pX gene products by human CD8 + cytotoxic T cells directed against HTLV-I-infected cells. *Int. Immunol.* 3, 761–767. doi: 10.1093/intimm/3.8.761

Kannagi, M., Harashima, N., Kurihara, K., Ohashi, T., Utsunomiya, A., Tanosaki, R., et al. (2005). Tumor immunity against adult T-cell leukemia. *Cancer Sci.* 96, 249–255. doi: 10.1111/j.1349-7006.2005.00050.x

Kattan, T., MacNamara, A., Rowan, A. G., Nose, H., Mosley, A. J., Tanaka, Y., et al. (2009). The avidity and lytic efficiency of the CTL response to HTLV-1. *J. Immunol.* 182, 5723–5729. doi: 10.4049/jimmunol.0900069

Kerkhofs, P., Adam, E., Droogmans, L., Portetelle, D., Mammerickx, M., Burny, A., et al. (1996). Cellular pathways involved in the ex vivo expression of bovine leukemia virus. *J. Virol.* 70, 2170–2177.

Kincaid, R. P., Burke, J. M., and Sullivan, C. S. (2012). RNA virus microRNA that mimics a B-cell oncomiR. *Proc. Natl. Acad. Sci. U.S.A.* 109, 3077–3082. doi: 10.1073/pnas.1116107109

Kulkarni, A., Mateus, M., Thinnes, C. C., McCullagh, J. S., Schofield, C. J., Taylor, G. P., et al. (2017). Glucose metabolism and oxygen availability govern reactivation from latency of the human retrovirus HTLV-1. *Cell Chem. Biol.* 24, 1377.e–1387.e. doi: 10.1016/j.chembiol.2017.08.016

Lagarias, D. M., and Radke, K. (1989). Transcriptional activation of bovine leukemia virus in blood cells from experimentally infected, asymptomatic sheep with latent infections. *J. Virol.* 63, 2099–2107.

Lezin, A., Gillet, N., Olindo, S., Signaté, A., Grandvaux, N., Verlaeten, O., et al. (2007). Histone deacetylase mediated transcriptional activation reduces proviral loads in HTLV-1 associated myelopathy/tropical spastic paraparesis patients. *Blood* 110, 3722–3728. doi: 10.1182/blood-2007-04-085076

Li, M., Kannian, P., Yin, H., Kesic, M., and Green, P. L. (2012). Human T lymphotropic virus type 1 regulatory and accessory gene transcript expression and export are not rex dependent. *AIDS Res. Hum. Retroviruses* 28, 405–410. doi: 10.1089/aid.2011.0130

Ma, G., Yasunaga, J.-I., and Matsuoka, M. (2016). Multifaceted functions and roles of HBZ in HTLV-1 pathogenesis. *Retrovirology* 13:16. doi: 10.1186/s12977-016-0249-x

MacNamara, A., Rowan, A., Hilburn, S., Kadolsky, U., Fujiwara, H., Suemori, K., et al. (2010). HLA class I binding of HBZ determines outcome in HTLV-1 infection. *PLOS Pathog.* 6:e1001117. doi: 10.1371/journal.ppat.1001117

Mahgoub, M., Yasunaga, J., Shimura, K., Furuta, R., and Matsuoka, M. (2017). "Transient tax expression is essential for ATL survival: insight based on single cell analysis," in *Proceedings of the 18th International Conference on Human Retrovirology HTLV and Related Viruses*, Tokyo.

Matsuoka, M., and Jeang, K.-T. (2007). Human T-cell leukaemia virus type 1 (HTLV-1) infectivity and cellular transformation. *Nat. Rev. Cancer* 7, 270–280. doi: 10.1038/nrc2111

Mitobe, Y., Yasunaga, J. I., Furuta, R., and Matsuoka, M. (2015). HTLV-1 bZIP Factor RNA and protein impart distinct functions on t-cell proliferation and survival. *Cancer Res.* 75, 4143–4152. doi: 10.1158/0008-5472.CAN-15-0942

Nagai, M., Kubota, R., Greten, T. F., Schneck, J. P., Leist, T. P., Jacobson, S., et al. (2001). Increased activated human T cell lymphotropic virus type I (HTLV-I) Tax11-19-specific memory and effector CD8 + cells in patients with HTLV-I-associated myelopathy/tropical spastic paraparesis: correlation with HTLV-I provirus load. *J. Infect. Dis.* 183, 197–205. doi: 10.1086/317932

Olindo, S., Belrose, G., Gillet, N., Rodriguez, S., Boxus, M., Verlaeten, O., et al. (2011). Brief report safety of long-term treatment of HAM/TSP patients with valproic acid. *Blood* 118, 6306–6309. doi: 10.1182/blood-2011-04-349910

Polat, M., Takeshima, S., and Aida, Y. (2017). Epidemiology and genetic diversity of bovine leukemia virus. *Virol. J.* 14, 209. doi: 10.1186/s12985-017-0876-4

Raval, G. U., Bidoia, C., Forlani, G., Tosi, G., Gessain, A., and Accolla, R. S. (2015). Localization, quantification and interaction with host factors of endogenous HTLV-1 HBZ protein in infected cells and ATL. *Retrovirology* 12:59. doi: 10.1186/s12977-015-0186-0

Rende, F., Cavallari, I., Corradin, A., Silic-Benussi, M., Toulza, F., Toffolo, G. M., et al. (2011). Kinetics and intracellular compartmentalization of HTLV-1 gene expression: nuclear retention of HBZ mRNAs. *Blood* 117, 4855–4859. doi: 10.1182/blood-2010-11-316463

Rodríguez, S. M., Florins, A., Gillet, N., de Brogniez, A., Sánchez-Alcaraz, M. T., Boxus, M., et al. (2011). Preventive and therapeutic strategies for bovine leukemia virus: lessons for HTLV. *Viruses* 3, 1210–1248. doi: 10.3390/v3071210

Rosewick, N., Momont, M., Durkin, K., Takeda, H., Caiment, F., Cleuter, Y., et al. (2013). Deep sequencing reveals abundant noncanonical retroviral microRNAs in B-cell leukemia/lymphoma. *Proc. Natl. Acad. Sci. U.S.A.* 110, 2306–2311. doi: 10.1073/pnas.1213842110

Rovnak, J., and Casey, J. W. (1999). Assessment of bovine leukemia virus transcripts in vivo. *J. Virol.* 73, 8890–8897.

Rowan, A. G., Suemori, K., Fujiwara, H., Yasukawa, M., Tanaka, Y., Taylor, G. P., et al. (2014). Cytotoxic T lymphocyte lysis of HTLV-1 infected cells is limited by weak HBZ protein expression, but non-specifically enhanced on induction of Tax expression. *Retrovirology* 11:116. doi: 10.1186/s12977-014-0116-6

Saito, M., Matsuzaki, T., Satou, Y., Yasunaga, J.-I., Saito, K., Arimura, K., et al. (2009). In vivo expression of the HBZ gene of HTLV-1 correlates with proviral load, inflammatory markers and disease severity in HTLV-1 associated myelopathy/tropical spastic paraparesis (HAM/TSP). *Retrovirology* 6:19. doi: 10.1186/1742-4690-6-19

Satou, Y., Yasunaga, J., Yoshida, M., and Matsuoka, M. (2006). HTLV-I basic leucine zipper factor gene mRNA supports proliferation of adult T cell leukemia cells. *Proc. Natl. Acad. Sci. U.S.A.* 103, 720–725. doi: 10.1073/pnas.0507631103

Satou, Y., Yasunaga, J., Zhao, T., Yoshida, M., Miyazato, P., Takai, K., et al. (2011). HTLV-1 bZIP factor induces T-cell lymphoma and systemic inflammation in vivo. *PLOS Pathog.* 7:e1001274. doi: 10.1371/journal.ppat.1001274

Shimizu-Kohno, K., Satou, Y., Arakawa, F., Kiyasu, J., Kimura, Y., Niino, D., et al. (2011). Detection of HTLV-1 by means of HBZ gene in situ hybridization in formalin-fixed and paraffin-embedded tissues. *Cancer Sci.* 102, 1432–1436. doi: 10.1111/j.1349-7006.2011.01946.x

Shiohama, Y., Naito, T., Matsuzaki, T., Tanaka, R., Tomoyose, T., Takashima, H., et al. (2016). Absolute quantification of HTLV-1 basic leucine zipper factor (HBZ) protein and its plasma antibody in HTLV-1 infected individuals with different clinical status. *Retrovirology* 13:29. doi: 10.1186/s12977-016-0263-z

Suemori, K., Fujiwara, H., Ochi, T., Ogawa, T., Matsuoka, M., Matsumoto, T., et al. (2009). HBZ is an immunogenic protein, but not a target antigen for human T-cell leukemia virus type 1-specific cytotoxic T lymphocytes. *J. Gen. Virol.* 90, 1806–1811. doi: 10.1099/vir.0.010199-0

Suzuki, T., Fujisawa, J. I., Toita, M., and Yoshida, M. (1993). The trans-activator tax of human T-cell leukemia virus type 1 (HTLV-1) interacts with cAMP-responsive element (CRE) binding and CRE modulator proteins that bind to the 21-base-pair enhancer of HTLV-1. *Proc. Natl. Acad. Sci. U.S.A.* 90, 610–614. doi: 10.1073/pnas.90.2.610

Tajima, S., and Aida, Y. (2005). Induction of expression of bovine leukemia virus (BLV) in blood taken from BLV-infected cows without removal of plasma. *Microbes Infect.* 7, 1211–1216. doi: 10.1016/j.micinf.2005.04.010

Tosi, G., Forlani, G., Andresen, V., Turci, M., Bertazzoni, U., Franchini, G., et al. (2011). Major histocompatibility complex class II transactivator CIITA is a viral restriction factor that targets human T-cell lymphotropic virus type 1 Tax-1 function and inhibits viral replication. *J. Virol.* 85, 10719–10729. doi: 10.1128/JVI.00813-11

Twizere, J.-C., Kruys, V., Lefèbvre, L., Vanderplasschen, A., Collete, D., Debacq, C., et al. (2003). Interaction of retroviral Tax oncoproteins with tristetraprolin and regulation of tumor necrosis factor-alpha expression. *J. Natl. Cancer Inst.* 95, 1846–1859. doi: 10.1093/jnci/djg118

Usui, T., Yanagihara, K., Tsukasaki, K., Murata, K., Hasegawa, H., Yamada, Y., et al. (2008). Characteristic expression of HTLV-1 basic zipper factor (HBZ) transcripts in HTLV-1 provirus-positive cells. *Retrovirology* 5:34. doi: 10.1186/1742-4690-5-34

Van Driessche, B., Rodari, A., Delacourt, N., Fauquenoy, S., Vanhulle, C., Burny, A., et al. (2016). Characterization of new RNA polymerase III and RNA polymerase II transcriptional promoters in the Bovine Leukemia Virus genome. *Sci. Rep.* 6:31125. doi: 10.1038/srep31125

Watanabe, T. (2017). Adult T-cell leukemia (ATL): molecular basis for clonal expansion and transformation of HTLV-1-infected T cells. *Blood* 129, 1071–1081. doi: 10.1182/blood-2016-09-692574

Willems, L., Hasegawa, H., Accolla, R., Bangham, C., Bazarbachi, A., Bertazzoni, U., et al. (2017). Reducing the global burden of HTLV-1 infection: an agenda for research and action. *Antiviral Res.* 137, 41–48. doi: 10.1016/j.antiviral.2016.10.015

Yoshida, M., and Seiki, M. (1987). Recent advances in the molecular biology of HTLV-1: trans-activation of viral and cellular genes. *Annu. Rev. Immunol.* 5, 541–557. doi: 10.1146/annurev.iy.05.040187.002545

3

Role of HTLV-1 Tax and HBZ in the Pathogenesis of HAM/TSP

*Yoshimi Enose-Akahata, Ashley Vellucci and Steven Jacobson**

Viral Immunology Section, Division of Neuroimmunology and Neurovirology, National Institute of Neurological Disorders and Stroke, National Institutes of Health, Bethesda, MD, United States

Correspondence:
Steven Jacobson
jacobsons@ninds.nih.gov

Human T cell lymphotropic virus type 1 (HTLV-1) infection can lead to development of adult T cell leukemia/lymphoma (ATL) or HTLV-1-associated myelopathy/tropical spastic paraparesis (HAM/TSP) in a subset of infected subjects. Understanding the interaction between host and HTLV-1 and the molecular mechanisms associated with disease pathogenesis is critical for development efficient therapies. Two HTLV-1 genes, *tax* and *HTLV-1 basic leucine zipper factor* (*HBZ*), have been demonstrated to play important roles in HTLV-1 infectivity and the growth and survival of leukemic cells. Increased HTLV-1 Tax expression induces the expression of various cellular genes such as IL-2 and IL-15, which directly contributes to lymphocyte activation and immunopathogenesis in HAM/TSP patients. However, little is known about the molecular and cellular mechanism of HBZ in development of HAM/TSP. It has been reported that *HBZ* mRNA expression was detected in HAM/TSP patients higher than in asymptomatic carriers and correlated with proviral load and disease severity. Unlike HTLV-1 tax, HBZ escapes efficient anti-viral immune responses and therefore these reactivities are difficult to detect. Thus, it is important to focus on understanding the function and the role of HTLV-1 tax and HBZ in disease development of HAM/TSP and discuss the potential use of these HTLV-1 viral gene products as biomarkers and therapeutic targets for HAM/TSP.

Keywords: HTLV-1, HAM/TSP, Tax, HBZ, neuroinflammatory disease

INTRODUCTION

Human T lymphotropic virus type 1 (HTLV-1) is the first human retrovirus discovered belonging to the deltaretrovirus family and is thought to infect approximately 10–20 million people worldwide (Poiesz et al., 1980; de The and Bomford, 1993). Several highly endemic areas for HTLV-1 are known in the world such as southern part of Japan, the Caribbean, North and South America, Central and West Africa, and foci in Middle East, Australia and Melanesia (Gessain and Cassar, 2012). HTLV-1 has been demonstrated to be the etiological agent of an aggressive mature T cell malignancy termed adult T cell leukemia (ATL) (Uchiyama, 1988) and a chronic, progressive neurological disease termed HTLV-1-associated myelopathy/tropic spastic paraparesis (HAM/TSP) (Gessain et al., 1985; Osame et al., 1986). Although the majority of infected individuals are asymptomatic carriers of the virus, approximately 0.25–3.8% of infected individuals develop HAM/TSP. HAM/TSP is clinically characterized by progressive lower extremity weakness, spasticity, and bladder/bowel sphincter dysfunction (Umehara et al., 1993). The disease development of HAM/TSP mainly occurs in adults, with a mean age at onset of 40–50 years, which is more common in women than in men (Yamano and Sato, 2012). The disease usually progresses slowly without remission, but the clinical course and rate of progression may vary greatly among patients (Yamano and Sato, 2012). HAM/TSP is also characterized by perivascular inflammatory infiltrates in the brain and spinal cord (Umehara et al., 1993; Aye et al., 2000). In early stages of HAM/TSP, infiltrating $CD4^+$ and $CD8^+$ lymphocytes are present in the inflammatory lesions in the spinal cord while $CD8^+$ T cells are predominantly detected in the chronic inflammatory lesions of patients with longer duration of disease (Umehara et al., 1993). The spinal cord is associated with a chronic inflammatory process with marked parenchymal exudation of inflammatory cells in both the gray and white matter, initially causing lower thoracic cord atrophy with extensive lateral and posterior column involvement (Iwasaki, 1990). As the disease progresses, axonal loss and degeneration in the form of myelin pallor is severe in the white matter and increases throughout the entire cord (Iwasaki, 1990). Recent studies of measuring spinal cord cross-sectional areas in HAM/TSP patients by magnetic resonance imaging (MRI) revealed that HAM/TSP patients had more atrophic cords compared to ACs and healthy normal donors, which was associated with disease duration (Liu et al., 2014; Azodi et al., 2017). In HAM/TSP patients who present with rapid progression, spinal cord atrophy has been demonstrated to begin in the thoracic cord and progress to the cervical cord (Azodi et al., 2017). A higher HTLV-1 proviral load (PVL) is frequently observed in the blood and the cells from cerebrospinal fluid (CSF) of HAM/TSP patients than ACs (Nagai et al., 2001b). High levels of antibodies against HTLV-1 antigens are present in blood and CSF (Gessain et al., 1985; Osame et al., 1986). Intrathecal HTLV-1-specific antibody production provides additional data to support the diagnosis of HAM/TSP. Thus, chronically activated immune responses against HTLV-1 and infiltration of inflammatory cells including HTLV-1 infected cells into the central nervous system (CNS) have been suggested to underlie the pathogenesis of HAM/TSP.

Like other retroviruses, the HTLV-1 proviral genome has structural genes, *gag*, *pol*, and *env* flanked by long terminal repeat at both ends. HTLV-1 genome also contains a *pX* region between *env* and 3′ LTR encoded several accessory genes including *tax*, *rex*, *p12*, *p21*, *p30*, *p13* and *HTLV-1 basic leucine zipper factor* (*HBZ*) (Matsuoka and Jeang, 2007). The viral genes are transcribed from the 5′ LTR, but only *HBZ* encoded on the minus strand of the provirus is transcribed from the 3′ LTR. HTLV-1 expresses a transcriptional *trans*-activator protein, Tax, which induce the expression of a various cellular genes. Increased expressions of critical immune mediators directly contribute to cell activation and proliferation observed in HAM/TSP patients, suggesting that chronically activated immune responses underlie the pathogenesis of this disorder (Matsuura et al., 2010). There is also increasing evidence that HBZ also plays a critical part in inflammation and pathogenesis of HAM/TSP. In this review, we discuss the understanding of the function and the role of HTLV-1 Tax and HBZ in disease development of HAM/TSP and the potential use of these HTLV-1 viral gene products as biomarkers and therapeutic targets for HAM/TSP.

HTLV-1 INFECTION AND HAM/TSP

HTLV-1 causes a persistent infection in humans and replicates mainly through clonal expansion of the infected cells rather than cell-free virus infection. HTLV-1 PVLs are significantly elevated in HAM/TSP patients, compared to ACs, and is strongly correlated with disease pathogenesis of HAM/TSP (Nagai et al., 1998). Higher PVLs are detected in CSF than in peripheral blood mononuclear cells (PBMCs) of HAM/TSP patients (Nagai et al., 2001b; Araya et al., 2014; Brunetto et al., 2014). In addition, PVL was significantly higher in CSF of HAM/TSP patients than in ACs and HTLV-1-infected subjects with other neurologic diseases (Puccioni-Sohler et al., 2007). HTLV-1 can infect a wide range of human cell types including T cells, monocytes, macrophages and dendritic cells (Hoffman et al., 1992; Koralnik et al., 1992; Koyanagi et al., 1993; Makino et al., 1999; Hanon et al., 2000c; Enose-Akahata et al., 2008). HTLV-1 provirus is predominantly detected in $CD4^+$ T cells *in vivo* (Richardson et al., 1990) and has been associated with leukemogenesis and reduced regulatory function of $CD4^+$ T cells (Uchiyama, 1988; Yamano et al., 2005). In HAM/TSP patients, $CD4^+CD25^+$ T cells are the main reservoir for HTLV-1 (Yamano et al., 2004). Further studies revealed that high HTLV-1 PVL was detected in $CD4^+CD25^+CCR4^+$ T cells and the frequency of IFN-γ-producing $CD4^+CD25^+CCR4^+$ T cells was dramatically increased in HAM/TSP patients, which was found to be correlated with disease activity and severity (Yamano et al., 2009; Araya et al., 2014). In addition, it has been demonstrated that abundant $CD4^+CCR4^+$ T cells which coexpressed the Th1 marker CXCR3 and produced T-bet and IFN-γ were present in CSF and spinal cord lesions of HAM/TSP patients (Araya et al., 2014). High CSF PVLs is a strong biomarker of HAM/TSP and HTLV-1-infected cells recruited into the CNS may alter the inflammatory milieu in the CNS of HAM/TSP patients.

HTLV-1 PVL varies widely among HTLV-1-infected subjects, but the PVL remains relatively stable within each subject. The HTLV-1 genome sequence is also stable within and between infected individuals. Integration of HTLV-1 was found to favor genes, transcriptional start sites, and CpG islands (Doi et al., 2005; Derse et al., 2007). Comparison of proviral integration sites between HTLV-1-infected subjects demonstrated that HTLV-1 integration might be more frequent in transcriptionally active areas of the genome in HAM/TSP patients than in ACs and that frequent integration into transcriptionally active area of the genome was associated with an increased rate of Tax expression (Meekings et al., 2008). Moreover, a larger number of unique insertion sites, but not a difference of clonality, was detected in HAM/TSP patients than in ACs (Gillet et al., 2011). Interestingly, the majority of spontaneous Tax expressing cells corresponded to the large number of low abundance clones, rather than a small number of high abundance clones (Melamed et al., 2013), suggesting that clonal expansion of infected cells might be controlled by host immune response to Tax or by other viral factor such as HBZ in HAM/TSP patients. These reports demonstrated that cells with HTLV-1 provirus integrated near transcriptionally active areas could establish and expand more frequently in HAM/TSP patients, which would influence expression of HTLV-1 gene products and further contribute to the development and pathogenesis of HAM/TSP.

MOLECULAR PATHOGENESIS OF TAX IN HAM/TSP

Tax is a transforming and transactivating protein of HTLV-1 and induces the expression of a variety of cellular genes by activation of the NF-κB and CREB/ATF pathways (Matsuoka and Jeang, 2011). HAM/TSP patients showed the spontaneous increase of *tax* mRNA and Tax protein expression in PBMCs after *ex vivo* culture without any exogenous stimulators that peaks at 12–24 h (Hanon et al., 2000a; Yamano et al., 2002). The expression of *tax* mRNA was significantly higher in HAM/TSP patients than in ACs (Yamano et al., 2002). Tax is prominently associated with dysregulation in immune cells of HAM/TSP patients, underlying many of the characteristic immune abnormalities.

Regulatory CD4$^+$ T Cells (Tregs)

In HAM/TSP patients, CD4$^+$CD25$^+$ T cells contain high frequency of HTLV-1 proviral DNA, express HTLV-1 *tax* mRNA at significantly higher levels than in CD4$^+$CD25$^-$ cells and induces various cytokines including IFN-γ (Yamano et al., 2004, 2009). CD4$^+$CD25$^+$ T cells, termed Tregs, that constitutively express CD25 (the IL-2 receptor α chain) and are engaged in the maintenance of immunologic self-tolerance by suppressing the activation and expansion of self-reactive lymphocytes that may cause autoimmune diseases (Sakaguchi et al., 2001). Although CD4$^+$CD25$^+$ T cells have an important role in suppression of T cell activation both *in vivo* and *in vitro*, HTLV-1-infected CD4$^+$CD25$^+$ T cells were not functionally suppressive but rather were shown to stimulate the proliferation of HTLV-1 Tax-specific CD8$^+$ T cells (Yamano et al., 2004). In HAM/TSP patients, the forkhead box P3 (FoxP3), which is critical for the function of Tregs, was decreased in CD4$^+$CD25$^+$ T cells (Yamano et al., 2005). When Tax was transduced in CD4$^+$CD25$^+$ T cells isolated from healthy volunteers, *foxp3* mRNA expression as well as regulatory function of the CD4$^+$CD25$^+$ T cells was inhibited (Yamano et al., 2005).

Other immune molecules related to Tregs were also dysregulated by Tax in HAM/TSP patients. The pleiotropic cytokine, transforming growth factor-β (TGF-β), play critical roles in suppressing the immune response, such as inhibition of inflammatory responses and promotion of Treg generation and function (Wan and Flavell, 2007). Dysregulation of TGF-β signaling has been reported in HAM/TSP patients, due to inhibition of TGF-β RII and Smad7 expression by Tax (Grant et al., 2008). OX40 is a member of the TNF co-stimulatory receptor family and is expressed on activated T cells. Costimulatory signals from OX40 promote proliferation and survival of effector and memory T cell population and also suppresses the differentiation and activity of Tregs (Croft et al., 2009). Tax has been demonstrated to transactivate OX40 (Higashimura et al., 1996). In HAM/TSP patients, OX40 was expressed in CD4$^+$ T cells depending on Tax expression after the culture (Saito et al., 2013). Higher levels of soluble OX40 was detected in the CSF of HAM/TSP patients with rapid progression, and OX40 was overexpressed in spinal cord infiltrating mononuclear cells in a clinically progressive HAM/TSP patient with a short duration of illness (Saito et al., 2013). Thus, decreased Treg function may cause immune dysregulation in HAM/TSP patients.

The Common γ Chain Family of Cytokines

The common γ chain family of cytokines including IL-2, IL-7, IL-9, IL-15, and IL-21 play an important role in lymphocyte proliferation, survival and function during immune responses and homeostasis. Tax has been shown to transactivate a number of the common γ chain family of cytokines and the receptors, such as IL-2/IL-2R, IL-9, IL-15/IL-15R, and IL-21/IL-21R (Cross et al., 1987; Siekevitz et al., 1987; Azimi et al., 1998; Mariner et al., 2001; Mizuguchi et al., 2009). One of the most striking features of the cellular immune response in HAM/TSP patients is the increased numbers of memory and/or effector CD8$^+$ T cells and also HTLV-1 Tax-specific cytotoxic CD8$^+$ T cells (Jacobson et al., 1990; Nagai et al., 2001a). Since both IL-2 and IL-15 induce the proliferation and increase the cytolytic activity of NK and CD8$^+$ T cells, it has been suggested that IL-2/IL-2R and IL-15/IL-15R autocrine loop may contribute to the pathogenesis of HAM/TSP (Azimi et al., 1999). In particular, IL-15 is critical for the development of NK cells and antigen-specific memory CD8$^+$ T cells and is well characterized for its role in maintaining memory pools of CD8$^+$ T cells. In HAM/TSP patients, *IL-15* mRNA level is up-regulated in non-T cells isolated from HAM/TSP patients and DC treated with HTLV-1 Tax protein *in vitro* (Azimi et al., 1999; Ahuja et al., 2006). IL-15 expression was also rapidly enhanced on the surface of CD14$^+$ cells in HAM/TSP patients after the PBMC culture, more than those in ACs (Enose-Akahata et al., 2008). In addition, high expression of IL-15Rα has been reported in HTLV-1 Tax-specific CD8$^+$ T cells, compared with CMV pp65-specific CD8$^+$ T cells (Azimi et al., 2001). IL-15 stimulated HTLV-1 Tax-specific CD8$^+$ T cells, but not CMV pp65-specific CD8$^+$ T cells, to induce degranulation and IFN-γ expression (Enose-Akahata et al., 2008). Thus, the increase of the common g chain family of cytokines and receptors in HAM/TSP patients may be involved in increased proliferation and enhanced cytolytic activity and inflammatory cytokine production of HTLV-1-specific CD8$^+$ T cells.

Increased levels of cytokines and the receptors driven by Tax further lead to upregulated protein expression and activated signaling cascades, such as JAK/STAT (Waldmann, 2015). IL-2/IL-15 binding to the β/γ chain complex results in heterodimerization of their cytoplasmic domains with activation of the Janus family tyrosine kinases, JAK1 and JAK3 (in association with β and γ chain, respectively). Activated JAK1 and JAK3 then phosphorylate signal transducer and activator of transcription proteins STAT3 and STAT5, respectively, to mediate IL-2 and IL-15 effects in immune cells (Waldmann, 2015). HAM/TSP patients showed increased STAT5 phosphorylation in T cells, which was inhibited by the blockade of IL-2Rα and IL-2/IL-15Rβ (Oh et al., 2011).

These reports demonstrated that continuous stimulation driven through HTLV-1 Tax may be associated with the pathogenesis of HAM/TSP.

The Blood–Brain Barrier (BBB)

The CNS is normally protected from infectious agents by a physiological structure called the blood–brain barrier (BBB), which consists primarily of a continuous endothelium with tight junctions. HAM/TSP develops upon infiltration of HTLV-1-infected lymphocytes into the CNS, mostly within the thoracic spinal cord. The tight junctions of the BBB endothelium in HAM/TSP patients are locally disorganized (Afonso et al., 2007, 2008). Matrix metalloproteinase (MMP) is known as a proteolytic enzyme, which is involved in degradation of many different components of extracellular matrix and is critical role in migration of leukocytes and damage of the BBB. It has been reported that high level of MMP-2 and MMP-9 was detected in the CSF and in infiltrating mononuclear cells of HAM/TSP patients, suggesting that MMP-2 and MMP-9 may cause disruption of BBB in HAM/TSP patients (Umehara et al., 1998). Later, MMP-9, but not MMP-2, was reported to be transactivated by Tax in HTLV-1-infected T cells (Mori et al., 2002). HTLV-1 infection has been also detected in astrocytes which interact with endothelial cells to form the BBB and may potentially also function as antigen-presenting cells (Lehky et al., 1995; Mendez et al., 1997). The expression of Tax in primary human astrocytomas and oligodendrogliomas resulted in robust induction of IL-1α, IL-1β, TNF-α, TNF-β, and IL-6 expression (Banerjee et al., 2007). These results suggested that increased inflammatory responses may cause disruption of BBB and the alteration of the BBB integrity may allow T cells to transmigrate into the CNS, resulting in neuroinflammation of HTLV-1-infected subjects.

Recruitment and extravasation of T cells through the BBB are favored by adhesion molecule-mediated interactions of circulating T cells with endothelial cells. Tax has been demonstrated to regulate cell adhesion molecules, such as intercellular adhesion molecule-1 (ICAM-1), vascular cell adhesion molecule 1 (VCAM-1) and cell adhesion molecule 1 (CADM1/TSLC1) (Valentin et al., 1997; Nejmeddine et al., 2009; Manivannan et al., 2016). Activated leukocyte cell adhesion molecule (ALCAM/CD166), a member of the immunoglobulin superfamily, is overexpressed on the surface of HTLV-1-infected lymphocytes, both in chronically infected cell lines and in primary $CD4^{+}CD25^{+}$ T cells from HAM/TSP patients (Curis et al., 2016). ALCAM expression was enhanced by Tax through the activation of the canonical NF-κB pathway (Curis et al., 2016). These results demonstrated that increase of ALCAM expression might facilitate the migration of HTLV-1-infected lymphocytes across the BBB endothelium.

Tax in CSF of HAM/TSP Patients

Although *tax* mRNA and Tax protein are rarely or undetectable directly in fresh uncultured PBMCs of HAM/TSP patients, it has been reported that HTLV-1 *tax* mRNA was detected in cells of spinal cord and cerebellar sections, and HTLV-1 Tax protein could be also detected in CSF cells of HAM/TSP patients (Lehky et al., 1995; Moritoyo et al., 1999; Cartier and Ramirez, 2005). The increased expression of HTLV-1 Tax protein in the CSF cells was more frequent in HAM/TSP patients with shorter duration of illness (Cartier and Ramirez, 2005). The presence of Tax protein in CSF might cause direct cell damage and loss in the CNS or immune cells to activate and generate Tax-specific immune responses in HAM/TSP patients. Axonal degeneration in HAM/TSP patients occurs without HTLV-1 infection of neurons, suggesting that secreted Tax protein might be involved. Tax from culture media of MT-2 and PBMCs of HAM/TSP patients caused retraction of differentiated human neuroblastoma cells (Maldonado et al., 2011; Medina et al., 2014). Tax expression also sensitized primary astrocytomas to apoptosis (Banerjee et al., 2007). These observations suggest that the chronic Tax secretion from infected cells could be sufficient for producing a neurotoxic effect on the long axons of corticospinal tracts involved in progressive neurological disease. Recent increasing evidences revealed that extracellular vesicles called exosomes play critical roles in viral pathogenesis and control of host immune responses to viral infection that deliver these microvesicles that contain host and viral components, including proteins, mRNA, and microRNA (Anderson et al., 2016). HTLV-1 has been shown to incorporate viral products into shed exosomes. Jaworski et al. (2014) found exosomes derived from HTLV-1-infected cells to contain Tax protein and proinflammatory mediators as well as viral mRNA transcripts, including Tax, HBZ, and Env. Once viral proteins or viral mRNAs enter or are released from exosomes in the CNS, these viral products might be able to stimulate or damage resident cells in the CNS or sensitize uninfected target cells for lysis by HTLV-1-specific $CD8^{+}$ T cells. These new findings suggest that incorporation of viral proteins and mRNAs into exosomes or alteration of host contents of immune cell derived exosomes may represent a mechanism by which viral antigens could be transported to the CNS and be associated with axonal degeneration and virus-specific immune responses in HAM/TSP.

MOLECULAR PATHOGENESIS OF HBZ IN HAM/TSP

HBZ encoded by the minus strand of the HTLV-1 proviral genome has been identified (Gaudray et al., 2002). While about 60% of ATL patients do not express the *tax* gene transcript in freshly isolated leukemic cells (Takeda et al., 2004), *HBZ* mRNA is ubiquitously expressed in HTLV-1-infected cells, ATL cells and PBMC of HTLV-1-infected individuals and promotes the growth and survival of the leukemic cells (Satou et al., 2006; Usui et al., 2008). HBZ interacts with CREB/ATF pathway, suppress Tax-mediated viral transcription, and selectively inhibits the classical NF-κB pathway (Matsuoka and Jeang, 2011). Previous *in vivo* studies also demonstrated that HBZ expression enhanced HTLV-1 infectivity, T cell proliferation and lymphoma (Arnold et al., 2006, 2008; Satou et al., 2011). While it has been reported that HBZ plays a crucial role in

HTLV-1 infectivity and the generation and maintenance of the oncogenic process, there is less information about the role of HBZ on the molecular and cellular mechanisms leading to HAM/TSP.

Localization of HBZ

The expression of *HBZ* mRNA was detected in PBMCs of HAM/TSP patients, which was significantly lower than in ATL patients but higher than in AC (Saito et al., 2009). Intriguingly, *HBZ* mRNA expression was correlated with PVL and disease severity in HAM/TSP patients (Saito et al., 2009). It remains unclear how expression of Tax (the plus strand transcription is minimized) and HBZ (constitutively expressed from the minus strand) is regulated in HTLV-1-infected cells, but there is increasing evidence that HBZ expression is detected in a different pattern or subset of $CD4^+$ T cells from Tax-expressing $CD4^+$ T cells. It has been demonstrated that high frequency of $CD39^+CD4^+$ T cells regardless of CD25 expression were detected in HAM/TSP patients compared to AC (Leal et al., 2013). When Tax and HBZ expression was examined in $CD4^+$ T cells expressing CD25 and CD39 of HAM/TSP patients, *HBZ* mRNA expression was significantly correlated with $CD39^+CD4^+$ T cells while Tax expression was restricted to CD25-expressing $CD4^+$ T cells regardless of CD39 expression (Leal et al., 2013). Moreover, a recent study using single molecule RNA fluorescent *in situ* hybridization targeted to the transcripts of *tax* or *HBZ* genes revealed that *tax* mRNA expression was enhanced in the absence of *HBZ* mRNA in patient-derived HTLV-1^+ T cell clones while *HBZ* mRNA expression was increased in cells with high *tax* mRNA expression (Billman et al., 2017). In addition, it has been demonstrated that at the single cell level, *HBZ* mRNA was expressed at much lower level, detected in more uniformly across cells compared to *tax* mRNA, and was not expressed in all T cell clones (Billman et al., 2017). These studies suggested that HBZ expression might be compartmentalized or cooperate with Tax expression in HTLV-1-infected T cells. A recent study demonstrated that CTCF, a zinc-finger protein and a key regulator of chromatin structure and function, bound to HTLV-1 and formed loops between proviral and host genes to regulate HTLV-1 proviral transcription and RNA splicing (Satou et al., 2016). Further molecular virological studies would be necessary to understand how the pattern of proviral transcription and latency is regulated in HTLV-1-infected individuals.

Unlike *HBZ* mRNA, HBZ protein was rarely detected in HAM/TSP patients (Shiohama et al., 2016). Previously, higher nuclear retention of *HBZ* mRNA and localization of HBZ in the nucleus have been demonstrated (Hivin et al., 2005; Rende et al., 2011), suggesting that HTLV-1 might favor viral persistence by reducing HBZ translation to escape the infected cells from HBZ-specific immune responses. However, HBZ protein has been recently reported to be localized in the cytoplasm of T cells depending on the expression of THEMIS (thymocyte-expressed molecule involved in selection), a recently identified T lineage-restricted protein (Kinosada et al., 2017). HBZ interfered complex formation of THEMIS with Grb2 and SHP-2, which resulted in inhibition of suppressive function of co-inhibitory receptors, such as T cell immunoglobulin and ITIM domain (TIGIT) and program death-1 (PD-1), and enhanced activation of T cells (Kinosada et al., 2017). Localization of HBZ protein was also reported in the cytoplasm of PBMCs of HAM/TSP patients (Baratella et al., 2017). Interestingly, HBZ protein is almost exclusively in $CD4^+$ T cells irrespective of co-expression of CD25 (Baratella et al., 2017). Thus, HBZ expression may be compartmentalized or cooperate with Tax expression in HTLV-1-infected $CD4^+$ T cells of HAM/TSP patients and may allow the virus to evade the host immune system. These findings of HBZ might shed light on a new molecular basis for a role of HBZ in the pathogenesis of HAM/TSP.

Inflammation by HBZ

As HBZ closely cooperates with Tax in many molecular mechanisms (Matsuoka and Jeang, 2011), opposing functions between HBZ and Tax have been also demonstrated in both *in vitro* and *in vivo* studies. HBZ enhanced TGF-β signaling and FoxP3 expression to induce Tregs from naïve $CD4^+$ T cells while Tax reduced TGF-β signaling and FoxP3 expression in $CD4^+$ T cells (Grant et al., 2008; Zhao et al., 2011). HBZ suppresses the *IFN-γ* gene transcription through inhibition of AP-1 and NFAT while Tax activates the *IFN-γ* gene promoter (Sugata et al., 2012). Interestingly, HBZ transgenic (HBZ-Tg) mice develop both T-cell lymphomas and chronic inflammation in lung and skin (Satou et al., 2011). Currently, there is no report that HBZ-Tg mice develop inflammatory neurologic diseases and it remains unknown how tissue specificity of HTLV-1-associated inflammatory diseases is determined, but similar immunological features with HAM/TSP patients have been demonstrated in HBZ-Tg mice. HBZ-Tg mice has been reported to show the increased effector/memory $CD4^+$ T cells while effector/memory $CD4^+$ T cells with high Tax-expression have been exhibited in HAM/TSP patients (Hanon et al., 2001; Satou et al., 2011). Further studies have demonstrated that Tregs of HBZ-Tg mice tend to lose expression of FoxP3 and Helios, leading to increased IFN-γ-expressing proinflammatory cells, associated with enhanced cell adhesion and migration of $CD4^+$ T cells of HBZ-Tg mice (Yamamoto-Taguchi et al., 2013). Importantly, decreased Helios expression and enhanced cell adhesion molecules observed in HBZ-Tg mice were also detected in $CD4^+$ T cells of HAM/TSP patients (Yamamoto-Taguchi et al., 2013). In addition, the conserved non-coding sequence 2 region of the *Foxp3* gene was hypermethylated in Tregs of HBZ-Tg mice, which is a characteristic of induced Tregs (Yamamoto-Taguchi et al., 2013). HAM/TSP patients also showed decreased demethylation of the *Foxp3* gene in $CD4^+CD25^+$ T cells, compared to NDs, which correlated with the decreased suppressive capacity of $CD4^+CD25^+$ T cells in HAM/TSP patients (Anderson et al., 2014). These results suggested that HBZ may be able to convert into the proinflammatory phenotype of HBZ-expressing T cells, suggesting that HBZ plays an important role in the disease process of HAM/TSP.

IMMUNE RESPONSE AGAINST TAX AND HBZ IN HAM/TSP

Tax-Specific Immune Responses

Tax is an immunodominant antigen recognized by HTLV-1-specific cytotoxic CD8$^+$ T cells (CTLs) (Jacobson et al., 1990). CD8$^+$ T cells play a crucial role in immunity against HTLV-1 through their ability to secrete various factors that suppress viral replication and kill infected target cells in HTLV-1-infected subjects (Hanon et al., 2000b; Vine et al., 2004). However, in HAM/TSP patients, the frequency of HTLV-1 Tax-specific CD8$^+$ T cells were even higher in CSF than in peripheral blood and were correlated with HTLV-1 PVL (Greten et al., 1998; Kubota et al., 1998; Nagai et al., 2001b). A recent study demonstrated that a more atrophic spinal cord in HAM/TSP was associated with higher percentage of inflammatory CD8$^+$ T cells and HTLV-1 PVL in CSF of HAM/TSP patients (Azodi et al., 2017). Moreover, it has been demonstrated that HTLV-1 Tax-specific CD8$^+$ T cells as well as CD4$^+$ T cells expressing HTLV-1 proteins were detected in the parenchyma of the spinal cords, suggesting that the interaction between HTLV-1-specific CTLs and HTLV-1-infected CD4$^+$ T cells may cause bystander damages to resident cells in the CNS (Matsuura et al., 2015).

While virus-specific antibodies play an important role in the control of viral infections in the CNS, intrathecal antibody synthesis appear to be associated with both protective and pathogenic function in chronic infection and immune-mediated diseases of the CNS. Intrathecal antibody synthesis against HTLV-1 has been also confirmed by the presence of HTLV-1-specific antibodies and oligoclonal IgG band in CSF of HAM/TSP patients (Gessain et al., 1988; Grimaldi et al., 1988; Link et al., 1989). Intrathecal antibody response to HTLV-1 inversely correlates with higher PVL and a worse prognostic outcome (Puccioni-Sohler et al., 1999). Moreover, antibodies against two HTLV-1 viral products, Tax and Gag p24, have been reported to cross-react with host antigens, heterogeneous ribonucleoprotein A1 (hnRNP A1) and peroxiredoxin-1 (PrX-1), respectively, suggesting that molecular mimicry may play a role in the pathogenesis of HAM/TSP, suggesting a role for molecular mimicry between an infectious agent and the CNS (Levin et al., 2002; Lee et al., 2008). Therefore, HTLV-1 Tax specific immune responses might be immunopathogenic, rather than protective, in HAM/TSP patients, due to high cytotoxicity, the production of inflammatory cytokines such as IFN-γ and TNF-α, associate with damage to the CNS.

HBZ-Specific Immune Responses

HBZ is also an immunogenic protein recognized by HBZ-specific CTL clones (Suemori et al., 2009; Macnamara et al., 2010). HBZ-specific CD8$^+$ T cells are detected in AC and HAM/TSP patients, and HBZ-specific CTL clones were able to lyse naturally infected cells isolated from AC and HAM/TSP patients, but not ATL patients (Suemori et al., 2009; Macnamara et al., 2010). However, the binding affinity of HBZ peptides to HLA class I molecules was found to be significantly weaker than that of peptides from Tax, and the frequency of HBZ-specific CD8$^+$ T cells was very low in peripheral blood (Macnamara et al., 2010; Hilburn et al., 2011).

Antibody response against HBZ was detected in serum/plasma of HTLV-1-infected subjects, but the frequency of the subjects with anti-HBZ antibody was about 10–16% in ACs, ATL, and HAM/TSP patients and did not discriminate between clinical status (Enose-Akahata et al., 2013; Shiohama et al., 2016). In addition, antibody responses against HBZ was detectable in the CSF of HAM/TSP patients, but was not dramatically elevated, suggesting that HBZ-specific antibody is not intrathecally synthesized (Enose-Akahata et al., 2013).

It has been suggested that the low frequency and affinity of HBZ-specific immune responses may be consequence of the low expression and antigenicity of HBZ in infected cells. Unlike Tax-specific immune responses, it might be difficult to induce effective HBZ-specific immune responses in HAM/TSP patients although the *HBZ* gene is constantly expressed while the *tax* gene is sporadically transcribed. Thus, HTLV-1-infected cells might be able to escape from host immune responses for long periods and result in persistence of HTLV-1 in infected individuals.

CONCLUSION

Both Tax and HBZ play critical roles in immune dysregulation in HAM/TSP. However, it is still unknown how expression of Tax and HBZ are regulated and how Tax and HBZ cooperate in naturally infected cells. Therapies that control the expression of HTLV-1 gene products might be effective in preventing the development of HAM/TSP. In addition, effective regulation or induction of HTLV-1 specific immune responses might improve the prognosis of patients with this disorder. Further studies will be necessary to identify the mechanism by which HTLV-1 gene products or other factors including HTLV-1-specific immune responses contribute to the pathogenesis of HAM/TSP.

AUTHOR CONTRIBUTIONS

YE-A contributed to paper writing. AV contributed to discussion. SJ supervised and contributed to discussion and writing.

ACKNOWLEDGMENT

This research was supported by the Intramural Research Program of the NIH, NINDS.

REFERENCES

Afonso, P. V., Ozden, S., Cumont, M. C., Seilhean, D., Cartier, L., Rezaie, P., et al. (2008). Alteration of blood-brain barrier integrity by retroviral infection. *PLOS Pathog.* 4:e1000205. doi: 10.1371/journal.ppat.1000205

Afonso, P. V., Ozden, S., Prevost, M. C., Schmitt, C., Seilhean, D., Weksler, B., et al. (2007). Human blood-brain barrier disruption by retroviral-infected lymphocytes: role of myosin light chain kinase in endothelial tight-junction disorganization. *J. Immunol.* 179, 2576–2583. doi: 10.4049/jimmunol.179.4.2576

Ahuja, J., Kampani, K., Datta, S., Wigdahl, B., Flaig, K. E., and Jain, P. (2006). Use of human antigen presenting cell gene array profiling to examine the effect of human T-cell leukemia virus type 1 Tax on primary human dendritic cells. *J. Neurovirol.* 12, 47–59. doi: 10.1080/13550280600614981

Anderson, M. R., Enose-Akahata, Y., Massoud, R., Ngouth, N., Tanaka, Y., Oh, U., et al. (2014). Epigenetic modification of the FoxP3 TSDR in HAM/TSP decreases the functional suppression of Tregs. *J. Neuroimmune Pharmacol.* 9, 522–532. doi: 10.1007/s11481-014-9547-z

Anderson, M. R., Kashanchi, F., and Jacobson, S. (2016). Exosomes in viral disease. *Neurotherapeutics* 13, 535–546. doi: 10.1007/s13311-016-0450-6

Araya, N., Sato, T., Ando, H., Tomaru, U., Yoshida, M., Coler-Reilly, A., et al. (2014). HTLV-1 induces a Th1-like state in CD4+CCR4+ T cells. *J. Clin. Invest.* 124, 3431–3442. doi: 10.1172/JCI75250

Arnold, J., Yamamoto, B., Li, M., Phipps, A. J., Younis, I., Lairmore, M. D., et al. (2006). Enhancement of infectivity and persistence in vivo by HBZ, a natural antisense coded protein of HTLV-1. *Blood* 107, 3976–3982. doi: 10.1182/blood-2005-11-4551

Arnold, J., Zimmerman, B., Li, M., Lairmore, M. D., and Green, P. L. (2008). Human T-cell leukemia virus type-1 antisense-encoded gene, Hbz, promotes T-lymphocyte proliferation. *Blood* 112, 3788–3797. doi: 10.1182/blood-2008-04-154286

Aye, M. M., Matsuoka, E., Moritoyo, T., Umehara, F., Suehara, M., Hokezu, Y., et al. (2000). Histopathological analysis of four autopsy cases of HTLV-I-associated myelopathy/tropical spastic paraparesis: inflammatory changes occur simultaneously in the entire central nervous system. *Acta Neuropathol.* 100, 245–252. doi: 10.1007/s004019900170

Azimi, N., Brown, K., Bamford, R. N., Tagaya, Y., Siebenlist, U., and Waldmann, T. A. (1998). Human T cell lymphotropic virus type I Tax protein trans-activates interleukin 15 gene transcription through an NF-kappaβ site. *Proc. Natl. Acad. Sci. U.S.A.* 95, 2452–2457. doi: 10.1073/pnas.95.5.2452

Azimi, N., Jacobson, S., Leist, T., and Waldmann, T. A. (1999). Involvement of IL-15 in the pathogenesis of human T lymphotropic virus type I-associated myelopathy/tropical spastic paraparesis: implications for therapy with a monoclonal antibody directed to the IL-2/15R beta receptor. *J. Immunol.* 163, 4064–4072.

Azimi, N., Nagai, M., Jacobson, S., and Waldmann, T. A. (2001). IL-15 plays a major role in the persistence of Tax-specific CD8 cells in HAM/TSP patients. *Proc. Natl. Acad. Sci. U.S.A.* 98, 14559–14564. doi: 10.1073/pnas.251540598

Azodi, S., Nair, G., Enose-Akahata, Y., Charlip, E., Vellucci, A., Cortese, I., et al. (2017). Imaging spinal cord atrophy in progressive myelopathies: HTLV-I-associated neurological disease (HAM/TSP) and multiple sclerosis (MS). *Ann. Neurol.* 82, 719–728. doi: 10.1002/ana.25072

Banerjee, P., Rochford, R., Antel, J., Canute, G., Wrzesinski, S., Sieburg, M., et al. (2007). Proinflammatory cytokine gene induction by human T-cell leukemia virus type 1 (HTLV-1) and HTLV-2 Tax in primary human glial cells. *J. Virol.* 81, 1690–1700. doi: 10.1128/JVI.01513-06

Baratella, M., Forlani, G., Raval, G. U., Tedeschi, A., Gout, O., Gessain, A., et al. (2017). Cytoplasmic localization of HTLV-1 HBZ protein: a biomarker of HTLV-1-associated myelopathy/tropical spastic paraparesis (HAM/TSP). *PLOS Negl. Trop. Dis.* 11:e0005285. doi: 10.1371/journal.pntd.0005285

Billman, M. R., Rueda, D., and Bangham, C. R. M. (2017). Single-cell heterogeneity and cell-cycle-related viral gene bursts in the human leukaemia virus HTLV-1. *Wellcome Open Res.* 2:87. doi: 10.12688/wellcomeopenres.12469.1

Brunetto, G. S., Massoud, R., Leibovitch, E. C., Caruso, B., Johnson, K., Ohayon, J., et al. (2014). Digital droplet PCR (ddPCR) for the precise quantification of human T-lymphotropic virus 1 proviral loads in peripheral blood and cerebrospinal fluid of HAM/TSP patients and identification of viral mutations. *J. Neurovirol.* 20, 341–351. doi: 10.1007/s13365-014-0249-3

Cartier, L., and Ramirez, E. (2005). Presence of HTLV-I Tax protein in cerebrospinal fluid from HAM/TSP patients. *Arch. Virol.* 150, 743–753. doi: 10.1007/s00705-004-0443-3

Croft, M., So, T., Duan, W., and Soroosh, P. (2009). The significance of OX40 and OX40L to T-cell biology and immune disease. *Immunol. Rev.* 229, 173–191. doi: 10.1111/j.1600-065X.2009.00766.x

Cross, S. L., Feinberg, M. B., Wolf, J. B., Holbrook, N. J., Wong-Staal, F., and Leonard, W. J. (1987). Regulation of the human interleukin-2 receptor alpha chain promoter: activation of a nonfunctional promoter by the transactivator gene of HTLV-I. *Cell* 49, 47–56. doi: 10.1016/0092-8674(87)90754-9

Curis, C., Percher, F., Jeannin, P., Montange, T., Chevalier, S. A., Seilhean, D., et al. (2016). Human T-lymphotropic virus type 1-induced overexpression of activated leukocyte cell adhesion molecule (ALCAM) facilitates trafficking of infected lymphocytes through the blood-brain barrier. *J. Virol.* 90, 7303–7312. doi: 10.1128/JVI.00539-16

de The, G., and Bomford, R. (1993). An HTLV-I vaccine: Why, how, for whom? *AIDS Res. Hum. Retroviruses* 9, 381–386. doi: 10.1089/aid.1993.9.381

Derse, D., Crise, B., Li, Y., Princler, G., Lum, N., Stewart, C., et al. (2007). Human T-cell leukemia virus type 1 integration target sites in the human genome: comparison with those of other retroviruses. *J. Virol.* 81, 6731–6741. doi: 10.1128/JVI.02752-06

Doi, K., Wu, X., Taniguchi, Y., Yasunaga, J., Satou, Y., Okayama, A., et al. (2005). Preferential selection of human T-cell leukemia virus type I provirus integration sites in leukemic versus carrier states. *Blood* 106, 1048–1053. doi: 10.1182/blood-2004-11-4350

Enose-Akahata, Y., Abrams, A., Massoud, R., Bialuk, I., Johnson, K. R., Green, P. L., et al. (2013). Humoral immune response to HTLV-1 basic leucine zipper factor (HBZ) in HTLV-1-infected individuals. *Retrovirology* 10:19. doi: 10.1186/1742-4690-10-19

Enose-Akahata, Y., Oh, U., Grant, C., and Jacobson, S. (2008). Retrovirally induced CTL degranulation mediated by IL-15 expression and infection of mononuclear phagocytes in patients with HTLV-I-associated neurologic disease. *Blood* 112, 2400–2410. doi: 10.1182/blood-2008-02-138529

Gaudray, G., Gachon, F., Basbous, J., Biard-Piechaczyk, M., Devaux, C., and Mesnard, J. M. (2002). The complementary strand of the human T-cell leukemia virus type 1 RNA genome encodes a bZIP transcription factor that down-regulates viral transcription. *J. Virol.* 76, 12813–12822. doi: 10.1128/JVI.76.24.12813-12822.2002

Gessain, A., Barin, F., Vernant, J. C., Gout, O., Maurs, L., Calender, A., et al. (1985). Antibodies to human T-lymphotropic virus type-I in patients with tropical spastic paraparesis. *Lancet* 2, 407–410. doi: 10.1016/S0140-6736(85)92734-5

Gessain, A., and Cassar, O. (2012). Epidemiological aspects and world distribution of HTLV-1 infection. *Front. Microbiol.* 3:388. doi: 10.3389/fmicb.2012.00388

Gessain, A., Caudie, C., Gout, O., Vernant, J. C., Maurs, L., Giordano, C., et al. (1988). Intrathecal synthesis of antibodies to human T lymphotropic virus type I and the presence of IgG oligoclonal bands in the cerebrospinal fluid of patients with endemic tropical spastic paraparesis. *J. Infect. Dis.* 157, 1226–1234. doi: 10.1093/infdis/157.6.1226

Gillet, N. A., Malani, N., Melamed, A., Gormley, N., Carter, R., Bentley, D., et al. (2011). The host genomic environment of the provirus determines the abundance of HTLV-1-infected T-cell clones. *Blood* 117, 3113–3122. doi: 10.1182/blood-2010-10-312926

Grant, C., Oh, U., Yao, K., Yamano, Y., and Jacobson, S. (2008). Dysregulation of TGF-beta signaling and regulatory and effector T-cell function in virus-induced neuroinflammatory disease. *Blood* 111, 5601–5609. doi: 10.1182/blood-2007-11-123430

Greten, T. F., Slansky, J. E., Kubota, R., Soldan, S. S., Jaffee, E. M., Leist, T. P., et al. (1998). Direct visualization of antigen-specific T cells: HTLV-1 Tax11-19- specific CD8$^+$ T cells are activated in peripheral blood and accumulate in cerebrospinal fluid from HAM/TSP patients. *Proc. Natl. Acad. Sci. U.S.A.* 95, 7568–7573. doi: 10.1073/pnas.95.13.7568

Grimaldi, L. M., Roos, R. P., Devare, S. G., Casey, J. M., Maruo, Y., Hamada, T., et al. (1988). HTLV-I-associated myelopathy: oligoclonal immunoglobulin G bands contain anti-HTLV-I p24 antibody. *Ann. Neurol.* 24, 727–731. doi: 10.1002/ana.410240606

Hanon, E., Asquith, R. E., Taylor, G. P., Tanaka, Y., Weber, J. N., and Bangham, C. R. (2000a). High frequency of viral protein expression in human T cell lymphotropic virus type 1-infected peripheral blood mononuclear cells. *AIDS Res. Hum. Retroviruses* 16, 1711–1715. doi: 10.1089/08892220050193191

Hanon, E., Goon, P., Taylor, G. P., Hasegawa, H., Tanaka, Y., Weber, J. N., et al. (2001). High production of interferon gamma but not interleukin-2 by human T-lymphotropic virus type I-infected peripheral blood mononuclear cells. *Blood* 98, 721–726. doi: 10.1182/blood.V98.3.721

Hanon, E., Hall, S., Taylor, G. P., Saito, M., Davis, R., Tanaka, Y., et al. (2000b). Abundant tax protein expression in CD4+ T cells infected with human T-cell lymphotropic virus type I (HTLV-I) is prevented by cytotoxic T lymphocytes. *Blood* 95, 1386–1392.

Hanon, E., Stinchcombe, J. C., Saito, M., Asquith, B. E., Taylor, G. P., Tanaka, Y., et al. (2000c). Fratricide among $CD8^+$ T lymphocytes naturally infected with human T cell lymphotropic virus type I. *Immunity* 13, 657–664.

Higashimura, N., Takasawa, N., Tanaka, Y., Nakamura, M., and Sugamura, K. (1996). Induction of OX40, a receptor of gp34, on T cells by trans-acting transcriptional activator, Tax, of human T-cell leukemia virus type I. *Jpn. J. Cancer Res.* 87, 227–231. doi: 10.1111/j.1349-7006.1996.tb00210.x

Hilburn, S., Rowan, A., Demontis, M. A., MacNamara, A., Asquith, B., Bangham, C. R., et al. (2011). In vivo expression of human T-lymphotropic virus type 1 basic leucine-zipper protein generates specific $CD8^+$ and $CD4^+$ T-lymphocyte responses that correlate with clinical outcome. *J. Infect. Dis.* 203, 529–536. doi: 10.1093/infdis/jiq078

Hivin, P., Frederic, M., Arpin-Andre, C., Basbous, J., Gay, B., Thebault, S., et al. (2005). Nuclear localization of HTLV-I bZIP factor (HBZ) is mediated by three distinct motifs. *J. Cell Sci.* 118(Pt 7), 1355–1362. doi: 10.1242/jcs.01727

Hoffman, P. M., Dhib-Jalbut, S., Mikovits, J. A., Robbins, D. S., Wolf, A. L., Bergey, G. K., et al. (1992). Human T-cell leukemia virus type I infection of monocytes and microglial cells in primary human cultures. *Proc. Natl. Acad. Sci. U.S.A.* 89, 11784–11788. doi: 10.1073/pnas.89.24.11784

Iwasaki, Y. (1990). Pathology of chronic myelopathy associated with HTLV-I infection (HAM/TSP). *J. Neurol. Sci.* 96, 103–123. doi: 10.1016/0022-510X(90)90060-Z

Jacobson, S., Shida, H., McFarlin, D. E., Fauci, A. S., and Koenig, S. (1990). Circulating $CD8^+$ cytotoxic T lymphocytes specific for HTLV-I pX in patients with HTLV-I associated neurological disease. *Nature* 348, 245–248. doi: 10.1038/348245a0

Jaworski, E., Narayanan, A., Van Duyne, R., Shabbeer-Meyering, S., Iordanskiy, S., Saifuddin, M., et al. (2014). Human T-lymphotropic virus type 1-infected cells secrete exosomes that contain Tax protein. *J. Biol. Chem.* 289, 22284–22305. doi: 10.1074/jbc.M114.549659

Kinosada, H., Yasunaga, J. I., Shimura, K., Miyazato, P., Onishi, C., Iyoda, T., et al. (2017). HTLV-1 bZIP factor enhances T-cell proliferation by impeding the suppressive signaling of co-inhibitory receptors. *PLOS Pathog.* 13:e1006120. doi: 10.1371/journal.ppat.1006120

Koralnik, I. J., Lemp, J. F. Jr., Gallo, R. C., and Franchini, G. (1992). In vitro infection of human macrophages by human T-cell leukemia/lymphotropic virus type I (HTLV-I). *AIDS Res. Hum. Retroviruses* 8, 1845–1849. doi: 10.1089/aid.1992.8.1845

Koyanagi, Y., Itoyama, Y., Nakamura, N., Takamatsu, K., Kira, J., Iwamasa, T., et al. (1993). In vivo infection of human T-cell leukemia virus type I in non-T cells. *Virology* 196, 25–33. doi: 10.1006/viro.1993.1451

Kubota, R., Kawanishi, T., Matsubara, H., Manns, A., and Jacobson, S. (1998). Demonstration of human T lymphotropic virus type I (HTLV-I) tax-specific $CD8^+$ lymphocytes directly in peripheral blood of HTLV-I-associated myelopathy/tropical spastic paraparesis patients by intracellular cytokine detection. *J. Immunol.* 161, 482–488.

Leal, F. E., Ndhlovu, L. C., Hasenkrug, A. M., Bruno, F. R., Carvalho, K. I., Wynn-Williams, H., et al. (2013). Expansion in $CD39^+$ $CD4^+$ immunoregulatory T cells and rarity of Th17 cells in HTLV-1 infected patients is associated with neurological complications. *PLOS Negl. Trop. Dis.* 7:e2028. doi: 10.1371/journal.pntd.0002028

Lee, S., Shin, Y., Marler, J., and Levin, M. C. (2008). Post-translational glycosylation of target proteins implicate molecular mimicry in the pathogenesis of HTLV-1 associated neurological disease. *J. Neuroimmunol.* 204, 140–148. doi: 10.1016/j.jneuroim.2008.07.020

Lehky, T. J., Fox, C. H., Koenig, S., Levin, M. C., Flerlage, N., Izumo, S., et al. (1995). Detection of human T-lymphotropic virus type I (HTLV-I) tax RNA in the central nervous system of HTLV-I-associated myelopathy/tropical spastic paraparesis patients by in situ hybridization. *Ann. Neurol.* 37, 167–175. doi: 10.1002/ana.410370206

Levin, M. C., Lee, S. M., Kalume, F., Morcos, Y., Dohan, F. C. Jr., Hasty, K. A., et al. (2002). Autoimmunity due to molecular mimicry as a cause of neurological disease. *Nat. Med.* 8, 509–513. doi: 10.1038/nm0502-509

Link, H., Cruz, M., Gessain, A., Gout, O., de The, G., and Kam-Hansen, S. (1989). Chronic progressive myelopathy associated with HTLV-I: oligoclonal IgG and anti-HTLV-I IgG antibodies in cerebrospinal fluid and serum. *Neurology* 39, 1566–1572. doi: 10.1212/WNL.39.12.1566

Liu, W., Nair, G., Vuolo, L., Bakshi, A., Massoud, R., Reich, D. S., et al. (2014). In vivo imaging of spinal cord atrophy in neuroinflammatory diseases. *Ann. Neurol.* 76, 370–378. doi: 10.1002/ana.24213

Macnamara, A., Rowan, A., Hilburn, S., Kadolsky, U., Fujiwara, H., Suemori, K., et al. (2010). HLA class I binding of HBZ determines outcome in HTLV-1 infection. *PLOS Pathog.* 6:e1001117. doi: 10.1371/journal.ppat.1001117

Makino, M., Shimokubo, S., Wakamatsu, S. I., Izumo, S., and Baba, M. (1999). The role of human T-lymphotropic virus type 1 (HTLV-1)-infected dendritic cells in the development of HTLV-1-associated myelopathy/tropical spastic paraparesis. *J. Virol.* 73, 4575–4581.

Maldonado, H., Ramirez, E., Utreras, E., Pando, M. E., Kettlun, A. M., Chiong, M., et al. (2011). Inhibition of cyclin-dependent kinase 5 but not of glycogen synthase kinase 3-beta prevents neurite retraction and tau hyperphosphorylation caused by secretable products of human T-cell leukemia virus type I-infected lymphocytes. *J. Neurosci. Res.* 89, 1489–1498. doi: 10.1002/jnr.22678

Manivannan, K., Rowan, A. G., Tanaka, Y., Taylor, G. P., and Bangham, C. R. (2016). CADM1/TSLC1 identifies HTLV-1-infected cells and determines their susceptibility to CTL-mediated lysis. *PLOS Pathog.* 12:e1005560. doi: 10.1371/journal.ppat.1005560

Mariner, J. M., Lantz, V., Waldmann, T. A., and Azimi, N. (2001). Human T cell lymphotropic virus type I Tax activates IL-15R alpha gene expression through an NF-kappa B site. *J. Immunol.* 166, 2602–2609. doi: 10.4049/jimmunol.166.4.2602

Matsuoka, M., and Jeang, K. T. (2007). Human T-cell leukaemia virus type 1 (HTLV-1) infectivity and cellular transformation. *Nat. Rev. Cancer* 7, 270–280. doi: 10.1038/nrc2111

Matsuoka, M., and Jeang, K. T. (2011). Human T-cell leukemia virus type 1 (HTLV-1) and leukemic transformation: viral infectivity, Tax, HBZ and therapy. *Oncogene* 30, 1379–1389. doi: 10.1038/onc.2010.537

Matsuura, E., Kubota, R., Tanaka, Y., Takashima, H., and Izumo, S. (2015). Visualization of HTLV-1-specific cytotoxic T lymphocytes in the spinal cords of patients with HTLV-1-associated myelopathy/tropical spastic paraparesis. *J. Neuropathol. Exp. Neurol.* 74, 2–14. doi: 10.1097/NEN.0000000000000141

Matsuura, E., Yamano, Y., and Jacobson, S. (2010). Neuroimmunity of HTLV-I infection. *J. Neuroimmune Pharmacol.* 5, 310–325. doi: 10.1007/s11481-010-9216-9

Medina, F., Quintremil, S., Alberti, C., Barriga, A., Cartier, L., Puente, J., et al. (2014). Tax posttranslational modifications and interaction with calreticulin in MT-2 cells and human peripheral blood mononuclear cells of human T cell lymphotropic virus type-I-associated myelopathy/tropical spastic paraparesis patients. *AIDS Res. Hum. Retroviruses* 30, 370–379. doi: 10.1089/AID.2013.0036

Meekings, K. N., Leipzig, J., Bushman, F. D., Taylor, G. P., and Bangham, C. R. (2008). HTLV-1 integration into transcriptionally active genomic regions is associated with proviral expression and with HAM/TSP. *PLOS Pathog.* 4:e1000027. doi: 10.1371/journal.ppat.1000027

Melamed, A., Laydon, D. J., Gillet, N. A., Tanaka, Y., Taylor, G. P., and Bangham, C. R. (2013). Genome-wide determinants of proviral targeting, clonal abundance and expression in natural HTLV-1 infection. *PLOS Pathog.* 9:e1003271. doi: 10.1371/journal.ppat.1003271

Mendez, E., Kawanishi, T., Clemens, K., Siomi, H., Soldan, S. S., Calabresi, P., et al. (1997). Astrocyte-specific expression of human T-cell lymphotropic virus type 1 (HTLV-1) Tax: induction of tumor necrosis factor alpha and susceptibility to lysis by $CD8^+$ HTLV-1-specific cytotoxic T cells. *J. Virol.* 71, 9143–9149.

Mizuguchi, M., Asao, H., Hara, T., Higuchi, M., Fujii, M., and Nakamura, M. (2009). Transcriptional activation of the interleukin-21 gene and its receptor gene by human T-cell leukemia virus type 1 Tax in human T-cells. *J. Biol. Chem.* 284, 25501–25511. doi: 10.1074/jbc.M109.010959

Mori, N., Sato, H., Hayashibara, T., Senba, M., Hayashi, T., Yamada, Y., et al. (2002). Human T-cell leukemia virus type I Tax transactivates the matrix metalloproteinase-9 gene: potential role in mediating adult T-cell leukemia invasiveness. *Blood* 99, 1341–1349. doi: 10.1182/blood.V99.4.1341

Moritoyo, T., Izumo, S., Moritoyo, H., Tanaka, Y., Kiyomatsu, Y., Nagai, M., et al. (1999). Detection of human T-lymphotropic virus type I p40tax protein in cerebrospinal fluid cells from patients with human T-lymphotropic virus type I-associated myelopathy/tropical spastic paraparesis. *J. Neurovirol.* 5, 241–248. doi: 10.3109/13550289909015810

Nagai, M., Kubota, R., Greten, T. F., Schneck, J. P., Leist, T. P., and Jacobson, S. (2001a). Increased activated human T cell lymphotropic virus type I (HTLV-I) Tax11-19-specific memory and effector $CD8^+$ cells in patients with HTLV-I-associated myelopathy/tropical spastic paraparesis: correlation with HTLV-I provirus load. *J. Infect. Dis.* 183, 197–205. doi: 10.1086/317932

Nagai, M., Usuku, K., Matsumoto, W., Kodama, D., Takenouchi, N., Moritoyo, T., et al. (1998). Analysis of HTLV-I proviral load in 202 HAM/TSP patients and 243 asymptomatic HTLV-I carriers: high proviral load strongly predisposes to HAM/TSP. *J. Neurovirol.* 4, 586–593. doi: 10.3109/13550289809114225

Nagai, M., Yamano, Y., Brennan, M. B., Mora, C. A., and Jacobson, S. (2001b). Increased HTLV-I proviral load and preferential expansion of HTLV-I Tax-specific $CD8^+$ T cells in cerebrospinal fluid from patients with HAM/TSP. *Ann. Neurol.* 50, 807–812.

Nejmeddine, M., Negi, V. S., Mukherjee, S., Tanaka, Y., Orth, K., Taylor, G. P., et al. (2009). HTLV-1-Tax and ICAM-1 act on T-cell signal pathways to polarize the microtubule-organizing center at the virological synapse. *Blood* 114, 1016–1025. doi: 10.1182/blood-2008-03-136770

Oh, U., McCormick, M. J., Datta, D., Turner, R. V., Bobb, K., Monie, D. D., et al. (2011). Inhibition of immune activation by a novel nuclear factor-kappa B inhibitor in HTLV-I-associated neurologic disease. *Blood* 117, 3363–3369. doi: 10.1182/blood-2010-10-306571

Osame, M., Usuku, K., Izumo, S., Ijichi, N., Amitani, H., Igata, A., et al. (1986). HTLV-I associated myelopathy, a new clinical entity. *Lancet* 1, 1031–1032. doi: 10.1016/S0140-6736(86)91298-5

Poiesz, B. J., Ruscetti, F. W., Gazdar, A. F., Bunn, P. A., Minna, J. D., and Gallo, R. C. (1980). Detection and isolation of type C retrovirus particles from fresh and cultured lymphocytes of a patient with cutaneous T-cell lymphoma. *Proc. Natl. Acad. Sci. U.S.A.* 77, 7415–7419. doi: 10.1073/pnas.77.12.7415

Puccioni-Sohler, M., Rios, M., Bianco, C., Zhu, S. W., Oliveira, C., Novis, S. A., et al. (1999). An inverse correlation of HTLV-I viral load in CSF and intrathecal synthesis of HTLV-I antibodies in TSP/HAM. *Neurology* 53, 1335–1339. doi: 10.1212/WNL.53.6.1335

Puccioni-Sohler, M., Yamano, Y., Rios, M., Carvalho, S. M., Vasconcelos, C. C., Papais-Alvarenga, R., et al. (2007). Differentiation of HAM/TSP from patients with multiple sclerosis infected with HTLV-I. *Neurology* 68, 206–213. doi: 10.1212/01.wnl.0000251300.24540.c4

Rende, F., Cavallari, I., Corradin, A., Silic-Benussi, M., Toulza, F., Toffolo, G. M., et al. (2011). Kinetics and intracellular compartmentalization of HTLV-1 gene expression: nuclear retention of HBZ mRNAs. *Blood* 117, 4855–4859. doi: 10.1182/blood-2010-11-316463

Richardson, J. H., Edwards, A. J., Cruickshank, J. K., Rudge, P., and Dalgleish, A. G. (1990). In vivo cellular tropism of human T-cell leukemia virus type 1. *J. Virol.* 64, 5682–5687.

Saito, M., Matsuzaki, T., Satou, Y., Yasunaga, J., Saito, K., Arimura, K., et al. (2009). In vivo expression of the HBZ gene of HTLV-1 correlates with proviral load, inflammatory markers and disease severity in HTLV-1 associated myelopathy/tropical spastic paraparesis (HAM/TSP). *Retrovirology* 6:19. doi: 10.1186/1742-4690-6-19

Saito, M., Tanaka, R., Arishima, S., Matsuzaki, T., Ishihara, S., Tokashiki, T., et al. (2013). Increased expression of OX40 is associated with progressive disease in patients with HTLV-1-associated myelopathy/tropical spastic paraparesis. *Retrovirology* 10:51. doi: 10.1186/1742-4690-10-51

Sakaguchi, S., Sakaguchi, N., Shimizu, J., Yamazaki, S., Sakihama, T., Itoh, M., et al. (2001). Immunologic tolerance maintained by $CD25^+$ $CD4^+$ regulatory T cells: their common role in controlling autoimmunity, tumor immunity, and transplantation tolerance. *Immunol. Rev.* 182, 18–32. doi: 10.1034/j.1600-065X.2001.1820102.x

Satou, Y., Miyazato, P., Ishihara, K., Yaguchi, H., Melamed, A., Miura, M., et al. (2016). The retrovirus HTLV-1 inserts an ectopic CTCF-binding site into the human genome. *Proc. Natl. Acad. Sci. U.S.A.* 113, 3054–3059. doi: 10.1073/pnas.1423199113

Satou, Y., Yasunaga, J., Yoshida, M., and Matsuoka, M. (2006). HTLV-I basic leucine zipper factor gene mRNA supports proliferation of adult T cell leukemia cells. *Proc. Natl. Acad. Sci. U.S.A.* 103, 720–725. doi: 10.1073/pnas.0507631103

Satou, Y., Yasunaga, J., Zhao, T., Yoshida, M., Miyazato, P., Takai, K., et al. (2011). HTLV-1 bZIP factor induces T-cell lymphoma and systemic inflammation in vivo. *PLOS Pathog.* 7:e1001274. doi: 10.1371/journal.ppat.1001274

Shiohama, Y., Naito, T., Matsuzaki, T., Tanaka, R., Tomoyose, T., Takashima, H., et al. (2016). Absolute quantification of HTLV-1 basic leucine zipper factor (HBZ) protein and its plasma antibody in HTLV-1 infected individuals with different clinical status. *Retrovirology* 13:29. doi: 10.1186/s12977-016-0263-z

Siekevitz, M., Feinberg, M. B., Holbrook, N., Wong-Staal, F., and Greene, W. C. (1987). Activation of interleukin 2 and interleukin 2 receptor (Tac) promoter expression by the trans-activator (tat) gene product of human T-cell leukemia virus, type I. *Proc. Natl. Acad. Sci. U.S.A.* 84, 5389–5393. doi: 10.1073/pnas.84.15.5389

Suemori, K., Fujiwara, H., Ochi, T., Ogawa, T., Matsuoka, M., Matsumoto, T., et al. (2009). HBZ is an immunogenic protein, but not a target antigen for human T-cell leukemia virus type 1-specific cytotoxic T lymphocytes. *J. Gen. Virol.* 90(Pt 8), 1806–1811. doi: 10.1099/vir.0.010199-0

Sugata, K., Satou, Y., Yasunaga, J., Hara, H., Ohshima, K., Utsunomiya, A., et al. (2012). HTLV-1 bZIP factor impairs cell-mediated immunity by suppressing production of Th1 cytokines. *Blood* 119, 434–444. doi: 10.1182/blood-2011-05-357459

Takeda, S., Maeda, M., Morikawa, S., Taniguchi, Y., Yasunaga, J., Nosaka, K., et al. (2004). Genetic and epigenetic inactivation of tax gene in adult T-cell leukemia cells. *Int. J. Cancer* 109, 559–567. doi: 10.1002/ijc.20007

Uchiyama, T. (1988). Adult T-cell leukemia. *Blood Rev.* 2, 232–238. doi: 10.1016/0268-960X(88)90012-4

Umehara, F., Izumo, S., Nakagawa, M., Ronquillo, A. T., Takahashi, K., Matsumuro, K., et al. (1993). Immunocytochemical analysis of the cellular infiltrate in the spinal cord lesions in HTLV-I-associated myelopathy. *J. Neuropathol. Exp. Neurol.* 52, 424–430. doi: 10.1097/00005072-199307000-00010

Umehara, F., Okada, Y., Fujimoto, N., Abe, M., Izumo, S., and Osame, M. (1998). Expression of matrix metalloproteinases and tissue inhibitors of metalloproteinases in HTLV-I-associated myelopathy. *J. Neuropathol. Exp. Neurol.* 57, 839–849. doi: 10.1097/00005072-199809000-00005

Usui, T., Yanagihara, K., Tsukasaki, K., Murata, K., Hasegawa, H., Yamada, Y., et al. (2008). Characteristic expression of HTLV-1 basic zipper factor (HBZ) transcripts in HTLV-1 provirus-positive cells. *Retrovirology* 5:34. doi: 10.1186/1742-4690-5-34

Valentin, H., Lemasson, I., Hamaia, S., Casse, H., Konig, S., Devaux, C., et al. (1997). Transcriptional activation of the vascular cell adhesion molecule-1 gene in T lymphocytes expressing human T-cell leukemia virus type 1 Tax protein. *J. Virol.* 71, 8522–8530.

Vine, A. M., Heaps, A. G., Kaftantzi, L., Mosley, A., Asquith, B., Witkover, A., et al. (2004). The role of CTLs in persistent viral infection: cytolytic gene expression in $CD8^+$ lymphocytes distinguishes between individuals with a high or low proviral load of human T cell lymphotropic virus type 1. *J. Immunol.* 173, 5121–5129. doi: 10.4049/jimmunol.173.8.5121

Waldmann, T. A. (2015). The shared and contrasting roles of IL2 and IL15 in the life and death of normal and neoplastic lymphocytes: implications for cancer therapy. *Cancer Immunol. Res.* 3, 219–227. doi: 10.1158/2326-6066.CIR-15-0009

Wan, Y. Y., and Flavell, R. A. (2007). 'Yin-Yang' functions of transforming growth factor-beta and T regulatory cells in immune regulation. *Immunol. Rev.* 220, 199–213. doi: 10.1111/j.1600-065X.2007.00565.x

Yamamoto-Taguchi, N., Satou, Y., Miyazato, P., Ohshima, K., Nakagawa, M., Katagiri, K., et al. (2013). HTLV-1 bZIP factor induces inflammation through labile Foxp3 expression. *PLOS Pathog.* 9:e1003630. doi: 10.1371/journal.ppat.1003630

Yamano, Y., Araya, N., Sato, T., Utsunomiya, A., Azakami, K., Hasegawa, D., et al. (2009). Abnormally high levels of virus-infected IFN-$gamma^+CCR4^+CD4^+CD25^+$ T cells in a retrovirus-associated neuroinflammatory disorder. *PLOS ONE* 4:e6517. doi: 10.1371/journal.pone.0006517

Yamano, Y., Cohen, C. J., Takenouchi, N., Yao, K., Tomaru, U., Li, H. C., et al. (2004). Increased expression of human T lymphocyte virus type I (HTLV-I) Tax11-19 peptide-human histocompatibility leukocyte antigen A*201 complexes on $CD4^+$ $CD25^+$ T Cells detected by peptide-specific, major histocompatibility complex-restricted antibodies in patients with HTLV-I-associated neurologic disease. *J. Exp. Med.* 199, 1367–1377. doi: 10.1084/jem.20032042

Yamano, Y., Nagai, M., Brennan, M., Mora, C. A., Soldan, S. S., Tomaru, U., et al. (2002). Correlation of human T-cell lymphotropic virus type 1 (HTLV-1) mRNA with proviral DNA load, virus-specific $CD8^+$ T cells, and disease severity in HTLV-1-associated myelopathy (HAM/TSP). *Blood* 99, 88–94. doi: 10.1182/blood.V99.1.88

Yamano, Y., and Sato, T. (2012). Clinical pathophysiology of human T-lymphotropic virus-type 1-associated myelopathy/tropical spastic paraparesis. *Front. Microbiol.* 3:389. doi: 10.3389/fmicb.2012.00389

Yamano, Y., Takenouchi, N., Li, H. C., Tomaru, U., Yao, K., Grant, C. W., et al. (2005). Virus-induced dysfunction of $CD4^+CD25^+$ T cells in patients with HTLV-I-associated neuroimmunological disease. *J. Clin. Invest.* 115, 1361–1368. doi: 10.1172/JCI23913

Zhao, T., Satou, Y., Sugata, K., Miyazato, P., Green, P. L., Imamura, T., et al. (2011). HTLV-1 bZIP factor enhances TGF-beta signaling through p300 coactivator. *Blood* 118, 1865–1876. doi: 10.1182/blood-2010-12-326199

4

Proposal of Classification Criteria for HTLV-1-Associated Myelopathy/Tropical Spastic Paraparesis Disease Activity

*Tomoo Sato[1], Naoko Yagishita[1], Keiko Tamaki[2], Eisuke Inoue[3], Daisuke Hasegawa[1], Misako Nagasaka[4,5], Hiroko Suzuki[1], Natsumi Araya[1], Ariella Coler-Reilly[1], Yasuhiro Hasegawa[6], Yoshio Tsuboi[2], Ayako Takata[7] and Yoshihisa Yamano[1,5]**

[1] Department of Rare Diseases Research, Institute of Medical Science, St. Marianna University School of Medicine, Kawasaki, Japan, [2] Department of Neurology, Fukuoka University, Fukuoka, Japan, [3] Medical Informatics, St. Marianna University School of Medicine, Kawasaki, Japan, [4] Department of Oncology, Karmanos Cancer Institute, Wayne State University, Detroit, MI, United States, [5] Department of Advanced Medical Innovation, St. Marianna University Graduate School of Medicine, Kawasaki, Japan, [6] Department of Neurology, St. Marianna University School of Medicine, Kawasaki, Japan, [7] Department of Preventive Medicine, St. Marianna University School of Medicine, Kawasaki, Japan

***Correspondence:**
Yoshihisa Yamano
yyamano@marianna-u.ac.jp

Human T-lymphotropic virus type 1 (HTLV-1)-associated myelopathy/tropical spastic paraparesis (HAM/TSP) is a rare chronic neuroinflammatory disease. While the disease usually progresses slowly without remission, there is a subgroup of patients with rapid progression and another subgroup with very slow progression. However, there have been no reports to date that have successfully determined the criteria to differentiate these subgroups. Therefore, we initially conducted a statistical modeling analysis to explore representative patterns of disease progression using data from our nationwide HAM/TSP patient registration system ("HAM-net"). The latent class mixed model analysis on the retrospective data (n = 205) of disease progression measured by the change in Osame Motor Disability Score from the onset of the disease to diagnosis demonstrated three representative progression patterns of HAM/TSP. Next, to test the effect of the progression rate at the initial phase of the disease on long-term prognosis, we divided 312 "HAM-net" registered patients into three groups (rapid, slow, and very slow progressors) based on the progression rate, then analyzed long-term functional prognosis of each group using the Kaplan–Meier method. Our data clearly demonstrated that the rapid progression at the early phase of the disease is an important poor prognostic factor. Moreover, to determine the biomarkers capable of discriminating the difference in disease activity, we compared the value of potential biomarkers of HAM/TSP among rapid (n = 15), slow (n = 74), very slow (n = 7), and controls (non-HAM/TSP patients, n = 18). The cerebrospinal fluid (CSF) levels of neopterin and C-X-C motif chemokine 10 (CXCL10) were the most valuable markers to discriminate among rapid, slow, and very slow progressors. To differentiate between rapid and slow progressors, the cut-off values of neopterin and CXCL10 were determined to be 44 pmol/mL and 4400 pg/mL, respectively. Furthermore, to differentiate between slow and very slow progressors, these values were determined to be 5.5 pmol/mL and 320 pg/mL, respectively. Notably, we found that CSF levels of these markers in very slow progressors were within the reference range. Thus, we propose a new classification criteria for disease activity of HAM/TSP that may contribute to improving the treatment algorithm for HAM/TSP.

Keywords: HTLV-1, HAM/TSP, classification criteria, disease activity, biomarker, neopterin, CXCL10

INTRODUCTION

Human T-lymphotropic virus type 1 (HTLV-1), the first human retrovirus discovered (Poiesz et al., 1980), causes a chronic neuroinflammatory disease termed HTLV-1-associated myelopathy/tropical spastic paraparesis (HAM/TSP) (Gessain et al., 1985; Osame et al., 1986). Infiltrated HTLV-1-infected cells cause chronic spinal inflammation and spinal cord tissue damage, which lead to HAM/TSP (Yamano and Sato, 2012; Bangham et al., 2015). In patients with HAM/TSP, the most frequent symptom is lower limb motor dysfunction, followed by bladder/bowel dysfunction and sensory disturbance (Osame, 1990; De Castro-Costa et al., 2006). Corticosteroids and interferon (IFN)-α are currently available for the treatment of HAM/TSP. However, there are no studies that support the need for stratified medicine based on disease activity, and there is no definite treatment algorithm for the use of therapeutic drugs (corticosteroids and IFN-α).

There exists a wide variation in the progression rates of HAM/TSP (Coler-Reilly et al., 2016), which suggests the need for stratified medicine. Some studies have reported a subgroup of patients with rapid progression and another subgroup with very slow progression. One example of rapid progressors has been reported in a Japanese study, in which 14 of 151 patients (9.3%) deteriorated more than three grades of a motor disability score within 2 years before initial examination (Nakagawa et al., 1995). A study from Peru demonstrates that 34 of 158 patients (21.5%) were rapid progressors who were unable to walk without two canes within 2 years of disease onset (Gotuzzo et al., 2004). A short communication from Brazil indicates that there were 7 of 88 patients (8%) with subacute progression, which is defined as the need to use a wheel chair during the first 2 years after the onset of symptoms (Lima et al., 2007). In an United Kingdom cohort, 3 of 48 patients (6%) were unable to walk within 2 years (Martin et al., 2010). A Martinique study also reports that a subgroup of patients have a faster rate of progression (Olindo et al., 2006). In the United Kingdom cohort, as an example of a subgroup of patients with very slow progression, 6 of 48 patients (12.5%) had little progression and deteriorated by only 0.3 s/10 m/year in a timed walk test (Martin et al., 2010). These studies put into context the wide variation in the clinical course of HAM/TSP patients and the need for a treatment algorithm to allow physicians to accurately determine disease activity and provide appropriate treatment. However, as mentioned above, the definitions representing the subgroups have differed from one researcher to another. There are no reports in which the clinical course from the disease onset was classified by subgroups defined by statistical methods.

In addition, because HAM/TSP progresses unremittingly, it is necessary to identify disease activity markers that can estimate the progression rate of HAM/TSP as early as possible. Several biomarker candidates for disease activity have already been identified. First, the level of proviral load in peripheral blood mononuclear cells (PBMCs) correlates with the progression of motor dysfunction and is associated with long-term prognosis (Matsuzaki et al., 2001; Olindo et al., 2005). Next, cell counts, anti-HTLV-1 antibody titer and protein levels in the cerebrospinal fluid (CSF) were elevated in rapid progressors, suggesting an association with progression of the disease (Matsuura et al., 2016). Furthermore, the CSF levels of C-X-C motif chemokine 10 (CXCL10) and neopterin, which are mainly produced in IFN-γ-stimulated astrocytes (Ando et al., 2013; de Paula Martins et al., 2018), strongly correlate with the rate of disease progression (Sato et al., 2013). However, there are no biomarker studies to date that have successfully determined the criteria for differentiating the subgroups with different disease activity.

In this study, we aimed to develop a new classification criteria for disease activity of HAM/TSP in order to provide appropriate treatment based on disease activity. First, we classified "clinical course from disease onset" of untreated HAM/TSP patients without any distribution assumption by applying a chronological change of motor disability to a mathematical model. Based on the results, we defined the criteria in which patients were divided into three groups. Next, we identified biomarkers and their cut-off values for discriminating the three groups. By summarizing the above results, we developed HAM/TSP classification criteria for disease activity based on clinical course and biomarkers.

MATERIALS AND METHODS

Ethical Considerations

The study was approved by the St. Marianna University School of Medicine Bioethics Committee (clinical course analysis using HAM/TSP patient registry data: Approval ID No. 2044, biomarker analysis: No. 1646) and the Fukuoka University Faculty of Medicine Bioethics Committee (biomarker analysis: No. 14-2-08). Prior to the collection of blood or CSF samples, all participants provided written informed consent permitting the analysis of their samples for research purposes as part of their clinical care.

Participants

For clinical course analysis, retrospective data from patients who had enrolled in the Japanese HAM/TSP patient registry called "HAM-net" were used (Coler-Reilly et al., 2016). There were 453 patients registered with "HAM-net" from March 2012 until December 2015 (**Figure 1**). Survey data, including sex, age at

TABLE 1 | Osame motor disability score.

Grade	Motor disability
0	No walking or running abnormalities
1	Normal gait but runs slowly
2	Abnormal gait (stumbling, stiffness)
3	Unable to run
4	Needs handrail to climb stairs
5	Needs a cane (unilateral support) to walk
6	Needs bilateral support to walk
7	Can walk 5–10 m with bilateral support
8	Can walk 1–5 m with bilateral support
9	Cannot walk, but able to crawl
10	Cannot crawl, but able to move using arms
11	Cannot move around, but able to turn over in bed
12	Cannot turn over in bed
13	Cannot even move toes

onset, age at the onset of motor symptoms, age at diagnosis, age at the time of the survey, Osame motor disability score (OMDS) (**Table 1**), age at the time of deterioration in OMDS, medical treatment history, Health assessment questionnaire-disability index (HAQ-DI), family history of HAM/TSP and adult T-cell leukemia/lymphoma (ATL), and blood transfusion history were used for analysis. Among the 453 participants, 28 cases for which the above data were missing were excluded. Therefore, 425 cases were included in the analysis (Arm A shown in **Figure 1**) for preparing classification criteria based on clinical course from onset. Among these cases, 312 HAM/TSP patients (Arm B) who had at least 2 observations at different time points were applied to Kaplan–Meier analysis, and 205 patients (Arm C) who had at least 3 observations at different time points were applied to latent class mixed model (LCMM) analysis. As shown in **Table 2**, there were no statistical differences in age, sex, age at onset, or age at diagnosis among these three groups. However, there were significant differences in OMDS between Arm A and Arm C. The participants for biomarker analysis were 96 HAM/TSP patients with clinical course data and blood and CSF marker data obtained before treatment and 18 control cases [asymptomatic carriers (n = 10) and non-HTLV-1-infected non-inflammatory neurological disease patients (n = 8)].

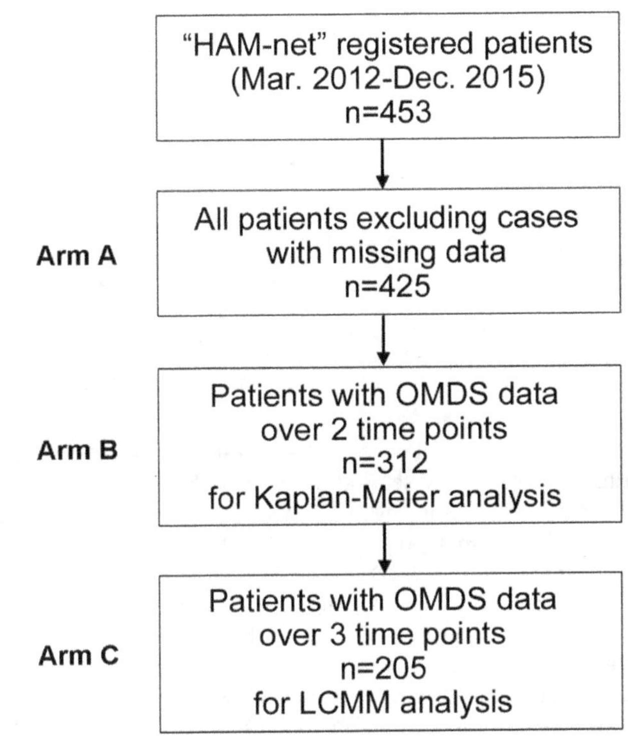

FIGURE 1 | Patient flow chart. There were 453 HAM/TSP patients registered with the HAM/TSP patient registry "HAM-net" from March 2012 to December 2015. After excluding 28 cases missing survey data, the participants in this study were 425 patients (Arm A). Among them, 312 HAM/TSP patients (Arm B) who had at least 2 observations at different time points were included in the Kaplan–Meier analysis, and 205 patients (Arm C) who had at least 3 observations at different time points were included in the latent class mixed model (LCMM) analysis. OMDS, Osame motor disability score.

Measurement of Biomarkers

Peripheral blood mononuclear cells, serum and CSF samples were prepared as described previously (Sato et al., 2013). Briefly, PBMCs were isolated with standard procedures using Pancoll® density gradient centrifugation (density 1.077 g/mL; PAN-Biotech GmbH, Aidenbach, Germany). Serum was obtained from venous blood samples by centrifugation after clotting. CSF was obtained by lumbar puncture. A small amount of CSF was used for routine laboratory tests, which included total protein, glucose, and cell counts. The remaining CSF was aliquoted into cryotubes and stored at −80°C until undergoing further analysis. The serum concentration of soluble IL-2 receptor (sIL-2R) was determined using a chemiluminescent enzyme immunoassay (LSI Medience Corporation, Tokyo, Japan). HTLV-1 proviral load was measured using real-time PCR, following DNA extraction from PBMCs, as previously described (Yamano et al., 2002). The anti-HTLV-1 antibody titer in CSF was determined using the gelatin particle agglutination test (Serodia-HTLV-1; Fujirebio, Tokyo, Japan). CSF neopterin level was measured using high-performance liquid chromatography at a commercial laboratory (SRL Inc., Tokyo, Japan). CXCL10 in CSF was measured using a cytometric bead array (BD Biosciences, Franklin Lakes, NJ, United States).

Statistical Analysis

In order to explore representative patterns of disease progression for HAM/TSP patients, the LCMM was applied to longitudinally collected OMDS data. This method assumes the existence of latent class, which affects the longitudinal trajectory of an outcome variable under the given number of latent classes. Specifically, we used the cumulative distribution function of Beta distribution as a link function of this model, and 2–5 latent classes were specified. The goodness of fit of the model was evaluated by the Bayesian Information Criterion

TABLE 2 | Demographics and clinical characteristics of "HAM-net"-registered HAM/TSP patients.

	Participants for LCMM analysis n = 205	Participants for Kaplan–Meier analysis n = 312	All registered patients n = 425	p-value
Sex: Female	160 (78.0%)	240 (76.9%)	319 (75.1%)	N.S.[(a)]
Age at onset*	41.7 ± 14.2	43.1 ± 14.3	44.5 ± 14.7	N.S.[(b)]
Age at diagnosis*	52.1 ± 12.7	52.4 ± 12.7	52.3 ± 12.9	N.S.[(b)]
Age (at present**)*	62.3 ± 10.1	61.7 ± 10.5	61.9 ± 10.5	N.S.[(b)]
OMDS (range: 0–13)*	6.1 ± 2.4†	6.4 ± 2.3	6.7 ± 2.3†	0.018[(b)]

Statistical methods: (a) By Fisher's exact test (b) by analysis of variance and Tukey post hoc tests.
**Data are expressed as mean ± SD.*
***Present is defined as the time of the subject's initial "HAM-net" interview.*
†There is significant difference between OMDS in participants for latent class analysis and OMDS in all registered patients.
LCMM, latent class mixed model; OMDS, Osame motor disability score.

(BIC), and the model with the smallest BIC was the most optimal. The analysis set was data from HAM/TSP patients registered to "HAM-net" who had at least three observations at different time points (Arm C shown in **Figure 1**). To exclude treatment effects on OMDS, OMDS from onset to diagnosis were analyzed. Additionally, time to deteriorating to OMDS grade 6 for each group determined by LCMM analysis was evaluated using Kaplan–Meier method followed by a log-rank test. Median time to OMDS grade 6 was calculated for each group. Calculations were performed using R[1] and the LCMM package[2]. Fisher's exact tests were used for comparison of categorical variables. Analyses of variance and Tukey *post hoc* tests were used for comparison of continuous variables. Kruskal–Wallis tests followed by Dunn's *post hoc* tests were used for comparison among the four groups for biomarker analysis. Jonckheere's trend test was used to investigate whether each marker had an increasing trend from the control group to the rapid progressor group. Receiver operating characteristic (ROC) analysis was performed to examine the sensitivity and specificity of individual biomarkers. Optimal sensitivity and specificity are defined as those yielding the minimal value for $(1 - \text{sensitivity})^2 + (1 - \text{specificity})^2$. Statistical analyses and graph composition were performed using R, IBM SPSS Statistics Version 22 (IBM Corp. Armonk, NY, United States), or GraphPad Prism 6 (GraphPad Software, Inc., San Diego, CA, United States). All p-values were two-tailed, and the threshold of significance was set at 0.05.

RESULTS

Classification of Clinical Course Patterns

LCCM was utilized to classify the patterns of "clinical course from disease onset" of untreated HAM/TSP patients. The analysis set for LCMM analysis consisted of data from 205 HAM/TSP patients (Arm C shown in **Figure 1**). The scores of chronological change of motor disability of each patient were used as input data (**Figure 2A**) and applied to a mathematical model without any distribution assumption. As shown in **Figure 2B**, when hypothesizing the clinical course patterns to 2, 3, 4, or 5, the BIC were calculated as 1672.7, 1666.6, 1679.9, or 1693.3, respectively. As the reliability of the model correlates to lower BIC, the clinical course patterns subdivided to 3, was considered the most appropriate (right upper graph in **Figure 2B**). To include all patients who showed rapid progression, we defined rapid progressors as those who developed OMDS grade 5 or above within 2 years from the onset of motor symptoms. Determining those who show little or no signs of progression despite a lack of treatment (very slow progressors) is also clinically important, as this group of patients could avoid unnecessary treatment. When patients were divided into three groups, there were patients who would eventually progress to OMDS grade 5 or above (data not shown) even in the most indolent group (right upper graph in **Figure 2B**). Therefore, we referred to the slowest progression group when dividing patients into four groups (left lower graph in **Figure 2B**), and defined the very slow progressor group as those who were at or lower than OMDS grade 3 at 10 years from the onset of motor symptoms. Those who did not meet the criteria for either rapid progressors or very slow progressors were defined as slow progressors.

Long-Term Functional Prognosis Based on the Three Groups

To investigate whether the long-term prognoses of these three groups were different, Kaplan–Meier analysis was used to evaluate the time to progression from OMDS grade 2 to grade 6 in all three groups (**Figure 3**). The longitudinal data (including treatment period) on the progression of OMDS in HAM/TSP patients (n = 312; Arm B shown in **Figure 1**) were utilized in the analysis. When Arm B was divided into the three groups based on previously provided definitions, 42 (13.5%) were rapid progressors, 249 (79.8%) were slow progressors, and 21 (6.7%) were very slow progressors (**Figure 3**). The difference between the time to progress from OMDS grade 2 to grade 6 in these three groups (rapid, slow, and very slow progressors) was statistically significant among the three groups ($p < 0.0001$). Furthermore, the median time to progression from OMDS grade 2 to grade 6 was 4, 19, and 35 years, respectively. This shows that the classification criteria based on the "clinical course from disease onset" was successful in defining three groups with different prognoses.

[1]http://cran.r-project.org
[2]https://CRAN.R-project.org/package=lcmm

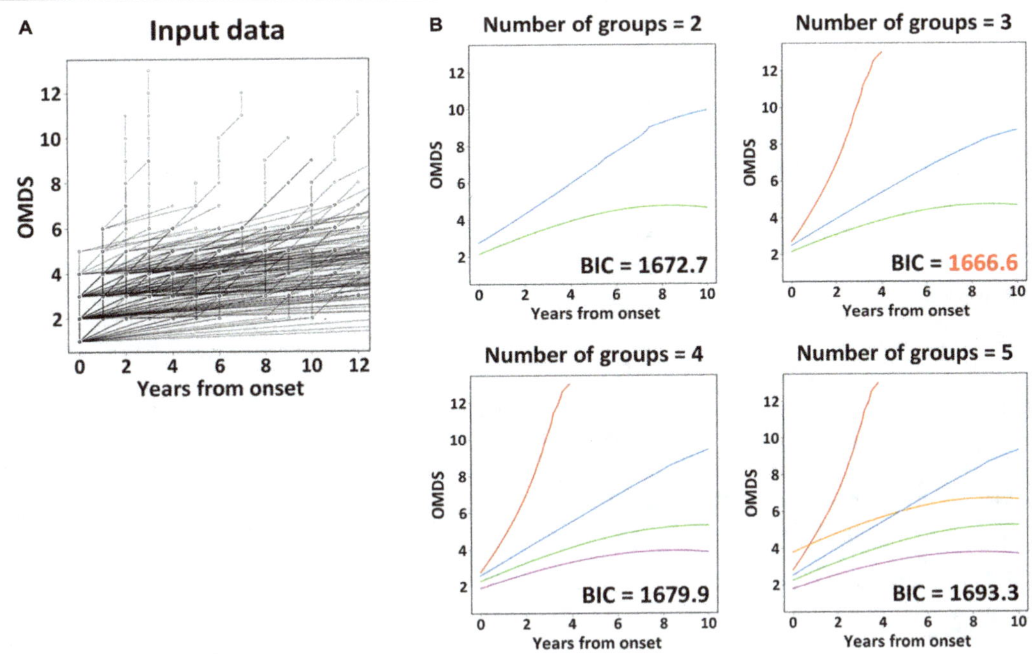

FIGURE 2 | Latent class mixed model analysis. The graph **(A)** shows the chronological change of Osame motor disability score (OMDS) from the onset to diagnosis in patients with HAM/TSP (n = 205), which was used as input data for latent class mixed model (LCMM) analysis. The graphs **(B)** show the representative progression pattern of OMDS when we set 2–5 latent classes. The horizontal axis represents the elapsed years from onset. The vertical axis represents OMDS. BIC, Bayesian Information Criterion.

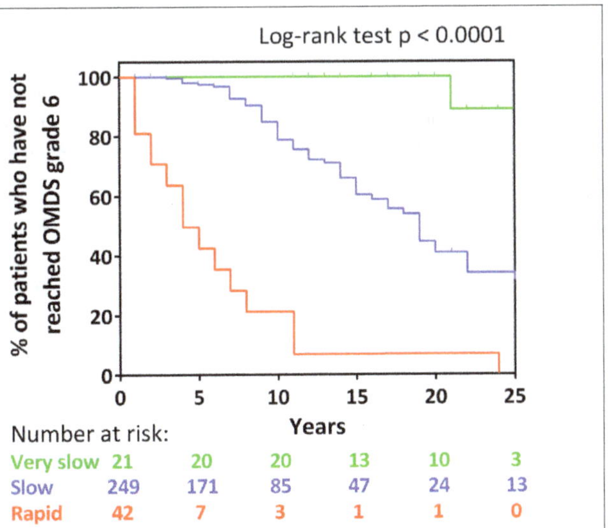

FIGURE 3 | Kaplan–Meier analysis. Kaplan–Meier analysis to evaluate time from Osame motor disability score (OMDS) grade 2 to deteriorating to OMDS grade 6 for the following patient group defined by the difference of the clinical course from onset: rapid progressors (red line), slow progressors (blue line), and very slow progressors (green line). The horizontal axis represents the elapsed years from OMDS grade 2. The vertical axis represents percentage of patients who have not reached OMDS grade 6. Numbers shown below the graph indicate number at risk. Differences in the degree of progress among the three groups were tested by log-rank test. Significance was defined as $p < 0.05$.

Characteristics of the Three Groups

Table 3 shows the characteristics of all "HAM-net" registered patients divided into the three groups (n = 425; Arm A shown in **Figure 1**). Specifically, 15.5, 79.5, and 4.9% of the patients composed the rapid, slow, and very slow progressor groups, respectively. There were no differences in sex among the three groups, but there were statistically significant differences in age of onset, time to diagnosis, and duration of disease among the three groups (all $p < 0.001$). To be specific, the more rapid the rate of progression, the older the age at onset, and the shorter the time to diagnosis and duration of disease. Furthermore, there were differences in OMDS and HAQ-DI among the three groups (p = 0.003 and 0.001, respectively), with the rapid progressor groups demonstrating significantly higher scores in both OMDS and HAQ-DI than the other two groups. There were no significant differences among the three groups with regard to initial symptoms and family histories of HAM/TSP and ATL, but rapid progressors tended to carry less such family history. There were also no significant differences among the three groups with respect to transfusion histories, but the rapid progressors tended to have higher rates of transfusion. This result does not directly imply that there are a high number of patients with HAM/TSP attributable to blood transfusion in the rapid progressors group because the patients may have already been infected before transfusion or they may have been infected by horizontal transmission, independent of blood transfusion.

TABLE 3 | Clinical attributes of rapid, slow, and very slow progressors who were defined based on the progression pattern after the onset of HAM/TSP.

	Group R Rapid progressor (*n* = 66, 15.5%)	Group S Slow progressor (*n* = 338, 79.5%)	Group VS Very slow progressor (*n* = 21, 4.9%)	*p*-value	Statistical methods	Groups with significant difference
Sex: Female	51 (77.3%)	255 (75.4%)	13 (61.9%)	N.S.	(a)	
Age at onset*	55.9 ± 11.7	43.0 ± 14.0	32.4 ± 14.5	<0.001	(b)	R > S > VS
Age at diagnosis*	58.7 ± 11.0	51.1 ± 12.9	51.0 ± 13.5	<0.001	(b)	R > S, R > VS
Age (at present**)*	65.6 ± 9.2	61.2 ± 10.6	62.1 ± 10.4	0.008	(b)	R > S
Diagnosis delay (time from onset to diagnosis)*	2.7 ± 4.3	8.1 ± 8.2	18.6 ± 12.9	<0.001	(b)	R > S > VS
Disease duration (time from onset to present)*	9.6 ± 7.7	18.2 ± 10.8	29.8 ± 10.0	<0.001	(b)	R > S > VS
Initial symptoms (inclusive)						
Gait disturbance	58 (87.9%)	277 (82.0%)	18 (85.7%)	N.S.	(a)	
Urinary disturbance	21 (31.8%)	145 (42.9%)	7 (33.3%)	N.S.	(a)	
Sensory disturbance (in legs)	10 (15.2%)	52 (15.4%)	2 (9.5%)	N.S.	(a)	
Others	17 (25.8%)	99 (29.3%)	7 (33.3%)	N.S.	(a)	
OMDS (range: 0–13)*	6.6 ± 2.3	5.6 ± 2.2	5.4 ± 1.5	0.003	(b)	R > S, R > VS
HAQ-DI (range: 0–3)*	1.4 ± 0.6	1.1 ± 0.7	1.0 ± 0.5	0.001	(b)	R > S, R > VS
Family history of HAM/TSP†: Yes	2 (3.0%)	36 (10.7%)	3 (14.3%)	N.S.	(a)	
Family history of ATL‡: Yes	2 (3.0%)	20 (5.9%)	1 (4.8%)	N.S.	(a)	
History of blood transfusion						
Yes (anytime)	19 (28.8%)	61 (18.0%)	2 (9.5%)	N.S.	(a)	
Yes, before 1986	16 (24.2%)	51 (15.1%)	2 (9.5%)	N.S.	(a)	

Statistical methods: (a) By Fisher's exact test, (b) by analysis of variance and Tukey post hoc tests.
**Data are expressed as mean ± SD, **present is defined as the time of the subject's initial "HAM-net" interview.*
†Family history of HAM/TSP indicates the subject has a first- or second-degree relative with HAM/TSP.
‡Family history of adult T-cell leukemia/lymphoma (ATL) indicates the subject has a first- or second degree relative with ATL.
N.S., not significant; OMDS, Osame motor disability score; HAQ-DI, health assessment questionnaire-disability index.

TABLE 4 | Trends for change in the values of candidate biomarkers from the control group to the rapid progressor group.

	Group R Rapid progressor*		Group S Slow progressor*		Group VS Very slow progressor*		Group C Control*		Jonckheere's trend test *p*-value
CSF neopterin pmol/mL	60	(46.5–75.5)	15.5	(7–29)	4	(3–4)	3	(2–3)	<0.0001
CSF CXCL10 pg/mL	6128.5	(4836–12217)	1807.5	(543.8–5000.8)	246	(138–263.5)	138	(115–160.8)	<0.0001
CSF cell count cells/μL	15	(11.5–29)	4	(2–6)	1	(1–2.5)	1	(0.5–1)	<0.0001
CSF glucose mg/dL	50	(48–60)	54	(51–60)	59	(57–68.5)	60	(53.3–63)	0.9976
CSF protein mg/dL	58	(45–92)	30	(27–36)	36	(27.5–41)	36.5	(26.75–48)	0.0081
CSF anti-HTLV-1 Ab titer	1536	(512–2048)	64	(32–256)	32	(32–32)	(–)**		<0.0001
Serum soluble IL-2R U/mL	761	(577.5–849)	568	(431–678)	349	(332.5–663)	208	(181–273)	<0.0001
Proviral load in PBMC copies/100 cells	7.95	(3.62–9.87)	4.04	(2.47–5.60)	4.44	(1.91–19.35)	0.25	(0.16–1.41)	0.0018

Data is expressed as median (interquartile range).(–) indicates that the CSF anti-HTLV-1-antibody test in the normal control group was all negative.*
CSF, cerebrospinal fluid; Ab, antibody; IL-2R, IL-2 receptor; PBMC, peripheral blood mononuclear cells.

Biomarker Analyses

Biomarker analyses were performed in 96 HAM/TSP cases in which blood and CSF marker data using pre-treatment samples were available and in 18 controls. The above 96 HAM/TSP patients were divided into the three patient groups (rapid, slow, and very slow progressors) defined by the difference of the clinical course from onset. Using a trend test, we investigated whether each of the eight candidate biomarkers had an increasing trend from the control group to the rapid progressor group (**Table 4**). All markers except CSF glucose had significantly increasing trends from the control group to the rapid progressor group. Next, we compared the eight candidate biomarkers among the three groups of HAM/TSP in addition to the control group (**Figure 4**). Consequently, CSF neopterin and CSF CXCL10 demonstrated significant differences in all pairwise comparisons among the three groups of rapid, slow, and very slow progressors (**Figures 4A,B**). The other six markers did not show statistically significant differences between slow and very slow progressors (**Figures 4C–H**). Notably, between very slow progressors and control participants, there were no statistically

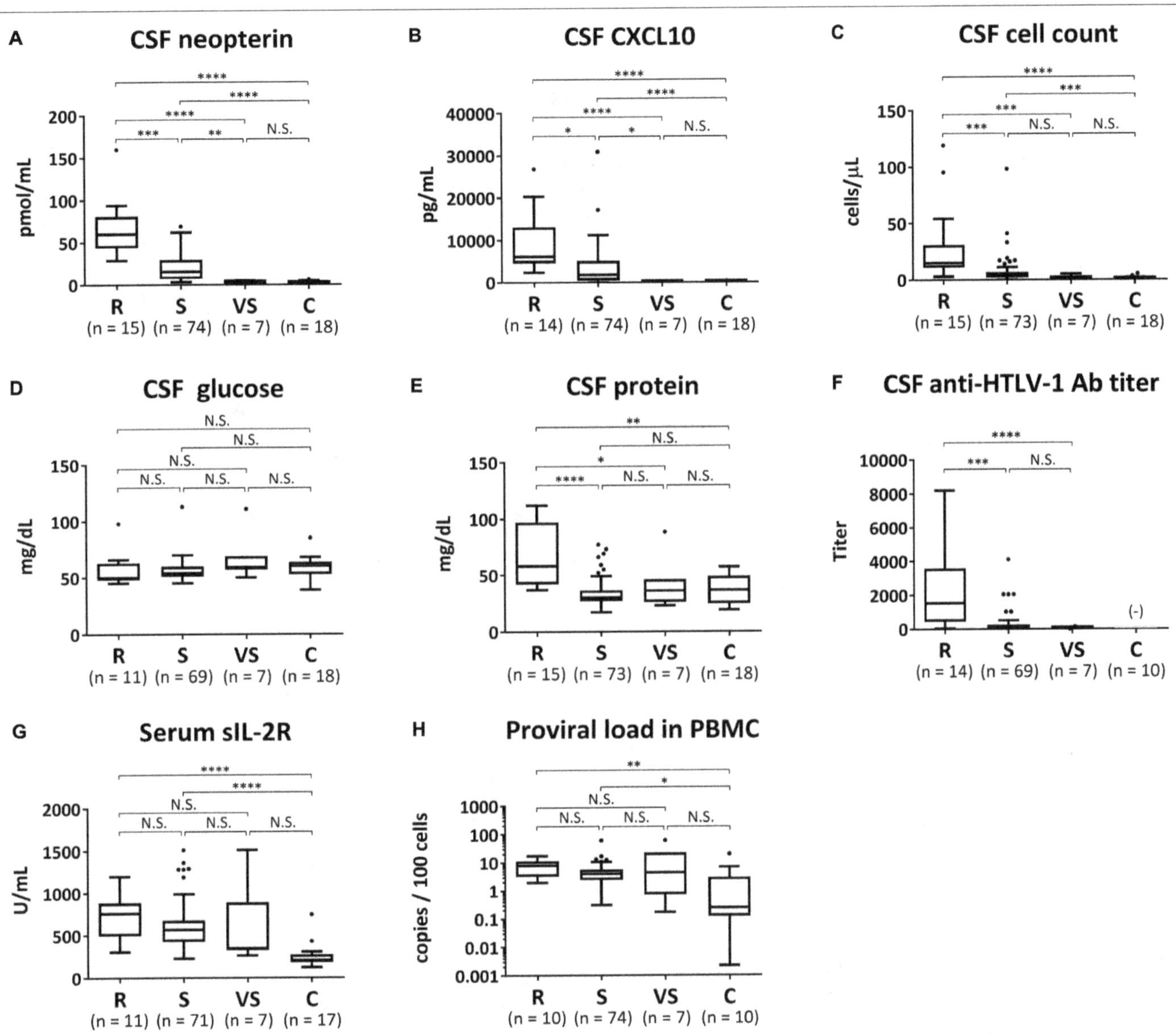

FIGURE 4 | Biomarker analyses. The following eight candidate biomarkers were compared among rapid progressors (R), slow progressors (S), very slow progressors (VS), and controls (C): **(A)** cerebrospinal fluid (CSF) levels of neopterin, **(B)** C-X-C motif chemokine 10 (CXCL10), **(C)** cell count, **(D)** glucose, **(E)** total protein, and **(F)** anti-HTLV-1 antibody titer; **(G)** serum level of soluble IL-2 receptor (sIL-2R); **(H)** proviral loads in peripheral blood mononuclear cells (PBMC). Data are shown as a Tukey box plots: data are presented as median (interquartile range; IQR) and whiskers represent 1.5 IQR and black dots represent outliers. Statistical analysis was performed using the Kruskal–Wallis test followed by Dunn's post hoc test: ****$p < 0.0001$, ***$p < 0.001$, **$p < 0.01$, *$p < 0.05$; N.S., not significant; Ab, antibody. (-) indicates that the CSF anti-HTLV-1-antibody test in normal control group was all negative.

significant differences in any of the markers, with the exception of the CSF anti-HTLV-1 antibody titer. Regarding the antibody titer, all patients in the very slow progressor group tested positive for this antibody, whereas all 10 asymptomatic carriers in the control group tested negative (**Figure 4F**). As shown in **Table 5**, there were no differences among sex, but rapid progressors had a tendency to be older than the other groups and controls had a tendency to be younger than the other groups. Age of disease onset and OMDS were compared among the three disease groups. Age at onset was higher in the rapid progressors than in the other groups, and OMDS was progressively lower in the order of rapid, slow, and very slow progressors. This is similar to the results of the overall analysis of all "HAM-net"-registered patients shown in **Table 3** and provides a good representation of the patients.

Determination of Cut-Off Values

Cerebrospinal fluid neopterin and CSF CXCL10 were both significantly different among the three groups. An ROC analysis was performed to evaluate the accuracy of the markers in differentiating rapid from slow progressors and to determine the cut-off values (**Figure 5A**). The areas under the curve (AUC) of CSF neopterin and CSF CXCL10 were 0.93 and 0.82, respectively.

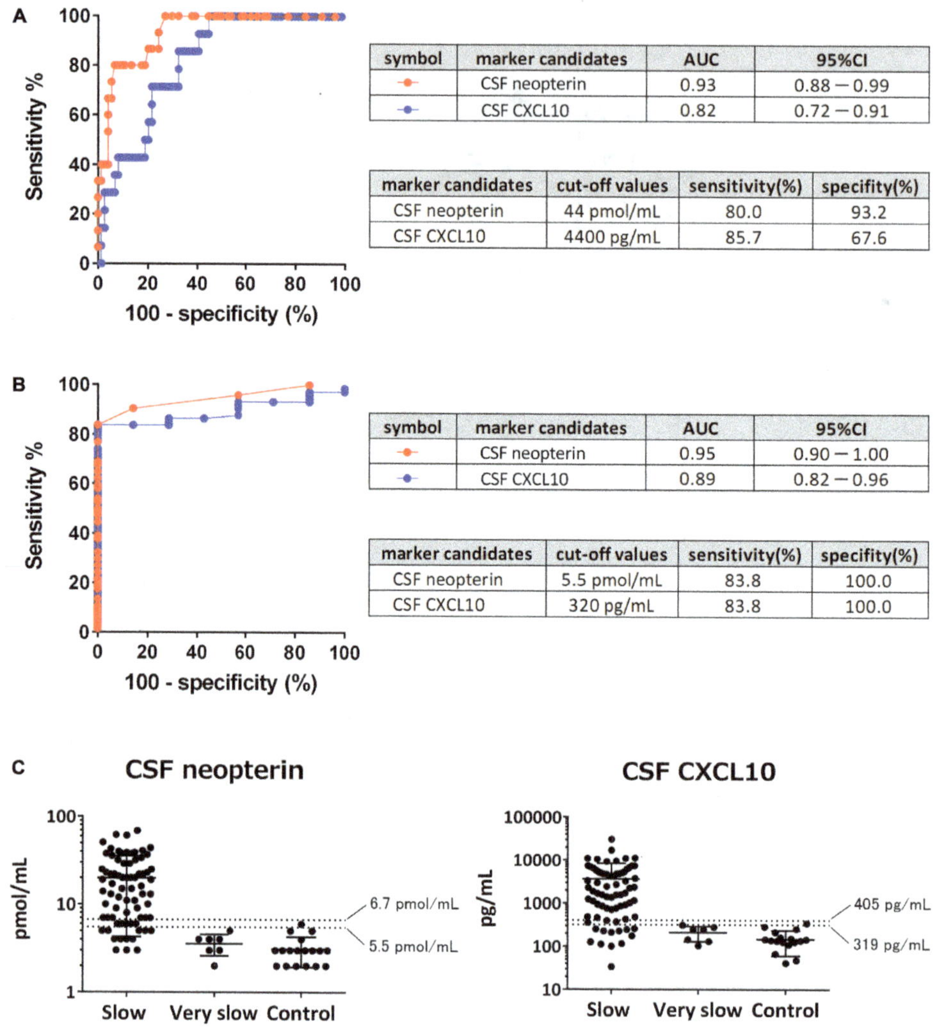

FIGURE 5 | Determination of cut-off values. **(A)** Receiver operating characteristic (ROC) analysis was employed to evaluate the sensitivities and specificities of cerebrospinal fluid (CSF) neopterin and CSF CXCL10 for discriminating rapid progressors from slow progressors. Greater proximity of the ROC curve to the upper left corner indicates higher sensitivity and specificity of the marker. AUC, area under the ROC curve; 95% CI, 95% confidence interval. **(B)** ROC analysis was employed to evaluate the sensitivities and specificities of CSF neopterin and CSF CXCL10 for discriminating slow progressors from very slow progressors. **(C)** CSF levels of neopterin and CXCL10 were compared among slow progressors, very slow progressors, and controls. The data are plotted on a logarithmic axis. Middle horizontal bars show the arithmetic mean and the error bars show plus or minus one standard deviation (SD). Horizontal dashed lines indicate mean + 3SD (upper) and mean + 2SD (lower) as reference values of each marker.

TABLE 5 | Demographics and clinical characteristics of HAM/TSP patients for biomarker analysis.

	Group R Rapid progressor (*n* = 15)	**Group S Slow progressor (*n* = 74)**	**Group VS Very slow progressor (*n* = 7)**	**Group C Control* (*n* = 18)**	***p*-value**	**Statistical methods**	**Groups with significant difference**
Sex: Female	12 (80.0%)	60 (81.1%)	3 (42.9%)	14 (77.8%)	0.1602	(a)	
Age (at present[†])**	67.0 ± 8.6	59.0 ± 10.3	59.1 ± 10.5	48.3 ± 13.1	<0.0001	(b)	R > S, R > C, S > C
Age at onset**	63.9 ± 9.4	44.0 ± 13.7	45.0 ± 12.3	–	<0.0001	(b)	R > S, R > VS
OMDS (range: 0–13)**	6.7 ± 1.9	5.4 ± 1.8	3.0 ± 0.8	–	<0.0001	(b)	R > S > VS

Statistical methods: (a) By Fisher's exact test, (b) by analysis of variance and Tukey post hoc tests.
**Normal control group consists of healthy HTLV-1 carrier (n = 10) and non-HTLV-1-infected non-inflammatory neurological disease patients (n = 8), **Data are expressed as mean ± SD, [†]Present is defined as the time of the sample collection.*
OMDS, Osame Motor Disability Score.

TABLE 6 | Proposed classification criteria for disease activity of HAM/TSP based on clinical course and biomarkers*.

Disease activity	Criteria based on clinical course after the onset of motor symptoms	Criteria based on the biomarkers	
		CSF neopterin (pmol/mL)	**CSF CXCL10 (pg/mL)**
High	**Rapid progressor:** Progression to OMDS grade 5 or greater within 2 years after the onset of motor symptoms	≥44	≥4400
Moderate	**Slow progressor:** Patients who does not meet the definition of either rapid- or very slow-progressor	6–43	320–4399
Low	**Very slow progressor:** Progression to OMDS grade 3 or less at least 10 years after the onset of motor symptoms	≤5	<320

*At the judgment of disease activity in individual patient, there is no need to meet all items, need to be judged comprehensively.

Both were higher than 0.8, therefore, demonstrating accuracy as markers to differentiate the two groups. The optimal cut-off value of CSF neopterin was 44 pmol/mL, which provided a sensitivity of 80.0% and specificity of 93.2% for the detection of rapid progressors. Similarly, the optimal cut-off for CSF CXCL10 was 4400 pg/mL with a sensitivity and specificity of 85.7 and 67.6%, respectively. Next, an ROC analysis to compare between slow and very slow progressors was performed (**Figure 5B**). The AUC were 0.95 and 0.89, respectively, again both higher than 0.8. The optimal cut-off for CSF neopterin was 5.5 pmol/mL with a sensitivity of 83.8% and specificity of 100% to detect slow progressors. Similarly, the optimal cut-off for CSF CXCL10 was 320 pg/mL, with sensitivity and specificity of 83.8 and 100%, respectively.

As neither marker has a pre-existing reference range, we attempted to determine a range using the values derived from controls. When the reference value was set at the mean + 3 standard deviations (SD), the upper limits of normal values of CSF neopterin and CSF CXCL10 were 6.7 pmol/mL and 405 pg/mL, respectively (**Figure 5C**). When these numbers were utilized as cut-off values to differentiate between slow and very slow progressors, CSF neopterin had a sensitivity of 77.0% and specificity of 100%, while CSF CXCL10 had a sensitivity of 79.7% and specificity of 100%. When the reference value was set at the mean + 2SD, the upper limits of normal values of CSF neopterin and CSF CXCL10 were determined to be at 5.5 pmol/mL and 319 pg/mL (≈320 pg/mL), consistent with previously derived cut-off value to distinguish between slow and very slow progressors. With the sensitivity of both markers at 83.8% and the specificity of them at 100%, the reference values set at mean + 2SD provided better sensitivity when compared to the reference value set at mean + 3SD. Therefore, we concluded that the reference value was best when set at the mean + 2SD, as fewer slow progressors would be included in the normal reference range.

DISCUSSION

In this study, we were able to classify HAM/TSP patients into three groups based on the disease activity, which is assessed by "progression rate of motor dysfunction after the onset of motor symptoms" and "the concentrations of neopterin and CXCL10 in CSF." Consequently, we propose a novel classification criteria in which HAM/TSP patients were divided into the three groups (High, Moderate, and Low) of the disease activity (**Table 6**).

The statistical pattern classification analysis using the chronological OMDS data from untreated 205 HAM/TSP patients demonstrated that the progression pattern after the onset of motor symptoms was divided into three patterns (**Figure 2**). This finding supported the existence of previously described rapid progressors (Nakagawa et al., 1995; Gotuzzo et al., 2004; Olindo et al., 2006; Lima et al., 2007; Martin et al., 2010) and very slow progressors (Martin et al., 2010). Importantly, the Kaplan–Meier analysis revealed that the long-term functional prognoses were different in the three groups (rapid, slow, and very slow) defined from results of pattern classification (**Figure 3**), suggesting that the clinical course at the early stage determines the subsequent prognosis. Rapid progressors had a significantly poorer prognosis than the other patients. This data clearly showed that the rapid progression at the early phase of the disease is an important poor prognostic factor in HAM/TSP patients. Such poor prognosis in rapid progressors might be caused by a delay in diagnosis, inappropriate treatment when diagnosed, ineffectiveness of treatment, and high disease activity. Therefore, it is imperative to survey the actual conditions at diagnosis and the treatment patterns for rapid progressors to determine methods to improve their prognosis.

This study revealed that there was a very slow progressor patient group among HAM/TSP patients. These patients were OMDS grade 3 or lower 10 years from the onset of motor symptoms, even in the absence of steroid treatment. This category contained patients with low disease activity whose CSF marker levels, which reflect spinal inflammation, were as low as those of the control group, even in the absence of treatment for HAM/TSP (**Figures 4, 5C**). They were also characterized by a younger age of onset than seen in the other patient groups (**Table 3**). Meanwhile, the rapid progressors were characterized by high levels of spinal inflammatory markers and old age at onset, consistent with previous reports (**Figure 4** and **Table 3**) (Nakagawa et al., 1995; Gotuzzo et al., 2004; Matsuura et al., 2016). Since there have been some case reports about the rapid progression of HAM/TSP due to organ transplantation (Gout et al., 1990; Toro et al., 2003), organ transplantation seems to be one of the background factors for the rapid progressors. In fact, of the 425 patients shown in **Table 3**, one with HAM/TSP that was caused by renal transplantation was a rapid progressor. In addition to transplantation, environmental factors (infection route, co-infection status), host factors (genetic factors such as HLA and gene mutation), and viral factors (viral gene expression and mutation) are candidate background factors that characterize the different disease activity of the three groups. In the future, we will clarify the importance of each factor.

There are various limitations to determine the disease activity based on the clinical course from onset. For example, there could be difficulties in judging the disease activity in patients with a short disease duration, or there may be a risk of overestimation or underestimation depending on the degree of the patients' complaints. To overcome these limitations, we further developed a more objective biomarker-based classification criteria. To date, several biomarker candidates to evaluate disease activity have been reported (Matsuzaki et al., 2001; Olindo et al., 2005; Sato et al., 2013; Matsuura et al., 2016). In the present study, among the eight candidate markers, both CSF neopterin and CSF CXCL10 were identified as markers that clearly distinguish the three groups with different disease activities. Furthermore, the cut-off values dividing the three groups were determined. The levels of two blood-derived markers (soluble IL-2 receptor and PBMC HTLV-1 proviral load) were significantly higher in HAM/TSP patients with rapid or slow progression compared with asymptomatic carriers (**Figures 4G,H**), but they could not distinguish among the three patient groups with different disease activity. Thus, at the moment, CSF tests are essential to determine disease activity. In addition, it has been reported that the cell counts, protein levels, and anti-HTLV-1 antibody titer in the CSF were elevated in rapid progressors (Matsuura et al., 2016). These markers showed significantly higher values in rapid progressors than in slow progressors in the present study as well. However, they showed no significant differences between slow and very slow progressors. Therefore, these markers were insufficient for classification criteria for disease activity. To be specific, when using 5 cells/μL as the reference value for CSF cell counts measurable in general practice, the sensitivity to detect slow progressors was extremely low at 28.8% (data not shown). This indicates that although CSF cell counts are within normal range, most HAM/TSP patients are in an active and progressive phase. This should be noted when evaluating the condition of these patients.

PBMC HTLV-1 proviral load is elevated in patients with rapid progression (Matsuzaki et al., 2001; Olindo et al., 2005) and is weakly but significantly correlated with CSF levels of neopterin and CXCL10 (Sato et al., 2013). Therefore, HTLV-1 proviral load was a candidate marker to divide patients into three groups with different disease activity. In the present study, HTLV-1 proviral load tended to be elevated in rapid progressors (**Table 4**). However, there were no statistically significant differences in HTLV-1 proviral load among the three groups (**Figure 4H**). This may be due to the small sample size and the fact that the slow and very slow progressor groups included patients with a high proviral load (**Figure 4H**). Interestingly, the number of IFN-γ-producing infected T cells is more correlated with disease activity than with the proviral load itself (Yamano et al., 2009). Additionally, it has recently been found that a certain percentage of HAM/TSP patients have increased ATL-like infected cells (CADM1+CD7- CD4 T-cells) (MN et al. unpublished data). This evidence suggests that HTLV-1-infected clones that reside in one HAM/TSP patient do not necessarily consist only of pro-inflammatory clones that contribute to the disease activity of HAM/TSP. Thus, to evaluate the disease activity using the number of infected cells, methods for evaluating the amount of proinflammatory clones rather than the proviral load itself must be created.

The results of this study suggest that CSF levels of neopterin and CXCL10 are not only biomarkers for evaluating disease activity, but also may be candidate biomarkers for drug response or surrogate markers reflecting a long-term functional prognosis that should be the true endpoint of HAM/TSP. This is likely, given that corticosteroid therapy decreases CSF neopterin levels (Nakagawa et al., 1996; Nagai et al., 2013) and appears to decrease CSF CXCL10 levels as well (unpublished data). Because levels of these two markers reflect disease activity and are related to the progression rate of HAM/TSP, the decrease in their levels after treatment is expected to provide a decrease in the progression rate of HAM/TSP and to predict an improved prognosis. Interestingly, a multicenter retrospective cohort study indicates that oral low-dose corticosteroid therapy reduces the progression rate of HAM/TSP and improves the long-term prognosis compared to untreated patients (Coler-Reilly et al., 2017). Therefore, the levels of these two markers reflect the therapeutic effect and long-term prognosis of HAM/TSP, indicating their potential as drug response or surrogate markers. Moreover, CSF levels of neopterin and CXCL10 were within the normal range in HAM/TSP patients with low disease activity as based on our developed classification criteria. This suggests that the level of spinal cord inflammation is comparable to that of the control group. Oral corticosteroids can be used to reduce the level of inflammation in HAM/TSP patients with high or medium disease activity (Nagai et al., 2013). However, in patients with low disease activity where the inflammation level is in the normal range in the untreated state, oral corticosteroid therapy may have little benefit and may result in an increased risk of side effects by corticosteroids. Therefore, evaluation of the disease activity could give us important information in determining the need for corticosteroid therapy. Because it is critical to prospectively investigate these hypotheses, we have been conducting a confirmatory multi-center Phase 2b study (UMIN trial number, UMIN000023798) on the efficacy and safety of stratified corticosteroid therapy for HAM/TSP patients. This clinical trial should provide significant evidence to test these hypotheses.

This study has several limitations. The first limitation is that we could not adjust for potential confounders (age and OMDS) because of the small sample size of each group after the patients with HAM/TSP were divided into the three groups to examine biomarkers. In this regard, both CSF neopterin and CSF CXCL10 have a strong correlation with the progression rate of HAM/TSP after adjustment for OMDS (Sato et al., 2013). This finding indicates that both markers do not simply reflect the severity of motor disability. Although a difference in age between slow and very slow progressors was not observed in this study, the levels of both markers were significantly different, suggesting that the influence of age is limited. Therefore, it is possible that the change in the level of both markers reflects the disease activity rather than age or disease severity. The second limitation is that the cut-off values and reference values of the biomarkers determined in this study may vary from other studies, depending on measurement methods and experimental environments. Thus, it is essential to standardize the measurement methods of neopterin and CXCL10. Thereafter, it will be necessary to determine the cut-off values and the reference values again. The third limitation is that the classification criteria were determined based on retrospective data. Therefore, it is necessary to prospectively verify the validity of this classification criteria in the future.

CONCLUSION

We herein propose a new classification criteria for disease activity of HAM/TSP. This classification criteria will enable clinicians to evaluate the disease activity early and support clinicians in providing appropriate treatment for each patient. This new classification criteria shall assist with establishing a novel therapeutic algorithm of HAM/TSP and incorporating tailored medicine based on the disease activity of each individual.

AUTHOR CONTRIBUTIONS

TS and YY contributed to the conception and design of the study. NY, EI, HS, AT, and YY run the patient registry "HAM-net." TS, KT, YH, YT, and YY contributed to the biomarker analysis. TS and EI performed the statistical analysis. TS, DH, NA, AC-R, MN, and YY drafted and corrected the manuscript. All authors contributed to manuscript revision, read, and approved the submitted version.

FUNDING

This study was supported by a grant from the Practical Research Project for Rare/Intractable Diseases of the Japan Agency for Medical Research and Development (Grant No. JP18ek0109356, JP18ek0109283, and JP18ek0109346), and a Health and Labour Sciences Research Grant on Rare and Intractable Diseases from the Ministry of Health, Labour and Welfare of Japan (Grant No. H28-nanchi-ippan-018).

ACKNOWLEDGMENTS

The authors would like to thank the support staff in the laboratory, including Katsunori Takahashi, Yasuo Kunitomo, Yumiko Hasegawa, Mikako Koike, Yuriko Hosokawa, Chihiro Sasa, Yumi Saito, Atsuko Kobayashi, and Miho Ishikawa. In addition, the authors thank the invaluable assistance of the other members of the "HAM-net" team. Employees of Accelight, Inc. provided data management and analysis services. Employees of Ata Life, Inc. (a site management organization) assisted in data collection by conducting telephone interviews and creating case report forms. The authors also would like to thank the staff of Rare Disease Data Registry of Japan (RADDAR-J) for their support for managing the data.

REFERENCES

Ando, H., Sato, T., Tomaru, U., Yoshida, M., Utsunomiya, A., Yamauchi, J., et al. (2013). Positive feedback loop via astrocytes causes chronic inflammation in virus-associated myelopathy. *Brain* 136(Pt 9), 2876–2887. doi: 10.1093/brain/awt183

Bangham, C. R., Araujo, A., Yamano, Y., and Taylor, G. P. (2015). HTLV-1-associated myelopathy/tropical spastic paraparesis. *Nat. Rev. Dis. Primers* 1:15012. doi: 10.1038/nrdp.2015.12

Coler-Reilly, A. L., Yagishita, N., Suzuki, H., Sato, T., Araya, N., Inoue, E., et al. (2016). Nation-wide epidemiological study of Japanese patients with rare viral myelopathy using novel registration system (HAM-net). *Orphanet. J. Rare Dis.* 11:69. doi: 10.1186/s13023-016-0451-x

Coler-Reilly, A. L. G., Sato, T., Matsuzaki, T., Nakagawa, M., Niino, M., Nagai, M., et al. (2017). Effectiveness of daily prednisolone to slow progression of human T-lymphotropic virus type 1-associated myelopathy/tropical spastic paraparesis: a multicenter retrospective cohort study. *Neurotherapeutics* 14, 1084–1094. doi: 10.1007/s13311-017-0533-z

de Paula Martins R, Ghisoni, K., Lim, C. K., Aguiar, A. S. Jr., Guillemin, G. J., and Latini, A. (2018). Neopterin preconditioning prevents inflammasome activation in mammalian astrocytes. *Free Radic. Biol. Med.* 115, 371–382. doi: 10.1016/j.freeradbiomed.2017.11.022

De Castro-Costa, C. M., Araujo, A. Q., Barreto, M. M., Takayanagui, O. M., Sohler, M. P., da Silva, E. L., et al. (2006). Proposal for diagnostic criteria of tropical spastic paraparesis/HTLV-I-associated myelopathy (TSP/HAM). *AIDS Res. Hum. Retroviruses* 22, 931–935. doi: 10.1089/aid.2006.22.931

Gessain, A., Barin, F., Vernant, J. C., Gout, O., Maurs, L., Calender, A., et al. (1985). Antibodies to human T-lymphotropic virus type-I in patients with tropical spastic paraparesis. *Lancet* 2, 407–410. doi: 10.1016/S0140-6736(85)92734-5

Gotuzzo, E., Cabrera, J., Deza, L., Verdonck, K., Vandamme, A. M., Cairampoma, R., et al. (2004). Clinical characteristics of patients in Peru with human T cell lymphotropic virus type 1-associated tropical spastic paraparesis. *Clin. Infect. Dis.* 39, 939–944. doi: 10.1086/423957

Gout, O., Baulac, M., Gessain, A., Semah, F., Saal, F., Peries, J., et al. (1990). Rapid development of myelopathy after HTLV-I infection acquired by transfusion during cardiac transplantation. *N. Engl. J. Med.* 322, 383–388. doi: 10.1056/NEJM199002083220607

Lima, M. A., Harab, R. C., Schor, D., Andrada-Serpa, M. J., and Araujo, A. Q. (2007). Subacute progression of human T-lymphotropic virus type I-associated myelopathy/tropical spastic paraparesis. *J. Neurovirol.* 13, 468–473. doi: 10.1080/13550280701510096

Martin, F., Fedina, A., Youshya, S., and Taylor, G. P. (2010). A 15-year prospective longitudinal study of disease progression in patients with HTLV-1 associated myelopathy in the UK. *J. Neurol. Neurosurg. Psychiatry* 81, 1336–1340. doi: 10.1136/jnnp.2009.191239

Matsuura, E., Nozuma, S., Tashiro, Y., Kubota, R., Izumo, S., and Takashima, H. (2016). HTLV-1 associated myelopathy/tropical spastic paraparesis (HAM/TSP): a comparative study to identify factors that influence disease progression. *J. Neurol. Sci.* 371, 112–116. doi: 10.1016/j.jns.2016.10.030

Matsuzaki, T., Nakagawa, M., Nagai, M., Usuku, K., Higuchi, I., Arimura, K., et al. (2001). HTLV-I proviral load correlates with progression of motor disability in

HAM/TSP: analysis of 239 HAM/TSP patients including 64 patients followed up for 10 years. *J. Neurovirol.* 7, 228–234. doi: 10.1080/13550280152403272

Nagai, M., Tsujii, T., Iwaki, H., Nishikawa, N., and Nomoto, M. (2013). Cerebrospinal fluid neopterin, but not osteopontin, is a valuable biomarker for the treatment response in patients with HTLV-I-associated myelopathy. *Intern. Med.* 52, 2203–2208. doi: 10.2169/internalmedicine.52.0869

Nakagawa, M., Izumo, S., Ijichi, S., Kubota, H., Arimura, K., Kawabata, M., et al. (1995). HTLV-I-associated myelopathy: analysis of 213 patients based on clinical features and laboratory findings. *J. Neurovirol.* 1, 50–61. doi: 10.3109/13550289509111010

Nakagawa, M., Nakahara, K., Maruyama, Y., Kawabata, M., Higuchi, I., Kubota, H., et al. (1996). Therapeutic trials in 200 patients with HTLV-I-associated myelopathy/ tropical spastic paraparesis. *J. Neurovirol.* 2, 345–355. doi: 10.3109/13550289609146899

Olindo, S., Cabre, P., Lezin, A., Merle, H., Saint-Vil, M., Signate, A., et al. (2006). Natural history of human T-lymphotropic virus 1-associated myelopathy: a 14-year follow-up study. *Arch. Neurol.* 63, 1560–1566. doi: 10.1001/archneur.63.11.1560

Olindo, S., Lezin, A., Cabre, P., Merle, H., Saint-Vil, M., Edimonana Kaptue, M., et al. (2005). HTLV-1 proviral load in peripheral blood mononuclear cells quantified in 100 HAM/TSP patients: a marker of disease progression. *J. Neurol. Sci.* 237, 53–59. doi: 10.1016/j.jns.2005.05.010

Osame, M. (1990). "Review of WHO kagoshima meeting and diagnostic guidelines for HAM/TSP," in *Human Retrovirology*, ed. W. A. Blattner (New York, NY: Raven Press), 191–197.

Osame, M., Usuku, K., Izumo, S., Ijichi, N., Amitani, H., Igata, A., et al. (1986). HTLV-I associated myelopathy, a new clinical entity. *Lancet* 1, 1031–1032. doi: 10.1016/S0140-6736(86)91298-5

Poiesz, B. J., Ruscetti, F. W., Gazdar, A. F., Bunn, P. A., Minna, J. D., and Gallo, R. C. (1980). Detection and isolation of type C retrovirus particles from fresh and cultured lymphocytes of a patient with cutaneous T-cell lymphoma. *Proc. Natl. Acad. Sci. U.S.A.* 77, 7415–7419. doi: 10.1073/pnas.77.12.7415

Sato, T., Coler-Reilly, A., Utsunomiya, A., Araya, N., Yagishita, N., Ando, H., et al. (2013). CSF CXCL10, CXCL9, and neopterin as candidate prognostic biomarkers for HTLV-1-associated myelopathy/tropical spastic paraparesis. *PLoS Negl. Trop. Dis.* 7:e2479. doi: 10.1371/journal.pntd.0002479

Toro, C., Rodes, B., Poveda, E., and Soriano, V. (2003). Rapid development of subacute myelopathy in three organ transplant recipients after transmission of human T-cell lymphotropic virus type I from a single donor. *Transplantation* 75, 102–104. doi: 10.1097/01.TP.0000040428.36030.FC

Yamano, Y., Araya, N., Sato, T., Utsunomiya, A., Azakami, K., Hasegawa, D., et al. (2009). Abnormally high levels of virus-infected IFN-gamma+ CCR4+ CD4+ CD25+ T cells in a retrovirus-associated neuroinflammatory disorder. *PLoS One* 4:e6517. doi: 10.1371/journal.pone.0006517

Yamano, Y., Nagai, M., Brennan, M., Mora, C. A., Soldan, S. S., Tomaru, U., et al. (2002). Correlation of human T-cell lymphotropic virus type 1 (HTLV-1) mRNA with proviral DNA load, virus-specific CD8(+) T cells, and disease severity in HTLV-1-associated myelopathy (HAM/TSP). *Blood* 99, 88–94. doi: 10.1182/blood.V99.1.88

Yamano, Y., and Sato, T. (2012). Clinical pathophysiology of human T-lymphotropic virus-type 1-associated myelopathy/tropical spastic paraparesis. *Front. Microbiol.* 3:389. doi: 10.3389/fmicb.2012.00389

5

Quantitating the Transfer of the HTLV-1 p8 Protein Between T-Cells by Flow Cytometry

Norbert Donhauser, Stefanie Heym and Andrea K. Thoma-Kress*

Institute of Clinical and Molecular Virology, Universitätsklinikum Erlangen, Friedrich-Alexander-Universität Erlangen-Nürnberg, Erlangen, Germany

Correspondence:
Andrea K. Thoma-Kress
andrea.thoma-kress@uk-erlangen.de

The Human T-cell leukemia virus type 1 (HTLV-1)-encoded accessory protein p8 is cleaved from the precursor protein p12 encoded by the HTLV-1 open reading frame I. Both p12 and p8 are thought to contribute to efficient viral persistence. Mechanistically, p8 induces T-cell conjugates and cellular conduits. The latter are considered to facilitate transfer of p8 to target cells and virus transmission. Transfer of p8 between p8-expressing T-cells and recipient cells has been analyzed by immunofluorescence and live imaging. However, automatic quantitation of p8-transfer between cells has not been studied yet. Here we developed a novel method allowing time saving quantitation of p8 transfer between cells by flow cytometry. After establishing a protocol for the detection of intracellular p8 by flow cytometry and validation of p8 protein expression by western blot and immunofluorescence, we set up a co-culture assay between p8-expressing donor Jurkat T-cells and recipient Jurkat T-cells that had been prestained with a well-retained live cell dye. Upon quantitating the amount of p8 positive recipient cells with regard to the percentage of p8 expressing donor cells, time course experiments confirmed that p8 is rapidly transferred between Jurkat T-cells. We found that p8 enters approximately 5% of recipient T-cells immediately upon co-culture for 5 min. Prolonged co-culture for up to 24 h revealed an increase of relative p8 transfer to approximately 23% of the recipient cells. Immunofluorescence analysis of co-culture experiments and manual quantitation of p8 expression in fluorescence images confirmed the validity of the flow cytometry based assay. Application of the new assay revealed that manipulation of actin polymerization significantly decreased p8 transfer between Jurkat T-cells suggesting an important role of actin dynamics contributing to p8 transfer. Further, transfer of p8 to co-cultured T-cells varies between different donor cell types since p8 transfer could hardly been detected in co-cultures of 293T donor cells with Jurkat acceptor cells. In summary, our novel assay allows automatic and rapid quantitation of p8 transfer to target cells and might thus contribute to a better understanding of cellular processes and dynamics regulating p8 transfer and HTLV-1 transmission.

Keywords: HTLV-1, p8, virus transmission, protein transport, flow cytometry

INTRODUCTION

Human T-cell leukemia virus type 1 (HTLV-1), a delta-retrovirus, infects ca. 5–10 million people worldwide and is the etiologic agent of adult T-cell leukemia/lymphoma (ATLL) (Poiesz et al., 1980; Yoshida et al., 1982; Gessain and Cassar, 2012). The virus is transmitted via cell-containing body fluids such as blood products, semen, and breast milk (Gross and Thoma-Kress, 2016). Upon integration into the host cell genome, HTLV-1 persists mainly in its provirus form. *In vivo*, integrated HTLV-1 is mainly detected in $CD4^+$ T-cells, and to a less extent in $CD8^+$ T-cells, dendritic cells (DC), or monocytes (Richardson et al., 1990; Macatonia et al., 1992; Nagai et al., 2001; Hishizawa et al., 2004; de Castro-Amarante et al., 2015; Melamed et al., 2015). Recent work shows that HTLV-1 infects hematopoietic stem cells and that these cells differentiate into diverse cell lineages (Furuta et al., 2017).

Replication of HTLV-1 occurs either by infection of new cells, or by mitotic division and clonal proliferation of infected $CD4^+$ T-cells (Carpentier et al., 2015; Turpin et al., 2017). Cell-cell contacts are a prerequisite for efficient infection of $CD4^+$ T-cells, while DC can be infected cell-free with viral biofilms (Alais et al., 2015). For infection of $CD4^+$ T-cells, the following mechanisms have been described: On the one hand, it is proposed that the virus is transmitted at tight cell-cell contacts via the virological synapse or via cell surface transfer of viral biofilms (Igakura et al., 2003; Pais-Correia et al., 2010). On the other hand, cellular conduits induced by the accessory protein p8 seem to be important for HTLV-1 transmission (Van Prooyen et al., 2010b).

p8 is a 70 amino acid (aa) protein that is proteolytically cleaved from the precursor protein p12 encoded by the HTLV-1 open reading frame I (Fukumoto et al., 2009): First, p12 is cleaved between amino acid (aa) 9/10 to remove an ER-retention signal. Second, a cleavage between aa 29/30 results in the p8 protein (Fukumoto et al., 2009). While the precursor protein p12 localizes to the endoplasmatic reticulum (ER) and to the *cis*-Golgi apparatus (Koralnik et al., 1993; Ding et al., 2001), p8 lacks an ER-retention signal and localizes to the cytoplasm and the cell membrane (Fukumoto et al., 2007, 2009). Both p12 and p8 are thought to contribute to efficient viral persistence (Valeri et al., 2010; Pise-Masison et al., 2014). Mechanistically, p8 induces T-cell conjugates and cellular conduits. The latter are considered to facilitate transfer of p8 to target cells and virus transmission (Van Prooyen et al., 2010b). The transfer of p8 to neighboring cells is supposed to induce T-cell anergy by decreasing T-cell receptor signaling (Fukumoto et al., 2007). Together, these potential functions of p8 could favor viral persistence in an immune competent host (Van Prooyen et al., 2010a,b; Edwards et al., 2011).

Thus far, little is known about the transfer of the mobile protein p8 between cells. Although immunofluorescence analysis and live imaging revealed that p8 transfer occurs within minutes (Van Prooyen et al., 2010b), p8 transfer between cells has not been automatically and quantitatively evaluated yet. Here, we developed a novel and simple method allowing a time saving quantitation of p8 transfer between co-cultured cells by flow cytometry. Manual quantitation of p8 expression in fluorescence images confirmed the validity of the assay. Application of the newly developed protocol indicated that transfer of p8 between cells depends on proper polymerization of the actin cytoskeleton and varies between different cell types. Summed up, use of this assay may contribute to a better understanding of cellular processes and dynamics regulating p8 transfer and HTLV-1 transmission.

METHODS

Cell Culture

$CD4^+$ Jurkat T-cells (from acute lymphoblastic leukemia) were cultured in RPMI 1640 with L-glutamine (0.35 g/l; GIBCO, Life Technologies, Darmstadt, Germany) supplemented with 45% Panserin 401 (PAN-Biotech, Aidenbach, Germany), 10% fetal calf serum (FCS; Sigma-Aldrich, Darmstadt, Germany) and penicillin/streptomycin (0.12 g/l each; Sigma-Aldrich). 293T cells were cultured in DMEM (GIBCO, Life Technologies), 10% FCS, L-glutamine and penicillin/streptomycin.

Expression Plasmids

Plasmids p8-HA (pME18S-p12I-Δ29) and the respective control pME (pME18S) were kindly provided by Genoveffa Franchini and have been described earlier (Fukumoto et al., 2009; Edwards et al., 2014). Constructs were checked for integrity by automated sequencing.

Transient Transfections

Jurkat T-cells ($1*10^7$) were transiently transfected by electroporation with a total amount of 100 μg of plasmid DNA using the *Gene Pulser Xcell™ Electroporation System* (BioRad, Munich, Germany) at 290 V and 1500 μF (exponential pulse). 293T cells were seeded at 5×10^5 cells per six-well. One day later, cells were transfected using *GeneJuice® Transfection Reagent* (Merck Millipore, Darmstadt, Germany) according to the manufacturer's protocol using a total amount of 2 μg DNA.

Western Blot

At day 2 post transfection, 293T or Jurkat T-cells were washed in phosphate-buffered saline (PBS without Ca^{2+} and Mg^{2+}) and lyzed in 150 mM NaCl, 10 mM Tris/HCl (pH 7.0), 10 mM ethylene-diamine tetra-acetic acid (EDTA), 1% Triton X-100, 2 mM dithiothreitol (DTT) supplemented with the protease inhibitors leupeptin, aprotinin (20 μg/ml each) and 1 mM phenyl-methylsulfonyl fluoride (PMSF; 1 mM) as described earlier (Mohr et al., 2014). Briefly, after repeated freeze-and-thaw cycles in liquid nitrogen, lysates were centrifuged at 14.000 rpm (15 min, 4°C), and supernatants containing cellular proteins were denatured in sodium dodecyl sulfate (SDS) loading dye [10 mM Tris/HCl (pH 6.8), 10% glycerine, 2% SDS, 0.1% bromphenole blue, 5% β-mercaptoethanol] for 10 min at 95°C. Subsequently, samples (50 μg) were subjected to SDS-polyacrylamide gel electrophoresis (SDS-PAGE) using the *XCell SureLock™ Mini-Cell Electrophoresis System* (Thermo Fisher Scientific, Waltham, MA, United States) and transferred to nitrocellulose membranes (Whatman®, Protran®, Whatman GmbH, Dassel,

Germany). Membranes were probed with rat monoclonal anti-HA-Peroxidase antibodies (clone 3F10; Roche, Mannheim, Germany), mouse monoclonal antibodies anti-β-actin (ACTB; Sigma-Aldrich/Merck, Darmstadt, Germany), or anti-glyceraldehyde-3-phosphate dehydrogenase (GAPDH; Sigma Aldrich/Merck). Secondary antibodies (anti-mouse) were conjugated with horseradish peroxidase (HRP; GE Healthcare, Little Chalfont, United Kingdom) and peroxidase activity was detected by enhanced chemoluminescence (ECL) using *INTAS Advanced Fluoreszenz und ECL Imager* (INTAS Science Imaging Instruments, Göttingen, Germany).

Flow Cytometry

To detect p8-HA expression, 293T cells or co-cultured cells were washed in PBS and fixed in 2% paraformaldehyde (PFA; 20 min, 20°C). After one washing step in wash buffer (PBS, 0.5% FCS and 2 mM EDTA), cells were permeabilized in wash buffer containing 0.5% saponin (Sigma-Aldrich/Merck) and stained in the same buffer using anti-HA-APC or the respective isotype-matched control antibody mouse IgG1-APC (both Milenty Biotech, Bergisch Gladbach, Germany; 1:40, 10 min, 20°C). After two washing steps in wash buffer containing 0.3% saponin, cells were resuspended in wash buffer and at least 3–5 × 10^5 events were analyzed using the *BD^{TM} LSR II* or the *BD LSRFortessa^{TM}* flow cytometer (Becton Dickinson GmbH, Heidelberg, Germany). Both devices were equipped with 405 and 633 nm laser. For evaluation of data, *FCS Express V3* (De Novo Software, Glendale, CA, United States) was used. In some experiments as indicated in the figure legend, cells were either stained without permeabilization in wash buffer, or cells were stained using *Inside Stain kit* (Miltenyi Biotec) according to the manufacturer's instructions.

To evaluate the vitality of Jurkat T-cells, cells were spun down, resuspended in PBS and analyzed using the *BD^{TM} LSR II* flow cytometer. The size of the cells (FSC, *forward scatter*) was plotted against the granularity (SSC, *side scatter*) and the percentage of living cells was assessed by gating.

Chemicals

To manipulate polymerization of the actin cytoskeleton, Jurkat T-cells transfected with p8-HA or pME (control) were pooled, respectively, at 24 h post transfection and incubated with increasing concentrations of cytochalasin D (Sigma-Aldrich/Merck; 0.5 μM, 1 μM, 2.5 μM, 5 μM), a potent inhibitor of actin polymerization, or the solvent control DMSO. Chemicals were added for 24 h and cells were incubated at 37°C. After one washing step in Jurkat T-cell culture medium (see Cell Culture), cells were used for co-culture assays (see Co-culture Assays).

Co-culture Assay Between p8-Expressing Donor Cells and Prestained Acceptor Jurkat T-Cells

Prestaining of Recipient Jurkat T-Cells

The live cell dye CellTracker^{TM} Blue CMAC (CMAC; 7-amino-4-chloromethylcoumarin; Thermo Fisher Scientific, Waltham, MA, United States; 10 mM in DMSO) was diluted in serum-free medium to a 20 μM working solution. To label acceptor T-cells, $1*10^6$ Jurkat T-cells per experimental condition were spun down (1200 rpm, 5 min, 20°C) and resuspended in 1 ml of prewarmed CellTracker^{TM} Blue CMAC (20 μM). After 45 min at 37°C, cells were spun down, washed in serum-free medium for three times, and directly used in co-culture assays (see Co-culture Assays).

Co-culture Assays

In time course experiments, $1*10^6$ Jurkat donor cells that had been transfected with p8-HA or pME (control) 48 h earlier were co-cultured with equal amounts of Jurkat acceptor cells prestained with CellTracker^{TM} Blue CMAC (Jurkat-CMAC) in 1 ml of Jurkat T-cell culture medium (see Cell Culture). For the time point 0 min, individual cell populations were fixed in 2% PFA prior to mixing of the donor and the acceptor cells. Co-cultures were incubated at 37° C for different time points (5, 30, 60 min, 24 h).

In experiments using cytochalasin D to manipulate cytoskeleton dynamics, $1*10^6$ transfected Jurkat T-cells pretreated with cytochalasin D or DMSO at one day post transfection for 24 h (see Chemicals) were co-cultured with equal amounts of Jurkat-CMAC in 1 ml fresh Jurkat T-cell culture medium (see Cell Culture) for another 24 h at 37°C. As a negative control for p8 transfer (time point 0 h), Jurkat donor cells transfected with p8-HA and pretreated with DMSO for 24 h were fixed in 2% PFA prior to mixing with fixed Jurkat-CMAC cells.

In co-cultures between 293T and Jurkat cells, transfected 293T cells were overlaid with equal amounts of Jurkat-CMAC acceptor cells ($3*10^6$) in 3 ml fresh culture medium at 37°C for either 5 min or for 24 h.

In experiments using fixed Jurkat donor cells, 24-well plates were coated with poly-L-lysine (Sigma-Aldrich) according to the manufacturer's instructions. Thereafter, $1*10^6$ transfected Jurkat T-cells were cultured in uncoated wells or they were allowed to adhere on poly-L-lysine-coated wells for 1 h (37°C). In the latter case, suspending cells were removed prior to adding equal amounts of Jurkat-CMAC for another 24 h at 37°C.

In all experiments, co-cultured cells were spun down at the end of the respective co-culture period (4°C, 1200 rpm, 5 min), washed in PBS, fixed in 2% PFA (20°C, 20 min) and subjected to intracellular staining using saponin (see Flow Cytometry).

Flow Cytometry and Quantitation of p8 Transfer After Flow Cytometry

After fixation and staining of co-cultured cells (see Flow Cytometry and Co-culture Assays), living cells were gated in dot plots displaying the forward scatter (FSC) plotted against the side scatter (SSC). Thereafter, CMAC-specific fluorescence was plotted against the SSC to discriminate between CMAC-negative donor cells (either Jurkat or 293T) and CMAC-positive Jurkat acceptor cells. Finally, HA-specific fluorescence was plotted against the SSC within the CMAC-negative Jurkat or 293T donor cells, representing the efficiency of transfection (E_T). In parallel, plotting of HA-specific fluorescence against the SSC within the CMAC-positive Jurkat acceptor cells, represented the transfer of p8 (T_{p8}) to the CMAC-positive Jurkat acceptor cells. To calculate the relative transfer of p8 [$T_{p8(relative)}$] between cells, T_{p8} was divided by E_T. To subtract background fluorescence of the HA-APC antibodies, gates were set according to labeling with isotype matched control antibodies and, additionally, co-cultures of cells transfected with the control plasmid pME were analyzed and considered in the evaluation. To quantitate the relative transfer of p8 $T_{p8(relative)}$, a mathematic equation was developed, which is described in more detail in the results section (see p8 Is Rapidly Transferred between Jurkat T-cells)

$$T_{p8(relative)} = \frac{T_{p8(p8_t)} - T_{p8(pME_t)}}{E_{T(p8_t)} - E_{T(pME_t)}}$$

T_{p8} shows the transfer of p8, which corresponds to the percentage of p8-HA positive cells within CMAC-positive acceptor cells ($T_{p8(p8_t)}$) at a given time point t and which was normalized on background fluorescence of the respective control cells transfected with pME ($T_{p8(pME_t)}$). E_T represents the efficiency of transfection at a given time point t and corresponds to the percentage of p8-HA positive cells within CMAC-negative donor cells ($E_{T(p8_t)}$), which is corrected by background fluorescence of the respective control cells transfected with pME ($E_{T(pME_t)}$).

Immunofluorescence and Confocal Laser Scanning Microscopy

At 48 h post transfection, p8-expressing donor Jurkat T-cells or control cells (Jurkat + pME) were co-cultured with acceptor Jurkat T-cells prestained with CellTracker™ Blue CMAC (see Prestaining of Recipient Jurkat T-cells). At different time points post co-culture (5, 30, 60 min, 24 h), cells were spotted on poly-L-lysine-coated glass slides and were fixed with 2% PFA (60 min, 20 °C). Cells were washed three times with PBS and permeabilized with 0.2% Triton X-100 (20 min, 4°C). After three washing steps, unspecific binding was prevented by 5% FCS/1% BSA in PBS (1 h, 20°C). Cells were stained with primary antibodies rabbit anti-HA (BioLegend, Eching, Germany; 1:100, 1 h, 20°C) followed by secondary antibodies anti-rabbit AlexaFluor 647 (Life Technologies/Thermo Fisher Scientific; 1:200, 30 min, 20°C). Slides were covered with ProLong Gold antifade reagent without DAPI (Molecular Probes/ Thermo Fisher Scientific) and analyzed by confocal laser scanning microscopy. All images were acquired using a Leica TCS SP5 confocal laser scanning microscope (Leica Microsystems GmbH, Wetzlar, Germany) equipped with a 63 × 1.4 HCX PL APO CS oil immersion objective lens. Images were analyzed using LAS AF software (Leica Microsystems GmbH). The numbers of cells expressing p8 in the CMAC-negative donor and CMAC-positive acceptor Jurkat T-cells were counted in 20 optical fields per experimental condition. $T_{p8(relative)}$ was calculated by normalizing the mean percentage of p8-HA positive cells within CMAC-positive acceptor cells on the mean percentage of p8-HA positive cells within CMAC-negative donor cells. In total, 6125 cells were analyzed manually.

To stain p8 in transfected Jurkat T-cells, p8-HA-expressing Jurkat T-cells or control cells (Jurkat + pME) were spotted on glass slides with marked rings (Medco, Munich, Germany), fixed with 2% PFA (60 min, 20°C) and washed with PBS for three times. The staining protocol was performed as described above, except that cells were either permeabilized with 0.2% Triton X-100 (20 min, 4°C) or they were stained without permeabilization. Upon staining, slides were covered with Vectashield® Antifade Mounting Medium with DAPI (Vector Laboratories/Biozol, Eching, Germany) and analyzed by confocal laser scanning microscopy.

Statistics

The means of at least three independently performed experiments were compared using t-tests as indicated in the figure legends. P-values were calculated using Microsoft Excel. $P < 0.05$ was considered to be significant (*), $P < 0.01$ was considered as highly significant (**).

RESULTS

p8 Protein Expression Is Detectable by Flow Cytometry

Expression of p8 has been analyzed by western blot, immunofluorescence or live cell imaging (Fukumoto et al., 2007, 2009; Van Prooyen et al., 2010b). To also establish a protocol to detect p8 expression by flow cytometry, we transfected 293T cells with p8-HA, a p8 expression plasmid C-terminally tagged with HA, and the respective control plasmid pME. At 48 h post transfection, cells were fixed and permeabilized with saponin, followed by staining with APC-labeled anti-HA-antibodies and the respective isotype-matched control antibodies. After setting gates that considered living cells and a marker that took into account the isotype control antibodies (**Figure 1A**, upper and lower left dot plots) and the low background fluorescence of pME-transfected control cells (**Figure 1A**, upper right dot plot), we found that ca. 61% of transfected 293T cells were p8-positive (**Figure 1A**, lower right dot plot). We tested several dilutions of the anti-HA-specific antibodies (ranging from 1:10 to 1:40), however, we could detect p8 expression at every concentration with a comparable frequency and sensitivity (data not shown). In control experiments of the same samples, we could also detect robust expression of p8-HA by western blot analysis (**Figure 1B**).

Next, we asked whether p8 can also be detected in T-cells, the natural cell type of HTLV-1 infection, and whether detection of p8 expression requires permeabilization of the cells. For this purpose, Jurkat T-cells were transfected with the C-terminally tagged p8-HA plasmid or the respective control plasmid pME. At 48 h post transfection, cells were subjected to flow cytometry (**Figure 1C**) or immunofluorescence (**Figure 1D**). For flow cytometry, cells were split and stained either in absence or presence of a permeabilization buffer. In the latter case, our homemade buffer with 0.5% saponin was compared to a commercially available intracellular staining kit (kit). In all experiments, markers were considered that took into account the isotype control antibodies and the low background fluorescence of pME-transfected control cells (data not shown). Flow cytometry of non-permeabilized cells revealed that ca. 15.5% (±3.6%) of Jurkat T-cells expressed p8-HA on their surface (**Figure 1C**, black bar). Permeabilization of cells using

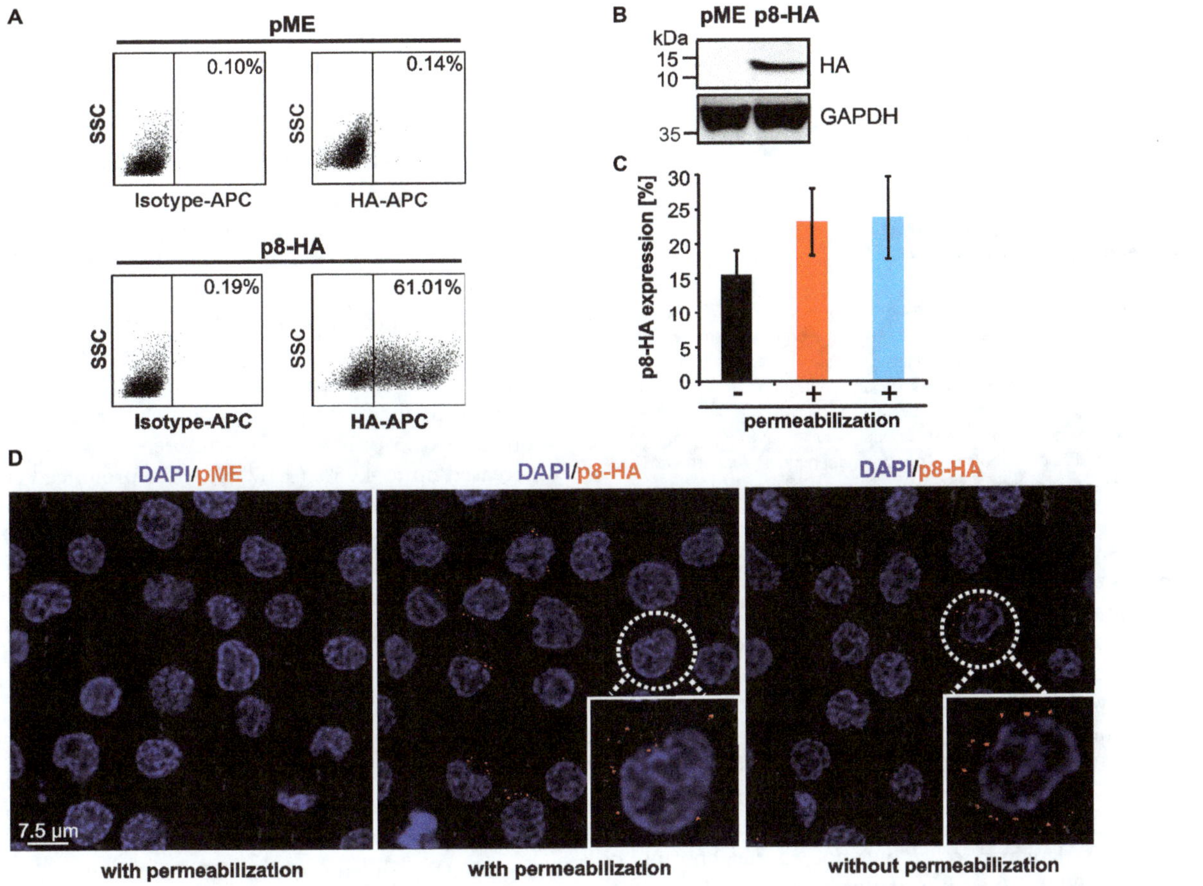

FIGURE 1 | Detection of p8 by flow cytometry compared to immunofluorescence. **(A,B)** 293T cells were transfected with p8-HA expression plasmids or the control plasmid pME for 48 h. **(A)** Flow cytometry after fixation and permeabilization of cells using HA-specific, APC-labeled antibodies. Dot plots display HA-APC-specific fluorescence plotted against the side scatter (SSC). **(B)** Immunoblot of p8-HA and the housekeeping gene glyceraldehyde-3-phosphate dehydrogenase (GAPDH). **(C)** Jurkat T-cells were transfected with p8-HA expression plasmids or the control plasmid pME for 48 h. Cells were split and stained either without permeablization buffer (black bar), or upon permeabilization with 0.5% saponin (red bar) or a commercially available intracellular staining kit (blue bar). The mean percentage of p8-HA expressing cells of three experiments ± standard deviation as detected by flow cytometry is shown. **(D)** Jurkat T-cells were transfected with expression plasmids p8-HA or pME for 48 h and spotted on glass slides. Cells were stained either with or without permeabilization with HA-specific antibodies and the respective secondary antibodies. Analysis by confocal microscopy shows p8-HA (red dots) and the nuclei (DAPI). Blow up: example of a p8-HA-expressing cell.

0.5% saponin (**Figure 1C**, red bar) or an intracellular staining kit (**Figure 1C**, blue bar) resulted in 23.1% (±4.8%) or 23.7% (±5.9%) (blue bar) of p8-HA-positive cells, respectively. First, these results show that transfection of T-cells is less efficient than transfection of 293T-cells. Second, the data suggest that the C-terminal part of p8-HA is oriented outside the cell membrane, otherwise surface staining of the HA-tagged p8 would not result in the detection of HA-positive cells. Third, these data support earlier observations that p8 is not only located intracellular, but also at the cell surface (Fukumoto et al., 2007, 2009; Edwards et al., 2014). Although staining upon permeabilization of cells resulted in a higher proportion of p8-positive cells than surface staining of p8, this difference was not statistically significant ($p > 0.05$). Next, we performed confocal microscopy of p8-transfected Jurkat T-cells either with or without permeabilization (**Figure 1D**). Imaging confirmed our results obtained by flow cytometry that p8 is detectable in both permeabilized and non-permeabilzed Jurkat T-cells. In more detail, p8 localized diffusely in dots in the cytoplasm and near the plasma membrane confirming earlier observations (Fukumoto et al., 2007, 2009; Edwards et al., 2014). Hence, to quantitate the total amount of p8 expression in cells by flow cytometry, we decided to stain p8 upon permeabilization of the cells with saponin in the next experiments. Taken together, p8 protein expression is not only detectable by immunofluorescence, but also by flow cytometry, thus, allowing automatic quantitation of p8 expression.

A Flow Cytometry-Based Co-culture Assay Allows to Quantitate the Transfer of p8 Between Cells

Since p8 is transferred rapidly to other cells, most likely via cellular protrusions (Van Prooyen et al., 2010b), we sought of establishing an experimental setup, which allows automatic quantification of p8 transfer between T-cells by flow cytometry (**Figure 2**). Moreover, we asked whether the cell-to-cell transfer of

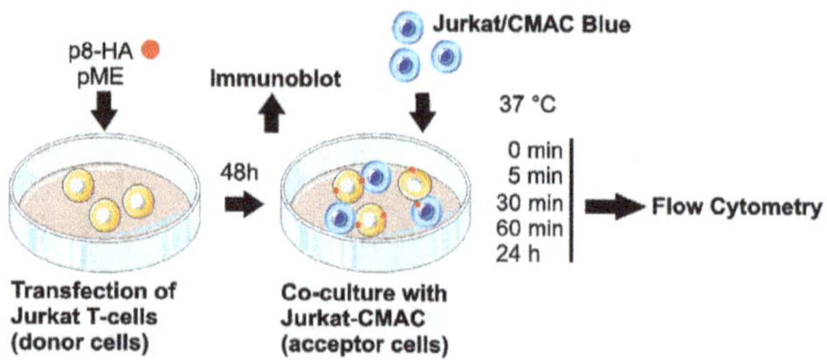

FIGURE 2 | Experimental setup of the co-culture assay. Jurkat T-cells were transfected with p8-HA expression plasmids or the control plasmid pME for 48 h. Transfected p8-donor or control (pME) cells were either subjected to immunoblot analysis or co-cultured with equal amounts of Jurkat acceptor cells ($1*10^6$) that had been prestained with Cell Tracker Blue CMAC Dye (Jurkat-CMAC). At different time points post co-culture at 37°C (0, 5, 30, 60 min, 24 h), cells were fixed in 2% paraformaldehyde (PFA), permeabilized, stained and analyzed by flow cytometry.

p8 increases upon prolonged co-culture. For this purpose, Jurkat T-cells were transfected with p8-HA or the control plasmids pME. After 2 days, a part of the transfected donor cells was lyzed for a control western blot, while the remaining cells were co-cultured with equal amounts of Jurkat acceptor T-cells that were prestained with the fluorescent CellTracker^{TM} Blue CMAC (Jurkat-CMAC). Briefly, after entering living cells, CellTracker^{TM} Blue CMAC is converted into a membrane-impermeable dye, which is well retained in living cells for several generations, transferred to daughter cells, but not to neighboring cells (King et al., 2004). Thus, use of this dye allows for a proper discrimination between p8-positive, CMAC-negative donor cells and Jurkat-CMAC acceptor cells. At different time points post co-culture (0, 5, 30, 60 min, 24 h), co-cultured cells were subjected to flow cytometry.

Co-cultured cells were analyzed by setting different gates, which are shown exemplarily for dot plots obtained at 60 min post co-culture (**Figure 3A**). First, the forward scatter (FSC) was plotted against the side scatter (SSC) and living cells were gated (**Figure 3A**, first line, red gates). Next, CMAC-specific fluorescence was plotted against the SSC (**Figure 3A**, second line) to discriminate between CMAC-negative Jurkat donor cells (purple gates) and CMAC-positive Jurkat acceptor cells (Jurkat-CMAC; blue gates). Finally, HA-specific fluorescence was plotted against the SSC to detect the efficiency of transfection (E_T) within the CMAC-negative donor cells (displayed on the left) or the transfer of p8 (T_{p8}) within the CMAC-positive acceptor cells Jurkat-CMAC (displayed on the right). Next to co-cultures of Jurkat-CMAC with p8-HA-expressing donor cells, co-cultures of Jurkat-CMAC with pME-transfected control cells were analyzed and background fluorescence of these cells was subtracted in the following evaluations. To quantitate the amount of p8 expressing recipient cells with regard to the percentage of p8 positive donor cells - the relative transfer of p8 [$T_{p8(relative)}$] between cells - a mathematic equation was developed (**Figure 3B**). Exemplarily, transfection of Jurkat T-cells with p8 resulted in 16.23% (16.59% – 0.36%) of p8-HA-expressing donor cells (E_T = 16.23%; **Figure 3A**), while transfer of p8-HA was detectable in 1.43% (1.85% - 0.42%) of all Jurkat-CMAC acceptor cells (T_{p8} = 1.43%) at 60 min post co-culture. Normalization of transferred p8 on the transfection efficiency using the equation (**Figure 3B**) resulted in $T_{p8(relative)}$) = 8.8%.

p8 Is Rapidly Transferred Between Jurkat T-Cells

Earlier work using real time live imaging techniques had shown that p8 is transferred within minutes between cells (Van Prooyen et al., 2010b). Using our flow cytometry-based approach, we analyzed the transfer of p8 between cells over time (**Figure 3C**). Cells that were fixed before co-culture served as negative control (0 min). Compared to the negative control, $T_{p8(relative)}$ significantly increased post co-culture compared to the negative control. At 5 min post co-culture, $T_{p8(relative)}$ significantly increased up to 4.95% (±1.73%) while prolonged co-culture for 30 and 60 min only led to a gradual enhancement of $T_{p8(relative)}$ ranging from 7.34% (±2.44%) to 8.48 % (±4.85%), which was not significantly increased when compared to the value obtained at 5 min post co-culture. This suggests that p8 is rapidly transferred within the first minutes of co-culture between cells confirming earlier findings (Van Prooyen et al., 2010b). However, prolonged co-culture of cells for 24 h resulted in a sharp and significant increase of $T_{p8(relative)}$ reaching values of approximately 23.17% (±5.55%). Thus, p8 is transferred over longer periods and it accumulates over time in co-cultured cells. The expression of p8-HA was also checked by western blotting using lysates of transfected donor cells that were obtained before co-culture (**Figure 3D**). Taken together, flow cytometry allows automatic quantitation of p8 transfer between Jurkat T-cells, revealing that p8 is rapidly transferred between cells and accumulates over time.

Comparison of Flow Cytometry and Immunofluorescence to Quantitate p8-Transfer

To evaluate the validity of the flow cytometry-based assay, we performed an immunofluorescence analysis of the co-culture experiments in parallel to flow cytometry similar to the experimental setup shown earlier (**Figure 2**). Thereafter, we manually quantitated p8 expression in fluorescence images at different time points post co-culture (5, 30, 60 min, 24 h; **Figure 4**). Briefly, co-cultures of donor Jurkat T-cells transfected with expression plasmids p8-HA or pME and acceptor cells Jurkat-CMAC were spotted on poly-L-lysine coated glass slides and stained with HA-specific antibodies and the respective secondary antibodies. A cutout of a fluorescent image depicting cells co-cultured for 24 h is shown exemplarily (**Figure 4A**). Imaging revealed that the p8 protein localized diffusely in dots in the cytoplasm and near the plasma membrane in Jurkat T-cells confirming earlier observations (Fukumoto et al., 2007, 2009; Edwards et al., 2014). Moreover, we could also confirm that p8 is transferred to other cells by immunofluorescence (Van Prooyen et al., 2010b) since we detected p8 (**Figure 4A**; red dots) not only in transfected donor cells, but also in co-cultured Jurkat-CMAC acceptor cells (**Figure 4A**; blue cells; *see* blow up for details). To calculate the relative transfer of p8 [$T_{p8(relative)}$] based on the

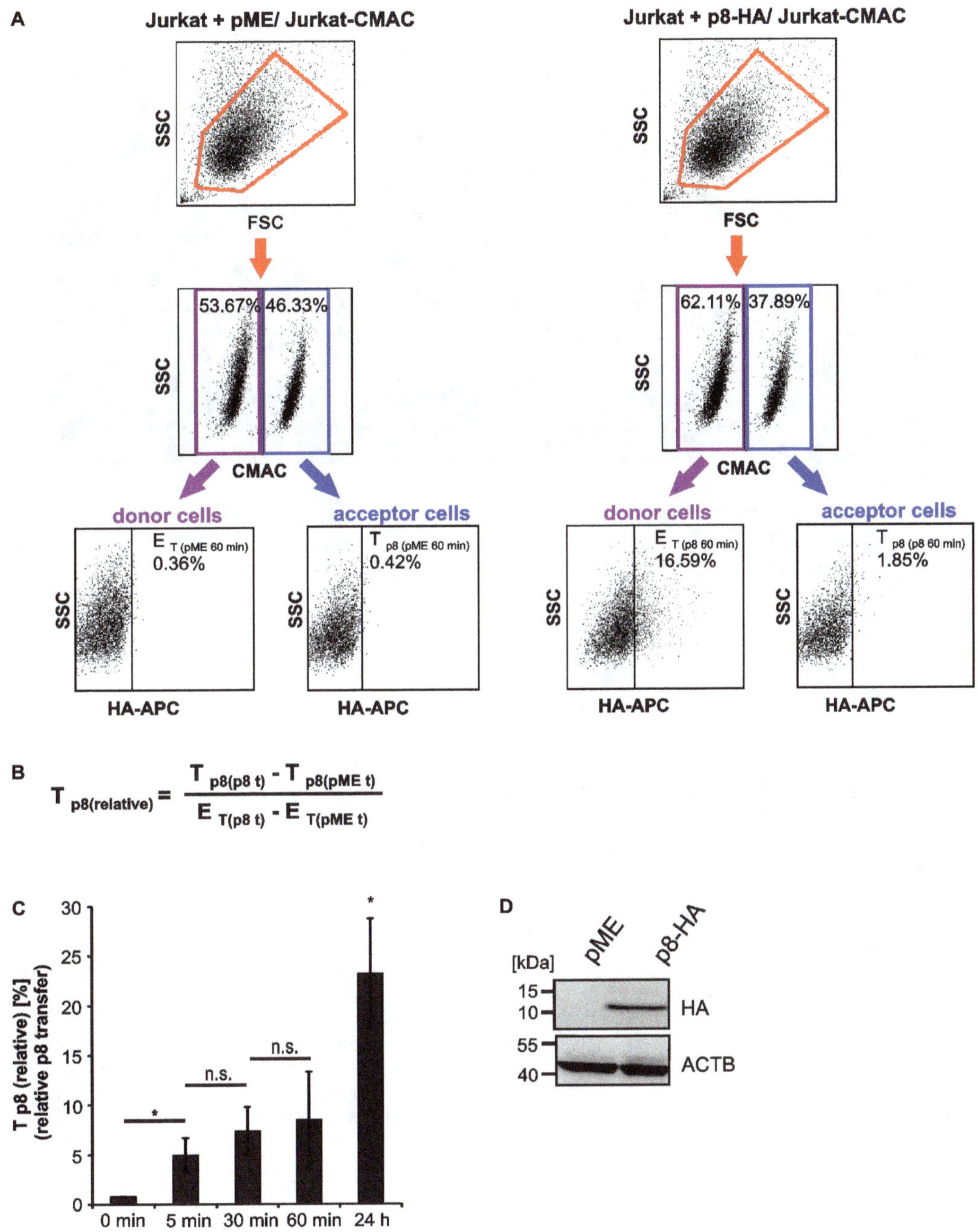

FIGURE 3 | Detection of rapid p8 transfer between Jurkat T-cells by flow cytometry. Jurkat donor T-cells were transfected with p8-HA or the control plasmid pME and co-cultured with prestained acceptor T-cells Jurkat-CMAC according to the experimental setup displayed in **Figure 2**. **(A)** Flow cytometry. At 48 h post transfection, equal amounts of donor and acceptor cells ($1*10^6$ cells each) were either directly fixed in 2% PFA and mixed (time point: 0 min), or they were co-cultured at 37°C for 5, 30, 60 min, or 24 h before fixation. After intracellular staining using HA-specific, APC-labeled antibodies or the respective isotype-matched control antibodies, flow cytometry was performed. Representative dot plots at 60 min post co-culture are shown. First line: Dot plots display the forward scatter (FSC) plotted against the side scatter (SSC) and living cells are gated (red gate). Second line: CMAC-specific fluorescence is plotted against the SSC, which allows discrimination between CMAC-negative donor (purple gate) and CMAC-positive acceptor (blue gate) cells. Third line: HA-specific fluorescence is plotted against the SSC and numbers represent the efficiency of transfection (E_T) within the CMAC-negative donor cells (displayed on the left) or the transfer of p8 (T_{p8}) within the CMAC-positive acceptor cells (displayed on the right). **(B)** Equation to calculate the relative transfer of p8 [$T_{p8(relative)}$] between cells. T_{p8} shows the transfer of p8, which corresponds to the percentage of p8-HA positive cells within CMAC-positive acceptor cells ($T_{p8(p8_t)}$) at a given time point t and which was normalized on background fluorescence of the respective control cells transfected with pME ($T_{p8(pME_t)}$). E_T represents the efficiency of transfection at a given time point t and corresponds to the percentage of p8-HA positive cells within CMAC-negative donor cells ($E_{T(p8_t)}$), which is corrected by background fluorescence of the respective control cells transfected with pME ($E_{T(pME_t)}$). **(C)** Time course analysis of $T_{p8(relative)}$ as measured by flow cytometry. The means of 3–4 independent experiments ±SE are shown and were compared as indicated using a paired *t*-test. *Indicates $p < 0.05$; * at 24 h, comparison between 24 h and all other times of measurement; n.s., not significant. **(D)** Representative immunoblot of p8-HA expression in Jurkat T-cells at 48 h post transfection. ACTB (β-Actin) served as housekeeping gene.

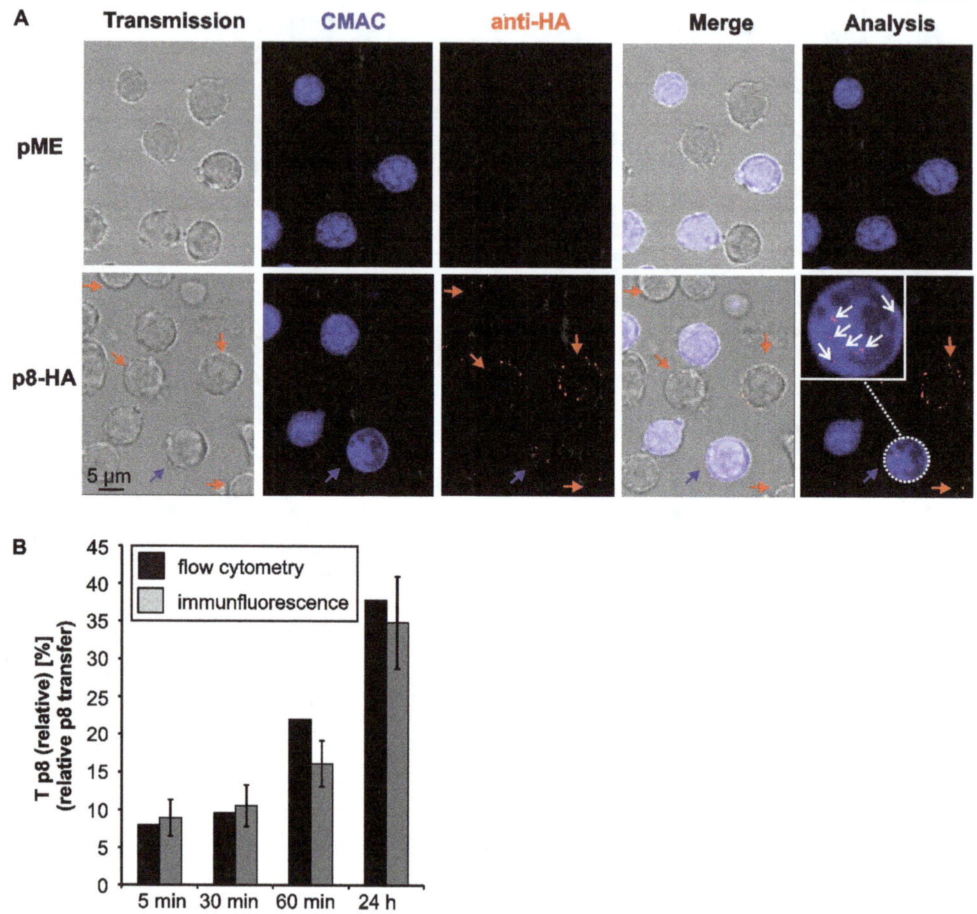

FIGURE 4 | Detection of p8 transfer between Jurkat T-cells by immunofluorescence. **(A)** Jurkat T-cells were transfected with expression plasmids p8-HA or pME for 48 h and co-cultivated with equal amounts of acceptor Jurkat T-cells prestained with Cell Tracker Blue CMAC (Jurkat-CMAC) on poly-L-lysine coated glass slides for 24 h at 37°C. Thereafter, cells were permeabilized and stained with HA-specific antibodies and the respective secondary antibodies. Slides were covered with ProLong Gold antifade reagent and analyzed by confocal microscopy. A cutout of an optical field shows cells expressing p8-HA (red) within the donor Jurkat T-cells (not stained) and the acceptor Jurkat T-cells (blue). The numbers of p8-positive cells (red) within the acceptor Jurkat T-cells (blue) were counted (white circles). Red arrows: p8-positive donor cells; blue arrows: p8-positive acceptor cell; blow up: example of a p8-expressing acceptor cell; white arrows: p8-HA. **(B)** Comparison of p8 transfer between flow cytometry (black bars) and immunofluorescence (gray bars). At 48 h post transfection with p8-HA, equal amounts of p8-donor Jurkat T- cells and Jurkat-CMAC acceptor cells ($1*10^6$ cells each) were co-cultured at 37°C for 5, 30, 60 min or 24 h before fixation. One representative time course experiment of relative p8 transfer [$T_{p8(relative)}$] as measured by flow cytometry (as shown for $n = 4$ in **Figure 3C**) is compared to the manual quantitation of relative p8 transfer within the same sample by immunofluorescence. $T_{p8(relative)}$ as measured by immunofluorescence was calculated by normalizing the mean percentage of p8-HA positive cells within CMAC-positive acceptor cells on the mean percentage of p8-HA positive cells within CMAC-negative donor cells in 20 optical fields. SE, standard error.

imaging data, we evaluated twenty optical fields per experimental condition and counted the number of p8-positive cells (red dots, **Figure 4A**) in the CMAC-negative donor and CMAC-positive acceptor T-cells (blue; white circles; **Figure 4A**). In total, 6125 cells were analyzed. Calculation of the relative p8 transfer revealed that $T_{p8(relative)}$ increased to 7.95% already at 5 min post co-culture (**Figure 4B**, gray bars). Prolonged co-culture led to a gradual increase of $T_{p8(relative)}$ over time with a sharp increase at 24 h post co-culture [$T_{p8(relative)} = 37.75\%$]. Comparison of the imaging data with the respective data obtained by flow cytometry (**Figure 4B**, black bars) revealed that $T_{p8(relative)}$ was comparable between both methods independent of the time point of co-culture. Thus, contrary to the manual counting of immunofluorescence images, flow cytometry allows an automatic and fast quantitation of p8 protein expression in a larger number of cells.

Transfer of p8 Between Cells Dependents on Proper Polymerization of the Cytoskeleton

Since p8 traffics to the cell membrane and increases T-cell conjugation depending on actin polymerization (Van Prooyen et al., 2010b), we now made use of our new assay to ask whether chemical manipulation of the actin cytoskeleton affects the transfer of p8 to target cells. Therefore, Jurkat T-cells were transfected with p8-HA or pME expression plasmids (**Figure 5A**). One day later, the cells were pre-treated with increasing concentrations of cytochalasin D, an inhibitor of actin

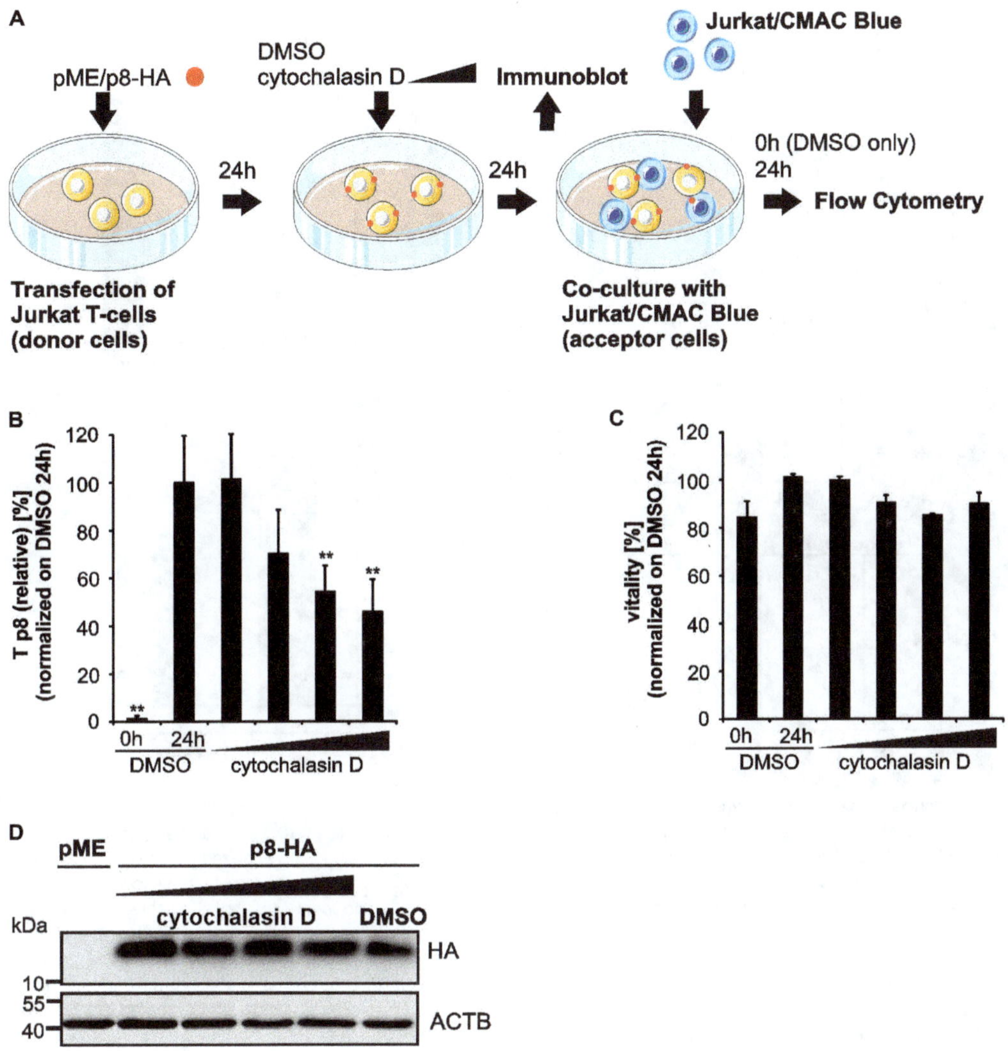

FIGURE 5 | Impairment of p8 transfer between T-cells after inhibition of actin polymerization. **(A)** Experimental setup. At 24 h post transfection of Jurkat T-cells with p8-HA or pME (control) expression plasmids (100 μg each), cells were cultured with increasing concentrations of an inhibitor of actin polymerization, cytochalasin D (0.5, 1, 2.5, 5 μM), or the solvent control dimethylsulfoxide (DMSO) for 24 h. Cells were either subjected to immunoblot analysis or co-cultured in fresh medium (without chemicals) with equal amounts of prestained Jurkat acceptor cells ($1*10^6$ cells, labeled with Cell Tracker Blue CMAC Dye) for 24 h at 37°C and analyzed by flow cytometry. DMSO-treated donor cells were also taken at 0 h post co-culture and served as negative control for p8 transfer. **(B)** The relative transfer of p8 [$T_{p8(relative)}$] was calculated as explained in **Figure 3**. Values display the means of at least four independent experiments (±SE) and were normalized on and compared to those of cells treated with DMSO and co-cultured for 24 h using an unpaired *t*-test. **Indicates $p < 0.01$. **(C)** Viability of Jurkat T-cells upon pretreatment with increasing amounts of cytochalasin D (0.5, 1, 2.5, 5 μM) or the solvent control DMSO for 24 h and co-culture for 0 h (DMSO) or 24 h (DMSO and all other samples) determined by forward-side scatter (FSC/SSC) analysis in flow cytometry. DMSO-treated cells (24 h co-culture) were set as 100%. The means of four independent experiments ±SE were compared to DMSO-treated cells using an unpaired *t*-test. **(D)** Representative immunoblot of p8-HA expression in Jurkat T-cells treated with the indicated inhibitors. ACTB served as housekeeping gene.

polymerization, or the solvent control DMSO for 24 h. Thereafter, a part of the transfected and pretreated donor cells was lyzed for a control western blot, while the remaining cells were co-cultured in fresh medium with equal amounts of acceptor T-cells Jurkat-CMAC for another 24 h. DMSO-treated donor cells were also taken at 0 h post co-culture and served as negative control for p8 transfer. Flow cytometry and evaluation of the results as described in **Figures 3A,B** revealed that $T_{p8(relative)}$ is significantly increased between DMSO-treated donor cells and Jurkat-CMAC after 24 h of co-culture compared to the control (0 h of co-culture; **Figure 5B**). Increasing concentrations of cytochalasin D (2.5 μM, 5 μM) led to a significant and dose-dependent decline of $T_{p8(relative)}$, while low concentrations did not (0.5 μM) or did only moderately (1 μM) affect $T_{p8(relative)}$. To check whether cytochalasin D affected the vitality of the cells, the percentage of living cells was assessed by flow cytometry of the co-cultured cells (**Figure 5C**). Cytochalasin D did not significantly reduce the vitality of the cells in any of the concentrations tested. In parallel to flow cytometry, we also performed immunoblots, which confirmed that p8 is properly expressed under all experimental

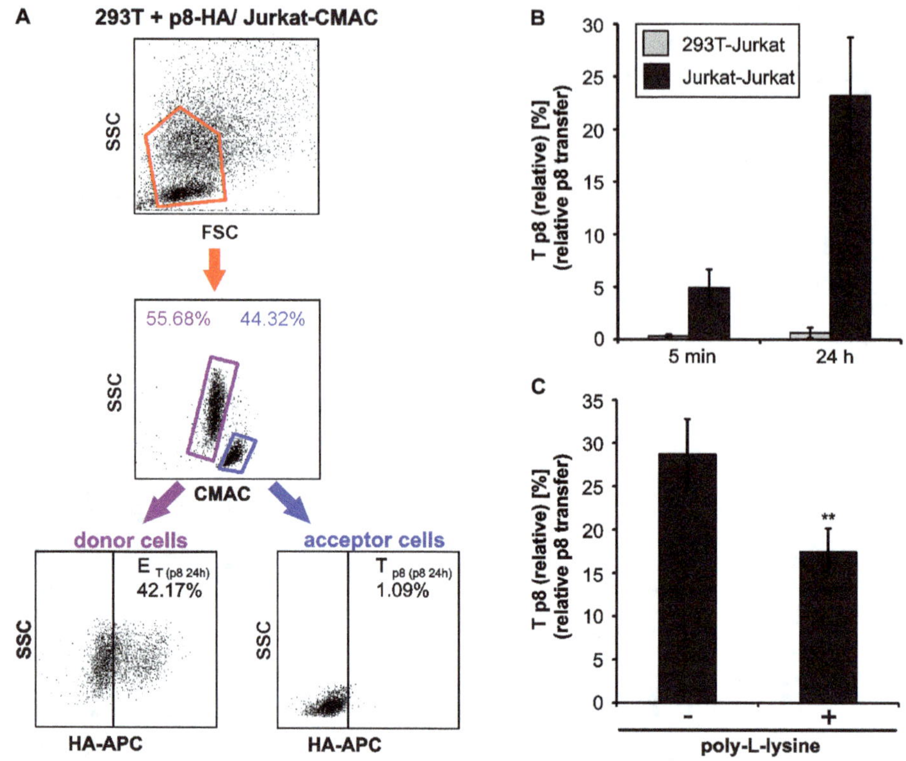

FIGURE 6 | Cell type-dependence of an efficient p8 transfer. **(A,B)** 293T were transfected with p8-HA expression plasmids or the control plasmid pME for 48h. Transfected p8-donor or control (pME) cells were co-cultured with equal amounts of Jurkat acceptor cells ($3*10^6$) that had been prestained with Cell Tracker Blue CMAC Dye. At 5 min or at 24 h post co-culture at 37°C, cells were fixed in 2% PFA, permeabilized, stained and analyzed by flow cytometry. **(A)** Representative dot plots of 293T cells transfected with p8-HA expression plasmids after co-culture with Jurkat-CMAC acceptor cells for 24 h are shown. First line: Dot plots display the forward scatter (FSC) plotted against the side scatter (SSC) and living cells are gated (red gate). Second line: CMAC-specific fluorescence is plotted against the SSC, which allows discrimination between CMAC-negative 293T donor (purple gate) and CMAC-positive Jurkat acceptor (blue gate) cells. Third line: HA-specific fluorescence is plotted against the SSC and numbers represent the efficiency of transfection (E_T) within the CMAC-negative 293T donor cells (displayed on the left) or the transfer of p8 (T_{p8}) within the CMAC-positive Jurkat acceptor cells (displayed on the right). **(B)** Time course analysis of $T_{p8(relative)}$ as measured by flow cytometry in co-cultures of 293T and Jurkat T-cells. The means of 4 independent experiments ±SE are shown and were compared as indicated using a *t*-test. For comparison, $T_{p8(relative)}$ between co-cultured Jurkat T-cells as shown in **Figure 3C** is displayed. **(C)** Jurkat donor T-cells were transfected with p8-HA or the control plasmid pME. At 48 h post transfection, $1*10^6$ donor cells were either fixed on poly-L-lysine coated culture plates for 1 h or they were left untreated. Thereafter, Jurkat donor T-cells were co-cultured with equal amounts of prestained acceptor T-cells Jurkat-CMAC at 37°C for 24 h and $T_{p8(relative)}$ was analyzed by flow cytometry. The means of 3 independent experiments ±SE are shown and were compared as indicated using an unpaired *t*-test. **Indicates $p < 0.01$.

conditions (**Figure 5D**). Summed up, as an application of our new flow cytometry assay, we could now show that transfer of p8 to target cells is actin-dependent.

Transfer of p8 Varies Between Cell Types

As another application for our new assay, we asked whether p8 transfer varies between different cell types. For this purpose, p8-expressing 293T donor cells and Jurkat-CMAC acceptor cells were co-cultured in analogy to the experimental setup described in **Figure 2**. At 5 min or 24 h post co-culture, cells were fixed, permeabilized and stained for flow cytometry (**Figure 6A**). Co-cultured cells were analyzed by setting gates, which are shown exemplarily for dot plots obtained at 24 h post co-culture. First, the forward scatter (FSC) was plotted against the side scatter (SSC) and living cells were gated (**Figure 6A**, first line, red gates). Next, CMAC-specific fluorescence was plotted against the SSC (**Figure 6A**, second line) to discriminate between CMAC-negative 293T donor cells (purple gates) and CMAC-positive Jurkat acceptor cells (Jurkat-CMAC; blue gates). Contrary to co-cultures in between Jurkat T-cells (**Figure 3A**), co-cultured 293T-cells and Jurkat T-cells could also be discriminated due to differences in the SSC (**Figure 6A**, second line). These data confirm that the dye CMAC is not leaky since only Jurkat T-cells (blue gate, lower SSC values) were CMAC-positive while 293T cells (purple gate, higher SSC values) remained CMAC-negative. Finally, HA-specific fluorescence was plotted against the SSC to detect E_T within the CMAC-negative 293T donor cells (displayed on the left) or T_{p8} within the CMAC-positive acceptor cells Jurkat-CMAC (displayed on the right). To calculate $T_{p8(relative)}$, we also analyzed co-cultures of Jurkat-CMAC with pME-transfected control 293T cells (data not shown), and background fluorescence of these cells were subtracted using the mathematic equation (**Figure 3B**). Evaluation of the relative p8 transfer revealed that despite high and robust expression of p8 in 293T donor cells, p8 could only be detected at a very low frequency in co-cultured Jurkat

T-cells reaching values of $T_{p8(relative)}$ = 0.29% (5 min) and $T_{p8(relative)}$ = 0.65% (24 h) (**Figure 6B**, gray bars). In some of the experimental replicates even none of the Jurkat acceptor T-cells was p8-positive. For comparison, $T_{p8(relative)}$ within co-cultures of Jurkat T-cells was plotted (**Figure 6B**, black bars), showing that p8 transfer was much higher [$T_{p8(relative)}$ = 23.17 % (24 h)]. To check whether this difference in p8-transfer between 293T and Jurkat T-cells is due to the fact that 293T cells are adherent while Jurkat are suspension cells, we allowed p8- and pME-transfected Jurkat T-cells to adhere on poly-L-lysine coated wells (1 h, 37°C) prior to addition of Jurkat-CMAC acceptor cells. After 24 h of co-culture, $T_{p8(relative)}$ was obtained by flow cytometry (**Figure 6C**). In comparison to co-cultures between non-adherent Jurkat T-cells (without poly-L-lysine), $T_{p8(relative)}$ was significantly reduced by 41.3% if donor Jurkat T-cells were fixed on poly-L-lysine-coated wells ($p < 0.01$). Despite this significant reduction of p8-transfer upon adherence of Jurkat donor cells, $T_{p8(relative)}$ was still higher between fixed Jurkat T-cells and Jurkat-CMAC (**Figure 6C**; $T_{p8(relative)}$ = 17.5%) than between 293T and Jurkat-CMAC (**Figure 6B**; 0.65%). Summed up, these data show that transfer of p8 depends on motility of the donor cells, and, to a greater extent, on the cell type. Thus, our novel protocol might be useful to study p8 transfer between cell types that are naturally infected with HTLV-1.

DISCUSSION

Quantitation of protein transport based on the evaluation of immunofluorescence images is a labor and time intensive effort. Dependent on the experimental setup, either manual or software-assisted evaluation of images is possible. Under each condition, high numbers of images with sufficient numbers of cells in several biological replicates have to be evaluated. Further, to localize and quantitate certain proteins within a cell, 3D evaluation of cells, including the time consuming analysis of Z-stacks, may be required. Certain circumstances allow the individual establishment of algorithms that may bypass the aforementioned difficulties (Wiesmann et al., 2017). To circumvent the stated problems of image evaluation, we developed a simple and fast flow cytometry-based assay to quantitate the transport of the HTLV-1-encoded protein p8 between cells. Contrary to the manual counting of immunofluorescence images, flow cytometry allows an automatic and fast quantitation of fluorescent protein expression in a large number of cells.

The viral protein p8 encoded by HTLV-1 is a mobile protein that is transferred rapidly between cells (Van Prooyen et al., 2010b). Despite low expression levels of p8 *in vivo*, it is required for establishing persistent infections (Valeri et al., 2010; Pise-Masison et al., 2014). Earlier work has shown that p8 enhances the formation of cellular conduits between T-cells, is transferred through these conduits to target T-cells and increases HTLV-1 transmission. Mechanistically, it was suggested that p8 dampens T-cell responses in target T-cells, thus facilitating HTLV-1 infection (Fukumoto et al., 2007; Van Prooyen et al., 2010a,b; Edwards et al., 2011). Thus, a better understanding of p8 transport between cells might broaden our understanding of HTLV-1 infectivity. Yet, p8 transfer between cells was analyzed by imaging techniques only (Van Prooyen et al., 2010b). However, flow cytometry based quantitation of p8-transfer provides several advantages compared to immunofluorescence-based quantitation of p8 transfer: Flow cytometry saves time, allows automatic quantitation of large number of cells, and thus, analysis of larger data sets. Further, use of CellTracker^{TM} Blue CMAC for staining of the acceptor cells provides a reliable discrimination between acceptor and donor cells since the dye is well retained in living cells (King et al., 2004), which we could also confirm in our co-culture experiments between 293T cells and Jurkat-CMAC (**Figure 6A**). This is an advantage over older dyes like Calcein-AM, which is cleaved by intracellular esterases to a fluorescent, membrane–impermeable, but gap junction–permeable form (Czyz et al., 2000). Further, the method is transferrable to be used with other cell-penetrating live cell dyes with different fluorescence. However, it is recommended to test several dilutions in co-culture experiments to ensure the dye is not leaky. Since further colors could be analyzed by flow cytometry, the method described here could be combined with methods measuring HTLV-1 virus transmission. It had been nicely shown by real-time live imaging that Gag-YFP expressed from an HTLV-1 construct and p8-mCherry are transported via protrusions to neighboring cells within minutes (Van Prooyen et al., 2010b). Combining the measurements of p8 transfer as described here and Gag transfer by flow cytometry (Gross et al., 2016) opens up the possibility to quantitatively evaluate p8's role in virus transmission.

Despite several advantages, the flow cytometry-based assay has also some limitations since it underestimates the real transfer of p8 between cells. First, the method does not consider transfer of p8 within transfected cells since all transfected cells are counted as donor cells. Second, it is also possible that p8 is secondarily transferred within the population of CMAC-blue stained acceptor cells and from CMAC-blue stained acceptor cells back to donor cells. However, all these circumstances cannot be excluded when evaluating immunofluorescence images, too, except, using live cell imaging of individual cells. A strategy to solve this issue would be the development of a fluorescently labeled p8 protein, which changes fluorescence when entering a new cell to discriminate between donor and acceptor cells.

Making use of the method, we performed time course experiments which confirmed earlier work based on real time live cell imaging that p8 is transferred between cells within few minutes after co-culture (Van Prooyen et al., 2010b). Extending these previous findings, we also observed both by flow cytometry and imaging that p8 transfer gradually increases over time. Assuming that every donor cell is p8 positive, p8 was transported to more than every fifth co-cultured T-cell at 24 h post co-culture, suggesting that p8 is efficiently spreading and accumulating over time.

Thus far, host factors regulating the transport of p8 between cells are unknown. Earlier work has shown that p8 traffics to the cell membrane and increases T-cell conjugation depending on actin polymerization (Van Prooyen et al., 2010b). We could now extend these earlier findings and show that transfer of p8 to target cells is actin-dependent, too. In unpublished work from our group (Donhauser et al., 2018, in preparation), we found a novel interaction between p8 and a modulator of actin-filament elongation, thus supporting an important role of proper actin polymerization for efficient transport of p8 to other cells. Taken together, the flow cytometry-based assay might be useful in determining the impact of various compounds on p8's trafficking between cells.

Yet, it is not understood whether p8 is transferred between all cell types that are HTLV-1-infected *in vivo*, or whether certain, cell-type specific host factors or co-culture conditions modulate the efficiency of p8 transfer between individual cell types. Our novel method may help to solve these questions in future studies. Data presented here clearly show that p8 transfer varies between different cell types. Despite high and robust expression of p8 in 293T donor cells, p8 could only be detected at a very low frequency in co-cultured Jurkat T-cells; in some of the experimental replicates even none of the Jurkat acceptor T-cells was p8-positive. First, host factors interacting with p8 and modulating p8 transfer may be missing in certain cell types like 293T cells. Second, the nature of cell–cell contact, the formation of cellular protrusions, and the ratio of donor and acceptor cells may be critical for p8 transfer. Third, it cannot be excluded that signals from the acceptor cell and respective receptors on the donor cell are critical for the transfer of p8, which is properly working between Jurkat T-cells but not between 293T and Jurkat T-cells. Fourth, the motility of the donor cell seems to be critical for p8-transfer since Jurkat T-cells fixed on poly-L-lysine transfer less p8 than suspension Jurkat T-cells. Lastly, it is unclear whether soluble p8 protein contributes to p8-transfer, which could be excluded by the use of filters (Chevalier et al., 2014). Hitherto, transfer of p8 has been shown by imaging techniques in co-cultures of p8-expressing Jurkat or MT-2 donor T-cells with Jurkat acceptor T-cells, or resting and activated PBMC acceptor cells (Van Prooyen et al., 2010b). Thus, further work applying our method is required to systematically and quantitatively evaluate the time course of p8 transfer in different cell types, including primary cell types like $CD4^+$ T-cells, $CD8^+$ T-cells, monocytes, dendritic cells and plasmacytoid dendritic cells, which are naturally infected with HTLV-1 *in vivo* (Richardson et al., 1990; Macatonia et al., 1992; Koyanagi et al., 1993; Nagai et al., 2001; Hishizawa et al., 2004; de Castro-Amarante et al., 2015; Melamed et al., 2015).

CONCLUSION

Use of the novel method described in this manuscript allows automatic and rapid quantitation of p8 transfer to target cells and might thus contribute to a better understanding of cellular processes and dynamics regulating p8 transfer and HTLV-1 transmission.

AUTHOR CONTRIBUTIONS

ND and SH performed the experiments and analyzed the data. AT-K conceived of the study, designed the experiments, analyzed the data, and wrote the manuscript.

FUNDING

This work was funded by Deutsche Forschungsgemeinschaft (DFG) (project C6, SFB796 and project TH2166/1-1). AT-K was further supported by the Exploration Grant of the Boehringer Ingelheim Foundation (BIS) and by the BFHZ (Bayerisch-Französisches Hochschulzentrum).

ACKNOWLEDGMENTS

We are grateful to Genoveffa Franchini, NIH, Bethesda, MD, United States, for providing p8 expression plasmids. We thank Klaus Überla and Bernhard Fleckenstein (Institute of Clinical and Molecular Virology, FAU) for continuous support. We acknowledge support by Deutsche Forschungsgemeinschaft and Friedrich-Alexander-Universität Erlangen-Nürnberg (FAU) within the funding program Open Access Publishing. Further, we acknowledge support by the Core Unit Cell Sorting and Immunomonitoring (Translational Research Center, Universitätsklinikum Erlangen, FAU, Erlangen, Germany). **Figures 2, 5A** were designed using the medical image bank Servier Medical Art, which is available under the Creative Commons license CC-BY.

REFERENCES

Alais, S., Mahieux, R., and Dutartre, H. (2015). Viral source-independent high susceptibility of dendritic cells to human T-cell leukemia virus type 1 infection compared to that of T lymphocytes. *J. Virol.* 89, 10580–10590. doi: 10.1128/JVI.01799-15

Carpentier, A., Barez, P. Y., Hamaidia, M., Gazon, H., de, B. A., Perike, S., et al. (2015). Modes of human T cell leukemia virus type 1 transmission, replication and persistence. *Viruses* 7, 3603–3624. doi: 10.3390/v7072793

Chevalier, S. A., Turpin, J., Cachat, A., Afonso, P. V., Gessain, A., Brady, J. N., et al. (2014). Gem-induced cytoskeleton remodeling increases cellular migration of HTLV-1-infected cells, formation of infected-to-target T-cell conjugates and viral transmission. *PLoS Pathog.* 10:e1003917. doi: 10.1371/journal.ppat.1003917

Czyz, J., Irmer, U., Schulz, G., Mindermann, A., and Hulser, D. F. (2000). Gap-junctional coupling measured by flow cytometry. *Exp. Cell Res.* 255, 40–46. doi: 10.1006/excr.1999.4760

de Castro-Amarante, M. F., Pise-Masison, C. A., McKinnon, K., Washington, P. R., Galli, V., Omsland, M., et al. (2015). HTLV-1 infection of the three monocyte subsets contributes to viral burden in humans. *J. Virol.* 90, 2195–2207. doi: 10.1128/JVI.02735-15

Ding, W., Albrecht, B., Luo, R., Zhang, W., Stanley, J. R., Newbound, G. C., et al. (2001). Endoplasmic reticulum and cis-Golgi localization of human T-lymphotropic virus type 1 p12(I): association with calreticulin and calnexin. *J. Virol.* 75, 7672–7682. doi: 10.1128/JVI.75.16.7672-7682.2001

Edwards, D., Fenizia, C., Gold, H., de Castro-Amarante, M. F., Buchmann, C., Pise-Masison, C. A., et al. (2011). Orf-I and orf-II-encoded proteins in HTLV-1 infection and persistence. *Viruses* 3, 861–885. doi: 10.3390/v3060861

Edwards, D., Fukumoto, R., de Castro-Amarante, M. F., Alcantara, L. C., Galvao-Castro, B., Washington, P. R., et al. (2014). Palmitoylation and p8-mediated human T-cell leukemia virus type 1 transmission. *J. Virol.* 88, 2319–2322. doi: 10.1128/JVI.03444-13

Fukumoto, R., Andresen, V., Bialuk, I., Cecchinato, V., Walser, J. C., Valeri, V. W., et al. (2009). In vivo genetic mutations define predominant functions of the human T-cell leukemia/lymphoma virus p12I protein. *Blood* 113, 3726–3734. doi: 10.1182/blood-2008-04-146928

Fukumoto, R., Dundr, M., Nicot, C., Adams, A., Valeri, V. W., Samelson, L. E., et al. (2007). Inhibition of T-cell receptor signal transduction and viral expression by the linker for activation of T cells-interacting p12(I) protein of human T-cell leukemia/lymphoma virus type 1. *J. Virol.* 81, 9088–9099. doi: 10.1128/JVI.02703-06

Furuta, R., Yasunaga, J. I., Miura, M., Sugata, K., Saito, A., Akari, H., et al. (2017). Human T-cell leukemia virus type 1 infects multiple lineage hematopoietic cells in vivo. *PLoS Pathog.* 13:e1006722. doi: 10.1371/journal.ppat.1006722

Gessain, A., and Cassar, O. (2012). Epidemiological aspects and world distribution of HTLV-1 infection. *Front. Microbiol.* 3:388. doi: 10.3389/fmicb.2012.00388

Gross, C., and Thoma-Kress, A. K. (2016). Molecular mechanisms of HTLV-1 cell-to-cell transmission. *Viruses* 8:74. doi: 10.3390/v8030074

Gross, C., Wiesmann, V., Millen, S., Kalmer, M., Wittenberg, T., Gettemans, J., et al. (2016). The tax-inducible actin-bundling protein fascin is crucial for release and cell-to-cell transmission of human T-cell leukemia virus type 1 (HTLV-1). *PLoS Pathog.* 12:e1005916. doi: 10.1371/journal.ppat.1005916

Hishizawa, M., Imada, K., Kitawaki, T., Ueda, M., Kadowaki, N., and Uchiyama, T. (2004). Depletion and impaired interferon-alpha-producing capacity of blood plasmacytoid dendritic cells in human T-cell leukaemia virus type I-infected individuals. *Br. J. Haematol.* 125, 568–575. doi: 10.1111/j.1365-2141.2004.04956.x

Igakura, T., Stinchcombe, J. C., Goon, P. K., Taylor, G. P., Weber, J. N., Griffiths, G. M., et al. (2003). Spread of HTLV-I between lymphocytes by virus-induced polarization of the cytoskeleton. *Science* 299, 1713–1716. doi: 10.1126/science.1080115

King, N., Korolchuk, S., McGivan, J. D., and Suleiman, M. S. (2004). A new method of quantifying glutathione levels in freshly isolated single superfused rat cardiomyocytes. *J. Pharmacol. Toxicol. Methods* 50, 215–222. doi: 10.1016/j.vascn.2004.05.003

Koralnik, I. J., Fullen, J., and Franchini, G. (1993). The p12I, p13II, and p30II proteins encoded by human T-cell leukemia/lymphotropic virus type I open reading frames I and II are localized in three different cellular compartments. *J. Virol.* 67, 2360–2366.

Koyanagi, Y., Itoyama, Y., Nakamura, N., Takamatsu, K., Kira, J., Iwamasa, T., et al. (1993). In vivo infection of human T-cell leukemia virus type I in non-T cells. *Virology* 196, 25–33. doi: 10.1006/viro.1993.1451

Macatonia, S. E., Cruickshank, J. K., Rudge, P., and Knight, S. C. (1992). Dendritic cells from patients with tropical spastic paraparesis are infected with HTLV-1 and stimulate autologous lymphocyte proliferation. *AIDS Res. Hum. Retroviruses* 8, 1699–1706. doi: 10.1089/aid.1992.8.1699

Melamed, A., Laydon, D. J., Al, K. H., Rowan, A. G., Taylor, G. P., and Bangham, C. R. (2015). HTLV-1 drives vigorous clonal expansion of infected $CD8^+$ T cells in natural infection. *Retrovirology* 12:91. doi:10.1186/s12977-015-0221-1

Mohr, C. F., Kalmer, M., Gross, C., Mann, M. C., Sterz, K. R., Kieser, A., et al. (2014). The tumor marker Fascin is induced by the Epstein-Barr virus-encoded oncoprotein LMP1 via NF-kappaB in lymphocytes and contributes to their invasive migration. *Cell Commun. Signal.* 12:46. doi: 10.1186/s12964-014-0046-x

Nagai, M., Brennan, M. B., Sakai, J. A., Mora, C. A., and Jacobson, S. (2001). $CD8^+$ T cells are an in vivo reservoir for human T-cell lymphotropic virus type I. *Blood* 98, 1858–1861. doi: 10.1182/blood.V98.6.1858

Pais-Correia, A. M., Sachse, M., Guadagnini, S., Robbiati, V., Lasserre, R., Gessain, A., et al. (2010). Biofilm-like extracellular viral assemblies mediate HTLV-1 cell-to-cell transmission at virological synapses. *Nat. Med.* 16, 83–89. doi: 10.1038/nm.2065

Pise-Masison, C. A., de Castro-Amarante, M. F., Enose-Akahata, Y., Buchmann, R. C., Fenizia, C., Washington, P. R., et al. (2014). Co-dependence of HTLV-1 p12 and p8 functions in virus persistence. *PLoS Pathog.* 10:e1004454. doi: 10.1371/journal.ppat.1004454

Poiesz, B. J., Ruscetti, F. W., Gazdar, A. F., Bunn, P. A., Minna, J. D., and Gallo, R. C. (1980). Detection and isolation of type C retrovirus particles from fresh and cultured lymphocytes of a patient with cutaneous T-cell lymphoma. *Proc. Natl. Acad. Sci. U.S.A.* 77, 7415–7419. doi: 10.1073/pnas.77.12.7415

Richardson, J. H., Edwards, A. J., Cruickshank, J. K., Rudge, P., and Dalgleish, A. G. (1990). In vivo cellular tropism of human T-cell leukemia virus type 1. *J. Virol.* 64, 5682–5687.

Turpin, J., Alais, S., Marcais, A., Bruneau, J., Melamed, A., Gadot, N., et al. (2017). Whole body clonality analysis in an aggressive STLV-1 associated leukemia (ATLL) reveals an unexpected clonal complexity. *Cancer Lett.* 389, 78–85. doi: 10.1016/j.canlet.2016.12.022

Valeri, V. W., Hryniewicz, A., Andresen, V., Jones, K., Fenizia, C., Bialuk, I., et al. (2010). Requirement of the human T-cell leukemia virus p12 and p30 products for infectivity of human dendritic cells and macaques but not rabbits. *Blood* 116, 3809–3817. doi: 10.1182/blood-2010-05-284141

Van Prooyen, N., Andresen, V., Gold, H., Bialuk, I., Pise-Masison, C., and Franchini, G. (2010a). Hijacking the T-cell communication network by the human T-cell leukemia/lymphoma virus type 1 (HTLV-1) p12 and p8 proteins. *Mol. Aspects Med.* 31, 333–343. doi: 10.1016/j.mam.2010.07.001

Van Prooyen, N., Gold, H., Andresen, V., Schwartz, O., Jones, K., Ruscetti, F., et al. (2010b). Human T-cell leukemia virus type 1 p8 protein increases cellular conduits and virus transmission. *Proc. Natl. Acad. Sci. U.S.A.* 107, 20738–20743. doi: 10.1073/pnas.1009635107

Wiesmann, V., Bergler, M., Palmisano, R., Prinzen, M., Franz, D., and Wittenberg, T. (2017). Using simulated fluorescence cell micrographs for the evaluation of cell image segmentation algorithms. *BMC Bioinformatics* 18:176. doi: 10.1186/s12859-017-1591-2

Yoshida, M., Miyoshi, I., and Hinuma, Y. (1982). Isolation and characterization of retrovirus from cell lines of human adult T-cell leukemia and its implication in the disease. *Proc. Natl. Acad. Sci. U.S.A.* 79, 2031–2035. doi: 10.1073/pnas.79.6.2031

6

The Exceptional Oncogenicity of HTLV-1

Yutaka Tagaya * *and Robert C. Gallo*

Division of Basic Science, Institute of Human Virology, University of Maryland School of Medicine, Baltimore, MD, United States

Correspondence:
Yutaka Tagaya
ytagaya@ihv.umaryland.edu

Human T-cell leukemia virus-1 (HTLV-1) is the first pathogenic human retrovirus identified in 1979 by the Gallo group. HTLV-1 causes fatal T-cell leukemia (adult T cell leukemia) and a progressive myelopahy (HTLV-1-associated myelopathy/ tropical spastic paraparesis, HAM/TSP) and other disorders. Since the discovery of HTLV-1, several other microorganisms are demonstrated to cause cancer in humans. In this article, we investigated the oncogenic capacity of HTLV-1, in comparison with those of other oncoviruses and one oncobacterium (Helicobacter pylori, H. Pylori) based on published literature. We conclude here that HTLV-1 is one of the most and may be the most carcinogenic among them and arguably one of the most potent of the known human carcinogens. This fact has not been noted before and is particularly important to justify why we need to study HTLV-1 as an important model of human viral oncogenesis.

Keywords: HTLV-1, retrovirus, oncogenicity, T-cell leukemia, oncoviruses

At the time of discovery of HTLV-1 (Poiesz et al., 1980a,b; Hinuma et al., 1981), the concept that viruses could cause cancer (ATL, adult T-cell leukemia/lymphoma by HTLV-1) in humans was a far-fetched idea that the medical community eventually accepted, despite the earlier identification of Epstein-Barr virus in cultured lymphoblasts from Burkitt's lymphoma (Epstein et al., 1964). After 35 years, the concept of viral oncogenesis has become a textbook fact and several viruses have joined the group of human oncoviruses. In addition, there is now one oncogenic bacterium (Helicobacter pylori) and a strong suspicion of more. The list of these oncogenic microbes, their global burden, and the incidence of malignant disorders involving them are provided in **Table 1** and summarized below.

The mechanism by which a virus causes tumor differs from virus to virus. While some viruses seem to require co-factors, and/or work indirectly such as by micro-environmental alterations, others directly transform host cells. One thing in common is that there are not yet effective cures for viral malignancies in humans. We conducted a literature survey in order to compare the carcinogenic potency of known oncoviruses, and present the risk of malignancy development following the infection of each oncovirus.

EPSTEIN BARR VIRUS (EBV)

In developing countries, the prevalence of EBV infection reaches over 90% before the age of 20. But even in high-risk area (East Asia), the incidence of nasopharyngeal carcinoma in association with EBV is at best 0.05% or less. Numbers for endemic Burkitt's lymphoma is more imprecise but estimated as 3~4/100,000 among Ugandan children (Hjalgrim et al., 2007), one of the most endemic area of EBV-related Burkitt's lymphoma (BL). It is also of note that the EBV-mediated malignancy development generally requires a co-factor, especially malaria in the case of BL and

TABLE 1 | Summary of the statistics on malignancy development associated with oncogenic microorganisms.

Infectious organism	Global burden (% in population)	Lifetime risk of malignancy development following infection
H. Pylori	5.5	3%
HPV	5.2	0.29%
HCV + HBV	4.9	HCV, 1∼3%/HBV, <1%
HHV-8	2∼5	Minimally oncogenic by itself
EBV	>90	0.3∼0.4%
HTLV-1	0.3% (∼40% in central Australian aboriginals)	5∼10%
MCV	0.06	??
AAV2	??	??

dietary factors in nasopharyngeal carcinoma. Elsewhere, the role of EBV as an oncogenic agent remain unclear. Thus, EBV is an oncogenic virus but one with very low level of oncogenicity and appears to be only one factor in a multi-factorial cause.

HUMAN HERPESVIRUS-8 (HHV-8)

Though it is a cause of Kaposi's sarcoma (HHV-8 is thus also known as the Kaposi Sarcoma-associated herpesvirus) and a more special and very atypical form of lymphoma in collaboration with HIV, HHV-8 is very minimally oncogenic in the absence of HIV. We also note that in most of primary effusion lymphoma cases associated with HHV-8, EBV is detected along with HHV-8, another support that HHV-8 is not potently oncogenic on its own (Bhutani et al., 2015).

HUMAN PAPILLOMA VIRUS (HPV)

More than 100 HPVs exist, and at least 13 of them are associated with human cancers (zur Hausen, 2009). The high rise types 16 and 18 are particularly responsible for 60∼70% of cervical cancer worldwide (Khan et al., 2005). More than 80∼90% of sexually active men and women will be infected with at least one type of HPV at some point in life and nearly 50% of these infections are with a high risk HPV type (Hariri et al., 2011). In the US, these high-risk HPV causes 2∼3% of all cancer cases among men and women (Jemal et al., 2013). This translates to 0.29% of the newly infected individuals with high-risk HPVs eventually developing HPV-related cancer. HPV infection can cause cervical, vaginal, and vulvar cancers, cancers of the penis or anus, as well as some head and neck cancers. Vaccines against HPV 16 and 18 have received approval in many countries.

HEPATITIS C VIRUS (HCV)/HEPATITIS B VIRUS (HBV)

[HCV] 85% of acute HCV infections will be unresolved, developing into chronic infection. Importantly, there is no evidence to support the conclusion that HCV is directly oncogenic, nonetheless it is clearly associated with hepatocellular carcinoma (HCC). That said, one study estimates that the HCC develops in 1∼3% of HCV infected persons after 30 years and the mechanisms are obscure (Goodgame et al., 2003). A chronic infection with HBV establishes when infection occurs in infants and children.

[HBV] <10% of healthy adults over 19 years will develop chronic hepatitis following HBV infection. A study conducted in Taiwan, a hyperendemic area of chronic HBV, shows that lifetime risk of HCC with HBsAg seropositive is 6.96% among women, but is an amazing 27.4% among men with HBsAg seropositivity (Huang et al., 2011). In a study conducted in the US, the 5 year HCC cumulative incidences with HBV-associated cirrhosis was 10% (Fattovich et al., 2008). Considering that <10% of HBV infected healthy adults develop chronic infection, the percentages of HCC development by HBV infection should be much lower than 1%.

ADENO-ASSOCIATED VIRUS TYPE 2 (AAV2)

This is the newest member to the family of oncogenic microorganisms (Nault et al., 2015). The oncogenesis of AAV2 is associated with HCC without cirrhosis (thus distinguishing then from those caused by hepatitis virus B or C). A clonal integration of AAV2 was found in 11 of 193 HCC cases in which AAV2 integrations occurred in cancer driver genes such as cyclin A2, E1, Telomerase reverse transcriptase etc. These insertions of AAV2 led to the over-expression of target genes. There is still very limited information available for the epidemiological facts about HCC caused by AAV2.

MERKEL CELL POLYOMAVIRUS (MCV)

MCV was discovered in 2008 (Feng et al., 2008) and causes a rare but aggressive neuroendocrine tumor (Merkel cell carcinoma, MCC) of the skin and its incidence seems rising. A 5 year mortality rate is as high as 46%. The annual incidence of MCC is 0.6 per 100,000 persons (0.06%) and about 1,600 new cases annually appear in the US. This rise in incidence may be partly because of the increased awareness and improved diagnostic methodology. The median age at diagnosis is >70 years. Only 4% of patients are diagnosed at 50 year or younger and it is extremely rare in children (Hughes et al., 2014). The oncogenic mechanism of MCV is still under investigation. No viral elements have been shown to directly cause cellular transformation. Nor does MCV seem to exploit cellular oncogenic mechanisms (p53, PTEN, Raf, Ras, etc.) (Lemos and Nghiem, 2007). Increase of MCC upon UV irradiation is reported, suggesting a potential involvement of defective DNA repair mechanism in the carcinogenesis (Lunder and Stern, 1998). In addition, a weakened immune function may be related to the occurrence of MCC. For example, the incidence of MCC is 5∼10 fold higher in immune-compromised individuals with AIDS or solid organ transplant (Becker, 2010). Some anecdotal regression of MCC has been observed following improvement in immune function (Wooff et al., 2010). As with most other oncogenic viruses, the presence of MCC seems insufficient to induce MCC and additional cellular events together with the loss of immune-surveillance are postulated. In summary, there is still missing information to evaluate the oncogenic potential of MCV in comparison with other oncogenic viruses, however MCC caused by MCV may represent another case of aggressive viral oncogenesis.

HUMAN IMMUNODEFICIENCY VIRUS-1 (HIV-1)

HIV is not generally listed with the oncogenic viruses by most virologists. However, in some reports it is analogous to HCV in that from an epidemiological view it is related to the frequency of developing cancer, notably non-Hodgkin lymphomas, Kaposi sarcoma, and cervical cancer that are known as AIDS-defining malignancies (reviewed in Levine, 1993). The mechanism(s) of cancer is indirect and relates to the microenvironment and perhaps to the diminished immune surveillance because the virus is rarely found in the tumor cells. The incidence of cancer development in HIV infected persons is about 40% (Levine, 1993) and usually accompanies co-infection with other oncoviruses.

HUMAN T-CELL LEUKEMIA VIRUS-1 (HTLV-1)

The oncogenic nature of HTLV-1 is solid; (1) Epidemiology studies (recently reviewed by Gessain and Cassar (Gessain and Cassar, 2012), (2) Clonal integration of the HTLV-1 in ATL cells (Hahn et al., 1983). This indicates that the virus was present at the level of the progenitor cell that gave rise to the leukemia. (3) Animal models that recapitulate human leukemia/lymphoma development by HTLV-1 or its genetic components (Tax-1, HBZ) of HTLV-1 (Hasegawa et al., 2006; Satou et al., 2011), (4) Immortalization of T cells by the Tax-1 gene of HTLV-1 (Grassmann et al., 1989), (5) Reproduction of the leukemia/lymphoma by the whole virus. The prevalence of ATL is 3–5% among infected persons and HTLV-1 infection is 0.1% as global average but the epidemiological data are still incomplete because those from highly populated areas such as China, India, the Maghreb, and East Africa are still unavailable. ATL is extremely difficult to treat, to the point that patients diagnosed with aggressive forms (acute and lymphomatous phases) were estimated to have <1 year of life left. However, recent years have seen promising developments of novel treatment modalities including arsenic trioxide (Bazarbachi et al., 2011), allogeneic stem cell transplant (Utsunomiya et al., 2001; Katsuya et al., 2015) and antibody therapy involving humanized anti-CCR4mAb (Ishida and Ueda, 2011; Katsuya et al., 2015). However, the establishment of a curative treatment for ATL probably needs more creative ideas.

HELICOBACTER PYLORI (H. PYLORI)

Helicobacter is a bacterium, and represents the first example of what may be emerging examples of bacterial oncogenesis. It has been somewhat controversial how infection with helocobacter pylori would increase or decrease various type of cancer in humans. A recent study in Japan (involving 1,526 patients) reported that gastric cancer develops in approximately 3% of H. pylori-infected patients, compared to none of the uninfected patients (Uemura et al., 2001). This number matches the development of Adult T cell Leukemia/lymphoma (ATL) in HTLV-1 infected individuals.

In summary, we note here that HTLV-1 is one of the most oncogenic entities known among human viruses and even among most known human carcinogens. The oncogenic mechanism of HTLV-1 is more direct compared with other entities and to date there is no evidence for any particular co-factor requirement. This striking characteristic of HTLV-1 seems overlooked because of its low prevalence in the U.S. and Europe. As discussed in an accompanying paper, we propose to revert the name of HTLV-1 back to the original, Human T-cell leukemia virus. This is also supported by a recent survey among 21 HTLV-1 experts from all over the world who belong to the HTLV-1 task-force of the Global Virus Network (GVN, http://gvn.org), a coalition of virologists on a global scale, as the result of the vote (16 vs. 5) supported the original name "leukemia". Another vote conducted at the 18th International HTLV-1 meeting (Tokyo, 2017) again showed majority support (78 vs. 26) on the "leukemia" name. We believe that this most carcinogenic virus deserves the "leukemia" name.

AUTHOR CONTRIBUTIONS

YT and RG conceptualized the idea, conducted literature search and investigation, and wrote the manuscript.

FUNDING

This work was supported by an internal funding of the Institute of Human Virology.

REFERENCES

Bazarbachi, A., Suarez, F., Fields, P., and Hermine, O. (2011). How i treat adult T-cell leukemia/lymphoma. *Blood* 118, 1736–1745. doi: 10.1182/blood-2011-03-345702

Becker, J. C. (2010). Merkel cell carcinoma. *Ann. Oncol.* 21(Suppl. 7), vii81–vii85. doi: 10.1093/annonc/mdq366

Bhutani, M., Polizzotto, M. N., Uldrick, T. S., and Yarchoan, R. (2015). Kaposi sarcoma-associated herpesvirus-associated malignancies: epidemiology, pathogenesis, and advances in treatment. *Semin. Oncol.* 42, 223–246. doi: 10.1053/j.seminoncol.2014.12.027

Epstein, M. A., Achong, B. G., and Barr, Y. M. (1964). Virus particles in cultured Lymphoblasts from Burkitt's Lymphoma. *Lancet* 1, 702–703. doi: 10.1016/S0140-6736(64)91524-7

Fattovich, G., Bortolotti, F., and Donato, F. (2008). Natural history of chronic hepatitis B: special emphasis on disease progression and prognostic factors. *J. Hepatol.* 48, 335–352. doi: 10.1016/j.jhep.2007.11.011

Feng, H., Shuda, M., Chang, Y., and Moore, P. S. (2008). Clonal integration of a polyomavirus in human Merkel cell carcinoma. *Science* 319, 1096–1100. doi: 10.1126/science.1152586

Gessain, A., and Cassar, O. (2012). Epidemiological aspects and world distribution of HTLV-1 infection. *Front. Microbiol.* 3:388. doi: 10.3389/fmicb.2012.00388

Goodgame, B., Shaheen, N. J., Galanko, J., and El-Serag, H. B. (2003). The risk of end stage liver disease and hepatocellular carcinoma among persons infected with hepatitis C virus: publication bias? *Am. J. Gastroenterol.* 98, 2535–2542. doi: 10.1111/j.1572-0241.2003.07678.x

Grassmann, R., Dengler, C., Muller-Fleckenstein, I., Fleckenstein, B., McGuire, K., Dokhelar, M., et al. (1989). Transformation to continuous growth of primary human T lymphocytes by human T-cell leukemia virus type I X-region genes transduced by a Herpesvirus saimiri vector. *Proc. Natl. Acad. Sci. U.S.A.* 86, 3351–3355. doi: 10.1073/pnas.86.9.3351

Hahn, B., Manzari, V., Colombini, S., Franchini, G., Gallo, R. C., and Wong-Staal, F. (1983). Common site of integration of HTLV in cells of three patients with mature T-cell leukaemia-lymphoma: a retraction. *Nature* 305:340. doi: 10.1038/305340a0

Hariri, S., Unger, E. R., Sternberg, M., Dunne, E. F., Swan, D., Patel, S., et al. (2011). Prevalence of genital human papillomavirus among females in the United States, the National Health And Nutrition Examination Survey, 2003–2006. *J. Infect. Dis.* 204, 566–573. doi: 10.1093/infdis/jir341

Hasegawa, H., Sawa, H., Lewis, M. J., Orba, Y., Sheehy, N., Yamamoto, Y., et al. (2006). Thymus-derived leukemia-lymphoma in mice transgenic for the Tax gene of human T-lymphotropic virus type I. *Nat. Med.* 12, 466–472. doi: 10.1038/nm1389

Hjalgrim, H., Friborg, J., and Melbye, M. (2007). *The Epidemiology of EBV and its Association with Malignant Disease*. Cambridge: Cambridge University Press.

Hinuma, Y., Nagata, K., Hanaoka, M., Nakai, M., Matsumoto, T., Kinoshita, K., et al. (1981). Adult T-cell leukemia: antigen in an ATL cell line and detection of antibodies to the antigen in human sera. *Proc. Natl. Acad. Sci. U.S.A.* 78, 6476–6480. doi: 10.1073/pnas.78.10.6476

Huang, Y. T., Jen, C. L., Yang, H. I., Lee, M. H., Su, J., Lu, S. N. et al. (2011). Lifetime risk and sex difference of hepatocellular carcinoma among patients with chronic hepatitis B and C. *J. Clin. Oncol.* 29, 3643–3650. doi: 10.1200/JCO.2011.36.2335

Hughes, M. P., Hardee, M. E., Cornelius, L. A., Hutchins, L. F., Becker, J. C., and Gao, L. (2014). Merkel Cell Carcinoma: epidemiology, target, and therapy. *Curr. Dermatol. Rep.* 3, 46–53. doi: 10.1007/s13671-014-0068-z

Ishida, T., and Ueda, R. (2011). Antibody therapy for adult T-cell leukemia-lymphoma. *Int. J. Hematol.* 94, 443–452. doi: 10.1007/s12185-011-0941-5

Jemal, A., Simard, E. P., Dorell, C., Noone, A., M., Markowitz, L. E., Edwards, B., K., et al. (2013). Annual Report to the Nation on the Status of Cancer, 1975-2009, featuring the burden and trends in human papillomavirus(HPV)-associated cancers and HPV vaccination coverage levels. *J. Natl. Cancer Inst.* 105, 175–201. doi: 10.1093/jnci/djs491

Katsuya, H., Ishitsuka, K., Utsunomiya, A., Hanada, S., Eto, T., Moriuchi, Y., et al. (2015). Treatment and survival among 1594 patients with ATL. *Blood* 126, 2570–2577. doi: 10.1182/blood-2015-03-632489

Khan, M. J., Castle, P. E., Lorincz, A. T., Wacholder, S., Sherman, M., Scott, D. R. et al. (2005). The elevated 10-year risk of cervical precancer and cancer in women with human papillomavirus (HPV) type 16 or 18 and the possible utility of type-specific HPV testing in clinical practice. *J. Natl. Cancer Inst.* 97, 1072–1079. doi: 10.1093/jnci/dji187

Lemos, B., and Nghiem, P. (2007). Merkel cell carcinoma: more deaths but still no pathway to blame. *J. Invest. Dermatol.* 127, 2100–2103. doi: 10.1038/sj.jid.5700925

Levine, A. M. (1993). AIDS-related malignancies: the emerging epidemic. *J. Natl. Cancer Inst.* 85, 1382–1397. doi: 10.1093/jnci/85.17.1382

Lunder, E. J., and Stern, R. S. (1998). Merkel-cell carcinomas in patients treated with methoxsalen and ultraviolet A radiation. *N. Engl. J. Med.* 339, 1247–1248. doi: 10.1056/NEJM199810223391715

Nault, J. C., Datta, S., Imbeaud, S., Franconi, A., Mallet, M., Couchy, G., et al. (2015). Recurrent AAV2-related insertional mutagenesis in human hepatocellular carcinomas. *Nat. Genet.* 47, 1187–1193. doi: 10.1038/ng.3389

Poiesz, B. J., Ruscetti, F. W., Gazdar, A. F., Bunn, P. A., Minna, J. D., and Gallo, R. C. (1980a). Detection and isolation of type C retrovirus particles from fresh and cultured lymphocytes of a patient with cutaneous T-cell lymphoma. *Proc. Natl. Acad. Sci. U.S.A.* 77, 7415–7419. doi: 10.1073/pnas.77.12.7415

Poiesz, B. J., Ruscetti, F. W., Mier, J. W., Woods, A. M., and Gallo, R. C. (1980b). T-cell lines established from human T-lymphocytic neoplasias by direct response to T-cell growth factor. *Proc. Natl. Acad. Sci. U.S.A.* 77, 6815–6819. doi: 10.1073/pnas.77.11.6815

Satou, Y., Yasunaga, J., Zhao, T., Yoshida, M., Miyazato, P., Takai, K., et al. (2011). HTLV-1 bZIP factor induces T-cell lymphoma and systemic inflammation *in vivo*. *PLoS Pathog.* 7:e1001274. doi: 10.1371/journal.ppat.1001274

Uemura, N., Okamoto, S., Yamamoto, S., Matsumura, N., Yamaguchi, S., Yamakido, M., et al. (2001). Helicobacter pylori infection and the development of gastric cancer. *N. Engl. J. Med.* 345, 784–789. doi: 10.1056/NEJMoa001999

Utsunomiya, A., Miyazaki, Y., Takatsuka, Y., Hanada, S., Uozumi, K., Yashiki, S., et al. (2001). Improved outcome of adult T cell leukemia/lymphoma with allogeneic hematopoietic stem cell transplantation. *Bone Marrow Transpl.* 27, 15–20. doi: 10.1038/sj.bmt.1702731

Wooff, J. C., Trites, J. R., Walsh, N. M., and Bullock, M. J. (2010). Complete spontaneous regression of metastatic merkel cell carcinoma: a case report and review of the literature. *Am. J. Dermatopathol.* 32, 614–617. doi: 10.1097/DAD.0b013e3181cd3158

zur Hausen, H. (2009). Papillomaviruses in the causation of human cancers - a brief historical account. *Virology* 384, 260–265. doi: 10.1016/j.virol.2008.11.046

7

Stability of the HTLV-1 Antisense-Derived Protein, HBZ, is Regulated by the E3 Ubiquitin-Protein Ligase, UBR5

*Amanda R. Panfil[1], Jacob Al-Saleem[1], Cory M. Howard[1], Nikoloz Shkriabai[2], Mamuka Kvaratskhelia[2] and Patrick L. Green[1,3]**

[1] *Department of Veterinary Biosciences, Center for Retrovirus Research, The Ohio State University, Columbus, OH, United States,* [2] *Division of Infectious Diseases, School of Medicine, University of Colorado Denver, Aurora, CO, United States,* [3] *Comprehensive Cancer Center and Solove Research Institute, The Ohio State University, Columbus, OH, United States*

Correspondence:
Patrick L. Green
green.466@osu.edu

Human T-cell leukemia virus type 1 (HTLV-1) encodes a protein derived from the antisense strand of the proviral genome designated HBZ (HTLV-1 basic leucine zipper factor). HBZ is the only viral gene consistently expressed in infected patients and adult T-cell leukemia/lymphoma (ATL) tumor cell lines. It functions to antagonize many activities of the Tax viral transcriptional activator, suppresses apoptosis, and supports proliferation of ATL cells. Factors that regulate the stability of HBZ are thus important to the pathophysiology of ATL development. Using affinity-tagged protein and shotgun proteomics, we identified UBR5 as a novel HBZ-binding partner. UBR5 is an E3 ubiquitin-protein ligase that functions as a key regulator of the ubiquitin proteasome system in both cancer and developmental biology. Herein, we investigated the role of UBR5 in HTLV-1-mediated T-cell transformation and leukemia/lymphoma development. The UBR5/HBZ interaction was verified *in vivo* using over-expression constructs, as well as endogenously in T-cells. shRNA-mediated knockdown of UBR5 enhanced HBZ steady-state levels by stabilizing the HBZ protein. Interestingly, the related HTLV-2 antisense-derived protein, APH-2, also interacted with UBR5 *in vivo*. However, knockdown of UBR5 did not affect APH-2 protein stability. Co-immunoprecipitation assays identified ubiquitination of HBZ and knockdown of UBR5 resulted in a decrease in HBZ ubiquitination. MS/MS analysis identified seven ubiquitinated lysines in HBZ. Interestingly, UBR5 expression was upregulated in established T lymphocytic leukemia/lymphoma cell lines and the later stage of T-cell transformation *in vitro*. Finally, we demonstrated loss of UBR5 decreased cellular proliferation in transformed T-cell lines. Overall, our study provides evidence for UBR5 as a host cell E3 ubiquitin-protein ligase responsible for regulating HBZ protein stability. Additionally, our data suggests UBR5 plays an important role in maintaining the proliferative phenotype of transformed T-cell lines.

Keywords: HTLV-1, HBZ, UBR5, ubiquitin, proliferation, T-cell

INTRODUCTION

Categorized as a tumorigenic virus, human T-cell leukemia virus type 1 (HTLV-1) is a deltaretrovirus that infects and transforms CD4+ T-cells. Approximately 10–15 million people worldwide are infected, with areas of endemic infection in southwestern Japan, Africa, South America, and the Pacific Islands (DeThe and Bomford, 1993; Matsuoka and Jeang, 2005; Gessain and Cassar, 2012). HTLV-1 is responsible for a highly aggressive and chemotherapy-resistant peripheral T-cell malignancy called adult T-cell leukemia/lymphoma (ATL) (Uchiyama et al., 1977; Poiesz et al., 1980; Yoshida et al., 1982) and a chronic progressive neurodegenerative disease termed HTLV-1-associated myelopathy/tropical spastic paraparesis (HAM/TSP) (Gessain et al., 1985; Osame et al., 1986). As a blood borne pathogen, HTLV-1 is primarily spread through breastfeeding, blood transfusions, and sexual contact. Not all HTLV-1 carriers will go on to develop disease. Disease penetrance is roughly 5% for the lifetime of an infected individual (Ishitsuka and Tamura, 2014). The discrepancy between individuals with disease development and lifelong healthy carriers is still not well understood.

Studies have identified at least two viral genes, *tax* and *hbz*, that are linked to oncogenic transformation and involved in the pathogenic process (Cheng et al., 2012; Ma et al., 2016). Derived from the sense strand of the proviral genome, Tax serves as a transcriptional activator of both viral and cellular gene expression (Bex and Gaynor, 1998). In addition, it deregulates the cell cycle, which ultimately leads to the accumulation of cellular genetic mutations (Marriott and Semmes, 2005). Although essential for viral transformation, Tax is frequently absent from ATL tumor cells through epigenetic silencing, 5′ LTR deletion, or abortive protein mutations in the *tax* gene (Takeda et al., 2004).

Conversely, a viral transcript that is consistently found in ATL cells is the antisense-derived *hbz* transcript (Satou et al., 2006). *Hbz* transcription initiates in the mostly epigenetically unmodified 3′ LTR (Larocca et al., 1989; Satou et al., 2006). Viral cAMP-responsive elements (CRE) and several SP1 binding sites help regulate transcription of *hbz* (Yoshida et al., 2008). *Hbz* mRNA exists in both a spliced and unspliced transcript variant (Satou et al., 2006). The proteins encoded by these transcripts have nearly identical amino acid sequence (with the exception of the first several amino acids) and demonstrate several functional differences in cells (Yoshida et al., 2008). Spliced HBZ is more abundant in infected cells (Usui et al., 2008) and therefore most research to date has focused on this isoform. The spliced *hbz* transcript encodes a 206-amino acid nuclear protein comprised of 3 domains: an N-terminal activation domain, a central basic region, and a C-terminal bZIP domain (Gaudray et al., 2002; Zhao and Matsuoka, 2012). Within the activation domain are two well-characterized LXXLL-like motifs. These motifs have been shown to bind the KIX domain of CBP/p300 and are also required for HBZ to activate TGF-β signaling (Clerc et al., 2008; Zhao et al., 2011). Through its bZIP domain, HBZ is able to hetero-dimerize with cellular bZIP proteins and affect their binding to DNA recognition sites (Matsuoka and Green, 2009).

Deletion of HBZ expression in the context of the virus has been studied using an HTLV-1 infectious molecular clone with a premature stop codon in HBZ, termed HTLV-1 ΔHBZ (Arnold et al., 2006). HBZ knockout had little effect on viral infectivity and transformation of T-cells in cellular immortalization assays *in vitro*. However, when the HTLV-1 ΔHBZ virus was injected in rabbits, there was a significant decrease in antibody response to the virus and proviral load, suggesting a positive role of HBZ in viral infectivity or proliferation of infected cells *in vivo*. In monkeys, the HBZ-KO clone gradually reverts to wild-type during prolonged infection (Valeri et al., 2010), confirming the importance of HBZ in viral infectivity and persistence *in vivo*.

A popular theory within the HTLV-1 field has been that Tax is responsible for initiating transformation, while HBZ provides the maintenance or cell survival signals later during transformation. However, HBZ expression is found both in early and late stages of viral infection (Li et al., 2009). In addition, it has been shown that HBZ is able to sustain proliferative cellular signaling, evade growth suppressors, resist cell death, cause genomic instability and mutations, enable replicative immortality, induce tumor-promoting inflammation, and has low immunogenicity (Ma et al., 2016). Therefore, the view is shifting to appreciate the role of HBZ throughout HTLV-1-mediated oncogenesis.

To date, little is known regarding the regulation of HBZ protein expression. We therefore sought to identify cellular proteins that regulate HBZ stability using affinity capture coupled with shotgun proteomics. These experiments have identified UBR5 as a novel HBZ-interacting partner. UBR5 is an E3 ubiquitin-protein ligase that functions as a key regulator of the ubiquitin proteasome system in both cancer and developmental biology. Through shRNA-mediated knockdowns and cycloheximide pulse chase experiments, we found that UBR5 regulates HBZ protein stability. MS/MS analysis enabled us to discover ubiquitination at seven lysine residues throughout the HBZ protein. Interestingly, UBR5 protein is over-expressed in T-cell lymphoma cell lines and using our HTLV-1-mediated model of T-cell transformation, we determined this dysregulation occurs late during the transformation process. In addition, we found UBR5 positively regulates cellular proliferation in T-cell lymphoma cell lines. Taken together, our data identifies UBR5 as an E3 ubiquitin-protein ligase responsible for the regulation of HBZ protein stability. In addition, it suggests UBR5 may play a role in maintaining the proliferative phenotype of transformed T-cell lines.

MATERIALS AND METHODS

Cell Lines and Culture

HEK293T cells were maintained in Dulbecco's modified Eagle's medium (DMEM) supplemented with 10% fetal bovine serum (FBS) (Gemini Bio-Products, Broderick, CA, United States), 2 mM glutamine, penicillin (100 U/mL), and streptomycin (100 μg/mL). PBL-ACH (early passage HTLV-1-immortalized human T-cells) were maintained in RPMI 1640 supplemented with 20% FBS, 20 U/mL recombinant human interleukin-2 (rhIL-2; Roche Applied Biosciences, Indianapolis, IN,

United States), 2 mM glutamine, penicillin (100 U/mL), and streptomycin (100 μg/mL). SLB-1 cells (HTLV-1-transformed T-cell line) were maintained in Iscove's medium supplemented with 10% FBS, 2 mM glutamine, penicillin (100 U/mL), and streptomycin (100 μg/mL). C8166, MT-1, MT-2, Hut-102 (HTLV-1-transformed T-cell lines), Hut-78, Jurkat (HTLV-1-negative transformed T-cell lines), TL-Om1, ATL-43T, and ATL-ED cells (ATL-derived T-cell lines) were maintained in RPMI 1640 supplemented with 10% FBS, 2 mM glutamine, penicillin (100 U/mL), and streptomycin (100 μg/mL). ATL-55T (ATL-derived T-cell line) were maintained in RPMI 1640 supplemented with 10% FBS, 20 U/mL rhIL-2, 2 mM glutamine, penicillin (100 U/mL), and streptomycin (100 μg/mL). The parental 729.B (uninfected) and derivative 729.ACH (HTLV-1 producing) cell lines were maintained in Iscove's medium supplemented with 10% FBS, 2 mM glutamine, penicillin (100 U/mL), and streptomycin (100 μg/mL). All cells were grown at 37°C in a humidified atmosphere of 5% CO_2 and air. Human PBMCs were isolated using Ficoll-Paque PLUS (GE Healthcare Life Sciences, Pittsburgh, PA, United States) and naïve T-cells were enriched using a Pan T-Cell Isolation Kit (Miltenyi Biotec., Inc., Gaithersburg, MD, United States), while memory CD4 T-cells were isolated using a Memory CD4+ T Cell Isolation Kit (Miltenyi Biotec, Inc.)

Plasmids and Cloning

Plasmid DNA was purified on maxi-prep columns according to the manufacturer's protocol (Qiagen, Valencia, CA, United States). The S-tagged APH-2 and HBZ expression vectors and the pME-APH-2 and HBZ expression vectors were generated and described previously (Arnold et al., 2006; Yin et al., 2012; Panfil et al., 2016). pCMV-Tag2B-EDD (Flag-UBR5) was a gift from Darren Saunders & Charles Watts (Addgene plasmid #37188) (Henderson et al., 2002). Flag-tagged ubiquitin was provided by F. Nina Papavasiliou (The Rockefeller University, New York, NY, United States) (Delker et al., 2013; Chesarino et al., 2014). S-tagged-BR, bZIP, and BR/bZIP were cloned by PCR amplifying the respective fragments from pME-HBZ and inserting them into pTriEx^{TM}-4 Neo (Millipore Sigma, Burlington, MA, United States) using BamHI and EcoRI restriction sites. HBZ plasmids S-tagged-AD, AD-BR2, AD-BR2-BR1, AD-BR2-BR1-BR3, and ΔbZIP were constructed using PCR-based site-directed mutagenesis to create stop codons with the following primer sets: HBZ AD forward (5′- C GCATCGTGATCGGTAGCGACGGGCTGAGGAG -3′) and reverse (5′- CTCCTCAGCCCGTCGCTACCGATCACGATG CG -3′); HBZ AD-BR2 forward (5′- GAGCGGGAGAAATAGGA GGAAAAGCAGATTG -3′) and reverse (5′- CAATCTGCTTT TCCTCCTATTTCTCCCGCTC -3′); HBZ AD-BR2-BR1 forward (5′- GTCGCCAGGAGAAAGTAGGAAGAGCAGGAG CGC -3′) and reverse (5′- GCGCTCCTGCTCTTCCTACT TTCTCCTGGCGAC -3′); HBZ-AD-BR2-BR1-BR3 forward (5′- GCAGGAGTTGGGGTAGGATGGCTATACTAGACAGTTGG -3′) and reverse (5′- CCAACTGTCTAGTATAGCCATCCTACCC CAACTCCTGC -3′); and HBZ-ΔbZIP forward (5′- GGAAG GCGAGGTGTAGTCCTTGGAGGCTG -3′) and reverse (5′- CA GCCTCCAAGGACTACACCTCGCCTTCC -3′). HBZ cDNA was cloned into the pCDH-EF1-MCS-T2A-copGFP lentiviral expression vector (SBI, Mountain View, CA, United States) to create an HBZ-GFP chimeric protein and used to generate the Jurkat-HBZ cell line.

S-Tag Affinity Pulldown Assays

HEK293T cells were transfected with the indicated expression vectors using Lipofectamine®2000 (Life Technologies, Carlsbad, CA, United States) according to the manufacturer's instructions. Twenty-four hours post-transfection, cells were treated with 10 μM MG-132 (Sigma–Aldrich, St. Louis, MO, United States) for 20 h. Cells were washed with 1× PBS and lysates were prepared with NP-40 lysis buffer in the presence of protease inhibitor (Roche Applied Biosciences) and 25 mM *N*-Ethylmaleimide (Sigma–Aldrich). Cells were centrifuged at maximum speed for 10 min at 4°C. S-tag purification was performed by rocking cell lysates with S beads (Millipore Sigma) overnight at 4°C. The S beads were washed twice with NP-40 lysis buffer. An equal volume of 2× SDS-sample buffer was added and proteins were extracted by heating at 95°C for 10 min.

Mass Spectrometry and Proteomic Analysis

The pulled-down proteins were subjected to SDS-PAGE and visualized by GelCode Blue staining. For identification of cellular binding partners of S-tagged HBZ and S-tagged APH-2 entire lanes were excised and cut into small pieces. For identification of ubiquitination sites the prominent S-tagged HBZ bands were excised. The gel pieces were then de-stained with 50% acetonitrile and then subjected to in-gel proteolysis using sequencing grade modified trypsin (Promega, Madison, WI, United States). The resulting peptides were extracted in acetonitrile by vortexing for 10 min and then desiccated in Eppendorf® Vacufuge® Plus Vacuum Concentrator. The samples were run on Thermo Scientific Q Exactive mass spectrometer and analyzed with Mascot software (Matrix Science, Boston, MA, United States). The data were visualized using Scaffold Viewer (Proteome Software, Inc., Portland, OR, United States) using 1.0% FDR Protein Threshold, 2 Min. #Peptides, and 1.0% Peptide Threshold.

Flag Immunoprecipitation Assays

HEK293T cells were transfected with the indicated expression vectors using Lipofectamine®2000 (Life Technologies) according to the manufacturer's instructions. Twenty-four hours post-transfection, cells were treated with 10 μM MG-132 (Sigma–Aldrich) for 20 h. Cells were washed with 1× PBS and then incubated with gentle rocking at 4°C for 30 min in 0.5% NP-40 lysis buffer (150 mM NaCl, 0.5% NP-40, 50 mM Tris-HCl, pH 8.0) in the presence of protease inhibitor (Roche Applied Biosciences) and 25 mM NEM (Sigma–Aldrich). Cells were centrifuged at maximum speed for 10 min at 4°C. The lysates were then diluted to 0.1% NP-40 lysis buffer. Flag immunoprecipitation was performed by rocking cell lysates with ANTI-FLAG® M2 Affinity Gel (Sigma–Aldrich) overnight at 4°C. The FLAG resin was washed twice with 0.1% NP-40 lysis buffer. An equal volume of 2× SDS-sample buffer was added and proteins were extracted by heating at 95°C for 10 min.

Immunoprecipitation Assays

Lymphocytes were treated with 10 μM MG-132 (Sigma–Aldrich) for 20 h. Cells were washed with 1× PBS and then incubated with gentle rocking at 4°C for 30 min in NP-40 lysis buffer in the presence of protease inhibitor (Roche Applied Biosciences) and 25 mM NEM (Sigma–Aldrich). Cells were centrifuged at maximum speed for 10 min at 4°C. Antibody (1 to 2 μg or no antibody for the direct load) was added to each sample, and the samples were rocked overnight at 4°C. The antibodies used were as follows: control rabbit IgG (Santa Cruz Biotechnology, Dallas, TX, United States), rabbit anti-HBZ antiserum, and rabbit anti-UBR5 (Bethyl Laboratories, Inc., Montgomery, TX, United States). Protein G Dynabeads (Fisher Scientific, Hampton, NH, United States) were added and the mixture was rocked at 4°C for 2 h. The beads were washed twice in NP-40 lysis buffer. An equal volume of 2× SDS-sample buffer was added and proteins were extracted by heating at 95°C for 10 min.

Immunoblotting

Cell lysates were harvested in NP-40 lysis buffer containing protease inhibitor cocktail (Roche Applied Bioscience) and quantitated using a Pierce bicinchoninic acid protein assay kit (Fisher Scientific). Equivalent amounts of protein were separated in Mini-Protean TGX precast 4 to 20% gels (Bio-Rad Laboratories, Hercules, CA, United States) and transferred to nitrocellulose membranes. Membranes were blocked in phosphate-buffered saline (PBS) containing 5% milk and 0.1% Tween 20 and incubated with primary antibody. The following antibodies were used: anti-S-tag (1:1000; Abcam, Cambridge, MA), anti-UBR5 (1:1000; Cell Signaling Technology, Danvers, MA, United States), anti-HBZ (1:1,000), anti-APH-2 (1:1,000), anti-FLAG clone M2 (1:1,000; Agilent Technologies, Santa Clara, CA, United States), anti-ubiquitin (1:250; Santa Cruz Biotechnology), anti-β-actin (1:5,000; Sigma–Aldrich), and anti-α-Tubulin (1:250; Santa Cruz Biotechnology). The secondary antibodies used were horseradish peroxidase-labeled goat anti-rabbit and goat anti-mouse immunoglobulin antibodies (1:5,000; Santa Cruz Biotechnology). The blots were developed using an ECL Western Blotting Substrate (Fisher Scientific). Images were taken using an Amersham Imager 600 imaging system (GE Healthcare Life Sciences), and densitometric data were calculated using the ImageQuant TL program (GE Healthcare Life Sciences).

Co-culture Immortalization Assays

Long-term immortalization assays were performed as detailed previously (Green et al., 1995). Briefly, 2 × 10^6 freshly isolated human PBMCs were co-cultivated at a 2:1 ratio with lethally irradiated cells (729.B uninfected parental; 729.ACH HTLV-1-producing) in 24-well culture plates (media was supplemented with 10 U/mL rhIL-2).

Quantitative RT-PCR

Total RNA was isolated using an RNeasy Mini kit (Qiagen) according to the manufacturer's instructions. Isolated total RNA was quantitated and DNase-treated using recombinant DNase I (Roche Applied Biosciences). Reverse transcription (RT) was performed using a SuperScript first-strand synthesis system for RT-PCR (Life Technologies) according to the manufacturer's instructions. The instrumentation and general principles of the CFX96 Touch real-time PCR detection system (Bio-Rad Laboratories) are described in detail in the operator's manual. PCR amplification was carried out in 96-well plates with optical caps. The final reaction volume was 20 μl and consisted of 10 μl iQ SYBR green Supermix (Bio-Rad Laboratories), 300 nM each specific primer, and 1.5 μl of cDNA template. For each run, standard cDNA, sample cDNA, and a no-template control were all assayed in triplicate. The reaction conditions were 95°C for 5 min, followed by 40 cycles of 94°C for 30 s, 56°C for 30 s, and 72°C for 45 s. Primer pairs for the specific detection of *ubr5* and human glyceraldehyde-3-phosphate dehydrogenase (*hgapdh*) were described previously (Subbaiah et al., 2016). Data from triplicate experiments are presented in histogram form as means with standard deviations. The total *ubr5* copy number for each cell line was determined using a plasmid DNA standard curve and normalized to 10^6 copies of hGAPDH mRNA.

Cycloheximide Pulse-Chase Experiments

HEK293T cells were transiently transfected with empty or untagged (pME) HBZ or APH-2 expression vectors using Lipofectamine®2000 (Life Technologies) according to the manufacturer's instructions. Forty-eight hours later, the cells were treated with 100 μg/ml cycloheximide (a translation elongation inhibitor; Sigma–Aldrich) and then harvested at different time points. Jurkat-HBZ cells were synchronized by serum starvation in 0.1% FBS overnight prior to treatment with 100 μg/ml cycloheximide and then harvested at different time points.

Infection and Packaging of Lentiviral Vectors

Lentiviral vectors expressing five different UBR5-directed short hairpin RNAs (shRNAs) (target set RHS4533-EG51366) and the universal negative control pLKO.1 (RHS4080) were purchased from Open Biosystems (Fisher Scientific) and propagated according to the manufacturer's instructions. HEK293T cells were transfected with lentiviral vector(s) plus DNA vectors encoding HIV Gag/Pol and vesicular stomatitis virus G in 10-cm dishes using Lipofetamine®2000 reagent according to the manufacturer's instructions. Media containing the lentiviral particles were collected 72 h later and filtered through 0.45-μm-pore-size filters (Fisher Scientific). Lentiviral particles were then concentrated using ultracentrifugation in a Sorvall SW-41 swinging bucket rotor at 90,000 × *g* for 1.5 h at 4°C. Target cells were infected with the indicated lentivirus by spininoculation at 2,000 × *g* for 2 h at room temperature. Three-days post-transduction, the cells were selected with puromycin for 7–10 days.

Proliferation Assays

Cell Titer 96 Aqueous One Solution Cell Proliferation Assays (Promega) were performed according to the manufacturer's protocol. Briefly, cells were counted and plated at 1,000 cells/well in 96-well round-bottom plates on day 0 and monitored over a 7-day time course. Cell Titer 96 reagent was added to each well, agitated slightly, and incubated at 37°C, 5% CO_2 for 2 h. The optical density absorbance at 490 nm was collected on an enzyme-linked immunosorbent assay (ELISA) plate reader. For each cell line, data represent three independent experiments performed in triplicate.

Annexin V Staining

Cells were stained using the FITC Annexin V Apoptosis Detection Kit I (BD Biosciences, San Jose, CA, United States) according to the manufacturer's instructions. Cells were analyzed for apoptosis via flow cytometry using a Guava EasyCyte Mini machine (Millipore Sigma).

RESULTS

HBZ Interacts with the Cellular E3 Ubiquitin-Protein Ligase, UBR5

HBZ has been linked to HTLV-1-mediated oncogenic transformation and plays a role in the pathogenic process (Ma et al., 2016). HBZ regulates several cellular functions and signaling pathways (NF-κB, CBP/p300, TGF-β/Smad, CREB2, c-Jun, JunB, JunD, CREB, MafB, ATF3, etc.) within the host cell (Matsuoka and Green, 2009). However, less is known concerning the regulation of HBZ expression. We therefore sought to identify cellular factors that interact with HBZ using affinity capture coupled with shotgun proteomics. For this, HBZ was cloned into a CMV driven pTriEx4-Neo plasmid, which created an amino terminal S-tagged HBZ expression vector. In addition, the HTLV-2 equivalent to HBZ, APH-2, was also cloned into pTriEx4-Neo. HTLV-1 and -2 are highly related retroviruses (Ciminale et al., 2014). While both viruses transform T-cells *in vitro* (Shoji et al., 2009; Imai et al., 2013; Romanelli et al., 2013), they have distinct pathological outcomes *in vivo* (Arnold et al., 2006; Yin et al., 2012). Consequently, comparisons of cellular interacting partners of HBZ (HTLV-1) and APH-2 (HTLV-2) may provide a better understanding of how HTLV-1, but not HTLV-2, infection is associated with disease. S-tagged-HBZ, APH-2, or empty expression vectors were ectopically expressed in HEK293T cells. Proteins were purified using S-tag affinity pull down assays. Samples were then run on a denaturing SDS-PAGE gel (**Figure 1A**). Entire lanes were subjected to trypsin digestion and analyzed by MS/MS. An abbreviated summary of the identified interacting cellular partners is depicted (**Figure 1A**). Several known cellular binding partners of both HBZ and APH-2 were identified, as well as the novel E3 ubiquitin-protein ligase UBR5. The interaction of HBZ and APH-2 with UBR5 was confirmed using S-tag affinity pulldowns in HEK293T cells followed by immunoblot for endogenous UBR5 (**Figure 1B**). Similar results were obtained using reciprocal coIPs in HEK293T cells with Flag-tagged UBR5 and either S-tagged or untagged HBZ expression constructs (**Figure 1C**). The interaction between UBR5 and HBZ was also confirmed in the more physiologically relevant cell lines Jurkat and SLB-1 (**Figures 1D,E**). Jurkat-HBZ cells are a transformed HTLV-1-negative T-cell line that ectopically expresses HBZ (see section "Materials and Methods"). SLB-1 cells are an HTLV-1-transformed T-cell line and therefore express physiologically relevant levels of HBZ protein. Taken together, these results identify and confirm UBR5 as a novel cellular binding partner of HBZ.

UBR5 Regulates HBZ Protein Stability

UBR5 is a cellular E3 ubiquitin-protein ligase involved in targeting proteins for ubiquitin-mediated proteolysis. We therefore examined the steady-state levels of both HBZ and APH-2 in the presence and with reduced amounts of endogenous UBR5. HEK293T cells were transduced with three separate shUBR5 or control lentiviral vectors. After a brief selection in puromycin (7–10 days), the cells were transfected with either HBZ or APH-2 expression vectors. When the level of UBR5 is decreased, the steady-state levels of HBZ, but not APH-2, are increased compared to control infected cells (**Figure 2A**). Given the disparity between HBZ expression levels in the presence and absence of UBR5, we next utilized cycloheximide pulse chase experiments to examine HBZ protein half-life. shControl and shUBR5 HEK293T cells were transfected with an untagged HBZ expression construct and then treated with the translation elongation inhibitor cycloheximide. Cell lysates were collected at defined time points, and the expression level of HBZ was examined by immunoblot analysis (**Figure 2B**). The calculated half-life of HBZ in shControl cells was approximately 6.7 h, consistent with previous reports (Panfil et al., 2016), whereas the calculated half-life in shUBR5 cells was greater than 24 h. As expected, based on the results in **Figure 2A**, the loss of UBR5 had no effect on the half-life of APH-2 in HEK293T cells (data not shown). The half-life of HBZ was also measured in Jurkat-HBZ cells transduced with either shControl or shUBR5 lentiviral vectors (**Figure 2C**). The calculated half-life of HBZ in Jurkat-HBZ shControl cells was 2.7 h and this increased roughly twofold to 5.8 h in Jurkat-HBZ shUBR5 cells. These results indicate that UBR5 regulates HBZ protein stability.

UBR5 Interacts with the Central Basic Region of HBZ

HBZ is a nuclear protein with an N-terminal activation domain, a central basic region, and a C-terminal basic leucine zipper domain. In order to determine the region of HBZ which interacts with UBR5, a series of mutants were created in the S-tagged HBZ expression construct as depicted in **Figure 3A**. HEK293T cells were transfected with FLAG-tagged UBR5 and either S-tagged HBZ wild-type, S-tagged HBZ mutant, or empty expression vector, as indicated. S-tag affinity pulldown assays were performed followed by immunoblot for UBR5 expression (**Figure 3B**). In addition to wild-type HBZ, the HBZ AD-BR2-BR1, HBZ AD-BR2-BR1-BR3, HBZ ΔbZIP, HBZ BR, and HBZ

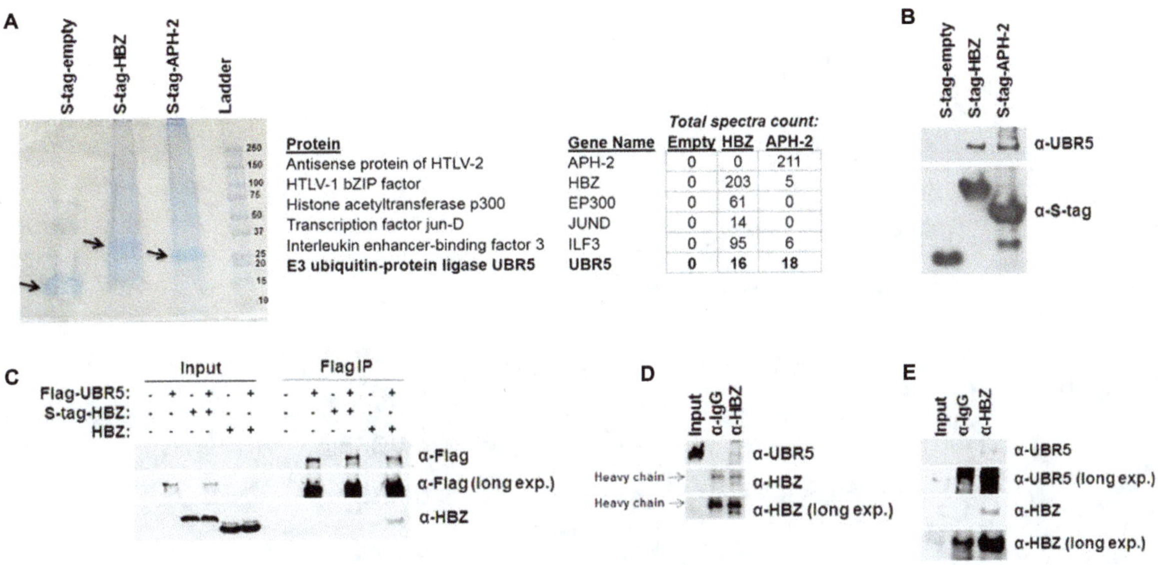

FIGURE 1 | HBZ interacts with the cellular E3 ubiquitin-protein ligase, UBR5. **(A)** (Left) S-tagged HBZ and APH-2 were purified from lysates of transfected HEK293T cells using S beads. The purified APH-2 or HBZ and associated protein complexes were resolved by SDS-PAGE and visualized with GelCode Blue staining prior to MS/MS analysis. Arrow indicates enriched S-tagged protein. (Right) Brief summary of HBZ and APH-2 interacting proteins identified by MS/MS. **(B)** HEK293T cells were transfected with empty, HBZ, or APH-2 S-tagged vectors. Tagged proteins were purified by S-tag affinity purification 48 h after transfection. Pulldowns were examined by immunoblot analysis using anti-S-tag and anti-UBR5 antibodies, as indicated. **(C)** HEK293T cells were transfected with FLAG-tagged UBR5 and empty, S-tagged HBZ or untagged HBZ expression vectors, as specified. FLAG IPs were performed 48 h after transfection as described in the Section "Materials and Methods." Immunoprecipitated proteins were examined by immunoblot analysis using anti-FLAG antibody and anti-HBZ antisera, as indicated. Five percent of the direct load was used for immunoblot analysis. **(D)** Jurkat-HBZ and **(E)** SLB-1 (HTLV-1-transformed; Right panel) cells were immunoprecipitated with a control rabbit antibody or HBZ rabbit antisera. Immunoprecipitated proteins were then examined by immunoblot analysis using anti-UBR5 antibody or anti-HBZ antisera as indicated. Five percent of the direct load was used for immunoblot analysis.

BR/bZIP mutants interacted with UBR5. These results indicate the basic region, specifically BR1 of HBZ, is important for interaction with UBR5.

HBZ Is Ubiquitinated

As an E3 ubiquitin-protein ligase, UBR5 acts as a scaffold between its protein substrate and an E2 ubiquitin-conjugating enzyme that is loaded with ubiquitin. To determine whether UBR5 acts as an E3 ligase for HBZ, we first examined whether HBZ is ubiquitinated (**Figure 4A**). HEK293T cells were co-transfected with either FLAG-tagged ubiquitin and/or S-tagged HBZ expression constructs. FLAG co-IPs were then performed to enrich all ubiquitinated proteins. Upon ubiquitin enrichment, we observed HBZ expression by immunoblot, indicating HBZ ubiquitination. Consistent with its role as an E3 ligase, we also found UBR5 expression in the co-IP samples upon ubiquitin enrichment. Next, FLAG-tagged ubiquitin and/or S-tagged HBZ expression constructs were co-transfected into shControl and shUBR5 HEK293T cell lines. S-tag affinity pulldowns were performed to enrich total HBZ protein, followed by immunoblot for ubiquitin (**Figure 4B**). HBZ is pulled down with equal efficiency in both control and UBR5 knockdown cells as indicated by arrow 1. However, in UBR5 knockdown cells, there is less ubiquitinated HBZ present as indicated by arrow 2. Taken together, our results confirm HBZ is ubiquitinated and UBR5 serves as its E3 cellular ubiquitin-protein ligase.

MS/MS Analysis of HBZ Ubiquitination Sites

Using mass spectrometry, we analyzed the ubiquitination site(s) within the HBZ protein. HEK293T cells were transfected with FLAG-tagged ubiquitin, FLAG-tagged UBR5, and S-tagged HBZ expression constructs. S-tag affinity pulldowns were then performed to enrich all HBZ protein in the cells. The sample was then run on a denaturing SDS-PAGE gel followed by staining with GelCode Blue. The predominant HBZ bands (35, 45 kDa) were excised and digested with trypsin and subjected to MS/MS analysis, which identified seven ubiquitinated lysine residues (K37, K119, K120, K153, K155, K181, K186) in the HBZ protein (**Figure 5C**). The representative MS/MS fragmentation spectra are shown in **Figures 5A,B**, demonstrating the ubiquitination of the indicated lysines.

UBR5 Is Dysregulated in T-Cell Leukemia/Lymphoma Cells

UBR5 dysregulation has been implicated in various aspects of cancer biology. UBR5 is overexpressed in both breast and ovarian cancer due to an allelic imbalance, which results in an increase in *ubr5* mRNA levels (Clancy et al., 2003). In mantle cell lymphoma, UBR5 protein is recurrently mutated (Meissner et al., 2013). To determine whether UBR5 is important to HTLV-1 biology and pathogenesis, we examined the level of UBR5 protein

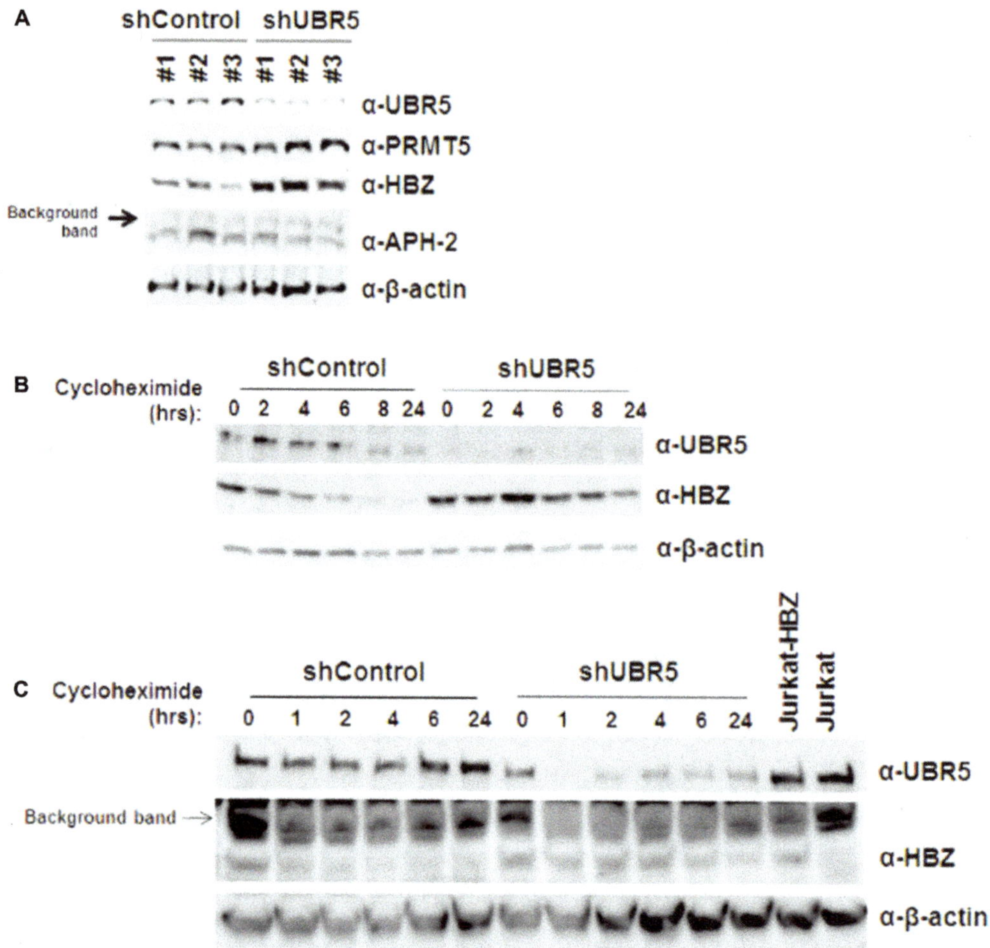

FIGURE 2 | UBR5 regulates HBZ protein stability. **(A)** HEK293T cells were infected with three different lentiviral vectors expressing shRNA directed against UBR5, or control shRNAs. After a brief puromycin selection, the cells were then transfected with HBZ or APH-2 expression plasmid. After 48 h, immunoblot analysis was performed to detect HBZ, APH-2, and UBR5 expression levels. β-actin was used as a loading control while PRMT5 was used as an internal control. **(B)** HEK293T and **(C)** Jurkat-HBZ cells were infected with either a lentiviral vector directed against UBR5 or a control shRNA. After a brief puromycin selection, the HEK293T cells were transfected with HBZ expression plasmid and used for cycloheximide pulse chase experiments after 48 h. Cells were treated with 100 μg/ml cycloheximide for the indicated times. Immunoblot analysis was performed to detect HBZ, UBR5, and β-actin (loading control) expression levels.

in a variety of transformed T-cell leukemia/lymphoma cell lines (**Figure 6A**). We included naïve primary T-cells, memory CD4 T-cells, HTLV-1-negative T-cell lines (Jurkat, Hut-78), HTLV-1-transformed T-cell lines (SLB-1, PBL-ACH, Hut-102, C8166, MT-1, MT-2), and ATL-derived T-cell lines (TL-Om1, ATL-55T, ATL-43T, ATL-ED). UBR5 protein was dramatically upregulated in all T-cell leukemia/lymphoma cell lines compared to both naïve and memory CD4 T-cells. Interestingly, UBR5 RNA levels were decreased in all T-cell leukemia/lymphoma cell lines compared to naïve and memory CD4 T-cells (**Figure 6B**), suggesting a post-transcriptional method of regulation. Using an HTLV-1-based *in vitro* T-cell immortalization co-culture assay, we determined whether UBR5 becomes dysregulated and over-expressed during the T-cell transformation process. Briefly, freshly isolated human PBMCs were co-cultured with lethally irradiated HTLV-1-producer cells. As a control, PBMCs were co-cultured with lethally irradiated HTLV-1-negative cells. Over the course of 10–14 weeks, the PBMCs co-cultured with HTLV-1-producer cells showed progressive growth indicative of HTLV-1 infection and transformation, while the PBMCs co-cultured with HTLV-1-negative cells were unable to sustain progressive growth. Using this *in vitro* transformation assay, T-cells were isolated at weekly time points and the levels of UBR5 protein were examined. We were unable to detect UBR5 protein expression at any of the time points from weeks 1–12 post-infection (data not shown) suggesting UBR5 is not upregulated early during T-cell transformation or as a result of HTLV-1 viral infection. HTLV-1-infected T-cells that became immortalized (termed HTLV-1 PBLs) continued to proliferate and grow in culture indefinitely. Several PBL clones from our *in vitro* transformation assay were kept in culture for up to 40 weeks. We examined the level of UBR5 protein expression in these cells and found upregulated UBR5 protein expression in 2 of the clones (clones 13 and 15) at 20 weeks post-infection

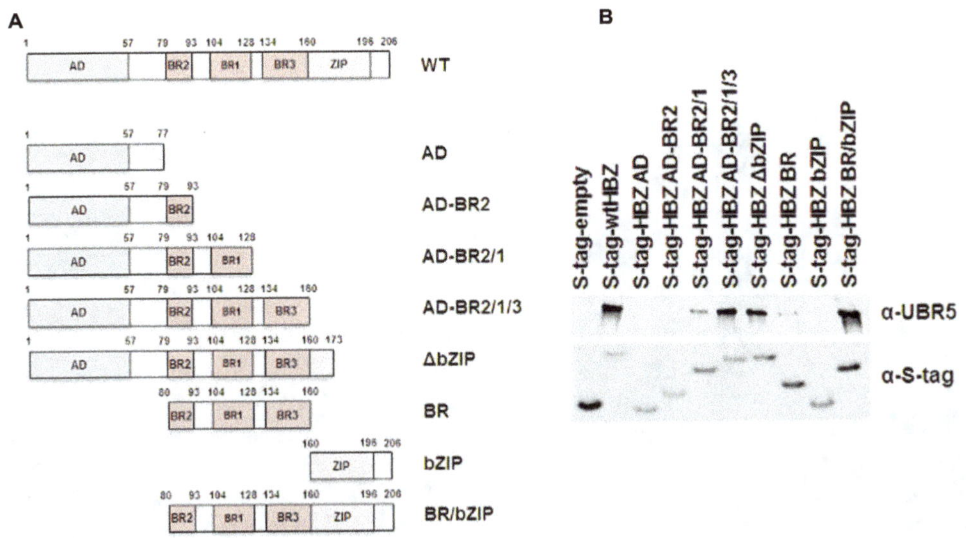

FIGURE 3 | UBR5 interacts with the central basic region of HBZ. **(A)** Schematic representation of the various HBZ mutant constructs. **(B)** HEK293T cells were transfected with FLAG-tagged UBR5 and empty, wtHBZ, or HBZ mutant S-tagged expression vectors, as indicated. S-tagged proteins were purified by S-tag affinity purification 48 h after transfection. Pulldowns were examined by immunoblot analysis using anti-S-tag and anti-UBR5 antibodies, as indicated.

and in 5 of the clones at 40 weeks post-infection (**Figure 6C**). Similar to T-cell leukemia/lymphoma cell lines, the level of UBR5 RNA was decreased in all PBL clones compared to naïve and memory CD4 T-cells (**Figure 6D**). These results suggest that UBR5 upregulation in T-cell lymphomas occurs late during transformation.

UBR5 Enhances Cellular Proliferation in T-Cell Leukemia/Lymphoma Cells

Recently, UBR5 has been shown to regulate proliferation and colony formation of gastric cancer cells (Yang et al., 2016). To examine if UBR5 affects cellular proliferation of T-cell lymphomas, we subjected both Jurkat (HTLV-1-negative) and SLB-1 (HTLV-1-transformed) cells to shRNA-mediated knockdown of UBR5. Knockdown of UBR5 inhibited cellular proliferation of Jurkat cells, but not SLB-1 cells (**Figure 7A**). Immunoblotting revealed knockdown of UBR5 in SLB-1 cells caused an increase in HBZ steady-state levels. The level of cellular apoptosis was also measured in Jurkat and Jurkat-HBZ cells infected with shControl and shUBR5 lentivirus. The number of apoptotic cells increased 2.5-fold in shUBR5 Jurkat cells compared to the lentiviral control (**Figure 7B**). However, shUBR5 Jurkat-HBZ cells had no difference in cellular apoptosis compared to the shControl cells. These results suggest UBR5 enhances cellular proliferation in T-cell leukemia/lymphoma cell lines.

DISCUSSION

HTLV-1 is a tumorigenic retrovirus that preferentially infects and transforms CD4+ T-cells both *in vitro* and *in vivo*. The virus is responsible for both an aggressive T-cell neoplasm termed ATL (Uchiyama et al., 1977; Poiesz et al., 1980; Yoshida et al., 1982) and a debilitating neurodegenerative disease called HAM/TSP (Gessain et al., 1985; Osame et al., 1986). Disease development occurs in roughly 5% of infected individuals after a prolonged clinical latency period (Ishitsuka and Tamura, 2014). During this time, it is believed that both genetic and epigenetic events accumulate in the cellular environment that contributes to disease development. The virus encodes an antisense-derived protein termed HBZ, which has been shown to play a role in oncogenic transformation and proliferation (Ma et al., 2016). HBZ is the only viral gene consistently found in ATL tumors (Satou et al., 2006). While many functions of HBZ protein have been revealed, little is known regarding the regulation and stability of HBZ protein. Given the importance of HBZ during ATL development, factors that regulate the stability of HBZ are of critical value.

Using an affinity capture coupled with shotgun proteomics approach, we identified cellular factors that interact with and regulate the stability of HBZ protein. These experiments have identified a cellular E3 ubiquitin-protein ligase called UBR5 as a novel cellular interacting partner of HBZ (**Figure 1A**). UBR5 is a member of the HECT domain family of E3 ubiquitin ligases and has widespread expression in various cell types (Mansfield et al., 1994). We confirmed interactions between HBZ and UBR5 using both expression vectors in HEK293T cells and endogenous levels of HBZ/UBR5 in physiologically relevant T-cell lines, Jurkat (HTLV-1 negative) and the HTLV-1 positive, SLB-1 (**Figures 1B–E**). Using HBZ domain mutants, we identified the central basic region, specifically BR1, as essential for UBR5 interaction (**Figure 3**). This region is the site of a dominant HBZ epitope for antibody recognition (Raval et al., 2015), therefore, predominantly exposed on the HBZ protein. Of interest, UBR5 was also identified and

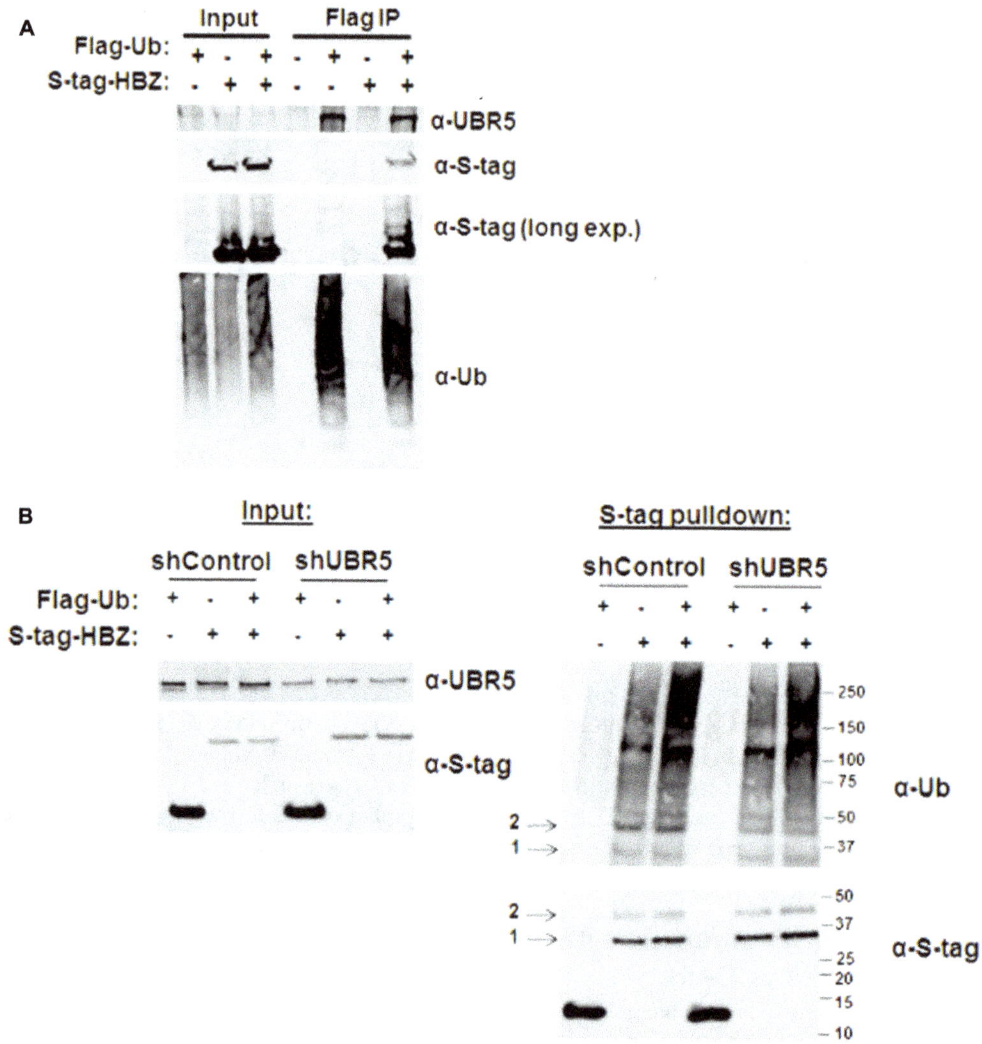

FIGURE 4 | HBZ is ubiquitinated. **(A)** HEK293T cells were co-transfected with FLAG-tagged ubiquitin (UB) and/or S-tagged HBZ expression vectors. FLAG IPs were performed 48 h after transfection. Immunoprecipitated proteins were examined by immunoblot analysis using anti-ubiquitin, anti-S-tag, and anti-UBR5 antibodies as indicated. Five percent of the direct load was used for immunoblot analysis. **(B)** HEK293T cells were infected with a lentiviral vector directed against UBR5 or control lentiviral vectors. After a brief selection with puromycin, the HEK293T cells were transfected with FLAG-tagged Ubiquitin (UB) and/or S-tagged HBZ expression vectors. S-tagged proteins were purified by S-tag affinity purification 48 h after transfection. Pulldowns were examined by immunoblot analysis using anti-S-tag and anti-ubiquitin antibodies, as indicated. Five percent of the direct load was used for immunoblot analysis and probed using anti-S-tag and anti-UBR5 antibodies, as indicated. Arrow 1 indicates total HBZ, while arrow 2 indicates ubiquitinated HBZ.

confirmed to be an interacting partner of the closely related HTLV-2 equivalent to HBZ, called APH-2 (**Figures 1A,B**). HTLV-1 and HTLV-2 are closely related retroviruses with drastically different pathological outcomes *in vivo* (Arnold et al., 2006; Yin et al., 2012). While both viruses share many similarities, HTLV-1 is associated with disease whereas HTLV-2 has not been associated with any disease to date (Ciminale et al., 2014).

A major function of UBR5 is to target cellular proteins for ubiquitination and eventual proteasome-mediated degradation. Using shRNA-mediated lentiviral vectors, we knocked down UBR5 expression in HEK293T cells and found HBZ steady state levels were increased in response to decreased UBR5 expression (**Figure 2A**). Conversely, APH-2 steady state levels were unchanged regardless of UBR5 expression levels. As an internal control, we also examined the expression level of a cellular protein, PRMT5, and found loss of UBR5 did not result in global increases in protein expression. We previously reported the HBZ half-life to be approximately 6.4 h in HEK293T cells and 2–3 h in Jurkat cells (Panfil et al., 2016). Using shRNA-mediated UBR5 knockdown coupled with cycloheximide pulse chase experiments, we found loss of UBR5 extended the HBZ half-life to greater than 24 h in HEK293T cells and approximately 6 h in Jurkat cells (**Figures 2B,C**). Of note, loss of UBR5 had no effect on the half-life of APH-2 protein (data not shown). It is possible there are additional E3 ligases that regulate

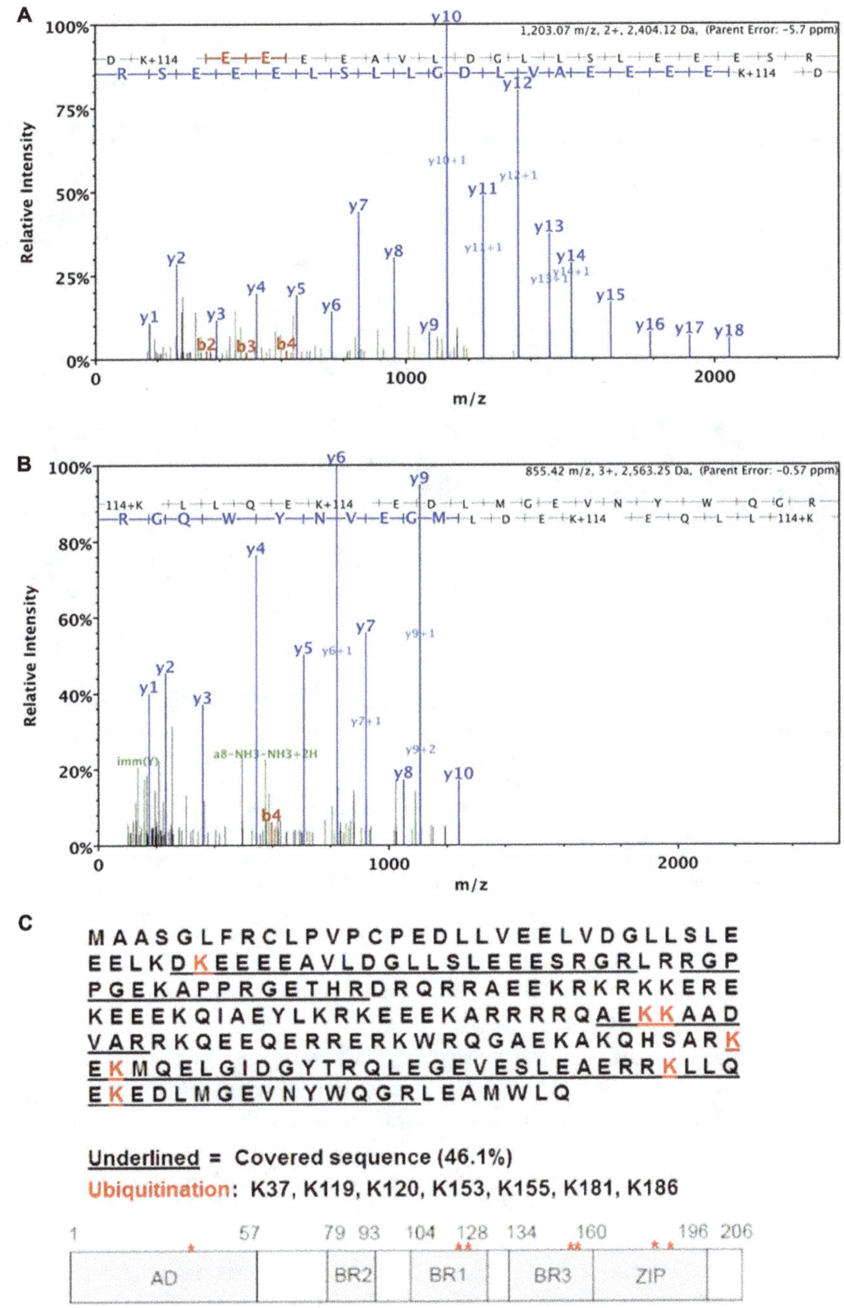

FIGURE 5 | MS/MS analysis of HBZ ubiquitination sites. HEK293T cells were co-transfected with S-tagged HBZ, FLAG-UB, and FLAG-UBR5 expression vectors. S-tagged HBZ was purified from lysates of transfected HEK293T cells using S beads. The purified HBZ was resolved by SDS-PAGE prior to MS/MS analysis. Representative MS/MS fragmentation spectra are shown for the following two HBZ peptides: **(A)** DK*EEEAVLDGLLSLEEES where lysine 37 is ubiquitinated. Mascot Ion score = 144.5. and **(B)** K*LLQEK*EDLMGEVNYWQ where lysines 181 and 186 are ubiquitinated. Mascot Ion score = 52.2. **(C)** Summary of MS results. The complete HBZ sequence is shown and the protein segments detected by MS/MS analysis are underlined. A schematic is provided to indicate ubiquitination sites on HBZ with respect to individual protein domains.

APH-2 and APH-2 protein stability does not solely depend on UBR5.

Given the difference in HBZ half-life in the presence and absence of UBR5, we examined HBZ ubiquitination using both coIPs (**Figure 4**) and MS/MS analysis (**Figure 5**). Upon ubiquitin enrichment, we identified both HBZ and UBR5 interaction (**Figure 4A**). Likewise, upon HBZ enrichment, we identified several higher molecular weight bands using ubiquitin antibody (**Figure 4B**). The predominant ubiquitin band, indicated with arrow 2, is decreased in UBR5 knockdown cells, indicating

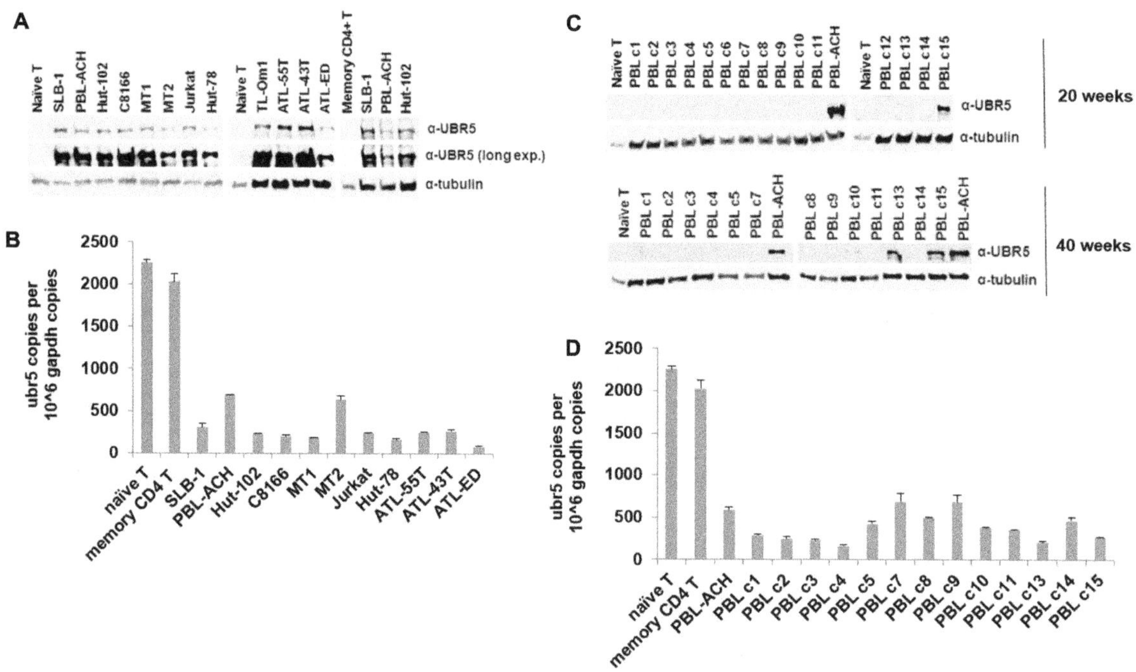

FIGURE 6 | UBR5 is dysregulated in T-cell leukemia/lymphoma cells. **(A)** Total cell lysates of HTLV-1 transformed cell lines (SLB-1, PBL-ACH, Hut-102, C8166, MT-1, MT-2), HTLV-1-negative transformed cell lines (Jurkat, Hut-78), ATL-derived cell lines (TL-Om1, ATL-55T, ATL-43T, ATL-ED), and naïve and memory CD4 T-cells were subjected to immunoblot analysis to compare the levels of endogenous UBR5 expression. α-Tubulin expression was used as a loading control. **(B)** Quantitative RT-PCR for *ubr5* and *gapdh* was performed on mRNA isolated from cells in **(A)**. The total numbers of UBR5 and GAPDH copies were determined using plasmid DNA standards and normalized to 10^6 copies of *gapdh* mRNA. **(C)** Total cell lysates of newly immortalized (20 weeks; upper panel and 40 weeks; lower panel) HTLV-1 transformed cell lines (clones 1–15), an established HTLV-1 transformed cell line (PBL-ACH) and naïve T-cells were subjected to immunoblot analysis to compare the levels of endogenous UBR5 expression. α-Tubulin expression was used as a loading control. **(D)** Quantitative RT-PCR for *ubr5* and *gapdh* was performed on mRNA isolated from cells in lower **(C)**. The total numbers of UBR5 and GAPDH copies were determined using plasmid DNA standards and normalized to 10^6 copies of *gapdh* mRNA.

UBR5 is responsible for HBZ ubiquitination. The principal ubiquitin bands, indicated with arrow 1 and 2, were subjected to MS/MS analysis and 7 ubiquitinated lysine residues were identified throughout the HBZ protein (K37, K119, K120, K153, K155, K181, K186) (**Figure 5**). A high percentage of the protein (45%) was covered in our analysis, especially the important C-terminal bZIP region. shRNA-mediated knockdown of UBR5 decreased the amount of ubiquitinated HBZ, again confirming UBR5 ubiquitinates HBZ. In a recent report, we found acetylation, phosphorylation, and methylation of the HBZ protein (Dissinger et al., 2014). However, mutational analysis of the modified residues failed to identify any effects on known HBZ functions. Surprisingly, several of the methylated (K37, K181, K186) and acetylated (K155) lysine residues in HBZ were also identified as ubiquitinated. It is possible that methylation or acetylation at these residues may inhibit the deposition of other PTMs, such as ubiquitination (Caron et al., 2005; Lanouette et al., 2014). Therefore, the role of HBZ methylation and acetylation could be to prevent ubiquitination and stabilize HBZ. Conversely, ubiquitination could prevent methylation and/or acetylation of these residues. In addition to promoting degradation, protein ubiquitination can also alter cellular localization, affect protein activity, and promote or prevent other protein interactions. The effect of HBZ ubiquitination on these various activities is of interest and will require further study.

UBR5 has been implicated in several cancers such as lung, breast and ovarian. UBR5 was upregulated in both breast and ovarian cancer due to an allelic imbalance (Clancy et al., 2003). The same report found a survival disadvantage for breast cancer patients with UBR5 dysregulation. A recent study for mantle cell lymphoma showed non-synonymous mutations in UBR5 in roughly 18% of MCL cases (Meissner et al., 2013). Most of these mutations are near the carboxy-terminus of UBR5, which is associated with the E3 ubiquitin ligase function. Data from The Cancer Genome Atlas (TCGA) shows UBR5 amplification as a common alteration in many cancer types[1]. However, the level of UBR5 in T-cell lymphomas was not included. Here, we report an upregulation of UBR5 in T-cell leukemia/lymphoma cell lines (**Figure 6**). Surprisingly, UBR5 mRNA is decreased, suggesting a post-transcriptional method of regulation. Upon further examination, we found a steady increase in UBR5 protein expression during *in vitro* T-cell transformation by HTLV-1. It will be important to follow-up and determine whether UBR5 expression is correlated to disease progression in ATL and HAM/TSP patients, as UBR5 may serve as a diagnostic marker.

[1]http://cancergenome.nih.gov/

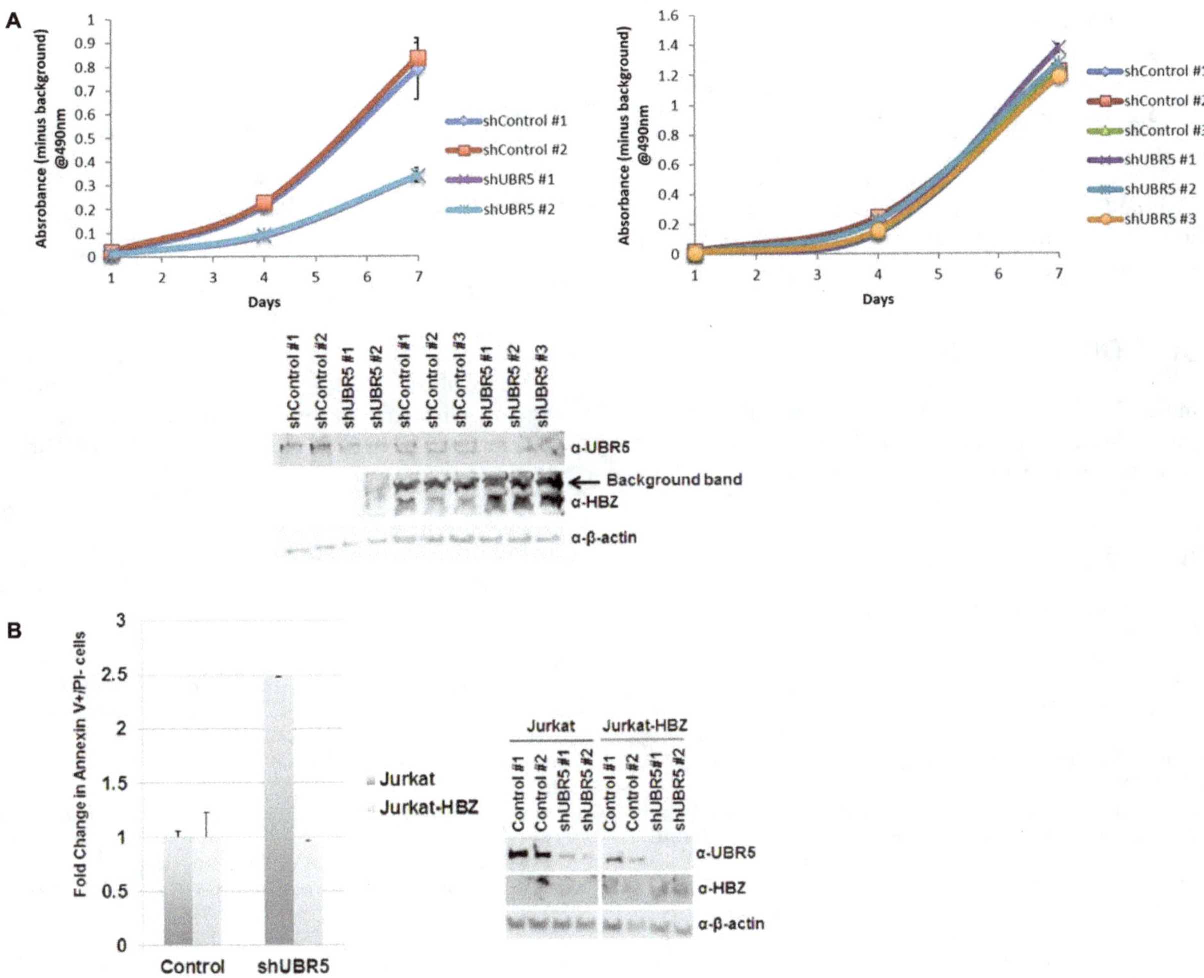

FIGURE 7 | UBR5 enhances cellular proliferation in T-cell leukemia/lymphoma cells. **(A)** 10^3 stable (Left) Jurkat and (Right) SLB-1 lentiviral control or shUBR5 infected cells were plated in normal growth medium in 96-well plates and MTS assays were performed on triplicate wells at 24-h intervals for a total of 7 days. The average absorbance numbers are plotted and error bars denote SD. (Below) Immunoblot analysis was performed on total cell lysates to compare the levels of endogenous UBR5 and HBZ expression. β-actin was used as a loading control. **(B)** Cellular apoptosis was measured using a FITC Annexin V Apoptosis Detection Kit as described in the Section "Materials and Methods." The percentage of cells undergoing apoptosis in lentiviral control infected cells was set at 1. The fold increase in apoptosis was measured in Jurkat and Jurkat-HBZ cells infected with lentiviral shUBR5. (Right) Immunoblot analysis was performed on total cell lysates to compare the levels of endogenous UBR5 and HBZ expression. β-actin was used as a loading control.

Recent work has shown HBZ can localize in different subcellular compartments depending upon the pathology associated to HTLV-1 infection (Baratella et al., 2017). Although UBR5 is mainly localized to the nucleoplasm, it can also be localized to the cytosol (The Human Protein Atlas). It will be of interest to determine if UBR5 also affects the stability of HBZ when it is segregated in the cytoplasm.

Recently, UBR5 was found to regulate gastric cancer cell growth both *in vitro* and *in vivo* (Yang et al., 2016). Using shRNA-mediated lentiviral knockdown, we found loss of UBR5 decreased T-cell lymphoma proliferation *in vitro* (**Figure 7A**, Jurkat cells). However, since UBR5 regulates the stability of HBZ, loss of UBR5 resulted in an increase in HBZ and therefore no difference in cell proliferation in HTLV-1-transformed cells (**Figure 7A**, SLB-1 cells). Upon further examination, we found that loss of UBR5 caused an increase in the level of apoptosis in Jurkat cells. Using Jurkat-HBZ cells we were able to rescue this difference due to the anti-apoptotic effects of HBZ (**Figure 7B**).

HBZ and Tax frequently have opposing effects on signaling pathways and various cellular processes. Tax is able to enhance viral transcription, whereas HBZ inhibits Tax-mediated transcription (Gaudray et al., 2002; Lemasson et al., 2007). Tax induces cell senescence whereas HBZ promotes cellular proliferation (Arnold et al., 2006, 2008; Satou et al., 2006; Ho et al., 2012). Tax activates NF-κB signaling whereas HBZ represses p65-mediated signaling (Ballard et al., 1988; Zhao et al., 2009). By having opposing effects, the virus is ensuring homeostasis within the cell. Thus far, the field has focused on regulation of Tax functions without regard for regulation of HBZ protein. Regulation of HBZ protein levels could very well have significant implications on Tax functional effects. A recent report details the occurrence of Tax gene bursts in HTLV-1-infected T-cells from patients (Billman et al., 2017). This result suggests that Tax may indeed be expressed throughout infection and highlights the relevance of HBZ regulation and the Tax/HBZ interplay.

We now show HBZ is ubiquitinated, presumably by the cellular E3 ubiquitin-protein ligase, UBR5. In addition, our work demonstrates the importance of UBR5 on the proliferative phenotype of transformed T-cell lines.

AUTHOR CONTRIBUTIONS

AP, PG, and MK conceived and planned the experiments. AP, JA-S, CH, and NS carried out the experiments and analyzed the data. AP took the lead in writing the manuscript. All authors provided the critical feedback and helped to shape the research, analysis, and the manuscript.

FUNDING

This work was supported by a grant from the National Institutes of Health (PO1CA100730) to PG.

ACKNOWLEDGMENTS

The authors thank Brandon Murphy for his preliminary work, which helped direct the focus of this manuscript, and Monika Dzieciatkowska for her technical expertise and assistance with MS/MS analysis. They also kindly thank Wessel Dirksen for editorial comments.

REFERENCES

Arnold, J., Yamamoto, B., Li, M., Phipps, A. J., Younis, I., Lairmore, M. D., et al. (2006). Enhancement of infectivity and persistence in vivo by HBZ, a natural antisense coded protein of HTLV-1. *Blood* 107, 3976–3982. doi: 10.1182/blood-2005-11-4551

Arnold, J., Zimmerman, B., Li, M., Lairmore, M. D., and Green, P. L. (2008). Human T-cell leukemia virus type-1 antisense-encoded gene, *Hbz*, promotes T-lymphocyte proliferation. *Blood* 112, 3788–3797. doi: 10.1182/blood-2008-04-154286

Ballard, D. W., Bohnlein, E., Lowenthal, J. W., Wano, Y., Franza, B. R., and Greene, W. C. (1988). HTLV-I tax induces cellular proteins that activate the κB element in the IL-2 receptor α gene. *Science* 241, 1652–1655. doi: 10.1126/science.2843985

Baratella, M., Forlani, G., Raval, G. U., Tedeschi, A., Gout, O., Gessain, A., et al. (2017). Cytoplasmic localization of HTLV-1 HBZ protein: a biomarker of HTLV-1-associated myelopathy/tropical spastic paraparesis (HAM/TSP). *PLOS Negl. Trop. Dis.* 11:e0005285. doi: 10.1371/journal.pntd.0005285

Bex, F., and Gaynor, R. B. (1998). Regulation of gene expression by HTLV-I Tax protein. *Methods* 16, 83–94. doi: 10.1006/meth.1998.0646

Billman, M. R., Rueda, D., and Bangham, C. R. M. (2017). Single-cell heterogeneity and cell-cycle-related viral gene bursts in the human leukaemia virus HTLV-1. *Wellcome Open Res.* 2:87. doi: 10.12688/wellcomeopenres.12469.2

Caron, C., Boyault, C., and Khochbin, S. (2005). Regulatory cross-talk between lysine acetylation and ubiquitination: role in the control of protein stability. *Bioessays* 27, 408–415. doi: 10.1002/bies.20210

Cheng, H., Ren, T., and Sun, S. C. (2012). New insight into the oncogenic mechanism of the retroviral oncoprotein Tax. *Protein cell* 3, 581–589. doi: 10.1007/s13238-012-2047-0

Chesarino, N. M., McMichael, T. M., Hach, J. C., and Yount, J. S. (2014). Phosphorylation of the antiviral protein interferon-inducible transmembrane protein 3 (IFITM3) dually regulates its endocytosis and ubiquitination. *J. Biol. Chem.* 289, 11986–11992. doi: 10.1074/jbc.M114.557694

Ciminale, V., Rende, F., Bertazzoni, U., and Romanelli, M. G. (2014). HTLV-1 and HTLV-2: highly similar viruses with distinct oncogenic properties. *Front. Microbiol.* 5:398. doi: 10.3389/fmicb.2014.00398

Clancy, J. L., Henderson, M. J., Russell, A. J., Anderson, D. W., Bova, R. J., Campbell, I. G., et al. (2003). EDD, the human orthologue of the *hyperplastic discs* tumour suppressor gene, is amplified and overexpressed in cancer. *Oncogene* 22, 5070–5081. doi: 10.1038/sj.onc.1206775

Clerc, I., Polakowski, N., Andre-Arpin, C., Cook, P., Barbeau, B., Mesnard, J. M., et al. (2008). An interaction between the human T cell leukemia virus type 1 basic leucine zipper factor (HBZ) and the KIX domain of p300/CBP contributes to the down-regulation of tax-dependent viral transcription by HBZ. *J. Biol. Chem.* 283, 23903–23913. doi: 10.1074/jbc.M803116200

Delker, R. K., Zhou, Y., Strikoudis, A., Stebbins, C. E., and Papavasiliou, F. N. (2013). Solubility-based genetic screen identifies RING finger protein 126 as an E3 ligase for activation-induced cytidine deaminase. *Proc. Natl. Acad. Sci. U.S.A.* 110, 1029–1034. doi: 10.1073/pnas.1214538110

DeThe, G., and Bomford, R. (1993). An HTLV-I vaccine: why, how, for whom? *AIDS Res. Hum. Retroviruses* 9, 381–386. doi: 10.1089/aid.1993.9.381

Dissinger, N., Shkriabai, N., Hess, S., Al-Saleem, J., Kvaratskhelia, M., and Green, P. L. (2014). Identification and characterization of HTLV-1 HBZ post-translational modifications. *PLOS ONE* 9:e112762. doi: 10.1371/journal.pone.0112762

Gaudray, G., Gachon, F., Basbous, J., Biard-Piechaczyk, M., Devaux, C., and Mesnard, J. (2002). The complementary strand of the human T-cell leukemia virus type 1 RNA genome encodes a bZIP transcription factor that down-regulates viral transcription. *J. Virol.* 76, 12813–12822. doi: 10.1128/JVI.76.24.12813-12822.2002

Gessain, A., Barin, F., Vernant, J. C., Gout, O., Maurs, L., Calender, A., et al. (1985). Antibodies to human T-lymphotropic virus type-I in patients with tropical spastic paraparesis. *Lancet* 2, 407–410. doi: 10.1016/S0140-6736(85)92734-5

Gessain, A., and Cassar, O. (2012). Epidemiological aspects and world distribution of HTLV-1 infection. *Front. Microbiol.* 3:388. doi: 10.3389/fmicb.2012.00388

Green, P. L., Ross, T. M., Chen, I. S. Y., and Pettiford, S. (1995). Human T-cell leukemia virus type II nucleotide sequences between *env* and the last exon of *tax/rex* are not required for viral replication or cellular transformation. *J. Virol.* 69, 387–394.

Henderson, M. J., Russell, A. J., Hird, S., Munoz, M., Clancy, J. L., Lehrbach, G. M., et al. (2002). EDD, the human hyperplastic discs protein, has a role in progesterone receptor coactivation and potential involvement in DNA damage response. *J. Biol. Chem.* 277, 26468–26478. doi: 10.1074/jbc.M203527200

Ho, Y. K., Zhi, H., DeBiaso, D., Philip, S., Shih, H. M., and Giam, C. Z. (2012). HTLV-1 tax-induced rapid senescence is driven by the transcriptional activity of NF-kappaB and depends on chronically activated IKKalpha and p65/RelA. *J. Virol.* 86, 9474–9483. doi: 10.1128/JVI.00158-12

Imai, M., Higuchi, M., Kawamura, H., Yoshita, M., Takahashi, M., Oie, M., et al. (2013). Human T cell leukemia virus type 2 (HTLV-2) Tax2 has a dominant activity over HTLV-1 Tax1 to immortalize human CD4+ T cells. *Virus Genes* 46, 39–46. doi: 10.1007/s11262-012-0831-9

Ishitsuka, K., and Tamura, K. (2014). Human T-cell leukaemia virus type I and adult T-cell leukaemia-lymphoma. *Lancet Oncol.* 15, e517–e526. doi: 10.1016/S1470-2045(14)70202-5

Lanouette, S., Mongeon, V., Figeys, D., and Couture, J. F. (2014). The functional diversity of protein lysine methylation. *Mol. Syst. Biol.* 10, 724. doi: 10.1002/msb.134974

Larocca, D., Chao, L. A., Seto, M. H., and Brunck, T. K. (1989). Human T-cell leukemia virus minus strand transcription in infected cells. *Biochem. Biophys. Res. Commun.* 163, 1006–1013. doi: 10.1016/0006-291X(89)92322-X

Lemasson, I., Lewis, M. R., Polakowski, N., Hivin, P., Cavanagh, M. H., Thebault, S., et al. (2007). Human T-cell leukemia virus type 1 (HTLV-1) bZIP protein interacts with the cellular transcription factor CREB to inhibit HTLV-1 transcription. *J. Virol.* 81, 1543–1553. doi: 10.1128/JVI.00480-06

Li, M., Kesic, M., Yin, H., Lianbo, Y., and Green, P. (2009). Kinetic analysis of human T-cell leukemia virus type 1 gene expression in cell culture and infected animals. *J. Virol.* 83, 3788–3797. doi: 10.1128/JVI.02315-08

Ma, G., Yasunaga, J., and Matsuoka, M. (2016). Multifaceted functions and roles of HBZ in HTLV-1 pathogenesis. *Retrovirology* 13:16. doi: 10.1186/s12977-016-0249-x

Mansfield, E., Hersperger, E., Biggs, J., and Shearn, A. (1994). Genetic and molecular analysis of *hyperplastic discs*, a gene whose product is required for regulation of cell proliferation in *Drosophila melanogaster* imaginal discs and germ cells. *Dev. Biol.* 165, 507–526. doi: 10.1006/dbio.1994.1271

Marriott, S. J., and Semmes, O. J. (2005). Impact of HTLV-I Tax on cell cycle progression and the cellular DNA damage repair response. *Oncogene* 24, 5986–5995. doi: 10.1038/sj.onc.1208976

Matsuoka, M., and Green, P. L. (2009). The HBZ gene, a key player in HTLV-1 pathogenesis. *Retrovirology* 6:71. doi: 10.1186/1742-4690-6-71

Matsuoka, M., and Jeang, K. T. (2005). Human T-cell leukemia virus type I at age 25: a progress report. *Cancer Res.* 65, 4467–4470. doi: 10.1158/0008-5472.CAN-05-0559

Meissner, B., Kridel, R., Lim, R. S., Rogic, S., Tse, K., Scott, D. W., et al. (2013). The E3 ubiquitin ligase *UBR5* is recurrently mutated in mantle cell lymphoma. *Blood* 121, 3161–3164. doi: 10.1182/blood-2013-01-478834

Osame, M., Usuku, K., Izumo, S., Ijichi, N., Amitani, H., Igata, A., et al. (1986). HTLV-I associated myelopathy, a new clinical entity. *Lancet* 1, 1031–1032. doi: 10.1016/S0140-6736(86)91298-5

Panfil, A. R., Dissinger, N. J., Howard, C. M., Murphy, B. M., Landes, K., Fernandez, S. A., et al. (2016). Functional comparison of HBZ and the related APH-2 protein provides insight into human T-Cell leukemia virus type 1 pathogenesis. *J. Virol.* 90, 3760–3772. doi: 10.1128/JVI.03113-15

Poiesz, B. J., Ruscetti, F. W., Gazdar, A. F., Bunn, P. A., Minna, J. D., and Gallo, R. C. (1980). Detection and isolation of type C retrovirus particles from fresh and cultured lymphocytes of a patient with cutaneous T-cell lymphoma. *Proc. Natl. Acad. Sci. U.S.A.* 77, 7415–7419. doi: 10.1073/pnas.77.12.7415

Raval, G. U., Bidoia, C., Forlani, G., Tosi, G., Gessain, A., and Accolla, R. S. (2015). Localization, quantification and interaction with host factors of endogenous HTLV-1 HBZ protein in infected cells and ATL. *Retrovirology* 12:59. doi: 10.1186/s12977-015-0186-0

Romanelli, M. G., Diani, E., Bergamo, E., Casoli, C., Ciminale, V., Bex, F., et al. (2013). Highlights on distinctive structural and functional properties of HTLV Tax proteins. *Front. Microbiol.* 4:271. doi: 10.3389/fmicb.2013.00271

Satou, Y., Yasunaga, J., Yoshida, M., and Matsuoka, M. (2006). *HTLV-I basic leucine zipper factor* gene mRNA supports proliferation of adult T cell leukemia cells. *Proc. Natl. Acad. Sci. U.S.A.* 103, 720–725. doi: 10.1073/pnas.0507631103

Shoji, T., Higuchi, M., Kondo, R., Takahashi, M., Oie, M., Tanaka, Y., et al. (2009). Identification of a novel motif responsible for the distinctive transforming activity of human T-cell leukemia virus (HTLV) type 1 Tax1 protein from HTLV-2 Tax2. *Retrovirology* 6:83. doi: 10.1186/1742-4690-6-83

Subbaiah, V. K., Zhang, Y., Rajagopalan, D., Abdullah, L. N., Yeo-Teh, N. S., Tomaic, V., et al. (2016). E3 ligase EDD1/UBR5 is utilized by the HPV E6 oncogene to destabilize tumor suppressor TIP60. *Oncogene* 35, 2062–2074. doi: 10.1038/onc.2015.268

Takeda, S., Maeda, M., Morikawa, S., Taniguchi, Y., Yasunaga, J., Nosaka, K., et al. (2004). Genetic and epigenetic inactivation of tax gene in adult T-cell leukemia cells. *Int. J. Cancer* 109, 559–567. doi: 10.1002/ijc.20007

Uchiyama, T., Yodoi, J., Sagawa, K., Takatsuki, K., and Uchino, H. (1977). Adult T-cell leukemia: clinical and hematologic features of 16 cases. *Blood* 50, 481–492.

Usui, T., Yanagihara, K., Tsukasaki, K., Murata, K., Hasegawa, H., Yamada, Y., et al. (2008). Characteristic expression of HTLV-1 basic zipper factor (HBZ) transcripts in HTLV-1 provirus-positive cells. *Retrovirology* 5:34. doi: 10.1186/1742-4690-5-34

Valeri, V. W., Hryniewicz, A., Andresen, V., Jones, K., Fenizia, C., Bialuk, I., et al. (2010). Requirement of the human T-cell leukemia virus p12 and p30 products for infectivity of human dendritic cells and macaques but not rabbits. *Blood* 116, 3809–3817. doi: 10.1182/blood-2010-05-284141

Yang, M., Jiang, N., Cao, Q. W., Ma, M. Q., and Sun, Q. (2016). The E3 ligase UBR5 regulates gastric cancer cell growth by destabilizing the tumor suppressor GKN1. *Biochem. Biophys. Res. Commun.* 478, 1624–1629. doi: 10.1016/j.bbrc.2016.08.170

Yin, H., Kannian, P., Dissinger, N., Haines, R., Niewiesk, S., and Green, P. L. (2012). HTLV-2 APH-2 is dispensable for *in vitro* immortalization, but functions to repress early viral replication *in vivo*. *J. Virol.* 86, 8412–8421. doi: 10.1128/JVI.00717-12

Yoshida, M., Miyoshi, I., and Hinuma, Y. (1982). Isolation and characterization of retrovirus from cell lines of human adult T-cell leukemia and its implication in the disease. *Proc. Natl. Acad. Sci. U.S.A.* 79, 2031–2035. doi: 10.1073/pnas.79.6.2031

Yoshida, M., Satou, Y., Yasunaga, J., Fujisawa, J., and Matsuoka, M. (2008). Transcriptional control of spliced and unspliced human T-cell leukemia virus type 1 bZIP factor (*HBZ*) gene. *J. Virol.* 82, 9359–9368. doi: 10.1128/JVI.00242-08

Zhao, T., and Matsuoka, M. (2012). HBZ and its roles in HTLV-1 oncogenesis. *Front. Microbiol.* 3:247. doi: 10.3389/fmicb.2012.00247

Zhao, T., Satou, Y., Sugata, K., Miyazato, P., Green, P. L., Imamura, T., et al. (2011). HTLV-1 bZIP factor enhances TGF-beta signaling through p300 coactivator. *Blood* 118, 1865–1876. doi: 10.1182/blood-2010-12-326199

Zhao, T., Yasunaga, J., Satou, Y., Nakao, M., Takahashi, M., Fujii, M., et al. (2009). Human T-cell leukemia virus type 1 bZIP factor selectively suppresses the classical pathway of NF-kappaB. *Blood* 113, 2755–2764. doi: 10.1182/blood-2008-06-161729

8

HTLV-1: Regulating the Balance Between Proviral Latency and Reactivation

*Anurag Kulkarni and Charles R. M. Bangham**

Section of Virology, Division of Infectious Diseases, Department of Medicine, Imperial College London, London, United Kingdom

Correspondence:
Charles R. M. Bangham
c.bangham@imperial.ac.uk

HTLV-1 plus-strand transcription begins with the production of doubly-spliced *tax/rex* transcripts, the levels of which are usually undetectable in freshly isolated peripheral blood mononuclear cells (PBMCs) from HTLV-1-infected individuals. However, the presence of a sustained chronically active cytotoxic T-cell response to HTLV-1 antigens in virtually all HTLV-1-infected individuals, regardless of their proviral load, argues against complete latency of the virus *in vivo*. There is an immediate burst of plus-strand transcription when blood from infected individuals is cultured *ex vivo*. How is the HTLV-1 plus strand silenced in PBMCs? Is it silenced in other anatomical compartments within the host? What reactivates the latent provirus in fresh PBMCs? While plus-strand transcription of the provirus appears to be intermittent, the minus-strand *hbz* transcripts are present in a majority of cells, albeit at low levels. What regulates the difference between the 5′- and 3′-LTR promoter activities and thereby the *tax-hbz* interplay? Finally, T lymphocytes are a migratory population of cells that encounter variable environments in different compartments of the body. Could these micro-environment changes influence the reactivation kinetics of the provirus? In this review we discuss the questions raised above, focusing on the early events leading to HTLV-1 reactivation from latency, and suggest future research directions.

Keywords: HTLV-1, latency, virus, metabolism, hypoxia, glucose, epigenetic regulation, transcription

INTRODUCTION

Human T-lymphotropic virus type 1 (HTLV-1), also known as Human T-cell leukemia virus type 1, is a retrovirus that mainly infects CD4+ T-cells *in vivo.* In ~5–10% of the infected individuals, HTLV-1 infection leads to either an aggressive T-cell malignancy, adult T-cell leukemia/lymphoma (ATL), or a chronic progressive neuro-inflammatory condition called HTLV-1-associated myelopathy/tropical spastic paraparesis (HAM/TSP). Both these conditions have limited treatment options, and ATL in particular carries a very poor prognosis.

Initial infection is asymptomatic and occurs through breast milk, blood, or semen of infected individuals (Bangham and Matsuoka, 2017). HTLV-1 replicates in the host through two distinct routes: (i) Infectious spread: this mode of spread involves productive replication from the integrated provirus followed by the transfer of newly produced virions through the virological synapse (Igakura et al., 2003; Pais-Correia et al., 2010). This is the major route of viral spread in the initial stages of infection when the proviral load – the percentage of infected PBMCs – is low (Bangham et al., 2014). Infectious spread results in the formation of distinct T-cell clones, each clone carrying a single-copy HTLV-1 provirus integrated in a unique genomic location within the host genome. (ii) Mitotic spread: proliferation of HTLV-1-infected host cells results in passive replication of the integrated HTLV-1 provirus within their genome. The two daughter cells resulting from mitosis of an HTLV-1-infected parent cell carry the provirus in the same genomic integration site. In contrast to infectious spread, the contribution of this mode to the proviral load in infected individuals may be small in the early stages of infection but gradually increases during the chronic stage of infection (Bangham et al., 2014). Although the proviral load in each host can fluctuate by a small factor (2- to 5-fold) over time, the proviral loads can vary between infected individuals by over 1000-fold (Nagai et al., 1998; Demontis et al., 2013). Individuals with a higher proviral load are at greater risk of developing either ATL or HAM/TSP (Matsuzaki et al., 2001; Iwanaga et al., 2010). An infected individual typically carries about 10^4 to 10^5 different T-cell clones, each with a unique proviral integration site (Bangham et al., 2014). The extent of proliferation of HTLV-1-infected T-cell clones, and thus their respective contribution to an individual's proviral load, both vary greatly from one clone to another.

A peculiar characteristic of HTLV-1 is the absence of detectable cell-free virions in infected individuals (Demontis et al., 2015). HTLV-1 was previously considered to be latent in infected individuals because there are no detectable plus-strand viral structural RNA or protein products in the peripheral blood mononuclear cells (PBMCs) freshly obtained from HTLV-1 infected individuals. Also, HTLV-1 is genetically stable, with minimal sequence variation over evolutionary time (Gessain et al., 1992), suggesting that *de novo* viral replication, which generates sequence variation, contributes little to the long-term persistence of HTLV-1 *in vivo*. However, a sustained chronically active CTL response to HTLV-1 can be observed in virtually all infected individuals (Jacobson et al., 1990; Bangham, 2009). This observation suggests that the host immune system routinely encounters viral antigens *in vivo*. Since the HTLV-1 plus-strand appears to be latent in peripheral blood, the implication is that plus-strand expression occurs either in intermittent bursts, or outside the circulation, or both. Thus it is necessary to understand the determinants of HTLV-1 latency and reactivation *in vivo* (**Figure 1**) to devise effective preventive and therapeutic approaches against HTLV-1-associated diseases such as HAM/TSP and ATL.

Proviral Genomic Integration Site

HTLV-1 cannot be looked at in isolation because it is an inseparable part of the chromatinized host genomic DNA. Can we predict the behavior of the integrated provirus in the context of its flanking host genome? HTLV-1 integration favors transcriptionally active regions of the host genome (Melamed et al., 2013). Recently, the host enzyme Protein Phosphatase 2A (PP2A) was identified as a major host co-factor for the HTLV-1 integrase, which could influence the selection of genomic integration sites (Maertens, 2016). Furthermore, plus-strand transcription is silenced when the viral DNA is integrated downstream of a host gene promoter in the same-sense orientation, possibly by transcriptional interference (Melamed et al., 2013). Similarly, the presence of the SWI/SNF-associated ATPase BRG-1 (identified by chromatin immunoprecipitation) upstream of the integrated HTLV-1 provirus is associated with silencing of plus-strand transcription, but its presence downstream of the HTLV-1 sequence is associated with proviral plus-strand expression (Melamed et al., 2013). Recently, it was reported that integration of HTLV-1 in the vicinity of cancer driver genes causes either premature interruption of transcription or antisense-dependent *cis*-perturbation of these genes, perhaps contributing to leukemogenesis (Rosewick et al., 2017).

The discovery of a functional CTCF binding site within the pX region of the HTLV-1 genome adds a new dimension to the importance of the flanking genome to proviral transcription (Satou et al., 2016). CTCF is a critical host protein which binds to an insulator motif in the DNA and appears to limit the spread of epigenetic modifications (Merkenschlager and Nora, 2016). Consistent with this function, the CTCF binding region in HTLV-1 is associated with the presence of a sharp epigenetic border in several histone and DNA modifications. It is possible that the epigenetic border formed by CTCF binding within the HTLV-1 provirus allows unhindered minus-strand transcription while reversibly inhibiting the plus-strand activity. Another central function of CTCF is the formation of chromatin loops, which regulate contacts between enhancers and promoters. It is thus possible that the CTCF bound to the provirus produces abnormal chromatin loops, resulting in aberrant cellular gene expression (Cook et al., 2017). Experiments are now in progress to test these hypotheses. Ultimately, this aberrant gene expression may be an important factor in the development of ATL.

The Tax-HBZ Interplay

Gene transcription in HTLV-1 progresses in both the sense and anti-sense directions, governed by promoters in the identical long terminal repeats (LTRs) situated at each end of the genome.

Plus-Strand Transcription

This is regulated by the TATA-box-containing inducible promoter in the 5′-LTR (Miyazato et al., 2016). TATA boxes are frequently observed in promoters of cellular genes which are highly regulated and responsive to stress and extracellular signals (Basehoar et al., 2004; Bahrami and Drablos, 2016). The 5′-LTR contains three tandem 21-bp imperfect nucleotide repeats called tax-responsive elements (TREs) which bind to the ATF/CREB family of proteins to activate HTLV-1 plus-strand transcription (Armstrong et al., 1993). The transactivator protein Tax, encoded by the pX region of the HTLV-1 genome by a double-splicing mechanism, is the most potent activator of plus-strand transcription and initiates a positive-feedback loop which induces transcription and promotes replication (Beimling and Moelling, 1992; Anderson and Dynan, 1994). Tax enhances the binding of CREB to the TRE, thereby promoting the transactivation of the proviral plus-strand (Suzuki et al., 1993). In addition to ATF/CREB family, the 5′-LTR also has binding sites for several major transcription factors, namely NF-κB, SRF and Sp1, which play direct or indirect roles in

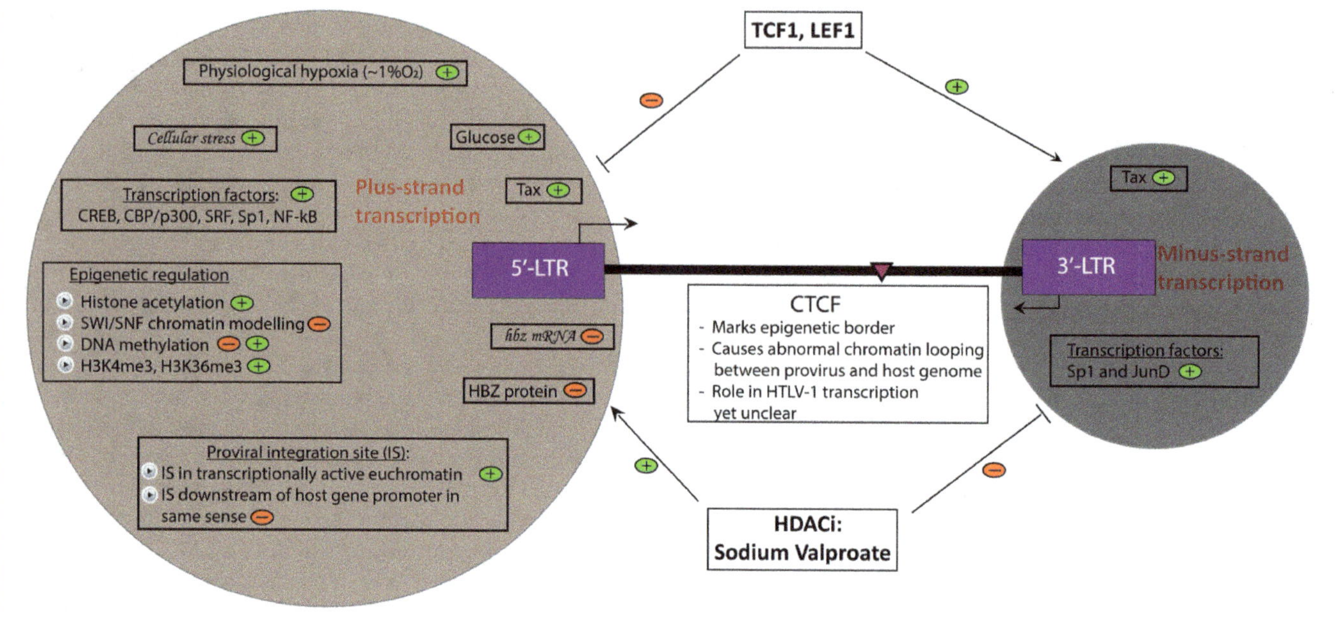

FIGURE 1 | Determinants of HTLV-1 plus- and minus-strand transcription. A summary of currently known factors that induce (depicted by ⊕) and inhibit (depicted by ⊖) plus- and minus-strand transcription of the HTLV-1 provirus. Italicized words denote hypothetical factors. The pink arrowhead marks the position of the CTCF binding site in the provirus.

activating plus-strand transcription (Armstrong et al., 1993). In addition to Tax, the HTLV-1 pX region encodes another regulatory protein, Rex, which regulates the stabilization of unspliced and singly spliced viral mRNAs, their nuclear export and subsequent effective translation of the viral proteins (Nakano and Watanabe, 2012). The other HTLV-1 pX region plus-strand products play accessory roles *in vivo*: p30 and p13 promote infectivity in a rabbit model (Bartoe et al., 2000; Silverman et al., 2004; Hiraragi et al., 2006), and p30 and p12/p8 enhance persistent infection in macaques (Valeri et al., 2010; Pise-Masison et al., 2014). p13 increases the production of reactive oxygen species, selectively killing transformed cells, and may favor persistence *in vivo* (Silic-Benussi et al., 2010). However, these genes are not essential for viral replication or T-cell immortalization (Derse et al., 1997; Robek et al., 1998), and may not be required for transmission (Furukawa et al., 2004).

Minus-Strand Transcription

This is initiated at the 3′-end of the provirus and progresses in the anti-sense direction (Gaudray et al., 2002). It is governed by the TATA-less promoter in the 3′-LTR (Yoshida et al., 2008), which contains three TREs and three Sp1 binding sites (Yoshida et al., 2008; Ma et al., 2016). *sHBZ* is the major transcript driven by the 3′-LTR.

The interplay between the plus-strand and minus-strand transcription is a critical and poorly understood aspect of the HTLV-1 life cycle. We infer that Tax expression *in vivo* is intermittent and highly regulated as it is a very immunogenic protein (Jacobson et al., 1990; Parker et al., 1992). There is no detectable plus-strand gene expression in freshly isolated PBMCs from HTLV-1 infected individuals (Kinoshita et al., 1989; Rende et al., 2011). HBZ, in contrast, is poorly immunogenic and is expressed at a very low level; however, the presence of a detectable cytotoxic T-lymphocyte (CTL) response to HBZ is associated with a lower proviral load (MacNamara et al., 2010). This constant, low-level *hbz* mRNA expression is thought to be necessary for the persistence of HTLV-1 *in vivo*: *hbz* expression can be observed in all HTLV-1-infected individuals, including those with HAM/TSP and ATL (Yoshida et al., 2008). Tax and HBZ have strong effects on a number of cellular processes and signaling cascades (Ma et al., 2016); the effects of the two proteins frequently act in opposition. Certain transcription factors including TCF1 and LEF1 inhibit plus-strand transcription from the 5′-LTR while simultaneously enhancing minus-strand transcription from the 3′-LTR (Ma et al., 2015). In contrast, the histone deacetylase inhibitor (HDACi) valproate activates plus-strand transcription while inhibiting expression of the minus-strand (Lezin et al., 2007; Belrose et al., 2011; Olindo et al., 2011). It is also known that Tax activates minus-strand transcription through binding of the TREs in the 3′-LTR (Landry et al., 2009). Recently, it was shown by single-molecule RNA-FISH that, at a single-cell level, both plus- and minus-strand transcription occurs in bursts, albeit at different times and intensities. HTLV-1 plus-strand mRNA transcription is enhanced in the absence of minus-strand transcripts, while high-level plus-strand transcription promotes transcription from the minus-strand (Billman et al., 2017), generating a putative negative feedback inhibition of plus-strand expression. Until now, it was believed that all HTLV-1-infected cells express *hbz* mRNA at all times; however, this study showed that, in naturally-infected T-cell clones isolated by limiting dilution from HTLV-1-infected individuals, not all HTLV-1-infected cells in each clone are *hbz* positive (Billman et al., 2017) at a given time. Many questions remain unanswered. Most *hbz* transcripts were found in the nuclei of cells, consistent with previous observations (Rende et al., 2011) and with the low level of protein expression of HBZ (Shiohama et al., 2016). HBZ protein inhibits *tax* expression by competing for binding of the transcription co-factor CREB (Gaudray et al., 2002; Lemasson et al., 2007; Clerc et al., 2008). However, since HBZ protein is expressed at such a low level in naturally-infected cells, the importance of this mechanism is uncertain (Shiohama et al., 2016).

These observations raise important questions: does *hbz* RNA also inhibit *tax* expression? What are the factors that regulate minus-strand transcriptional bursts? It has been shown that Sp1 co-operates with JunD to activate minus-strand transcription, but how and when this happens in the context of the cell cycle and viral replication are still unknown (Gazon et al., 2012). Recently, Billman et al. (2017) showed at the single cell level, in naturally-infected HTLV-1 positive T-cell clones, that cells in S and G2/M phase of the cell cycle have elevated *tax* and *hbz* mRNA levels, suggesting that the cell cycle plays a role in regulating HTLV-1 transcription. It is now becoming clear that the results of population-level analyses can mask strong single-cell heterogeneity within the population.

Epigenetic Regulation of HTLV-1 Transcription

Small-molecule inhibitors targeting epigenetic enzymes have been widely used in HIV latency research. Indeed, reactivation of the latent HIV-1 provirus occurs in response to HDACi [e.g., suberanilohydroxamic acid (SAHA)], DNA methylation inhibitors (e.g., 5-azacytidine) and histone methyltransferase inhibitors (e.g., the EZH1/2 inhibitor GSK343) (Spina et al., 2013; Tripathy et al., 2015). Similarly, in the case of HTLV-1, the HDACi valproate and DNA methyltransferase inhibitor 5-azacytidine have been shown to activate the latent HTLV-1 provirus (Koiwa et al., 2002; Belrose et al., 2011). These observations suggest that epigenetic modulations in the promoters, enhancers and gene bodies of the integrated provirus play an important role in proviral transcription. Cytosine methylation of the 5′-LTR of HTLV-1 is associated with silencing of plus-strand transcription; the degree of this 5′-LTR methylation varies from individual to individual, whereas the 3′-LTR is unmethylated in most individuals (Taniguchi et al., 2005), consistent with the constitutive activation of the promoter in the 3′-LTR. However, it is unclear how dynamic the DNA methylation pattern of the provirus is. It is unlikely that the highly regulated promoter in the 5′-LTR of HTLV-1 would rely primarily on DNA demethylation for its reactivation. Data are needed on the extent of DNA methylation during bursts of gene expression from the plus-strand and minus-strand, and whether this methylation is a cause or an effect of promoter activation. Activation of plus-strand transcription is associated with Tax-mediated recruitment of pCREB and CBP/p300, resulting in histone acetylation and nucleosomal remodeling through BRG-1-containing SWI/SNF complexes (Sharma and Nyborg, 2008; Easley et al., 2010). Similarly, HTLV-1 plus-strand transcription is associated with a significant increase in the activatory histone epigenetic modifications H3K4me3 and H3K36me3 in different regions of the HTLV-1 provirus (Kulkarni et al., 2017). It is likely that, as in host gene expression, many epigenetic modifications are a cause as well as consequence of viral reactivation. Most of the epigenetic studies carried out so far have been in *in vitro* cellular models where Tax expression is the primary mechanism of 5′-LTR activation through a positive feedback loop. However, evidence from freshly isolated PBMCs from HTLV-1-infected individuals indicates that there is little or no Tax protein expression *in vivo* at steady state. Hence, it is important to examine the earliest epigenetic changes that occur at the provirus in freshly isolated PBMCs from HTLV-1-infected individuals to provide clues to the changes that occur *in vivo*. It is still unclear what primarily reactivates plus-strand transcription in *ex vivo* PBMCs from HTLV-1-infected individuals. The answer to this conundrum may elucidate the key question – what regulates the rapidly reversible silencing of plus-strand transcription *in vivo*?

Metabolic Regulation of HTLV-1 Transcription

The metabolic and functional activities of circulating PBMCs depend directly on their micro-environment. What effects do metabolic changes have on the integrated HTLV-1 provirus within these cells? We recently showed that HTLV-1 reactivation from latency is severely limited in the absence of glucose in the surrounding medium (Kulkarni et al., 2017). In support of this conclusion, we found that PBMCs treated with glycolysis inhibitors – and to a lesser extent with mitochondrial electron transport chain (ETC) inhibitors – show a significant impairment in HTLV-1 plus-strand transcription when they are cultured *ex vivo*. The importance of glucose in the HTLV-1 life cycle is emphasized by the fact that the glucose receptor GLUT-1 is a cellular receptor for HTLV-1 (Manel et al., 2003). Surprisingly, drugs that inhibit or stimulate the TCA cycle do not alter HTLV-1 transcription (Kulkarni et al., 2017). Indeed, it is known that different subsets of T-cells use distinct metabolic pathways to satisfy their energy requirements (Dimeloe et al., 2017). How glucose metabolism influences HTLV-1 reactivation from latency is not known. A plausible explanation is that inhibition of glycolysis or the mitochondrial ETC reduces the availability of Acetyl CoA and ATP. Acetyl CoA is an important acetyl group donor in the process of histone acetylation, which is associated with transcriptional activation. Similarly, the SWI/SNF family of chromatin-remodeling complexes such as BAF and PBAF rely on energy in the form of ATP for their nucleosomal remodeling function. It has been shown before that HTLV-1 plus-strand activation is intimately linked with histone acetylation and chromatin remodeling at the 5′-LTR promoter (Easley et al., 2010; Nyborg et al., 2010). Another important question that arises from these results is the possible impact of high blood glucose levels on HTLV-1 replication. Do HTLV-1-infected individuals with poorly controlled diabetes have a higher proviral load of HTLV-1, resulting in a significantly higher risk of HTLV-1-associated diseases such as HAM/TSP or ATL? Further work is needed to clarify the cause and effect relationships between metabolic activity and proviral activation in HTLV-1-infected primary $CD4^+$ T-cells.

Anatomical Regulation of HTLV-1 Transcription

Lymphocytes are constantly exposed to highly variable stresses during circulation through different body compartments. As mentioned above, HTLV-1 appears to be latent in freshly isolated PBMCs from the peripheral blood of HTLV-1 infected individuals. The presence of a chronically active CTL response to HTLV-1 in all infected individuals raises the possibility that HTLV-1 expression is more frequent in different compartments like the bone marrow, lymph, and lymph nodes (Bangham, 2008). Also, Yasunaga et al. (2016) found that *tax* transcription in the bone marrow of STLV-1-infected Japanese macaques was significantly higher than that in other tissues. One factor that differs significantly between these compartments is the oxygen tension. While the lymph, lymph nodes, and bone marrow are highly hypoxic (~1% oxygen), PBMCs in peripheral venous blood are exposed to significantly higher levels of oxygen (~6–10%) (Tsai et al., 2004; Hangai-Hoger et al., 2007). We recently reported that, in primary PBMCs isolated from HTLV-1 infected individuals, physiological hypoxia (~1% oxygen) enhances plus-strand HTLV-1 transcription (Kulkarni et al., 2017). Again, further work is needed to identify the molecular mechanism of this effect, and the consequences of this observation on the establishment and spread of viral infection within and between individuals.

SUMMARY

The transcriptional regulation of the plus- and minus-strands of HTLV-1 is affected by a number of virus-specific and host-specific factors. Advances in single-cell genomics and transcriptomics have made it possible to study HTLV-1 infection at the single-cell level. We anticipate that these approaches will shed light on the effect of cellular heterogeneity on viral replication and help identify novel mechanisms that were missed by earlier studies which depended on population-level analyses. Precise identification of the molecular mechanisms involved in HTLV-1 reactivation from latency would help us design and test novel therapeutic strategies to reduce the disease burden associated with this infection.

AUTHOR CONTRIBUTIONS

Both authors have made a substantial, direct and intellectual contribution to the work, and approved it for publication.

FUNDING

This work was supported by a Project Grant from the Medical Research Council, United Kingdom (MR/K019090/1), a Senior Investigator award from the Wellcome Trust, United Kingdom (WT100291MA) to CB, and the Imperial National Institute for Health Research Biomedical Research Centre (https://imperialbrc.nihr.ac.uk/).

ACKNOWLEDGMENTS

We thank Jocelyn Turpin and Goedele Maertens for comments on the manuscript.

REFERENCES

Anderson, M. G., and Dynan, W. S. (1994). Quantitative studies of the effect of HTLV-I tax protein on CREB protein-DNA binding. *Nucleic Acids Res.* 22, 3194–3201. doi: 10.1093/nar/22.15.3194

Armstrong, A. P., Franklin, A. A., Uittenbogaard, M. N., Giebler, H. A., and Nyborg, J. K. (1993). Pleiotropic effect of the human T-cell leukemia virus Tax protein on the DNA binding activity of eukaryotic transcription factors. *Proc. Natl. Acad. Sci. U.S.A.* 90, 7303–7307. doi: 10.1073/pnas.90.15.7303

Bahrami, S., and Drablos, F. (2016). Gene regulation in the immediate-early response process. *Adv. Biol. Regul.* 62, 37–49. doi: 10.1016/j.jbior.2016.05.001

Bangham, C. R. (2008). HTLV-1 infection: role of CTL efficiency. *Blood* 112, 2176–2177. doi: 10.1182/blood-2008-06-163071

Bangham, C. R. (2009). CTL quality and the control of human retroviral infections. *Eur. J. Immunol.* 39, 1700–1712. doi: 10.1002/eji.200939451

Bangham, C. R., Cook, L. B., and Melamed, A. (2014). HTLV-1 clonality in adult T-cell leukaemia and non-malignant HTLV-1 infection. *Semin. Cancer Biol.* 26, 89–98. doi: 10.1016/j.semcancer.2013.11.003

Bangham, C. R. M., and Matsuoka, M. (2017). Human T-cell leukaemia virus type 1: parasitism and pathogenesis. *Philos. Trans. R. Soc. Lond. B Biol. Sci.* 372:20160272. doi: 10.1098/rstb.2016.0272

Bartoe, J. T., Albrecht, B., Collins, N. D., Robek, M. D., Ratner, L., Green, P. L., et al. (2000). Functional role of pX open reading frame II of human T-lymphotropic virus type 1 in maintenance of viral loads in vivo. *J. Virol.* 74, 1094–1100. doi: 10.1128/JVI.74.3.1094-1100.2000

Basehoar, A. D., Zanton, S. J., and Pugh, B. F. (2004). Identification and distinct regulation of yeast TATA box-containing genes. *Cell* 116, 699–709. doi: 10.1016/S0092-8674(04)00205-3

Beimling, P., and Moelling, K. (1992). Direct interaction of CREB protein with 21 bp Tax-response elements of HTLV-ILTR. *Oncogene* 7, 257–262.

Belrose, G., Gross, A., Olindo, S., Lezin, A., Dueymes, M., Komla-Soukha, I., et al. (2011). Effects of valproate on Tax and HBZ expression in HTLV-1 and HAM/TSP T lymphocytes. *Blood* 118, 2483–2491. doi: 10.1182/blood-2010-11-321364

Billman, M. R., Rueda, D., and Bangham, C. R. M. (2017). Single-cell heterogeneity and cell-cycle-related viral gene bursts in the human leukaemia virus HTLV-1. *Wellcome Open Res.* 2:87. doi: 10.12688/wellcomeopenres.12469.2

Clerc, I., Polakowski, N., Andre-Arpin, C., Cook, P., Barbeau, B., Mesnard, J. M., et al. (2008). An interaction between the human T cell leukemia virus type 1 basic leucine zipper factor (HBZ) and the KIX domain of p300/CBP contributes to the down-regulation of tax-dependent viral transcription by HBZ. *J. Biol. Chem.* 283, 23903–23913. doi: 10.1074/jbc.M803116200

Cook, L., Melamed, A., Yaguchi, H., and Bangham, C. R. (2017). The impact of HTLV-1 on the cellular genome. *Curr. Opin. Virol.* 26, 125–131. doi: 10.1016/j.coviro.2017.07.013

Demontis, M. A., Hilburn, S., and Taylor, G. P. (2013). Human T cell lymphotropic virus type 1 viral load variability and long-term trends in asymptomatic carriers and in patients with human T cell lymphotropic virus type 1-related diseases. *AIDS Res. Hum. Retroviruses* 29, 359–364. doi: 10.1089/AID.2012.0132

Demontis, M. A., Sadiq, M. T., Golz, S., and Taylor, G. P. (2015). HTLV-1 viral RNA is detected rarely in plasma of HTLV-1 infected subjects. *J. Med. Virol.* 87, 2130–2134. doi: 10.1002/jmv.24264

Derse, D., Mikovits, J., and Ruscetti, F. (1997). X-I and X-II open reading frames of HTLV-I are not required for virus replication or for immortalization of primary T-cells in vitro. *Virology* 237, 123–128. doi: 10.1006/viro.1997.8781

Dimeloe, S., Burgener, A. V., Grahlert, J., and Hess, C. (2017). T-cell metabolism governing activation, proliferation and differentiation; a modular view. *Immunology* 150, 35–44. doi: 10.1111/imm.12655

Easley, R., Carpio, L., Guendel, I., Klase, Z., Choi, S., Kehn-Hall, K., et al. (2010). Human T-lymphotropic virus type 1 transcription and chromatin-remodeling complexes. *J. Virol.* 84, 4755–4768. doi: 10.1128/JVI.00851-09

Furukawa, Y., Usuku, K., Izumo, S., and Osame, M. (2004). Human T cell lymphotropic virus type I (HTLV-I) p12I is dispensable for HTLV-I transmission and maintenance of infection in vivo. *AIDS Res. Hum. Retroviruses* 20, 1092–1099. doi: 10.1089/aid.2004.20.1092

Gaudray, G., Gachon, F., Basbous, J., Biard-Piechaczyk, M., Devaux, C., and Mesnard, J. M. (2002). The complementary strand of the human T-cell leukemia virus type 1 RNA genome encodes a bZIP transcription factor that down-regulates viral transcription. *J. Virol.* 76, 12813–12822. doi: 10.1128/JVI.76.24.12813-12822.2002

Gazon, H., Lemasson, I., Polakowski, N., Cesaire, R., Matsuoka, M., Barbeau, B., et al. (2012). Human T-cell leukemia virus type 1 (HTLV-1) bZIP factor requires cellular transcription factor JunD to upregulate HTLV-1 antisense transcription from the 3' long terminal repeat. *J. Virol.* 86, 9070–9078. doi: 10.1128/JVI.00661-12

Gessain, A., Gallo, R. C., and Franchini, G. (1992). Low degree of human T-cell leukemia/lymphoma virus type I genetic drift in vivo as a means of monitoring viral transmission and movement of ancient human populations. *J. Virol.* 66, 2288–2295.

Hangai-Hoger, N., Tsai, A. G., Cabrales, P., and Intaglietta, M. (2007). Terminal lymphatics: the potential "lethal corner" in the distribution of tissue pO2. *Lymphat. Res. Biol.* 5, 159–168. doi: 10.1089/lrb.2007.5303

Hiraragi, H., Kim, S. J., Phipps, A. J., Silic-Benussi, M., Ciminale, V., Ratner, L., et al. (2006). Human T-lymphotropic virus type 1 mitochondrion-localizing protein p13(II) is required for viral infectivity in vivo. *J. Virol.* 80, 3469–3476. doi: 10.1128/JVI.80.7.3469-3476.2006

Igakura, T., Stinchcombe, J. C., Goon, P. K., Taylor, G. P., Weber, J. N., Griffiths, G. M., et al. (2003). Spread of HTLV-I between lymphocytes by virus-induced polarization of the cytoskeleton. *Science* 299, 1713–1716. doi: 10.1126/science.1080115

Iwanaga, M., Watanabe, T., Utsunomiya, A., Okayama, A., Uchimaru, K., Koh, K. R., et al. (2010). Human T-cell leukemia virus type I (HTLV-1) proviral load and disease progression in asymptomatic HTLV-1 carriers: a nationwide prospective study in Japan. *Blood* 116, 1211–1219. doi: 10.1182/blood-2009-12-257410

Jacobson, S., Shida, H., Mcfarlin, D. E., Fauci, A. S., and Koenig, S. (1990). Circulating CD8$^+$ cytotoxic T lymphocytes specific for HTLV-I pX in patients with HTLV-I associated neurological disease. *Nature* 348, 245–248. doi: 10.1038/348245a0

Kinoshita, T., Shimoyama, M., Tobinai, K., Ito, M., Ito, S., Ikeda, S., et al. (1989). Detection of mRNA for the tax1/rex1 gene of human T-cell leukemia virus type I in fresh peripheral blood mononuclear cells of adult T-cell leukemia patients and viral carriers by using the polymerase chain reaction. *Proc. Natl. Acad. Sci. U.S.A.* 86, 5620–5624. doi: 10.1073/pnas.86.14.5620

Koiwa, T., Hamano-Usami, A., Ishida, T., Okayama, A., Yamaguchi, K., Kamihira, S., et al. (2002). 5'-long terminal repeat-selective CpG methylation of latent human T-cell leukemia virus type 1 provirus in vitro and in vivo. *J. Virol.* 76, 9389–9397. doi: 10.1128/JVI.76.18.9389-9397.2002

Kulkarni, A., Mateus, M., Thinnes, C. C., Mccullagh, J. S., Schofield, C. J., Taylor, G. P., et al. (2017). Glucose metabolism and oxygen availability govern reactivation of the latent human retrovirus HTLV-1. *Cell Chem. Biol.* 24, 1377.e3–1387.e3. doi: 10.1016/j.chembiol.2017.08.016

Landry, S., Halin, M., Vargas, A., Lemasson, I., Mesnard, J. M., and Barbeau, B. (2009). Upregulation of human T-cell leukemia virus type 1 antisense transcription by the viral tax protein. *J. Virol.* 83, 2048–2054. doi: 10.1128/JVI.01264-08

Lemasson, I., Lewis, M. R., Polakowski, N., Hivin, P., Cavanagh, M. H., Thebault, S., et al. (2007). Human T-cell leukemia virus type 1 (HTLV-1) bZIP protein interacts with the cellular transcription factor CREB to inhibit HTLV-1 transcription. *J. Virol.* 81, 1543–1553. doi: 10.1128/JVI.00480-06

Lezin, A., Gillet, N., Olindo, S., Signate, A., Grandvaux, N., Verlaeten, O., et al. (2007). Histone deacetylase-mediated transcriptional activation reduces proviral loads in HTLV-1-associated myelopathy/tropical spastic paraparesis patients. *Blood* 110, 3722–3728. doi: 10.1182/blood-2007-04-085076

Ma, G., Yasunaga, J., Akari, H., and Matsuoka, M. (2015). TCF1 and LEF1 act as T-cell intrinsic HTLV-1 antagonists by targeting Tax. *Proc. Natl. Acad. Sci. U.S.A.* 112, 2216–2221. doi: 10.1073/pnas.1419198112

Ma, G., Yasunaga, J., and Matsuoka, M. (2016). Multifaceted functions and roles of HBZ in HTLV-1 pathogenesis. *Retrovirology* 13:16. doi: 10.1186/s12977-016-0249-x

MacNamara, A., Rowan, A., Hilburn, S., Kadolsky, U., Fujiwara, H., Suemori, K., et al. (2010). HLA class I binding of HBZ determines outcome in HTLV-1 infection. *PLoS Pathog.* 6:e1001117. doi: 10.1371/journal.ppat.1001117

Maertens, G. N. (2016). B'-protein phosphatase 2A is a functional binding partner of delta-retroviral integrase. *Nucleic Acids Res.* 44, 364–376. doi: 10.1093/nar/gkv1347

Manel, N., Kim, F. J., Kinet, S., Taylor, N., Sitbon, M., and Battini, J. L. (2003). The ubiquitous glucose transporter GLUT-1 is a receptor for HTLV. *Cell* 115, 449–459. doi: 10.1016/S0092-8674(03)00881-X

Matsuzaki, T., Nakagawa, M., Nagai, M., Usuku, K., Higuchi, I., Arimura, K., et al. (2001). HTLV-I proviral load correlates with progression of motor disability in HAM/TSP: analysis of 239 HAM/TSP patients including 64 patients followed up for 10 years. *J. Neurovirol.* 7, 228–234. doi: 10.1080/13550280152403272

Melamed, A., Laydon, D. J., Gillet, N. A., Tanaka, Y., Taylor, G. P., and Bangham, C. R. (2013). Genome-wide determinants of proviral targeting, clonal abundance and expression in natural HTLV-1 infection. *PLoS Pathog.* 9:e1003271. doi: 10.1371/journal.ppat.1003271

Merkenschlager, M., and Nora, E. P. (2016). CTCF and cohesin in genome folding and transcriptional gene regulation. *Annu. Rev. Genomics Hum. Genet.* 17, 17–43. doi: 10.1146/annurev-genom-083115-022339

Miyazato, P., Matsuo, M., Katsuya, H., and Satou, Y. (2016). Transcriptional and epigenetic regulatory mechanisms affecting HTLV-1 provirus. *Viruses* 8:E171. doi: 10.3390/v8060171

Nagai, M., Usuku, K., Matsumoto, W., Kodama, D., Takenouchi, N., Moritoyo, T., et al. (1998). Analysis of HTLV-I proviral load in 202 HAM/TSP patients and 243 asymptomatic HTLV-I carriers: high proviral load strongly predisposes to HAM/TSP. *J. Neurovirol.* 4, 586–593. doi: 10.3109/13550289809114225

Nakano, K., and Watanabe, T. (2012). HTLV-1 Rex: the courier of viral messages making use of the host vehicle. *Front. Microbiol.* 3:330. doi: 10.3389/fmicb.2012.00330

Nyborg, J. K., Egan, D., and Sharma, N. (2010). The HTLV-1 Tax protein: revealing mechanisms of transcriptional activation through histone acetylation and nucleosome disassembly. *Biochim. Biophys. Acta* 1799, 266–274. doi: 10.1016/j.bbagrm.2009.09.002

Olindo, S., Belrose, G., Gillet, N., Rodriguez, S., Boxus, M., Verlaeten, O., et al. (2011). Safety of long-term treatment of HAM/TSP patients with valproic acid. *Blood* 118, 6306–6309. doi: 10.1182/blood-2011-04-349910

Pais-Correia, A. M., Sachse, M., Guadagnini, S., Robbiati, V., Lasserre, R., Gessain, A., et al. (2010). Biofilm-like extracellular viral assemblies mediate HTLV-1 cell-to-cell transmission at virological synapses. *Nat. Med.* 16, 83–89. doi: 10.1038/nm.2065

Parker, C. E., Daenke, S., Nightingale, S., and Bangham, C. R. (1992). Activated, HTLV-1-specific cytotoxic T-lymphocytes are found in healthy seropositives as well as in patients with tropical spastic paraparesis. *Virology* 188, 628–636. doi: 10.1016/0042-6822(92)90517-S

Pise-Masison, C. A., Castro-Amarante, M. F. D., Enose-Akahata, Y., Buchmann, R. C., Fenizia, C., Washington, P. R., et al. (2014). Co-dependent of HTLV-1 p12 and p8 function in virus persistence. *PLoS Pathog.* 10:e1004454. doi: 10.1371/journal.ppat.1004454

Rende, F., Cavallari, I., Corradin, A., Silic-Benussi, M., Toulza, F., Toffolo, G. M., et al. (2011). Kinetics and intracellular compartmentalization of HTLV-1 gene expression: nuclear retention of HBZ mRNAs. *Blood* 117, 4855–4859. doi: 10.1182/blood-2010-11-316463

Robek, M. D., Wong, F. H., and Ratner, L. (1998). Human T-cell leukemia virus type 1 pX-I and pX-II open reading frames are dispensable for the immortalization of primary lymphocytes. *J. Virol.* 72, 4458–4462.

Rosewick, N., Durkin, K., Artesi, M., Marcais, A., Hahaut, V., Griebel, P., et al. (2017). Cis-perturbation of cancer drivers by the HTLV-1/BLV proviruses is an early determinant of leukemogenesis. *Nat. Commun.* 8:15264. doi: 10.1038/ncomms15264

Satou, Y., Miyazato, P., Ishihara, K., Yaguchi, H., Melamed, A., Miura, M., et al. (2016). The retrovirus HTLV-1 inserts an ectopic CTCF-binding site into the human genome. *Proc. Natl. Acad. Sci. U.S.A.* 113, 3054–3059. doi: 10.1073/pnas.1423199113

Sharma, N., and Nyborg, J. K. (2008). The coactivators CBP/p300 and the histone chaperone NAP1 promote transcription-independent nucleosome eviction at the HTLV-1 promoter. *Proc. Natl. Acad. Sci. U.S.A.* 105, 7959–7963. doi: 10.1073/pnas.0800534105

Shiohama, Y., Naito, T., Matsuzaki, T., Tanaka, R., Tomoyose, T., Takashima, H., et al. (2016). Absolute quantification of HTLV-1 basic leucine zipper factor (HBZ) protein and its plasma antibody in HTLV-1 infected individuals with different clinical status. *Retrovirology* 13:29. doi: 10.1186/s12977-016-0263-z

Silic-Benussi, M., Cavallari, I., Vajente, N., Vidali, S., Chieco-Bianchi, L., Di Lisa, F., et al. (2010). Redox regulation of T-cell turnover by the p13 protein of human T-cell leukemia virus type 1: distinct effects in primary versus transformed cells. *Blood* 116, 54–62. doi: 10.1182/blood-2009-07-235861

Silverman, L. R., Phipps, A. J., Montgomery, A., Ratner, L., and Lairmore, M. D. (2004). Human T-cell lymphotropic virus type 1 open reading frame II-encoded p30II is required for in vivo replication: evidence of in vivo reversion. *J. Virol.* 78, 3837–3845. doi: 10.1128/JVI.78.8.3837-3845.2004

Spina, C. A., Anderson, J., Archin, N. M., Bosque, A., Chan, J., Famiglietti, M., et al. (2013). An in-depth comparison of latent HIV-1 reactivation in multiple cell model systems and resting CD4+ T cells from aviremic patients. *PLoS Pathog.* 9:e1003834. doi: 10.1371/journal.ppat.1003834

Suzuki, T., Fujisawa, J. I., Toita, M., and Yoshida, M. (1993). The trans-activator tax of human T-cell leukemia virus type 1 (HTLV-1) interacts with cAMP-responsive element (CRE) binding and CRE modulator proteins that bind to the 21-base-pair enhancer of HTLV-1. *Proc. Natl. Acad. Sci. U.S.A.* 90, 610–614. doi: 10.1073/pnas.90.2.610

Taniguchi, Y., Nosaka, K., Yasunaga, J., Maeda, M., Mueller, N., Okayama, A., et al. (2005). Silencing of human T-cell leukemia virus type I gene transcription by epigenetic mechanisms. *Retrovirology* 2:64. doi: 10.1186/1742-4690-2-64

Tripathy, M. K., Mcmanamy, M. E. M., Burch, B. D., Archin, N. M., and Margolis, D. M. (2015). H3K27 demethylation at the proviral promoter sensitizes latent HIV to the effects of vorinostat in *ex vivo* cultures of resting $CD4^+$ T cells. *J. Virol.* 89, 8392–8405. doi: 10.1128/JVI.00572-15

Tsai, A. G., Cabrales, P., Hangai-Hoger, N., and Intaglietta, M. (2004). Oxygen distribution and respiration by the microcirculation. *Antioxid. Redox Signal.* 6, 1011–1018. doi: 10.1089/ars.2004.6.1011

Valeri, V. W., Hryniewicz, A., Andresen, V., Jones, K., Fenizia, C., Bialuk, I., et al. (2010). Requirement of the human T-cell leukemia virus p12 and p30 products for infectivity of human dendritic cells and macaques but not rabbits. *Blood* 116, 3809–3817. doi: 10.1182/blood-2010-05-284141

Yasunaga, J.-I., Furuta, R., Miura, M., Sugata, K., Saito, A., Hirofumi, A., et al. (2016). Hematopoietic stem cell infected with HTLV-1 functions as a viral reservoir in vivo. *Blood* 128, 1343–1343.

Yoshida, M., Satou, Y., Yasunaga, J., Fujisawa, J., and Matsuoka, M. (2008). Transcriptional control of spliced and unspliced human T-cell leukemia virus type 1 bZIP factor (HBZ) gene. *J. Virol.* 82, 9359–9368. doi: 10.1128/JVI.00242-08

9

IFNG +874A/T Polymorphism Among Asymptomatic HTLV-1-Infected Individuals is Potentially Related to a Worse Prognosis

Maria A. F. Queiroz, Vânia N. Azevedo, Ednelza da S. G. Amoras, Tuane C. F. Moura, Marluísa de O. Guimarães Ishak, Ricardo Ishak, Antonio C. R. Vallinoto and Rosimar N. Martins Feitosa*

Laboratory of Virology, Institute of Biological Sciences, Federal University of Pará, Belém, Brazil

***Correspondence:**
Maria A. F. Queiroz
alicefarma@hotmail.com

HTLV-1 infections are persistent and frequently latent; however, productive infections trigger different types of immunological responses that utilize cytokines to control infection. The present study investigated the role of *IFNG* +874A/T polymorphisms among 153 HTLV-1-infected individuals (33 clinically diagnosed with TSP/HAM, 22 with rheumatologic manifestations, 2 with dermatitis, 1 with uveitis, and 95 asymptomatic patients) and 300 healthy control individuals. Genotyping and proviral HTLV-1 load assessment were performed using real-time PCR assays, and the plasma levels of IFN-γ were measured using an enzyme immunoassay (ELISA). Genotype frequencies were not significantly different, but the presence of the T allele was higher ($p < 0.0142$) among the asymptomatic patients. Plasma levels of IFN-γ were significantly higher ($p < 0.0137$) among those with the TT genotype. Their proviral load was also higher, although this elevation did not reach statistical significance. There was no difference in the IFN-γ plasma levels among the symptomatic patients, even when ranked according to disease severity (TSP/HAM or rheumatologic manifestations). However, the difference among asymptomatic patients with the T allele was significantly higher ($p < 0.0016$) and similar to the plasma levels observed among symptomatic individuals. These results suggest that the *IFNG* +874A/T polymorphism may modulate the plasma levels of IFN-γ during HTLV-1 infection. Asymptomatic carriers of the polymorphic genotypes appear to develop an inflammatory response in a shorter timeframe, triggering progression to HTLV-1-related symptoms and disorders. These results further suggest that HTLV-1-infected asymptomatic individuals expressing the *IFNG* +874A/T polymorphism should be monitored more closely in order to readily detect the increase in clinical symptoms, as these patients are potentially at risk of a poor prognosis and should therefore start available treatment procedures earlier.

Keywords: HTLV-1, IFN-γ, polymorphism, plasma dosage, clinical symptoms

INTRODUCTION

Human T-cell lymphotropic virus type 1 (HTLV-1) infects approximately 5–10 million people worldwide (Cassar and Gessain, 2017). The most relevant endemic regions are located in southeastern Japan, sub-Saharan Africa, the Caribbean, the Middle East, the Austro-Melanesia region, and South America (Gessain and Cassar, 2012). Brazil is an important endemic area for the virus, and a great diversity of diseases associated with HTLV-1 have been observed there (Pombo de Oliveira et al., 1999; Rathsam-Pinheiro et al., 2009; Grassi et al., 2011; Okajima et al., 2013). The prevalence of infection varies among different regions of the country: the southern region usually shows the lowest rates, and the northeast region shows the highest rates (Catalan-Soares et al., 2005).

Most HTLV-1 infections are asymptomatic; however, under certain conditions not yet fully understood, the virus may lead to the development of associated diseases, including HTLV-1-associated myelopathy/tropical spastic paraparesis (HAM/TSP), adult T-cell leukemia/lymphoma (ATLL), and inflammatory syndromes such as rheumatoid arthritis, dermatitis, and uveitis (Yakova et al., 2005; Okajima et al., 2013; Quaresma et al., 2015). The development of symptoms, particularly those of HAM/TSP and rheumatoid arthritis, are associated with a high proviral load in the peripheral blood and dysregulation of the immunological response against the virus (Grassi et al., 2011; Coutinho et al., 2014; da Silva Dias et al., 2016). Several studies have investigated the mechanisms underlying how immunological factors may change the course of HTLV-1 infection (Goon et al., 2004; Kato et al., 2004; Yakova et al., 2005; Best et al., 2009; Araya et al., 2014; Coutinho et al., 2014). The virus induces spontaneous proliferation of T $CD4^+$ and T $CD8^+$ lymphocytes and natural killer cells (Goon et al., 2004; Pinto et al., 2011). The increases in the numbers of these cells may lead to the development of a hyperimmune response and the marked production of proinflammatory cytokines, contributing to the pathogenesis of inflammatory disorders associated with HTLV-1 (Montanheiro et al., 2009; Yamano et al., 2009).

Genetic variations in important components of the immunological system are associated with the presence of symptoms in HTLV-1 infection (Vallinoto et al., 2012; Shirdel et al., 2013; de Sá et al., 2016). IFN-γ is the main proinflammatory cytokine associated with clinical symptoms etiologically linked to HTLV-1 infection (Montanheiro et al., 2009; da Silva Dias et al., 2016). The *IFNG* gene expresses CA-repeat microsatellite polymorphisms, and the major one—*IFNG* +874 A/T—is associated with the increased production of IFN-γ (Pravica et al., 1999). The presence of this polymorphism has been associated with multiple viral infections, including HIV-1, hepatitis B virus (Freitas et al., 2015; Al Kadi and Monem, 2017) and susceptibility to HTLV-1 infection (Rocha-Júnior et al., 2012).

The direct influence of this polymorphism on IFN-γ levels in HTLV-1-infected individuals has not reportedly been characterized. The marked inflammatory response observed with HTLV-1-associated diseases was therefore investigated in order to determine the influence of the *IFNG* +874 A/T polymorphism on the plasma level of IFN-γ and its relationship with the progression of HTLV-1 infection to symptomatic disease.

MATERIALS AND METHODS

Study Population

This study examined 153 HTLV-1-infected individuals (33 clinically diagnosed with HAM/TSP, 22 with rheumatic manifestations, 2 with dermatitis, 1 with uveitis, and 95 asymptomatic individuals) of both sexes, older than 18 years, not currently being treated with glucocorticoids, who were followed in the outpatient clinic of the Tropical Medicine Division of the Federal University of Pará. Clinical and laboratory criteria were used to diagnose the diseases associated with HTLV-1 according to the Brazilian Guidelines for HTLV-1 diseases from the Brazilian Ministry of Health (Brasil. Ministério da Saúde, 2013). The control group consisted of 300 blood donors from the Center for Hemotherapy and Hematology of Pará (HEMOPA) who were used to compare the genotype and allele frequencies of the *IFNG* +874 A/T polymorphism. The control group was matched by age and sex with the HTLV-1-infected individuals.

Collection and Storage of Samples

Blood samples were collected (10 mL) by intravenous puncture using a vacuum collection system containing ethylenediaminetetraacetic acid (EDTA) as an anticoagulant. The samples were centrifuged at 3,000 rpm for 10 min to separate the leukocytes. Leukocytes were used to extract genomic DNA for analysis of the *IFNG* +874 A/T polymorphism and quantification of the proviral load. Plasma samples were used for the quantification of IFN-γ. Samples were stored at −70°C until use.

Laboratory Tests

DNA Extraction

DNA was extracted from peripheral blood leukocytes using a Puregene kit (Puregene, Gentra Systems, Inc., United States) according to the manufacturer's protocol, which included cell lysis, protein precipitation, DNA precipitation, and hydration. After extraction, the DNA was quantified using a Qubit® 2.0 fluorometer (Life Technologies, Carlsbad, CA, United States) and the Qubit™ DNA Assay Kit (Life Technologies, Carlsbad, CA, United States) reagents, following the manufacturer's protocol.

Quantification of the Proviral Load of HTLV-1

Proviral load was quantified by qPCR using three target sequences synthesized using the TaqMan® system (Life Technologies, Foster City, CA, United States), according to a previously described protocol by Tamegão-Lopes et al. (2006), namely, collection of 5 mL of whole blood for DNA extraction from leukocytes, followed by relative quantification using real-time PCR. The obtained results were further adjusted to an absolute proviral quantification by considering the leukocyte counts per mm^3, and the results were expressed as DNA proviral copies/mm^3.

Genotyping of *IFNG*+874 A/T (rs2430561)

IFNG +874 A/T polymorphism located in the first intron of the gene encoding IFN-γ was analyzed by real-time PCR using a StepOnePLUS™ Real-Time PCR system. The specific primers (IFNG-F: 5′-TTC AGA CAT TCA CAA TTG ATT TTA TTC T-3′ and IFNG-R: 5′-CCC CCA ATG GTA CAG GTT TC-3′) and probes (FAM-AAAATCAAATCTCACACACACA-MGB and VIC-AAAATCAAATCACACACACACA-MGB) were previously described (Tso et al., 2005). The reaction followed a program of 10 min at 95°C, 40 cycles of 15 s at 95°C and 1 min at 60°C.

Quantification of Plasma IFN-γ Levels

Plasma IFN-γ levels were measured by the Ready-SET-Go® enzyme-linked immunosorbent assay (ELISA) (eBioscience, San Diego, CA, United States), which uses specific monoclonal antibodies to detect the cytokine following the manufacturer's instructions.

Statistical Analysis

Genotype and allele frequencies were estimated by direct counting. Significant differences between groups were determined using the chi-squared test. Hardy–Weinberg equilibrium was calculated to evaluate whether the distribution of the genotype frequencies observed was in agreement with the expected frequencies. Plasma IFN-γ levels were compared between groups using the non-parametric Mann–Whitney test. All tests were performed using the software BioEstat 5.3 (Ayres et al., 2011). Statistical associations at *p-values* < 0.05 were considered statistically significant.

Ethical Considerations

The project was approved by the Research Ethics Committee of the João de Barros Barreto University Hospital of the Federal University of Pará (protocol no. 2061/2005). All study participants were fully informed of the research objectives, and those who agreed to participate signed an informed consent form.

RESULTS

The majority of HTLV-1-infected individuals were female (69.9%, 107/153), with a mean age of 50.3 years. The frequency of the wild-type genotype was higher in the infected group, but there was no significant difference between the genotype and allele frequencies when compared to the control group (**Table 1**).

Comparison of genotype frequencies between infected (asymptomatic) and diseased (symptomatic) individuals showed no significant differences. However, the presence of the T allele was significantly higher ($p = 0.0142$) among asymptomatic individuals than among diseased patients (**Table 2**).

Plasma IFN-γ levels were significantly higher ($p = 0.0137$) in the group with the TT genotype (**Figure 1A**); they also showed a higher proviral load, although this difference was not statistically significant when compared with the other genotypes (**Figure 1B**). The proviral load was significantly higher ($p = 0.0002$) among the HAM/TSP patients than among asymptomatic individuals (**Figure 1C**). No significant difference in plasma levels was observed between patients with the wild-type genotype and those with the polymorphic allele in the symptomatic group (**Figure 2A**), and no significant difference was observed when comparing patients with the wild-type genotype and those with TSP/HAM or those with rheumatic disease (**Figures 2C,D**). However, there was a significant difference among the asymptomatic HTLV-1-infected individuals: IFN-γ plasma levels among patients with the T allele were significantly higher ($p = 0.0016$) and reached values that were similar to those exhibited by symptomatic individuals (**Figure 2B**).

DISCUSSION

The immunological response elicited by HTLV-1-infected individuals appears to be influenced and modulated by the virus, as its presence may promote the spontaneous proliferation of T $CD4^+$ and T $CD8^+$ lymphocytes and induce the production of proinflammatory cytokines responsible for the symptoms of various diseases, such as HAM/TSP (Goon et al., 2002). Along with other cytokines involved in this process, IFN-γ is likely the most important in the immunological pathogenesis of HAM/TSP (Montanheiro et al., 2009), and genetic alterations in its gene sequence lead to an exacerbated inflammatory process and an increase in the severity of the disease.

In the present study, the *IFNG* +874 A/T polymorphism was not associated with susceptibility to HTLV-1 infection; there was no significant difference in genotype or allele frequencies between HTLV-1-infected individuals and those in the control group. However, the polymorphic allele was associated with the absence of symptoms in HTLV-1-infected individuals. The wild-type allele was already associated as a risk factor for HIV-1 infection (Freitas et al., 2015) and for the disease progression of hepatitis B (Al Kadi and Monem, 2017). The results obtained herein with the allele frequencies suggest that the T allele would

TABLE 1 | Genotype and allele frequencies of the *IFNG* +874 A/T polymorphism among HTLV-1-infected individuals and the control group.

Genotypes and alleles	HTLV	Control	p^*
	n (%)	*n* (%)	
AA	83 (54.25)	174 (58.00)	0.5153
AT	58 (37.91)	110 (36.67)	
TT	12 (7.84)	16 (5.33)	
*A	0.73	0.76	0.7456
*T	0.27	0.24	

*n, number of individuals; *chi-squared (χ^2) test.*

TABLE 2 | Genotype and allele frequencies of the *IFNG* +874 A/T polymorphism among symptomatic and asymptomatic HTLV-1-infected individuals.

Genotypes and alleles	Symptomatic	Asymptomatic	p^*
	n (%)	*n* (%)	
AA	34 (58.63)	49 (51.58)	0.6689
AT	19 (32.75)	39 (41.05)	
TT	5 (8.62)	7 (7.37)	
*A	0.87	0.72	0.0142
*T	0.13	0.28	

*n, number of individuals; *chi-squared (χ^2) test.*

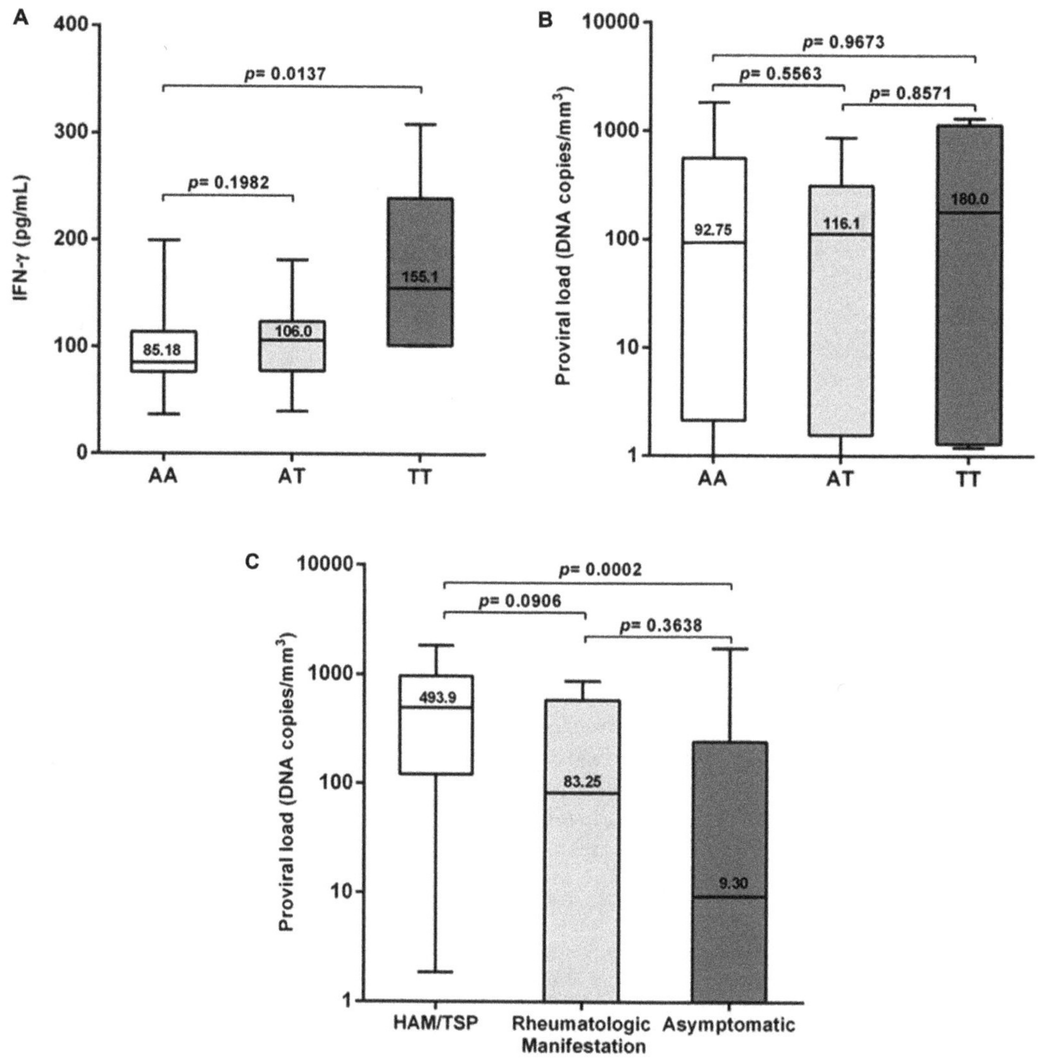

FIGURE 1 | (A) IFN-γ plasma levels classified according to *IFNG* +874 A/T genotypes, **(B)** proviral load among HTLV-1-infected individuals expressing *IFNG* +874 A/T genotypes, and **(C)** proviral load classified according to the presence or absence of symptoms. Mann–Whitney test.

act as a protective factor against the progression to disease among HTLV-1-infected individuals.

These results were not correlated with IFN-γ levels measured in plasma. Plasma IFN-γ levels were significantly higher among individuals carrying the *IFNG* +874 T allele (genotypes AT and TT); these individuals also showed a higher proviral load, although the differences were not statistically significant. The presence of the polymorphism may increase the levels of inflammation and lead to disease progression. Elevated IFN-γ levels in patients with HAM/TSP have previously been associated with central nervous system inflammatory disorders (Yamano et al., 2009).

Proviral load was higher in the patient group; however, in contrast to a previous report from Yakova et al. (2005), the plasma proviral loads of patients with rheumatic disease were higher than those of asymptomatic individuals but lower than those of TSP/HAM patients. Statistical significance was achieved only when HAM/TSP patients were compared with asymptomatic carriers of the virus.

The polymorphism did not affect plasma IFN-γ levels among symptomatic patients when they were assessed according to the clinical presentation of the disease or in the presence of the enhanced inflammatory process involving the nervous tissue and joints, as observed among patients with HAM/TSP and rheumatoid arthritis. Several other factors have been associated with the severity of disease, which may be more relevant than the presence of the *IFNG* +874 A/T polymorphism in this situation (Ahuja et al., 2007; Quaresma et al., 2015).

A significantly different situation was observed with the comparison of IFN-γ levels in the asymptomatic group. Individuals carrying the polymorphic allele showed significantly higher levels of IFN-γ than those carrying the wild-type allele. This may represent a previously unidentified risk factor for disease progression, as these infected individuals show IFN-γ

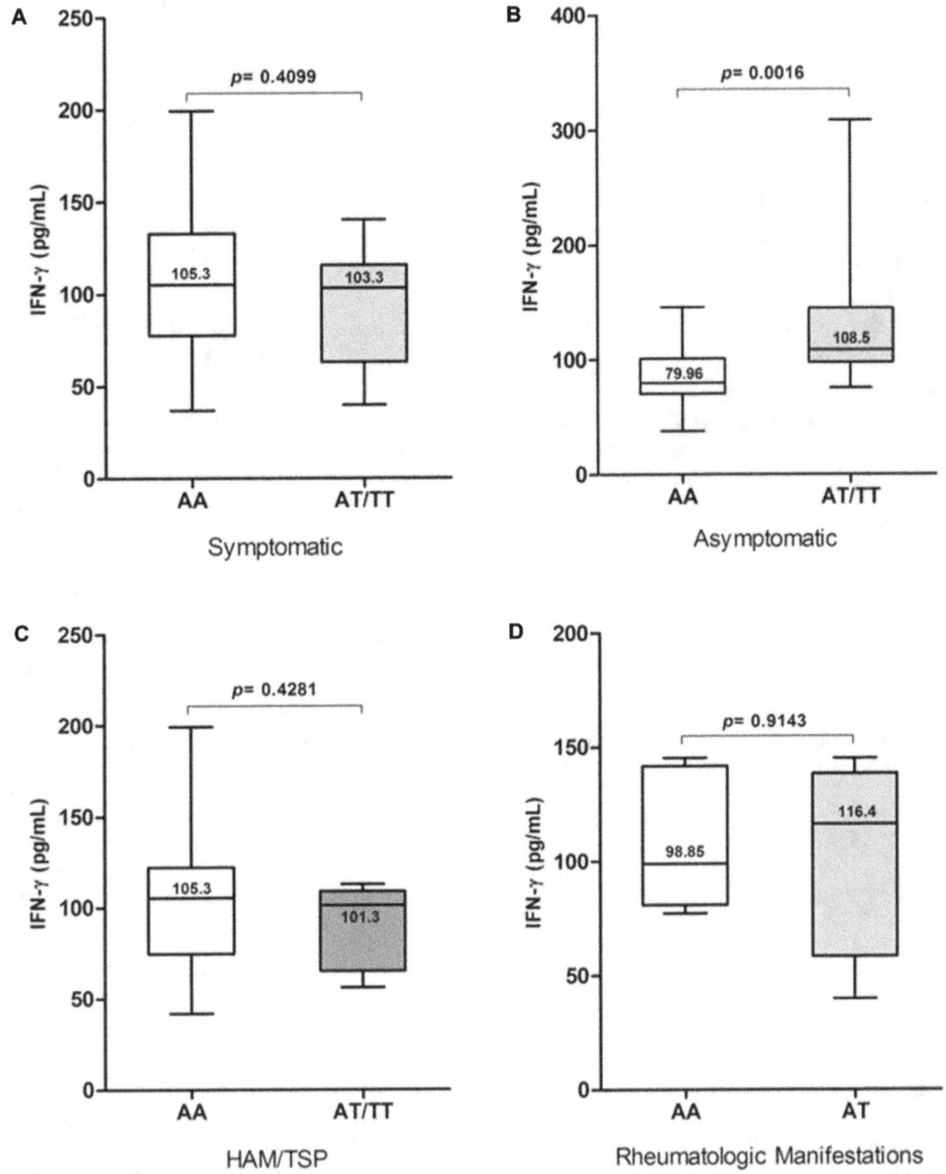

FIGURE 2 | IFN-γ plasma levels among **(A)** symptomatic and **(B)** asymptomatic HTLV-1-infected individuals carrying *IFNG* +874 A/T genotypes, as well as **(C,D)** according to the presence and presentation of disease. Mann–Whitney test.

levels similar to those of symptomatic individuals. The proviral load was occasionally detected, and the increase in IFN-γ levels may serve as a safe, reliable immunological marker of a poor prognosis and the evolution of disease pathogenesis, which may indicate clinical progression to HAM/TSP (Montanheiro et al., 2009; da Silva Dias et al., 2016).

Although the proviral load is reportedly elevated among HTLV-1-infected individuals with neurological dysfunction (Silva et al., 2007; Grassi et al., 2011), several asymptomatic individuals may present a pattern of immunological response that is associated with a high proviral load (Coutinho et al., 2014) and an inflammatory response similar to those with HAM/TSP (Santos et al., 2004).

Another investigation of the *IFNG* +874 A/T polymorphism, albeit one that did not measure the intensity of the inflammatory process (using the IFN-γ levels as a marker of inflammation), showed that the AT genotype was associated with a higher proviral load (Rocha-Júnior et al., 2012). This suggests that asymptomatic HTLV-1-infected individuals with the T allele of *IFNG* +874 A/T and a high proviral load have a high probability of developing HTLV-1-associated inflammatory diseases.

The immunological response during HTLV-1 infection is complex, and although IFN-γ is crucial in fighting intracellular viral agents by blocking their replication (Schoenborn and Wilson, 2007), elevated levels of this cytokine are harmful to the host; in the case of HTLV-1, this may lead to the progression of severe diseases, including HAM/TSP.

CONCLUSION

The present results suggest that the *IFNG* +874 A/T polymorphism may influence IFN-γ plasma levels upon HTLV-1 infection. Asymptomatic individuals carrying the T allele appear to be more likely to develop inflammation more rapidly, which could lead to the onset of associated diseases. The present investigation identifies IFN-γ levels and its sequence polymorphism as an important biomarker to be further investigated and monitored among HTLV-1-infected patients as part of the routine follow-up of asymptomatic individuals, as they are potentially at risk of developing a worse prognosis of disease and should start available treatment procedures earlier.

AUTHOR CONTRIBUTIONS

MQ, RI, AV, and RM designed the study, analyzed and interpreted the data. MQ, VA, EA, and TM performed the experiments. MQ wrote the manuscript. MG, RI, and AV oversaw the experiments and edited the manuscript. MQ, VA, EA, TM, MG, RI, AV and RM reviewed the manuscript.

FUNDING

The present study was fully supported by grants from the Conselho Nacional de Desenvolvimento Científico e Tecnológico - CNPq (Grant No. 302582/2013-4) and the Universidade Federal do Pará. The funding source(s) had no involvement in conducting the research and/or preparing the article.

REFERENCES

Ahuja, J., Lepoutre, V., Wigdahl, B., Khan, Z. K., and Jain, P. (2007). Induction of pro-inflammatory cytokines by human T-cell leukemia virus type-1 Tax protein as determined by multiplexed cytokine protein array analyses of human dendritic cells. *Biomed. Pharmacother.* 61, 201–208. doi: 10.1016/j.biopha.2007.02.006

Al Kadi, M., and Monem, F. (2017). Polymorphism of IFN-γ (+874 T/A) in Syrian patients wit/h chronic hepatitis B. *Gastroenterol. Hepatol. Bed Bench* 10, 34–38.

Araya, N., Sato, T., Ando, H., Tomaru, U., Yoshida, M., Coler-Reilly, A., et al. (2014). HTLV-1 induces a Th1-like state in CD4+CCR4+ T cells. *J. Clin. Invest.* 124, 3431–3442. doi: 10.1172/JCI75250

Ayres, M., Ayres, J. R., Ayres, D. L., and Santos, A. S. (2011). *BioEstat 5.3: Aplicações Estatísticas nas Areas das Ciências Biológicas e Médicas*. Belém: Editora Sociedade Civil Mamirauá, 364.

Best, I., López, G., Verdonck, K., González, E., Tipismana, M., Gotuzzo, E., et al. (2009). IFN-gamma production in response to Tax 161-233, and frequency of CD4+ Foxp3+ and Lin HLA-DRhigh CD123+ cells, discriminate HAM/TSP patients from asymptomatic HTLV-1-carriers in a Peruvian population. *Immunology* 128, e777–e786. doi: 10.1111/j.1365-2567.2009.03082.x

Brasil. Ministério da Saúde (2013). *Guia de Manejo Clínico da Infecção pelo HTLV*. Available at: http://www.sierj.org.br/artigos/htlv_manual_final_pdf_25082.pdf

Cassar, O., and Gessain, A. (2017). Serological and molecular methods to study epidemiological aspects of human T-cell lymphotropic virus type 1 infection. *Methods Mol. Biol.* 1582, 3–24. doi: 10.1007/978-1-4939-6872-5_1

Catalan-Soares, B., Carneiro-Proietti, A. B. F., and Proietti, F. A. (2005). Heterogeneous geographic distribution of human T-cell lymphotropic viruses I and II (HTLV-I/II): serological screening prevalence rates in blood donors from large urban areas in Brazil. *Cad. Saúde Pública* 21, 926–931. doi: 10.1590/S0102-311X2005000300027

Coutinho, R. Jr., Grassi, M. F., Korngold, A. B., Olavarria, V. N., Galvão-Castro, B., and Mascarenhas, R. E. (2014). Human T lymphotropic virus type 1 (HTLV-1) proviral load induces activation of T-lymphocytes in asymptomatic carriers. *BMC Infect. Dis.* 14:453. doi: 10.1186/1471-2334-14-453

da Silva Dias, G. A., Sousa, R. C. M., Gomes, L. F., Caldas, C. A. M., Nassiri, R., Quaresma, J. A. S., et al. (2016). Correlation between clinical symptoms and peripheral immune response in HAM/TSP. *Microb. Pathog.* 92, 72–75. doi: 10.1016/j.micpath.2015.11.018

de Sá, K. S., Santana, B. B., de Souza Ferreira, T. C., Sousa, R. C., Caldas, C. A., Azevedo, V. N., et al. (2016). IL28B gene polymorphisms and Th1/Th2 cytokine levels might be associated with HTLV-associated arthropathy. *Cytokine* 77, 79–87. doi: 10.1016/j.cyto.2015.11.004

Freitas, F. B., Lima, S. S., Feitosa, R. N., Azevedo, V. N., Ishak, M. O., Ishak, R., et al. (2015). Polymorphisms in the IFNγ, IL-10, and TGFβ genes may be associated with HIV-1 infection. *Dis. Markers* 2015:248571. doi: 10.1155/2015/248571

Gessain, A., and Cassar, O. (2012). Epidemiological aspects and world distribution of HTLV-1 infection. *Front. Microbiol.* 3:388. doi: 10.3389/fmicb.2012.00388

Goon, P. K., Hanon, E., Igakura, T., Tanaka, Y., Weber, J. N., Taylor, G. P., et al. (2002). High frequencies of Th1-type CD4(+) T cells specific to HTLV-1 Env and Tax proteins in patients with HTLV-1-associated myelopathy/tropical spastic paraparesis. *Blood* 99, 3335–3341. doi: 10.1182/blood.V99.9.3335

Goon, P. K., Igakura, T., Hanon, E., Mosley, A. J., Barfield, A., Barnard, A. L., et al. (2004). Human T cell lymphotropic virus type I (HTLV-I)-specific CD4+ T cells: immunodominance hierarchy and preferential infection with HTLV-I. *J. Immunol.* 172, 1735–1743. doi: 10.4049/jimmunol.172.3.1735

Grassi, M. F., Olavarria, V. N., Kruschewsky, R. d. A, Mascarenhas, R. E., Dourado, I., Correia, L. C., et al. (2011). Human T cell lymphotropic virus type 1 (HTLV-1) proviral load of HTLV-associated myelopathy/tropical spastic paraparesis (HAM/TSP) patients according to new diagnostic criteria of HAM/TSP. *J. Med. Virol.* 83, 1269–1274. doi: 10.1002/jmv.22087

Kato, T., Asahara, H., Kurokawa, M. S., Fujisawa, K., Hasunuma, T., Inoue, H., et al. (2004). HTLV-I env protein acts as a major antigen in patients with HTLV-I-associated arthropathy. *Clin. Rheumatol.* 23, 400–409. doi: 10.1007/s10067-004-0901-z

Montanheiro, P. A., Penalva de Oliveira, A. C., Smid, J., Fukumori, L. M., Olah, I., da S Duarte, A. J., et al. (2009). The elevated interferon gamma production is an important immunological marker in HAM/TSP pathogenesis. *Scand. J. Immunol.* 70, 403–407. doi: 10.1111/j.1365-3083.2009.02291.x

Okajima, R., Oliveira, A. C., Smid, J., Casseb, J., and Sanches, J. A. Jr. (2013). High prevalence of skin disorders among HTLV-1 infected individuals independent of clinical status. *PLoS Negl. Trop. Dis.* 7:e2546. doi: 10.1371/journal.pntd.0002546

Pinto, L. A., Galvão Castro, B., Soares, M. B., and Grassi, M. F. (2011). An evaluation of the spontaneous proliferation of peripheral blood mononuclear cells in HTLV-1-infected individuals using flow cytometry. *ISRN Oncol.* 2011:326719. doi: 10.5402/2011/326719

Pombo de Oliveira, M. S., Loureiro, P., Bittencourt, A., Chiattone, C., Borducchi, D., De Carvalho, S. M., et al. (1999). Geographic diversity of adult T-cell leukemia lymphoma in Brazil. The Brazilian ATLL Study Group. *Int. J. Cancer* 83, 291–298. doi: 10.1002/(SICI)1097-0215(19991029)83:3<291::AID-IJC1>3.0.CO;2-P

Pravica, V., Asderakis, A., Perrey, C., Hajeer, A., Sinnott, P. J., and Hutchinson, I. V. (1999). In vitro production of IFN-gamma correlates with CA repeat polymorphism in the human IFN-gamma gene. *Eur. J. Immunogenet.* 26, 1–3. doi: 10.1046/j.1365-2370.1999.00122.x

Quaresma, J. A., Yoshikawa, G. T., Koyama, R. V., Dias, G. A., Fujihara, S., and Fuzii, H. T. (2015). HTLV-1, immune response and autoimmunity. *Viruses* 8:5. doi: 10.3390/v8010005

Rathsam-Pinheiro, R. H., Boa-Sorte, N., Castro-Lima-Vargens, C., Pinheiro, C. A., Castro-Lima, H., and Galvão-Castro, B. (2009). Ocular lesions in HTLV-1 infected patients from Salvador, State of Bahia: the city with the highest prevalence of this infection in Brazil. *Rev. Soc. Bras. Med. Trop.* 42, 633–637. doi: 10.1590/S0037-86822009000600004

Rocha-Júnior, M. C., Haddad, R., Cilião Alves, D. C., de Deus Wagatsuma, V. M., Mendes-Junior, C. T., Deghaide, N. H., et al. (2012). Interleukin-18 and interferon-gamma polymorphisms are implicated on proviral load and susceptibility to human T-lymphotropic virus type 1 infection. *Tissue Antigens* 80, 143–150. doi: 10.1111/j.1399-0039.2012.01887.x

Santos, S. B., Porto, A. F., Muniz, A. L., de Jesus, A. R., Magalhães, E., Melo, A., et al. (2004). Exacerbated inflammatory cellular immune response characteristics of HAM/TSP is observed in a large proportion of HTLV-I asymptomatic carriers. *BMC Infect. Dis.* 4:7. doi: 10.1186/1471-2334-4-7

Schoenborn, J. R., and Wilson, C. B. (2007). Regulation of interferon-gamma during innate and adaptive immune responses. *Adv. Immunol.* 96, 41–101. doi: 10.1016/S0065-2776(07)96002-2

Shirdel, A., Azarpazhooh, M. R., Sahebari, M., Ghanbari, M., Mirfeizi, S. Z., Hutchinson, I., et al. (2013). Association of IL-10 gene polymorphisms and human T lymphotropic virus type I-associated myelopathy/tropical spastic paraparesis in North-East of Iran (Mashhad). *Iran J. Basic Med. Sci.* 16, 258–263.

Silva, M. T., Harab, R. C., Leite, A. C., Schor, D., Araújo, A., and Andrada-Serpa, M. J. (2007). Human T lymphotropic virus type 1 (HTLV-1) proviral load in asymptomatic carriers, HTLV-1-associated myelopathy/tropical spastic paraparesis, and other neurological abnormalities associated with HTLV-1 infection. *Clin. Infect. Dis.* 44, 689–692. doi: 10.1086/510679

Tamegão-Lopes, B. P., Rezende, P. R., Maradei-Pereira, L. M. C., and Lemos, J. A. R. (2006). HTLV-1 and HTLV-2 proviral load: a simple method using quantitative real-time PCR. *Rev. Soc. Bras. Med. Trop.* 39, 548–552. doi: 10.1590/S0037-86822006000600007

Tso, H. W., Ip, W. K., Chong, W. P., Tam, C. M., Chiang, A. K., and Lau, Y. L. (2005). Association of interferon gamma and interleukin 10 genes with tuberculosis in Hong Kong Chinese. *Genes Immun.* 6, 358–363. doi: 10.1038/sj.gene.6364189

Vallinoto, A. C., Santana, B. B., dos Santos, E. L., Santo, R. R., Hermes, R. B., Sousa, R. C., et al. (2012). FAS-670A/G single nucleotide polymorphism may be associated with human T lymphotropic virus-1 infection and clinical evolution to TSP/HAM. *Virus Res.* 163, 78–82. doi: 10.1016/j.virusres.2011.09.015

Yakova, M., Lézin, A., Dantin, F., Lagathu, G., Olindo, S., Jean-Baptiste, G., et al. (2005). Increased proviral load in HTLV-1-infected patients with rheumatoid arthritis or connective tissue disease. *Retrovirology* 2:4. doi: 10.1186/1742-4690-2-4

Yamano, Y., Araya, N., Sato, T., Utsunomiya, A., Azakami, K., Hasegawa, D., et al. (2009). Abnormally high levels of virus-infected IFN-gamma+ CCR4+ CD4+ CD25+ T cells in a retrovirus-associated neuroinflammatory disorder. *PLoS One* 4:e6517. doi: 10.1371/journal.pone.0006517

10

HTLV-1 Alters T Cells for Viral Persistence and Transmission

Azusa Tanaka[1] *and Masao Matsuoka*[2,3]*

[1] *Department of Drug Discovery Medicine, Graduate School of Medicine, Kyoto University, Kyoto, Japan,*
[2] *Department of Hematology, Rheumatology and Infectious Diseases, Faculty of Life Sciences, Kumamoto University, Kumamoto, Japan,* [3] *Institute for Frontier Life and Medical Sciences, Kyoto University, Kyoto, Japan*

***Correspondence:**
Masao Matsuoka
mamatsu@kumamoto-u.ac.jp;
mmatsuok@virus.kyoto-u.ac.jp

Human T-cell leukemia virus type 1 (HTLV-1) was the first retrovirus to be discovered as a causative agent of adult T-cell leukemia-lymphoma (ATL) and chronic inflammatory diseases. Two viral factors, Tax and HTLV-1 bZIP factor (HBZ), are thought to be involved in the leukemogenesis of ATL. Tax expression is frequently lost due to DNA methylation in the promoter region, genetic changes to the *tax* gene, and deletion of the 5′ long terminal repeat (LTR) in approximately half of all ATL cases. On the other hand, *HBZ* is expressed in all ATL cases. HBZ is known to function in both protein form and mRNA form, and both forms play an important role in the oncogenic process of HTLV-1. HBZ protein has a variety of functions, including the suppression of apoptosis, the promotion of proliferation, and the impairment of anti-viral activity, through the interaction with several host cellular proteins including p300/CBP, Foxp3, and Foxo3a. These functions dramatically modify the transcriptional profiling of host T cells. *HBZ* mRNA also promotes T cell proliferation and viability. HBZ changes infected T cells to $CCR4^{+}TIGIT^{+}CD4^{+}$ effector/memory T cells. This unique immunophenotype enables T cells to migrate into various organs and tissues and to survive *in vivo*. In this review, we summarize how HBZ hijacks the transcriptional networks and immune systems of host T cells to contribute to HTLV-1 pathogenesis on the basis of recent new findings about *HBZ* and *tax*.

Keywords: HTLV-1, HBZ, tax, viral oncogenesis, Regulatory T Cell

INTRODUCTION

The burden of viral infection in cancer is high, with estimates having more than 20% of cancer cases caused by infection (Bouvard et al., 2009). One oncogenic virus, human T-cell leukemia virus type 1 (HTLV-1), was identified in the United States and Japan almost 40 years ago (Poiesz et al., 1980; Yoshida et al., 1982). Thereafter, HTLV-1 was found to be a causative agent of adult T-cell leukemia-lymphoma (ATL) and HTLV-1-associated myelopathy/tropical spastic paraparesis (HAM/TSP) (Hinuma et al., 1981; Gessain et al., 1985; Gallo, 2005). Today, it is estimated that HTLV-1 infects approximately 10 million people worldwide (Gessain and Cassar, 2012). HTLV-1 transmits through cell-to-cell contacts (Igakura et al., 2003; Pais-Correia et al., 2010), while the free virus shows poor infectivity with the exception of dendritic cells (DCs) that can be infected without cell-to-cell contact (Derse et al., 2001; Jones et al., 2008; Mazurov et al., 2010; Alais et al., 2015). Accordingly, HTLV-1 increases its copy number primarily by triggering the proliferation of infected cells to facilitate its transmission (Etoh et al., 1997; Cavrois et al., 1998).

These properties distinguish HTLV-1 from another well-known human retrovirus, human immunodeficiency virus type 1 (HIV-1). The chemokine receptors CXCR4 and CCR5 function as co-receptors for HIV-1 to infect host $CD4^+$ T cells. Unlike HIV-1, HTLV-1 can infect a variety of cell types, but more than 90% of infected cells are $CD4^+$ memory T cells *in vivo* (Richardson et al., 1990; Rizkallah et al., 2017).

EXPRESSION AND FUNCTION OF TAX

The HTLV-1 provirus is 9 kb in length and contains multiple coding regions for Gag, Pol, Env, p12, p30, p13, Rex, Tax, and HBZ. Among the viral proteins of HTLV-1, Tax can activate various signal pathways including the NF-κB and AP-1 pathways (Grassmann et al., 2005; Gazon et al., 2018). It also can induce T-cell leukemia or lymphoma *in vivo* (Grossman et al., 1995; Hasegawa et al., 2006). However, Tax expression is frequently undetectable in ATL cases due to genetic and epigenetic aberrations (Takeda et al., 2004). Importantly, nonsense mutations in the *tax* gene are often observed not only in ATL cases but also in infected cells of asymptomatic HTLV-1 carriers (Furukawa et al., 2001; Fan et al., 2010). Tax was initially discovered as the viral trans-activator for HTLV-1 RNA transcription from a promoter located in the 5′ LTR (Felber et al., 1985), thus its expression is essential for viral replication. However, it is a major target antigen recognized by cytotoxic T lymphocytes (CTL) (Kannagi et al., 1991). Therefore, Tax expression is tightly controlled for the survival of HTLV-1 infected cells in order to evade host immunosurveillance. Tax expression is usually silenced in ATL cells, but a single cell transcript analysis has revealed that Tax expression is not completely suppressed and that a small percentage of MT-1 cells transiently express Tax (Mahgoub et al., 2018). Because Tax expression is necessary for *de novo* infection, it may play a key role in the spreading of HTLV-1. Taken together, Tax expression is usually suppressed in order to escape from CTL, but at the same time, Tax is transiently expressed to maintain and expand HTLV-1 infected cells. These findings suggest that another key regulator may contribute to the onset of ATL.

HBZ AND ITS ROLE IN THE ONCOGENESIS

The HTLV-1 bZIP factor (HBZ) was first identified in 2002 as a novel viral protein that contains a N-terminal transcription activation domain and a leucine zipper motif in its C-terminus (Gaudray et al., 2002). It has been reported that *HBZ* mRNA is expressed in all ATL cases. Its mRNA form promotes the proliferation of T-cells, and its protein form induces the development of T-cell lymphomas in transgenic mice, indicating that HBZ is critical for the proliferation of ATL cells and leukemogenesis (Satou et al., 2006, 2011).

THE LOCALIZATION OF HBZ AND ITS FUNCTION IN THE NUCLEUS AND CYTOPLASM

HTLV-1 bZIP factor contains nuclear localization signals in its central/bZIP domain and nuclear export signals in its N terminus (Hivin et al., 2005). HBZ is mainly localized in the cytoplasm of peripheral blood mononuclear cells (PBMCs) in HAM/TSP patients, suggesting cytoplasmic HBZ as a possible biomarker of the disorder (Baratella et al., 2017). In addition, cytoplasmic HBZ interacts with GADD34 to suppress GADD34 function and positively regulate the mechanistic target of rapamycin (mTOR) signaling pathway (Mukai and Ohshima, 2014) (**Figure 1**). Finally, the cytoplasmic localization of HBZ protein in T cells depends on the host factor THEMIS (Kinosada et al., 2017). Since THEMIS is expressed only in T cells, this function might explain why HTLV-1 promotes the proliferation of T cells.

Compared to cytoplasmic HBZ, far more studies have focused on the function of nuclear HBZ. HBZ in the nucleus makes a complex with several important transcription factors including p300, p65, LEF1, AP-1 transcription factors and forkhead family proteins (Basbous et al., 2003; Thebault et al., 2004; Satou et al.,

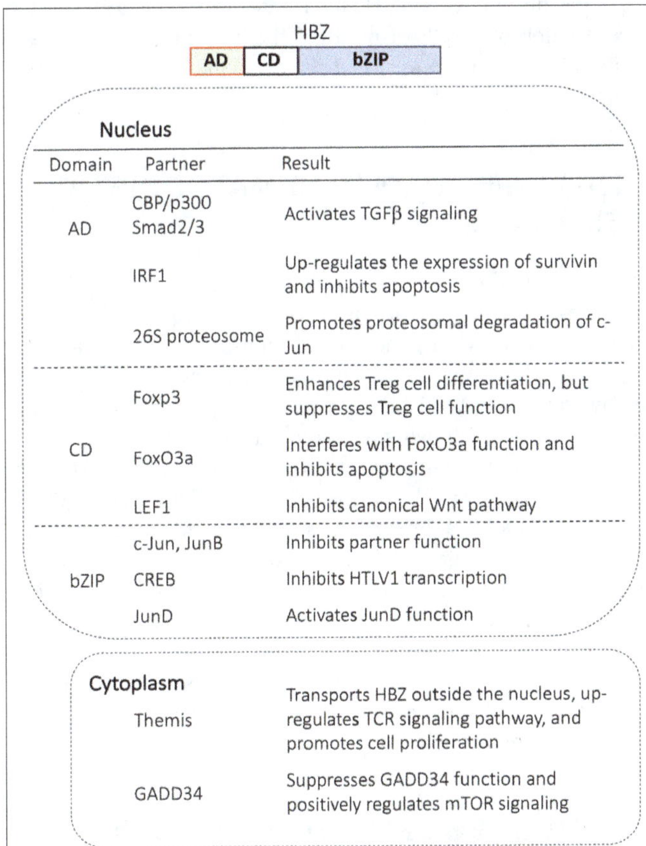

FIGURE 1 | Cellular proteins that interact with HBZ in the nucleus and in the cytoplasm. HBZ has three domains: activation domain (AD), central domain (CD), and basic ZIP domain (bZIP). Each domain interacts with important regulators and modulates cellular function.

2011; Zhao et al., 2011; Ma et al., 2013; Tanaka-Nakanishi et al., 2014) (**Figure 1**). For example, HBZ binds to p65 to diminish the p65 affinity for DNA and thus suppresses the classical pathway of NF-κB. Also, the interaction with c-Jun leads to a transcriptional repression of AP-1 regulated genes. Overall, nuclear HBZ has various functions necessary for the onset of ATL.

TARGET CELLS OF HTLV-1 INFECTION

HTLV-1 provirus is mainly detected in T cells *in vivo*. In particular, most provirus (~90%) was found in CD4$^+$effector/memory T cells, while the remaining 10% of provirus was detected in CD8$^+$ T cells (Yasunaga et al., 2001). HTLV-1 infected CD4$^+$ T cells tend to express cell adhesion molecule 1 (CADM1), C-C chemokine receptor type 4 (CCR4), T cell immunoreceptor with Ig and ITIM domains (TIGIT), CD45RO, and sometimes CD25, indicating that these cells are effector/memory T cells (Bangham and Matsuoka, 2017). These findings suggest two scenarios for the transmission of HTLV-1. First, HTLV-1 preferentially infects these subpopulations of CD4$^+$ T cells. It has been reported that Tax induces the expression of CCL22, one ligand of CCR4 (Hieshima et al., 2008), which might attract CCR4$^+$ T cells and transmit the virus in the periphery. Second, HTLV-1 infection itself induces the differentiation of T cells or modifies the immunophenotype to a specific phenotype. To answer this question, we focus on HBZ below.

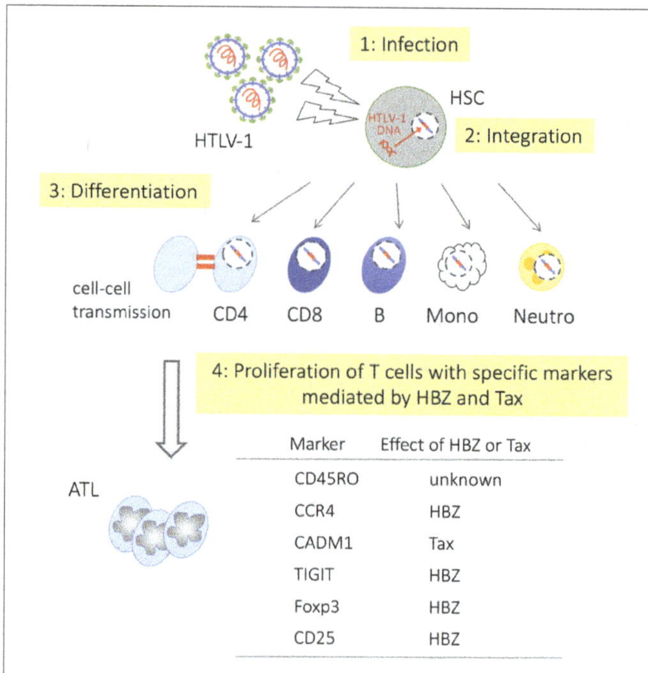

FIGURE 2 | Differentiation of hematopoietic cells and HTLV-1 infection HTLV-1 infected HSCs differentiate into diverse hematopoietic cell lineages, but only memory CD4$^+$ T cells become predominant. HSC, hematopoietic stem cell; CD4, CD4$^+$ T cell; CD8, CD8$^+$ T cell; B, B cell; Mono, monocyte; Neutro, neutrophil, respectively.

HTLV-1 INFECTION IN HEMATOPOIETIC STEM CELLS

Tax is essential for *de novo* infection since transcription of the sense-strand of the provirus, which is responsible for the generation of the viral genome and viral proteins such as Gag, Env, and Pol, is Tax dependent. Tax expression in the bone marrow cells of HAM/TSP patients was reported (Levin et al., 1997), suggesting *de novo* infection occurs in this area. However, it remains unknown whether hematopoietic stem cells (HSCs) are truly infected with HTLV-1 or not. Recently, Furuta et al. (2017) reported that identical integration sites were detected in multiple hematopoietic lineage cells. This result indicates that HTLV-1 infects HSCs *in vivo* and that HTLV-1 infected HSCs differentiate into diverse hematopoietic cell lineages (**Figure 2**). Since HTLV-1 infected cells possess a similar immunophenotype, viral factors are implicated to acquire the phenotype. Because HBZ is the only viral gene that is consistently expressed in infected cells and ATL cells, we consider its role.

HBZ INDUCES EFFECTOR/MEMORY PHENOTYPE

Here we summarize how HBZ influences hematopoietic cell differentiation to effector/memory CD4$^+$ T cells by modulating the function of key regulators (**Figure 2**).

Foxp3

As mentioned above, although HTLV-1 can infect many types of cells including HSCs, most ATL cells and HTLV-1 infected cells are CD4$^+$CD45RO$^+$ conventional memory T cells. It is speculated that HBZ influences the differentiation of host T cells. For instance, HBZ affects the expression and function of Foxp3, a critical regulator for the development of Treg cells. In HBZ-Tg mice, the number of CD4$^+$Foxp3$^+$ T cells is increased, suggesting that HBZ induces Foxp3 expression (Satou et al., 2011). HBZ promotes Foxp3 expression by activating the TGF-β/Smad pathway (Zhao et al., 2011). Although HBZ induces Foxp3 expression, HBZ directly interacts with Foxp3 to interfere with Foxp3 DNA binding activity and function. In addition, HBZ increases the number of induced Treg cells, in which Foxp3 expression is unstable, and converts them to Foxp3$^-$ T cells with increased interferon-γ production (Yamamoto-Taguchi et al., 2013). These results suggest that HBZ induces Foxp3$^+$ T cells but hijacks their transcriptional network, which leads to inflammatory disease in the host.

CCR4

It is well known that CCR4 is a hallmark of ATL cells (Yoshie et al., 2002; Ishida et al., 2003; Iqbal et al., 2010). CCR4 is not only a marker of Th2 cells, but it is also involved in the suppressive function of Treg cells (Yuan et al., 2007). CCR4 mutations were frequently observed in ATL samples, with most mutations being nonsense or frame-shift (Nakagawa et al., 2014; Kataoka et al., 2015). A truncated mutant of the cytoplasmic region of CCR4 enhances PI3K/Akt signaling, the migration and proliferation of ATL cells, but still many questions remain about the mechanisms for the increased expression of CCR4 and enhanced proliferation of cells even in cells expressing wild-type CCR4. Sugata et al. (2016) studied the mechanisms and effects of CCR4 expression on ATL cells. They found that both *HBZ* RNA and HBZ protein increased the expression of CCR4 through the induction of GATA3 expression, thereby activating transcription of the *CCR4* gene promoter. Upregulated CCR4 expression is associated with enhanced T cell migration and proliferation, which are implicated in the infiltration and proliferation of HTLV-1 infected cells. They further showed that wild-type CCR4 also promotes the proliferation of T cells and that a CCR4 antagonist inhibits cell migration and proliferation effectively. Thus, CCR4 induced by HBZ is implicated in proliferation and migration of infected cells.

TIGIT AND THEMIS

One gene upregulated by HBZ is the co-inhibitory receptor T cell immunoglobulin and ITIM domain (*TIGIT*) (Yasuma et al., 2016). ATL cells and HTLV-1 infected cells highly express TIGIT on their surfaces. Enhanced expression of TIGIT induces IL-10 production, leading to the suppression of host immune responses.

Although TIGIT is a co-inhibitory receptor that suppresses T-cell activation, it does not inhibit the proliferation of ATL cells or HTLV-1 infected cells (Kinosada et al., 2017). TIGIT exerts inhibitory signal through the intracytoplasmic immunoreceptor tyrosine inhibitory motif (ITIM) that interacts with a complex of SHP-2, Grb2, and THEMIS. HBZ interacts with THEMIS and hinders the inhibitory signal from THEMIS. Cytoplasmic HBZ almost disappears after knock-down of THEMIS expression while HBZ protein exists in the nucleus, suggesting that THEMIS is responsible for cytoplasmic localization of HBZ. THEMIS is a T cell-specific protein, this finding could explain why HTLV-1 induces the proliferation of only T cells despite the fact that HTLV-1 can infect many cell types.

Taken together, these studies suggest that HTLV-1 might influence earlier stages of hematopoietic cell differentiation and induce infected HSCs or other infected progenitor cells to abnormally differentiate into $CD4^+$ Treg-like cells.

ROLE OF TAX IN T CELL DIFFERENTIATION

No direct evidence has associated Tax with T-cell differentiation. Indeed, several HTLV-1 associated molecules, including Foxp3, CCR4, and TIGIT, are not induced in Tax transgenic mice, indicating that HBZ is closely linked to the expression of these molecules (Satou et al., 2011). HTLV-1 targets $CD4^+$ effector/memory T cells. As described above, HBZ is mainly responsible for immunophenotypes of HTLV-1 infected cells, but Tax has also been implicated. Two T cell specific proteins, TCF-1 and LEF-1, are highly expressed in thymocytes. It has been reported that TCF1 and LEF1 perturb Tax function (Ma et al., 2015), leading to the inhibition of HTLV-1 replication in the thymus. On the other hand, TCF-1/LEF-1 expression is suppressed in peripheral T cells. Therefore, Tax is fully functional in peripheral T cells.

GENETIC INSTABILITY INDUCED BY HBZ

Some individuals infected with HTLV-1 succumb to ATL after a long latency period (Grassmann et al., 2005). This effect suggests that along with the viral proteins Tax and HBZ, the accumulations of genetic or epigenetic alterations are required for ATL onset. Indeed, gain-of-function alterations were frequently observed in TCR/NF-κB signaling from integrated genetic and transcriptomic analysis (Kataoka et al., 2015). Recently, it has been reported that HBZ induces genomic instability in an oncogenic miRNA-dependent manner. HBZ increases the expression of miR17 and miR21, and these oncomiRs suppress the expression of *OBFC2A*, which codes a single-stranded DNA-binding protein that protects genome stability (Vernin et al., 2014).

HBZ INTERFERES WITH THE HOST TRANSCRIPTIONAL AND TRANSLATIONAL MACHINERY

Detailed analysis of RNA-seq data has revealed that host genes undergo frequent and provirus-dependent transcription termination. The expression level of host-genome exons located downstream of HTLV-1 was halved on average (Rosewick et al., 2017). At the same time, systematic interaction between the HTLV-1 antisense transcript corresponding to HBZ and host genes located upstream of the provirus is observed. This fusion is classified into four types depending on the location and the direction of viral insertion (genic or intergenic integration, and concordant or discordant gene-provirus orientation). In this way, HBZ transcription perturbs host gene transcription around HTLV-1 integration sites.

Both RNA and DNA viruses are heavily dependent on the host translational machinery to produce the polypeptides that are necessary for replication. Most viral mRNAs recruit 40S ribosomal subunit to initiate their translation. For example, RPS25, a component of 40S, is essential for the internal ribosome entry sites (IRESs) of hepatits C virus (Landry et al., 2009). This recruitment not only promotes viral protein synthesis, it also impedes the host innate defenses that inhibit viral protein production. It has been reported that HBZ disrupts the host translational mechanism by negatively modulating RPS25. JunD, a functional component of AP-1 has two isoforms, a full-length isoform containing 341 amino acids (JunD-FL) and a truncated isoform lacking 48 amino acids at the N terminus (ΔJunD) (Yazgan and Pfarr, 2002; Terol et al., 2017). There is a functional difference between the two isoforms. JunD-FL is the more potent transcriptional activator, but only ΔJunD induces the proliferation and transformation of cells. HBZ induces the loss of RPS25 protein, which leads to the down-regulation of JunD-FL and up-regulation of ΔJunD. As a result, HBZ shifts the function of JunD from growth suppressor to tumor promoter, a condition advantageous for HTLV-1 infected cells to proliferate. Thus, HBZ influences host transcriptional and translational machinery.

HBZ SUPPRESSES APOPTOSIS

To increase the number of infected cells, both HBZ and Tax play a key role in maintaining clonal longevity. One mechanism for this maintenance is achieved by inhibiting the apoptosis of HTLV-1 infected cells. HBZ protein perturbs the localization and function of FOXO3a to down-regulate the pro-apoptotic genes *Bim* and *FasL* (Tanaka-Nakanishi et al., 2014). Similarly, the N-terminal region of HBZ interacts with IRF-1 to inhibit IRF-1 DNA binding/transcriptional activity and reduce the number of cells undergoing apoptosis (Chopra et al., 1990; Panfil et al., 2016). Finally, in addition to HBZ protein, *HBZ* mRNA inhibits apoptosis by upregulating the expression of survivin by enhancing its promoter activity (Mitobe et al., 2015).

CONCLUSION AND FUTURE DIRECTION

HTLV-1 bZIP factor affects many aspects of host immunity to promote the survival and proliferation of the infected cells through its mRNA form and protein form. In this review, we mainly focused on recent findings about the functions of HBZ. As described above, HBZ positively or negatively regulates various host transcription factors, modulates multiple signaling pathways and perturbs T cell function and differentiation. One of the characteristics of HBZ is that it functions both in protein form and RNA form, both of which play crucial roles in cell proliferation and cell survival. Additionally, HBZ activates oncomiRs to induce genetic instability, a hallmark of cancer.

Owing to the effort of many researchers around the world, the mystery of HBZ has been gradually uncovered. Yet, it is still unknown why most ATL cells are CD4 positive memory T cells and whether HBZ contributes to this phenomenon. Answering these questions will contribute to new therapeutic targets.

AUTHOR CONTRIBUTIONS

All authors listed have made a substantial, direct and intellectual contribution to the work, and approved it for publication.

FUNDING

This work was supported by JSPS KAKENHI (15J04853 to AT and JP16H05336 to MM), the Project for Cancer Research and Therapeutic Evolution (P-CREATE) (17cm0106306h0002 to MM), the Research Program on Emerging and Re-emerging Infectious Diseases (17fk0108227h0002 to MM), and a grant from The Naito Foundation to AT.

REFERENCES

Alais, S., Mahleux, R., and Dutarte, H. (2015). Viral source-independent high susceptibility of dendritic cells to human T-cell leukemia virus type 1 infection compared to that of T lymphocytes. *J. Virol.* 89, 10580–10590. doi: 10.1128/JVI.01799-15

Bangham, C. R. M., and Matsuoka, M. (2017). Human T-cell leukaemia virus type 1: parasitism and pathogenesis. *Philos. Trans. R. Soc. Lond. B Biol. Sci.* 372:20160272. doi: 10.1098/rstb.2016.0272

Baratella, M., Forlani, G., Raval, G. U., Tedeschi, A., Gout, O., Gessain, A., et al. (2017). Cytoplasmic localization of HTLV-1 HBZ protein: a biomarker of HTLV-1-associated myelopathy/tropical spastic paraparesis (HAM/TSP). *PLoS Negl. Trop. Dis.* 11:e0005285. doi: 10.1371/journal.pntd.0005285

Basbous, J., Arpin, C., Gaudray, G., Piechaczyk, M., Devaux, C., and Mesnard, J. M. (2003). The HBZ factor of human T-cell leukemia virus type I dimerizes with transcription factors JunB and c-Jun and modulates their transcriptional activity. *J. Biol. Chem.* 278, 43620–43627. doi: 10.1074/jbc.M307275200

Bouvard, V., Baan, R., Straif, K., Grosse, Y., Secretan, B., El Ghissassi, F., et al. (2009). A review of human carcinogens–Part B: biological agents. *Lancet Oncol.* 10, 321–322. doi: 10.1016/S1470-2045(09)70096-8

Cavrois, M., Leclercq, I., Gout, O., Gessain, A., Wain-Hobson, S., and Wattel, E. (1998). Persistent oligoclonal expansion of human T-cell leukemia virus type 1-infected circulating cells in patients with Tropical spastic paraparesis/HTLV-1 associated myelopathy. *Oncogene* 17, 77–82. doi: 10.1038/sj.onc.1201906

Chopra, I., Hacker, K., Misulovin, Z., and Rothstein, D. M. (1990). Sensitive biological detection method for tetracyclines using a tetA-lacZ fusion system. *Antimicrob. Agents Chemother.* 34, 111–116. doi: 10.1128/AAC.34.1.111

Derse, D., Hill, S. A., Lloyd, P. A., Chung, H., and Morse, B. A. (2001). Examining human T-lymphotropic virus type 1 infection and replication by cell-free infection with recombinant virus vectors. *J. Virol.* 75, 8461–8468. doi: 10.1128/JVI.75.18.8461-8468.2001

Etoh, K., Tamiya, S., Yamaguchi, K., Okayama, A., Tsubouchi, H., Ideta, T., et al. (1997). Persistent clonal proliferation of human T-lymphotropic virus type I-infected cells in vivo. *Cancer Res.* 57, 4862–4867

Fan, J., Ma, G., Nosaka, K., Tanabe, J., Satou, Y., Koito, A., et al. (2010). APOBEC3G generates nonsense mutations in human T-cell leukemia virus type 1 proviral genomes *in vivo*. *J. Virol.* 84, 7278–7287. doi: 10.1128/JVI.02239-09

Felber, B. K., Paskalis, H., Kleinman-Ewing, C., Wong-Staal, F., and Pavlakis, G. N. (1985). The pX protein of HTLV-I is a transcriptional activator of its long terminal repeats. *Science* 229, 675–679. doi: 10.1126/science.2992082

Furukawa, Y., Kubota, R., Tara, M., Izumo, S., and Osame, M. (2001). Existence of escape mutant in HTLV-I tax during the development of adult T-cell leukemia. *Blood* 97, 987–993. doi: 10.1182/blood.V97.4.987

Furuta, R., Yasunaga, J. I., Miura, M., Sugata, K., Saito, A., Akari, H., et al. (2017). Human T-cell leukemia virus type 1 infects multiple lineage hematopoietic cells in vivo. *PLoS Pathog.* 13:e1006722. doi: 10.1371/journal.ppat.1006722

Gallo, R. C. (2005). History of the discoveries of the first human retroviruses: HTLV-1 and HTLV-2. *Oncogene* 24, 5926–5930. doi: 10.1038/sj.onc.1208980

Gaudray, G., Gachon, F., Basbous, J., Biard-Piechaczyk, M., Devaux, C., and Mesnard, J. M. (2002). The complementary strand of the human T-cell leukemia virus type 1 RNA genome encodes a bZIP transcription factor that down-regulates viral transcription. *J. Virol.* 76, 12813–12822. doi: 10.1128/JVI.76.24.12813-12822.2002

Gazon, H., Barbeau, B., Mesnard, J. M., and Peloponese, J. M. Jr. (2018). Hijacking of the AP-1 signaling Pathway during development of ATL. *Front. Microbiol.* 15:2686. doi: 10.3389/fmicb.2017.02686

Gessain, A., Barin, F., Vernant, J. C., Gout, O., Maurs, L., Calender, A., et al. (1985). Antibodies to human T-lymphotropic virus type-I in patients with tropical spastic paraparesis. *Lancet* 2, 407–410. doi: 10.1016/S0140-6736(85)92734-5

Gessain, A., and Cassar, O. (2012). Epidemiological aspects and world distribution of HTLV-1 infection. *Front. Microbiol.* 3:388. doi: 10.3389/fmicb.2012.00388

Grassmann, R., Aboud, M., and Jeang, K. T. (2005). Molecular mechanisms of cellular transformation by HTLV-1 Tax. *Oncogene* 24, 5976–5985. doi: 10.1038/sj.onc.1208978

Grossman, W. J., Kimata, J. T., Wong, F. H., Zutter, M., Ley, T. J., and Ratner, L. (1995). Development of leukemia in mice transgenic for the tax gene of human T-cell leukemia virus type I. *Proc. Natl. Acad. Sci. U.S.A.* 92, 1057–1061. doi: 10.1073/pnas.92.4.1057

Hasegawa, H., Sawa, H., Lewis, M. J., Orba, Y., Sheehy, N., Yamamoto, Y., et al. (2006). Thymus-derived leukemia-lymphoma in mice transgenic for the Tax gene of human T-lymphotropic virus type I. *Nat. Med.* 12, 466–472. doi: 10.1038/nm1389

Hieshima, K., Nagakubo, D., Nakayama, T., Shirakawa, A. K., Jin, Z., and Yoshie, O. (2008). Tax-inducible production of CC chemokine ligand 22 by human T cell leukemia virus type 1 (HTLV-1)-infected T cells promotes preferential transmission of HTLV-1 to CCR4-expressing CD4+ T cells. *J. Immunol.* 180, 931–939. doi: 10.4049/jimmunol.180.2.931

Hinuma, Y., Nagata, K., Hanaoka, M., Nakai, M., Matsumoto, T., Kinoshita, K. I., et al. (1981). Adult T-cell leukemia: antigen in an ATL cell line and detection of antibodies to the antigen in human sera. *Proc. Natl. Acad. Sci. U.S.A.* 78, 6476–6480. doi: 10.1073/pnas.78.10.6476

Hivin, P., Frederic, M., Arpin-Andre, C., Basbous, J., Gay, B., Thebault, S., et al. (2005). Nuclear localization of HTLV-I bZIP factor (HBZ) is mediated by three distinct motifs. *J. Cell Sci.* 118, 1355–1362. doi: 10.1242/jcs.01727

Igakura, T., Stinchcombe, J. C., Goon, P. K., Taylor, G. P., Weber, J. N., Griffiths, G. M., et al. (2003). Spread of HTLV-I between lymphocytes by virus-induced polarization of the cytoskeleton. *Science* 299, 1713–1716. doi: 10.1126/science.1080115

Iqbal, J., Weisenburger, D. D., Greiner, T. C., Vose, J. M., McKeithan, T., Kucuk, C., et al. (2010). Molecular signatures to improve diagnosis in peripheral T-cell lymphoma and prognostication in angioimmunoblastic T-cell lymphoma. *Blood* 115, 1026–1036. doi: 10.1182/blood-2009-06-227579

Ishida, T., Utsunomiya, A., Iida, S., Inagaki, H., Takatsuka, Y., Kusumoto, S., et al. (2003). Clinical significance of CCR4 expression in adult T-cell leukemia/lymphoma: its close association with skin involvement and unfavorable outcome. *Clin. Cancer Res.* 9, 3625–3634.

Jones, K. S., Petrow-Sadowski, C., Huang, Y. K., Bertolette, D. C., and Ruscetti, F. W. (2008). Cell-free HTLV-1 infects dendritic cells leading to transmission and transformation of CD4(+) T cells. *Nat. Med.* 14, 429–436. doi: 10.1038/nm1745

Kannagi, M., Harada, S., Maruyama, I., Inoko, H., Igarashi, H., Kuwashima, G., et al. (1991). Predominant recognition of human T cell leukemia virus type I (HTLV-I) pX gene products by human CD8+ cytotoxic T cells directed against HTLV-I-infected cells. *Int. Immunol.* 3, 761–767. doi: 10.1093/intimm/3.8.761

Kataoka, K., Nagata, Y., Kitanaka, A., Shiraishi, Y., Shimamura, T., Yasunaga, J., et al. (2015). Integrated molecular analysis of adult T cell leukemia/lymphoma. *Nat. Genet.* 47, 1304–1315. doi: 10.1038/ng.3415

Kinosada, H., Yasunaga, J. I., Shimura, K., Miyazato, P., Onishi, C., Iyoda, T., et al. (2017). HTLV-1 bZIP factor enhances T-Cell proliferation by impeding the suppressive signaling of co-inhibitory receptors. *PLoS Pathog.* 13:e1006120. doi: 10.1371/journal.ppat.1006120

Landry, D. M., Hertz, M. I., and Thompson, S. R. (2009). RPS25 is essential for translation initiation by the Dicistroviridae and hepatitis C viral IRESS. *Genes Dev.* 23, 2753–2764. doi: 10.1101/gad.1832209

Levin, M. C., Krichavsky, M., Fox, R. J., Lehky, T., Jacobson, S., Fox, C., et al. (1997). Extensive latent retroviral infection in bone marrow of patients with HTLV-I-associated neurologic disease. *Blood* 89, 346–348.

Ma, G., Yasunaga, J., Akari, H., and Matsuoka, M. (2015). TCF1 and LEF1 act as T-cell intrinsic HTLV-1 antagonists by targeting Tax. *Proc. Natl. Acad. Sci. U.S.A.* 112, 2216–2221. doi: 10.1073/pnas.1419198112

Ma, G., Yasunaga, J., Fan, J., Yanagawa, S., and Matsuoka, M. (2013). HTLV-1 bZIP factor dysregulates the Wnt pathways to support proliferation and migration of adult T-cell leukemia cells. *Oncogene* 32, 4222–4230. doi: 10.1038/onc.2012.450

Mahgoub, M., Yasunaga, J., Iwami, S., Nakaoka, S., Koizumi, Y., Shimura, K., et al. (2018). Sporadic on/off switching of HTLV-1 Tax expression is crucial to maintain the whole population of virus-induced leukemic cells. *Proc. Natl. Acad. Sci. U.S.A.* 115, E1269–E1278. doi: 10.1073/pnas.1715724115

Mazurov, D., Ilinskaya, A., Heidecker, G., Lloyd, P., and Derse, D. (2010). Quantitative comparison of HTLV-1 and HIV-1 cell-to-cell infection with new replication dependent vectors. *PLoS Pathog.* 6:e1000788. doi: 10.1371/journal.ppat.1000788

Mitobe, Y., Yasunaga, J., Furuta, R., and Matsuoka, M. (2015). HTLV-1 bZIP factor RNA and protein impart distinct functions on T-cell proliferation and survival. *Cancer Res.* 75, 4143–4152. doi: 10.1158/0008-5472.CAN-15-0942

Mukai, R., and Ohshima, T. (2014). HTLV-1 HBZ positively regulates the mTOR signaling pathway via inhibition of GADD34 activity in the cytoplasm. *Oncogene* 33, 2317–2328. doi: 10.1038/onc.2013.181

Nakagawa, M., Schmitz, R., Xiao, W., Goldman, C. K., Xu, W., Yang, Y., et al. (2014). Gain-of-function CCR4 mutations in adult T cell leukemia/lymphoma. *J. Exp. Med.* 211, 2497–2505. doi: 10.1084/jem.20140987

Pais-Correia, A. M., Sachse, M., Guadagnini, S., Robbiati, V., Lasserre, R., Gessain, A., et al. (2010). Biofilm-like extracellular viral assemblies mediate HTLV-1 cell-to-cell transmission at virological synapses. *Nat. Med.* 16, 83–89. doi: 10.1038/nm.2065

Panfil, A. R., Dissinger, N. J., Howard, C. M., Murphy, B. M., Landes, K., Fernandez, S. A., et al. (2016). Functional comparison of HBZ and the related APH-2 protein provides insight into human T-Cell leukemia virus type 1 pathogenesis. *J. Virol.* 90, 3760–3772. doi: 10.1128/JVI.03113-15

Poiesz, B. J., Ruscetti, F. W., Gazdar, A. F., Bunn, P. A., Minna, J. D., and Gallo, R. C. (1980). Detection and isolation of type C retrovirus particles from fresh and cultured lymphocytes of a patient with cutaneous T-cell lymphoma. *Proc. Natl. Acad. Sci. U.S.A.* 77, 7415–7419. doi: 10.1073/pnas.77.12.7415

Richardson, J. H., Edwards, A. J., Cruickshank, J. K., Rudge, P., and Dalgleish, A. G. (1990). In vivo cellular tropism of human T-cell leukemia virus type 1. *J. Virol.* 64, 5682–5687.

Rizkallah, G., Alais, S., Futsch, N., Tanaka, Y., Journo, C., Mahieux, R., et al. (2017). Dendritic cell maturation, but not type I interferon exposure, restricts infection by HTLV-1, and viral transmission to T-cells. *PLoS Pathog.* 4:e1006353. doi: 10.1371/journal.ppat.1006353

Rosewick, N., Durkin, K., Artesi, M., Marcais, A., Hahaut, V., Griebel, P., et al. (2017). Cis-perturbation of cancer drivers by the HTLV-1/BLV proviruses is an early determinant of leukemogenesis. *Nat. Commun.* 8:15264. doi: 10.1038/ncomms15264

Satou, Y., Yasunaga, J., Yoshida, M., and Matsuoka, M. (2006). HTLV-I basic leucine zipper factor gene mRNA supports proliferation of adult T cell leukemia cells. *Proc. Natl. Acad. Sci. U.S.A.* 103, 720–725. doi: 10.1073/pnas.0507631103

Satou, Y., Yasunaga, J., Zhao, T., Yoshida, M., Miyazato, P., Takai, K., et al. (2011). HTLV-1 bZIP factor induces T-cell lymphoma and systemic inflammation *in vivo*. *PLoS Pathog.* 7:e1001274. doi: 10.1371/journal.ppat.1001274

Sugata, K., Yasunaga, J., Kinosada, H., Mitobe, Y., Furuta, R., Mahgoub, M., et al. (2016). HTLV-1 viral factor HBZ induces CCR4 to promote T-cell migration and proliferation. *Cancer Res.* 76, 5068–5079. doi: 10.1158/0008-5472.CAN-16-0361

Takeda, S., Maeda, M., Morikawa, S., Taniguchi, Y., Yasunaga, J., Nosaka, K., et al. (2004). Genetic and epigenetic inactivation of tax gene in adult T-cell leukemia cells. *Int. J. Cancer* 109, 559–567. doi: 10.1002/ijc.20007

Tanaka-Nakanishi, A., Yasunaga, J., Takai, K., and Matsuoka, M. (2014). HTLV-1 bZIP factor suppresses apoptosis by attenuating the function of FoxO3a and altering its localization. *Cancer Res.* 74, 188–200. doi: 10.1158/0008-5472.CAN-13-0436

Terol, M., Gazon, H., Lemasson, I., Duc-Dodon, M., Barbeau, B., Cesaire, R., et al. (2017). HBZ-mediated shift of JunD from growth suppressor to tumor promoter in leukemic cells by inhibition of ribosomal protein S25 expression. *Leukemia* 31, 2235–2243. doi: 10.1038/leu.2017.74

Thebault, S., Basbous, J., Hivin, P., Devaux, C., and Mesnard, J. M. (2004). HBZ interacts with JunD and stimulates its transcriptional activity. *FEBS Lett.* 562, 165–170. doi: 10.1016/S0014-5793(04)00225-X

Vernin, C., Thenoz, M., Pinatel, C., Gessain, A., Gout, O., Delfau-Larue, M. H., et al. (2014). HTLV-1 bZIP factor HBZ promotes cell proliferation and genetic instability by activating OncomiRs. *Cancer Res.* 74, 6082–6093. doi: 10.1158/0008-5472.CAN-13-3564

Yamamoto-Taguchi, N., Satou, Y., Miyazato, P., Ohshima, K., Nakagawa, M., Katagiri, K., et al. (2013). HTLV-1 bZIP factor induces inflammation through labile Foxp3 expression. *PLoS Pathog.* 9:e1003630. doi: 10.1371/journal.ppat.1003630

Yasuma, K., Yasunaga, J., Takemoto, K., Sugata, K., Mitobe, Y., Takenouchi, N., et al. (2016). HTLV-1 bZIP factor impairs anti-viral immunity by inducing co-inhibitory molecule, T Cell immunoglobulin and ITIM domain (TIGIT). *PLoS Pathog.* 12:e1005372. doi: 10.1371/journal.ppat.1005372

Yasunaga, J., Sakai, T., Nosaka, K., Etoh, K., Tamiya, S., Koga, S., et al. (2001). Impaired production of naive T lymphocytes in human T-cell leukemia virus type I-infected individuals: its implications in the immunodeficient state. *Blood* 97, 3177–3183. doi: 10.1182/blood.V97.10.3177

Yazgan, O., and Pfarr, C. M. (2002). Regulation of two JunD isoforms by Jun N-terminal kinases. *J. Biol. Chem.* 277, 29710–29718. doi: 10.1074/jbc.M204552200

Yoshida, M., Miyoshi, I., and Hinuma, Y. (1982). Isolation and characterization of retrovirus from cell lines of human adult T-cell leukemia and its implication in the disease. *Proc. Natl. Acad. Sci. U.S.A.* 79, 2031–2035. doi: 10.1073/pnas.79.6.2031

Yoshie, O., Fujisawa, R., Nakayama, T., Harasawa, H., Tago, H., Izawa, D., et al. (2002). Frequent expression of CCR4 in adult T-cell leukemia and human T-cell leukemia virus type 1-transformed T cells. *Blood* 99, 1505–1511. doi: 10.1182/blood.V99.5.1505

Yuan, Q., Bromley, S. K., Means, T. K., Jones, K. J., Hayashi, F., Bhan, A. K., et al. (2007). CCR4-dependent regulatory T cell function in inflammatory bowel disease. *J. Exp. Med.* 204, 1327–1334. doi: 10.1084/jem.20062076

Zhao, T., Satou, Y., Sugata, K., Miyazato, P., Green, P. L., Imamura, T., et al. (2011). HTLV-1 bZIP factor enhances TGF-beta signaling through p300 coactivator. *Blood* 118, 1865–1876. doi: 10.1182/blood-2010-12-326199

11

HTLV-1-Mediated Epigenetic Pathway to Adult T-Cell Leukemia–Lymphoma

Makoto Yamagishi[1], Dai Fujikawa[1], Toshiki Watanabe[2] and Kaoru Uchimaru[1]**

[1] *Department of Computational Biology and Medical Sciences, Graduate School of Frontier Sciences, The University of Tokyo, Tokyo, Japan,* [2] *The Institute of Medical Science, The University of Tokyo, Tokyo, Japan*

***Correspondence:**
Makoto Yamagishi
myamagishi@edu.k.u-tokyo.ac.jp
Kaoru Uchimaru
uchimaru@cbms.k.u-tokyo.ac.jp

Human T-cell leukemia virus type 1 (HTLV-1), the first reported human oncogenic retrovirus, is the etiologic agent of highly aggressive, currently incurable diseases such as adult T-cell leukemia–lymphoma (ATL) and HTLV-1-associated myelopathy/tropical spastic paraparesis (HAM/TSP). HTLV-1 proteins, including Tax and HBZ, have been shown to have critical roles in HTLV-1 pathogenicity, yet the underlying mechanisms of HTLV-1-driven leukemogenesis are unclear. The frequent disruption of genetic and epigenetic gene regulation in various types of malignancy, including ATL, is evident. In this review, we illustrate a focused range of topics about the establishment of HTLV-1 memory: (1) genetic lesion in the Tax interactome pathway, (2) gene regulatory loop/switch, (3) disordered chromatin regulation, (4) epigenetic lock by the modulation of epigenetic factors, (5) the loss of gene fine-tuner microRNA, and (6) the alteration of chromatin regulation by HTLV-1 integration. We discuss the persistent influence of Tax-dependent epigenetic changes even after the disappearance of HTLV-1 gene expression due to the viral escape from the immune system, which is a remaining challenge in HTLV-1 research. The summarized evidence and conceptualized description may provide a better understanding of HTLV-1-mediated cellular transformation and the potential therapeutic strategies to combat HTLV-1-associated diseases.

Keywords: HTLV-1, ATLL, epigenetics, EZH2, gene expression, gene mutations

INTRODUCTION

Human T-cell leukemia virus type 1 (HTLV-1) infection (Poiesz et al., 1980; Hinuma et al., 1981; Yoshida et al., 1982) is associated with the development of adult T-cell leukemia–lymphoma (ATL) and HTLV-1-associated myelopathy/tropical spastic paraparesis (HAM/TSP), although most virus carriers remain asymptomatic throughout their lifespan. ATL is a highly aggressive T-cell malignancy refractory to the currently available combination chemotherapies (Uchiyama et al., 1977; Tsukasaki et al., 2007; Katsuya et al., 2012). HAM/TSP, a debilitating neuro-inflammatory disease, expresses chronic spinal cord inflammation and progressive myelopathic symptoms (Gessain et al., 1985; Osame et al., 1986).

Accumulating evidence has shown that HTLV-1 exhibits complicated involvement in the pathogenesis (Matsuoka and Jeang, 2007; Yamagishi and Watanabe, 2012). In particular, HTLV-1 Tax significantly affects host gene expression and interacts with multiple partner proteins (Boxus et al., 2008; Chevalier et al., 2012; Simonis et al., 2012). Moreover, Tax-transgenic mice develop malignant lymphoma, suggesting that Tax is an oncoprotein (Hasegawa et al., 2006; Ohsugi et al., 2007). The evolution of viral genes with virus expansion indicates that leukemogenesis by Tax is selectively advantageous for viral replication and cell proliferation. Transgenic expression of *HBZ* in CD4$^+$ T-cells also induces T-cell lymphomas and systemic inflammation in mice (Satou et al., 2011). Tax and HBZ certainly contribute to leukemogenesis in HTLV-1-infected T-cells. However, considering the low rate of incidence, clinical observation implies that HTLV-1 lacks a strong capacity to induce leukemogenesis, in contrast to other animal leukemia viruses.

Notably, most leukemic cells do not express viral genes, excluding HBZ (Gaudray et al., 2002; Taniguchi et al., 2005; Satou et al., 2006). Tax, a highly immunogenic protein, is not expressed in most aggressive-type ATL cases because HTLV-1 provirus is substantially silenced by proviral defect and/or an epigenetic mechanism (Tamiya et al., 1996; Koiwa et al., 2002; Taniguchi et al., 2005). It is assumed that this is one of the strategies that viruses use to evade host immune defense.

However, leukemic cells possess similar traits as Tax-expressing cells (Yamagishi and Watanabe, 2012). Although the reason for this seemingly paradoxical observation is yet to be determined, it is suggested that the acquired cellular characteristics, including promoting cell proliferation and apoptotic resistance, is conferred by viral genes in early-phase infected cells and by genetic/epigenetic abnormalities in late-phase, highly malignant ATL cells (**Figure 1**). Although Tax has already disappeared at the time of ATL onset, Tax and its interactome (described in later chapters) have already left multiple genetic and epigenetic memories, contributing ATL onset. This switch during leukemogenesis is indeed supported by transcriptome data; the changes in gene expression in infected cells are dominated by disordered homeostasis and the characteristics of ATL.

Although cancer is typically considered to be a genetic disease, chromatin and epigenetic aberrations as well as active roles of HBZ play important roles in tumor potentiation, initiation, and progression in ATL and HTLV-1-associated diseases. Based on recent findings, we introduce a hypothesis with important implications that might explain the underlying mechanism of the issue: the molecular memories inherited from HTLV-1.

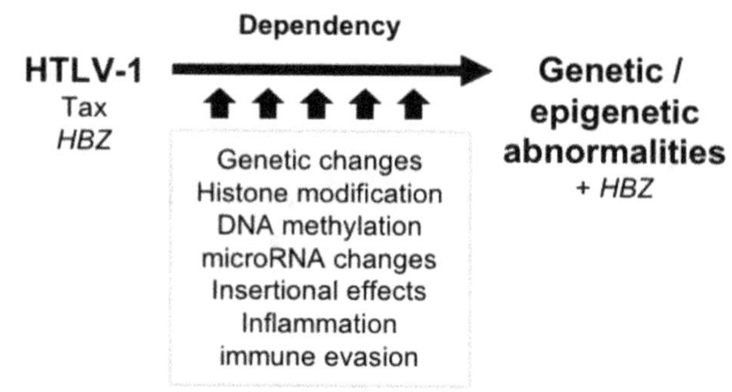

FIGURE 1 | Transition of the molecular characteristics during latent period. The acquired cellular characteristics such as promoting cell proliferation and apoptotic resistance are conferred by viral genes in early-phase infected cells and by genetic/epigenetic abnormalities in late-phase ATL cells. The aberrant characteristics are acquired and imprinted, nevertheless Tax disappears. The consequent genotype and epigenotype support the differential phenotypes and the disease entities.

TRANSCRIPTOME ABNORMALITY IN ATL

Cellular characteristics (i.e., phenotype) are strictly defined by the regulation of gene expression. HTLV-1 Tax directly affects host gene expression through multiple mechanisms, including the binding with host transcription factors and the perturbation of multiple signaling pathways (Ballard et al., 1988; Ruben et al., 1988; Kim et al., 1990; Migone et al., 1995; Good et al., 1996; Takemoto et al., 1997; Boxus et al., 2008). Intriguingly, the molecular hallmarks of aggressive ATL cells at the final stage of progression still comprise pronounced dysregulation of the signaling pathways that control the cell cycle, the resistance to apoptosis, and the proliferation of leukemic cells without Tax expression.

Cell cycle regulation is a typical example of the correlation between gene expression and phenotypic changes. The oncogenic function of Tax was first demonstrated in a study of cell cycle regulation. Tax inhibits cyclin-dependent kinase (CDK) inhibitor, *CDKN2A* (p16^{INK4A}), via physical interaction (Suzuki et al., 1996). The mitogenic activity of Tax is exerted through the stimulation of G_1-to-S-phase transition. Additionally, Tax affects a cohort of cell cycle-related proteins, including *CDKs*, *CDKN1A*, *CDKN1B*, and *CDKN2A*, via the regulation of their expression or physical interaction (Akagi et al., 1996; Neuveut et al., 1998; Schmitt et al., 1998; Santiago et al., 1999; Suzuki et al., 1999; de La Fuente et al., 2000; Iwanaga et al., 2001; Haller et al., 2002; Liang et al., 2002).

Comprehensive gene expression profiling revealed that several positive regulators of the cell cycle process are overexpressed in acute-type ATL, in most of which HTLV-1 sense-transcripts and the virus replication is silenced. Diverse abnormalities were also found in each of these comprehensive studies; however, several gene alterations and other critical events have been commonly implicated as the determinants of gene expression pattern. The abnormalities in the expression of different cytokines, their receptors, and various proteins that act as anti-apoptotic factors or proliferating agents are the cellular hallmarks responsible for malignant phenotypes (Tsukasaki et al., 2004; Sasaki et al., 2005; Watanabe et al., 2010; Yamagishi et al., 2012). These notable traits in the transcriptome may be genetically and epigenetically established during long-term latency periods.

ESTABLISHMENT OF HTLV-1 MEMORY

Genetic, metabolic, and environmental stimuli can induce overly restrictive or permissive epigenetic landscapes that contribute to the pathogenesis of cancer and other diseases. The restrictive chromatin states prevent the appropriate expression of tumor suppressors or block differentiation. In contrast, the permissive states allow the stochastic activation of oncogenic genes and stochastic silencing of tumor suppressor genes. The abnormal restriction or plasticity may also affect other processes mediated through factors such as chromatin–DNA repair and telomere maintenance.

Chromatin homeostasis, a basis of molecular memory (Flavahan et al., 2017), is disrupted by genetic and epigenetic stimuli (e.g., inflammation, aging, hypoxia, cell stress, developmental cues, metabolism, and pathogens). The heritable, selective adaptive changes are the hallmarks of cancers. Herein, we introduce the abnormality contributing to the molecular pathogenesis of HTLV-1 infection by tracing the function(s) of Tax and the characteristics of ATL cells.

Genetic Lesion in Tax Interactome Pathway

Tax directly participates in genetic damage (Jeang et al., 1990; Saggioro et al., 1994; Kao and Marriott, 1999; Haoudi et al., 2003). In parallel with this, persistent proliferation, which is boosted by cell cycle progression, may cause genetic instability and create stochastic genetic lesions; ≥ 1 lesions may then act as "drivers," allowing clonal evolution.

The recent advanced technology-based comprehensive characterization of genetic abnormalities delineated the spectrum of genetic alterations in ATL (Kataoka et al., 2015). Genomic data from a total of 426 patients with ATL identified 6,404 mainly age-related somatic mutations (2.3 mutations/Mb/sample) by whole-exome sequencing, including 6,096 single-nucleotide variants and 308 insertions–deletions, strongly suggesting that the clonal expansion of aggressive ATL cells is driven by multiple genetic abnormalities. One of the remarkable indications is that some of the somatically altered genes in ATL (mutation and copy number variation) encode the pivotal molecules that Tax physically interacts with and/or deregulates, including the components of TCR–NF-κB pathway [activated by Tax (Yamaoka et al., 1998; reviewed in Sun and Yamaoka, 2005)] and p53 and p16 tumor suppressors [inactivated by Tax (Suzuki et al., 1996; Grassmann et al., 2005)]; this strongly suggests that ATL cells still depend on the dysregulated Tax interactome even after the disappearance of Tax expression in most ATL cases, i.e., the influence of Tax is genetically imprinted in ATL cells.

Gene Regulatory Loop/Switch

Depending on the cellular status, a transient cue such as an inflammatory cytokine can induce stable malignant transformation through a positive feedback network that is normally held in check by a host defense mechanism (Iliopoulos et al., 2009; Barabási et al., 2011; Yosef and Regev, 2011). In this manner, network motifs, including a coherent feedforward, mutual negative feedback, and positive feedback loops, may switch the cell fate in some cases.

Tax can activate several signaling pathways and lead to an abnormal gene expression pattern. For instance, it can activate NF-κB and NFAT pathways responsible for the predominant expression of IL-2 and its receptor IL2R (Ballard et al., 1988; Ruben et al., 1988; Hoyos et al., 1989; McGuire et al., 1993; Good et al., 1996), whose activation leads to a positive feedback loop. The target transcriptome of NF-κB pathway includes the genes encoding the members of the Rel family, p100/p105, NF-κB-inducing kinase (NIK), and several cytokines that stimulate the same pathway.

Negative regulators within the network are critical for the homeostasis of the regulatory motif. In the developmental process of HTLV-1-infected cells, some NF-κB negative regulators are diminished or inactivated, leading to chronic activation of the signaling pathway. For example, miR-31, a new class of negative regulator of the non-canonical NF-κB pathway, acts by regulating NIK. One mechanism of NF-κB activation without Tax is the epigenetic silencing of miR-31 in HTLV-1-infected cells and aggressive ATL cells (Yamagishi et al., 2012).

Another NF-κB negative regulator, p47, which is essential for Golgi membrane fusion, associates with the NEMO subunit of IκB kinase (IKK) complex upon TNF-α or IL-1 stimulation and inhibits IKK activation. Tax inhibits the interaction between p47 and the IKK complex. In contrast, a significant reduction of p47 expression has been reported in ATL cells, which show a high-level constitutive NF-κB activation that protects ATL cells from apoptosis in a Tax-independent manner (Shibata et al., 2012). These findings indicate that defenseless signaling may cause automatically and chronically activated signaling pathways (Yamagishi et al., 2015), possibly even after the loss of Tax.

Chromatin Regulation

Chromatin is the fundamental medium through which transcription factors, signaling pathways, and various other cues influence gene activity. A dynamic change of the chromatin conformation reinforces regulatory activity or repression at each locus and causes reorganization in response to appropriate intrinsic and extrinsic stimuli.

Genes encoding epigenetic factors, including SWI/SNF complex members and DNA methylation modifiers, are among the most frequently mutated genes in human cancers (Lawrence et al., 2014). However, the genetic changes of such epigenetic factors are less common in ATL, although epigenetic dysregulation such as DNA methylation and histone acetylation is observed at each investigated locus (Nosaka et al., 2000; Tsuchiya et al., 2000; Hofmann et al., 2001; Yasunaga et al., 2004; Yoshida et al., 2004; Yang et al., 2005; Daibata et al., 2007; Taniguchi et al., 2008).

The ATL is also characterized by prominent CpG island DNA hypermethylation, leading to transcriptional silencing (Kataoka et al., 2015). Approximately 40% of the cases showed the CpG island methylator phenotype without any mutation at *TET2*,

IDH2, and *DNMT3A*. Additionally, C2H2-type zinc finger genes (implicated in the suppression of endogenous and exogenous retroviruses) were hypermethylated and silenced. Furthermore, the hypermethylation of MHC-I expression may contribute to immune evasion.

When we consider the chromatin aberrations that confer plasticity, the polycomb family and its substrate histone, H3K27, are of particular interest. EZH2 can repress a wide range of genes by catalyzing the trimethylation of H3K27 (H3K27me3). Regarding the cellular function, EZH2 and H3K27me3 act in a highly context-dependent manner. EZH2 gain-of-function mutations may be oncogenic in a B-cell lineage (Morin et al., 2010; Yap et al., 2011). In addition, an aberrant activation of polycomb repressive complex 2 (PRC2) mainly based on the overexpression of EZH2 is frequently observed in hematological malignancies and solid tumors (Yamagishi and Uchimaru, 2017). In contrast, EZH2 is genetically inactivated in myelodysplastic syndromes (Ernst et al., 2010) and T-cell acute lymphoblastic leukemia (Ntziachristos et al., 2012).

We recently analyzed the pattern of ATL histone modification and integrated it with the transcriptome from primary ATL cells to decipher the ATL-specific "epigenetic code" (Kobayashi et al., 2014; Fujikawa et al., 2016). PRC2-mediated H3K27me3 is significantly and frequently reprogrammed at half of genes in ATL cells. A large proportion of abnormal gene downregulation is observed at an early stage of disease progression, which is explained by H3K27me3 accumulation. Global H3K27me3 alterations involve ATL-specific gene expression changes that include several tumor suppressors, transcription factors, epigenetic modifiers, miRNAs, and developmental genes (Fujikawa et al., 2016), suggesting the diverse outcomes of the PRC2-dependent hierarchical regulation.

Importantly, the Tax-dependent immortalized cells also show significantly similar H3K27me3 reprogramming as that of ATL cells. A majority of the epigenetic silencing occurs in leukemic cells from indolent ATL and in HTLV-1-infected premalignant T-cells from asymptomatic HTLV-1 carriers.

The important implications for deciphering the triggers of the specific histone code are physical interaction and other influences on the host epigenetic machinery by Tax, including the key histone modifiers HDAC1 (Ego et al., 2002), SUV39H1 (Kamoi et al., 2006), SMYD3 (Yamamoto et al., 2011), and EZH2 (Fujikawa et al., 2016).

Epigenetic Lock by Modulation of Epigenetic Factors

The functional classification of genes has revealed that genes epigenetically suppressed by H3K27me3 are enriched in certain biological processes, including transcriptional regulation and histone modifiers, in ATL. Among these, the expression of *KDM6B*, encoding a JMJD3 demethylase of H3K27me3, is significantly downregulated upon H3K27me3 gain (Fujikawa et al., 2016). Because JMJD3 downregulation causes the global accumulation of H3K27me3, ATL cells seemingly acquire a coherent pattern that produces and maintains the systematic abnormality of H3K27me3.

Another coherent pattern is observed in EZH2 regulation. EZH2 is sensitive to promiscuous signaling networks, including NF-κB pathway. Upregulated EZH2 causes excessive PRC2 activity and suppresses multiple target genes such as NF-κB negative regulators (Yamagishi et al., 2012); this forms a positive feedback loop. HTLV-1 Tax is significantly involved in this motif by interacting with EZH2 and activating NF-κB pathway (Fujikawa et al., 2016). Regarding the chronic activation of PRC2 without Tax, an initial triggering event is unnecessary for the maintenance of epigenetic loop.

Loss of Gene Fine-Tuner microRNA

Among the regulators of gene expression, microRNAs are recognized as "buffers" and/or "fine-tuners." MicroRNA can reduce the noise in gene expression; thus, the loss of microRNA may create perturbed gene expression at the post-transcriptional level (Huntzinger and Izaurralde, 2011; Ebert and Sharp, 2012).

One of the key characteristics of ATL is the global downregulation of microRNA (Yamagishi et al., 2012). Although it has not been experimentally demonstrated, the loss of functional small RNA may cause disordered gene expression through transcriptional and post-transcriptional levels. Notably, this global loss is caused and imprinted by HTLV-1-induced H3K27me3 accumulation, suggesting that the global loss of microRNA is one of the processes required for the developmental pathway leading to ATL.

Alteration of Chromatin Regulation by HTLV-1 Integration

Recent advances regarding insertional effects by HTLV-1 have provided critical implications. Satou et al. (2016) found that CTCF (a key regulator of chromatin structure and function) binds to the provirus in the provirus pX region and acts as an enhancer blocker, leading to long-distance interactions with flanking host chromatin. Indeed, HTLV-1 was reported to alter local higher-order chromatin structure and gene expression in the host genome (Satou et al., 2016).

Rosewick et al. (2017) employed stranded RNA-seq data in combination with improved DNA-seq-based high-throughput mapping of integration sites and found that HTLV-1/BLV proviruses are integrated near cancer drivers, which they affect via provirus-dependent transcription termination or as a result of viral antisense RNA-dependent *cis*-perturbation. Remarkably, a similar result was observed at polyclonal non-malignant stages, indicating that provirus-dependent host gene perturbation contributes to the initial selection of the multiple clones characterizing the asymptomatic stage, requiring additional alterations in the clone that will evolve into aggressive leukemia/lymphoma. Although the hotspots of proviral integration sites and the influence of their insertion into the host genome/epigenome are still being discussed, the previously unrecognized mechanisms may be complementary to viral gene products and the acquisition of somatic alterations in the host genome.

EPIGENETIC LANDSCAPE OF HTLV-1-INFECTED CELLS

The biologist Conrad Waddington first conceptualized developmental fate decision as an epigenetic landscape wherein differentiating cells proceed downhill along the branching canals separated by the walls that restrict cell identity (Waddington, 1957).

Decades of research have revealed that transcription factors are the predominant specifiers of cellular identity (Zaret and Mango, 2016; Bradner et al., 2017). However, the topography of this "hill" seems to be determined by the chromatin pattern, which is directly regulated by epigenetic mechanisms in response to the intrinsic and extrinsic (environmental) stimuli exemplified in this review.

Therefore, as a hypothesis, we propose, in agreement with the established developmental pathway of HTLV-1-infected cells, that disease progression fits with the epigenetic landscape, wherein the height of the walls between the canals is determined by several molecular events (**Figure 2**).

The initial trigger for restricting gene expression is HTLV-1 infection. This violent event significantly affects cell fate, primarily by Tax and HBZ. Then, the immortalized cells possibly undergo several molecular events, as described above (including genetic and epigenetic alterations). During a long period, several aberrant characteristics are acquired and fixed, nevertheless Tax disappears. The consequent genotype and epigenotype support the differential phenotypes and the disease entities of ATL and HAM/TSP.

HTLV-1 provirus is frequently defective or silenced in ATL. However, the lesions recurrently detected in ATL cells imitate the function of Tax and would be stably inherited in the progeny of the malignant cells. This raises several critical possibilities such as that the active imprinting of the viral function into the host genome and epigenome is one of the critical steps of leukemogenesis. Furthermore, the features of ATL cells are not accidental but are the products of HTLV-1 infection. In addition to the sustained roles of HBZ (reviewed in Ma et al., 2016), some crucial outcomes (including gene mutations in the components of TCR–NF-κB pathway and abnormal H3K27me3 accumulation) and many other stochastic events shape ATL cells and their characteristics.

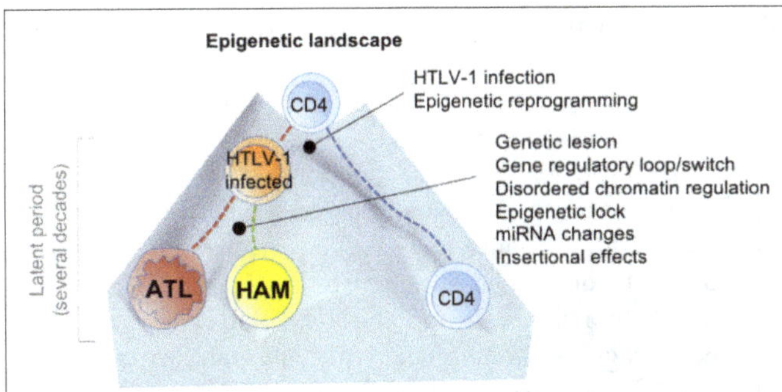

FIGURE 2 | A hypothetical model of developmental pathway in HTLV-1-associated diseases [adapted from Waddington (1957)]. The height of the walls between the valleys (or canals) is determined by several molecular events. The first HTLV-1 infection and the accompanied epigenetic alterations change the cell fate. The permissive state induced by HTLV-1-infection allows following stochastic perturbations such as genetic mutations and dysregulation of the signaling pathways and clonal selection, paralleled by a decrease in transcriptional noise, and the stabilization of cell states (deepening of the valleys).

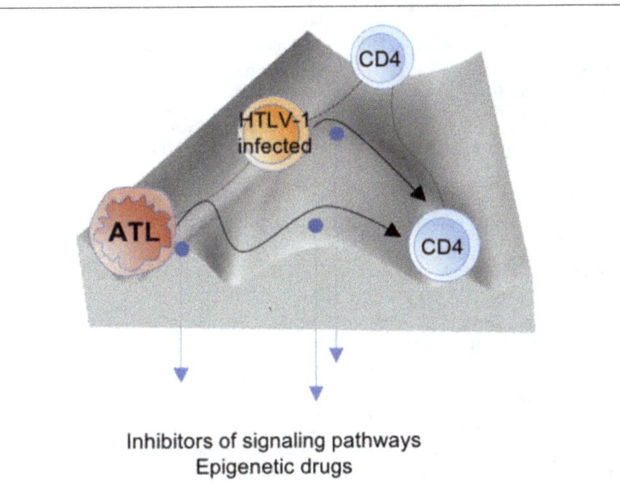

FIGURE 3 | The mechanism-based medicines such as epigenetic drugs and inhibitors of signaling pathways could reprogram the fate of HTLV-1-infected cells (conceptualized as a reduction or elevation of the walls (blue arrows), which promote crossing or bypassing within the epigenetic landscape (black curved arrows) into the normal state).

FUTURE DIRECTION

At present, researchers and hematologists are sharing their findings on the characteristics of ATL cells. Additionally, the phenotypic characteristics of HAM/TSP have been studied. Considering the therapeutics for the HTLV-1-associated diseases and the need to eliminate the premalignant cell population, the establishment of a precise understanding of disease developmental pathways (routes, branch points, and the events that influence the landscape, as shown in **Figure 2**) is an urgent requirement. Therefore, there is a need to investigate the abnormalities contributing to the molecular pathogenesis, including those in master transcription factors and chromatin regulators. Furthermore, in addition to cellular traits, environmental parameters such as aging, cellular stress, and immune response should be integrated into our model of this process. The order of the molecular events is just a pathway of disease development. HTLV-1 infection and following epigenetic reprogramming may be an initial step of fate changes.

Intentional regulation such as by inhibitor treatment will reprogram the fate of HTLV-1-infected cells, which can be conceptualized as a reduction or elevation of the walls between the canals in the epigenetic landscape, in line with the analogy mentioned above (**Figure 3**). Realizing the potential of such mechanism-based medicines and advanced diagnostic tools for the detection and evaluation of tumor stage and heterogeneity will require a deeper understanding of epigenetic plasticity and restriction. The road ahead is long but must be challenged to capture this major component of HTLV-1 biology and its associated diseases.

AUTHOR CONTRIBUTIONS

MY conceived and supervised the project, summarized and conceptualized the evidence, and wrote the paper. DF, TW, and KU discussed the new concept.

FUNDING

This research was supported by AMED under Grant Nos. JP18ak0101086 (MY, KU, and TW), JP17fk0108112 (KU and TW), JP17im0210101 (KU), and JP17fk0410208 (KU), and JSPS KAKENHI Grant Nos. JP16H05323 (MY), JP18K08317 (MY), and JP16H0627 (TW).

REFERENCES

Akagi, T., Ono, H., and Shimotohno, K. (1996). Expression of cell-cycle regulatory genes in HTLV-I infected T-cell lines: possible involvement of Taxl in the altered expression of cyclin D2, p18Ink4 and p21Waf1/Cip1/Sdi1. *Oncogene* 12, 1645–1652.

Ballard, D. W., Böhnlein, E., Lowenthal, J. W., Wano, Y., Franza, B. R., and Greene, W. C. (1988). HTLV-I Tax induces cellular proteins that activate the kappa B element in the IL- 2 receptor alpha gene. *Science* 241, 1652–1655. doi: 10.1126/science.2843985

Barabási, A. L., Gulbahce, N., and Loscalzo, J. (2011). Network medicine: a network-based approach to human disease. *Nat. Rev. Genet.* 12, 56–68. doi: 10.1038/nrg2918

Boxus, M., Twizere, J. C., Legros, S., Dewulf, J. F., Kettmann, R., and Willems, L. (2008). The HTLV-1 Tax interactome. *Retrovirology* 5:76. doi: 10.1186/1742-4690-5-76

Bradner, J. E., Hnisz, D., and Young, R. A. (2017). Transcriptional addiction in cancer. *Cell* 168, 629–643. doi: 10.1016/j.cell.2016.12.013

Chevalier, S. A., Durand, S., Dasgupta, A., Radonovich, M., Cimarelli, A., Brady, J. N., et al. (2012). The Transcription profile of Tax-3 is more similar to Tax-1 than Tax-2: insights into HTLV-3 potential leukemogenic properties. *PLoS One* 7:e41003. doi: 10.1371/journal.pone.0041003

Daibata, M., Nemoto, Y., Bandobashi, K., Kotani, N., Kuroda, M., Tsuchiya, M., et al. (2007). Promoter hypermethylation of the bone morphogenetic protein-6 gene in malignant lymphoma. *Clin. Cancer Res.* 13, 3528–3535. doi: 10.1158/1078-0432.CCR-06-2766

de La Fuente, C., Santiago, F., Chong, S. Y., Deng, L., Mayhood, T., Fu, P., et al. (2000). Overexpression of p21waf1 in human T-cell lymphotropic virus type 1-infected cells and its association with cyclin A/cdk2. *J. Virol.* 74, 7270–7283. doi: 10.1128/JVI.74.16.7270-7283.2000

Ebert, M. S., and Sharp, P. A. (2012). Roles for MicroRNAs in conferring robustness to biological processes. *Cell* 149, 515–524. doi: 10.1016/j.cell.2012.04.005

Ego, T., Ariumi, Y., and Shimotohno, K. (2002). The interaction of HTLV-1 Tax with HDAC1 negatively regulates the viral gene expression. *Oncogene* 21, 7241–7246. doi: 10.1038/sj.onc.1205701

Ernst, T., Chase, A. J., Score, J., Hidalgo-Curtis, C. E., Bryant, C., Jones, A. V., et al. (2010). Inactivating mutations of the histone methyltransferase gene EZH2 in myeloid disorders. *Nat. Genet.* 42, 722–726. doi: 10.1038/ng.621

Flavahan, W. A., Gaskell, E., and Bernstein, B. E. (2017). Epigenetic plasticity and the hallmarks of cancer. *Science* 357:eaal2380. doi: 10.1126/science.aal2380

Fujikawa, D., Nakagawa, S., Hori, M., Kurokawa, N., Soejima, A., Nakano, K., et al. (2016). Polycomb-dependent epigenetic landscape in adult T-cell leukemia. *Blood* 127, 1790–1802. doi: 10.1182/blood-2015-08-662593

Gaudray, G., Gachon, F., Basbous, J., Biard-Piechaczyk, M., Devaux, C., and Mesnard, J. M. (2002). The complementary strand of the human T-cell leukemia virus type 1 RNA genome encodes a bZIP transcription factor that downregulates viral transcription. *J. Virol.* 76, 12813–12822. doi: 10.1128/JVI.76.24.12813-12822.2002

Gessain, A., Barin, F., Vernant, J. C., Gout, O., Maurs, L., Calender, A., et al. (1985). Antibodies to human T-lymphotropic virus type-I in patients with tropical spastic paraparesis. *Lancet* 2, 407–410. doi: 10.1016/S0140-6736(85)92734-5

Good, L., Maggirwar, S. B., and Sun, S. C. (1996). Activation of the IL-2 gene promoter by HTLV-I tax involves induction of NF-AT complexes bound to the CD28-responsive element. *EMBO J.* 15, 3744–3750.

Grassmann, R., Aboud, M., and Jeang, K. T. (2005). Molecular mechanisms of cellular transformation by HTLV- 1 Tax. *Oncogene* 24, 5976–5985. doi: 10.1038/sj.onc.1208978

Haller, K., Wu, Y., Derow, E., Schmitt, I., Jeang, K. T., and Grassmann, R. (2002). Physical interaction of human T-cell leukemia virus type 1 Tax with cyclin-dependent kinase 4 stimulates the phosphorylation of retinoblastoma protein. *Mol. Cell. Biol.* 22, 3327–3338. doi: 10.1128/MCB.22.10.3327-3338.2002

Haoudi, A., Daniels, R. C., Wong, E., Kupfer, G., and Semmes, O. J. (2003). Human T-cell leukemia virus-I tax oncoprotein functionally targets a subnuclear complex involved in cellular DNA damage-response. *J. Biol. Chem.* 278, 37736–37744. doi: 10.1074/jbc.M301649200

Hasegawa, H., Sawa, H., Lewis, M. J., Orba, Y., Sheehy, N., Yamamoto, Y., et al. (2006). Thymus-derived leukemia-lymphoma in mice transgenic for the Tax gene of human T-lymphotropic virus type I. *Nat. Med.* 12, 466–472. doi: 10.1038/nm1389

Hinuma, Y., Nagata, K., Hanaoka, M., Nakai, M., Matsumoto, T., Kinoshita, K. I., et al. (1981). Adult T-cell leukemia: antigen in an ATL cell line and detection of antibodies to the antigen in human sera. *Proc. Natl. Acad. Sci. U.S.A.* 78, 6476–6480. doi: 10.1073/pnas.78.10.6476

Hofmann, W. K., Tsukasaki, K., Takeuchi, N., Takeuchi, S., and Koeffler, H. P. (2001). Methylation analysis of cell cycle control genes in adult T-cell leukemia/lymphoma. *Leuk. Lymphoma* 42, 1107–1109. doi: 10.3109/10428190109097731

Hoyos, B., Ballard, D. W., Böhnlein, E., Siekevitz, M., and Greene, W. C. (1989). Kappa B-specific DNA binding proteins: role in the regulation of human interleukin-2 gene expression. *Science* 244, 457–460. doi: 10.1126/science.2497518

Huntzinger, E., and Izaurralde, E. (2011). Gene silencing by microRNAs: contributions of translational repression and mRNA decay. *Nat. Rev. Genet.* 12, 99–110. doi: 10.1038/nrg2936

Iliopoulos, D., Hirsch, H. A., and Struhl, K. (2009). An Epigenetic switch involving NF-(B, Lin28, Let-7 MicroRNA, and IL6 links inflammation to cell transformation. *Cell* 139, 693–706. doi: 10.1016/j.cell.2009.10.014

Iwanaga, R., Ohtani, K., Hayashi, T., and Nakamura, M. (2001). Molecular mechanism of cell cycle progression induced by the oncogene product Tax of human T-cell leukemia virus type I. *Oncogene* 20, 2055–2067. doi: 10.1038/sj.onc.1204304

Jeang, K. T., Widen, S. G., Semmes, O. J. T., and Wilson, S. H. (1990). HTLV-1 trans-activator protein, tax, is a trans-repressor of the human beta-polymerase gene. *Science* 247, 1082–1084. doi: 10.1126/science.2309119

Kamoi, K., Yamamoto, K., Misawa, A., Miyake, A., Ishida, T., Tanaka, Y., et al. (2006). SUV39H1 interacts with HTLV-1 Tax and abrogates Tax transactivation of HTLV-1 LTR. *Retrovirology* 3:5. doi: 10.1186/1742-4690-3-5

Kao, S. Y., and Marriott, S. J. (1999). Disruption of nucleotide excision repair by the human T-cell leukemia virus type 1 Tax protein. *J. Virol.* 73, 4299–4304.

Kataoka, K., Nagata, Y., Kitanaka, A., Shiraishi, Y., Shimamura, T., Yasunaga, J., et al. (2015). Integrated molecular analysis of adult T cell leukemia/lymphoma. *Nat. Genet.* 47, 1304–1315. doi: 10.1038/ng.3415

Katsuya, H., Yamanaka, T., Ishitsuka, K., Utsunomiya, A., Sasaki, H., Hanada, S., et al. (2012). Prognostic index for acute- and lymphoma-type adult T-cell leukemia/lymphoma. *J. Clin. Oncol.* 30, 1635–1640. doi: 10.1200/JCO.2011.38.2101

Kim, S. J., Kehrl, J. H., Burton, J., Tendler, C. L., Jeang, K. T., Danielpour, D., et al. (1990). Transactivation of the transforming growth factor beta 1 (TGF-beta 1) gene by human T lymphotropic virus type 1 tax: a potential mechanism for the increased production of TGF-beta 1 in adult T cell leukemia. *J. Exp. Med.* 172, 121–129. doi: 10.1084/jem.172.1.121

Kobayashi, S., Nakano, K., Watanabe, E., Ishigaki, T., Ohno, N., Yuji, K., et al. (2014). CADM1 expression and stepwise downregulation of CD7 are closely associated with clonal expansion of HTLV-I-infected cells in adult T-cell leukemia/lymphoma. *Clin. Cancer Res.* 20, 2851–2861. doi: 10.1158/1078-0432.CCR-13-3169

Koiwa, T., Hamano-Usami, A., Ishida, T., Okayama, A., Yamaguchi, K., Kamihira, S., et al. (2002). 5'-Long terminal repeat-selective CpG methylation of latent human T-cell leukemia virus type 1 provirus in vitro and in vivo. *J. Virol.* 76, 9389–9397. doi: 10.1128/JVI.76.18.9389-9397.2002

Lawrence, M. S., Stojanov, P., Mermel, C. H., Robinson, J. T., Garraway, L. A., Golub, T. R., et al. (2014). Discovery and saturation analysis of cancer genes across 21 tumour types. *Nature* 505, 495–501. doi: 10.1038/nature12912

Liang, M. H., Geisbert, T., Yao, Y., Hinrichs, S. H., and Giam, C. Z. (2002). Human T-lymphotropic virus type 1 oncoprotein tax promotes S-phase entry but blocks mitosis. *J. Virol.* 76, 4022–4033. doi: 10.1128/JVI.76.8.4022-4033.2002

Ma, G., Yasunaga, J., and Matsuoka, M. (2016). Multifaceted functions and roles of HBZ in HTLV-1 pathogenesis. *Retrovirology* 13:16. doi: 10.1186/s12977-016-0249-x

Matsuoka, M., and Jeang, K. T. (2007). Human T-cell leukaemia virus type 1 (HTLV-1) infectivity and cellular transformation. *Nat. Rev. Cancer* 7, 270–280. doi: 10.1038/nrc2111

McGuire, K. L., Curtiss, V. E., Larson, E. L., and Haseltine, W. A. (1993). Influence of human T-cell leukemia virus type I tax and rex on interleukin-2 gene expression. *J. Virol.* 67, 1590–1599.

Migone, T. S., Lin, J. X., Cereseto, A., Mulloy, J. C., O'Shea, J. J., Franchini, G., et al. (1995). Constitutively activated JAK-STAT pathway in T cells transformed with HTLV-I. *Science* 269, 79–81. doi: 10.1126/science.7604283

Morin, R. D., Johnson, N. A., Severson, T. M., Mungall, A. J., An, J., Goya, R., et al. (2010). Somatic mutations altering EZH2 (Tyr641) in follicular and diffuse large B-cell lymphomas of germinal-center origin. *Nat. Genet.* 42, 181–185. doi: 10.1038/ng.518

Neuveut, C., Low, K. G., Maldarelli, F., Schmitt, I., Majone, F., Grassmann, R., et al. (1998). Human T-cell leukemia virus type 1 Tax and cell cycle progression: role of cyclin D-cdk and p110Rb. *Mol. Cell. Biol.* 18, 3620–3632. doi: 10.1128/MCB.18.6.3620

Nosaka, K., Maeda, M., Tamiya, S., Sakai, T., Mitsuya, H., and Matsuoka, M. (2000). Increasing methylation of the CDKN2A gene is associated with the progression of adult T-cell leukemia. *Cancer Res.* 60, 1043–1048.

Ntziachristos, P., Tsirigos, A., Van Vlierberghe, P., Nedjic, J., Trimarchi, T., Flaherty, M. S., et al. (2012). Genetic inactivation of the polycomb repressive complex 2 in T cell acute lymphoblastic leukemia. *Nat. Med.* 18, 298–303. doi: 10.1038/nm.2651

Ohsugi, T., Kumasaka, T., Okada, S., and Urano, T. (2007). The Tax protein of HTLV-1 promotes oncogenesis in not only immature T cells but also mature T cells. *Nat. Med.* 13, 527–528. doi: 10.1038/nm0507-527

Osame, M., Usuku, K., Izumo, S., Ijichi, N., Amitani, H., Igata, A., et al. (1986). HTLV-I associated myelopathy, a new clinical entity. *Lancet* 1, 1031–1032. doi: 10.1016/S0140-6736(86)91298-5

Poiesz, B. J., Ruscetti, F. W., Gazdar, A. F., Bunn, P. A., Minna, J. D., and Gallo, R. C. (1980). Detection and isolation of type C retrovirus particles from fresh and cultured lymphocytes of a patient with cutaneous T cell lymphoma. *Proc. Natl. Acad. Sci. U.S.A.* 77, 7415–7419. doi: 10.1073/pnas.77.12.7415

Rosewick, N., Durkin, K., Artesi, M., Marçais, A., Hahaut, V., Griebel, P., et al. (2017). Cis-perturbation of cancer drivers by the HTLV-1/BLV proviruses is an early determinant of leukemogenesis. *Nat. Commun.* 8:15264. doi: 10.1038/ncomms15264

Ruben, S., Poteat, H., Tan, T. H., Kawakami, K., Roeder, R., Haseltine, W., et al. (1988). Cellular transcription factors and regulation of IL-2 receptor gene expression by HTLV-I tax gene product. *Science* 241, 89–92. doi: 10.1126/science.2838905

Saggioro, D., Majone, F., Forino, M., Turchetto, L., Leszl, A., and Chieco-Bianchi, L. (1994). Tax protein of human T-lymphotropic virus type I triggers DNA damage. *Leuk. Lymphoma* 12, 281–286. doi: 10.3109/10428199409059600

Santiago, F., Clark, E., Chong, S., Molina, C., Mozafari, F., Mahieux, R., et al. (1999). Transcriptional Up-regulation of the cyclin D2 gene and acquisition of new cyclin-dependent kinase partners in human T-cell leukemia virus type 1-infected cells. *J. Virol.* 73, 9917–9927.

Sasaki, H., Nishikata, I., Shiraga, T., Akamatsu, E., Fukami, T., Hidaka, T., et al. (2005). Overexpression of a cell adhesion molecule, TSLC1, as a possible molecular marker for acute-type adult T- cell leukemia. *Blood* 105, 1204–1213. doi: 10.1182/blood-2004-03-1222

Satou, Y., Miyazato, P., Ishihara, K., Yaguchi, H., Melamed, A., Miura, M., et al. (2016). The retrovirus HTLV-1 inserts an ectopic CTCF-binding site into the human genome. *Proc. Natl. Acad. Sci. U.S.A.* 113, 3054–3059. doi: 10.1073/pnas.1423199113

Satou, Y., Yasunaga, J., Zhao, T., Yoshida, M., Miyazato, P., Takai, K., et al. (2011). HTLV-1 bZIP factor induces T-cell lymphoma and systemic inflammation in vivo. *PLoS Pathog.* 7:e1001274. doi: 10.1371/journal.ppat.1001274

Satou, Y., Yasunaga, J. I., Yoshida, M., and Matsuoka, M. (2006). HTLV-I basic leucine zipper factor gene mRNA supports proliferation of adult T cell leukemia cells. *Proc. Natl. Acad. Sci. U.S.A.* 103, 720–725. doi: 10.1073/pnas.0507631103

Schmitt, I., Rosin, O., Rohwer, P., Gossen, M., and Grassmann, R. (1998). Stimulation of cyclin-dependent kinase activity and G1- to S-phase transition in human lymphocytes by the human T-cell leukemia/lymphotropic virus type 1 Tax protein. *J. Virol.* 72, 633–640.

Shibata, Y., Oyama, M., Kozuka-Hata, H., Han, X., Tanaka, Y., Gohda, J., et al. (2012). p47 negatively regulates IKK activation by inducing the lysosomal degradation of polyubiquitinated NEMO. *Nat. Commun.* 3:1061. doi: 10.1038/ncomms2068

Simonis, N., Rual, J. F., Lemmens, I., Boxus, M., Hirozane-Kishikawa, T., Gatot, J. S., et al. (2012). Host-pathogen interactome mapping for HTLV-1 and -2 retroviruses. *Retrovirology* 9:26. doi: 10.1186/1742-4690-9-26

Sun, S. C., and Yamaoka, S. (2005). Activation of NF-kappaB by HTLV-I and implications for cell transformation. *Oncogene* 24, 5952–5964. doi: 10.1038/sj.onc.1208969

Suzuki, T., Kitao, S., Matsushime, H., and Yoshida, M. (1996). Tax protein interacts with cyclin-dependent kinase inhibitor p16INK4A and counteracts its inhibitory activity towards CDK4. *EMBO J.* 15, 1607–1614.

Suzuki, T., Narita, T., Uchida-Toita, M., and Yoshida, M. (1999). Down-regulation of the INK4 family of cyclin-dependent kinase inhibitors by tax protein of HTLV-1 through two distinct mechanisms. *Virology* 259, 384–391. doi: 10.1006/viro.1999.9760

Takemoto, S., Mulloy, J. C., Cereseto, A., Migone, T. S., Patel, B. K., Matsuoka, M., et al. (1997). Proliferation of adult T cell leukemia/lymphoma cells is associated with the constitutive activation of JAK/STAT proteins. *Proc. Natl. Acad. Sci. U.S.A.* 94, 13897–13902. doi: 10.1073/pnas.94.25.13897

Tamiya, S., Matsuoka, M., Etoh, K., Watanabe, T., Kamihira, S., Yamaguchi, K., et al. (1996). Two types of defective human T-lymphotropic virus type I provirus in adult T-cell leukemia. *Blood* 88, 3065–3073.

Taniguchi, A., Nemoto, Y., Yokoyama, A., Kotani, N., Imai, S., Shuin, T., et al. (2008). Promoter methylation of the bone morphogenetic protein-6 gene in association with adult T-cell leukemia. *Int. J. Cancer* 123, 1824–1831. doi: 10.1002/ijc.23749

Taniguchi, Y., Nosaka, K., Yasunaga, J., Maeda, M., Mueller, N., Okayama, A., et al. (2005). Silencing of human T-cell leukemia virus type I gene transcription by epigenetic mechanisms. *Retrovirology* 2:64. doi: 10.1186/1742-4690-2-64

Tsuchiya, T., Tamura, G., Sato, K., Endoh, Y., Sakata, K., Jin, Z., et al. (2000). Distinct methylation patterns of two APC gene promoters in normal and cancerous gastric epithelia. *Oncogene* 19, 3642–3646. doi: 10.1038/sj.onc.1203704

Tsukasaki, K., Tanosaki, S., DeVos, S., Hofmann, W. K., Wachsman, W., Gombart, A. F., et al. (2004). Identifying progression-associated genes in adult T-cell leukemia/lymphoma by using oligonucleotide microarrays. *Int. J. Cancer* 109, 875–881. doi: 10.1002/ijc.20028

Tsukasaki, K., Utsunomiya, A., Fukuda, H., Shibata, T., Fukushima, T., Takatsuka, Y., et al. (2007). VCAP-AMP-VECP compared with biweekly CHOP for adult T-cell leukemia-lymphoma: Japan clinical oncology group study JCOG9801. *J. Clin. Oncol.* 25, 5458–5464. doi: 10.1200/JCO.2007.11.9958

Uchiyama, T., Yodoi, J., Sagawa, K., Takatsuki, K., and Uchino, H. (1977). Adult T-cell leukemia: clinical and hematologic features of 16 cases. *Blood* 50, 481–492.

Waddington, C. H. (1957). *The Strategy of the Genes. A Discussion of some Aspects of Theoretical Biology*. London: George Allen & Unwin.

Watanabe, M., Nakahata, S., Hamasaki, M., Saito, Y., Kawano, Y., Hidaka, T., et al. (2010). Down-regulation of CDKN1A in adult T-cell leukemia/lymphoma despite overexpression of CDKN1A in human T-lymphotropic virus 1- infected cell lines. *J. Virol.* 84, 6966–6977. doi: 10.1128/JVI.00073-10

Yamagishi, M., Katano, H., Hishima, T., Shimoyama, T., Ota, Y., Nakano, K., et al. (2015). Coordinated loss of microRNA group causes defenseless signaling in malignant lymphoma. *Sci. Rep.* 5:17868. doi: 10.1038/srep17868

Yamagishi, M., Nakano, K., Miyake, A., Yamochi, T., Kagami, Y., Tsutsumi, A., et al. (2012). Polycomb-mediated loss of miR-31 activates NIK-dependent NF-(B pathway in adult T cell leukemia and other cancers. *Cancer Cell* 21, 121–135. doi: 10.1016/j.ccr.2011.12.015

Yamagishi, M., and Uchimaru, K. (2017). Targeting EZH2 in cancer therapy. *Curr. Opin. Oncol.* 29, 375–381. doi: 10.1097/CCO.0000000000000390

Yamagishi, M., and Watanabe, T. (2012). Molecular hallmarks of adult T cell leukemia. *Front. Microbiol.* 3:334. doi: 10.3389/fmicb.2012.00334

Yamamoto, K., Ishida, T., Nakano, K., Yamagishi, M., Yamochi, T., Tanaka, Y., et al. (2011). SMYD3 interacts with HTLV-1 Tax and regulates subcellular localization of Tax. *Cancer Sci.* 102, 260–266. doi: 10.1111/j.1349-7006.2010.01752.x

Yamaoka, S., Courtois, G., Bessia, C., Whiteside, S. T., Weil, R., Agou, F., et al. (1998). Complementation cloning of NEMO, a component of the IkappaB kinase complex essential for NF-kappaB activation. *Cell* 93, 1231–1240. doi: 10.1016/S0092-8674(00)81466-X

Yang, Y., Takeuchi, S., Tsukasaki, K., Yamada, Y., Hata, T., Mori, N., et al. (2005). Methylation analysis of the adenomatous polyposis coli (APC) gene in adult T-cell leukemia/lymphoma. *Leuk. Res.* 29, 47–51. doi: 10.1016/j.leukres.2004.05.004

Yap, D. B., Chu, J., Berg, T., Schapira, M., Cheng, S. W., Moradian, A., et al. (2011). Somatic mutations at EZH2 Y641 act dominantly through a mechanism of selectively altered PRC2 catalytic activity, to increase H3K27 trimethylation. *Blood* 117, 2451–2459. doi: 10.1182/blood-2010-11-321208

Yasunaga, J., Taniguchi, Y., Nosaka, K., Yoshida, M., Satou, Y., Sakai, T., et al. (2004). Identification of aberrantly methylated genes in association with adult T-cell leukemia. *Cancer Res.* 64, 6002–6009. doi: 10.1158/0008-5472.CAN-04-1422

Yosef, N., and Regev, A. (2011). Impulse control: temporal dynamics in gene transcription. *Cell* 144, 886–896. doi: 10.1016/j.cell.2011.02.015

Yoshida, M., Miyoshi, I., and Hinuma, Y. (1982). Isolation and characterization of retrovirus from cell lines of human adult T-cell leukemia and its implication in the disease. *Proc. Natl. Acad. Sci. U.S.A.* 79, 2031–2035. doi: 10.1073/pnas.79.6.2031

Yoshida, M., Nosaka, K., Yasunaga, J., Nishikata, I., Morishita, K., and Matsuoka, M. (2004). Aberrant expression of the MEL1S gene identified in association with hypomethylation in adult T-cell leukemia cells. *Blood* 103, 2753–2760. doi: 10.1182/blood-2003-07-2482

Zaret, K. S., and Mango, S. E. (2016). Pioneer transcription factors, chromatin dynamics, and cell fate control. *Curr. Opin. Genet. Dev.* 37, 76–81. doi: 10.1016/j.gde.2015.12.003

12

Hijacking of the AP-1 Signaling Pathway during Development of ATL

Hélène Gazon[1]*, Benoit Barbeau[2], Jean-Michel Mesnard[3] and Jean-Marie Peloponese Jr.[3]*

[1] Belgium Molecular and Cellular Epigenetics, Interdisciplinary Cluster for Applied Genoproteomics, University of Liège, Liège, Belgium, [2] Département des Sciences Biologiques and Centre de Recherche BioMed, Université du Québec à Montréal, Montréal, QC, Canada, [3] Institut de Recherche en Infectiologie de Montpellier, Centre National de la Recherche Scientifique, Université de Montpellier, Montpellier, France

***Correspondence:**
Jean-Marie Peloponese Jr.
jean-marie.peloponese@irim.cnrs.fr
Hélène Gazon
helene.gazon@ulg.ac.be

Human T-cell leukemia virus type 1 (HTLV-1) is the causative agent of a fatal malignancy known as adult T-cell leukemia (ATL). One way to address the pathology of the disease lies on conducting research with a molecular approach. In addition to the analysis of ATL-relevant signaling pathways, understanding the regulation of important and relevant transcription factors allows researchers to reach this fundamental objective. HTLV-1 encodes for two oncoproteins, Tax and HTLV-1 basic leucine-zipper factor, which play significant roles in the cellular transformation and the activation of the host's immune responses. Activating protein-1 (AP-1) transcription factor has been linked to cancer and neoplastic transformation ever since the first representative members of the Jun and Fos gene family were cloned and shown to be cellular homologs of viral oncogenes. AP-1 is a dimeric transcription factor composed of proteins belonging to the Jun (c-Jun, JunB, and JunD), Fos (c-Fos, FosB, Fra1, and Fra2), and activating transcription factor protein families. Activation of AP-1 transcription factor family by different stimuli, such as inflammatory cytokines, stress inducers, or pathogens, results in innate and adaptive immunity. AP-1 is also involved in various cellular events including differentiation, proliferation, survival, and apoptosis. Deregulated expression of AP-1 transcription factors is implicated in various lymphomas such as classical Hodgkin lymphomas, anaplastic large cell lymphomas, diffuse large B-cell lymphomas, and adult T-cell leukemia. Here, we review the current thinking behind deregulation of the AP-1 pathway and its contribution to HTLV-induced cellular transformation.

Keywords: AP-1, HTLV-1, antisense transcription, leukemia, HBZ, JunD

INTRODUCTION

The human T-cell leukemia virus type 1 (HTLV-1) was the first pathogenic retrovirus identified in human (Matsuoka and Jeang, 2010). It is estimated that 10–15 million individuals are infected with HTLV-1 around the world, with endemic areas in the Caribbean, southern Japan, Central and South America, Iran, Melanesia, and sub-Saharan Africa (Sonoda et al., 2011; Gessain and Cassar, 2012). While the vast majority of HTLV-1-infected individuals remain clinically asymptomatic, around 5% of them will develop a highly aggressive T-cell malignancy, termed adult T-cell leukemia/lymphoma (ATL) (Matsuoka and Jeang, 2007; Kogure and Kataoka, 2017). ATL presents four distinct clinical stages ranging from smoldering to acute leukemia. It generally occurs in individuals infected around the time of birth eventually and it develops only after prolonged incubation periods ranging from 20 to 60 years (Matsuoka et al., 1997). Although several studies have reported that the proviral DNA load is a critical factor for promoting disease progression in infected individuals (Olindo et al., 2005; Iwanaga et al., 2010; Yoshida, 2010), 30 years after its characterization in T-lymphocytes from leukemic patients, it is still not fully understood how HTLV-1 transforms human CD4+ T cells in a stepwise fashion. The current view is that pleiotropic functions of the HTLV-1 viral transcriptional transactivator Tax (Peloponese et al., 2007; Journo et al., 2009), such as deregulation of the signaling pathways AP-1 pathway (Fujii et al., 2000) and NF-kB (Rosin et al., 1998; Peloponese et al., 2006; Chan and Greene, 2012), and inactivation of tumor suppressors (Tabakin-Fix et al., 2006) are promitotic events, which drive CD4+ T-cell proliferation during the preleukemic stage (Matsuoka and Jeang, 2007). Paradoxically, fresh ATL cells lack Tax expression, due to genetic and epigenetic modifications in the HTLV-1 provirus (Tamiya et al., 1996; Kataoka et al., 2015). In contrast, HTLV-1 basic leucine-zipper factor (HBZ) mRNA which is encoded by the complementary strand of the HTLV-1 genome is expressed in all ATL cells (Mesnard et al., 2006; Matsuoka and Green, 2009). Recent studies have provided striking evidence for the important role played by of HBZ and the AP-1 pathway in HTLV-1 pathogenesis. In this review, we will limit our focus to the role of AP-1 activation by Tax and HBZ and discuss, in a non-exhaustive manner, how this activation relates to oncogenesis and inflammation.

THE AP-1 PATHWAY, A KEY REGULATOR OF CELLULAR TRANSFORMATION

Activating protein-1 (AP-1) transcription factor has been linked to cancer and neoplastic transformation since the first cloning of *jun* and *fos* proto-oncogenes were cloned following their identification as cellular homologs of avian sarcoma virus 17 (ASV 17)-encoded oncogenes *vjun* and *vfos* 30 years ago (Curran and Franza, 1988). AP-1 is composed of 18 dimeric complexes which included members of four families of DNA-binding proteins: Jun family (c-Jun, JunB, v-Jun, JunD), Fos family (c-Fos, FosB, Fra-1, and Fra-2,) ATF/cyclic AMP-responsive element-binding (CREB) (activating transcription factor: ATF1–4, ATF-6, b-ATF, ATFx), and Maf family (musculoaponeurotic fibrosarcoma c-Maf, MafA, MafB, MafG/F/K, and Nrl) (Eferl and Wagner, 2003; Milde-Langosch, 2005; Hernandez et al., 2008; **Figure 1**). Transcriptional activity of AP-1 is regulated by a wide array of cellular stimuli including growth factors, bacterial and viral infection, cytokines, UV radiation, and cellular stress (Eferl and Wagner, 2003; Milde-Langosch, 2005; Hernandez et al., 2008; **Figure 1**). These transcription factors have critical functions in wide variety of cellular processes, including inflammation, proliferation, differentiation, and apoptosis (Eferl and Wagner, 2003; Milde-Langosch, 2005; Hernandez et al., 2008). The activity of the different AP-1 dimer also depends on the cell type and its differentiation state. In response to external stimuli, MAPK activity increases and regulates both the abundance and transactivating capacities of Jun, Fos, and ATF (**Figure 2**). MAPKs are a serine/threonine kinase superfamily that comprises extracellular signal-regulated kinases (ERK), c-Jun NH2-terminal kinases (JNK), p38, and c-Fos-regulating kinases (FRK) (Cavigelli et al., 1995; Karin, 1995; Srivastava et al., 1999). The regulation of AP-1 is complex and occurs at multiple levels, ranging from dimer composition, to transcriptional and post-translational events, and to specific interactions between AP-1 proteins and other transcription cofactors (**Figure 2**).

AP-1 Transcriptional Regulation

Activating protein-1 activity is modulated through its dimer composition which is determined by the differential expression of Jun, Fos, ATF, and Maf families (**Figure 1**) and through the sequence of the AP-1 DNA-binding sites (**Table 1**). The abundance of the subunits can be controlled either via the regulation of the synthesis and stability of respective mRNAs or via the regulation of protein stability (for example, stimulus-dependent degradation via the ubiquitin pathway) (Musti et al., 1996). Most of the genes that encode AP-1 subunits behave as "immediate-early" genes. Indeed, they are rapidly but transiently transcribed in response to extracellular stimuli, such as growth factor (Ryder and Nathans, 1988; Karin et al., 1997) and cellular stress (Angel and Karin, 1991; Zhou et al., 2007; **Figure 2**). Among these, the transcriptional regulation of *c-jun* and *c-fos* is well studied and characterized (Abate and Curran, 1990). The transcription of *c-fos* is induced in response to a diverse spectrum of extracellular stimuli and its promoter is composed of several transcription factor-binding sites, such as a cAMP-response element (CRE), which can drive transcriptional activation in response to elevation of intracellular Ca^{2+} or cAMP concentrations under stimulation from neurotransmitters and polypeptide hormones (**Figure 2**; Lucibello et al., 1993; Cavigelli et al., 1995; Tulchinsky, 2000). It also contains a serum-response element (SRE), which can drive transcription in response to growth factors, cytokines, UV irradiation, and other stimuli. SRE is recognized by a dimer composed of serum-response factor (SRF) and Elk-1, the major component of ternary complex factor (TCF) in human cells (Lucibello et al., 1993; Cavigelli et al., 1995; Tulchinsky, 2000; **Figure 2**). The third major element of c-Fos promoter is the v-Sis-inducible element (SIE) (Wagner et al., 1990). SIE is mostly recognized by homodimers and heterodimers of STAT1 and STAT3, two members of the signal transducers and activators of transcription (STAT) family (Sadowski et al., 1993). Tyrosine phosphorylation of these factors in the cytoplasm is mediated by janus kinase/tyrosine kinase (JAK/TYKs) and drives their dimerization. The dimerized factor can then translocates to the nucleus, binds the SIE, and participates in promoter activation. Finally, the c-Fos promoter also contains a 12-*O*-tetradecanoyl-phorbol-13-acetate-response element (TRE) (**Figure 2**; Lucibello et al., 1993; Cavigelli et al., 1995; Tulchinsky, 2000).

The *c-jun* promoter is simpler, being mostly induced through the TRE element that preferentially binds c-Jun/ATF2 heterodimers (**Figure 2**; Van Dam et al., 1993). Despite its inducible expression, most cell types prior to their stimulation contain basal levels of c-Jun protein. Like the *c-fos* SRE, the

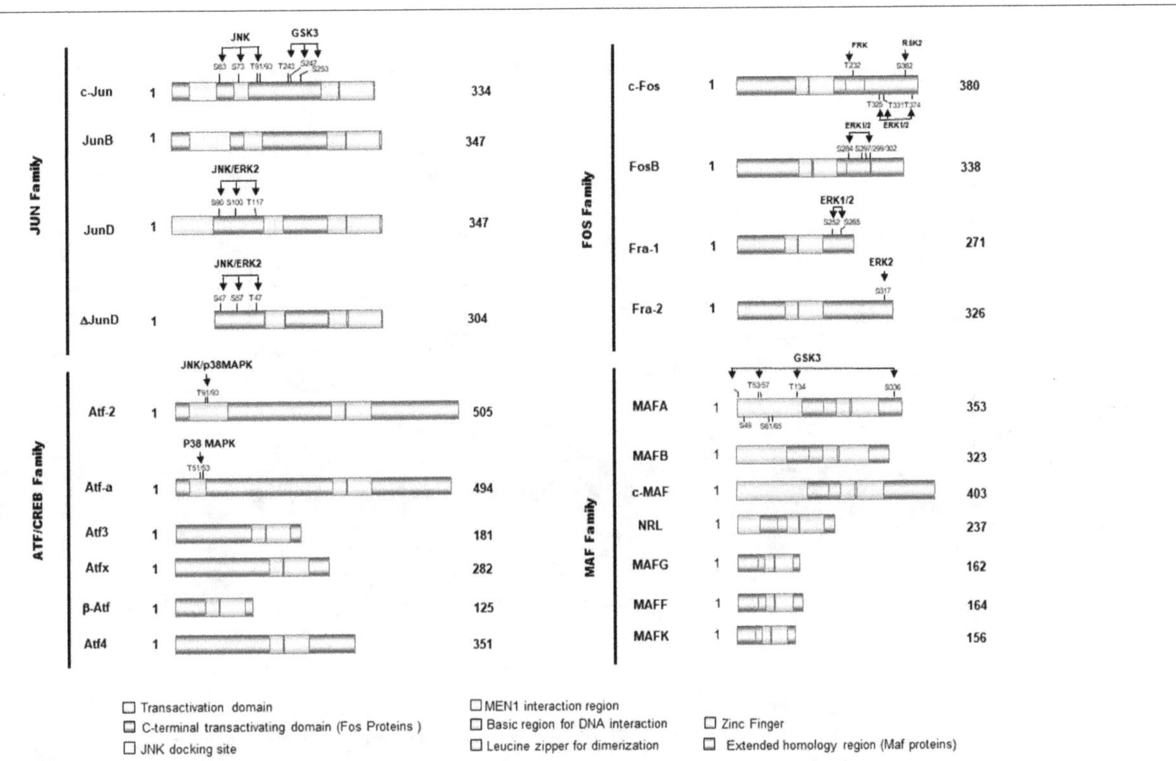

FIGURE 1 | Schematical presentation of the structure of AP-1 proteins. Activator protein 1 (AP-1) proteins include the JUN, FOS, activating transcription factor (ATF), and musculoaponeurotic fibrosarcoma (MAF) protein families, which can form homodimers and heterodimers through their leucine-zipper domains. The AP-1 proteins exhibit several domains, including the bZIP domain (leucine zipper plus basic domain), transactivation domains, and docking sites for several kinases, such as JNK or ERK. These kinases modulate the activity of those transcription factors by phosphorylation of serine and threonine residues.

c-jun TRE is constitutively occupied *in vivo* (Van Dam et al., 1993). Thus, the expression of more than one AP-1 component is under positive and negative AP-1 (auto-)regulation. For example, *c-jun* and *atf3* promoters can be activated by c-Jun/ATF2 and/or ATF2/ATF2 via TRE-binding sites, whereas the *atf3* promoter is inhibited by ATF3 (Angel et al., 1988; Hai and Curran, 1991; Van Dam et al., 1993). The *c-jun* promoter can be inhibited by JunB, c-Jun itself, and c-Fos. This feedback control allows fine-tuned regulation of AP-1 heterodimer activity over longer periods of time (Chiu et al., 1988).

Post-transcriptional Regulation of AP-1 Transcription Factors

Phosphorylation of AP-1 components modulates the dimers transcriptional activities (Karin and Hunter, 1995). Serum and growth factors stimulation induces AP-1 by activating the ERK which then directly phosphorylate c-Jun, Fra-1, and Fra-2 (**Figure 2**). While phosphorylation of c-Jun by ERK on one serine located next to the C-terminal DNA-binding domain inhibits c-Jun DNA-binding activity, phosphorylation of Fra-1 and Fra-2 enhances their DNA binding in conjunction with c-Jun (Woodgett et al., 1993; Punga et al., 2006).

The induction of AP-1 by pro-inflammatory cytokines and genotoxic stress is mostly mediated by the JNK and p38MAPK pathways (Chang and Karin, 2001; **Figure 2**). Once activated, JNKs translocate to the nucleus, where they phosphorylate c-Jun on Ser 63/73 and Thr 91/93 and thereby potentiates its ability to activate transcription upon homodimerization or a heterodimerization with c-Fos (Woodgett et al., 1993; Deng and Karin, 1994; Punga et al., 2006). The molecular mechanisms underlying the capacity of JNK to control c-Jun activity involve the modulation of interactions with histone deacetylase complexes, sub-nuclear localization of AP-1 proteins, and related factors required for c-Jun-dependent activity. JNKs also phosphorylate ATF2 on Thr69/71 and potentiate its activity after heterodimerization with c-Jun, leading to its binding to divergent AP-1 sites in the *c-jun* promoter (Van Dam et al., 1993; Shaulian and Karin, 2002). Transactivation by ATF2 is also potentiated by binding of retinoblastoma (Rb) or E1A, to the DNA-bound ATF2 dimer (Lopez-Bergami et al., 2010). Both E1A and Rb act in concert with phosphorylation of ATF2. Although E1A induces c-jun transcription (Van Dam et al., 1993), it concomitantly represses AP-1 activity through competition for CREB-binding protein (CBP) in a similar manner to the competitive effect of E1A binding on p300 (Offringa et al., 1990; Arany et al., 1994).

The contribution of p38 to AP-1 induction can be mediated by the direct phosphorylation and activation of ATF2 and TCFs (**Figure 2**; Mendelson et al., 1996; Whitmarsh and Davis, 1996). The PI3K/AKT pathway is activated in response to cytokine receptors and T-cell receptor activation in normal T cells. Akt is

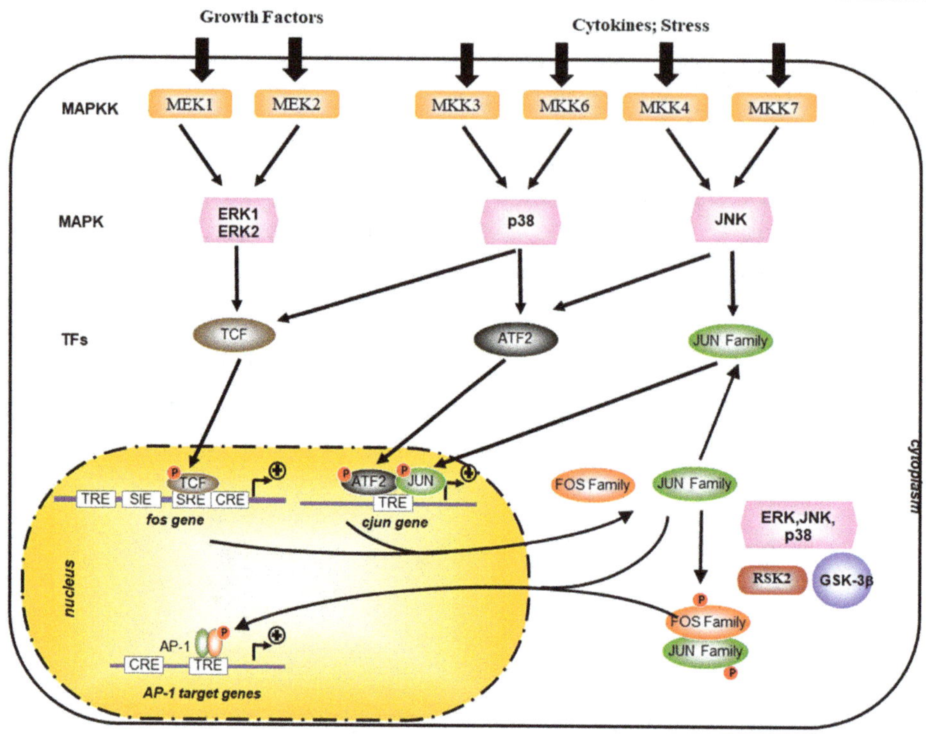

FIGURE 2 | Transcriptional and post-translational activation of AP-1. AP-1 activity is stimulated by external stimuli like growth factors or inflammatory cytokines and a complex network of kinase such as mitogen-activated protein kinases (MAPKs) of the extracellular-signal regulated kinase (ERK), p38, and JUN amino-terminal kinase (JNK) families. Posttranslational phosphorylation by various kinases regulates AP-1 activity, which includes its transactivating potential, DNA-binding capacity, and the stability of AP-1 components. GSK-3β, glycogen synthase kinase-3β; MAPKK, MAPK kinase; RSK2, ribosomal S6 kinase 2; TCF, ternary complex factor; SRE, serum-response element; TRE, TPA-responsive element; CRE, cAMP-response element; SIE, Sis-inducible element.

a serine/threonine protein kinase activated by PI3K through phosphorylation of Ser473, which acts as a regulator of cell survival and proliferation (Warfel and Kraft, 2015). In addition, glycogen synthase kinase-3 (GSK3), an effector kinase of the PI3K pathway, has the capacity to negatively regulate AP-1 transcriptional activity (Koul et al., 2007; Warfel and Kraft, 2015). GSK3 is a ubiquitously expressed serine/threonine kinase normally active in unstimulated cells. Upon stimulation by growth factors, GSK3 is phosphorylated at Ser9 and Ser21 (for GSK3β and GSK3a, respectively) by Akt and other kinases of the AGC family (protein kinase A, protein kinase G, protein kinase C) thus leading to an important decrease of its activity (Koul et al., 2007; Venkatesan et al., 2010). GSK3 activity is controlled mainly through the PI3K/AKT pathway upon AKT phosphorylation on Ser473 (Koul et al., 2007). This complex network of signaling pathways reveals that a particular stimulus can evoke a specific "spectrum" of AP-1 activity and thereby activate and/or repress distinct subset of AP-1-targetted genes.

TABLE 1 | The different AP-1-binding sites.

	AP-1-binding sequence
TRE	TGACTCA
MAREI	TGCTGACTCAGCA
CRE	TGACGTCA
MARE II	TGCTGACGTCAGCA
ARE	a/gtGACnnnGC

The main DNA response element is the TPA-responsive element (TRE), but different dimers preferentially bind to elements such as the cAMP-response element (CRE), the MAF-recognition elements (MAREs), and the antioxidant-response elements (AREs).

The Janus (Dual) Role of AP-1 in Cancer Development

A large amount of studies have shown that AP-1 components play an important role in oncogenesis. *c-jun* and *c-fos* were first identified as retrovirus-activated genes with oncogenic potential in avian and mammalian cells (Abate and Curran, 1990; Verma et al., 1990). Chronic exposure to carcinogens can promote tumorigenesis through the activation of a wide array of signaling pathways, ranging from inflammatory to pro-proliferative and survival pathways. Furthermore, environmental or dietary carcinogens have been shown to induce increased AP1 activity (Abate and Curran, 1990; Verma et al., 1990).

Many human cancers exhibit overexpression of Jun family members (Neyns et al., 1996; Langer et al., 2006; Kharman-Biz et al., 2013). Consistent with the idea that c-Jun can promote tumorigenicity, overexpression of this transcription factor is observed in some of the more aggressive CD30-positive lymphomas (Drakos et al., 2007; Mao et al., 2007). In breast cancer, alteration of RB, VEGF, and EGFR pathways has been shown to induce c-Jun overexpression (Kharman-Biz et al., 2013). Interestingly, increased c-Fos expression is associated with poor clinical outcome in osteosarcoma and endometrial carcinoma, while loss of c-Fos expression is associated with tumor progression and adverse outcome in ovarian carcinoma and gastric carcinoma (Tulchinsky, 2000). On the other side, Fra-1 overexpression is associated with the development of thyroid, breast, lung, brain, nasopharyngeal, esophageal, endometrial, prostate, and colon carcinomas, along with glioblastomas and mesotheliomas (Tulchinsky, 2000; Young and Colburn, 2006). The studies are strongly suggesting that the role of Fos family in tumors development depends on the tissue of origin.

Several studies have shown that AP-1 activity is crucial for tumorigenesis, as its inhibition by dominant-negative c-Jun mutants or AP-1 decoys strongly inhibits the growth of various tumor cell lines both *in vitro* and *in vivo* (Angel and Karin, 1991; Kajanne et al., 2009; Eckert et al., 2013; Kharman-Biz et al., 2013). These studies have also led to the identification of AP1 target genes involved in carcinogenesis (Eferl and Wagner, 2003; Lopez-Bergami et al., 2010; Nakayama et al., 2012). In addition, chronic exposure to environmental and dietary carcinogens such as cigarette smoke or nicotine or ethanol, activates AP1 activity in mouse brain or epithelial cell lines or neuroblastoma cells (Fried et al., 2001; Jochum et al., 2001; Manna et al., 2006). Interestingly, increase in AP1 activity has been also reported in drug-resistant cancer, suggesting that some chemotherapeutic agents can elicit AP1 activation and favor tumor cell survival by making them refractory to long-term treatments (Malorni et al., 2016; Fan et al., 2017; Liou et al., 2017).

Overexpression of Jun and Fos proteins can also suppress tumor formation (Eferl and Wagner, 2003; Hess et al., 2004; Shaulian, 2010), thus revealing the double-edged activity of AP-1 transcription factors. These dual properties depend on the genetic background of the tumor, its differentiation state, and tumor stage (Eferl and Wagner, 2003; Hess et al., 2004; Shaulian, 2010). Several studies have shown that increased AP-1 activity can lead to apoptosis in human tumor cells but it can also antagonize apoptosis in specific cell types, such as liver tumors (Eferl and Wagner, 2003; Hess et al., 2004; Shaulian, 2010). This dual effect of AP-1 on apoptosis can be further exemplified. Indeed, increased c-Jun activity promotes apoptosis in neuronal cells *in vitro* (Ham et al., 2000). When the activation of c-Jun is impaired, either in Jnk3-null or JunAA mice, which express a c-Jun insensitive to JNK-mediated activation, neurons are protected from apoptosis (Behrens et al., 1999). In contrast, c-Jun is required for the survival of fetal hepatocytes, which undergo apoptosis in c-Jun-deficient mouse embryos (Eferl et al., 1999; Hasselblatt et al., 2007). The cell-specific consequence of AP-1 activity over apoptosis is likely due to its differential regulation of pro-apoptotic and anti-apoptotic target genes. In neurons, c-Jun regulates the expression of Bim, a pro-apoptotic Bcl-2 family member that is crucial for neuronal apoptosis. While in T cells, c-Jun regulates the expression of Fas ligand (FasL), which upon binding to the Fas receptor triggers apoptosis (Eferl and Wagner, 2003; Hess et al., 2004; Shaulian, 2010). Bcl-2 family members are also part of the list of anti-apoptotic targets that are regulated by AP-1 (Kirkin et al., 2004). In T cells, Jun members exert a protective signal through the induction of Bcl-3, while in myeloid cells, inactivation of JunB leads to reduced apoptosis with increased expression of the anti-apoptotic Bcl-2 gene (Kyriakis, 1999; Srivastava et al., 1999; Kirkin et al., 2004) have shown that hepatocytes deficient for c-Jun or JunD are highly sensitive to tumor necrosis factor-α-induced apoptosis, thus suggesting that c-Jun and JunD might regulate genes that protect cells from TNF-α-induced-cell death. Those studies have demonstrated that depending on the type of extracellular stimuli and on the cellular context, activation of AP-1 can have a different outcome on the cell fate. Therefore, the role and function of AP-1 in cancer development should be examined within the context of a complex network of simultaneously triggered signaling pathways.

REGULATION OF AP-1 DURING HTLV-1 INFECTION

Human T-cell leukemia virus type 1 can infect a variety of cell types *in vivo*, including T cells, B cells, and macrophages (Jones et al., 2008; Pique and Jones, 2012; Gross and Thoma-Kress, 2016; Rizkallah et al., 2017). The HTLV-1 provirus is detected mainly in CD4+ T cells and to lesser extent in CD8+ T cells (Richardson et al., 1990; Iwahashi et al., 1991; Melamed et al., 2015). This asymmetry in detection may be caused by recruitment of CD4+ T cells and induction of their proliferation following HTLV-1 infection in contrast to a delayed cell death in CD8+ T cells (Sibon et al., 2006; Alais et al., 2015). A subset of HTLV-1-infected individuals will develop ATL after an extended period of time (Matsuoka and Jeang, 2007; Kannian and Green, 2011). Infected cells, however, must initiate proliferation and evade apoptosis as a prelude to immortalization and transformation. Virally encoded oncogenic proteins are known to dysregulate various cellular pathways or processes through the regulation of the activity of target proteins. HTLV-1 transforms T cells via its transactivator Tax, which interferes with pathways regulating cell growth control through activation of various cellular transcription factors (NF-κB, E2F, and AP-1) (Fujii et al., 2000; Peloponese et al., 2006; Journo et al., 2009) and inactivation of p53 (Tabakin-Fix et al., 2006). Since AP-1 has been implicated in transformation of T cells, it has been hypothesized that inappropriate activation of AP-1 could contribute to the dysregulated phenotype of HTLV-1-infected cells or to the development of ATL. Indeed, ATL leukemic cells exhibit increased levels of mRNAs encoding JunD and Fra-2 (Nakayama et al., 2008; Terol et al., 2017) and high levels of AP-1-binding activity (Fujii et al., 2000; Iwai et al., 2001), with in addition, AP-1, more specifically c-Jun/c-Fos heterocomplex, induces HTLV-1 promoter activity and thus participates in HTLV-1 basal transcription (Jeang et al., 1991).

Activation of the AP-1 Pathway by Tax during Acute HTLV-1 Infection

The phosphoprotein Tax is encoded within the pX region of HTLV-1 genome and is a viral regulatory protein. This protein mainly localizes to the nucleus and is a well-known trans-activator of the HTLV-1 long terminal repeat (LTR) operating through three 21-bp repeats called Tax-responsive elements (TxREs) (Meertens et al., 2004; Grassmann et al., 2005). Its pleiotropic properties confer a pivotal role for Tax toward viral pathogenicity and immortalization/transformation of infected cells, causing the onset of HTLV-1-associated diseases (Giam and Jeang, 2007; Peloponese et al., 2007). Indeed, the protein is sufficient for immortalizing primary human T cells and rodent fibroblasts as well as inducing tumors in nude mice inoculated with Tax-transformed cells (Grassmann et al., 2005). Furthermore, Tax interferes with important functions, leading to cell cycle dysregulation and promoting *in vivo* clonal expansion (Giam and Jeang, 2007; Peloponese et al., 2007). Moreover, Tax can alter the expression of cellular proteins involved in cell growth and proliferation such as cytokines (Grassmann et al., 2005; Matsuoka and Jeang, 2010). Importantly, this trans-acting factor also acts as a transcriptional regulator of gene expression by recruiting or modifying the activity of cellular transcription factors (such as CREB protein, SRF, NF-κB, and notably AP-1) through direct or indirect interactions (Grassmann et al., 2005; Matsuoka and Jeang, 2010).

Tax activates the transcription of cellular genes by activating AP-1 DNA binding to promoter elements in T cells (Fujii et al., 2000; Iwai et al., 2001; **Figure 3**). Thus, activation of AP-1 by Tax is thought to contribute to the deregulated phenotypes and leukemogenesis of T cells infected with HTLV-I. Among these factors, c-Fos, Fra-1, c-Jun, JunB, and JunD genes have been shown to be activated by Tax at the transcriptional level (Fujii et al., 2000; Iwai et al., 2001; **Figure 3**). Interestingly, exogenous expression of Tax in Jurkat cells induced AP-1-dependent transcription of a reporter gene more efficiently than any combinations of AP-1 proteins (Fujii et al., 2000; Iwai et al., 2001). Thus, the well-known induction of expression of multiple Fos and Jun family members by Tax is known to be essential, but may not be sufficient, for the transcriptional activation of AP-1 sites mediated by Tax (Fujii et al., 2000; Iwai et al., 2001). Since DNA-binding activity and transcriptional activation of AP-1 are regulated at a post-translational level, Tax might be involved in this regulation (Fujii et al., 2000; Iwai et al., 2001). Tax contributes to the high activity of AP-1 in HTLV-1-infected cells through different mechanisms with intricate ramifications in cascade signaling (**Figure 3**). A constitutive activation of JNK had initially been reported in HTLV-1-infected cells, such as Tax-expressing MT-2 cells and Tax-ATL primary cells. Furthermore, this activity seemed to depend on the status of infection (Xu et al., 1996). Interestingly, Tax is responsible for the constitutive activation of PI3K/Akt by impairing the association between the catalytic (p110) and the regulatory subunit (p85) leading to Akt Ser ^{473} phosphorylation in HTLV-1-infected cell lines (Peloponese and Jeang, 2006; **Figure 3**).

Pathways lying upstream of AP-1 are normally activated in response to external stimuli whereas Tax overrides this requirement (**Figure 3**). Additionally, the PI3K/Akt-AP-1 pathway had been involved in survival and is likely to be required for the immortalization of HTLV-1-infected cells (Jeong et al., 2005). Indeed, Tax even in the absence of NF-κB signaling is able to activate the Akt/PI3K pathway, which upon inhibition by dominant-negative mutants for Akt or for c-Jun abolishes the proliferation of Tax-transfected cell lines as well as the transformed phenotype (Peloponese and Jeang, 2006). AP-1 sites are Tax-inducible elements in different cellular genes, which promote cell proliferation and are associated with clinical characteristic features of ATL (**Figure 3**). Among target genes regulated by AP-1-binding sites, Tax activates growth-promoting cytokine genes, such as IL-2, IL-5, IL-13, as well as proinflammatory cytokines, such as IL-8 and TNF-a, and further induces the expression of immunosuppressors, like TGF-b1 and proenkephalin (PENK). This deregulation implies different combinations of dimerized AP-1 complex although they have not all been characterized as to their targeted promoters (Brady, 1992; Waldmann, 1996; Yamada et al., 1996; Jeang, 2001; Hall and Fujii, 2005).

Tax does not always activate cellular promoters bearing AP-1-binding site (for example, the collagenase gene) (Hall and Fujii, 2005). Rather, another mechanism, by which promoter could be regulated by Tax through AP1, is linked to the binding of Tax and c-Jun to an overlapping region corresponding to the KIX domain of CBP. This could induce a competition between these two proteins for CBP interaction and lead to the repression of c-Jun transcription activity by Tax (Van Orden et al., 1999; Van Orden and Nyborg, 2000). Although Tax induces the transactivation of the TGF-b1 promoter through AP-1 sites, Tax inhibits TGF-b1 signaling by reducing DNA-binding activity of Smad3 through a Smad3/c-Jun complex, which might be involved in the resistance of TGF-b1-induced growth inhibition observed in ATL cells, an important step in the pathogenesis of ATL (Arnulf et al., 2002). An opposite effect could be imputed to HBZ which could overcome the repressing effect of Tax on TGF-b1 signaling (Zhao et al., 2011). Tax might not solely account for the constitutive activity of AP-1, since a high activity of AP-1 has been detected in primary Tax^{-} leukemia cells of ATL patients (Fujii et al., 2000). These observations indicate that AP-1 is activated in HTLV-1-infected T cells through Tax-dependent and Tax-independent mechanisms.

Hijacking of the AP-1 Signaling Pathway by HBZ

Within its antisense strand, HTLV-1 codes for a bZIP factor, which was appropriately named HBZ (HTLV-I bZIP factor) (Gaudray et al., 2002; Mesnard et al., 2006). HBZ can express under three isoforms: one unspliced form (usHBZ) and two alternatively spliced forms (HBZ-SP1 and HBZ-SP2) (Cavanagh et al., 2006; Murata et al., 2006) with HBZ-SP1 (or sHBZ) being the most abundant spliced variant. sHBZ is a 31-kDa protein with an N-terminal transcriptional activation domain, a central domain involved in nuclear localization, and a C-terminal bZIP domain (Gaudray et al., 2002; Mesnard et al., 2006). Sequence comparison between HBZ bZIP region and several bZIP factors clearly indicates that HBZ possesses a c-Fos-like bZIP domain although its DNA-binding domain lacks the consensus amino acid sequence bb-bN–AA-b(C/S)R-bb thought to be critical for DNA binding. HBZ interacts with all the

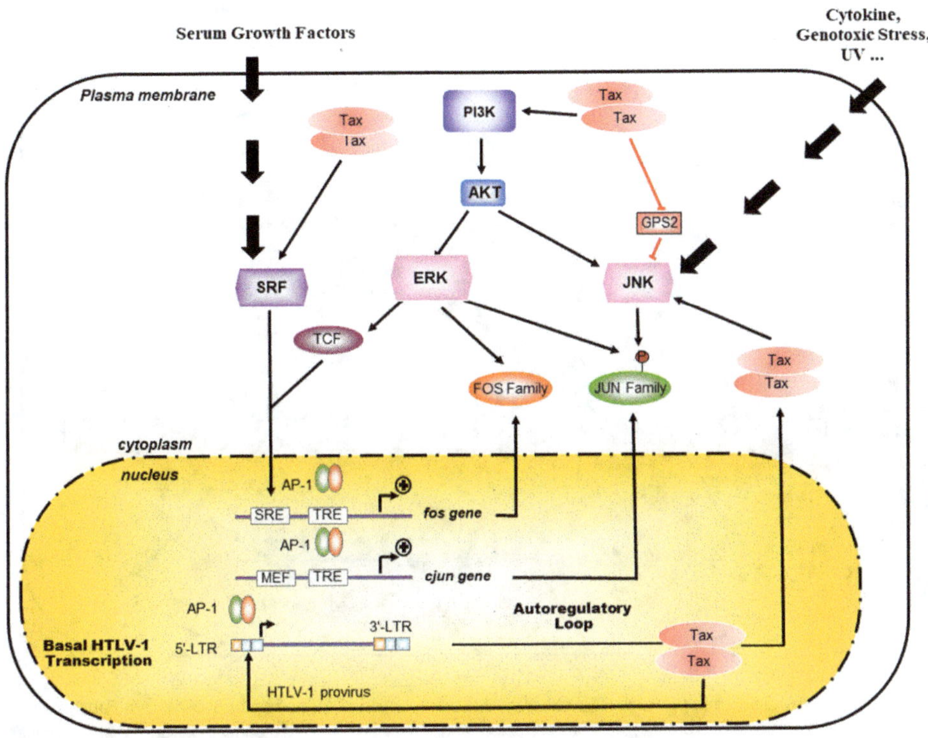

FIGURE 3 | Mechanisms of Tax activation of AP-1 pathway. Pathways upstream of AP-1 are normally activated in response to external stimuli whereas presence of Tax overrides this requirement. Indeed Tax is able to activate the Akt/PI3K pathway as well as the SRF pathways thus activating at the transcription of c-Fos, Fra-1, c-Jun, JunB, and JunD genes. By interacting with the JNK inhibitor G-protein pathway suppressor 2 (GPS2), Tax also participate to the exhibited highly activity of AP-1 in HTLV-1 infected cells through a constitutive activation of JNK (Jin et al., 1997). Interestingly, AP-1 proteins such as c-Jun and c-Fos activate the transcription through the 21 bp repeat in HTLV-1 LTR.

members of the Jun family (JunB, c-Jun, and JunD), and differently regulates the transcriptional properties of the Jun family (Basbous et al., 2003; Hivin et al., 2005, 2007; Clerc et al., 2009).

Sequestration of JunB and c-Jun in HBZ Nuclear Bodies Inhibits Their Transcriptional Activities

HTLV-1 basic leucine-zipper factor is a nuclear protein, which not only accumulates in specific nuclear bodies (called here HBZ-NBs) but is targeted to nucleoli (Hivin et al., 2005). Using a fluorescence recovery after photobleaching approach (FRAP) and an EGFP-tagged-HBZ, Hivin et al. (2007) have observed that the deletion of its leucine-zipper domain altered the rate of nuclear flux of HBZ, suggesting that HBZ heterodimerization partners are involved in controlling its own nuclear trafficking. Indeed, HBZ modifies the localization of JunB and targets JunB to the HBZ-NBs. Moreover, the relocalization of JunB into HBZ-NBs inhibits its transcriptional activity (Hivin et al., 2007; Clerc et al., 2009; **Figure 4**).

Although HBZ and c-Fos can both interact with c-Jun, they differ greatly in their abilities to activate transcription of AP-1-regulated genes. Indeed, the interaction of HBZ with c-Jun prevents this transcription factor from activating transcription of AP-1-dependent promoters by decreasing its DNA-binding activity (Clerc et al., 2009). The generation of different c-Fos/HBZ chimeras by region swapping indicates that the HBZ DNA-binding motif has an important impact on the transcriptional activity of both transcription factors in the presence of c-Jun (Hivin et al., 2007; Clerc et al., 2009). Indeed, the mutant HBZ-mutMD/DBD, for which specific residues present in the MD and DBD regions of HBZ were substituted for corresponding amino acids of c-Fos, showed a significant *in vitro* affinity for the AP-1-binding site TRE but remained unable to stimulate promoter activity of the AP-1-dependent collagenase gene *in vivo*. Like JunB, c-Jun is also relocalized to HBZ-NBs in the presence of HBZ-mutMD/DBD, while this transcription factor is diffusely distributed throughout the nucleus in the presence of HBZ-H14F (a construction in which the bZIP domain of HBZ-mutMD/DBD was replaced by the corresponding ZIP domain of c-Fos), suggesting that HBZ- inhibits c-Jun DNA-binding capacity *in vivo* mainly by its sequestration to the HBZ-NBs (Hivin et al., 2007; Clerc et al., 2009) (**Figure 4** blue circle).

HBZ Promotes the Proteosomal Degradation of c-Jun

HTLV-1 basic leucine-zipper factor can also decrease the stability of c-Jun in cells and promote its degradation (Matsumoto et al.,

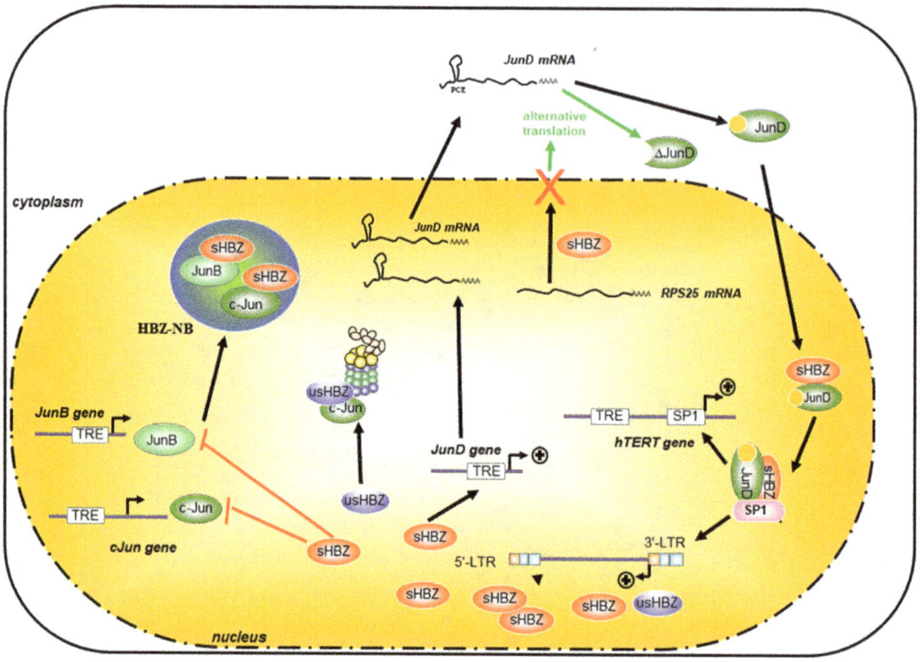

FIGURE 4 | Differential effects of HBZ on the Jun family proteins. HBZ specifically interacted with all three members of the Jun Family (JunB, c-Jun, and JunD), but it regulates differently the transcriptional properties of the Jun family. HBZ dramatically suppressed c-Jun- and JunB-induced transcriptional activation from the AP-1 element by sequestering c-Jun and JunB into HBZ-NB and by decreasing the steady-state level of c-Jun and the stability of c-Jun protein in cells through a proteasome-dependent pathway. It is particularly interesting to note that HBZ has a different and opposite action on JunD expression. HBZ can stimulate the transcription of JunD and by nuclear retention of RPS25, HBZ allows the expression of an alternative isoform of JunD called ΔJunD. Furthermore, HBZ cooperates with JunD and sp1 to enhance transcription of the 3′-LTR and also the human telomerase reverse transcriptase gene (hTERT).

2005). Indeed, in cells transfected with usHBZ, Matsumoto et al. (2005) observed that treatment with proteasome inhibitors but not with calpain inhibitors prevented the reduction in the steady state of c-Jun, suggesting that this HBZ-mediated reduction in c-Jun abundance could also occur through a proteasome-dependent pathway. It has also been suggested by Isono et al. (2008) that HBZ could act as a tethering factor between the 26S proteasome and c-Jun (**Figure 4**). However, c-Jun is less degraded by sHBZ (also called HBZ-SP1) than by usHBZ (Isono et al., 2008) and it remains unclear how both isoforms of HBZ could inhibit c-Jun through two different mechanisms.

HBZ Activates the Transcriptional Activity of JunD

It is particularly interesting to note that HBZ has a different and opposite effects on c-Jun- and JunD-dependent transcription. Indeed, these two proteins belong to the same family of transcription factors, but they are very different proteins. In JunD expressing cells, HBZ is diffusely distributed throughout the nucleoplasm, while no HBZ-NBs are formed (Hivin et al., 2007). Interestingly, JunD is the only Jun family member, which can be activated by HBZ (Thebault et al., 2004; Kuhlmann et al., 2007). It is worth noting that the presence of the EQERRE motif in HBZ modulates JunD activity (Hivin et al., 2005). When the HBZ DNA-binding motif is substituted by the c-Fos modulatory domain, HBZ is no longer able to stimulate the transcriptional activity of JunD, although no alteration in the JunD DNA-binding activity is observed (Hivin et al., 2005). It is important to note that the abundance and activity of JunD increase in freshly isolated ATL cells concomitantly with an increase of HBZ expression (Nakayama et al., 2008; Terol et al., 2017). These observations suggest that HBZ modulates its own expression through a positive-feedback loop in resting cells that involves cooperation with JunD (**Figure 4**). Indeed, in HTLV-1-infected cells, HBZ enhances expression of JunD, which leads to the association of JunD and HBZ to Sp-1 bound to the $3''$-LTR-containing antisense promoter and, ultimately, to the activation of *hbz* transcription (Gazon et al., 2012).

In association with JunD and Sp-1, HBZ also activates the transcription of the human telomerase catalytic subunit gene (hTERT) (Kuhlmann et al., 2007). Telomerase, a ribonucleoprotein complex that extends telomeres which are essential for protecting chromosomal ends against end-end fusions or degradation (Cong et al., 2002; Brunori et al., 2005; Segal-Bendirdjian and Gilson, 2008). While mouse telomerase (mTERT) is activated in many normal tissues, human cells rarely spontaneously reactivate expression of the telomerase gene, as its expression is tightly regulated (Cong et al., 2002; Brunori et al., 2005; Segal-Bendirdjian and Gilson, 2008). However, 75–85% of cancer cells including ATL cells present an increase in telomerase expression and activity (Uchida et al., 1999; Brunori and Gilson, 2004; Brunori et al., 2005; Shay and Wright, 2011). Human telomerase is composed of a structural

RNA component (hTERC), which contains an 11-base sequence complementary to the telomeric single-stranded overhang acting as a template for the synthesis of telomeric DNA. The other main component of the hTERT is its enzymatic reverse transcriptase subunit (Cong et al., 2002; Brunori et al., 2005; Segal-Bendirdjian and Gilson, 2008). Expression of *hTERT* is regulated mainly at the transcriptional level. The proximal 180 bp of the *hTERT* promoter, which does not contain any AP-1-binding site, is important for maintaining basal transcriptional activity and is thought to be the essential component for its regulation. Interestingly, Kuhlmann et al. (2007) have observed an increase in hTERT transcripts in cells co-expressing HBZ and JunD. Chromatin immunoprecipitation (ChIP) assays revealed that HBZ/JunD heterodimers interact with Sp1 and that activation of hTERT transcription by this trimer is mediated through Sp-1-binding sites present in the core region of the *hTERT* promoter (Kuhlmann et al., 2007).

We recently uncovered an additional mechanism used by HBZ to turn JunD from a growth suppressor to a tumor promoter (Terol et al., 2017). *JunD* is an intronless gene and produces two predominant isoforms by alternative initiation of translation, a 39-kDa protein (JunD-FL) through initiation from the first AUG codon and a shorter, 34-kDa JunD protein (ΔJunD) through the use of the second in-frame AUG codon (Hirai et al., 1989; Berger and Shaul, 1994; Short and Pfarr, 2002). Several studies indicated that JunD-FL and ΔJunD are differentially regulated through interactions with other nuclear proteins (Hirai et al., 1989; Berger and Shaul, 1994; Short and Pfarr, 2002). For example, menin, the product of the tumor-suppressor MEN-1 gene, represses JunD-FL transcriptional activity by interacting through its first 48 amino acids (Agarwal et al., 1999; Yazgan and Pfarr, 2001). Loss of menin expression or lost the ability of menin to bind JunD confers JunD with growth-promoting capabilities (Agarwal et al., 2003). ΔJunD does not bind menin, and its transcriptional activity is unaffected by menin overexpression (Yazgan and Pfarr, 2001). In addition, JNK binds and activates JunD-FL more efficiently than ΔJunD, even though both JunD isoforms contain a JNK-docking domain and three JNK phosphorylation target sites. It is interesting to note that freshly isolated ATL cells and HBZ-expressing T lymphocytes express both JunD isoforms (Yazgan and Pfarr, 2002).

JunD mRNA contains a third functional out-of-frame ORF (uORFs) positioned between the ATG of JunD-FL and ATG of ΔJunD (Short and Pfarr, 2002). Translation of downstream ORFs by uORF appears to be a common translational regulatory mechanism, as uORFs are present in two-third of mRNAs encoding oncoproteins and proteins that regulate important cellular processes. Alteration of protein expression levels by disruption or creation of uORF has been associated with the development of several human disease such as Alzheimer's disease, acute myeloid leukemia, and breast cancer (Short and Pfarr, 2002; Zhou and Song, 2006; Wethmar et al., 2010; Barbosa et al., 2013). HBZ relieves uORF translational control by reducing the cellular abundance of RPS25, a ribosomal protein known to play a key role in several alternative translation mechanisms (Nishiyama et al., 2007). Using an immortalized fibroblast cell line model, we found that ΔJunD exhibits growth-promoting and -transforming activities that are enhanced in presence of HBZ (Terol et al., 2017). In summary, it has been proposed that HBZ/JunD heterodimers induce down-regulation of lymphocyte activation and viral transcription to favor viral latency and persistence of the infected cells. Future studies will aim to clarify how HBZ/JunD and/or HBZ/ΔJunD coordinately drive cell fate toward cellular transformation.

PERSPECTIVES AND CONCLUSION

Activating protein-1 family members have both overlapping and unique roles, and the transcriptional activity of the AP-1 dimer functions in a tissue-specific fashion (Shaulian, 2010). With respect to this important fact, recent studies have included the analysis of expression and/or activity of all Jun and Fos family members. Thus, it has been demonstrated that malignant transformation and tumor progression is accompanied by a cell-type-specific shift in AP-1 dimer composition (Verma et al., 1990; Radler-Pohl et al., 1993; Karin et al., 1997; Tulchinsky, 2000; Shaulian and Karin, 2002; Eferl and Wagner, 2003; Milde-Langosch, 2005; Hernandez et al., 2008; Shaulian, 2010). Those studies support a model, in which a shift in the expression pattern of the Fos family members is a crucial step in carcinogenesis and/or tumor progression (Tulchinsky, 2000; Milde-Langosch, 2005). Indeed, while uninfected CD4+ T-lymphocytes express mainly c-Jun and c-Fos proteins, ATL-leukemic cells are expressing JunD and Fra-2 (Nakayama et al., 2008; Terol et al., 2017). Due to their lack of a trans-activating domain, it has been suggested that Fra-1 and Fra-2 might exert anti-tumor effect and inhibit tumor cell growth. Yet, recent studies point to a positive effect of Fra-1, and partly Fra-2, on tumor growth (Young and Colburn, 2006; Milde-Langosch et al., 2008; Nakayama et al., 2008; Davies et al., 2011; Higuchi et al., 2013; Wang et al., 2014; Gupta et al., 2015).

Indeed, gene substitution experiments in mice have shown that growth retardation and osteoporosis observed in c-Fos null-mice were rescued by Fra-1 overexpression, although Fra-1 failed to induce expression of c-Fos target genes, such as MMP13 and vimentin (Young and Colburn, 2006). This observation is in line with results obtained in various cancer cell types, in which Fra-1 alters the biological behavior of the cells without directly activating AP-1-responsive promoters (Zerbini et al., 2003; Young and Colburn, 2006). Surprisingly, in most of the clinical tumor samples analyzed so far, Fra-1 expression is lower than in normal cells, and the protein is poorly phosphorylated (Zerbini et al., 2003; Young and Colburn, 2006). These observations raised several interesting questions on the true role of Fra-1 in oncogenesis and tumor progression. Indeed, whether the low level Fra-1 expression is due to tumor heterogeneity and if expression Fra-1 in specific clones within the tumors have a similar effect to what is seen in experimental systems and contributes to local invasion and metastasis should be further studied. In contrast to the bulk of data available on the function of c-Fos and Fra-1 in carcinogenesis, far less is known on the role of other Fos family members (FosB, FosB2, deltaFosB2, and Fra-2), which are often found expressed in high levels in cancer tissues. Further study of the role of all Fos proteins in carcinogenesis will be of great importance.

AUTHOR CONTRIBUTIONS

J-MP and HG wrote the first draft of the manuscript. HG, BB, J-MM, and J-MP wrote sections of the manuscript. All authors contributed to manuscript revision, read, and approved the submitted version.

FUNDING

This work was supported by the Fondation pour la Recherche Médicale (Equipe FRM DEQ20161136701) and the Ligue Contre le Cancer, Comité Hérault.

REFERENCES

Abate, C., and Curran, T. (1990). Encounters with Fos and Jun on the road to AP-1. *Semin. Cancer Biol.* 1, 19–26.

Agarwal, S. K., Guru, S. C., Heppner, C., Erdos, M. R., Collins, R. M., and Park, S. Y. (1999). Menin interacts with the AP1 transcription factor JunD and represses JunD-activated transcription. *Cell* 96, 143–152. doi: 10.1016/S0092-8674(00)80967-8

Agarwal, S. K., Novotny, E. A., Crabtree, J. S., Weitzman, J. B., Yaniv, M., Burns, A. L., et al. (2003). Transcription factor JunD, deprived of menin, switches from growth suppressor to growth promoter. *Proc. Natl. Acad. Sci. U.S.A.* 100, 10770–10775. doi: 10.1073/pnas.1834524100

Alais, S., Mahieux, R., and Dutartre, H. (2015). Viral source-independent high susceptibility of dendritic cells to human T-cell leukemia virus type 1 infection compared to that of T Lymphocytes. *J. Virol.* 89, 10580–10590. doi: 10.1128/JVI.01799-15

Angel, P., Hattori, K., Smeal, T., and Karin, M. (1988). The jun proto-oncogene is positively autoregulated by its product, Jun/AP-1. *Cell* 55, 875–885. doi: 10.1016/0092-8674(88)90143-2

Angel, P., and Karin, M. (1991). The role of Jun, Fos and the AP-1 complex in cell-proliferation and transformation. *Biochim. Biophys. Acta* 1072, 129–157. doi: 10.1016/0304-419X(91)90011-9

Arany, Z., Sellers, W. R., Livingston, D. M., and Eckner, R. (1994). E1A-associated p300 and CREB-associated CBP belong to a conserved family of coactivators. *Cell* 77, 799–800. doi: 10.1016/0092-8674(94)90127-9

Arnulf, B., Villemain, A., Nicot, C., Mordelet, E., Charneau, P., Kersual, J., et al. (2002). Human T-cell lymphotropic virus oncoprotein Tax represses TGF-beta 1 signaling in human T cells via c-Jun activation: a potential mechanism of HTLV-I leukemogenesis. *Blood* 100, 4129–4138. doi: 10.1182/blood-2001-12-0372

Barbosa, C., Peixeiro, I., and Romão, L. (2013). Gene expression regulation by upstream open reading frames and human disease. *PLOS Genet.* 9:e1003529. doi: 10.1371/journal.pgen.1003529

Basbous, J., Arpin, C., Gaudray, G., Piechaczyk, M., Devaux, C., and Mesnard, J. M. (2003). The HBZ factor of human T-cell leukemia virus type I dimerizes with transcription factors JunB and c-Jun and modulates their transcriptional activity. *J. Biol. Chem.* 278, 43620–43627. doi: 10.1074/jbc.M307275200

Behrens, A., Sibilia, M., and Wagner, E. F. (1999). Amino-terminal phosphorylation of c-Jun regulates stress-induced apoptosis and cellular proliferation. *Nat. Genet.* 21, 326–329. doi: 10.1038/6854

Berger, I., and Shaul, Y. (1994). The human junD gene is positively and selectively autoregulated. *DNA Cell Biol.* 13, 249–255. doi: 10.1089/dna.1994.13.249

Brady, J. N. (1992). Extracellular Tax1 protein stimulates NF-kB and expression of NF-kB-responsive Ig kappa and TNF beta genes in lymphoid cells. *AIDS Res. Hum. Retroviruses* 8, 724–727.

Brunori, M., and Gilson, E. (2004). The opposite roles of telomerase during initiation and progression of cancers. *J. Soc. Biol.* 198, 105–111. doi: 10.1051/jbio/2004198020105

Brunori, M., Luciano, P., Gilson, E., and Geli, V. (2005). The telomerase cycle: normal and pathological aspects. *J. Mol. Med. (Berl)* 83, 244–257. doi: 10.1007/s00109-004-0616-2

Cavanagh, M. H., Landry, S., Audet, B., Arpin-Andre, C., Hivin, P., and Pare, M. E. (2006). HTLV-I antisense transcripts initiating in the 3'LTR are alternatively spliced and polyadenylated. *Retrovirology* 3:15. doi: 10.1186/1742-4690-3-15

Cavigelli, M., Dolfi, F., Claret, F. X., and Karin, M. (1995). Induction of c-fos expression through JNK-mediated TCF/Elk-1 phosphorylation. *EMBO J.* 14, 5957–5964.

Chan, J. K., and Greene, W. C. (2012). Dynamic roles for NF-kappaB in HTLV-I and HIV-1 retroviral pathogenesis. *Immunol. Rev.* 246, 286–310. doi: 10.1111/j.1600-065X.2012.01094.x

Chang, L., and Karin, M. (2001). Mammalian MAP kinase signalling cascades. *Nature* 410, 37–40. doi: 10.1038/35065000

Chiu, R., Boyle, W. J., Meek, J., Smeal, T., Hunter, T., and Karin, M. (1988). The c-Fos protein interacts with c-Jun/AP-1 to stimulate transcription of AP-1 responsive genes. *Cell* 54, 541–552. doi: 10.1016/0092-8674(88)90076-1

Clerc, I., Hivin, P., Rubbo, P. A., Lemasson, I., Barbeau, B., and Mesnard, J. M. (2009). Propensity for HBZ-SP1 isoform of HTLV-I to inhibit c-Jun activity correlates with sequestration of c-Jun into nuclear bodies rather than inhibition of its DNA-binding activity. *Virology* 391, 195–202. doi: 10.1016/j.virol.2009.06.027

Cong, Y.-S., Wright, W. E., and Shay, J. W. (2002). Human telomerase and its regulation. *Microbiol. Mol. Biol. Rev.* 66, 407–425. doi: 10.1128/MMBR.66.3.407-425.2002

Curran, T., and Franza, B. R. (1988). Fos and jun: the AP-1 connection. *Cell* 55, 395–397. doi: 10.1016/0092-8674(88)90024-4

Davies, J. S., Klein, D. C., and Carter, D. A. (2011). Selective genomic targeting by FRA-2/FOSL2 transcription factor: regulation of the Rgs4 gene is mediated by a variant activator protein 1 (AP-1) promoter sequence/CREB-binding protein (CBP) mechanism. *J. Biol. Chem.* 286, 15227–15239. doi: 10.1074/jbc.M110.201996

Deng, T., and Karin, M. (1994). c-Fos transcriptional activity stimulated by H-Ras-activated protein kinase distinct from JNK and ERK. *Nature* 371, 171–175. doi: 10.1038/371171a0

Drakos, E., Leventaki, V., Schlette, E. J., Jones, D., Lin, P., Medeiros, L. J., et al. (2007). c-Jun expression and activation are restricted to CD30+ lymphoproliferative disorders. *Am. J. Surg. Pathol.* 31, 447–453. doi: 10.1097/01.pas.0000213412.25935.e4

Eckert, R. L., Adhikary, G., Young, C. A., Jans, R., Crish, J. F., Xu, W., et al. (2013). AP1 transcription factors in epidermal differentiation and skin cancer. *J. Skin Cancer* 2013:537028. doi: 10.1155/2013/537028

Eferl, R., Sibilia, M., Hilberg, F., Fuchsbichler, A., Kufferath, I., Guertl, B., et al. (1999). Functions of c-Jun in Liver and Heart Development. *J. Cell Biol.* 145, 1049–1061. doi: 10.1083/jcb.145.5.1049

Eferl, R., and Wagner, E. F. (2003). AP-1: a double-edged sword in tumorigenesis. *Nat. Rev. Cancer* 3, 859–868. doi: 10.1038/nrc1209

Fan, F., Bashari, M. H., Morelli, E., Tonon, G., Malvestiti, S., Vallet, S., et al. (2017). The AP-1 transcription factor JunB is essential for multiple myeloma cell proliferation and drug resistance in the bone marrow microenvironment. *Leukemia* 31, 1570–1581. doi: 10.1038/leu.2016.358

Fried, U., Kotarsky, K., and Alling, C. (2001). Chronic ethanol exposure enhances activating protein-1 transcriptional activity in human neuroblastoma cells. *Alcohol* 24, 189–195. doi: 10.1016/S0741-8329(01)00151-3

Fujii, M., Iwai, K., Oie, M., Fukushi, M., Yamamoto, N., Kannagi, M., et al. (2000). Activation of oncogenic transcription factor AP-1 in T cells infected with human T cell leukemia virus type 1. *AIDS Res. Hum. Retroviruses* 16, 1603–1606. doi: 10.1089/08892220050193029

Gaudray, G., Gachon, F., Basbous, J., Biard-Piechaczyk, M., Devaux, C., and Mesnard, J. M. (2002). The complementary strand of the human T-cell leukemia virus type 1 RNA genome encodes a bZIP transcription factor that down-regulates viral transcription. *J. Virol.* 76, 12813–12822. doi: 10.1128/JVI.76.24.12813-12822.2002

Gazon, H., Lemasson, I., Polakowski, N., Cesaire, R., Matsuoka, M., Barbeau, B., et al. (2012). Human T-cell leukemia virus type 1 (HTLV-1) bZIP factor requires cellular transcription factor JunD to upregulate HTLV-1 antisense transcription from the 3' long terminal repeat. *J. Virol.* 86, 9070–9078. doi: 10.1128/JVI.00661-12

Gessain, A., and Cassar, O. (2012). Epidemiological aspects and world distribution of HTLV-1 infection. *Front. Microbiol.* 3:388. doi: 10.3389/fmicb.2012.00388

Giam, C. Z., and Jeang, K. T. (2007). HTLV-1 Tax and adult T-cell leukemia. *Front. Biosci.* 12, 1496–1507. doi: 10.2741/2163

Grassmann, R., Aboud, M., and Jeang, K. T. (2005). Molecular mechanisms of cellular transformation by HTLV-1 Tax. *Oncogene* 24, 5976–5985. doi: 10.1038/sj.onc.1208978

Gross, C., and Thoma-Kress, A. K. (2016). Molecular mechanisms of HTLV-1 Cell-to-Cell transmission. *Viruses* 8:74. doi: 10.3390/v8030074

Gupta, S., Kumar, P., Kaur, H., Sharma, N., Saluja, D., Bharti, A. C., et al. (2015). Selective participation of c-Jun with Fra-2/c-Fos promotes aggressive tumor phenotypes and poor prognosis in tongue cancer. *Sci. Rep.* 5:16811. doi: 10.1038/srep16811

Hai, T., and Curran, T. (1991). Cross-family dimerization of transcription factors Fos/Jun and ATF/CREB alters DNA binding specificity. *Proc. Natl. Acad. Sci. U.S.A.* 88, 3720–3724. doi: 10.1073/pnas.88.9.3720

Hall, W. W., and Fujii, M. (2005). Deregulation of cell-signaling pathways in HTLV-1 infection. *Oncogene* 24, 5965–5975. doi: 10.1038/sj.onc.1208975

Ham, J., Eilers, A., Whitfield, J., Neame, S. J., and Shah, B. (2000). c-Jun and the transcriptional control of neuronal apoptosis. *Biochem. Pharmacol.* 60, 1015–1021. doi: 10.1016/S0006-2952(00)00372-5

Hasselblatt, P., Rath, M., Komnenovic, V., Zatloukal, K., and Wagner, E. F. (2007). Hepatocyte survival in acute hepatitis is due to c-Jun/AP-1-dependent expression of inducible nitric oxide synthase. *Proc. Natl. Acad. Sci. U.S.A.* 104, 17105–17110. doi: 10.1073/pnas.0706272104

Hernandez, J. M., Floyd, D. H., Weilbaecher, K. N., Green, P. L., and Boris-Lawrie, K. (2008). Multiple facets of junD gene expression are atypical among AP-1 family members. *Oncogene* 27, 4757–4767. doi: 10.1038/onc.2008.120

Hess, J., Angel, P., and Schorpp-Kistner, M. (2004). AP-1 subunits: quarrel and harmony among siblings. *J. Cell Sci.* 117, 5965–5973. doi: 10.1242/jcs.01589

Higuchi, T., Nakayama, T., Arao, T., Nishio, K., and Yoshie, O. (2013). SOX4 is a direct target gene of FRA-2 and induces expression of HDAC8 in adult T-cell leukemia/lymphoma. *Blood* 121, 3640–3649. doi: 10.1182/blood-2012-07-441022

Hirai, S. I., Ryseck, R. P., Mechta, F., Bravo, R., and Yaniv, M. (1989). Characterization of junD: a new member of the jun proto-oncogene family. *EMBO J.* 8, 1433–1439.

Hivin, P., Basbous, J., Raymond, F., Henaff, D., Arpin-Andre, C., Robert-Hebmann, V., et al. (2007). The HBZ-SP1 isoform of human T-cell leukemia virus type I represses JunB activity by sequestration into nuclear bodies. *Retrovirology* 4:14. doi: 10.1186/1742-4690-4-14

Hivin, P., Frederic, M., Arpin-Andre, C., Basbous, J., Gay, B., Thebault, S., et al. (2005). Nuclear localization of HTLV-I bZIP factor (HBZ) is mediated by three distinct motifs. *J. Cell Sci.* 118, 1355–1362. doi: 10.1242/jcs.01727

Isono, O., Ohshima, T., Saeki, Y., Matsumoto, J., Hijikata, M., Tanaka, K., et al. (2008). Human T-cell leukemia virus type 1 HBZ protein bypasses the targeting function of ubiquitination. *J. Biol. Chem.* 283, 34273–34282. doi: 10.1074/jbc.M802527200

Iwahashi, M., Nakahara, K., Takeshita, T., Shimotakahara, S., Uozumi, K., and Hanada, S. (1991). Surface phenotypic diversity of CD4 or CD8 in ATL. *Rinsho Ketsueki* 32, 142–146.

Iwai, K., Mori, N., Oie, M., Yamamoto, N., and Fujii, M. (2001). Human T-cell leukemia virus type 1 tax protein activates transcription through AP-1 site by inducing DNA binding activity in T cells. *Virology* 279, 38–46. doi: 10.1006/viro.2000.0669

Iwanaga, M., Watanabe, T., Utsunomiya, A., Okayama, A., Uchimaru, K., Koh, K. R., et al. (2010). Human T-cell leukemia virus type I (HTLV-1) proviral load and disease progression in asymptomatic HTLV-1 carriers: a nationwide prospective study in Japan. *Blood* 116, 1211–1219. doi: 10.1182/blood-2009-12-257410

Jeang, K. T. (2001). Functional activities of the human T-cell leukemia virus type I Tax oncoprotein: cellular signaling through NF-kappa B. *Cytokine Growth Factor Rev.* 12, 207–217. doi: 10.1016/S1359-6101(00)00028-9

Jeang, K. T., Chiu, R., Santos, E., and Kim, S. J. (1991). Induction of the HTLV-I LTR by Jun occurs through the Tax-responsive 21-bp elements. *Virology* 181, 218–227. doi: 10.1016/0042-6822(91)90487-V

Jeong, S. J., Pise-Masison, C. A., Radonovich, M. F., Park, H. U., and Brady, J. N. (2005). Activated AKT regulates NF-kappaB activation, p53 inhibition and cell survival in HTLV-1-transformed cells. *Oncogene* 24, 6719–6728. doi: 10.1038/sj.onc.1208825

Jin, D. Y., Teramoto, H., Giam, C. Z., Chun, R. F., Gutkind, J. S., and Jeang, K. T. (1997). A human suppressor of c-Jun N-terminal kinase 1 activation by tumor necrosis factor alpha. *J. Biol. Chem.* 272, 25816–25823. doi: 10.1074/jbc.272.41.25816

Jochum, W., Passegue, E., and Wagner, E. F. (2001). AP-1 in mouse development and tumorigenesis. *Oncogene* 20, 2401–2412. doi: 10.1038/sj.onc.1204389

Jones, K. S., Petrow-Sadowski, C., Huang, Y. K., Bertolette, D. C., and Ruscetti, F. W. (2008). Cell-free HTLV-1 infects dendritic cells leading to transmission and transformation of CD4(+) T cells. *Nat. Med.* 14, 429–436. doi: 10.1038/nm1745

Journo, C., Douceron, E., and Mahieux, R. (2009). HTLV gene regulation: because size matters, transcription is not enough. *Future Microbiol.* 4, 425–440. doi: 10.2217/fmb.09.13

Kajanne, R., Miettinen, P., Tenhunen, M., and Leppa, S. (2009). Transcription factor AP-1 promotes growth and radioresistance in prostate cancer cells. *Int. J. Oncol.* 35, 1175–1182.

Kannian, P., and Green, P. L. (2011). Human T lymphotropic virus type 1 (HTLV-1): molecular biology and oncogenesis. *Viruses* 2, 2037–2077. doi: 10.3390/v2092037

Karin, M. (1995). The regulation of AP-1 activity by mitogen-activated protein kinases. *J. Biol. Chem.* 270, 16483–16486. doi: 10.1074/jbc.270.28.16483

Karin, M., and Hunter, T. (1995). Transcriptional control by protein phosphorylation: signal transmission from the cell surface to the nucleus. *Curr. Biol.* 5, 747–757. doi: 10.1016/S0960-9822(95)00151-5

Karin, M., Liu, Z., and Zandi, E. (1997). AP-1 function and regulation. *Curr. Opin. Cell Biol.* 9, 240–246. doi: 10.1016/S0955-0674(97)80068-3

Kataoka, K., Nagata, Y., Kitanaka, A., Shiraishi, Y., Shimamura, T., Yasunaga, J., et al. (2015). Integrated molecular analysis of adult T cell leukemia/lymphoma. *Nat. Genet.* 47, 1304–1315. doi: 10.1038/ng.3415

Kharman-Biz, A., Gao, H., Ghiasvand, R., Zhao, C., Zendehdel, K., and Dahlman-Wright, K. (2013). Expression of activator protein-1 (AP-1) family members in breast cancer. *BMC Cancer* 13:441. doi: 10.1186/1471-2407-13-441

Kirkin, V., Joos, S., and Zörnig, M. (2004). The role of Bcl-2 family members in tumorigenesis. *Biochim. Biophys. Acta* 1644, 229–249. doi: 10.1016/j.bbamcr.2003.08.009

Kogure, Y., and Kataoka, K. (2017). Genetic alterations in adult T-cell leukemia/lymphoma. *Cancer Sci.* 108, 1719–1725. doi: 10.1111/cas.13303

Koul, D., Shen, R., Shishodia, S., Takada, Y., Bhat, K. P., Reddy, S. A., et al. (2007). PTEN down regulates AP-1 and targets c-fos in human glioma cells via PI3-kinase/Akt pathway. *Mol. Cell. Biochem.* 300, 77–87. doi: 10.1007/s11010-006-9371-8

Kuhlmann, A. S., Villaudy, J., Gazzolo, L., Castellazzi, M., Mesnard, J. M., and Duc Dodon, M. (2007). HTLV-1 HBZ cooperates with JunD to enhance transcription of the human telomerase reverse transcriptase gene (hTERT). *Retrovirology* 4:92. doi: 10.1186/1742-4690-4-92

Kyriakis, J. M. (1999). Activation of the AP-1 transcription factor by inflammatory cytokines of the TNF family. *Gene Expr.* 7, 217–231.

Langer, S., Singer, C. F., Hudelist, G., Dampier, B., Kaserer, K., Vinatzer, U., et al. (2006). Jun and Fos family protein expression in human breast cancer: correlation of protein expression and clinicopathological parameters. *Eur. J. Gynaecol. Oncol.* 27, 345–352.

Liou, J. T., Lin, C. S., Liao, Y. C., Ho, L. J., Yang, S. P., and Lai, J. H. (2017). JNK/AP-1 activation contributes to tetrandrine resistance in T-cell acute lymphoblastic leukaemia. *Acta Pharmacol. Sin.* 38, 1171–1183. doi: 10.1038/aps.2017.26

Lopez-Bergami, P., Lau, E., and Ronai, Z. E. (2010). Emerging roles of ATF2 and the dynamic AP1 network in cancer. *Nat. Rev. Cancer* 10, 65–76. doi: 10.1038/nrc2681

Lucibello, F. C., Brusselbach, S., Sewing, A., and Muller, R. (1993). Suppression of the growth factor-mediated induction of c-fos and down-modulation of AP-1-binding activity are not required for cellular senescence. *Oncogene* 8, 1667–1672.

Malorni, L., Giuliano, M., Migliaccio, I., Wang, T., Creighton, C. J., Lupien, M., et al. (2016). Blockade of AP-1 potentiates endocrine therapy and overcomes resistance. *Mol. Cancer Res.* 14, 470–481. doi: 10.1158/1541-7786.MCR-15-0423

Manna, S. K., Rangasamy, T., Wise, K., Sarkar, S., Shishodia, S., Biswal, S., et al. (2006). Long term environmental tobacco smoke activates nuclear transcription factor-kappa B, activator protein-1, and stress responsive kinases in mouse brain. *Biochem. Pharmacol.* 71, 1602–1609. doi: 10.1016/j.bcp.2006.02.014

Mao, X., Orchard, G., Russell-Jones, R., and Whittaker, S. (2007). Abnormal activator protein 1 transcription factor expression in CD30-positive cutaneous large-cell lymphomas. *Br. J. Dermatol.* 157, 914–921. doi: 10.1111/j.1365-2133.2007.08150.x

Matsumoto, J., Ohshima, T., Isono, O., and Shimotohno, K. (2005). HTLV-1 HBZ suppresses AP-1 activity by impairing both the DNA-binding ability and the stability of c-Jun protein. *Oncogene* 24, 1001–1010. doi: 10.1038/sj.onc.1208297

Matsuoka, M., and Green, P. L. (2009). The HBZ gene, a key player in HTLV-1 pathogenesis. *Retrovirology* 6:71. doi: 10.1186/1742-4690-6-71

Matsuoka, M., and Jeang, K. T. (2007). Human T-cell leukaemia virus type 1 (HTLV-1) infectivity and cellular transformation. *Nat. Rev. Cancer* 7, 270–280. doi: 10.1038/nrc2111

Matsuoka, M., and Jeang, K. T. (2010). Human T-cell leukemia virus type 1 (HTLV-1) and leukemic transformation: viral infectivity, Tax, HBZ and therapy. *Oncogene* 30, 1379–1389. doi: 10.1038/onc.2010.537

Matsuoka, M., Tamiya, S., Takemoto, S., Yamaguchi, K., and Takatsuki, K. (1997). HTLV-I provirus in the clinical subtypes of ATL. *Leukemia* 11(Suppl. 3), 67–69.

Meertens, L., Chevalier, S., Weil, R., Gessain, A., and Mahieux, R. (2004). A 10-amino acid domain within human T-cell leukemia virus type 1 and type 2 tax protein sequences is responsible for their divergent subcellular distribution. *J. Biol. Chem.* 279, 43307–43320. doi: 10.1074/jbc.M400497200

Melamed, A., Laydon, D. J., Al Khatib, H., Rowan, A. G., Taylor, G. P., and Bangham, C. R. (2015). HTLV-1 drives vigorous clonal expansion of infected CD8(+) T cells in natural infection. *Retrovirology* 12:91. doi: 10.1186/s12977-015-0221-1

Mendelson, K. G., Contois, L. R., Tevosian, S. G., Davis, R. J., and Paulson, K. E. (1996). Independent regulation of JNK/p38 mitogen-activated protein kinases by metabolic oxidative stress in the liver. *Proc. Natl. Acad. Sci. U.S.A.* 93, 12908–12913. doi: 10.1073/pnas.93.23.12908

Mesnard, J. M., Barbeau, B., and Devaux, C. (2006). HBZ, a new important player in the mystery of adult T-cell leukemia. *Blood* 108, 3979–3982. doi: 10.1182/blood-2006-03-007732

Milde-Langosch, K. (2005). The Fos family of transcription factors and their role in tumourigenesis. *Eur. J. Cancer* 41, 2449–2461. doi: 10.1016/j.ejca.2005.08.008

Milde-Langosch, K., Janke, S., Wagner, I., Schroder, C., Streichert, T., Bamberger, A. M., et al. (2008). Role of Fra-2 in breast cancer: influence on tumor cell invasion and motility. *Breast Cancer Res. Treat.* 107, 337–347. doi: 10.1007/s10549-007-9559-y

Murata, K., Hayashibara, T., Sugahara, K., Uemura, A., Yamaguchi, T., Harasawa, H., et al. (2006). A novel alternative splicing isoform of human T-cell leukemia virus type 1 bZIP factor (HBZ-SI) targets distinct subnuclear localization. *J. Virol.* 80, 2495–2505. doi: 10.1128/JVI.80.5.2495-2505.2006

Musti, A. M., Treier, M., Peverali, F. A., and Bohmann, D. (1996). Differential regulation of c-Jun and JunD by ubiquitin-dependent protein degradation. *Biol. Chem.* 377, 619–624. doi: 10.1515/bchm3.1996.377.10.619

Nakayama, T., Hieshima, K., Arao, T., Jin, Z., Nagakubo, D., Shirakawa, A. K., et al. (2008). Aberrant expression of Fra-2 promotes CCR4 expression and cell proliferation in adult T-cell leukemia. *Oncogene* 27, 3221–3232. doi: 10.1038/sj.onc.1210984

Nakayama, T., Higuchi, T., Oiso, N., Kawada, A., and Yoshie, O. (2012). Expression and function of FRA2/JUND in cutaneous T-cell lymphomas. *Anticancer Res.* 32, 1367–1373.

Neyns, B., Katesuwanasing Vermeij, J., Bourgain, C., Vandamme, B., and Amfo, K. (1996). Expression of the jun family of genes in human ovarian cancer and normal ovarian surface epithelium. *Oncogene* 12, 1247–1257.

Nishiyama, T., Yamamoto, H., Uchiumi, T., and Nakashima, N. (2007). Eukaryotic ribosomal protein RPS25 interacts with the conserved loop region in a dicistroviral intergenic internal ribosome entry site. *Nucleic Acids Res.* 35, 1514–1521. doi: 10.1093/nar/gkl1121

Offringa, R., Gebel, S., Van Dam, H., Timmers, M., Smits, A., Zwart, R., et al. (1990). A novel function of the transforming domain of E1a: repression of AP-1 activity. *Cell* 62, 527–538. doi: 10.1016/0092-8674(90)90017-9

Olindo, S., Lezin, A., Cabre, P., Merle, H., Saint-Vil, M., Edimonana Kaptue, M., et al. (2005). HTLV-1 proviral load in peripheral blood mononuclear cells quantified in 100 HAM/TSP patients: a marker of disease progression. *J. Neurol. Sci.* 237, 53–59. doi: 10.1016/j.jns.2005.05.010

Peloponese, J. M. Jr., and Jeang, K. T. (2006). Role for Akt/protein kinase B and activator protein-1 in cellular proliferation induced by the human T-cell leukemia virus type 1 tax oncoprotein. *J. Biol. Chem.* 281, 8927–8938. doi: 10.1074/jbc.M510598200

Peloponese, J. M. Jr., Kinjo, T., and Jeang, K. T. (2007). Human T-cell leukemia virus type 1 Tax and cellular transformation. *Int. J. Hematol.* 86, 101–106. doi: 10.1532/IJH97.07087

Peloponese, J. M., Yeung, M. L., and Jeang, K. T. (2006). Modulation of nuclear factor-kappaB by human T cell leukemia virus type 1 Tax protein: implications for oncogenesis and inflammation. *Immunol. Res.* 34, 1–12. doi: 10.1385/IR:34:1:1

Pique, C., and Jones, K. S. (2012). Pathways of cell-cell transmission of HTLV-1. *Front. Microbiol.* 3:378. doi: 10.3389/fmicb.2012.00378

Punga, T., Bengoechea-Alonso, M. T., and Ericsson, J. (2006). Phosphorylation and ubiquitination of the transcription factor sterol regulatory element-binding protein-1 in response to DNA binding. *J. Biol. Chem.* 281, 25278–25286. doi: 10.1074/jbc.M604983200

Radler-Pohl, A., Gebel, S., Sachsenmaier, C., Konig, H., Kramer, M., Oehler, T., et al. (1993). The activation and activity control of AP-1 (fos/jun). *Ann. N. Y. Acad. Sci.* 684, 127–148. doi: 10.1111/j.1749-6632.1993.tb32277.x

Richardson, J. H., Edwards, A. J., Cruickshank, J. K., Rudge, P., and Dalgleish, A. G. (1990). In vivo cellular tropism of human T-cell leukemia virus type 1. *J. Virol.* 64, 5682–5687.

Rizkallah, G., Alais, S., Futsch, N., Tanaka, Y., Journo, C., Mahieux, R., et al. (2017). Dendritic cell maturation, but not type I interferon exposure, restricts infection by HTLV-1, and viral transmission to T-cells. *PLOS Pathog.* 13:e1006353. doi: 10.1371/journal.ppat.1006353

Rosin, O., Koch, C., Schmitt, I., Semmes, O. J., Jeang, K. T., and Grassmann, R. (1998). A human T-cell leukemia virus Tax variant incapable of activating NF-kappaB retains its immortalizing potential for primary T-lymphocytes. *J. Biol. Chem.* 273, 6698–6703. doi: 10.1074/jbc.273.12.6698

Ryder, K., and Nathans, D. (1988). Induction of protooncogene c-jun by serum growth factors. *Proc. Natl. Acad. Sci. U.S.A.* 85, 8464–8467. doi: 10.1073/pnas.85.22.8464

Sadowski, H. B., Shuai, K., Darnell, J. E. Jr., and Gilman, M. Z. (1993). A common nuclear signal transduction pathway activated by growth factor and cytokine receptors. *Science* 261, 1739–1744. doi: 10.1126/science.8397445

Segal-Bendirdjian, E., and Gilson, E. (2008). Telomeres and telomerase: from basic research to clinical applications. *Biochimie* 90, 1–4. doi: 10.1016/j.biochi.2007.10.003

Shaulian, E. (2010). AP-1–The Jun proteins: Oncogenes or tumor suppressors in disguise? *Cell Signal* 22, 894–899. doi: 10.1016/j.cellsig.2009.12.008

Shaulian, E., and Karin, M. (2002). AP-1 as a regulator of cell life and death. *Nat. Cell Biol.* 4, E131–E136. doi: 10.1038/ncb0502-e131

Shay, J. W., and Wright, W. E. (2011). Role of telomeres and telomerase in cancer. *Semin. Cancer Biol.* 21, 349–353. doi: 10.1016/j.semcancer.2011.10.001

Short, J. D., and Pfarr, C. M. (2002). Translational regulation of the JunD messenger RNA. *J. Biol. Chem.* 277, 32697–32705. doi: 10.1074/jbc.M204553200

Sibon, D., Gabet, A. S., Zandecki, M., Pinatel, C., Thete, J., Delfau-Larue, M. H., et al. (2006). HTLV-1 propels untransformed CD4 lymphocytes into the cell cycle while protecting CD8 cells from death. *J. Clin. Invest.* 116, 974–983. doi: 10.1172/JCI27198

Sonoda, S., Li, H. C., and Tajima, K. (2011). Ethnoepidemiology of HTLV-1 related diseases: ethnic determinants of HTLV-1 susceptibility and its worldwide dispersal. *Cancer Sci.* 102, 295–301. doi: 10.1111/j.1349-7006.2010.01820.x

Srivastava, S., Weitzmann, M. N., Cenci, S., Ross, F. P., Adler, S., and Pacifici, R. (1999). Estrogen decreases TNF gene expression by blocking JNK activity and the resulting production of c-Jun and JunD. *J. Clin. Invest.* 104, 503–513. doi: 10.1172/JCI7094

Tabakin-Fix, Y., Azran, I., Schavinky-Khrapunsky, Y., Levy, O., and Aboud, M. (2006). Functional inactivation of p53 by human T-cell leukemia virus type 1 Tax protein: mechanisms and clinical implications. *Carcinogenesis* 27, 673–681. doi: 10.1093/carcin/bgi274

Tamiya, S., Matsuoka, M., Etoh, K., Watanabe, T., Kamihira, S., Yamaguchi, K., et al. (1996). Two types of defective human T-lymphotropic virus type I provirus in adult T-cell leukemia. *Blood* 88, 3065–3073.

Terol, M., Gazon, H., Lemasson, I., Duc-Dodon, M., Barbeau, B., Cesaire, R., et al. (2017). HBZ-mediated shift of JunD from growth suppressor to tumor promoter in leukemic cells by inhibition of ribosomal protein S25 expression. *Leukemia* 31, 2235–2243. doi: 10.1038/leu.2017.74

Thebault, S., Basbous, J., Hivin, P., Devaux, C., and Mesnard, J. M. (2004). HBZ interacts with JunD and stimulates its transcriptional activity. *FEBS Lett.* 562, 165–170. doi: 10.1016/S0014-5793(04)00225-X

Tulchinsky, E. (2000). Fos family members: regulation, structure and role in oncogenic transformation. *Histol. Histopathol.* 15, 921–928.

Uchida, N., Otsuka, T., Arima, F., Shigematsu, H., Fukuyama, T., Maeda, M., et al. (1999). Correlation of telomerase activity with development and progression of adult T-cell leukemia. *Leuk. Res.* 23, 311–316. doi: 10.1016/S0145-2126(98)00170-2

Van Dam, H., Duyndam, M., Rottier, R., Bosch, A., De Vries-Smits, L., Herrlich, P., et al. (1993). Heterodimer formation of cJun and ATF-2 is responsible for induction of c-jun by the 243 amino acid adenovirus E1A protein. *EMBO J.* 12, 479–487.

Van Orden, K., and Nyborg, J. K. (2000). Insight into the tumor suppressor function of CBP through the viral oncoprotein tax. *Gene Expr.* 9, 29–36. doi: 10.3727/000000001783992678

Van Orden, K., Yan, J. P., Ulloa, A., and Nyborg, J. K. (1999). Binding of the human T-cell leukemia virus Tax protein to the coactivator CBP interferes with CBP-mediated transcriptional control. *Oncogene* 18, 3766–3772. doi: 10.1038/sj.onc.1202703

Venkatesan, B., Valente, A. J., Prabhu, S. D., Shanmugam, P., Delafontaine, P., and Chandrasekar, B. (2010). EMMPRIN activates multiple transcription factors in cardiomyocytes, and induces interleukin-18 expression via Rac1-dependent PI3K/Akt/IKK/NF-kappaB andMKK7/JNK/AP-1 signaling. *J. Mol. Cell Cardiol.* 49, 655–663. doi: 10.1016/j.yjmcc.2010.05.007

Verma, I. M., Ransone, L. J., Visvader, J., Sassone-Corsi, P., and Lamph, W. W. (1990). fos-jun conspiracy: implications for the cell. *Ciba Found. Symp.* 150, 128–137;discussion 137–146.

Wagner, B. J., Hayes, T. E., Hoban, C. J., and Cochran, B. H. (1990). The SIF binding element confers sis/PDGF inducibility onto the c-fos promoter. *EMBO J.* 9, 4477–4484.

Waldmann, T. A. (1996). The promiscuous IL-2/IL-15 receptor: a target for immunotherapy of HTLV-I-associated disorders. *J. Acquir. Immune Defic. Syndr. Hum. Retrovirol.* 13(Suppl. 1), S179–S185. doi: 10.1097/00042560-199600001-00027

Wang, J., Sun, D., Wang, Y., Ren, F., Pang, S., Wang, D., et al. (2014). FOSL2 positively regulates TGF-beta1 signalling in non-small cell lung cancer. *PLOS ONE* 9:e112150. doi: 10.1371/journal.pone.0112150

Warfel, N. A., and Kraft, A. S. (2015). PIM kinase (and Akt) biology and signaling in tumors. *Pharmacol. Ther.* 151, 41–49. doi: 10.1016/j.pharmthera.2015.03.001

Wethmar, K., Smink, J. J., and Leutz, A. (2010). Upstream open reading frames: molecular switches in (patho)physiology. *Bioessays* 32, 885–893. doi: 10.1002/bies.201000037

Whitmarsh, A. J., and Davis, R. J. (1996). Transcription factor AP-1 regulation by mitogen-activated protein kinase signal transduction pathways. *J. Mol. Med. (Berl)* 74, 589–607. doi: 10.1007/s001090050063

Woodgett, J. R., Pulverer, B. J., Nikolakaki, E., Plyte, S., Hughes, K., Franklin, C. C., et al. (1993). Regulation of jun/AP-1 oncoproteins by protein phosphorylation. *Adv. Second Messenger Phosphoprotein Res.* 28, 261–269.

Xu, X., Heidenreich, O., Kitajima, I., Mcguire, K., Li, Q., Su, B., et al. (1996). Constitutively activated JNK is associated with HTLV-1 mediated tumorigenesis. *Oncogene* 13, 135–142.

Yamada, Y., Ohmoto, Y., Hata, T., Yamamura, M., Murata, K., Tsukasaki, K., et al. (1996). Features of the cytokines secreted by adult T cell leukemia (ATL) cells. *Leuk. Lymphoma* 21, 443–447. doi: 10.3109/10428199609093442

Yazgan, O., and Pfarr, C. M. (2001). Differential binding of the Menin tumor suppressor protein to JunD isoforms. *Cancer Res.* 61, 916–920.

Yazgan, O., and Pfarr, C. M. (2002). Regulation of two JunD isoforms by Jun N-terminal kinases. *J. Biol. Chem.* 277, 29710–29718. doi: 10.1074/jbc.M204552200

Yoshida, M. (2010). Molecular approach to human leukemia: isolation and characterization of the first human retrovirus HTLV-1 and its impact on tumorigenesis in adult T-cell leukemia. *Proc. Jpn. Acad. Ser. B Phys. Biol. Sci.* 86, 117–130. doi: 10.2183/pjab.86.117

Young, M. R., and Colburn, N. H. (2006). Fra-1 a target for cancer prevention or intervention. *Gene* 379, 1–11. doi: 10.1016/j.gene.2006.05.001

Zerbini, L. F., Wang, Y., Cho, J. Y., and Libermann, T. A. (2003). Constitutive activation of nuclear factor kappaB p50//p65 and Fra-1 and JunD is essential for deregulated interleukin 6 expression in prostate cancer. *Cancer Res.* 63, 2206–2215.

Zhao, T., Satou, Y., Sugata, K., Miyazato, P., Green, P. L., Imamura, T., et al. (2011). HTLV-1 bZIP factor enhances TGF-beta signaling through p300 coactivator. *Blood* 118, 1865–1876. doi: 10.1182/blood-2010-12-326199

Zhou, H., Gao, J., Lu, Z. Y., Lu, L., Dai, W., and Xu, M. (2007). Role of c-Fos//JunD in protecting stress-induced cell death. *Cell Prolif.* 40, 431–444. doi: 10.1111/j.1365-2184.2007.00444.x

Zhou, W., and Song, W. (2006). Leaky scanning and reinitiation regulate BACE1 gene expression. *Mol. Cell. Biol.* 26, 3353–3364. doi: 10.1128/MCB.26.9.3353-3364.2006

13

Time to Go Back to the Original Name

*Robert C. Gallo [1,2], Luc Willems [2,3] and Yutaka Tagaya [1,2]**

[1] Division of Basic Science, Institute of Human Virology, University of Maryland School of Medicine, Baltimore, MD, United States, [2] HTLV-1 Task Force, Global Virus Network, Baltimore, MD, United States, [3] Research Director, National Fund for Scientific Research at University of Liège, Liège, Belgium

Keywords: HTLV-1, T-cell leukemia, myelopathy, oncogenic viruses, retrovirus

***Correspondence:**
Yutaka Tagaya
ytagaya@ihv.umaryland.edu

Almost 40 years ago, the first pathogenic human retrovirus was discovered and shown to cause a T-cell leukemia in humans (Poiesz et al., 1980a, 1981). Thus, a new paradigm was established showing that a virus can directly cause cancer in humans. At that time, the only other suspected case was the connection between the Epstein-Barr Virus (EBV) and Burkitt's Lymphoma and later nasopharyngeal carcinoma. Naturally, and in keeping with the names used for animal retroviruses associated with leukemia, we (the Gallo group) named this new virus as human T-cell leukemia virus (HTLV), which was formally accepted by the scientific community. As of today, our list of human oncogenic viruses has expanded to include hepatitis B/C virus (HBV, HCV), papillomaviruses (HPV), Herpes virus-8 (HHV-8), and Merkel cell polyomavirus (MCV). In parallel, the group of Delta-retroviruses that HTLV belongs to has expanded to accommodate additional members over the years, now consisting of HTLV-1~4, and STLV-1~4. In the meantime, the name of the original virus was changed to human T-lymphotropic virus-1, because of the addition of the viral causative agent of AIDS as HTLV-III to the same group (Coffin, 2015). Thus, one of us (RCG) agreed on this name change with several other retrovirologists at a Cold-Spring Harbor meeting in 1983. In retrospect, the name change has made it ambiguous as to which virus should enter this group, and since the AIDS virus was formally named HIV and separated from the HTLV group, making the need for the name of HTLV-1 as human T-"lymphotropic" virus no longer particularly useful or meaningful. Finally, we note here the greater oncogenicity of HTLV-1 compared to other viruses, or even other known carcinogens which is a point in keeping a name that describes their effect (discussed in an accompanying article). Therefore, we propose to restore the original name "leukemia virus."

LYMPHOTROPIC VS. LEUKEMIA; A HISTORICAL BACKGROUND

The original names for HTLV-1 came from two groups in association with an unusual type of human T-cell leukemia (adult T-cell leukemia) (Poiesz et al., 1980a,b; Hinuma et al., 1981). At first two names existed; "Human T-cell leukemia virus (-I)" and "Adult T-cell leukemia virus." A study in 1982 confirmed the identical nature of these two viruses (Popovic et al., 1982) and a meeting was held to unify "names." US and Japanese researchers agreed on acknowledging the US priority for the discovery of the virus and the Japanese priority in defining the very particular type of leukemia (Yodoi et al., 1974; Uchiyama et al., 1977). Hence, the virus adopted the name "Human T-cell leukemia" and the disease "adult T-cell leukemia." Around this time, a genetically related (70% structural similarity to HTLV-1) virus was identified in association with a CD8 T-cell variant of a hairy cell leukemia (Kalyanaraman et al., 1982) and this virus was added to the group as HTLV-II. A turn of event followed shortly that another virus originally thought to be closely related to the HTLV group was discovered in connection with AIDS (Popovic et al., 1984), a virus now known as HIV-1. HIV-1 was originally regarded as another member of HTLVs and added to the group as HTLV-III (Gallo et al., 1983). As mentioned above, these events led to the change of the naming from "Leukemia" to "Lymphotropic" (Coffin, 2015). The following years have seen explosive research on HIV-1 and soon HTLV-III was shown to be more substantially different from HTLV-I/II. Eventually a new name, HIV-1, was given to HTLV-III and HIV-1 was taxonomically separated from HTLVs. It has become also clear that HTLV-1 does not only infect T cells (Koyanagi et al., 1993). Thus, the first name change has lost its ground and perhaps the nomenclature should have been reversed to the original at much earlier point.

HTLV-1 AS A POTENT CARCINOGEN

HTLVs cause not only leukemia, but also HAM/TSP, a progressive and incurable myelopathy (Nagai and Osame, 2003). Disease manifestations caused by HTLV-1 also include inflammatory arthropathy and uveitis. Because of this diverse clinical impact, there have been arguments even in the HTLV-1 community if the "leukemia" name is the most appropriate. We note several points; (1) Precedent throughout animal retroviruses and notably these animal retroviruses also can cause non-neoplastic disorders as well as leukemias/lymphomas, (2) Precedent as the first formal name for HTLV, (3) the fact that leukemia is the most frequent severe outcome of HTLV-1 infection, and (4) the fact that viruses commonly cause more than one disease as noted above. There are some suggestions that the name "leukemia" might be overly intimidating, given that over 80% of the HTLV-1-infected individuals remain disease-free for their lifetime. We are of course aware of the challenges imposed on infected people and clinicians who are in charge of caring the infected individuals. However, we note that HTLV-1 is one, if not the most, potent oncogenic virus and therefore the name "leukemia virus" has scientific significance that outweighs other concerns. We have recently seen reports involving the Japanese Ministry of Health and Labor which declared to target HTLV-1 as one of the major focuses in 2010 that they revised the prevalence of HAM/TSP in Japan down to 0.3% whereas that of ATL remains 4–5% as proposed before, which suggests that ATL indeed represents more than 80% of the disease manifestations caused by HTLV-1 in Japan. Moreover, this argument flies in the face of names for most other disease-causing agents (e.g., influenza, polio, hepatitis, human immunodeficiency virus, Kaposi's sarcoma herpes virus, tuberculosis, etc.). Finally, in an age when patients have easy access to information, the point seems moot. It is of note that unlike animal oncoviruses, not necessarily all human known oncoviruses include the direct mention to cancer. Merkel cell carcinoma virus and Kaposi-sarcoma herpes virus (HHV-8) have direct words in their names. Hepatitis B and C viruses were identified as viruses causing hepatitis and later linked to carcinogenesis. Epstein-Barr virus and Papilloma viruses (HPV) were isolated from cancer cell/tissues, but took a long time to establish the link to cancer. The HTLV-1 represents a unique exception. The peculiar geographical distribution of adult-T cell leukemia (Yodoi et al., 1974; Uchiyama et al., 1977) promoted researchers to suspect that a virus might be the cause of it. Moreover, there are many precedent connecting retroviruses and leukemogenesis in animals (hence many animal retroviruses cause leukemia carry the name "leukemic" in their names). The discovery of HTLV-1 was thus a consequence of "hunting for a leukemia-causing virus." Moreover, the causative link of the discovered virus to ATL was quickly established within a few years after the discovery. This makes a pivotal success of modern virology and therefore, we feel compelled to commemorate the relevance of these events by renaming HTLV-1 as the original "leukemia virus."

The oncogenic association of HTLV-1 to ATL is unquestionable. Once diagnosed as an acute phase or lymphomatous phase of ATL, the patient has less than 12 months of life left, indicating the aggressive nature of the oncogenic property of HTLV-1 as well as the lack of effective treatment to this fatal disease caused by HTLV-1, though there are promising treatments emerging. In comparison, the oncogenic capacity were not as convincing for other members of the HTLV family. However, the original identification of HTLV-2, which is not highly oncogenic, came from a CD8 T-cell variant of Hairy Leukemia cells. STLV-associated leukemia among monkeys has been reported (Miura et al., 2013). There is no concrete evidence to strongly link HTLV-2, -3, and -4 with particular human malignant diseases. Indeed some of the earlier epidemiology studies failed to show strong correlation between HTLV-2 and the T-cell type hairy cell leukemia (Quesada et al., 1986; Rosenblatt et al., 1988). HTLV-2 encodes Tax-2 gene which is structurally highly similar to the oncogenic Tax-1 gene of HTLV-1. However, recent studies suggest that another gene called HBZ (HTLV-1 bZIP factor) which is encoded by the minus-strand of the HTLV-1 can cause an ATL-like CD4 leukemia in mice (Satou et al., 2011). Interestingly, HTLV-2 encodes a similar anti-sense gene (Anti-sense viral protein of HTLV-2, APH-2), but the structure and function of APH-2 seems very different and this may account for the differences of oncogenic properties of HTLV-1 and -2 (Panfil et al., 2016). However, an alternative to the above explanation might be connected to the tropism of these two viruses. HTLV-1 is clearly CD4 T-cell tropic whereas HTLV-2 is mainly detected in CD8 T cells (Ijichi et al., 1992; Xie and Green, 2005; Jones et al., 2006). HTLV-1 does not transform CD8 T cells though it infects CD8 T cells as well at the early phase of infection. This may reflect the fundamental differences of intracellular pathways required for cellular transformation in CD4 and CD8 T cells, rather than the differences between HTLV-1 and -2. Consistently, transgenic experiments over-expressing so-called T-cell growth cytokines such as IL-15 and IL-7 results in the development of T-cell leukemia in these mice but that is only limited to CD8 (and NK) lineages (Rich et al., 1993; Fehniger et al., 2001; Sato et al., 2011). It is of note that CD4 T cells in these mice show robust DNA synthesis (Marks-Konczalik et al., 2000) but never become leukemic. Conversely, we only see leukemia of CD4 T cells by HTLV-1. It is likely that that mechanism by which HTLVs transform host cells works efficiently in CD4 T cells, but not in CD8 T cells. If this were true, HTLV-2 can be leukemic, but not very efficient in CD8 T cells. This view contrasts with the former argument which regards HTLV-2 "non-oncogenic because of the lack of molecular machinery." In fact, recent studies demonstrate that Tax-2 of HTLV-2 transforms T cells as efficient as does Tax-1 *in vitro* (Ren and Cheng, 2013). Another recent publication demonstrates that HTLV-2 facilitates clonal expansion of CD4 or CD8 T cells in infected individuals. The same study showed that the resultant clone frequency resembles that in established ATL than the frequency in asymptomatic carriers of HTLV-1 (Melamed et al., 2014). Needless to say, clonal expansion is often associated with cancers. These pieces of argument may not provide a strong justification to call HTLV-2 an oncogenic virus, but surely suggest that the issue is still pending. We also need more data about HTLV-3 and 4 to determine their pathogenicity. In conclusion, we only limit the renaming proposal to one member, the HTLV-1, leaving other members for a final decision in the future.

In an accompanying manuscript (Tagaya and Gallo, 2017), we also discuss an intriguing finding from a survey of recent literature that HTLV-1 is the most oncogenic virus known. We therefore propose that the name of this virus should clearly represent the concept by including the word "leukemia" in it. We recently conducted a survey among the HTLV-1 experts who belong to the HTLV-1 task-force of the Global Virus Network (http://gvn.org). GVN represents a coalition of virologists on a global scale, and its HTLV-1 task-force includes major HTLV-1 researchers from all-over the world (member roster shown in Willems et al., 2017). A vote was made on the following four choices; (1) to keep the lymphotropic virus as its name (2) to re-adopt the leukemia virus name, (3) to change it to Primate T-lymphotropic virus-1, (4) to change it to Primate T-cell leukemia virus-1, and the result was 5 for (1), and 16 for (2) (none for 3 and 4). The vast majority therefore supported readopting the original name "Human T-cell leukemia virus-1." In addition, a similar poll conducted at the 18th International Conference on Human Retrovirology (Tokyo, 2017) showed 78 vs. 26 in support of the "leukemia" name. We consider these votes compelling because the former (the GVN vote) reflects the majority opinion of the HTLV experts in the research community and the latter (Tokyo, 2017) represents the community opinion of many HTLV-1 researchers from all over the world.

We thus propose the renaming of HTLV-1 as "human T-cell leukemia virus-1" to commemorate an illustrious event in the history of modern virology.

AUTHOR CONTRIBUTIONS

All three authors conceptualized and discussed the idea and accordingly LW and RG conducted the first poll among GVN's HTLV-1 task force members. Then LW organized a poll at the 18th International Retrovirology meeting in Tokyo (2017). The three wrote the manuscript together.

ACKNOWLEDGMENTS

The authors thank other members of the GVN (Global Virus Network) HTLV-1 taskforce for supporting this work. Also special thanks for the participants of the 18th International Conference on Human Retrovirology to have enabled the poll on this topic.

REFERENCES

Coffin, J. M. (2015). The discovery of HTLV-1, the first pathogenic human retrovirus. *Proc. Natl. Acad. Sci. U.S.A.* 112, 15525–15529. doi: 10.1073/pnas.1521629112

Fehniger, T. A., Suzuki, K., VanDeusen, J. B., Cooper, M. A., Freud, A. G., and Caligiuri, M. A. (2001). Fatal leukemia in interleukin-15 transgenic mice. *Blood Cells Mol. Dis.* 27, 223–230. doi: 10.1006/bcmd.2001.0379

Gallo, R. C., Sarin, P. S., Gelmann, E. P., Robert-Guroff, M., Richardson, E., Kalyanaraman, V. S., et al. (1983). Isolation of human T-cell leukemia virus in acquired immune deficiency syndrome (AIDS). *Science* 220, 865–867. doi: 10.1126/science.6601823

Hinuma, Y., Nagata, K., Hanaoka, M., Nakai, M., Matsumoto, T., Kinoshita, K. I., et al. (1981). Adult T-cell leukemia: antigen in an ATL cell line and detection of antibodies to the antigen in human sera. *Proc. Natl. Acad. Sci. U.S.A.* 78, 6476–6480. doi: 10.1073/pnas.78.10.6476

Ijichi, S., Ramundo, M. B., Takahashi, H., and Hall, W. W. (1992). *In vivo* cellular tropism of human T cell leukemia virus type II (HTLV-II). *J. Exp. Med.* 176, 293–296. doi: 10.1084/jem.176.1.293

Jones, K. S., Fugo, K., Petrow-Sadowski, C., Huang, Y., Bertolette, D. C., Lisinski, I., et al. (2006). Human T-cell leukemia virus type 1 (HTLV-1) and HTLV-2 use different receptor complexes to enter T cells. *J. Virol.* 80, 8291–8302. doi: 10.1128/JVI.00389-06

Kalyanaraman, V. S., Sarngadharan, M. G., Robert-Guroff, M., Miyoshi, I., Golde, D., and Gallo, R. C. (1982). A new subtype of human T-cell leukemia virus (HTLV-II) associated with a T-cell variant of hairy cell leukemia. *Science* 218, 571–573. doi: 10.1126/science.6981847

Koyanagi, Y., Itoyama, Y., Nakamura, N., Takamatsu, K., Kira, J., Iwamasa, T., et al. (1993). *In vivo* infection of human T-cell leukemia virus type I in non-T cells. *Virology* 196, 25–33. doi: 10.1006/viro.1993.1451

Marks-Konczalik, J., Dubois, S., Losi, J. M., Sabzevari, H., Yamada, N., Feigenbaum, L., et al. (2000). IL-2-induced activation-induced cell death is inhibited in IL-15 transgenic mice. *Proc. Natl. Acad. Sci. U.S.A.* 97, 11445–11450. doi: 10.1073/pnas.200363097

Melamed, A., Witkover, A. D., Laydon, D. J., Brown, R., Ladell, K., Miners, K., et al. (2014). Clonality of HTLV-2 in natural infection. *PLoS Pathog.* 10:e1004006. doi: 10.1371/journal.ppat.1004006

Miura, M., Yasunaga, J., Tanabe, J., Sugata, K., Zhao, T., Ma, G., et al. (2013). Characterization of simian T-cell leukemia virus type 1 in naturally infected Japanese macaques as a model of HTLV-1 infection. *Retrovirology* 10:118. doi: 10.1186/1742-4690-10-118

Nagai, M., and Osame, M. (2003). Human T-cell lymphotropic virus type I and neurological diseases. *J. Neurovirol.* 9, 228–235. doi: 10.1080/13550280390194028

Panfil, A. R., Dissinger, N. J., Howard, C. M., Murphy, B. M., Landes, K., Fernandez, S. A., et al. (2016). Functional comparison of HBZ and the related APH-2 protein provides insight into human T-cell leukemia virus type 1 pathogenesis. *J. Virol.* 90, 3760–3772. doi: 10.1128/JVI.03113-15

Poiesz, B. J., Ruscetti, F. W., Gazdar, A. F., Bunn, P. A., Minna, J. D., and Gallo, R. C. (1980a). Detection and isolation of type C retrovirus particles from fresh and cultured lymphocytes of a patient with cutaneous T-cell lymphoma. *Proc. Natl. Acad. Sci. U.S.A.* 77, 7415–7419. doi: 10.1073/pnas.77.12.7415

Poiesz, B. J., Ruscetti, F. W., Mier, J. W., Woods, A. M., and Gallo, R. C. (1980b). T-cell lines established from human T-lymphocytic neoplasias by direct response to T-cell growth factor. *Proc. Natl. Acad. Sci. U.S.A.* 77, 6815–6819. doi: 10.1073/pnas.77.11.6815

Poiesz, B. J., Ruscetti, F. W., Reitz, M. S., Kalyanaraman, V. S., and Gallo, R. C. (1981). Isolation of a new type C retrovirus (HTLV) in primary uncultured cells of a patient with Sezary T-cell leukaemia. *Nature* 294, 268–271. doi: 10.1038/294268a0

Popovic, M., Reitz, M. S. Jr., Sarngadharan, M. G., Robert-Guroff, M., Kalyanaraman, V. S., Nakao, Y., et al. (1982). The virus of Japanese adult T-cell leukaemia is a member of the human T-cell leukaemia virus group. *Nature* 300, 63–66. doi: 10.1038/300063a0

Popovic, M., Sarngadharan, M. G., Read, E., and Gallo, R. C. (1984). Detection, isolation, and continuous production of cytopathic retroviruses (HTLV-III) from patients with AIDS and pre-AIDS. *Science* 224, 497–500. doi: 10.1126/science.6200935

Quesada, J. R., Reuben, J., Hopfer, R. L., Mundon, F. K., and Hersh, E. M. (1986). Serologic studies in hairy cell leukemia: high prevalence of Epstein-Barr and cytomegalovirus antibodies and absence of human T-cell lymphotrophic viruses antibodies. *Leuk. Res.* 10, 1169–1173. doi: 10.1016/0145-2126(86)90234-1

Ren, T., and Cheng, H. (2013). Differential transforming activity of the retroviral Tax oncoproteins in human T lymphocytes. *Front. Microbiol.* 4:287. doi: 10.3389/fmicb.2013.00287

Rich, B. E., Campos-Torres, J., Tepper, R. I., Moreadith, R. W., and Leder, P. (1993). Cutaneous lymphoproliferation and lymphomas in interleukin 7 transgenic mice. *J. Exp. Med.* 177, 305–316. doi: 10.1084/jem.177.2.305

Rosenblatt, J. D., Chen, I. S., and Golde, D. W. (1988). HTLV-II and human lymphoproliferative disorders. *Clin. Lab. Med.* 8, 85–95.

Sato, N., Sabzevari, H., Fu, S., Ju, W., Petrus, M. N., Bamford, R. N., et al. (2011). Development of an IL-15-autocrine CD8 T-cell leukemia in IL-15-transgenic mice requires the cis expression of IL-15Ralpha. *Blood* 117, 4032–4040. doi: 10.1182/blood-2010-09-307504

Satou, Y., Yasunaga, J., Zhao, T., Yoshida, M., Miyazato, P., Takai, K., et al. (2011). HTLV-1 bZIP factor induces T-cell lymphoma and systemic inflammation *in vivo*. *PLoS Pathog.* 7:e1001274. doi: 10.1371/journal.ppat.1001274

Tagaya, Y., and Gallo, R. C. (2017). The exceptional oncogenicity of HTLV-1. *Front. Microbiol.* 8:1425. doi: 10.3389/fmicb.2017.01425

Uchiyama, T., Yodoi, J., Sagawa, K., Takatsuki, K., and Uchino, H. (1977). Adult T-cell leukemia: clinical and hematologic features of 16 cases. *Blood* 50, 481–492.

Willems, L., Hasegawa, H., Accolla, R., Bangham, C., Bazarbachi, A., Bertazzoni, U., et al. (2017). Reducing the global burden of HTLV-1 infection: an agenda for research and action. *Antiviral Res.* 137, 41–48. doi: 10.1016/j.antiviral.2016.10.015

Xie, L., and Green, P. L. (2005). Envelope is a major viral determinant of the distinct *in vitro* cellular transformation tropism of human T-cell leukemia virus type 1 (HTLV-1) and HTLV-2. *J. Virol.* 79, 14536–14545. doi: 10.1128/JVI.79.23.14536-14545.2005

Yodoi, J., Takatsuki, K., and Masuda, T. (1974). Letter: two cases of T-cell chronic lymphocytic leukemia in Japan. *N. Engl. J. Med.* 290, 572–573. doi: 10.1056/NEJM197403072901019

14

Mitochondrial Proteins Coded by Human Tumor Viruses

Ilaria Cavallari [1†], Gloria Scattolin [2†], Micol Silic-Benussi [1], Vittoria Raimondi [1], Donna M. D'Agostino [3] and Vincenzo Ciminale [1,2*]*

[1] Veneto Institute of Oncology IOV-IRRCS, Padova, Italy, [2] Department of Surgery, Oncology, and Gastroenterology, University of Padova, Padova, Italy, [3] Department of Biomedical Sciences, University of Padova, Padova, Italy

***Correspondence:**
Donna M. D'Agostino
dm.dagostino@unipd.it
Vincenzo Ciminale
v.ciminale@unipd.it

† These authors have contributed equally to this work.

Viruses must exploit the cellular biosynthetic machinery and evade cellular defense systems to complete their life cycles. Due to their crucial roles in cellular bioenergetics, apoptosis, innate immunity and redox balance, mitochondria are important functional targets of many viruses, including tumor viruses. The present review describes the interactions between mitochondria and proteins coded by the human tumor viruses human T-cell leukemia virus type 1, Epstein-Barr virus, Kaposi's sarcoma-associated herpesvirus, human hepatitis viruses B and C, and human papillomavirus, and highlights how these interactions contribute to viral replication, persistence and transformation.

Keywords: Mitochondria, EBV, HTLV-1, HPV, HBV, HCV, KSHV

INTRODUCTION

It is estimated that about 20% of human cancer cases world-wide are caused by infection with bacteria, parasites, or viruses (de Martel et al., 2012; IARC Monograph, 2012). Among these infectious agents, viruses have the greatest reliance on cellular pathways for completion of their life cycles, and have evolved complex mechanisms to manipulate the anabolic and proliferative capacity of the host cell while minimizing effects on cell death and destruction by the immune system.

Viruses that are linked to human cancer include human-T-cell leukemia virus type 1 (HTLV-1), Epstein-Barr virus (EBV), Kaposi's sarcoma-associated herpesvirus (KSHV), hepatitis viruses B and C (HBV and HCV, respectively), high-risk genotypes of human papillomavirus (HPV, e.g., HPV-16, HPV-18), and Merkel cell polyomavirus (MCPyV). In addition, the AIDS-causing retrovirus human immunodeficiency virus is indirectly linked to cancer through its immunosuppressive effects, which favor transformation by other tumor viruses, especially EBV and KSHV (IARC Monograph, 2012).

Many of the strategies used by viruses to replicate and persist in host cells intersect at the level of mitochondria (reviewed by Claus and Liebert, 2014). This is not surprising, given the range of essential roles of these organelles in multiple cellular processes. (i) Energy supply: as the sites of pyruvate- and fatty acid oxidation, the citric acid cycle and the electron transport chain, mitochondria are the main source of ATP in most differentiated eukaryotic cells. (ii) Ca^{2+} signaling: through their ability to take up Ca^{2+} from the cytoplasm, mitochondria act as key regulators of intracellular calcium homeostasis. (iii) ROS homeostasis: the mitochondrial electron transport chain is a major source of reactive oxygen species (ROS) that influence cell turnover. (iv) Apoptosis: proteins present within mitochondria (e.g., cytochrome *c*) and on the outer mitochondrial membrane (e.g., Bax, Bak) play a key role in the triggering of intrinsic apoptosis. (v) Innate immunity: MAVS (Mitochondrial antiviral signaling protein), a component of RIG-1 (retinoic acid-inducible gene **I**) signaling, is located on the outer mitochondrial membrane.

All of the human tumor viruses except MCPyV (the most recently identified human tumor virus, associated with a rare form of skin cancer Feng et al., 2008b) code for one or more proteins that affect mitochondrial function.

The present review summarizes current knowledge on the interplay between mitochondria and proteins produced by HTLV-1, EBV, KSHV, HBV, HCV, and HPV (**Table 1**), and how these proteins influence the replication strategies and oncogenic properties of these viruses (**Table 2**).

HTLV-1

HTLV-1 is a retrovirus that infects at least 10 million people worldwide, with most cases identified in southwestern Japan, the Caribbean basin, sub-Saharan Africa, and Brazil (reviewed by Gessain and Cassar, 2012). HTLV-1 is transmitted through transfer of infected cells during breast-feeding, sexual contact, and exposure to blood and results in persistent infection, mainly in CD4+ T-cells. About 3–5% of infected patients develop an aggressive neoplasm of mature CD4+ T-cells named adult T-cell leukemia/lymphoma (ATLL) or a neurological disease named tropical spastic paraparesis/HTLV-associated myelopathy (TSP/HAM) after a latency period of decades (ATLL) or years (TSP/HAM). Other diseases associated with HTLV-1 infection include uveitis, infective dermatitis, myositis and other pathologies with an important inflammatory component (reviewed by Goncalves et al., 2010). Almost 40 years after the discovery of HTLV-1, we still lack an accurate measure of the global burden of infection, there is no HTLV-1 vaccine, and biomarkers to predict clinical outcome remain to be identified (reviewed by Willems et al., 2017).

In addition to the gag, pol, pro and env gene products produced by all retroviruses, the genome of HTLV-1 codes for nonstructural proteins named Tax, Rex, p12, p13, p21Rex, p30/Tof, and HBZ (reviewed by Journo et al., 2009; Lairmore et al., 2012). Tax and Rex are essential for completion of the replication cycle, with Tax driving transcription from the viral 5′-LTR (long terminal repeat) promoter, and Rex promoting expression of incompletely spliced transcripts, including those coding for the virion proteins. Tax and HBZ are considered to be the principal viral factors that drive development of ATLL (reviewed by Panfil et al., 2016).

p13

p13 is an 87-amino acid protein that accumulates primarily in the inner mitochondrial membrane (**Table 1**; D'Agostino et al., 2002). p13 is considered to be an accessory protein, as its deletion from the viral genome does not abolish viral replication *in vitro* (Derse et al., 1997). However, studies carried out in a rabbit model of HTLV-1 infection indicated that p13 is important for establishing a persistent infection *in vivo* (Hiraragi et al., 2006).

Studies of transfected HeLa cells demonstrated that mitochondrial accumulation of p13 is directed by a 10-residue mitochondrial targeting signal near its amino terminus (MTS; amino acids 22-31) which, unlike most MTS, is not cleaved during mitochondrial import (Ciminale et al., 1999). The MTS contains 4 arginines and folds into an amphipathic α-helix (D'Agostino et al., 2002). The carboxy-terminal half of p13 contains a putative hinge (amino acids 42-48) and β-sheet (amino acids 65-75) (Silic-Benussi et al., 2010a) and a cluster of prolines that mediate binding to proteins containing SH3 domains (Ghorbel et al., 2006; Tibaldi et al., 2011).

Expression of p13 in HeLa cells alters mitochondrial morphology and produces isolated clusters of round-shaped, fragmented mitochondria (Ciminale et al., 1999). These effects depend on the presence of the 4 arginines that constitute the charged face of the amphipathic α-helix (Silic-Benussi et al., 2004). When added to isolated rat liver mitochondria, synthetic p13 protein induces an inward K^+ current, mitochondrial swelling and loss of mitochondrial membrane potential ($\Delta\Psi_m$). However, this depolarization triggers a compensatory increase in electron transport chain activity, which restores $\Delta\Psi_m$ but raises levels of mitochondrial reactive oxygen species (ROS) and lowers the threshold for opening of the permeability transition pore (PTP), a channel that regulates apoptosis (reviewed by Bernardi et al., 2015). These effects are dose-dependent, as low concentrations of p13 induce mitochondrial swelling and increased mitochondrial ROS, but not mitochondrial depolarization, while high levels of p13 cause irreversible swelling, depolarization and cytochrome *c* release (Silic-Benussi et al., 2009).

p13 corresponds to the carboxy-terminal portion of p30/Tof, a nucleolar/nuclear accessory protein whose activities include suppression of Tax and Rex expression and modulation of cellular transcription (reviewed by Anupam et al., 2013). Despite this sequence overlap, p13 and p30/Tof are expressed from distinct alternatively spliced mRNAs. Studies of the temporal regulation of HTLV-1 gene expression in infected cells demonstrated that the mRNA coding for p13 is expressed as a late gene (Cavallari et al., 2011, 2013; Rende et al., 2011) and is Rex-dependent (Cavallari et al., 2015).

The structural and functional properties of p13 suggest that it may act as a viroporin (reviewed by D'Agostino et al., 2005a,b). The viroporins are a family of small, hydrophobic viral proteins with one or more membrane-spanning domains that, upon oligomerization in membranes, alter membrane permeability to ions and small molecules through the formation of channels or pores, and change the trafficking, processing and lifespan of membrane-associated proteins. Although these alterations have diverse effects on infected cells, the principal role of viroporins is to promote the assembly and egress of virus particles (reviewed by Nieva et al., 2012; Scott and Griffin, 2015).

The activity of Ca^{2+} as a second messenger is crucial for the metabolism and function of all cells, including T-cells (reviewed by Fracchia et al., 2013), the main targets of HTLV-1 infection *in vivo*. Ca^{2+} messages are delivered in the form of transient elevations in cytosolic Ca^{2+} concentration upon its release from organelles, especially the ER, and upon entry from the extracellular environment; mitochondria participate in this process through their ability to take up Ca^{2+}. Experiments carried out in HeLa cells expressing organelle-targeted aequorins revealed that p13 specifically reduces mitochondrial Ca^{2+} uptake (Biasiotto et al., 2010).

TABLE 1 | Localization and main mitochondrial effects of viral proteins.

Virus	Viral Protein	Localization	Main effects on mitochondria	Key References
HTLV-1	p13	IMM[a] nucleus[f]	• Fragmentation[b] • mtROS production (low doses), irreversible swelling depolarization and cytochrome *c* release (high dose)[c] • Reduced mitochondrial Ca^{2+} uptake[d] • Recruitment of SFKs to the intermembrane space[e]	Ciminale et al., 1999[b]; D'Agostino et al., 2002[a]; Silic-Benussi et al., 2009[c]; Biasiotto et al., 2010[d]; Tibaldi et al., 2011[e]; Andresen et al., 2011[f]
EBV	BHRF1	OMM[a]	• Binds BH3-only proteins and inhibits Bak-Bax oligomerization in OMM[b]	Hickish et al., 1994[a]; Cross et al., 2008[b]; Flanagan and Letai, 2008[b]; Kvansakul et al., 2010[b]
KSHV	K7	Mitochondria[a], ER[b], nuclear membranes[c]	• Blocks apoptosis by forming bridge with Bcl-2 and Caspase-3[d]	Feng et al., 2002[a]; Wang et al., 2002[a,b,c,d]
	KS-Bcl-2	Mitochondria[a], nucleus[b]	• Blocks apoptosis by binding to BH3 domains of pro-apoptotic Bcl-2 family proteins[c] and by sequestration of Aven[d]	Kalt et al., 2010 [a,b]; Gallo et al., 2017[a,b]; Flanagan and Letai, 2008[c]; Chau et al., 2000[d]
HBV	HBx	Nucleus[a], cytoplasm[b], Mitochondria[c], in OMM[d]	• Change in $\Delta\psi_m^{e,f}$ • Increased mtROS[g] • Increased COXIII activity[h] • Increased mitochondrial uptake of Ca^{2+i} • Increased mitochondria fission through translocation of Drp1 and degration of Mfn2[j] • Targeting of Parkin to mitochondria and mitophagy[k] • Degradation of Mfn2[1]	Henkler et al., 2001[a,b,c]; Takada et al., 1999[c,f]; Huh and Siddiqui, 2002[d]; Rahmani et al., 2000[f]; Shirakata and Koike, 2003[f]; Clippinger and Bouchard, 2008[d,f]; Zheng et al., 2014[e,h]; Ren et al., 2016[g];Zou et al., 2015[h]; Yang and Bouchard, 2012[i]; Kim et al., 2013a,b[j,k]
	Pol	Mitochondria[a]	n. d.	Unchwaniwala et al., 2016[a]
HCV	Core	ER[a], OMM[b], MAMs[c], lipid droplets[d]	• Loss of $\Delta\psi_m^e$ • Increased mtROS[f] • Increased uptake of Ca^{2+g} • promotion[h] or inhibition[i] of mitophagy	Lo et al., 1995[a]; Santolini et al., 1995[a]; Barba et al., 1997[d]; Schwer et al., 2004[b,c]; Benali-Furet et al., 2005[e,g]; Suzuki et al., 2005[a,b]; Rouille et al., 2006[a,d]; Machida et al., 2006[e]; Wang et al., 2010[b,c];Chu et al., 2011[f];Kim et al., 2013b[h]; Hara et al., 2014[i]
	P7	ER[a], MAMs[b]	• Loss of $\Delta\psi_m^c$	Griffin et al., 2005[b]; Haqshenas et al., 2007[a]; Qi et al., 2017[c]
	NS4A	ER, Mitochondria	• Perinuclear clustering[b] • Loss of $\Delta\psi_m^c$	Nomura-Takigawa et al., 2006[a,b,c]
	NS3/4A	ER[a], MAMs[b], mitochondria[c]	• Cleavage of MAVS[d]	Wolk et al., 2000[a,b]; Nomura-Takigawa et al., 2006[a,c]; Horner et al., 2011[d]; Bender et al., 2015[d]
HPV	E1^E4	Cytokeratin network[a], Mitochondria[b]	• Dissociation from microtubules, perinuclear clustering[c] • Loss of $\Delta\psi_m^d$	Doorbar et al., 1991[a]; Raj et al., 2004[a,b,c,d]
	E2	Nucleus[a], cytoplasm[b], Mitochondria[c]	• Perinuclear clustering[d], loss of cristae structure[e] • Loss of $\Delta\psi_m^f$ • Increased mtROS[g]	Blachon et al., 2005[a,b]; Lai et al., 2013[a,b,c,d,e,f,g]; Chen et al., 2014[f]

ER, endoplasmic reticulum; MAM, mitochondria-associated membranes; OMM, outer mitochondrial membrane; IMM, inner mitochondrial membrane; $\Delta\Psi_m$, mitochondrial membrane potential; mtROS, mitochondrial reactive oxygen species; cyt c, cytochrome c. Letters in superscript indicate references for each property listed columns.

The impact of p13 on mitochondrial Ca^{2+} flux might intersect with that of p12, a 99-residue viral protein that accumulates in the endoplasmic reticulum (ER) and cis-Golgi (reviewed by Van Prooyen et al., 2010). p12 increases Ca^{2+} release from the ER, thus promoting activation of NFAT (nuclear factor of activated T-cells), a key mitogenic transcription factor whose activity is controlled by the calcium-dependent phosphatase calcineurin (Kim et al., 2003). p12-mediated release of ER calcium stores and p13-mediated interference with mitochondrial calcium uptake could be predicted to amplify Ca^{2+} signals in response to T-cell receptor activation with consequent enhancement of NFAT activation (Silic-Benussi et al., 2010a). In alternative, this combination of effects could convert a Ca^{2+} signal to a prolonged increase in cytosolic Ca^{2+}, which might trigger apoptosis.

Efforts to understand the function of p13 through identification of its binding partners indicated that it interacts with farnesyl pyrophosphate synthetase (FPPS) (Lefebvre et al., 2002); with members of the Src family of protein kinases (Tibaldi et al., 2011), and with the HTLV-1 regulatory protein Tax (Andresen et al., 2011). Yeast 2-hybrid screening assays

TABLE 2 | Role of viral mitochondrial proteins in replication and transformation.

Viral protein	Effects on virus replication	Role in transformation
HTLV-1 p13	Not required for replication *in vitro* (Derse et al., 1997), but necessary for virus persistence in *in vivo* rabbit model (Hiraragi et al., 2006); mechanism unknown	Proposed negative role: interferes with transformation of fibroblasts by Myc and Ras; favors death of transformed but not normal T-cells, possibly by raising mitochondrial ROS production (reviewed by Silic-Benussi et al., 2010a)
EBV BHRF1	Not required for virus replication (Altmann and Hammerschmidt, 2005)	Required for transformation of resting B-cells, but dispensable for transformation of activated B-cells (Altmann and Hammerschmidt, 2005)
KSHV K7	Not required for virus replication (Liang et al., 2015; Gallo et al., 2017)	Possible negative role through interference with transforming activity of vGPCR (Feng et al., 2008a)
KSHV KS-Bcl-2	Required for efficient virus replication (Gelgor et al., 2015; Liang et al., 2015; Gallo et al., 2017) Not required for establishment of latent infection (Gelgor et al., 2015)	Positive role likely, through inhibition of apoptosis in infected cells
HBV HBx	Favors viral replication; enhances viral polymerase activity through Ca^{2+} signaling (Lucifora et al., 2011)	Transforms cells *in vitro* and causes tumors in transgenic mice (reviewed by Levrero and Zucman-Rossi, 2016)
HBV Pol	Needed for packaging of pregenomic RNA and reverse transcription into dsDNA genome (Bartenschlager and Schaller, 1992)	Not defined
HCV Core	Forms virion capsid (reviewed by Scheel and Rice, 2013)	Transforms cells *in vitro* and causes tumors in transgenic mice (reviewed by Banerjee et al., 2010)
HCV p7	Viroporin required for virion assembly and release (reviewed by Madan and Bartenschlager, 2015)	Not defined
HCV NS3/4A	Required for viral RNA replication, polyprotein processing and virion assembly (reviewed by Morikawa et al., 2011); cleavage of MAVS proposed to favor immune evasion (Horner et al., 2011)	Not defined; NS3 alone transforms cells *in vitro* (Sakamuro et al., 1995)
NS4A	Forms complex with NS3 (see above)	Not defined
HPV E1^E4	May promote virion release through perturbation of the cytokeratin network (Raj et al., 2004)	Not defined
HPV E2	Essential for coordination of late events of viral replication (reviewed by Graham, 2016)	Negative role, through inhibition of E6 and E7 expression (reviewed by Woodman et al., 2007)

also indicated binding of p13 to a protein of the nucleoside monophosphate kinase superfamily and to actin-binding protein 280 (Hou et al., 2000), but these interactions were not explored in detail.

Binding of p13 to Src family kinases (SFKs) involves p13's proline-rich C-terminal domain and the kinases' SH3 domains (Tibaldi et al., 2011). This interaction promotes the accumulation of SFKs in the mitochondrial intermembrane space and enhances their tyrosine kinase activity, while interfering with targeting of p13 to the inner membrane and attenuating the effects of p13 on $\Delta\Psi_m$ (Tibaldi et al., 2011).

Experiments carried out in 293T cells and HeLa cells showed that when co-expressed with Tax, p13 undergoes ubiquitination, becomes more stable, and is partially rerouted to nuclear speckles containing Tax (Andresen et al., 2011). These effects are mediated by direct interaction of Tax with p13 through a disulphide bond involving cysteine 27 in p13. Results of co-immunoprecipitation and LTR-reporter assays showed that the p13-Tax interaction interferes with binding of Tax to the transcription cofactor CBP/p300 and leads to a decrease in Tax-mediated viral gene transcription, a crucial step in viral replication. The finding that substitution of the p13 start codon with isoleucine in an HTLV-1 proviral clone resulted in an increase in viral gene expression supports the proposal that p13 might acts as a negative regulator of HTLV-1 replication (Andresen et al., 2011).

The effect of ablation of p13 expression on the ability of the virus to transform primary T-cells needs to be investigated. However, results of studies of primary cells and cell lines forced to express p13 indicate that it may limit the oncogenic potential of HTLV-1. p13 interfered with the ability of Myc and Ras to transform rat embryo fibroblasts, and HeLa cells expressing p13 exhibited a proliferation defect *in vitro* and were less tumorigenic in nude mice compared to parental cells (Silic-Benussi et al., 2004). Subsequent comparisons of the effects of p13 in the T-cell line Jurkat and normal primary T-cells showed that the protein induced ROS production in both cell types, and slowed proliferation and increased apoptotic death in Jurkat cells, but activated normal primary T-cells (Silic-Benussi et al., 2010a,b). These observations suggest that p13 might expand the pool of untransformed infected cells and on the other hand favor the elimination of transformed cells, thus increasing HTLV-1's adaptation to the host and lifelong persistence of the infection. It will be interesting to determine whether the effects of p13 on T-cell activation are connected to its impact on Ca^{2+} homeostasis.

EBV AND KSHV

EBV and KSHV are members of the γ-herpesvirus subfamily of the Herpesviridae, a family of large, enveloped viruses with a linear double-stranded DNA genome. EBV targets mainly B-cells and epithelial cells, while KSHV targets mainly B-cells and endothelial cells. Like all herpesviruses, the life cycle of EBV and KSHV comprises latent states that favor viral persistence and a lytic phase that produces virus particles. EBV infection is present in more than 90% of adults in a uniform distribution worldwide. Although studies of global KSHV epidemiology are incomplete, it appears to be more restricted geographically to parts of sub-Saharan Africa, the Mediterranean basin, Brazil and China, and in certain subpopulations such as men who have sex with men (Morrison et al., 2015). Primary infection with EBV and KSHV occurs mainly through contact with saliva during early childhood and usually does not cause overt clinical symptoms, except for a self-limiting polyclonal B-lymphoproliferative disease termed infectious mononucleosis that arises in at least 25% of patients who become infected with EBV in adolescence or early adulthood. Neoplasias associated with EBV include an endemic form of Burkitt's lymphoma, post-transplant lymphoproliferative disorders, some cases of Hodgkin's lymphoma, and nasopharyngeal carcinoma (reviewed by Young et al., 2016). KSHV is linked to 3 types of tumors, with immunosuppression representing a risk factor: the endothelial-derived tumor Kaposi's sarcoma (KS, always KSHV-positive), a B-cell malignancy named primary effusion lymphoma (PEL; always KSHV-positive, sometimes also EBV-positive) and a plasmablastic form of the B-lymphoproliferative disorder multicentric Castleman disease (MCD, KSHV-positive in about half of cases) (reviewed by Bhutani et al., 2015). EBV and KSHV share 59 gene homologs whose similarity ranges from 41 to 75% (Russo et al., 1996); both viruses possess genes that closely resemble cellular genes. Proteins of EBV and KSHV with an impact on mitochondria include EBV BHRF1, BZLF1, and LMP2A, and KSHV K7 and KS-Bcl-2 (**Table 1**).

EBV: BHRF1, BZLF1 BALF1, LMP2A

BHRF1 is a 191-amino acid, early lytic-phase EBV protein with similarity to cellular Bcl-2 family proteins (Pearson et al., 1987) that blocks the mitochondrial arm of apoptosis mediated by pro-apoptotic Bcl-2 proteins (Henderson et al., 1993). In this manner, BHRF1 promotes the survival of EBV-infected cells, thus promoting viral persistence/replication. The production of viral Bcl-2 homologs (v-Bcl-2) is shared by several other viruses, including KSHV, which produces KS-Bcl-2 (see below; reviewed by Kvansakul et al., 2017) (**Figures 1A,B**).

BHRF1 accumulates in the outer mitochondrial membrane (OMM) of B-cells, a distribution similar to that of Bcl-2 (Hickish et al., 1994). Mitochondrial targeting of BHRF1 is directed by a C-terminal sequence that shares homology with the transmembrane domain (TM) present in some cellular Bcl-2 family members (Bcl-2 38%, Bcl-xl 32%, Bax 34%) (Hickish et al., 1994). As shown in **Figure 1A**, BHRF1 also has all of the Bcl-2 homology (BH) domains except for BH4, which is in general poorly conserved among Bcl-2 family members.

The anti-apoptotic effects of BHRF1 are mainly due to its ability to bind to BH3-only pro-apoptotic Bcl-2 proteins and to inhibit the formation of Bax/Bak oligomers in the OMM (Cross et al., 2008; Flanagan and Letai, 2008; Kvansakul et al., 2010; Milian et al., 2015). BHRF1 binds to the BH3-only proteins Bim, Bak, Bid and PUMA (Kvansakul et al., 2010). Structural studies of the BHRF1:Bim/Bak complexes indicated that an aspartic acid at position 100 of BHRF1 is essential for its association with the BH3 domains of Bim and Bak, in interactions similar to those occurring between Bim and the cellular anti-apoptotic protein Bcl-xL (Kvansakul et al., 2010). EBV also codes for BALF1, a cytoplasmic protein with homology to Bcl-2 family members that was shown by one group to interfere with the anti-apoptotic activity of BHRF1 (Bellows et al., 2002) but by others to inhibit apoptosis (Marshall et al., 1999) and promote transformation (Hsu et al., 2012).

BZLF1 (also known as ZEBRA, *Zta*) is a 245-amino acid EBV protein of the basic leucine zipper (b-zip) family that is essential for transcription of EBV lytic-phase genes and replication of viral genomes (reviewed by McKenzie and El-Guindy, 2015). Experiments carried out in EBV-positive B-cell lines indicated that BZLF1 can interact with the mitochondrial protein mtSSB, a single-stranded DNA-binding protein that is needed for replication of the mitochondrial genome. Cells in the lytic phase showed reduced mtDNA synthesis and contained fewer mitochondrial genomes, while silencing of mtSSB interfered with BZLF1-dependent entry into the lytic phase. Imaging analyses of cells expressing FLAG-tagged BZLF1 indicated that BZLF1 partially redirects mtSSB from mitochondria to the nucleus, which is the main site of BZLF1 accumulation (Wiedmer et al., 2008). Expression of BZLF1 in HeLa cells resulted in fused masses of mitochondria, some of which were localized near the nucleus (LaJeunesse et al., 2005); it is however unclear how BZLF1 induces these changes in mitochondrial morphology.

EBV LMP2A (latent membrane protein 2A) is a 497-amino acid protein with 12 transmembrane domains that accumulates mainly in plasma membrane rafts, where it influences signal transduction pathways affecting cell activation, proliferation, survival and migration (reviewed by Cen and Longnecker, 2015). Pal et al. (2014) showed that LMP2A has indirect effects on mitochondria. Expression of LMP2A in EBV-negative gastric- and breast cancer cell lines resulted in increased mitochondrial fission accompanied by an increase in migration and induction of the epithelial-mesenchymal transition. These effects were attributed to LMP2A's ability to stimulate the Notch pathway, which in turn upregulates Drp1 (dynamin-related protein 1), a protein that induces mitochondrial fission (Pal et al., 2014).

KSHV: K7, KS-Bcl-2

K7 is a 126-amino acid protein that shows homology to the cellular protein Survivin, a member of the family of inhibitor of apoptosis proteins (IAPs). K7 is detected in the endoplasmic reticulum, nucleus and in mitochondria (Feng et al., 2002; Wang et al., 2002) and is expressed early after induction of the lytic cycle (Wang et al., 2002). It contains an amino-terminal, atypical MTS, a putative transmembrane domain partially overlapping the MTS, a baculovirus IAP repeat (BIR), and a carboxy-terminal BH2 domain (Wang et al., 2002).

K7's BH2 and BIR domains mediate its direct interaction with Bcl-2 and activated caspase-3, respectively; the resulting bridge between K7, Bcl-2, and activated caspase-3 inhibits the apoptotic caspase cascade (Wang et al., 2002). Another

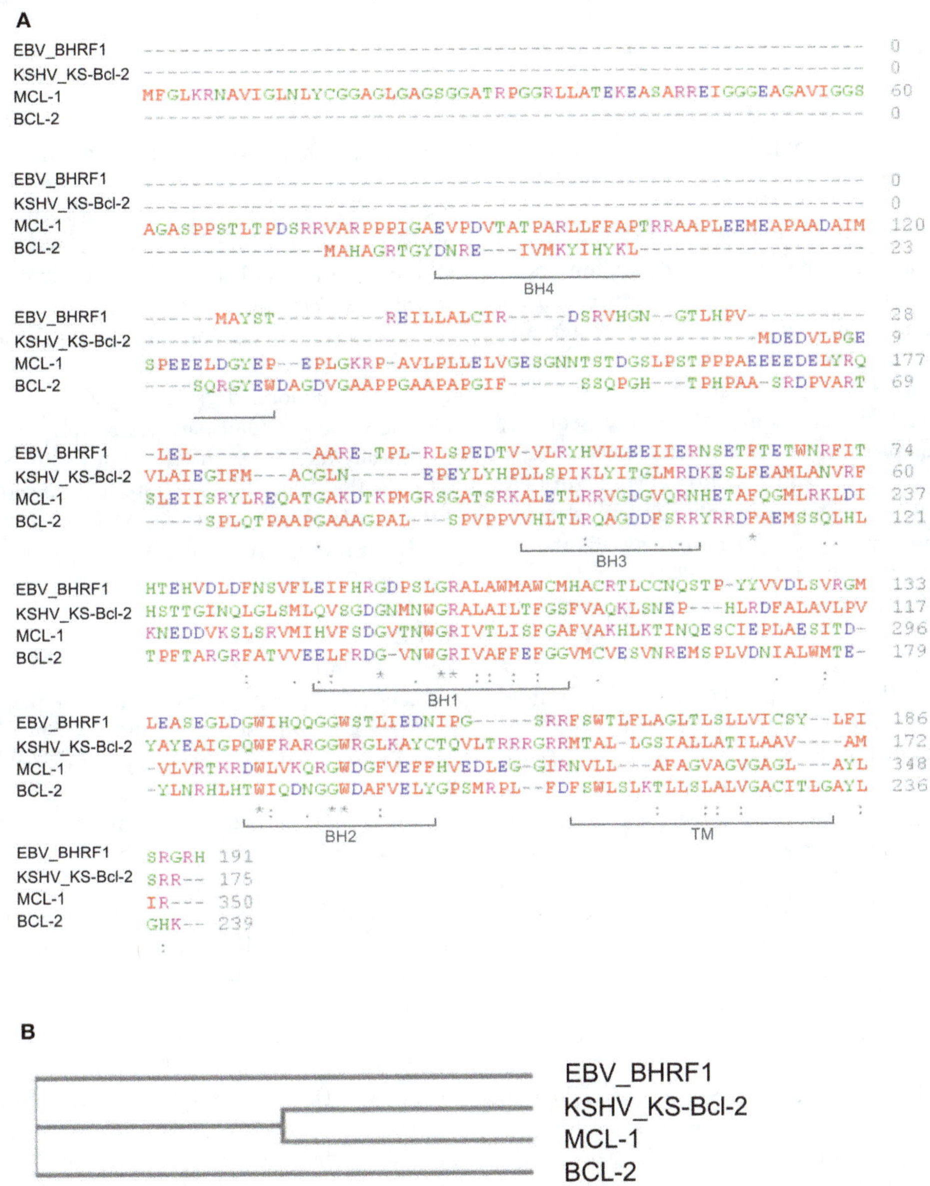

FIGURE 1 | Sequence comparison of human Bcl-2 family proteins and viral homologues. **(A)** Multiple sequence alignment. Protein sequences were obtained from the UniProt database (http://www.uniprot.org/) and analyzed with the online software Clustal Omega (http://www.ebi.ac.uk/Tools/msa/clustalo/). Amino acids are labeled in different colors according to their biochemical properties (red: small/hydrophobic; blue: acidic; magenta: basic; green: hydroxyl/sulfhydryl/amine/Glycine). Asterisks indicate single conserved residues; periods indicate residues with similar properties, and brackets indicate locations of the BH and TM domains in Bcl-2 (UniProtKB–P10415 (BCL2_HUMAN). Accession IDs: KS-Bcl-2 (>sp|F5HGJ3|ARBH_HHV8P); BHRF1 (>sp|P03182|EAR_EBVB9); Bcl-2 (>sp|P10415|BCL2_HUMAN)); Mcl-1 (>sp|Q07820|MCL1_HUMAN). **(B)** Cladogram representing the similarity between human Bcl-2 and Mcl-1 and the viral orthologs of EBV (EBV_BHRF1) and KSHV (KSHV_KS-Bcl-2). The cladrogram was generated with the online software Clustal Omega.

anti-apoptotic function of K7 involves its association with CAML (calcium-modulating cyclophilin ligand), an ER protein that controls intracellular Ca^{2+} homeostasis (Feng et al., 2002). The CAML-K7 interaction alters the changes in cytosolic Ca^{2+} induced by thapsigargin, an inhibitor of the sarcoplasmic/endoplasmic Ca^{2+}-ATPase (SERCA), whose role is to transport Ca^{2+} from the cytosol to the ER. In this manner, K7 protects thapsigargin-treated cells from Ca^{2+}-induced loss of $\Delta\Psi_m$ and apoptosis (Feng et al., 2002).

K7 also interacts with a helix-coiled region (HC) of Rubicon, an inhibitor of autophagosomal maturation that forms a complex with Beclin 1, UVRAG and Vps34 (Liang et al., 2013). K7 increases the expression of Rubicon and enhances its interaction with the autophagy machinery, resulting in a block in the autophagy process. Cells expressing a virus knocked out for K7 were able to produce virus, but expression of the viral proteins K3 and K8 was reduced (Liang et al., 2013). K7-mediated regulation of K3 and K8 expression would have an important effect on viral persistence, as K3 induces internalization and degradation of MHC-I complexes, thus favoring the escape of virus-infected cells from the control of the immune system, while K8 is part of the viral lytic DNA replication complex (Liang et al., 2013).

Another binding partner of K7 is vGPCR, a KSHV homolog of the human interleukin-8 receptor that induces tumors in nude mice. K7 induces the degradation of vGPCR through the ER-associated degradation pathway, resulting in a decrease in tumorigenicity (Feng et al., 2008a).

KS-Bcl-2 is a 175-amino acid protein that is expressed from KSHV ORF16. KS-Bcl-2 shares about 60% overall sequence homology with cellular proteins of the Bcl-2 family, with high conservation of the BH1, BH2 and TM domains (**Figure 1A**; Sarid et al., 1997). KS-Bcl-2 appears to be more closely related to MCL-1 than to Bcl-2 in terms of sequence homology (**Figure 1B**) and functional activity (Flanagan and Letai, 2008).

The first study of KS-Bcl-2 described a punctate cytoplasmic distribution, indicative of accumulation in organelle membranes, but did not identify these organelles (Sarid et al., 1997). More recent studies of KS-Bcl-2 indicated its partial localization in mitochondria and in the nucleus (Gallo et al., 2017) or nucleoli (Kalt et al., 2010); these studies employed tagged versions of KS-Bcl-2 and a fluorescent marker to distinguish the mitochondrial compartment in different cell lines. The combination of tagging and cell lines used in these studies likely influenced KS-Bcl-2's localization: in the study by Gallo et al., when expressed in 293A cells, FLAG-tagged KS-Bcl-2 colocalized with the fluorescent mitochondrial marker in compact crescent-shaped structures lying against the nucleus, while in HUVEC cells the protein colocalized with the marker in structures resembling the web-like mitochondrial network (Gallo et al., 2017).

The KS-Bcl-2 mRNA is detected in KS lesions and in PEL cells upon induction of the lytic cycle and is required for lytic reactivation and replication (Gelgor et al., 2015; Liang et al., 2015). The best-defined roles of KS-Bcl-2 are to inhibit both apoptosis and autophagy. KS-Bcl-2 impedes apoptosis through interactions with the BH3 domains of pro-apoptotic proteins of the Bcl-2 family (Bim, Bid, PUMA, Bik, NOXA, Bmf) (Flanagan and Letai, 2008) and through its association with Aven, a protein that interferes with the ability of Apaf-1 to activate caspase 9 (Chau et al., 2000). Inhibition of autophagy by KS-Bcl-2 is mediated by its binding to Beclin 1, a key component of the autophagy pathway (Pattingre et al., 2005). Interestingly, the role of KS-Bcl-2 in promoting viral reactivation and replication is apparently independent from its effects on apoptosis and autophagy (Gelgor et al., 2015; Liang et al., 2015).

The nuclear targeting of KS-Bcl-2 was mapped to its amino-terminal 17 amino acids and was found to be important for the protein's effects of viral replication (Gallo et al., 2017). Nucleolar accumulation of KS-Bcl-2 depended on its association with the cellular nucleolar protein GLTSCR2/PICT-1, and interfered with KS-Bcl-2's anti-apoptotic properties (Kalt et al., 2010).

HBV

HBV is a small, enveloped DNA virus of the Hepadnaviridae family that is transmitted parenterally through blood and other body fluids. Most HBV-infected patients are asymptomatic or present signs of acute liver disease and inflammation. Active viral replication and insufficient immune clearance may result in chronic hepatitis, with tissue injury, inflammation and regeneration that can progress to cirrhosis, liver failure and hepatocellular carcinoma (HCC), an aggressive neoplasm with a dismal prognosis. Despite the availability of a prophylactic vaccine since the 1980s, approximately 257 million persons are chronically infected with HBV worldwide, with high-prevalence areas in the African and Western Pacific regions (statistics for 2015, World Health Organization, 2017).

HBV produces two proteins that interact with mitochondria: HBx and Pol (**Table 1**).

HBx

The 154-amino acid HBx protein favors HBV replication, influences cellular transcription, signal transduction, cell proliferation and survival, and is considered to be a key contributor to the oncogenic potential of HBV (reviewed by Benhenda et al., 2009; Motavaf et al., 2013). Takada et al. provided the first evidence for mitochondrial accumulation of HBx, which was accompanied by aggregation of mitochondria near the nucleus, accumulation of p53 in mitochondria, loss of $\Delta\Psi_m$ and increased apoptotic death (Takada et al., 1999). A subsequent study revised HBx's targeting properties to include the nucleus and cytoplasm (Henkler et al., 2001). The mitochondrially-associated fraction of HBx was detected in the outer mitochondrial membrane in the hepatoma cell line Huh7, in the hepatocarcinoma cell line HepG2, and in primary rat hepatocytes (Huh and Siddiqui, 2002; Clippinger and Bouchard, 2008). Experiments carried out with GFP fusion proteins in Huh7 cells indicated that residues 68-117 of HBx confer mitochondrial targeting (Shirakata and Koike, 2003).

HBx was shown to induce loss of $\Delta\Psi_m$ when expressed in Huh7 cells (Takada et al., 1999; Rahmani et al., 2000; Shirakata and Koike, 2003) or HepG2 cells (Clippinger and Bouchard, 2008). However, primary rat hepatocytes expressing HBx were protected from loss of $\Delta\Psi_m$ induced by treatment with TNF-α, a property attributed to HBx's ability to induce the NF-κB pathway (Clippinger and Bouchard, 2008). More recently, Zheng et al. (2014) showed that HBx increases $\Delta\Psi_m$ in HepG2 cells, an effect that was correlated with increased levels of cytochrome *c* oxidase III (COXIII) and increased COX activity. HBx-induced upregulation of COXIII expression and activity were also observed in the immortalized hepatocyte cell line HL-7702 (Zou et al., 2015). Although no changes in $\Delta\Psi_m$ were observed in this system, HBx increased cellular ROS production (Zou et al., 2015).

An HBx-mediated increase in cellular ROS was associated with activation of the STAT3 and NF-κB pathways (Waris et al., 2001), both of which are important in inflammation and transformation. HBx-induced cellular ROS was also accompanied by induction of cyclooxygenase 2 (COX-2), an enzyme that catalyzes the production of mediators of the inflammatory response (Lim et al., 2010). The observation that ROS-enhancing agents such as the chemotherapeutic drug adriamycin increase the stability of HBx protein suggests the existence of a ROS-HBx positive feedback loop (Wang et al., 2003). Such a positive role for ROS would fit in with results of a recent study which indicated that HBV infection induces oxidative stress, in part through the ability of HBx to increase mitochondrial ROS, and that ROS favor replication of the virus (Ren et al., 2016).

Binding partners of HBx in mitochondria include the voltage-dependent anion channel 3 (VDAC3), which resides in the OMM (Rahmani et al., 2000), and heat shock protein 60 (HSP60), a multifunctional chaperone that is located mainly in the mitochondrial matrix (Tanaka et al., 2004). Studies carried out in Huh7 cells showed that HBx-induced ROS promotes the translocation of the serine/threonine kinase Raf-1 (C-Raf) from the cytoplasm to mitochondria, where it forms complexes with HBx (Chen and Siddiqui, 2007).

HBx-induced alterations in intracellular calcium signaling play an important role in HBV replication (Bouchard et al., 2001). HBx increases the basal levels of cytosolic Ca^{2+} in HepG2 cells (McClain et al., 2007) and augments the spike in cytosolic Ca^{2+} levels provoked by ATP (Chami et al., 2003). The HBx-mediated increase in cytosolic Ca^{2+} reflects influx of Ca^{2+} into cells due to store-operated calcium entry and is associated to greater uptake of Ca^{2+} by mitochondria (Yang and Bouchard, 2012).

HBx-expressing cells also show changes in mitochondrial dynamics with increased fission, which results both from phosphorylation and mitochondrial translocation of the fission protein Drp1 and from ubiquitination/degradation of Mitofusin 2 (Mfn2), a protein that mediates mitochondrial fusion (Kim et al., 2013a). The study by Kim et al. also showed that HBx promotes mitophagy, a specialized form of autophagy that eliminates dysfunctional mitochondria. This effect was attributed to increased expression and mitochondrial targeting of Parkin, a ubiquitin ligase whose substrates provide a signal for sequestration and degradation of damaged mitochondria (reviewed by Bernardini et al., 2017). Thus, by enhancing mitophagy, HBV could promote cell survival and possibly viral persistence. On the other hand, induction of mitochondrial fission may lead to mitochondrial injury, which might play a role in the pathogenesis of HBV-related liver disease (Kim et al., 2013a).

HBx can either directly affect apoptosis or modify the response of cells to apoptotic stimuli such as TNF-α (reviewed by Rawat et al., 2012). The varied effects of HBx on cell death likely depend on its expression levels and on the cell context.

Pol

HBV Pol is an 832-residue protein whose principal role is to direct packaging of viral pregenomic RNA (pgRNA) molecules into capsids and to reverse-transcribe the RNA into the dsDNA genome. In a study of the sites of HBV replication, Pol was observed to accumulate mainly in mitochondria (Unchwaniwala et al., 2016). A segment of Pol spanning residues 141-160 functioned as a mitochondrial targeting sequence (MTS) when attached to GFP and was important for pgRNA packaging, but could be deleted from Pol without abolishing its mitochondrial targeting, indicating that additional amino-terminal sequences contribute MTS properties. Based on the finding that neither the pgRNA nor the viral Core protein (Cp) localized to the mitochondria during replication, Unchwaniwala et al. (2016) suggested that the binding of pgRNA to the MTS may block mitochondrial localization, so that Pol bound to pgRNA would be incorporated into in capsids, while unbound Pol would accumulate in mitochondria. Although the role of Pol in mitochondria remains to be understood, the protein is known to have diverse activities besides HBV genome packaging and reverse transcription, including interference with the antiviral interferon response (Wu et al., 2007; Wang and Ryu, 2010), an effect that would favor viral persistence in chronically infected patients.

HCV

HCV belongs to the *Flaviviridae* family of small, enveloped viruses with a single-stranded RNA genome. About 80–85% of HCV-infected persons fail to eliminate the virus due to the ability of HCV to evade innate and adaptive immune surveillance. Persistent HCV infection is associated with liver pathologies (chronic hepatitis, hepatic steatosis, hepatic fibrosis, cirrhosis, and hepatocellular carcinoma) and lymphoproliferative disorders. Although new treatment regimens with direct-acting antivirals (DAAs) are capable of eliminating the infection, HCV remains an important clinical and social problem due to the high cost of DAAs, susceptibility of cured patients to reinfection, and the lack of a prophylactic vaccine (reviewed by Webster et al., 2015; Dustin et al., 2016). Approximately 71 million persons are chronically infected with HCV, in a heterogeneous world-wide distribution with higher prevalence in the Eastern Mediterranean and European regions (statistics for 2015, World Health Organization, 2017).

The 9.6 kb HCV genome codes for a polyprotein precursor that is cleaved by viral and cellular proteases into the structural proteins Core, E1 and E2, which make up the virion, and nonstructural proteins p7, NS2, NS3, NS4A, NS4B, NS5A, and NS5B, which are involved in polyprotein processing, genome replication and virion assembly (reviewed by Scheel and Rice, 2013). The viral life cycle takes place in the ER and an ER-derived membranous web associated with lipid droplets, and results in ER stress (Dash et al., 2016). Core, p7, NS3/4, NS4A, and NS5A show partial accumulation in mitochondria or MAM (**Table 1**).

Core

Core is a 177-amino acid protein that is derived from the amino-terminal end of the polyprotein precursor by two proteolytic cleavage events (Okamoto et al., 2008). In addition to forming the virion capsid, Core has multiple effects on viral and cellular pathways, and can protect against or sensitize cells to apoptotic stimuli (reviewed by Kao et al., 2016). Studies of cell lines ectopically expressing Core indicated its accumulation mainly in the ER (Lo et al., 1995; Santolini et al., 1995), in both the ER and outer mitochondrial membrane (Suzuki et al., 2005), or in MAMs and the OMM (Schwer et al., 2004). The association of Core with the ER/MAMs/OMM is mediated by an amphipathic alpha helical sequence near its carboxy terminus (Schwer et al., 2004; Suzuki et al., 2005).

The targeting properties of Core are influenced by its interactions with other viral proteins. When expressed with E1, Core was targeted mainly to lipid droplets (Barba et al., 1997). Analyses of HCV-infected Huh7 cells indicated accumulation of Core mainly in lipid droplets and associated membranes (Rouille et al., 2006) or mainly in the ER and to a lesser extent in mitochondria (Schwer et al., 2004). The fraction of Core associated with mitochondria in infected cells is restricted to the outer mitochondrial membrane (OMM) or possibly MAMs (Wang et al., 2010).

The phenotype of Core-transgenic mice includes hepatic steatosis and hepatocellular carcinoma (Moriya et al., 1997, 1998). The hepatocytes of these mice showed age-dependent accumulation of Core in the nucleus and in morphologically altered, dysfunctional mitochondria (Moriya et al., 1998), and signs of oxidative stress (Moriya et al., 2001), suggesting a direct role for Core in the oxidative stress and impaired mitochondrial function observed in patients with HCV-induced hepatitis (Farinati et al., 1995; Barbaro et al., 1999). Core was also detected in mitochondria of hepatocytes from mice co-expressing Core, E1, E2, and p7; the mitochondria contained reduced levels of glutathione and NADPH, indicating oxidative stress, and showed impaired Complex I activity, reduced respiration, and increased ROS (Korenaga et al., 2005).

Further studies of Core-expressing cell lines confirmed that it induces mitochondrial ROS (Okuda et al., 2002; Chu et al., 2011), as well as loss of $\Delta\Psi_m$ (Benali-Furet et al., 2005; Machida et al., 2006), and release of cytochrome *c* into the cytoplasm (Okuda et al., 2002; Benali-Furet et al., 2005). Other effects of Core include increased lipid peroxidation (Okuda et al., 2002; Machida et al., 2006), accumulation of lipid droplets (Chu et al., 2011), increased autophagy (Chu et al., 2011), and apoptosis (Benali-Furet et al., 2005). Studies in Huh7 cells and the B-cell line Raji linked the effects of Core on $\Delta\Psi_m$ and cellular ROS to activation of the STAT3 pathway and induction of DNA damage (Machida et al., 2006).

Core also induces an accumulation of Ca^{2+} in the cytoplasm due to a defect in SERCA function (Benali-Furet et al., 2005). The consequent loading of mitochondria with Ca^{2+} favors mitochondrial ROS production and triggers the mitochondrial permeability transition (Li et al., 2007). Augmented mitochondrial uptake of Ca^{2+} as well as inhibition of Complex I activity and increased ROS and glutathione oxidation are also observed when Core is added to isolated mitochondria (Korenaga et al., 2005). The effects of Core on intracellular Ca^{2+} likely contribute to the signs of ER stress observed in an osteosarcoma cell line expressing HCV (Piccoli et al., 2006, 2007, 2009). In Huh7 cells, HCV was found to favor ROS induction, loss of $\Delta\Psi_m$ and apoptosis triggered by treatment with the pro-oxidant t-BOOH; interestingly, these effects that were lost in the absence of Core, E1, E2, and p7 (Wang et al., 2010).

Analyses of Core deletion mutants indicated that its first 75 amino acids mediate binding to the mitochondrial matrix protein HSP60, and are required for Core-mediated induction of intracellular ROS (measured using DCF-DA) and sensitization to TNF-α-induced apoptosis (Kang et al., 2009); the location in which the Core-HSP60 interaction influences ROS production was not identified.

Core also interacts with Prohibitin, a multifunctional protein that influences mitochondrial function and apoptosis (reviewed by Peng et al., 2015). Core-expressing HepG2 cells expressed increased levels of Prohibitin, but reduced levels of COX subunits I and II, suggesting that the Core-Prohibitin interaction interferes with proper assembly and activity of respiratory chain complexes, which would be expected to result in increased mitochondrial ROS production (Tsutsumi et al., 2009).

Studies of the impact of HCV on mitophagy indicated either a promoting (Kim et al., 2013b) or blocking (Hara et al., 2014) effect. The inhibitory effect was attributed to the ability of Core to interact with and inhibit mitochondrial accumulation of Parkin (Hara et al., 2014). Persistence of damaged mitochondria due to the Core-Parkin interaction may thus contribute to HCV-associated liver damage and HCC.

Alcohol consumption worsens the clinical course of chronic hepatitis C. A study of Huh7 cells provided evidence that Core may contribute to oxidative stress that arises during the metabolism of ethanol by cytochrome p4502E1 (CYP2E1) (Otani et al., 2005).

Hepatitis C patients and Core-transgenic mice show hepatic accumulation of iron (Farinati et al., 1995; Moriya et al., 2010). In iron-overload experiments, hepatocytes from Core-transgenic mice and Core-expressing HepG2 cells showed impaired upregulation of the antioxidant heme-oxygenase-1, which is normally induced by excess iron (Moriya et al., 2010). Such a defect in the iron-overload response could contribute to deregulated ROS production and scavenging (Fujinaga et al., 2011).

p7, NS3/4A, and NS5A

There is also evidence that the HCV nonstructural proteins p7, NS3/4A, and NS5A influence mitochondrial function. p7 is a 63-amino acid hydrophobic protein that forms hexameric ion channels in membranes and is classified as a viroporin (reviewed by Nieva et al., 2012). p7 is required for the assembly and release of infectious virions, thus making it a potential target for antiviral therapy (reviewed by Madan and Bartenschlager, 2015). p7 accumulates mainly in the ER (Carrere-Kremer et al., 2002; Isherwood and Patel, 2005; Haqshenas et al., 2007; Vieyres et al., 2013) and in MAMs (Griffin et al., 2004, 2005). A recent study of the impact of HCV on the interferon response indicated that p7 induces mitochondrial membrane depolarization and binds to IFI6-16, an interferon response protein that stabilizes $\Delta\Psi_m$, an effect that is counteracted by p7 (Qi et al., 2017).

NS3 is a 631-amino acid protein with serine protease- and RNA helicase activities that, in association with the 54-amino acid accessory protein NS4A, forms a membrane-bound complex that is required for HCV RNA replication, polyprotein processing and virion assembly (reviewed by Morikawa et al., 2011), and is a target of DAAs. NS3/4A also cleaves several cellular proteins, including MAVS, a component of the innate immune response signaling pathway triggered by binding of pathogen-associated molecular patterns (PAMPs) to RIG-1 (Meylan et al., 2005; Belgnaoui et al., 2011). In Huh7 cells, MAVS is detected in peroxisomes, mitochondria and MAMs (Seth et al., 2005; Horner et al., 2011) and is susceptible to cleavage by NS3/4A in these compartments (Horner et al., 2011; Bender et al., 2015). Consistent with these observations, there is evidence that NS3/4A is partially targeted to mitochondria (Wolk et al., 2000; Nomura-Takigawa et al., 2006). Huh7 cells expressing both NS3 and NS4A showed increased sensitivity to an apoptosis-inducing drug, suggesting that NS3/4A may contribute to HCV-associated hepatic injury by priming cells to death stimuli (Nomura-Takigawa et al., 2006). Huh7 cells expressing NS4A alone showed accumulation of the protein in the ER and in mitochondria that were perinuclear, doughnut-shaped, and depolarized. NS4A-expressing cells released cytochrome *c* in the cytoplasm and showed increased spontaneous death compared to control cells or cells coexpressing NS3 and NS4A (Nomura-Takigawa et al., 2006).

NS5A is a 467-amino acid ER/membranous web-associated phosphoprotein that regulates viral replication and assembly, and is a target of DAA (reviewed by Ross-Thriepland and Harris, 2015). When expressed in Huh7 cells, NS5A activates the NF-κB and STAT3 pathways through Ca^{2+}-mediated induction of oxidative stress (Gong et al., 2001). Studies of a GFP-NS5A fusion protein in HEK293 cells indicated a role for NS5A in tethering the ER to mitochondria through a mechanism that involves binding of NS5A to phosphatidylinositol 4-kinase PI4KA III-α (PI4KA) (Siu et al., 2016). GFP-NS5A-expressing cells contained fragmented mitochondria but were more resistant to apoptosis induced by hydrogen peroxide compared to control cells, suggesting that NS5A may favor survival of HCV-infected cells under conditions of oxidative stress (Siu et al., 2016).

HPV

Human papillomaviruses (HPVs), members of the Papillomaviridae family, are small non-enveloped viruses that possess a double-stranded circular DNA genome and show a tropism for epithelial cells (reviewed by Doorbar et al., 2015). The HPV genome codes for 6 early genes named E1, E2, E4, E5, E6, and E7, which regulate virus replication, and 2 late genes named L1 and L2, which make up the virion capsid. The pattern of HPV gene expression is tightly controlled by the differentiation status of the infected cell: after penetration into the mucosal or cutaneous epithelium through microlesions, replication initiates with expression of early genes in the basal layer, while expression of late proteins making up the virus particles is restricted to the upper layers (Tomaic, 2016).

HPV is classified into more than 200 genotypes based on the sequence of the L1 (major capsid) gene. The genotypes are grouped into 5 genera, the 2 most numerous of which are named alpha and beta. Alpha-HPVs are further classified as low-risk or high-risk based on their connection with cancer (reviewed by Egawa et al., 2015). The low-risk viruses (e.g., HPV-6 and HPV-11) cause warts and other benign proliferative lesions of the skin and mucosa, while high-risk viruses (e.g., HPV-16 and HPV-18) are associated with cervical carcinoma and cancers of the anogenital and head-and neck regions. Beta-HPVs infect cutaneous epithelia and may contribute to the initiation of non-melanoma skin cancers, with immunosuppression representing a risk factor (Tomaic, 2016). HPV-mediated transformation is driven by the viral proteins E6 and E7 (reviewed by Doorbar et al., 2015; Tomaic, 2016). The frequency of HPV-associated pathologies should be reduced with the availability, since 2006, of vaccines based on virus-like particles containing the L1 protein of selected high-risk or high- plus low-risk genotypes (reviewed by Harper and DeMars, 2017).

E1^E4

Two HPV proteins, E1^E4 and E2, partially localize to mitochondria (**Table 1**, **Figure 2**). E1^E4 is produced by a spliced mRNA that joins the first few codons from the E1 ORF in frame to the E4 ORF, resulting in a fusion protein of about 90-120 amino acids, depending on the genotype. E1^E4 is the most abundantly expressed viral protein in the mid-upper epithelial layers of productively infected lesions (reviewed by Doorbar, 2013). Doorbar et al. (1991) showed that E1^E4 binds to and collapses the cytokeratin network in mature human keratinocytes. Collapse of the cytokeratin network leads to accumulation of E1^E4 in mitochondria, mediated by an N-terminal leucine-rich sequence (Raj et al., 2004). E1^E4-containing mitochondria dissociate from microtubules, are clustered in the perinuclear region and exhibit a severe reduction in $\Delta\Psi_m$; these changes are associated with increased apoptosis (Raj et al., 2004).

E2

E2 is a protein of about 365 amino acids (HPV genotype 16; its length varies among genotypes) whose expression gradually increases from the mid- to upper layers of infected lesions. This pattern of expression reflects the key role of E2 in the coordination of late events of viral replication through its influence on transcription, viral genome replication and partitioning in dividing cells, and RNA processing (reviewed by Graham, 2016).

A comparison of the properties of E2 from low-risk (genotypes 6, 11) and high-risk (genotypes 16, 18) viruses revealed stable nuclear compartmentalization of low-risk E2, and shuttling of high-risk E2 between the nucleus and cytoplasm, a property that was connected to the ability of high-risk E2 to induce apoptosis (Blachon et al., 2005). A more recent analysis of E2 trafficking provided evidence that the cytoplasmic fraction of E2 eventually accumulates in mitochondria. Lai et al. studied GFP-tagged E2 proteins from HPV-18 and HPV-6 (high- and low-risk genotypes, respectively) in a keratinocyte cell line (Lai et al., 2013). Time-lapse studies showed that HPV-18 GFP-E2 was gradually relocalized from the nucleus to mitochondria, whereas HPV-6 GFP-E2 remained predominantly nuclear and

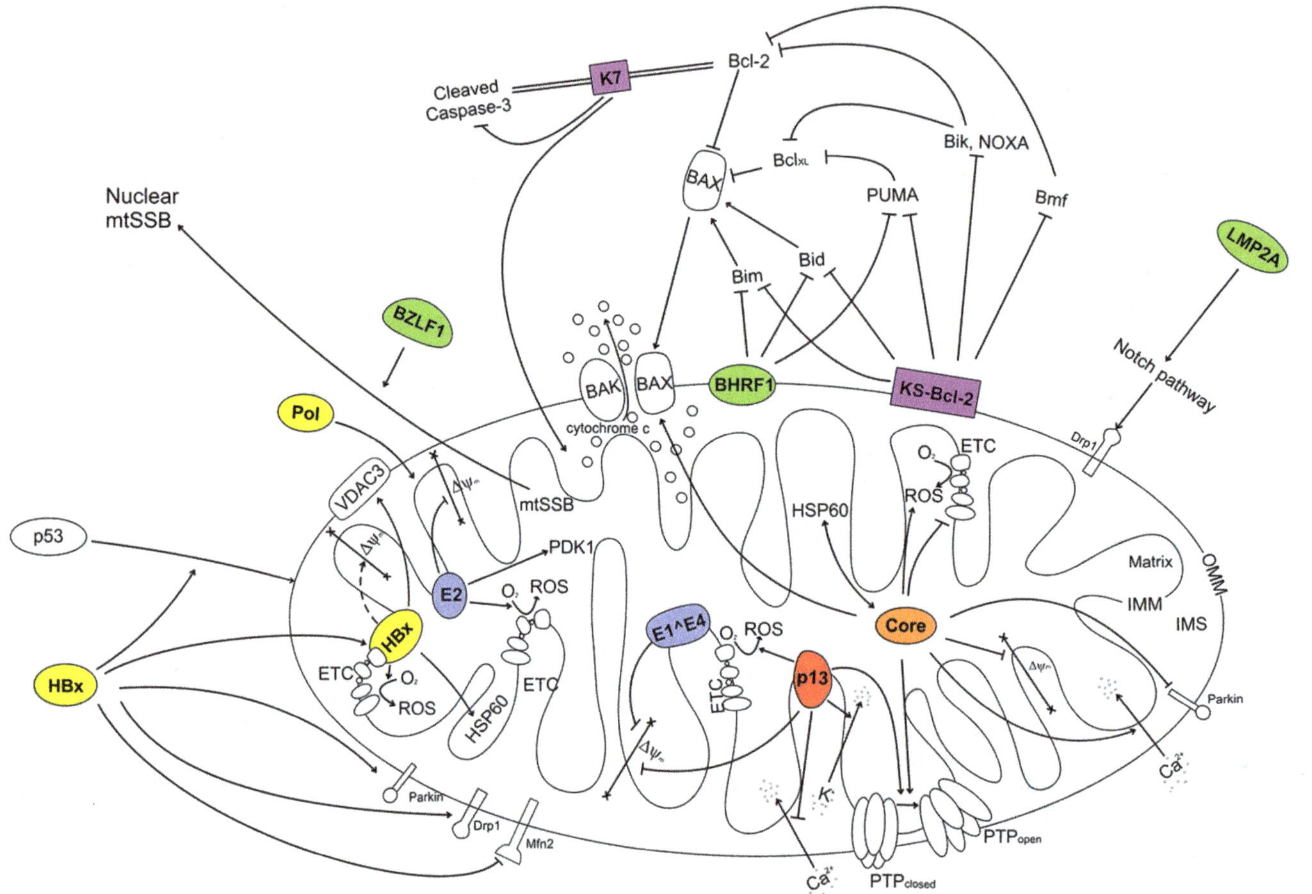

FIGURE 2 | Interactions of human tumor virus proteins with mitochondria. HTLV-1 (red): p13 causes an inward K^+ current that leads to mitochondrial swelling, depolarization and increased ROS production that lowers the PTP opening threshold. p13 also reduces mitochondrial Ca^{2+} uptake. EBV (green): The Bcl-2 homolog BHRF1 localizes at the OMM and binds to Bim, Bid and PUMA, resulting in the inhibition of Bax translocation to the OMM. BZLF1 interacts with mtSSB. LMP2A increases expression of Drp1 (dynamin-related protein 1) through stimulation of the Notch pathway. KSHV (purple): The Mcl-1 homolog KS-Bcl-2 localizes at the OMM and can bind and inhibit a variety of BH3-only proteins, resulting in the inhibition of Bax-Bak oligomerization at the OMM. The K7 protein forms a bridge between cellular Bcl-2 and cleaved Caspase-3, resulting in inhibition of Caspase-3 activity. HBV (yellow): HBx interacts with the Complex IV subunit COXIII and increases ROS generation by the ETC; HBx can interact with VDAC3 and HSP60. Furthermore, HBx can induce p53 translocation to mitochondria. HBx was also shown to influence mitochondrial dynamics through its interaction with Drp1 and Mnf2. Polymerase (Pol) contains an amino-terminal MTS that determines its mitochondrial targeting; it impact on mitochondria remains to be understood. HCV (orange): Core increases mitochondrial respiration, ROS generation, and uptake of Ca^{2+}, which sensitizes PTP opening. Core also inhibits translocation of Parkin to mitochondria, favors/facilitates/promotes Bax-Bak oligomerization, and interacts with the matrix chaperone HSP60. HPV (blue): The E2 protein interacts with IMM proteins and induce expression of the matrix protein PDK1 (pyruvate dehydrogenase kinase 1); E2 also increases ROS generation in mitochondria. The E1^E4 protein causes loss of $\Delta\Psi_m$. OMM, outer mitochondrial membrane; IMM, inner mitochondrial membrane; IMS, inter-membrane space; ROS, reactive oxygen species; MTS, mitochondrial targeting sequence; ETC, electron transport chain; PTP, permeability transition pore; not determined.

was rerouted to mitochondria only when expressed at high levels. Mass spectrometry analysis of proteins co-immunoprecipitated with HPV-18 E2 revealed the association of E2 with many mitochondrial proteins, mostly of the IMM, including subunits of Complexes III, IV, and V. Cells expressing HPV-18 GFP-E2 exhibited perinuclear clustering of the mitochondria and loss of the cristae structure. HPV-18 GFP-E2-expressing keratinocytes showed higher levels of mitochondrial ROS however, this effect did not induce cytochrome *c* release or apoptosis.

Lai et al. (2013) also showed that HPV-18 E2 stabilizes HIF-1α and induces the expression of known HIF-1α target genes, including PDK1 (pyruvate dehydrogenase kinase 1) and CAIX (carbonic anhydrase IX). These changes were accompanied by a modest but statistically significant increase in lactate production, an indication of a shift toward glycolytic metabolism. These effects were not observed in cells expressing low-risk HPV-6 E2. Although mitochondrial localization of HPV-18 E2 was evident Lai et al.'s transfection assays, their results of immunohistochemistry assays revealed cytoplasmic staining of HPV-18 E2 in a CINII lesion (grade II cervical intraepithelial neoplasia) and mainly nuclear accumulation of HPV-6 E2 in a benign condyloma.

A study by Chen et al. (2014) indicated that HPV-induced mitochondrial alterations also involve interplay between E2 and C1QBP (complement C1q-binding protein, also named gC1qR, receptor of the globular heads of complement C1q), a predominantly mitochondrial protein that has many roles, including regulation of mitochondrial and ER morphology and cell metabolism (Hu et al., 2013). Both HPV-16 E2 and C1QBP were less abundant in cervical carcinoma samples compared to non-neoplastic cervical tissue, suggesting a negative role for these proteins in the growth of cervical epithelial cells (Chen et al., 2014). Ectopic expression of E2 in keratinocytes led to upregulation of C1QBP and signs of mitochondrial dysfunction, including increased cellular ROS, augmented levels of cytosolic Ca^{2+}, loss of $\Delta\Psi_m$, and increased apoptosis; similar effects were observed in cells overexpressing C1QBP. These properties described for E2 contrast with previous observations made for the HPV oncoproteins E6 and E7, which downregulate C1QBP expression and protect cells from apoptosis (Gao et al., 2011; Chen et al., 2014). It is noteworthy that HCV Core protein also binds to C1QBP, resulting in impaired cytokine production (Song et al., 2016).

CONCLUDING REMARKS

Mitochondria play a central role in key biological processes, including energy conservation, cell death and Ca^{2+} signaling. These processes are involved both in the control of physiological tissue homeostasis and in the process of neoplastic transformation. It is thus not surprising that tumor viruses have developed common strategies to impinge on mitochondrial function. The experimental evidence collected so far suggests five major mechanisms by which proteins coded by tumor viruses interact with mitochondria (see **Figure 2**): (i) direct inhibition of apoptosis by viral Bcl-2 homologs; (ii) deregulation of cellular bioenergetics through the interaction with ETC components and inner mitochondrial membrane complexes; (iii) changes in the mitochondrial production of ROS, which indirectly influence cell turnover; (iv) changes in mitochondrial Ca^{2+} uptake, which influence Ca^{2+} homeostasis and Ca^{2+}-dependent signal transduction pathways; (v) interaction with OMM proteins that control innate immunity. Mitochondrial viral proteins can also alter mitochondrial morphology through interactions with the fission/fusion machinery.

Although much experimental information supports these points of interaction between tumor virus-encoded proteins and mitochondria, it must be emphasized that in most cases these results were obtained from experimental systems based on the ectopic overexpression of the viral protein of interest, in the absence of other viral proteins, and in cell types that are not the targets of natural infection by the viruses, conditions that may not recapitulate the situation in naturally infected cells. Further investigation is thus needed in order to validate and clarify the effects of viral-encoded mitochondrial proteins in the context of the viruses' life cycle and in connection to their pathogenic properties in the natural host.

In addition to providing information about the life cycle and pathogenesis of human tumor viruses, the examination of virus-mitochondrial interactions using more refined experimental systems will enhance our understanding of basic cellular processes controlled by mitochondria, thus paving the road to the identification of new strategies to clear viral infections and treat cancer patients.

AUTHOR CONTRIBUTIONS

IC prepared the paragraph on HPV and **Table 1**. GS prepared the paragraph on EBV and **Figures 1, 2**. MS-B prepared the paragraph on HSHV. VR prepared the paragraph on HTLV-1. VC prepared the paragraph on HBV and revised the final version of the paper. DD prepared the paragraph on HCV, **Table 2** and revised the final version of the paper.

ACKNOWLEDGMENTS

The authors thank Luigi Chieco-Bianchi for discussions. Research in VC laboratory is supported by an Investigator Grant from the Associazione Italiana per la Ricerca sul Cancro (AIRC IG # 17794), a University of Padova Ateneo grant, and a University of Verona-Veneto Institute of Oncology Joint Project grant. DD research is supported by the University of Padova. IC, VR, and MS-B are supported by fellowships from the Veneto Institute of Oncology and GS is a recipient of a Ph.D. fellowship from the Fondazione Cassa di Risparmio di Padova e Rovigo.

REFERENCES

Altmann, M., and Hammerschmidt, W. (2005). Epstein-Barr virus provides a new paradigm: a requirement for the immediate inhibition of apoptosis. *PLoS Biol.* 3:e404. doi: 10.1371/journal.pbio.0030404

Andresen, V., Pise-Masison, C. A., Sinha-Datta, U., Bellon, M., Valeri, V., Washington Parks, R., et al. (2011). Suppression of HTLV-1 replication by Tax-mediated re-routing of the p13 viral protein to nuclear speckles. *Blood* 11, 1549–1559.doi: 10.1182/blood-2010-06-293340

Anupam, R., Doueiri, R., and Green, P. L. (2013). The need to accessorize: molecular roles of HTLV-1 p30 and HTLV-2 p28 accessory proteins in the viral life cycle. *Front. Microbiol.* 4:275. doi: 10.3389/fmicb.2013.00275

Banerjee, A., Ray, R. B., and Ray, R. (2010). Oncogenic potential of hepatitis C virus proteins. *Viruses* 2, 2108–2133. doi: 10.3390/v2092108

Barba, G., Harper, F., Harada, T., Kohara, M., Goulinet, S., Matsuura, Y., et al. (1997). Hepatitis C virus core protein shows a cytoplasmic localization and associates to cellular lipid storage droplets. *Proc. Natl. Acad. Sci. U.S.A.* 94, 1200–1205.

Barbaro, G., Di Lorenzo, G., Asti, A., Ribersani, M., Belloni, G., Grisorio, B., et al. (1999). Hepatocellular mitochondrial alterations in patients with chronic hepatitis C: ultrastructural and biochemical findings. *Am. J. Gastroenterol.* 94, 2198–2205. doi: 10.1111/j.1572-0241.1999.01294.x

Bartenschlager, R., and Schaller, H. (1992). Hepadnaviral assembly is initiated by polymerase binding to the encapsidation signal in the viral RNA genome. *EMBO J* 11, 3413–3420.

Belgnaoui, S. M., Paz, S., and Hiscott, J. (2011). Orchestrating the interferon antiviral response through the mitochondrial antiviral signaling (MAVS) adapter. *Curr. Opin. Immunol.* 23, 564–572. doi: 10.1016/j.coi.2011.08.001

Bellows, D. S., Howell, M., Pearson, C., Hazlewood, S. A., and Hardwick, J. M. (2002). Epstein-Barr virus BALF1 is a BCL-2-like antagonist of the herpesvirus antiapoptotic BCL-2 proteins. *J. Virol.* 76, 2469–2479. doi: 10.1128/jvi.76.5.2469-2479.2002

Benali-Furet, N. L., Chami, M., Houel, L., De Giorgi, F., Vernejoul, F., Lagorce, D., et al. (2005). Hepatitis C virus core triggers apoptosis in liver cells by inducing ER stress and ER calcium depletion. *Oncogene* 24, 4921–4933. doi: 10.1038/sj.onc.1208673

Bender, S., Reuter, A., Eberle, F., Einhorn, E., Binder, M., and Bartenschlager, R. (2015). Activation of Type I and III interferon response by mitochondrial and peroxisomal MAVS and inhibition by Hepatitis C Virus. *PLoS Pathog.* 11:e1005264. doi: 10.1371/journal.ppat.1005264

Benhenda, S., Cougot, D., Buendia, M. A., and Neuveut, C. (2009). Hepatitis B virus X protein molecular functions and its role in virus life cycle and pathogenesis. *Adv. Cancer Res.* 103, 75–109. doi: 10.1016/S0065-230X(09)03004-8

Bernardi, P., Rasola, A., Forte, M., and Lippe, G. (2015). The mitochondrial permeability transition pore: channel formation by F-ATP synthase, integration in signal transduction, and role in pathophysiology. *Physiol. Rev.* 95, 1111–1155. doi: 10.1152/physrev.00001.2015

Bernardini, J. P., Lazarou, M., and Dewson, G. (2017). Parkin and mitophagy in cancer. *Oncogene* 36, 1315–1327. doi: 10.1038/onc.2016.302

Bhutani, M., Polizzotto, M. N., Uldrick, T. S., and Yarchoan, R. (2015). Kaposi sarcoma-associated herpesvirus-associated malignancies: epidemiology, pathogenesis, and advances in treatment. *Semin. Oncol.* 42, 223–246. doi: 10.1053/j.seminoncol.2014.12.027

Biasiotto, R., Aguiari, P., Rizzuto, R., Pinton, P., D'Agostino, D. M., and Ciminale, V. (2010). The p13 protein of human T cell leukemia virus type 1 (HTLV-1) modulates mitochondrial membrane potential and calcium uptake. *Biochim. Biophys. Acta* 1797, 945–951. doi: 10.1016/j.bbabio.2010.02.023

Blachon, S., Bellanger, S., Demeret, C., and Thierry, F. (2005). Nucleo-cytoplasmic shuttling of high risk human Papillomavirus E2 proteins induces apoptosis. *J. Biol. Chem.* 280, 36088–36098. doi: 10.1074/jbc.M505138200

Bouchard, M. J., Wang, L. H., and Schneider, R. J. (2001). Calcium signaling by HBx protein in hepatitis B virus DNA replication. *Science* 294, 2376–2378. doi: 10.1126/science.294.5550.2376

Carrere-Kremer, S., Montpellier-Pala, C., Cocquerel, L., Wychowski, C., Penin, F., and Dubuisson, J. (2002). Subcellular localization and topology of the p7 polypeptide of hepatitis C virus. *J. Virol.* 76, 3720–3730. doi: 10.1128/JVI.76.8.3720-3730.2002

Cavallari, I., Rende, F., Bender, C., Romanelli, M. G., D'Agostino, D. M., and Ciminale, V. (2013). Fine tuning of the temporal expression of HTLV-1 and HTLV-2. *Front. Microbiol.* 4:235. doi: 10.3389/fmicb.2013.00235

Cavallari, I., Rende, F., Bona, M. K., Sztuba-Solinska, J., Silic-Benussi, M., Tognon, M., et al. (2015). Expression of alternatively spliced human T-Cell Leukemia Virus Type 1 mRNAs is influenced by mitosis and by a novel cis-acting regulatory sequence. *J. Virol.* 90, 1486–1498. doi: 10.1128/JVI.02298-15

Cavallari, I., Rende, F., D'Agostino, D. M., and Ciminale, V. (2011). Converging strategies in expression of human complex retroviruses. *Viruses* 3, 1395–1414. doi: 10.3390/v3081395

Cen, O., and Longnecker, R. (2015). Latent Membrane Protein 2 (LMP2). *Curr. Top. Microbiol. Immunol.* 391, 151–180. doi: 10.1007/978-3-319-22834-1_5

Chami, M., Ferrari, D., Nicotera, P., Paterlini-Brechot, P., and Rizzuto, R. (2003). Caspase-dependent alterations of Ca^{2+} signaling in the induction of apoptosis by hepatitis B virus X protein. *J. Biol. Chem.* 278, 31745–31755. doi: 10.1074/jbc.M304202200

Chau, B. N., Cheng, E. H., Kerr, D. A., and Hardwick, J. M. (2000). Aven, a novel inhibitor of caspase activation, binds Bcl-xL and Apaf-1. *Mol. Cell* 6, 31–40. doi: 10.1016/S1097-2765(05)00021-3

Chen, J., and Siddiqui, A. (2007). Hepatitis B virus X protein stimulates the mitochondrial translocation of Raf-1 via oxidative stress. *J. Virol.* 81, 6757–6760. doi: 10.1128/JVI.00172-07

Chen, Z. L., Su, Y. J., Zhang, H. L., Gu, P. Q., and Gao, L. J. (2014). The role of the globular heads of the C1q receptor in HPV-16 E2-induced human cervical squamous carcinoma cell apoptosis via a mitochondria-dependent pathway. *J. Transl. Med.* 12:286. doi: 10.1186/s12967-014-0286-y

Chu, V. C., Bhattacharya, S., Nomoto, A., Lin, J., Zaidi, S. K., Oberley, T. D., et al. (2011). Persistent expression of hepatitis C virus non-structural proteins leads to increased autophagy and mitochondrial injury in human hepatoma cells. *PLoS ONE* 6:e28551. doi: 10.1371/journal.pone.0028551

Ciminale, V., Zotti, L., D'Agostino, D. M., Ferro, T., Casareto, L., Franchini, G., et al. (1999). Mitochondrial targeting of the p13II protein coded by the x-II ORF of human T-cell leukemia/lymphotropic virus type I (HTLV-I). *Oncogene* 18, 4505–4514.

Claus, C., and Liebert, U. G. (2014). A renewed focus on the interplay between viruses and mitochondrial metabolism. *Arch. Virol.* 159, 1267–1277. doi: 10.1007/s00705-013-1841-1

Clippinger, A. J., and Bouchard, M. J. (2008). Hepatitis B virus HBx protein localizes to mitochondria in primary rat hepatocytes and modulates mitochondrial membrane potential. *J. Virol.* 82, 6798–6811. doi: 10.1128/JVI.00154-08

Cross, J. R., Postigo, A., Blight, K., and Downward, J. (2008). Viral pro-survival proteins block separate stages in Bax activation but changes in mitochondrial ultrastructure still occur. *Cell Death Differ* 15, 997–1008. doi: 10.1038/cdd.2008.14

D'Agostino, D. M., Bernardi, P., Chieco-Bianchi, L., and Ciminale, V. (2005a). Mitochondria as Functional Targets of Proteins Coded by Human Tumor Viruses. *Adv. Cancer Res.* 94, 87–142. doi: 10.1016/S0065-230X(05)94003-7

D'Agostino, D. M., Ranzato, L., Arrigoni, G., Cavallari, I., Belleudi, F., Torrisi, M. R., et al. (2002). Mitochondrial alterations induced by the p13II protein of human T-cell leukemia virus type 1. Critical role of arginine residues. *J. Biol. Chem.* 277, 34424–34433. doi: 10.1074/jbc.M203023200

D'Agostino, D. M., Silic-Benussi, M., Hiraragi, H., Lairmore, M. D., and Ciminale, V. (2005b). The human T-cell leukemia virus type 1 p13II protein: effects on mitochondrial function and cell growth. *Cell Death Differ* 12(Suppl. 1), 905–915. doi: 10.1038/sj.cdd.4401576

Dash, S., Chava, S., Aydin, Y., Chandra, P. K., Ferraris, P., Chen, W., et al. (2016). Hepatitis C virus infection induces autophagy as a prosurvival mechanism to alleviate hepatic ER-stress response. *Viruses* 8:150. doi: 10.3390/v8050150

de Martel, C., Ferlay, J., Franceschi, S., Vignat, J., Bray, F., Forman, D., et al. (2012). Global burden of cancers attributable to infections in 2008: a review and synthetic analysis. *Lancet Oncol.* 13, 607–615. doi: 10.1016/S1470-2045(12)70137-7

Derse, D., Mikovits, J., and Ruscetti, F. (1997). X-I and X-II open reading frames of HTLV-I are not required for virus replication or for immortalization of primary T-cells *in vitro*. *Virology* 237, 123–128.

Doorbar, J. (2013). The E4 protein; structure, function and patterns of expression. *Virology* 445, 80–98. doi: 10.1016/j.virol.2013.07.008

Doorbar, J., Egawa, N., Griffin, H., Kranjec, C., and Murakami, I. (2015). Human papillomavirus molecular biology and disease association. *Rev. Med. Virol.* 25(Suppl. 1), 2–23. doi: 10.1002/rmv.1822

Doorbar, J., Ely, S., Sterling, J., McLean, C., and Crawford, L. (1991). Specific interaction between HPV-16 E1-E4 and cytokeratins results in collapse of the epithelial cell intermediate filament network. *Nature* 352, 824–827.

Dustin, L. B., Bartolini, B., Capobianchi, M. R., and Pistello, M. (2016). Hepatitis C virus: life cycle in cells, infection and host response, and analysis of molecular markers influencing the outcome of infection and response to therapy. *Clin. Microbiol. Infect.* 22, 826–832. doi: 10.1016/j.cmi.2016.08.025

Egawa, N., Egawa, K., Griffin, H., and Doorbar, J. (2015). Human Papillomaviruses; Epithelial Tropisms, and the Development of Neoplasia. *Viruses* 7, 3863–3890. doi: 10.3390/v7072802

Farinati, F., Cardin, R., De Maria, N., Della Libera, G., Marafin, C., Lecis, E., et al. (1995). Iron storage, lipid peroxidation and glutathione turnover in chronic anti-HCV positive hepatitis. *J. Hepatol.* 22, 449–456.

Feng, H., Dong, X., Negaard, A., and Feng, P. (2008a). Kaposi's sarcoma-associated herpesvirus K7 induces viral G protein-coupled receptor degradation and reduces its tumorigenicity. *PLoS Pathog* 4:e1000157. doi: 10.1371/journal.ppat.1000157

Feng, H., Shuda, M., Chang, Y., and Moore, P. S. (2008b). Clonal integration of a polyomavirus in human Merkel cell carcinoma. *Science* 319, 1096–1100. doi: 10.1126/science.1152586

Feng, P., Park, J., Lee, B. S., Lee, S. H., Bram, R. J., and Jung, J. U. (2002). Kaposi's sarcoma-associated herpesvirus mitochondrial K7 protein targets a cellular calcium-modulating cyclophilin ligand to modulate intracellular calcium concentration and inhibit apoptosis. *J. Virol.* 76, 11491–11504. doi: 10.1128/JVI.76.22.11491-11504.2002

Flanagan, A. M., and Letai, A. (2008). BH3 domains define selective inhibitory interactions with BHRF-1 and KSHV BCL-2. *Cell Death Differ* 15, 580–588. doi: 10.1038/sj.cdd.4402292

Fracchia, K. M., Pai, C. Y., and Walsh, C. M. (2013). Modulation of T cell metabolism and function through calcium signaling. *Front. Immunol.* 4:324. doi: 10.3389/fimmu.2013.00324

Fujinaga, H., Tsutsumi, T., Yotsuyanagi, H., Moriya, K., and Koike, K. (2011). Hepatocarcinogenesis in hepatitis C: HCV shrewdly exacerbates oxidative stress by modulating both production and scavenging of reactive oxygen species. *Oncol.* 81(Suppl. 1), 11–17. doi: 10.1159/000333253

Gallo, A., Lampe, M., Gunther, T., and Brune, W. (2017). The viral Bcl-2 homologs of kaposi's sarcoma-associated herpesvirus and rhesus rhadinovirus share an essential role for viral replication. *J. Virol.* 91:e01875-16. doi: 10.1128/JVI.01875-16

Gao, L. J., Gu, P. Q., Fan, W. M., Liu, Z., Qiu, F., Peng, Y. Z., et al. (2011). The role of gC1qR in regulating survival of human papillomavirus 16 oncogene-transfected cervical cancer cells. *Int. J. Oncol.* 39, 1265–1272. doi: 10.3892/ijo.2011.1108

Gelgor, A., Kalt, I., Bergson, S., Brulois, K. F., Jung, J. U., and Sarid, R. (2015). Viral Bcl-2 Encoded by the Kaposi's Sarcoma-Associated Herpesvirus Is Vital for Virus Reactivation. *J. Virol.* 89, 5298–5307. doi: 10.1128/JVI.00098-15

Gessain, A., and Cassar, O. (2012). Epidemiological Aspects and World Distribution of HTLV-1 Infection. *Front. Microbiol.* 3:388. doi: 10.3389/fmicb.2012.00388

Ghorbel, S., Sinha-Datta, U., Dundr, M., Brown, M., Franchini, G., and Nicot, C. (2006). Human T-cell leukemia virus type I p30 nuclear/nucleolar retention is mediated through interactions with RNA and a constituent of the 60 S ribosomal subunit. *J. Biol. Chem.* 281, 37150–37158. doi: 10.1074/jbc.M603981200

Goncalves, D. U., Proietti, F. A., Ribas, J. G., Araujo, M. G., Pinheiro, S. R., Guedes, A. C., et al. (2010). Epidemiology, treatment, and prevention of human T-cell leukemia virus type 1-associated diseases. *Clin. Microbiol. Rev.* 23, 577–589. doi: 10.1128/CMR.00063-09

Gong, G., Waris, G., Tanveer, R., and Siddiqui, A. (2001). Human hepatitis C virus NS5A protein alters intracellular calcium levels, induces oxidative stress, and activates STAT-3 and NF-kappa B. *Proc. Natl. Acad. Sci. U.S.A.* 98, 9599–9604. doi: 10.1073/pnas.171311298

Graham, S. V. (2016). Human Papillomavirus E2 protein: linking replication, transcription, and RNA processing. *J. Virol.* 90, 8384–8388. doi: 10.1128/JVI.00502-16

Griffin, S., Clarke, D., McCormick, C., Rowlands, D., and Harris, M. (2005). Signal peptide cleavage and internal targeting signals direct the hepatitis C virus p7 protein to distinct intracellular membranes. *J. Virol.* 79, 15525–15536. doi: 10.1128/JVI.79.24.15525-15536.2005

Griffin, S. D., Harvey, R., Clarke, D. S., Barclay, W. S., Harris, M., and Rowlands, D. J. (2004). A conserved basic loop in hepatitis C virus p7 protein is required for amantadine-sensitive ion channel activity in mammalian cells but is dispensable for localization to mitochondria. *J. Gen. Virol.* 85(Pt 2), 451–461. doi: 10.1099/vir.0.19634-0

Haqshenas, G., Mackenzie, J. M., Dong, X., and Gowans, E. J. (2007). Hepatitis C virus p7 protein is localized in the endoplasmic reticulum when it is encoded by a replication-competent genome. *J. Gen. Virol.* 88(Pt 1), 134–142. doi: 10.1099/vir.0.82049-0

Hara, Y., Yanatori, I., Ikeda, M., Kiyokage, E., Nishina, S., Tomiyama, Y., et al. (2014). Hepatitis C virus core protein suppresses mitophagy by interacting with parkin in the context of mitochondrial depolarization. *Am. J. Pathol.* 184, 3026–3039. doi: 10.1016/j.ajpath.2014.07.024

Harper, D. M., and DeMars, L. R. (2017). HPV vaccines - A review of the first decade. *Gynecol. Oncol.* 146, 196–204. doi: 10.1016/j.ygyno.2017.04.004

Henderson, S., Huen, D., Rowe, M., Dawson, C., Johnson, G., and Rickinson, A. (1993). Epstein-Barr virus-coded BHRF1 protein, a viral homologue of Bcl-2, protects human B cells from programmed cell death. *Proc. Natl. Acad. Sci. U.S.A.* 90, 8479–8483.

Henkler, F., Hoare, J., Waseem, N., Goldin, R. D., McGarvey, M. J., Koshy, R., et al. (2001). Intracellular localization of the hepatitis B virus HBx protein. *J. Gen. Virol.* 82(Pt 4), 871–882. doi: 10.1099/0022-1317-82-4-871

Hickish, T., Robertson, D., Clarke, P., Hill, M., di Stefano, F., Clarke, C., et al. (1994). Ultrastructural localization of BHRF1: an Epstein-Barr virus gene product which has homology with bcl-2. *Cancer Res.* 54, 2808–2811.

Hiraragi, H., Kim, S. J., Phipps, A. J., Silic-Benussi, M., Ciminale, V., Ratner, L., et al. (2006). Human T-lymphotropic virus type 1 mitochondrion-localizing protein p13(II) is required for viral infectivity *in vivo*. *J. Virol.* 80, 3469–3476. doi: 10.1128/JVI.80.7.3469-3476.2006

Horner, S. M., Liu, H. M., Park, H. S., Briley, J., and Gale, M. Jr. (2011). Mitochondrial-associated endoplasmic reticulum membranes (MAM) form innate immune synapses and are targeted by hepatitis C virus. *Proc. Natl. Acad. Sci. U.S.A.* 108, 14590–14595. doi: 10.1073/pnas.1110133108

Hou, X., Foley, S., Cueto, M., and Robinson, M. A. (2000). The human T-cell leukemia virus type I (HTLV-I) X region encoded protein p13(II) interacts with cellular proteins. *Virology* 277, 127–135. doi: 10.1006/viro.2000.0604

Hsu, W. L., Chung, P. J., Tsai, M. H., Chang, C. L., and Liang, C. L. (2012). A role for Epstein-Barr viral BALF1 in facilitating tumor formation and metastasis potential. *Virus Res.* 163, 617–627. doi: 10.1016/j.virusres.2011.12.017

Hu, M., Crawford, S. A., Henstridge, D. C., Ng, I. H., Boey, E. J., Xu, Y., et al. (2013). p32 protein levels are integral to mitochondrial and endoplasmic reticulum morphology, cell metabolism and survival. *Biochem. J.* 453, 381–391. doi: 10.1042/BJ20121829

Huh, K. W., and Siddiqui, A. (2002). Characterization of the mitochondrial association of hepatitis B virus X protein, HBx. *Mitochondrion* 1, 349–359. doi: 10.1016/S1567-7249(01)00040-X

IARC Monograph (2012). Biological agents. Volume 100 B. A review of human carcinogens. *IARC Monogr. Eval. Carcinog. Risks Hum.* 100(Pt B), 1–441.

Isherwood, B. J., and Patel, A. H. (2005). Analysis of the processing and transmembrane topology of the E2p7 protein of hepatitis C virus. *J. Gen. Virol.* 86(Pt 3), 667–676. doi: 10.1099/vir.0.80737-0

Journo, C., Douceron, E., and Mahieux, R. (2009). HTLV gene regulation: because size matters, transcription is not enough. *Future Microbiol.* 4, 425–440. doi: 10.2217/fmb.09.13

Kalt, I., Borodianskiy-Shteinberg, T., Schachor, A., and Sarid, R. (2010). GLTSCR2/PICT-1, a putative tumor suppressor gene product, induces the nucleolar targeting of the Kaposi's sarcoma-associated herpesvirus KS-Bcl-2 protein. *J. Virol.* 84, 2935–2945. doi: 10.1128/JVI.00757-09

Kang, S. M., Kim, S. J., Kim, J. H., Lee, W., Kim, G. W., Lee, K. H., et al. (2009). Interaction of hepatitis C virus core protein with Hsp60 triggers the production of reactive oxygen species and enhances TNF-alpha-mediated apoptosis. *Cancer Lett.* 279, 230–237. doi: 10.1016/j.canlet.2009.02.003

Kao, C. C., Yi, G., and Huang, H. C. (2016). The core of hepatitis C virus pathogenesis. *Curr. Opin. Virol.* 17, 66–73. doi: 10.1016/j.coviro.2016.01.009

Kim, S. J., Ding, W., Albrecht, B., Green, P. L., and Lairmore, M. D. (2003). A conserved calcineurin-binding motif in human T lymphotropic virus type 1 p12I functions to modulate nuclear factor of activated T cell activation. *J. Biol. Chem.* 278, 15550–15557. doi: 10.1074/jbc.M210210200

Kim, S. J., Khan, M., Quan, J., Till, A., Subramani, S., and Siddiqui, A. (2013a). Hepatitis B virus disrupts mitochondrial dynamics: induces fission and mitophagy to attenuate apoptosis. *PLoS Pathog.* 9:e1003722. doi: 10.1371/journal.ppat.1003722

Kim, S. J., Syed, G. H., and Siddiqui, A. (2013b). Hepatitis C virus induces the mitochondrial translocation of Parkin and subsequent mitophagy. *PLoS Pathog.* 9:e1003285. doi: 10.1371/journal.ppat.1003285

Korenaga, M., Wang, T., Li, Y., Showalter, L. A., Chan, T., Sun, J., et al. (2005). Hepatitis C virus core protein inhibits mitochondrial electron transport and increases reactive oxygen species (ROS) production. *J. Biol. Chem.* 280, 37481–37488. doi: 10.1074/jbc.M506412200

Kvansakul, M., Caria, S., and Hinds, M. G. (2017). The Bcl-2 family in host-virus interactions. *Viruses* 9:pii: E290. doi: 10.3390/v9100290

Kvansakul, M., Wei, A. H., Fletcher, J. I., Willis, S. N., Chen, L., Roberts, A. W., et al. (2010). Structural basis for apoptosis inhibition by Epstein-Barr virus BHRF1. *PLoS Pathog.* 6:e1001236. doi: 10.1371/journal.ppat.1001236

Lai, D., Tan, C. L., Gunaratne, J., Quek, L. S., Nei, W., Thierry, F., et al. (2013). Localization of HPV-18 E2 at mitochondrial membranes induces ROS release and modulates host cell metabolism. *PLoS ONE* 8:e75625. doi: 10.1371/journal.pone.0075625

Lairmore, M. D., Haines, R., and Anupam, R. (2012). Mechanisms of human T-lymphotropic virus type 1 transmission and disease. *Curr. Opin. Virol.* 2, 474–481. doi: 10.1016/j.coviro.2012.06.007

LaJeunesse, D. R., Brooks, K., and Adamson, A. L. (2005). Epstein-Barr virus immediate-early proteins BZLF1 and BRLF1 alter mitochondrial morphology during lytic replication. *Biochem. Biophys. Res. Commun.* 333, 438–442. doi: 10.1016/j.bbrc.2005.05.120

Lefebvre, L., Vanderplasschen, A., Ciminale, V., Heremans, H., Dangoisse, O., Jauniaux, J. C., et al. (2002). Oncoviral bovine leukemia virus G4 and human T-cell leukemia virus type 1 p13(II) accessory proteins interact with farnesyl pyrophosphate synthetase. *J. Virol.* 76, 1400–1414. doi: 10.1128/JVI.76.3.1400-1414.2002

Levrero, M., and Zucman-Rossi, J. (2016). Mechanisms of HBV-induced hepatocellular carcinoma. *J. Hepatol.* 64, S84–S101. doi: 10.1016/j.jhep.2016.02.021

Li, Y., Boehning, D. F., Qian, T., Popov, V. L., and Weinman, S. A. (2007). Hepatitis C virus core protein increases mitochondrial ROS production by stimulation of Ca^{2+} uniporter activity. *FASEB J.* 21, 2474–2485. doi: 10.1096/fj.06-7345com

Liang, Q., Chang, B., Brulois, K. F., Castro, K., Min, C. K., Rodgers, M. A., et al. (2013). Kaposi's sarcoma-associated herpesvirus K7 modulates Rubicon-mediated inhibition of autophagosome maturation. *J. Virol.* 87, 12499–12503. doi: 10.1128/JVI.01898-13

Liang, Q., Chang, B., Lee, P., Brulois, K. F., Ge, J., Shi, M., et al. (2015). Identification of the essential role of viral Bcl-2 for Kaposi's Sarcoma-Associated Herpesvirus Lytic Replication. *J. Virol.* 89, 5308–5317. doi: 10.1128/JVI.00102-15

Lim, W., Kwon, S. H., Cho, H., Kim, S., Lee, S., and Ryu, W. S. (2010). HBx targeting to mitochondria and ROS generation are necessary but insufficient for HBV-induced cyclooxygenase-2 expression. *J. Mol. Med.* 88, 359–369. doi: 10.1007/s00109-009-0563-z

Lo, S. Y., Masiarz, F., Hwang, S. B., Lai, M. M., and Ou, J. H. (1995). Differential subcellular localization of hepatitis C virus core gene products. *Virology* 213, 455–461. doi: 10.1006/viro.1995.0018

Lucifora, J., Arzberger, S., Durantel, D., Belloni, L., Strubin, M., Levrero, M., et al. (2011). Hepatitis B virus X protein is essential to initiate and maintain virus replication after infection. *J. Hepatol.* 55, 996–1003. doi: 10.1016/j.jhep.2011.02.015

Machida, K., Cheng, K. T., Lai, C. K., Jeng, K. S., Sung, V. M., and Lai, M. M. (2006). Hepatitis C virus triggers mitochondrial permeability transition with production of reactive oxygen species, leading to DNA damage and STAT3 activation. *J. Virol.* 80, 7199–7207. doi: 10.1128/JVI.00321-06

Madan, V., and Bartenschlager, R. (2015). Structural and Functional Properties of the Hepatitis C Virus p7 Viroporin. *Viruses* 7, 4461–4481. doi: 10.3390/v7082826

Marshall, W. L., Yim, C., Gustafson, E., Graf, T., Sage, D. R., Hanify, K., et al. (1999). Epstein-Barr virus encodes a novel homolog of the bcl-2 oncogene that inhibits apoptosis and associates with Bax and Bak. *J. Virol.* 73, 5181–5185.

McClain, S. L., Clippinger, A. J., Lizzano, R., and Bouchard, M. J. (2007). Hepatitis B virus replication is associated with an HBx-dependent mitochondrion-regulated increase in cytosolic calcium levels. *J. Virol.* 81, 12061–12065. doi: 10.1128/JVI.00740-07

McKenzie, J., and El-Guindy, A. (2015). Epstein-Barr Virus Lytic Cycle Reactivation. *Curr. Top. Microbiol. Immunol.* 391, 237–261. doi: 10.1007/978-3-319-22834-1_8

Meylan, E., Curran, J., Hofmann, K., Moradpour, D., Binder, M., Bartenschlager, R., et al. (2005). Cardif is an adaptor protein in the RIG-I antiviral pathway and is targeted by hepatitis C virus. *Nature* 437, 1167–1172. doi: 10.1038/nature04193

Milian, E., Prats, E., Cairo, J. J., Godia, F., and Vives, J. (2015). BHRF1 exerts an antiapoptotic effect and cell cycle arrest via Bcl-2 in murine hybridomas. *J. Biotechnol.* 209, 58–67. doi: 10.1016/j.jbiotec.2015.06.379

Morikawa, K., Lange, C. M., Gouttenoire, J., Meylan, E., Brass, V., Penin, F., et al. (2011). Nonstructural protein 3-4A: the Swiss army knife of hepatitis C virus. *J. Viral Hepat.* 18, 305–315. doi: 10.1111/j.1365-2893.2011.01451.x

Moriya, K., Fujie, H., Shintani, Y., Yotsuyanagi, H., Tsutsumi, T., Ishibashi, K., et al. (1998). The core protein of hepatitis C virus induces hepatocellular carcinoma in transgenic mice. *Nat. Med.* 4, 1065–1067. doi: 10.1038/2053

Moriya, K., Miyoshi, H., Shinzawa, S., Tsutsumi, T., Fujie, H., Goto, K., et al. (2010). Hepatitis C virus core protein compromises iron-induced activation of antioxidants in mice and HepG2 cells. *J. Med. Virol.* 82, 776–792. doi: 10.1002/jmv.21661

Moriya, K., Nakagawa, K., Santa, T., Shintani, Y., Fujie, H., Miyoshi, H., et al. (2001). Oxidative stress in the absence of inflammation in a mouse model for hepatitis C virus-associated hepatocarcinogenesis. *Cancer Res.* 61, 4365–4370.

Moriya, K., Yotsuyanagi, H., Shintani, Y., Fujie, H., Ishibashi, K., Matsuura, Y., et al. (1997). Hepatitis C virus core protein induces hepatic steatosis in transgenic mice. *J. Gen. Virol.* 78(Pt 7), 1527–1531. doi: 10.1099/0022-1317-78-7-1527

Morrison, B. J., Labo, N., Miley, W. J., and Whitby, D. (2015). Serodiagnosis for tumor viruses. *Semin. Oncol.* 42, 191–206. doi: 10.1053/j.seminoncol.2014.12.024

Motavaf, M., Safari, S., Saffari Jourshari, M., and Alavian, S. M. (2013). Hepatitis B virus-induced hepatocellular carcinoma: the role of the virus x protein. *Acta Virol.* 57, 389–396. doi: 10.4149/av_2013_04_389

Nieva, J. L., Madan, V., and Carrasco, L. (2012). Viroporins: structure and biological functions. *Nat. Rev. Microbiol.* 10, 563–574. doi: 10.1038/nrmicro2820

Nomura-Takigawa, Y., Nagano-Fujii, M., Deng, L., Kitazawa, S., Ishido, S., Sada, K., et al. (2006). Non-structural protein 4A of Hepatitis C virus accumulates on mitochondria and renders the cells prone to undergoing mitochondria-mediated apoptosis. *J. Gen. Virol.* 87(Pt 7), 1935–1945. doi: 10.1099/vir.0.81701-0

Okamoto, T., Omori, H., Kaname, Y., Abe, T., Nishimura, Y., Suzuki, T., et al. (2008). A single-amino-acid mutation in hepatitis C virus NS5A disrupting FKBP8 interaction impairs viral replication. *J. Virol.* 82, 3480–3489. doi: 10.1128/JVI.02253-07

Okuda, M., Li, K., Beard, M. R., Showalter, L. A., Scholle, F., Lemon, S. M., et al. (2002). Mitochondrial injury, oxidative stress, and antioxidant gene expression are induced by hepatitis C virus core protein. *Gastroenterology* 122, 366–375. doi: 10.1053/gast.2002.30983

Otani, K., Korenaga, M., Beard, M. R., Li, K., Qian, T., Showalter, L. A., et al. (2005). Hepatitis C virus core protein, cytochrome P450 2E1, and alcohol produce combined mitochondrial injury and cytotoxicity in hepatoma cells. *Gastroenterology* 128, 96–107. doi: 10.1053/j.gastro.2004.10.045

Pal, A. D., Basak, N. P., Banerjee, A. S., and Banerjee, S. (2014). Epstein-Barr virus latent membrane protein-2A alters mitochondrial dynamics promoting cellular migration mediated by Notch signaling pathway. *Carcinogenesis* 35, 1592–1601. doi: 10.1093/carcin/bgu069

Panfil, A. R., Martinez, M. P., Ratner, L., and Green, P. L. (2016). Human T-cell leukemia virus-associated malignancy. *Curr. Opin. Virol.* 20, 40–46. doi: 10.1016/j.coviro.2016.08.009

Pattingre, S., Tassa, A., Qu, X., Garuti, R., Liang, X. H., Mizushima, N., et al. (2005). Bcl-2 antiapoptotic proteins inhibit Beclin 1-dependent autophagy. *Cell* 122, 927–939. doi: 10.1016/j.cell.2005.07.002

Pearson, G. R., Luka, J., Petti, L., Sample, J., Birkenbach, M., Braun, D., et al. (1987). Identification of an Epstein-Barr virus early gene encoding a second component of the restricted early antigen complex. *Virology* 160, 151–161.

Peng, Y. T., Chen, P., Ouyang, R. Y., and Song, L. (2015). Multifaceted role of prohibitin in cell survival and apoptosis. *Apoptosis* 20, 1135–1149. doi: 10.1007/s10495-015-1143-z

Piccoli, C., Quarato, G., Ripoli, M., D'Aprile, A., Scrima, R., Cela, O., et al. (2009). HCV infection induces mitochondrial bioenergetic unbalance: causes and effects. *Biochim. Biophys. Acta* 1787, 539–546. doi: 10.1016/j.bbabio.2008.11.008

Piccoli, C., Scrima, R., D'Aprile, A., Ripoli, M., Lecce, L., Boffoli, D., et al. (2006). Mitochondrial dysfunction in hepatitis C virus infection. *Biochim. Biophys. Acta* 1757, 1429–1437. doi: 10.1016/j.bbabio.2006.05.018

Piccoli, C., Scrima, R., Quarato, G., D'Aprile, A., Ripoli, M., Lecce, L., et al. (2007). Hepatitis C virus protein expression causes calcium-mediated mitochondrial bioenergetic dysfunction and nitro-oxidative stress. *Hepatology* 46, 58–65. doi: 10.1002/hep.21679

Qi, H., Chu, V., Wu, N. C., Chen, Z., Truong, S., Brar, G., et al. (2017). Systematic identification of anti-interferon function on hepatitis C virus genome reveals p7 as an immune evasion protein. *Proc. Natl. Acad. Sci. U.S.A.* 114, 2018–2023. doi: 10.1073/pnas.1614623114

Rahmani, Z., Huh, K. W., Lasher, R., and Siddiqui, A. (2000). Hepatitis B virus X protein colocalizes to mitochondria with a human voltage-dependent anion channel, HVDAC3, and alters its transmembrane potential. *J. Virol.* 74, 2840–2846. doi: 10.1128/JVI.74.6.2840-2846.2000

Raj, K., Berguerand, S., Southern, S., Doorbar, J., and Beard, P. (2004). E1 empty set E4 protein of human papillomavirus type 16 associates with mitochondria. *J. Virol.* 78, 7199–7207. doi: 10.1128/JVI.78.13.7199-7207.2004

Rawat, S., Clippinger, A. J., and Bouchard, M. J. (2012). Modulation of apoptotic signaling by the hepatitis B virus X protein. *Viruses* 4, 2945–2972. doi: 10.3390/v4112945

Ren, J. H., Chen, X., Zhou, L., Tao, N. N., Zhou, H. Z., Liu, B., et al. (2016). Protective Role of Sirtuin3 (SIRT3) in Oxidative Stress Mediated by Hepatitis B Virus X Protein Expression. *PLoS ONE* 11:e0150961. doi: 10.1371/journal.pone.0150961

Rende, F., Cavallari, I., Corradin, A., Silic-Benussi, M., Toulza, F., Toffolo, G. M., et al. (2011). Kinetics and intracellular compartmentalization of HTLV-1 gene expression: nuclear retention of HBZ mRNA. *Blood* 117, 4855–4859. doi: 10.1182/blood-2010-11-316463

Ross-Thriepland, D., and Harris, M. (2015). Hepatitis C virus NS5A: enigmatic but still promiscuous 10 years on! *J. Gen. Virol.* 96(Pt 4), 727–738. doi: 10.1099/jgv.0.000009

Rouille, Y., Helle, F., Delgrange, D., Roingeard, P., Voisset, C., Blanchard, E., et al. (2006). Subcellular localization of hepatitis C virus structural proteins in a cell culture system that efficiently replicates the virus. *J. Virol.* 80, 2832–2841. doi: 10.1128/JVI.80.6.2832-2841.2006

Russo, J. J., Bohenzky, R. A., Chien, M. C., Chen, J., Yan, M., Maddalena, D., et al. (1996). Nucleotide sequence of the Kaposi sarcoma-associated herpesvirus (HHV8). *Proc. Natl. Acad. Sci. U.S.A.* 93, 14862–14867.

Sakamuro, D., Furukawa, T., and Takegami, T. (1995). Hepatitis C virus nonstructural protein NS3 transforms NIH 3T3 cells. *J. Virol.* 69, 3893–3896.

Santolini, E., Pacini, L., Fipaldini, C., Migliaccio, G., and Monica, N. (1995). The NS2 protein of hepatitis C virus is a transmembrane polypeptide. *J. Virol.* 69, 7461–7471.

Sarid, R., Sato, T., Bohenzky, R. A., Russo, J. J., and Chang, Y. (1997). Kaposi's sarcoma-associated herpesvirus encodes a functional bcl-2 homologue. *Nat. Med.* 3, 293–298.

Scheel, T. K., and Rice, C. M. (2013). Understanding the hepatitis C virus life cycle paves the way for highly effective therapies. *Nat. Med.* 19, 837–849. doi: 10.1038/nm.3248

Schwer, B., Ren, S., Pietschmann, T., Kartenbeck, J., Kaehlcke, K., Bartenschlager, R., et al. (2004). Targeting of hepatitis C virus core protein to mitochondria through a novel C-terminal localization motif. *J. Virol.* 78, 7958–7968. doi: 10.1128/JVI.78.15.7958-7968.2004

Scott, C., and Griffin, S. (2015). Viroporins: structure, function and potential as antiviral targets. *J. Gen. Virol.* 96, 2000–2027. doi: 10.1099/vir.0.000201

Seth, R. B., Sun, L., Ea, C. K., and Chen, Z. J. (2005). Identification and characterization of MAVS, a mitochondrial antiviral signaling protein that activates NF-kappaB and IRF 3. *Cell* 122, 669–682. doi: 10.1016/j.cell.2005.08.012

Shirakata, Y., and Koike, K. (2003). Hepatitis B virus X protein induces cell death by causing loss of mitochondrial membrane potential. *J. Biol. Chem.* 278, 22071–22078. doi: 10.1074/jbc.M301606200

Silic-Benussi, M., Biasiotto, R., Andresen, V., Franchini, G., D'Agostino, D. M., and Ciminale, V. (2010a). HTLV-1 p13, a small protein with a busy agenda. *Mol. Aspects Med.* 31, 350–358. doi: 10.1016/j.mam.2010.03.001

Silic-Benussi, M., Cannizzaro, E., Venerando, A., Cavallari, I., Petronilli, V., La Rocca, N., et al. (2009). Modulation of mitochondrial K^+ permeability and reactive oxygen species production by the p13 protein of human T-cell leukemia virus type 1. *Biochim. Biophys. Acta* 1787, 947–954. doi: 10.1016/j.bbabio.2009.02.001

Silic-Benussi, M., Cavallari, I., Zorzan, T., Rossi, E., Hiraragi, H., Rosato, A., et al. (2004). Suppression of tumor growth and cell proliferation by p13II, a mitochondrial protein of human T cell leukemia virus type 1. *Proc. Natl. Acad. Sci. U.S.A.* 101, 6629–6634. doi: 10.1073/pnas.0305502101

Silic-Benussi, M., Marin, O., Biasiotto, R., D'Agostino, D. M., and Ciminale, V. (2010b). Effects of human T-cell leukemia virus type 1 (HTLV-1) p13 on mitochondrial K^+ permeability: a new member of the viroporin family? *FEBS Lett.* 584, 2070–2075. doi: 10.1016/j.febslet.2010.02.030

Siu, G. K., Zhou, F., Yu, M. K., Zhang, L., Wang, T., Liang, Y., et al. (2016). Hepatitis C virus NS5A protein cooperates with phosphatidylinositol 4-kinase IIIalpha to induce mitochondrial fragmentation. *Sci. Rep.* 6:23464. doi: 10.1038/srep23464

Song, X., Yao, Z., Yang, J., Zhang, Z., Deng, Y., Li, M., et al. (2016). HCV core protein binds to gC1qR to induce A20 expression and inhibit cytokine production through MAPKs and NF-kappaB signaling pathways. *Oncotarget* 7, 33796–33808. doi: 10.18632/oncotarget.9304

Suzuki, R., Sakamoto, S., Tsutsumi, T., Rikimaru, A., Tanaka, K., Shimoike, T., et al. (2005). Molecular determinants for subcellular localization of hepatitis C virus core protein. *J. Virol.* 79, 1271–1281. doi: 10.1128/JVI.79.2.1271-1281.2005

Takada, S., Shirakata, Y., Kaneniwa, N., and Koike, K. (1999). Association of hepatitis B virus X protein with mitochondria causes mitochondrial aggregation at the nuclear periphery, leading to cell death. *Oncogene* 18, 6965–6973.

Tanaka, Y., Kanai, F., Kawakami, T., Tateishi, K., Ijichi, H., Kawabe, T., et al. (2004). Interaction of the hepatitis B virus X protein (HBx) with heat shock protein 60 enhances HBx-mediated apoptosis. *Biochem. Biophys. Res. Commun.* 318, 461–469. doi: 10.1016/j.bbrc.2004.04.046

Tibaldi, E., Venerando, A., Zonta, F., Bidoia, C., Magrin, E., Marin, O., et al. (2011). Interaction between the SH3 domain of Src family kinases and the proline-rich motif of HTLV-1 p13: a novel mechanism underlying delivery of Src family kinases to mitochondria. *Biochem. J.* 439, 505–516. doi: 10.1042/BJ20101650

Tomaic, V. (2016). Functional Roles of E6 and E7 Oncoproteins in HPV-Induced Malignancies at Diverse Anatomical Sites. *Cancers (Basel).* 8:95. doi: 10.3390/cancers8100095

Tsutsumi, T., Matsuda, M., Aizaki, H., Moriya, K., Miyoshi, H., Fujie, H., et al. (2009). Proteomics analysis of mitochondrial proteins reveals overexpression of a mitochondrial protein chaperon, prohibitin, in cells expressing hepatitis C virus core protein. *Hepatology* 50, 378–386. doi: 10.1002/hep.22998

Unchwaniwala, N., Sherer, N. M., and Loeb, D. D. (2016). Hepatitis B Virus Polymerase Localizes to the Mitochondria, and Its Terminal Protein Domain Contains the Mitochondrial Targeting Signal. *J. Virol.* 90, 8705–8719. doi: 10.1128/JVI.01229-16

Van Prooyen, N., Andresen, V., Gold, H., Bialuk, I., Pise-Masison, C., and Franchini, G. (2010). Hijacking the T-cell communication network by the human T-cell leukemia/lymphoma virus type 1 (HTLV-1) p12 and p8 proteins. *Mol. Aspects Med.* 31, 333–343. doi: 10.1016/j.mam.2010.07.001

Vieyres, G., Brohm, C., Friesland, M., Gentzsch, J., Wolk, B., Roingeard, P., et al. (2013). Subcellular localization and function of an epitope-tagged p7 viroporin in hepatitis C virus-producing cells. *J. Virol.* 87, 1664–1678. doi: 10.1128/JVI.02782-12

Wang, H., and Ryu, W. S. (2010). Hepatitis B virus polymerase blocks pattern recognition receptor signaling via interaction with DDX3: implications for immune evasion. *PLoS Pathog.* 6:e1000986. doi: 10.1371/journal.ppat.1000986

Wang, H. W., Sharp, T. V., Koumi, A., Koentges, G., and Boshoff, C. (2002). Characterization of an anti-apoptotic glycoprotein encoded by Kaposi's sarcoma-associated herpesvirus which resembles a spliced variant of human survivin. *EMBO J.* 21, 2602–2615. doi: 10.1093/emboj/21.11.2602

Wang, J. H., Yun, C., Kim, S., Lee, J. H., Yoon, G., Lee, M. O., et al. (2003). Reactive oxygen species modulates the intracellular level of HBx viral oncoprotein. *Biochem. Biophys. Res. Commun.* 310, 32–39. doi: 10.1016/j.bbrc.2003.08.113

Wang, T., Campbell, R. V., Yi, M. K., Lemon, S. M., and Weinman, S. A. (2010). Role of Hepatitis C virus core protein in viral-induced mitochondrial dysfunction. *J. Viral Hepat.* 17, 784–793. doi: 10.1111/j.1365-2893.2009.01238.x

Waris, G., Huh, K. W., and Siddiqui, A. (2001). Mitochondrially associated hepatitis B virus X protein constitutively activates transcription factors

STAT-3 and NF-kappa B via oxidative stress. *Mol. Cell. Biol.* 21, 7721–7730. doi: 10.1128/MCB.21.22.7721-7730.2001

Webster, D. P., Klenerman, P., and Dusheiko, G. M. (2015). Hepatitis C. *Lancet* 385, 1124–1135. doi: 10.1016/S0140-6736(14)62401-6

Wiedmer, A., Wang, P., Zhou, J., Rennekamp, A. J., Tiranti, V., Zeviani, M., et al. (2008). Epstein-Barr virus immediate-early protein Zta co-opts mitochondrial single-stranded DNA binding protein to promote viral and inhibit mitochondrial DNA replication. *J. Virol.* 82, 4647–4655. doi: 10.1128/JVI.02198-07

Willems, L., Hasegawa, H., Accolla, R., Bangham, C., Bazarbachi, A., Bertazzoni, U., et al. (2017). Reducing the global burden of HTLV-1 infection: an agenda for research and action. *Antiviral Res.* 137, 41–48. doi: 10.1016/j.antiviral.2016.10.015

Wolk, B., Sansonno, D., Krausslich, H. G., Dammacco, F., Rice, C. M., Blum, H. E., et al. (2000). Subcellular localization, stability, and trans-cleavage competence of the hepatitis C virus NS3-NS4A complex expressed in tetracycline-regulated cell lines. *J. Virol.* 74, 2293–2304. doi: 10.1128/JVI.74.5.2293-2304.2000

World Health Organization (2017). *Global Hepatitis Report 2017.* Geneva: World Health Organization.

Woodman, C. B., Collins, S. I., and Young, L. S. (2007). The natural history of cervical HPV infection: unresolved issues. *Nat. Rev. Cancer* 7, 11–22. doi: 10.1038/nrc2050

Wu, M., Xu, Y., Lin, S., Zhang, X., Xiang, L., and Yuan, Z. (2007). Hepatitis B virus polymerase inhibits the interferon-inducible MyD88 promoter by blocking nuclear translocation of Stat1. *J. Gen. Virol.* 88(Pt 12), 3260–3269. doi: 10.1099/vir.0.82959-0

Yang, B., and Bouchard, M. J. (2012). The hepatitis B virus X protein elevates cytosolic calcium signals by modulating mitochondrial calcium uptake. *J. Virol.* 86, 313–327. doi: 10.1128/JVI.06442-11

Young, L. S., Yap, L. F., and Murray, P. G. (2016). Epstein-Barr virus: more than 50 years old and still providing surprises. *Nat. Rev. Cancer* 16, 789–802. doi: 10.1038/nrc.2016.92

Zheng, B. Y., Fang, X. F., Zou, L. Y., Huang, Y. H., Chen, Z. X., Li, D., et al. (2014). The co-localization of HBx and COXIII upregulates COX-2 promoting HepG2 cell growth. *Int. J. Oncol.* 45, 1143–1150. doi: 10.3892/ijo.2014.2499

Zou, L. Y., Zheng, B. Y., Fang, X. F., Li, D., Huang, Y. H., Chen, Z. X., et al. (2015). HBx co-localizes with COXIII in HL-7702 cells to upregulate mitochondrial function and ROS generation. *Oncol. Rep.* 33, 2461–2467. doi: 10.3892/or.2015.3852

15

Mouse Models that Enhanced our Understanding of Adult T Cell Leukemia

Sara Moodad[1,2], Abdou Akkouche[1], Rita Hleihel[1], Nadine Darwiche[3], Marwan El-Sabban[2], Ali Bazarbachi[1,2] and Hiba El Hajj[1,4*]*

[1] Department of Internal Medicine, Faculty of Medicine, American University of Beirut, Beirut, Lebanon, [2] Department of Anatomy, Cell Biology and Physiological Sciences, Faculty of Medicine, American University of Beirut, Beirut, Lebanon, [3] Department of Biochemistry and Molecular Genetics, Faculty of Medicine, American University of Beirut, Beirut, Lebanon, [4] Department of Experimental Pathology, Immunology and Microbiology, Faculty of Medicine, American University of Beirut, Beirut, Lebanon

***Correspondence:**
Ali Bazarbachi
bazarbac@aub.edu.lb
Hiba El Hajj
he21@aub.edu.lb

Adult T cell Leukemia (ATL) is an aggressive lymphoproliferative malignancy secondary to infection by the human T-cell leukemia virus type I (HTLV-I) and is associated with a dismal prognosis. ATL leukemogenesis remains enigmatic. In the era of precision medicine in oncology, mouse models offer one of the most efficient *in vivo* tools for the understanding of the disease biology and developing novel targeted therapies. This review provides an up-to-date and comprehensive account of mouse models developed in the context of ATL and HTLV-I infection. Murine ATL models include transgenic animals for the viral proteins Tax and HBZ, knock-outs for key cellular regulators, xenografts and humanized immune-deficient mice. The first two groups provide a key understanding of the role of viral and host genes in the development of ATL, as well as their relationship with the immunopathogenic processes. The third group represents a valuable platform to test new targeted therapies against ATL.

Keywords: adult T cell leukemia, HTLV-I, mouse models, Tax, HBZ

INTRODUCTION

Human T Cell Leukemia Virus

Human T-cell leukemia virus type I (HTLV-I) retrovirus belongs to the deltaretroviridae family of viruses (reviewed in Matsuoka and Jeang, 2007). It is the first described oncogenic retrovirus and is responsible for a spectrum of diseases, the most aggressive of which is Adult T Cell Leukemia (ATL) (Poiesz et al., 1980; Hinuma et al., 1981; Yoshida et al., 1982, reviewed in Watanabe, 2017; Zhang et al., 2017). Approximately 5–20 million people are infected with HTLV-I worldwide (reviewed in Gessain and Cassar, 2012). However, the highest prevalence is reported in endemic areas that include Japan, the Caribbean, South America, inter-tropical Africa, Pacific islands, some areas in the Middle East, and Romania (Nosaka et al., 2017, reviewed in Edlich et al., 2003; Gessain and Cassar, 2012). The genome of this virus encodes for classical structural proteins required for retroviral replication and a series of accessory and regulatory proteins including the viral transcriptional activator Tax (Lee et al., 1984, reviewed in Azran et al., 2004) and the HTLV-I bZIP factor gene (HBZ), a viral protein encoded from the 3′ long terminal repeat (LTR) in the complementary strand of the proviral genome (Gaudray et al., 2002, reviewed in Matsuoka and Jeang, 2011; Giam and Semmes, 2016; Ma et al., 2016). Both Tax and HBZ are linked to HTLV-I pathogenesis (reviewed in Boxus and Willems, 2009; Kannian and Green, 2010; Giam and Semmes, 2016).

Adult T Cell Leukemia

Adult T cell leukemia is an aggressive hematological malignancy with very poor prognosis, high relapse rate, resistance to therapy, and a limited survival rate (Takatsuki et al., 1977; reviewed in Hermine et al., 1998; Bazarbachi et al., 2004, 2010; Goncalves et al., 2010; Marcais et al., 2013; Nasr et al., 2017; Watanabe, 2017). ATL develops in around 5% of infected carriers, following a long latency period exceeding 20 years (reviewed in Ishitsuka and Tamura, 2014; Bangham and Ratner, 2015).

Adult T cell leukemia is characterized by the presence of leukemic cells with atypical morphology and lobulated nucleus (Shimoyama et al., 1983). The majority of these cells are mature $CD3^+$ $CD4^+$ $CD25^+$ $CD7^-$ cells exhibiting an increased expression of the alpha chain of interleukin 2 receptor (IL-2R) but also an overexpression of Foxp3, a marker of T regulatory (T_{reg}) cells (Waldmann et al., 1984; Shimoyama, 1991; Okayama et al., 1997; Karube et al., 2004).

Tax as a Viral Oncoprotein

Tax is a 40 kDa viral transactivator protein that promotes viral transcription via the 5′-LTR. Tax functions as an oncogene resulting in leukemia (Grassmann et al., 1989; Tanaka et al., 1990). In rat fibroblasts, the expression of Tax is sufficient for induction of transformation and development of tumors (Grassmann et al., 1989, 1992; Tanaka et al., 1990). Tax transactivates transcription by activating promoters implicated in cell proliferation, activation, and survival leading to accumulation of diverse genetic and epigenetic mutations, genetic instability, cell cycle checkpoint disruption, and damage of DNA repair mechanisms (reviewed in Marriott and Semmes, 2005; Kfoury et al., 2012; Bazarbachi, 2016; Watanabe, 2017). Tax binds critical transcription factors including the cAMP response element binding protein (Zhao and Giam, 1992; Brauweiler et al., 1995; Giebler et al., 1997), AP-1 (Fujii et al., 2000; Iwai et al., 2001), and serum response element (SRF) (Fujii et al., 1991, 1992). In addition, Tax inactivates tumor suppressor genes including p53 (Pise-Masison et al., 1998; Portis et al., 2001) and p16 (Suzuki et al., 1996), represses the expression of cyclin A, and antagonizes apoptosis through inhibiting apoptotic genes expression such as Bax and promoting anti-apoptotic ones including Bcl-xL and BFI-1 (Brauweiler et al., 1997; Nicot et al., 2000; Marriott and Semmes, 2005; Macaire et al., 2012).

Importantly, Tax activates the NF-κB pathway (Sun et al., 1994; Good and Sun, 1996; Mori et al., 1999; Hironaka et al., 2004), after binding the regulatory subunit of the IkappaB kinase (IKK) complex called nemo or IKK-γ (Harhaj and Sun, 1999; Kfoury et al., 2008; Wang et al., 2016) resulting in the activation of downstream effector genes (reviewed in Kfoury et al., 2005). Furthermore, Tax induced NF-κB activation is highly dependent on its post translational modifications namely ubiquitylation and sumoylation (reviewed in Kfoury et al., 2012). Tax also affects the expression of various micro RNAs (mi-RNA) including mi-RNA31 known to inhibit the expression of the NF-κB non-canonical pathway components (Yamagishi et al., 2012). Tax-induced mi-RNA31 downregulation occurs via a deregulation of polycomb proteins, leading to a consequent activation of NF-κB, and inhibition of apoptosis in ATL cells (Yamagishi et al., 2012). Moreover, Tax modulates the microenvironment and increases ATL cells' invasion and extravasation through affecting gap junctions between endothelial cells and infected cells (El-Sabban et al., 2002; Bazarbachi et al., 2004).

Due to genetic/epigenetic alterations in the HTLV-I genome, including mutations and promoter methylation, most ATL cells lack detectable Tax expression (Takeda et al., 2004). Despite its undetectable levels in ATL patients (Matsuoka and Jeang, 2007), Tax is essential for ATL cells survival as its silencing results in cell death (Dassouki et al., 2015; Bazarbachi, 2016). Recent data revealed transient Tax expression in the form of Tax bursts (Tax expression switching on/off) occurring in a small fraction of ATL-derived or HTLV-I transformed cells (Mahgoub et al., 2018, reviewed in Bangham and Matsuoka, 2017). In a similar context, ATL epigenome was analyzed and an ATL-specific "epigenetic code" paramount for cell identity was deciphered (Fujikawa et al., 2016). In more details, Tax was shown to induce an epigenetic-dependent global alteration including increased polycomb complex 2 (PRC2) trimethylation at histone 3 (H3k27me3) resulting in cellular transformation, immortalization, and epigenome reprogramming, similar to that observed in ATL patients (Fujikawa et al., 2016). All these properties make Tax an ideal oncoprotein for *in vivo* investigation.

HBZ Biology in ATL

HBZ is a nuclear protein encoded by the complementary strand of HTLV-I RNA genome (Larocca et al., 1989; Gaudray et al., 2002). Unlike Tax that is often undetected in ATL cells, *Hbz* gene undergoes no abortive mutations and the protein is expressed in all ATL patients and HTLV-I infected carriers (Fan et al., 2010; Kataoka et al., 2015; reviewed in Satou et al., 2006; Matsuoka and Jeang, 2011). HBZ was found to be a negative regulator of Tax-mediated viral transcription (Gaudray et al., 2002). This opposite expression pattern of the two proteins may indicate a possible differential role in HTLV-I pathogenesis and suggests HBZ as a candidate for a possible HTLV-I vaccine (Mahieux, 2015; Sugata et al., 2015). The mRNA of HBZ positively correlates with the proviral load of HTLV-I in carriers, and ATL patients (Saito et al., 2009). *In vitro*, HBZ promotes the proliferation of ATL cells but its suppression by short hairpin RNA (shRNA) results in modest inhibition of ATL cells proliferation (Satou et al., 2006; Arnold et al., 2008). HBZ affects several cellular pathways implicated in cellular proliferation such as NF-κB (Zhao et al., 2009; Panfil et al., 2016), AP-1 (Matsumoto et al., 2005), JunD (Thebault et al., 2004; Kuhlmann et al., 2007), c-Jun, JunB (Basbous et al., 2003), and CREB (Lemasson et al., 2007). In contrast to Tax which constitutively activates both canonical and non-canonical NF-κB pathways, HBZ was shown to inhibit the canonical pathway of NF-κB via proteasomal degradation of p65 while the non-canonical pathway was not affected (Zhao et al., 2009; Panfil et al., 2016).

Animal Models in ATL

Due to the complexity of HTLV-I associated diseases and the enigmatic mechanisms dictating their occurrence, in particular in ATL, animal models have been instrumental in providing a platform for answering pivotal questions related to HTLV-I infection, disease progression, and importantly developing new effective therapeutic approaches (reviewed in El Hajj et al., 2012; Niewiesk, 2016). Among these models, rabbits, monkeys but also rats were useful to understand early HTLV-I viral infection and transmission as well as the induced host immune response against the virus (reviewed in El Hajj et al., 2012). More recently, transgenic *Drosophila* models expressing Tax in the compound eye and plasmatocytes were generated (Shirinian et al., 2015). However, mice remain by far one of the most efficient tools helping in understanding the biology of this affliction. Murine ATL models include transgenic animals for the viral proteins Tax and HBZ, xenografts inoculated with ATL cells (either cells lines or patient-derived cells) and humanized mouse models (reviewed in Panfil et al., 2013; Niewiesk, 2016). In this review, we attempt to provide an updated summary of these various mouse models, the key advances they offered in the understanding of HTLV-I infection, as well as their contribution to ATL research and drug development.

MOUSE MODELS OF ATL

Immunocompromised Mouse Models

Mice are relevant tools to study the molecular mechanisms of carcinogenesis and to develop new antitumor therapies. However, in immunocompetent mice, transplantation is often hindered by the functional host immune response resulting in low or no tumor engraftment. This problem was overcome after the discovery of the immunocompromised CB17 *scid/scid* (SCID) mouse model making a revolution in the cancer field. These mice harbor a spontaneous non-sense mutation in the *scid* gene, encoding for the protein kinase DNA activated catalytic polypeptide (Pkrdc), indispensable for efficient B and T lymphocytes recombination (Bosma et al., 1983). The loss of *Pkrdc* results in impaired adaptive immunity whereby B and T cells are both non-functional. Despite the lack of adaptive immunity, SCID mice retain a normal innate immunity in which macrophages, antigen-presenting cells, and natural killer (NK) cells carry normal functions (Bosma et al., 1983).

To further improve tumor engraftment, a non-obese diabetic (NOD/SCID) model exhibiting additional mutations resulting in further impairment of NK activity was generated (Shultz et al., 1995). This model was further immunosuppressed to generate the NOD/SCID β2-microglobulin^{null} mice in which the *β2-microglobulin* gene was deleted resulting in a complete abolishment of the NK cell activity (Koller and Smithies, 1989). Importantly, a NOD/SCID IL2-R$\gamma^{-/-}$ or NSG model was generated by deletion or truncation of the gamma chain of IL-2R (Ito et al., 2002), reviewed in (Ito et al., 2008). Therefore, in addition to all the abnormalities of their predecessors, NSG mice possess a defective production of IL-2, IL-4, IL-7, IL-9, IL-15, and IL-21 as well as a severe impairment of the dendritic cell (DC) and their capacity to produce interferon γ (IFN-γ) upon stimulation (Ito et al., 2002; Ishikawa et al., 2005). For further immunosuppression, the Rag2$^{-/-}\gamma c^{-/-}$ model was established. These mice have a deletion of the Recombination Activating Genes (*RAG2*), impairing the production of both T and B cells and NK cell-mediated immunity in murine hosts. Moreover, because the rag proteins are not involved in DNA repair, *RAG2*-deleted mice do no show the leakiness or radio-sensitivity observed in SCID mice (Traggiai et al., 2004; reviewed in Chicha et al., 2005). Because of these properties, these mice allowed to study human hematopoiesis (Hiramatsu et al., 2003; Ishikawa et al., 2005; reviewed in Pearson et al., 2008).

ATL Development in Xenograft Mouse Models

The use of xenograft mice has provided invaluable information pertaining to the tumorigenic and proliferative potential of ATL (Ohsugi et al., 1994). Under this section, we will provide an overview of most tested immunocompromised animals injected with HTLV-I-transformed or ATL-derived cell lines or those injected with patient-derived ATL cells. Initial xenograft studies investigating ATL development and progression were performed in the SCID mouse model (Ishihara et al., 1992; Feuer et al., 1993; Kondo et al., 1993). Later studies reported the use of NOD/SCID and SCID/Beige mice for injection of transformed or immortalized cell lines (Liu et al., 2002). More recently, an HBZ xenograft mouse model was generated by transplantation of retrovirally transduced T cells with Bcl-xL, AKT, and HBZ (Kasugai et al., 2016). Only mice groups transplanted with T cells triply transduced with plasmids encoding for the three proteins generated tumors, highlighting the need of key components from different cellular pathways along with the viral HBZ for cellular transformation (Kasugai et al., 2016) (**Table 1**).

Xenograft Mouse Models as Platforms for ATL Targeted Drug Development

Given its resistance to therapy and its high relapse rate, ATL remains an aggressive disease with an unfavorable prognosis (reviewed in Bazarbachi et al., 2011; Nasr et al., 2017). Despite remarkable progress in ATL therapies with the use of antiviral therapy and allogeneic stem cell transplantation (Bazarbachi et al., 2010, 2014), most patients relapse highlighting the necessity of novel therapeutic approaches (**Table 2**).

Targeting the NF-κB pathway in ATL therapy

Adult T cell leukemia development entails the deregulation of multiple cellular pathways, including the constitutive activation of the NF-κB pathway (reviewed in Kfoury et al., 2005), making this pathway an attractive therapeutic target against ATL. Accordingly, a wide array of NF-κB inhibitors including specific (Bay 11-7082), and non-specific inhibitors (such as bortezomib and Dehydroxymethylepoxyquinomicin DHMEQ), were tested in ATL xenograft mouse models.

Bay 11-7082, specifically inhibiting the NF-κB DNA binding activity, prevented primary tumor growth and leukemic organ infiltration in NSG mice xenografted with HTLV-I infected cell lines (Dewan et al., 2003). Bortezomib prevented tumor growth in an ED, ATL-derived T cell line, xenografted SCID model (Satou et al., 2004). DHMEQ, known to inhibit the activation of NF-κB by preventing the nuclear translocation of its active subunit p65

TABLE 1 | Summary of the contribution of mouse models to ATL biology.

Model	Contribution to ATL biology	Reference
Xenografts		
Xenograft mouse models injected with HTLV-I transformed, ATL-derived cell lines or ATL patients derived cells	Represented a platform for targeted drug development.	Zimmerman et al., 2011; Ishitsuka et al., 2012; Masaki et al., 2013; Saitoh et al., 2016
	Provided a potential tool to test therapeutic agents: – Targeting the constitutively active NF-κB pathway. – Specifically targeting leukemic cells.	Phillips et al., 2000; Zhang et al., 2003, 2005; Maeda et al., 2010, 2015
	Allowed a better understanding of the immune response against HTLV-I.	Ishihara et al., 1992; Feuer et al., 1995; Stewart et al., 1996; Uchiyama, 1996; Liu et al., 2002
	Was instrumental for studying early stages of primary HTLV-I infection and subsequent clonal proliferation.	Miyazato et al., 2006
	Allowed the assessment of early steps of HTLV-I infection, proviral load, and clonal proliferation.	Takajo et al., 2007
Humanized		
HTLV-I infected human $CD34^+$ in $Rag2^{-/-}$ gamma $c^{-/-}$ mice models	Allowed the understanding of early infection stages. Confirmed the *in vivo* correlation of Tax and NF-κB activation upon expansion of $CD4^+CD25^+$ malignant cells.	Villaudy et al., 2011
HTLV-I infected human $CD133^+$ in NSG	Generated a human adaptive immune system in immunodeficient mice. Was the closest model to recapitulate the *in vivo* ATL development. Assessed the initiated immune system against the virus and clonal selection.	Tezuka et al., 2014
***Tax* Transgenics** LTR-*Tax*-Tg	Established Tax as an oncoprotein and HTLV-I as transforming virus resulting in mesenchymal tumors and neurofibroma.	Hinrichs et al., 1987; Nerenberg et al., 1987
	Showed that Tax expression in oxidative fibers resulted in HTLV-I associated myopathies.	Nerenberg and Wiley, 1989
	Showed that Tax induced autoimmune like Sjogren like syndrome.	Green et al., 1989
	Showed that Tax induced skeletal abnormalities and fragile bones similar to ATL patients.	Ruddle et al., 1993
	Showed that Tax is arthrogenic and induced ankylotic arthropathy.	Habu et al., 1999
	Showed that Immune system activation contributes to ATL pathogenesis in infected carriers.	Swaims et al., 2010
Double transgenic -β*gal*-Tg model LTR-*Tax* HTLV-I LTR	Unveiled tissues supporting tax-mediated transcriptional transactivation. Provided a model system to study the mechanism of gene regulation by Tax.	Bieberich et al., 1993
huGMZB*Tax* GMZ-*Tax*-Tg Granzyme promoter	First model to generate leukemia (LGL), tumor infiltration, and splenomegaly partly resembling ATL.	Grossman et al., 1995
	Showed that Tax functionally inactivates P53 contributing to late stage tumor progression.	Portis et al., 2001
	Showed that innate immune system, specifically IFN-γ, is crucial for ATL development.	Mitra-Kaushik et al., 2004
	Revealed malignant hypercalcemia and osteolytic bone lesions resembling human ATL.	Gao et al., 2005
	Showed that Tax expression *in vivo* induced constitutive activation of HTLV-I.	Bernal-Mizrachi et al., 2006
	Revealed that Tax activation of lymphocytes recruits, activates, and transforms NK/T-cells.	Rauch et al., 2009b
CD3- *Tax* Tg CD3-epsilon promoter/enhancer	Model failed to develop leukemia. Tax expression closely associated with apoptosis *in vivo*.	Hall et al., 1998
tTA/*Tax* mice Bi-transgenic doxycycline inducible model, EmuSR alpha promoter-enhancer	Revealed ATL-like cutaneous lesions and splenomegaly via HTLV-I activation. Showed that Tax or HTLV-I suppression resolves cutaneous symptoms.	Kwon et al., 2005
lck-*Tax*-tg model Lck-proximal promoter	Showed diffuse large cell lymphoma after prolonged latency. Model exhibits acute ATL like symptoms and HTLV-I activation.	Hasegawa et al., 2006
	Provided a candidate ATL stem cells of $CD38^-/CD71^-/CD117^+$ phenotype and decreased expression of Tax, Notch, BMI1 were isolated.	Yamazaki et al., 2009

(Continued)

TABLE 1 | Continued

Model	Contribution to ATL biology	Reference
lck-*Tax*-tg model Lck-Distal promoter	Showed leukemia of mature $CD4^+$ cells resembling mature $CD4^+$ ATL cells	Ohsugi et al., 2007b
***HBZ* transgenic**		
HBZ-Tg model CD4-specific promoter/enhancer/silencer	Showed systemic inflammation and later lymphoma in 30% of mice upon aging.	Satou et al., 2011
	Increased effector/memory $CD4^+$ cells and functionally impaired $CD4^+$ $Foxp3^+$ T_{reg} cells.	Yamamoto-Taguchi et al., 2013
	Showed that HBZ promotes a pro-inflammatory phenotype via labile Foxp3 expression.	Sugata et al., 2012
	Showed that HBZ suppresses Th1 cytokines and impairs cell-mediated immunity.	Kuribayashi et al., 2016
	HBZ-Tg mice exhibit ATL stem cells of c-kit^+/$CD4^-$/$CD8^-$ phenotype.	
Double Transgenic *Tax/HBZ* *HBZ/Tax* model CD4 promoter/enhancer	This model failed to generate ATL-like leukemia. Skin lesions, T-cell lymphoma, and splenomegaly with increased $CD4^+$ memory and $Foxp3^+T_{reg}$ cells.	Zhao et al., 2014

TABLE 2 | ATL mouse models as platform for ATL targeted therapy.

Model	Drug	Drug target	ATL therapy	Reference
Xenograft model	Bay 11-7082	NF-κB	Prevented tumor growth and infiltration.	Dewan et al., 2003
	Bortezomib	NF-κB	Prevented tumor growth in an ED SCID.	Satou et al., 2004
	DHMEQ	NF-κB	Prolonged survival and Prevented tumor growth.	Ohsugi et al., 2007a
	9-aminoacridine and Campath-1H	NF-κB and CD25	Prolonged survival and induced P53-mediated apoptosis.	Ju et al., 2014
	Compound E, Bortezomib, and Romidepsin	γ-secretase, NF-κB, and HDAC	Assessed interaction between Notch-1 and HTLV-I pathways, combination exhibited synergy supporting clinical trials.	Yu et al., 2015
	AR-42	HDAC	Prolonged survival.	Zimmerman et al., 2011
	Campath-1H and HAT or MEDI 507	CD2 and CD25	Targeting different CD25 epitopes exhibited synergy.	Zhang et al., 2006
	Flavopiridol and HAT	cyclin-dependent kinase and CD25	Synergy enhancing antitumor effect and survival.	Chen et al., 2009
	HAT, MAT, and 7G7B6	CD25	Inhibited tumor growth.	Maeda et al., 2010
	7G7/B6 and daclizumab	CD25	Presented osteoponin-integrin interaction as novel therapeutic target for ATL.	Maeda et al., 2015
	Daclizumab and Depsipeptide	CD25 and HDAC inhibition	Improved survival and attenuated tumor infiltration and viral production.	Ikebe et al., 2013
	A20 ShRNA	A20, ubiquitin-editing Enzyme	Decreased tumor growth and revealed a novel role for ubiquitin-editing enzymes in ATL development.	Saitoh et al., 2016
	ABT-737	Bcl-2 and Bcl-xL inhibition	Inhibited tumor growth.	Ishitsuka et al., 2012
	Adoptive patient-autologous Tax-CTL	ATL cells	Decreased tumor infiltration and enhanced survival *in vivo*.	Masaki et al., 2013
Humanized model	Tinofovir and Azidothymidine	Reverse Transcriptase inhibition	Prophylactic potential by blocking primary infection *in vivo*.	Miyazato et al., 2006
	TARC-PE38	CCR-4	CCR4 is a potential ATL target.	Hiyoshi et al., 2015
Transgenic model	Arsenic/IFN	NF-κB, Tax, LIC	Cured ATL via LIC elimination.	El Hajj et al., 2010
	ST1926	NF-κB, Tax, LIC	Highlights retinoids as promising therapies by enhancing survival and decreasing tumor infiltration.	El Hajj et al., 2014

DHMEQ, dehydroxymethylepoxyquinomicin; HDAC, histone deacetylase; HAT, humanized anti-Tac; MAT, murin anti-Tac; arsenic, arsenic Trioxide; IFN, interferon-alpha; LIC, leukemia initiating cells.

(Ohsugi et al., 2006), showed a significantly prolonged survival and prevented tumor growth in an ATL NSG xenograft mouse model (Ohsugi et al., 2007a).

In a similar context, a small molecule, 9-aminoacridine (9AA) selectively induced *in vitro* ATL cell death through inhibition of the NF-κB pathway and induction of p53 responsive genes (Ju et al., 2014). The efficacy of 9AA alone or in combination with Campath-1H (a monoclonal antibody directed against CD52) was assessed using a xenograft NOD/SCID model, inoculated with MET-1 cells. MET-1 are activated T cells that express CD2,

CD3, CD4, CD25, CD122, and CD52. An enhanced survival of tumor bearing mice was reported upon treatment with both compounds as compared to either compound alone. This involved the induction of p53, PARP cleavage, and apoptosis of splenic cells from leukemia-bearing mice (Ju et al., 2014).

Finally, and in order to assess the physical and functional interaction between Notch-1 signaling (increased in some ATL patients) and NF-κB activation, the therapeutic efficacy of the γ-secretase inhibitor compound E along with bortezomib and romidepsin, a histone deacetylase inhibitor (HDACI), was explored in an NSG model injected with MT-1 cells (Yu et al., 2015). Each of the three reagents alone or in double combination inhibited tumor growth as monitored by tumor size, the level of tumor markers in the serum, and significantly prolonged the survival of tumor-bearing animals (Yu et al., 2015).

At the cellular and molecular levels, and to assess the *in vivo* implication of NF-κB in ATL pathogenesis, Nitta et al. (2008) investigated ATL development in a defective NF-κB setting. In this study, alymphoplasia *(aly/aly)* mice bearing an NF-κB inducing kinase (NIK) mutation were used. These mice harbored defects in lymphoid organs development and severe deficiencies in both humoral and cell-mediated immunity. In contrary to BALB/c and C57BL/6J control mice, *aly/aly* animals inoculated with HTLV-I producing MT-2 cells, did not maintain the provirus and antibodies against HTLV-I were not detected suggesting that NIK is required for the initial proliferation and maintenance of HTLV-I infected cells in mice (Nitta et al., 2008).

Targeted ATL drug development other than NF-κB inhibition

One of the suggested approaches in ATL treatment is virotherapy. Oncolytic virotherapy is a relatively new approach in cancer treatment, utilizing replication-competent viruses that selectively target cancer cells while sparing the normal healthy ones (Russell et al., 2012). MET-1 NOD/SCID model was used to evaluate the potential of measles-virus-virotherapy in treating ATL. Measles virus treatment of tumor cells lacking type I interferon (IFN-α) secretion was shown to be more efficient both *in vitro* and *in vivo* as compared to other tumor cells (Parrula et al., 2011). Using the same animal model, Zimmerman et al. (2011) investigated the therapeutic effect of AR-42, an HDACI, in alleviating HTLV-I-associated lymphoid malignancies. A dietary formulation of AR-42 was shown to prolong survival of ATL engrafted mice as compared to controls.

Non-obese diabetic/SCID xenograft mice were also used to test ABT-737, a small molecule inhibitor of Bcl-2 and Bcl-xL, whereby ABT-737 resulted in a significant inhibition of the tumor growth *in vivo* (Ishitsuka et al., 2012).

In another study, Ikebe et al. (2013) investigated the efficacy of 17-DMAG, an HSP90 inhibitor, as a therapeutic agent against ATL. Oral administration of 17-DMAG dramatically attenuated the aggressive infiltration of multiple organs in an ATL xenograft mouse model. It also inhibited the *de novo* viral production and improved the overall survival of ATL mice (Ikebe et al., 2013).

More recently, an unrecognized role of ubiquitin-editing enzyme, A20, in the survival of HTLV-I-infected cells was unveiled in a SCID model. In brief, the ATL-derived HuT-102 cell line was first transduced with A20 shRNA and then inoculated into SCID mice (Saitoh et al., 2016). Depletion of A20 induced apoptosis and affected the *in vivo* growth of HTLV-I infected cells, highlighting the importance of ubiquitin in ATL development.

Monoclonal antibodies as an ATL therapy in xenograft mouse models

The expression of specific cell surface markers on ATL cells implanted into mice, renders them an excellent model for testing the pre-clinical therapeutic potential of monoclonal antibodies. In this context, and given that ATL cells express CD2 and CD25 among other surface markers, the effect of monoclonal antibodies targeted against these two surface markers was performed in NOD/SCID mice intraperiteneally injected with MET-1 cells (Zhang et al., 2003). Furthermore, the effect of campath-1H (anti CD52), either alone or in combination with a humanized anti-Tac (HAT) or MEDI 507, a monoclonal antibody directed against CD2, was investigated. Campath-1H led to a striking prolongation of the survival of MET-1 ATL-bearing mice. This survival was significantly longer than that of the group receiving HAT. Moreover, the study revealed the anti-leukemic mechanism of action of Campath-1H which involves FcR-gamma-containing receptors (e.g., FcRgamma-III) present on polymorphonuclear leukocytes and macrophages, known to normally mediate antibody-dependent cellular cytotoxicity (ADCC) and/or trigger cross-linking induced apoptosis (Zhang et al., 2003). The same group explored the use of flavopiridol, a cyclin-dependent kinase inhibitor, alone or in combination with HAT. HAT/flavopiridol combination resulted in a prolonged survival and dramatic enhancement of the antitumor effect in MET-1 NOD/SCID mice as compared to the control group (Zhang et al., 2005).

A MET-1 NOD/SCID mouse model was also used to investigate the anti-leukemic effects of specific antibodies targeting IL-2R (Phillips et al., 2000) whereby HAT, murine anti-Tac (MAT), and 7G7/B6, all of which targeting IL-2Rα, significantly delayed leukemia progression resulting in enhanced survival (Phillips et al., 2000). To decipher the mechanism of action of these antibodies, comparison between treated-NOD/SCID and NSG mice was carried out (Zhang et al., 2004). In contrast to what was seen in NOD/SCID mice, treatment of NSG mice did not affect leukemia growth nor improved mice survival highlighting that the immune system difference, specifically polymorphonuclear cells, plays a crucial role in leukemia elimination by anti-IL-2R antibodies (Zhang et al., 2004).

Another promising therapeutic target is CC chemoreceptor 4 (CCR4). CCR4 is a chemokine receptor expressed by tumor cells in about 90% of ATL patients (Ishida et al., 2003). Several studies have investigated the effect of monoclonal Anti-CCR4 antibodies *in vivo* as potential therapies in ATL (Yano et al., 2008; Ito et al., 2009; Ishii et al., 2010; Hiyoshi et al., 2015). Using an ATL SCID xenograft mouse model, Yano et al. tried to augment the ADCC effect induced by defucosylated chimeric Anti-CCR4 IgG1 monoclonal antibody KM2760 *via* addition of granulocyte colony stimulating factor (G-CSF). Addition of G-CSF resulted in a more robust antitumor effect as compared to KM2670 alone (Yano et al., 2008). Using the same model, another monoclonal anti-CCR4 antibody KW-0761 was investigated (Ishii et al., 2010). KW-0761 also presented promising antitumor activity where tumor volume was significantly decreased (Ishii et al., 2010).

Anti-CD25 antibodies such as 7G7/B6 and daclizumab, directed against different epitopes of CD25 were also tested either alone or in combination (Zhang et al., 2006). Overall, 91% of the mice receiving the combination survived showing the promising synergistic effect of these two antibodies, especially when compared to the single antibody treatment (Zhang et al., 2006). In a similar study and using the same model, daclizumab was investigated in combination with an HDAI, depsipeptide, and showed an enhanced antitumor effect and survival of the leukemia-bearing mice, compared with those in the depsipeptide or daclizumab alone groups (Chen et al., 2009).

Given that some ATL cells express CD30 on their surface, the therapeutic effect of two anti-CD30 monoclonal antibodies was also investigated. SGN-30, a chimeric anti-CD30 mAb, and SGN-35, a monomethyl auristatin E-conjugated anti-CD30 mAb, were used. Maeda et al. (2010) treated NOD/SCID mice subcutaneously engrafted with HTLV-I-infected cells and reported a significant inhibition of the tumor growth upon treatment with either antibodies (Maeda et al., 2010).

In another context, the therapeutic efficacy of adoptive patient-autologous Tax-specific cytotoxic T cells (Tax-CTL) was assessed in NSG mice bearing primary ATL cells from three patients. Tax-CTL treatment resulted in a significant decrease of ATL cell infiltration into blood, spleen, and liver as well as a significant prolonged survival time in ATL NSG mice that received cells from two out of three patients (Masaki et al., 2013).

More recently, the NSG mice were used to investigate the physiological roles of osteopontin (OPN)-integrin interaction in ATL pathogenesis *in vivo*. ATL cell lines inoculated into NSG mice resulted in an increased OPN plasma levels. Treatment of these mice with anti-OPN mAbs inhibited not only tumor growth but also tumor invasion and metastasis suggesting a pivotal role of OPN in these processes (Maeda et al., 2015).

Altogether, the above studies highlight the importance of xenograft mice models in targeting and understanding ATL.

Humanized Mouse Models of ATL

Despite the importance of ATL xenograft mouse models, these models present with the limitation of being injected with ATL-derived or HTLV-I-transformed cell lines that were maintained for years in culture. This entails the potential changes in their genetic drift and the attenuation in their *in vivo* potential. Moreover, ATL xenograft mouse models do not provide answers to how HTLV-I induces ATL at early steps and how it maintains leukemogenesis (reviewed in Niewiesk, 2016). In this context, humanized mouse models were generated (reviewed in Panfil et al., 2013; Duc Dodon, 2014; Niewiesk, 2016) (**Table 1**).

Humanized Mouse Models: A Transformation in ATL Biology and Immune Responses

The lack of an adaptive and/or innate immune system is advantageous for HTLV-I replication and tumor engraftment. Injecting $CD34^+$ hematopoietic stem cells into NSG mice resulted in generation of human lymphocytes. The first humanized ATL model was developed by Miyazato et al. (2006) in which human peripheral blood mononuclear cell (PBMC) were first injected in NSG mice to establish a human-like setting followed by inoculation of HTLV-I virus producing MT-2 cells. MT-2 cells ensured cell-to-cell transmission required for HTLV-I infection. In this model, proviral load was increased in both $CD4^+$ and $CD8^+$ cells (Miyazato et al., 2006).

In a similar study, PBMC from HTLV-I infected carriers were injected in NSG mice to establish HTLV-I infection (Takajo et al., 2007). Despite the different methodology, both studies resulted in mice harboring human infected cells, which later undergo clonal proliferation (Miyazato et al., 2006; Takajo et al., 2007).

Another humanized model was developed by Villaudy et al. (2011) whereby $CD34^+$ human umbilical stem cells (HUSC) were intra-hepatically inoculated into newborn BALB/c/Rag22/2IL-2Rgc2/2 (also known as *BRG* mice) mice generating human lymphocytes. These lymphocytes were later infected by intraperitoneal injection of irradiated HTLV-producing MT-2 cell line. This study reported an alteration in T cell development alongside an increase in the proviral load and expansion of $CD4^+CD25^+$ cells. Mice also developed ATL- like splenomegaly as well as lymphoma (Villaudy et al., 2011).

In a different study, intra-bone marrow injection (IBMI) of cord blood $CD133^+$ stem cells intratibialy into sublethally irradiated NSG mice was performed (Tezuka et al., 2014). This study is based on the idea that $CD133^+$ cells are believed to be the ancestral of $CD34^+$ in hematopoiesis and carry the potential to differentiate to any hematopoietic cells including lymphocytes (Tezuka et al., 2014; reviewed in Duc Dodon, 2014). One month following $CD133^+$ inoculation, human $CD45^+$ leukocytes were found to completely reconstitute the murine bone marrow. Later, human B and T lymphocytes were detected and a balanced B/T lymphocyte ratio was attained and remained stable for up to 8 months (Tezuka et al., 2014). After establishing a human immune system in this model, HTLV-I infection involved the injection of sublethally irradiated MT-2 cells known to produce HTLV-I. Shortly after infection, an increase in $CD4^+$ cells was reported where $CD4^+$ $CD25^+$ clones gradually dominated indicating clonal selection. ATL like features including splenomegaly, hepatomegaly with ATL infiltration, and ATL-like "flower cells" were also reported (Tezuka et al., 2014). Screening of cytokines profiles demonstrated an initial elevation in the levels of IL-6, IL-8, IL-10, IL-12, IL- 13, IFN-γ, and TNF-α. In addition, granulocyte-macrophage colony-stimulating factor (GM-CSF) and chemokine (C-C motif) ligand 4 were also increased, clearly referring to an initiated immune response against the virus (Tezuka et al., 2014). One of the main findings of this study was the generation of a functional adaptive immune response in a humanized mouse model, summarized by detection of anti-HTLV-I antibodies as well as Tax-specific cytotoxic T cells (CTL). These CTL inversely correlated with proviral load of infected cells (Tezuka et al., 2014). Thus Tezuka's model was by far the closest one in recapitulating the development of ATL *in vivo*. While most of the other humanized models developed lymphoma and/or thymoma, Tezuka's model exclusively developed leukemia. This may be due to the fact that infection was carried out after a humanized immune system may be fully developed.

Another humanized ATL model was generated by Nakamura et al. (2015) where PBMC from ATL patients were injected into NOD/SCID/Jak3-null mice (NOJ mice). In Brief, the model involved subcutaneous injection of the ATL S1T cell line into mice, followed by subcutaneous, intraperitoneal, or intravenous transplantation of primary ATL cells. ATL cells successfully infiltrated various organs and could transplant into secondary mice establishing NOJ mice as successful models for primary ATL xenotransplantation (Nakamura et al., 2015).

Humanized Models as Platforms for ATL Therapy

After the revolution that humanized mouse models generated in the ATL field, they provided a strong platform for testing anti-ATL therapies (**Table 2**). Two antiviral reverse transcriptase inhibitors, Tenofovir and Azidothymidine, were tested in the first humanized ATL model developed by Miyazato et al. (2006). After the uncertain role of these agents in halting HTLV-I infection, their prophylactic potential was investigated. Both agents were reported to block primary infection in these mice (Miyazato et al., 2006).

Moreover, humanized models proved instrumental in investigating the anti-tumor potential of anti-CCR4 antibodies. NSG mice inoculated with primary ATL cells and autologous immune cells belonging to the same patient were used to assess the antitumor efficacy of KM2760 antibody (Ito et al., 2009). KM2760 decreased the number of ATL cells in blood, spleen, and liver, as well as ATL lesions and organ infiltration, and lowered IL-2R concentration in serum (Ito et al., 2009). Using a humanized mouse model similar to that developed by Villaudy et al. (2011), the efficacy of TARC-PE38 targeting CCR4 was investigated (Hiyoshi et al., 2015). TARC-PE38 is a complex of thymus and activation-regulated chemokine (TARC), CCR4 ligand, fused to a truncated *Pseudomonas aeruginosa* exotoxin A (PE38). TARC-PE38 efficiently killed HTLV-I-infected cell lines and shrank HTLV-I-associated solid tumors size. Moreover, TARC-PE38 markedly inhibited the proliferation of HTLV-I-infected human $CD4^+CD25^+$ or $CD4^+CD25^+CCR4^+$ cells and reduced the proviral loads in PBMC obtained from both patients and asymptomatic carriers (Hiyoshi et al., 2015).

Using the NOJ model generated by Nakamura et al. (2015) the antitumor effect of pyrrolidine dithiocarbamate (PDTC), an antioxidant agent, was tested in ATL and showed that PDTC significantly enhanced the survival of these mice (Nakamura et al., 2015).

Transgenic Mouse Models of HTLV-I

To decipher the oncogenic potential role of HTLV-I proteins *in vivo*, transgenic mice overexpressing the viral oncoproteins Tax or HBZ were generated. So far several *Tax* transgenic models expressing Tax under different promoters and two *HBZ* models were developed. Despite exhibiting many features of ATL, none of these models could exactly recapitulate HTLV-I-associated ATL. For instance, and as opposed to HTLV-I infection where HBZ- and Tax-specific CTLs and antibodies are generated, transgenics for both oncoproteins lack a host induced immune response against these viral proteins. the importance of transgenic models in the ATL context lies in confirming the oncogenic functions of HTLV-I proteins, Tax and HBZ, *in vivo* and in disclosing various host pathways manipulated by these proteins ultimately leading to tumor generation (**Table 1**).

Tax Transgenic Models Develop ATL-Like Leukemia/Lymphoma

Tax transgenic mice

The generation of *Tax* transgenic models relied on the earliest discoveries of the *in vitro* Tax oncogenic potential, from independent laboratories (Grassmann et al., 1989; Tanaka et al., 1990). To investigate the *in vivo* role of Tax in leukemogenesis, several mouse models have been generated (reviewed in Ohsugi, 2013; Niewiesk, 2016). Depending on the used promoter, Tax expression resulted in various tumors (Ohsugi, 2013). Except for models utilizing the lymphocyte-specific protein tyrosine kinase p56 (lck) or granzyme promoters, most of the models could not generate leukemia/lymphoma and rather resulted in less typical HTLV-I tumors and manifestations (Ohsugi, 2013; Panfil et al., 2013; Niewiesk, 2016). Nerenberg et al. (1987) established the first *Tax* transgenic mouse where *Tax* gene was under the control of the natural HTLV-I promoter, the LTR promoter [Tg (HIV-*tat*) 6-2Gja] (Nerenberg et al., 1987). Referred to initially as tat protein, Tax exhibited tissue-specific expression and LTR-*Tax* mice mainly developed mesenchymal tumors in the nose, ear, mouth, tail, as well as the foot. Despite not recapitulating human ATL, this research was novel in establishing Tax as an oncoprotein and HTLV-I as a transforming virus *in vivo* (Nerenberg et al., 1987). Beside mesenchymal tumors, LTR-*Tax* mice developed tumors at other multiple sites resembling neurofibroma making this model useful for studying neurofibromatosis (Hinrichs et al., 1987). Apart from tumors, mice were shown to develop myopathies similar to those associated with HTLV-I, which are due to the atrophy/degeneration of oxidative muscle fibers (Nerenberg and Wiley, 1989). Using the same model, Habu et al. (1999) reported high incidence of inflammatory polyarthropathy resembling arthritis. Profound skeletal alterations resulting in fragile bones with high turnover rate were also reported (Ruddle et al., 1993). Salivary and lacrimal glands showed exocrinopathy and lesions resembling Sjogren's syndrome due to the attack by immune cells (Green et al., 1989).

Afterwards, a bi-transgenic mouse was generated by crossing LTR-*Tax* mice with LTR-β *gal* mice (β-galactosidase) to better visualize the organ involvement in ATL (Bieberich et al., 1993). Tax acts on the LTR resulting in an increased β-gal expression; and this enzyme was detected in specific tissues including bones, muscles, exocrine glands, as well as mesenchymal tumors (Benvenisty et al., 1992; Bieberich et al., 1993). Using this same model, Swaims et al. (2010) assessed the interaction with the host immune system and demonstrated that the activation of infected $CD4^+$ T cells may induce Tax expression and thus may contribute to ATL pathogenesis in infected carriers. In addition, infected $CD4^+$ cells harboring Tax exhibited changes in the expression of surface markers and resulted in changes in $CD4^+$ subtype specifications (Swaims et al., 2010, reviewed in Kress et al., 2011). Despite not developing leukemia/lymphoma, the LTR-*Tax* model provided a clear-cut evidence that Tax expression is solely sufficient for tumor induction establishing Tax as an oncoprotein *in vivo*.

In another attempt, a *Tax* transgenic model of C57BL/6TgN mice (huGMZB*Tax*) was developed where *Tax* expression was controlled by the granzyme B promoter (Grossman et al., 1995). This promoter restricted the expression of Tax to the T cell compartment specifically $CD4^+$ and $CD8^+$, NK, and lymphokine activated killer cells. In this model, mice developed large granular lymphocytic leukemia (LGL), neutrophil dominated inflammation, as well as tumors on the ears, tail, and leg (Grossman et al., 1995). Mice also developed symptoms resembling human ATL such as high white blood cell count, splenomegaly, neutrophilia, and lymphadenopathy. LGL cells disseminated to distant organs such as lungs, bone marrow, and liver (Grossman et al., 1995). Despite not fully recapitulating the human disease, this model was instrumental in showing that Tax expression in the lymphocytes is sufficient to cause leukemia. Later, Gao et al. (2005) demonstrated that these mice also exhibited malignant hypercalcemia and symptoms associated with metastasis such as osteolytic bone lesions which again resemble the disease in ATL patients (Gao et al., 2005). Activated NK and T cells from this model demonstrated Tax-mediated constitutive activation of NF-κB in both its canonical and non-canonical pathways (Bernal-Mizrachi et al., 2006). Using the same model, the role of p53 inactivation in Tax-induced tumor development was assessed and showed that the p53 apoptotic pathway was functionally inactivated (Portis et al., 2001). P53 mutations in tumors were also detected but were associated with secondary organ infiltration. In the same context, mating of these *Tax*-transgenic mice with P53-deficient mice did not accelerate the initial tumor development (Portis et al., 2001). However, it significantly increased disease progression and mortality in *p53* heterozygous mice. This suggested that Tax functionally inactivates p53 which contributes to late stage tumor progression rather than initial tumor formation (Portis et al., 2001).

In a different study, the role of the innate immune system and inflammation in ATL development was assessed. The *Tax*-granzyme model was mated with an *IFN-γ* knock-out model. The resulting mice exhibited enhanced tumorigenesis with accelerated lesions development (Mitra-Kaushik et al., 2004). Using the same model, Rauch et al. generated *Tax*-LUC double transgenic mouse model whereby luciferase bioluminescent imaging techniques allowed to track tumor engraftment *in vivo* (Rauch et al., 2009a,b). The onset of peripheral subcutaneous tumors was preceded by the formation of microscopic intra-epithelial lesions. This suggests that Tax activates lymphocytes, which then recruit NK/T-cells to be activated and transformed (Rauch et al., 2009b). Another study from the same group suggested that in *Tax*-LUC model, lymphoma development is promoted by an inflammatory stimulus whereby T cell activation via T cell receptor (TCR) was shown to promote/exacerbate tumorigenesis (Rauch et al., 2009a). Afterwards, the same *Tax*-LUC model was utilized to investigate the role of IL-15 in spontaneous lymphoma development (Rauch et al., 2014). $IL15^{-/-}$ *Tax*-LUC mice were generated and resulted in an aggressive lymphoma development and accelerated mortality, suggesting that IL-15 contributes to the antitumor immunity in ATL. Knocking out IL-15 also resulted in elevation of IL-1α and IL-1α driven cytokines (Rauch et al., 2014). Treatment with anti-IL-1α antibodies resulted in decreased tumor growth. This study suggested IL-15 and IL-1α as potential therapeutic options in ATL.

Hall et al. (1998) established another *Tax* transgenic model where *Tax* gene was under the control of *CD3-epsilon* promoter-enhancer sequences. These mice, of C57/CBA origins, developed a variety of tumors including salivary and mammary adenomas as well as mesenchymal tumors, specifically at wound sites, yet failed to develop leukemia (Hall et al., 1998).

In an attempt to target Tax expression to the leukocyte compartment, a bi-transgenic doxycycline inducible model [Tg (*EmuSR-tTa*) 83Bop] was generated (Kwon et al., 2005). This conditional "tet off" *Tax* transgenic model targeted both wild-type Tax and Tax mutants that selectively compromise NF-κB or CREB pathways to the leukocyte compartment. Wild type *Tax* transgenic mice developed a lethal cutaneous disease with skin lesions infiltrated by $CD3^+CD4^+MHC\text{-}2^+$ T cells, similar to those seen in ATL patients. Mice also developed systemic lymphadenopathy and splenomegaly. Moreover, inflammatory cytokines including TNF-α, IL-6, IL1 α/β, IFN-γ were induced (Kwon et al., 2005). Of note, suppression of Tax by doxycycline administration resulted in disappearance of skin lesions directly linking Tax to this dermal pathogenesis (Kwon et al., 2005).

To further restrict Tax expression to the thymus compartment, being the site of maturation of T cells, the two *Lck* promoters, distal and proximal, were used. The proximal promoter drives gene expression in thymocytes while the distal one restricts gene expression specifically to mature T lymphocytes (Hasegawa et al., 2006; Ohsugi et al., 2007b; reviewed in Ohsugi, 2013). Hasegawa et al. (2006) generated a *Tax* transgenic model using the *lck* proximal promoter [C57BL/6-Tg (Lck-HTLV-I *Tax*)]. As compared to previous transgenic models, Hasegawa's model was the closest to recapitulate ATL. In this model, mice developed $CD4^-CD8^-CD44^+CD25^+$ diffuse large cell lymphoma and leukemia after a prolonged latency period of around 18 months resembling the latency period required in humans to generate tumors (Hasegawa et al., 2006). Interestingly, mice exhibited clinical and histological resemblance to acute ATL seen in patients, such as characteristic "flower cells" in blood smears, lymphadenopathy, splenomegaly with extensive infiltration by lymphomatous cells, as well as distant organ infiltration of bone marrow, liver, kidney, lung, skin, and meninges. Infiltrating T cells were of malignant phenotype with increased expression of $CD25^+$ cell surface marker (Hasegawa et al., 2006). In addition, mice exhibited further common features with ATL patients such as significant leukocytosis, hypercalcemia, elevated LDH, and constitutive NF-κB activation (Hasegawa et al., 2006). A major difference between Hasegawa's mice and human ATL is in the phenotype of leukemic cells; mice exhibited immature $CD4^+$ cells while human ATL involves leukemia generally of mature $CD4^+$ cells (reviewed in Ohsugi, 2013). Since the development of leukemia required a long time, splenocytes derived from this *Tax* transgenic model were intraperitoneally injected in SCID mice. This model recapitulated most of the ATL phenotypes, similar to the transgenic model and to ATL patients, in less than 1 month (Hasegawa et al., 2006; El Hajj et al., 2010). Moreover, this model was the first to be used for the isolation of candidate ATL stem cells of $CD38^-/CD71^-/CD117^+$ phenotype (Hasegawa et al., 2006; Yamazaki et al., 2009). These stem cells exhibited decreased levels of Notch, Tax, and BMI-1 expression all of which indicate their early hematopoietic cell origin (Yamazaki et al., 2009).

The distal promoter of *Lck* was also used to develop *Tax* transgenic mice (Ohsugi et al., 2007b). The expression of Tax under the distal promoter resulted in leukemia of mature $CD4^+$ cells unlike the immature ones obtained in the Hasegawa et al generated *Tax* transgenic model (Ohsugi et al., 2007b).

Tax transgenic models for ATL-targeted therapy

Despite the improvements in prolonging survival of the leukemic subtypes of ATL upon using zidovudine and IFN-α (IFN) (Bazarbachi et al., 2010), most patients relapse underlying the urgent need for novel therapies and targets. Different drugs and drug combinations were investigated, of which a very promising treatment is the combination of arsenic trioxide (arsenic) and IFN-α (**Table 2**). *In vitro* studies involving ATL cell lines and primary patients' leukemic cells showed the efficacy of the arsenic/IFN combination in triggering Tax degradation by the proteasome resulting in cell cycle arrest and apoptosis (Bazarbachi et al., 1999; El-Sabban et al., 2000; Nasr et al., 2003). To assess the efficacy of this combination *in vivo*, El Hajj et al. used spleen cells derived from the murine *Tax* transgenic ATL model (Hasegawa et al., 2006) and injected them into SCID mice (El Hajj et al., 2010). Approximately 1 month after transplantation, mice recapitulated ATL manifestations (Hasegawa et al., 2006; El Hajj et al., 2010). Strikingly, arsenic/IFN cured ATL in these mice. Although this combination did not rapidly decrease the tumor bulk, as ATL cells only underwent modest cell cycle arrest and apoptosis, the curative efficacy of arsenic/IFN occurred through clearance of leukemia initiating cells (LIC). Briefly, using the serial transplantation method, ATL cells derived from primary mice treated with the arsenic/IFN combination resulted in lower leukemia transplantation ability in untreated secondary mice and no transplantation in untreated tertiary mice (El Hajj et al., 2010). Addition of the proteasome inhibitor bortezomib to arsenic/IFN treatment of primary mice reversed all the observed phenotypes and led to normal ATL development in serial transplantation experiments, demonstrating that Tax degradation is the critical step for LIC exhaustion which further highlighted the oncogenic addiction of ATL cells to Tax (El Hajj et al., 2010). Later, Kchour et al. (2009) have investigated the effect of the triple combination of zidovudine, arsenic, and IFN, and translated the promising pre-clinical results to patients. Interestingly, the triple drug combination of zidovudine/arsenic/IFN resulted in 70% complete remission rate and 100% overall response rate in chronic ATL patients, as well as an unpreceded prolonged survival in some patients, strongly suggesting a similar mechanism of LIC eradication in patients (Kchour et al., 2009).

Using the same *in vivo* model, the preclinical efficacy of a synthetic retinoid ST1926 was investigated (El Hajj et al., 2014) (**Table 2**). Oral treatment of ST1926 induced massive apoptosis, prolonged survival and decreased tumor infiltration, leukocytosis, and splenomegaly as compared to untreated group of animals. This study highlights the potential of synthetic retinoids as a promising therapy for ATL (El Hajj et al., 2014). It remains to be determined whether ST1926 treatment alone targets LIC in ATL.

HBZ Transgenic Models Generate T-Cell Lymphoma but Not Leukemia

HBZ transgenic models

The first *in vivo HBZ* transgenic mouse model (*HBZ*-Tg mice) was generated by Satou et al. (2006), where the *HBZ* gene was expressed under the control of a murine *CD4*-specific promoter/enhancer/silencer (Satou et al., 2006). Restricted HBZ expression to $CD4^+$ T cells resulted in systemic inflammation and development of T cell lymphoma in only 30% of mice after a long latency period. *HBZ*-Tg spontaneously developed systemic dermatitis, alveolitis, and later lymphoma upon aging (Satou et al., 2011). At the cellular level, HBZ increased the generation and proliferation of $Foxp3^+$ T cells as well as the transcription of Foxp3 mRNA. *HBZ*-Tg mice exhibited an increase in the number of functionally impaired $CD4^+$ $Foxp3^+$ T_{reg} cells as well as effector/memory $CD4^+$ T cells (Satou et al., 2011). Recently, this model was used to investigate whether the proliferation of $CD4^+$ T cells is increased *in vivo*. Allergic encephalomyelitis was experimentally induced by immunization with myelin oligodendrocyte glycoprotein (MOG)/complete Freund's adjuvant. Disease severity was not increased but the number of $CD4^+$ T cells was increased only in the immunized *HBZ*-Tg mice suggesting that HBZ-expressing T cells have higher susceptibility to immune stimulation *in vivo* (Kinosada et al., 2017). Using the same model, Yamamoto-Taguchi et al. (2013) showed that HBZ promotes inflammation through labile Foxp3 expression; T_{reg} cells induced by HBZ have unstable Foxp3 expression and tend to convert to $Foxp3^-$T cells producing IFN-γ. This HBZ-induced pro-inflammatory phenotype of $CD4^+$ T cells was suggested to be involved in the HTLV-I-associated pathogenesis and inflammation (Yamamoto-Taguchi et al., 2013). Later, Mitagami et al. (2015) closely investigated HBZ-induced inflammation and revealed that in *HBZ*-Tg mice, inflammation severity significantly correlate with lymphoma development. The study suggested a link between HBZ inflammation and oncogenesis in $CD4^+$ T cells (Mitagami et al., 2015).

In another context, HBZ expression was found to impair the cell-mediated immunity of HBZ-Tg via suppression of Th-1 cytokine production (Sugata et al., 2012). Despite being of low immunogenicity, anti-HBZ antibodies can be detected in patient's serum. Sugata et al. (2012) have utilized the *HBZ*-Tg mouse model to investigate the possibility of generation of HBZ-targeted HTLV-I vaccine. Accordingly, splenocytes from HBZ mice, called HT-48, were inoculated into immunodeficient mice generating an ATL model. C57BL/6 mice immunized by a recombinant vaccinia virus-based HBZ vaccine generated HBZ-specific CD4 and CD8 T-cell response. Afterwards, inoculation of anti-HBZ cytotoxic T cells into the generated HT-48 mouse model increased survival of these mice suggesting that HBZ might be a candidate for vaccine production (Sugata et al., 2012). Recently, ATL stem cells were identified in an HBZ model (Kuribayashi et al., 2016). In this study, HT-48 cells from *HBZ*-Tg mice were injected intraperitoneally into C57BL/6 mice, then serial transplantation experiments were done to assess the presence of LIC where nine consecutive transplantations were done (Kuribayashi et al., 2016). Surprisingly, HT-48 cells were able to regenerate leukemia in all transplantations. In this model, ATL stem cells were identified as c-kit$^+$/CD4$^-$/CD8$^-$ cells. Compared to the T cell progenitors, reported ATL stem cells had a similar gene expression profile (Kuribayashi et al., 2016).

More recently, a transgenic *HBZ* mouse model using Granzyme B (Gzmb-*HBZ*) was generated (Esser et al., 2017). In addition to splenomegaly, abnormal white cell count, tumors developing after 18 months, in two thirds of the Gzmb-*HBZ* mice, as well as pathologic bone loss and hypercalcaemia were obtained (Kinosada et al., 2017) (**Table 1**).

The Double Transgenic HBZ/Tax Mouse Model Fails to Recapitulate ATL

Zhao et al. (2014) established a double transgenic mouse model expressing both Tax and HBZ viral proteins. In this model, both proteins were exclusively expressed in CD4$^+$T cells. The concomitant transgenic expression of both Tax and HBZ resulted in skin lesions, T-cell lymphoma, and splenomegaly resembling in part diseases observed in HTLV-I infected individuals (Zhao et al., 2014). In addition, HBZ/Tax double expression resulted in an increase in the number of CD4$^+$ memory T cells and Foxp3$^+$ T$_{reg}$ cells. Overall, they reported that little difference is seen between the phenotype produced by *HBZ*-Tg model and *HBZ/Tax* double transgenic model. However, in contrast to all published findings, this study reported that "Tax expression alone failed to generate major health problems" and did not result in tumor development (Zhao et al., 2014). In addition, this model resulted in lymphoma and thus failed to generate ATL-like leukemia (**Table 1**).

CONCLUSION

In an attempt to unveil the molecular mechanisms dictating ATL development and to advance novel therapeutic options, multiple animal models were utilized. The use of animal models to study HTLV-I infection and ATL development has been instrumental in providing valuable data concerning disease progression and potential therapy targets. Besides ATL patients, mice remain the most valuable and useful tools for *in vivo* investigation of ATL. Generation of ATL xenograft models strikingly advanced the search for targeted therapies for ATL. To assess ATL pathogenesis in a more human setting, humanized models were remarkably helpful. These models are critical for studying early steps of HTLV-I infection. Transgenic ATL models expressing HTLV-I proteins, Tax, HBZ, or both have disclosed the potential roles and contributions of either protein to ATL pathogenesis. In this context, *Tax* transgenic mice, specifically *Lck*-promoter-models, were shown to develop ATL-like leukemia with pathology and molecular changes resembling acute ATL including activation of NF-κB pathway. *Tax* transgenic mice also served as means of testing targeted drug therapies. On the other hand, *HBZ* transgenic mice showed manifestations of systemic inflammation establishing HBZ as a pro-inflammatory protein. Finally, despite the noted progress in disease understanding and its treatment strategies, an animal model that can fully recapitulate the human ATL disease has not been achieved yet.

AUTHOR CONTRIBUTIONS

All authors listed have made a substantial, direct and intellectual contribution to the work, and approved it for publication.

REFERENCES

Arnold, J., Zimmerman, B., Li, M., Lairmore, M. D., and Green, P. L. (2008). Human T-cell leukemia virus type-1 antisense-encoded gene, hbz, promotes T-lymphocyte proliferation. *Blood* 112, 3788–3797. doi: 10.1182/blood-2008-04-154286

Azran, I., Schavinsky-Khrapunsky, Y., and Aboud, M. (2004). Role of Tax protein in human T-cell leukemia virus type-I leukemogenicity. *Retrovirology* 1:20. doi: 10.1186/1742-4690-1-20

Bangham, C. R., and Matsuoka, M. (2017). Human T-cell leukaemia virus type 1: parasitism and pathogenesis. *Philos. Trans. R. Soc. B* 372:20160272. doi: 10.1098/rstb.2016.0272

Bangham, C. R. M., and Ratner, L. (2015). How does HTLV-I cause adult T-cell leukaemia/lymphoma (ATL)? *Curr. Opin. Virol.* 14, 93–100. doi: 10.1016/j.coviro.2015.09.004

Basbous, J., Arpin, C., Gaudray, G., Piechaczyk, M., Devaux, C., and Mesnard, J. M. (2003). The HBZ factor of human T-cell leukemia virus type I dimerizes with transcription factors JunB and c-jun and modulates their transcriptional activity. *J. Biol. Chem.* 278, 43620–43627. doi: 10.1074/jbc.M307275200

Bazarbachi, A. (2016). Tax fingerprint in adult T-cell leukemia. *Blood* 127, 1737–1738. doi: 10.1182/blood-2016-02-694885

Bazarbachi, A., Cwynarski, K., Boumendil, A., Finel, H., Fields, P., Raj, K., et al. (2014). Outcome of patients with HTLV-1-associated adult T-cell leukemia/lymphoma after SCT: a retrospective study by the EBMT LWP. *Bone Marrow Transplant.* 49, 1266–1268. doi: 10.1038/bmt.2014.143

Bazarbachi, A., El-Sabban, M. E., Nasr, R., Quignon, F., Awaraji, C., Kersual, J., et al. (1999). Arsenic trioxide and interferon-alpha synergize to induce cell cycle arrest and apoptosis in human T-cell lymphotropic virus type I-transformed cells. *Blood* 93, 278–283.

Bazarbachi, A., Ghez, D., Lepelletier, Y., Nasr, R., de The, H., El-Sabban, M. E., et al. (2004). New therapeutic approaches for adult T-cell leukaemia. *Lancet Oncol.* 5, 664–672. doi: 10.1016/S1470-2045(04)01608-0

Bazarbachi, A., Plumelle, Y., Carlos Ramos, J., Tortevoye, P., Otrock, Z., Taylor, G., et al. (2010). Meta-analysis on the use of zidovudine and interferon-alfa in adult T-cell leukemia/lymphoma showing improved survival in the leukemic subtypes. *J. Clin. Oncol.* 28, 4177–4183. doi: 10.1200/JCO.2010.28.0669

Bazarbachi, A., Suarez, F., Fields, P., and Hermine, O. (2011). How I treat adult T-cell leukemia/lymphoma. *Blood* 118, 1736–1745. doi: 10.1182/blood-2011-03-345702

Benvenisty, N., Ornitz, D. M., Bennett, G. L., Sahagan, B. G., Kuo, A., Cardiff, R. D., et al. (1992). Brain tumours and lymphomas in transgenic mice that carry HTLV-I LTR/c-myc and ig/tax genes. *Oncogene* 7, 2399–2405.

Bernal-Mizrachi, L., Lovly, C. M., and Ratner, L. (2006). The role of NF-κB-1 and NF-κB-2-mediated resistance to apoptosis in lymphomas. *Proc. Natl. Acad. Sci. U.S.A.* 103, 9220–9225. doi: 10.1073/pnas.0507809103

Bieberich, C. J., King, C. M., Tinkle, B. T., and Jay, G. (1993). A transgenic model of transactivation by the tax protein of HTLV-I. *Virology* 196, 309–318. doi: 10.1006/viro.1993.1481

Bosma, G. C., Custer, R. P., and Bosma, M. J. (1983). A severe combined immunodeficiency mutation in the mouse. *Nature* 301, 527–530. doi: 10.1038/301527a0

Boxus, M., and Willems, L. (2009). Mechanisms of HTLV-I persistence and transformation. *Br. J. Cancer* 101, 1497–1501. doi: 10.1038/sj.bjc.6605345

Brauweiler, A., Garl, P., Franklin, A. A., Giebler, H. A., and Nyborg, J. K. (1995). A molecular mechanism for human T-cell leukemia virus latency and tax transactivation. *J. Biol. Chem.* 270, 12814–12822. doi: 10.1074/jbc.270.21.12814

Brauweiler, A., Garrus, J. E., Reed, J. C., and Nyborg, J. K. (1997). Repression of bax gene expression by the HTLV-I tax protein: Implications for suppression of apoptosis in virally infected cells. *Virology* 231, 135–140. doi: 10.1006/viro.1997.8509

Chen, J., Zhang, M., Ju, W., and Waldmann, T. A. (2009). Effective treatment of a murine model of adult T-cell leukemia using depsipeptide and its combination with unmodified daclizumab directed toward CD25. *Blood* 113, 1287–1293. doi: 10.1182/blood-2008-04-149658

Chicha, L., Tussiwand, R., Traggiai, E., Mazzucchelli, L., Bronz, L., Piffaretti, J. C., et al. (2005). Human adaptive immune system Rag2-/-gamma(c)-/- mice. *Ann. N. Y. Acad. Sci.* 1044, 236–243. doi: 10.1196/annals.1349.029

Dassouki, Z., Sahin, U., El Hajj, H., Jollivet, F., Kfoury, Y., Lallemand-Breitenbach, V., et al. (2015). ATL response to arsenic/interferon therapy is triggered by SUMO/PML/RNF4-dependent tax degradation. *Blood* 125, 474–482. doi: 10.1182/blood-2014-04-572750

Dewan, M. Z., Terashima, K., Taruishi, M., Hasegawa, H., Ito, M., Tanaka, Y., et al. (2003). Rapid tumor formation of human T-cell leukemia virus type 1-infected cell lines in novel NOD-SCID/gammac(null) mice: Suppression by an inhibitor against NF-kappaB. *J. Virol.* 77, 5286–5294. doi: 10.1128/JVI.77.9.5286-5294.2003

Duc Dodon, M. D. (2014). Of mice, men, and HTLV-I. *Blood* 123, 303–304. doi: 10.1182/blood-2013-11-538272

Edlich, R. F., Hill, L. G., and Williams, F. M. (2003). Global epidemic of human T-cell lymphotrophic virus type-I (HTLV-I): an update. *J. Long Term Eff. Med. Implants* 13, 127–140. doi: 10.1615/JLongTermEffMedImplants.v13.i2.70

El Hajj, H., El-Sabban, M., Hasegawa, H., Zaatari, G., Ablain, J., Saab, S. T., et al. (2010). Therapy-induced selective loss of leukemia-initiating activity in murine adult T cell leukemia. *J. Exp. Med.* 207, 2785–2792. doi: 10.1084/jem.20101095

El Hajj, H., Khalil, B., Ghandour, B., Nasr, R., Shahine, S., Ghantous, A., et al. (2014). Preclinical efficacy of the synthetic retinoid ST1926 for treating adult T-cell leukemia/lymphoma. *Blood* 124, 2072–2080. doi: 10.1182/blood-2014-03-560060

El Hajj, H. E., Nasr, R., Kfoury, Y., Dassouki, Z., Nasser, R., Kchour, G., et al. (2012). Animal models on HTLV-I and related viruses: What did we learn? *Front. Microbiol.* 3:333. doi: 10.3389/fmicb.2012.00333

El-Sabban, M. E., Merhi, R. A., Haidar, H. A., Arnulf, B., Khoury, H., Basbous, J., et al. (2002). Human T-cell lymphotropic virus type 1-transformed cells induce angiogenesis and establish functional gap junctions with endothelial cells. *Blood* 99, 3383–3389. doi: 10.1182/blood.V99.9.3383

El-Sabban, M. E., Nasr, R., Dbaibo, G., Hermine, O., Abboushi, N., Quignon, F., et al. (2000). Arsenic-interferon-alpha-triggered apoptosis in HTLV-I transformed cells is associated with tax down-regulation and reversal of NF-kappa B activation. *Blood* 96, 2849–2855.

Esser, A. K., Rauch, D. A., Xiang, J., Harding, J. C., Kohart, N. A., Ross, M. H., et al. (2017). HTLV-I viral oncogene HBZ induces osteolytic bone disease in transgenic mice. *Oncotarget* 8, 69250–69263. doi: 10.18632/oncotarget.20565

Fan, J., Ma, G., Nosaka, K., Tanabe, J., Satou, Y., Koito, A., et al. (2010). APOBEC3G generates nonsense mutations in human T-cell leukemia virus type 1 proviral genomes *in vivo*. *J. Virol.* 84, 7278–7287. doi: 10.1128/JVI.02239-09

Feuer, G., Stewart, S. A., Baird S. M., Lee, F., Feuer, R., and Chen, I. S. (1995). Potential role of natural killer cells in controlling tumorigenesis by human T-cell leukemia viruses. *J. Virol.* 69, 1328–1333.

Feuer, G., Zack, J. A., Harrington, W. J. Jr., W. J., Valderama, R., Rosenblatt, J. D., Wachsman, W., et al. (1993). Establishment of human T-cell leukemia virus type I T-cell lymphomas in severe combined immunodeficient mice. *Blood* 82, 722–731.

Fujii, M., Iwai, K., Oie, M., Fukushi, M., Yamamoto, N., Kannagi, M., et al. (2000). Activation of oncogenic transcription factor AP-1 in T cells infected with human T cell leukemia virus type 1. *AIDS Res. Hum. Retroviruses* 16, 1603–1606. doi: 10.1089/08892220050193029

Fujii, M., Niki, T., Mori, T., Matsuda, T., Matsui, M., Nomura, N., et al. (1991). HTLV-I tax induces expression of various immediate early serum responsive genes. *Oncogene* 6, 1023–1029.

Fujii, M., Tsuchiya, H., Chuhjo, T., Akizawa, T., and Seiki, M. (1992). Interaction of HTLV-I Tax1 with p67SRF causes the aberrant induction of cellular immediate early genes through CArG boxes. *Genes Dev.* 6, 2066–2076. doi: 10.1101/gad.6.11.2066

Fujikawa, D., Nakagawa, S., Hori, M., Kurokawa, N., Soejima, A., Nakano, K., et al. (2016). Polycomb-dependent epigenetic landscape in adult T-cell leukemia. *Blood* 127, 1790–1802. doi: 10.1182/blood-2015-08-662593

Gao, L., Deng, H., Zhao, H., Hirbe, A., Harding, J., Ratner, L., et al. (2005). HTLV-I tax transgenic mice develop spontaneous osteolytic bone metastases prevented by osteoclast inhibition. *Blood* 106, 4294–4302. doi: 10.1182/blood-2005-04-1730

Gaudray, G., Gachon, F., Basbous, J., Biard-Piechaczyk, M., Devaux, C., and Mesnard, J. M. (2002). The complementary strand of the human T-cell leukemia virus type 1 RNA genome encodes a bZIP transcription factor that down-regulates viral transcription. *J. Virol.* 76, 12813–12822. doi: 10.1128/JVI.76.24.12813-12822.2002

Gessain, A., and Cassar, O. (2012). Epidemiological aspects and world distribution of HTLV-I infection. *Front. Microbiol.* 3:388. doi: 10.3389/fmicb.2012.00388

Giam, C. Z., and Semmes, O. J. (2016). HTLV-I infection and adult T-cell leukemia/lymphoma-A tale of two proteins: Tax and HBZ. *Viruses* 8:E161. doi: 10.3390/v8060161

Giebler, H. A., Loring, J. E., van Orden, K., Colgin, M. A., Garrus, J. E., Escudero, K. W., et al. (1997). Anchoring of CREB binding protein to the human T-cell leukemia virus type 1 promoter: a molecular mechanism of tax transactivation. *Mol. Cell. Biol.* 17, 5156–5164. doi: 10.1128/MCB.17.9.5156

Goncalves, D. U., Proietti, F. A., Ribas, J. G., Araujo, M. G., Pinheiro, S. R., Guedes, A. C., et al. (2010). Epidemiology, treatment, and prevention of human T-cell leukemia virus type 1-associated diseases. *Clin. Microbiol. Rev.* 23, 577–589. doi: 10.1128/CMR.00063-09

Good, L., and Sun, S. C. (1996). Persistent activation of NF-kappa B/rel by human T-cell leukemia virus type 1 tax involves degradation of I kappa B beta. *J. Virol.* 70, 2730–2735.

Grassmann, R., Berchtold, S., Radant, I., Alt, M., Fleckenstein, B., Sodroski, J. G., et al. (1992). Role of human T-cell leukemia virus type 1 X region proteins in immortalization of primary human lymphocytes in culture. *J. Virol.* 66, 4570–4575.

Grassmann, R., Dengler, C., Muller-Fleckenstein, I., Fleckenstein, B., McGuire, K., Dokhelar, M. C., et al. (1989). Transformation to continuous growth of primary human T lymphocytes by human T-cell leukemia virus type I X-region genes transduced by a herpesvirus saimiri vector. *Proc. Natl. Acad. Sci. U.S.A.* 86, 3351–3355. doi: 10.1073/pnas.86.9.3351

Green, J. E., Hinrichs, S. H., Vogel, J., and Jay, G. (1989). Exocrinopathy resembling Sjogren's syndrome in HTLV-I tax transgenic mice. *Nature* 341, 72–74. doi: 10.1038/341072a0

Grossman, W. J., Kimata, J. T., Wong, F. H., Zutter, M., Ley, T. J., and Ratner, L. (1995). Development of leukemia in mice transgenic for the tax gene of human T-cell leukemia virus type I. *Proc. Natl. Acad. Sci. U.S.A.* 92, 1057–1061. doi: 10.1073/pnas.92.4.1057

Habu, K., Nakayama-Yamada, J., Asano, M., Saijo, S., Itagaki, K., Horai, R., et al. (1999). The human T cell leukemia virus type I-tax gene is responsible for the development of both inflammatory polyarthropathy resembling rheumatoid arthritis and noninflammatory ankylotic arthropathy in transgenic mice. *J. Immunol.* 162, 2956–2963.

Hall, A. P., Irvine, J., Blyth, K., Cameron, E. R., Onions, D. E., and Campbell, M. E. (1998). Tumours derived from HTLV-I tax transgenic mice are characterized by enhanced levels of apoptosis and oncogene expression. *J. Pathol.* 186, 209–214. doi: 10.1002/(SICI)1096-9896(1998100)186:2<209::AID-PATH162>3.0.CO;2-I

Harhaj, E. W., and Sun, S. C. (1999). IKKgamma serves as a docking subunit of the IkappaB kinase (IKK) and mediates interaction of IKK with the human T-cell leukemia virus tax protein. *J. Biol. Chem.* 274, 22911–22914. doi: 10.1074/jbc.274.33.22911

Hasegawa, H., Sawa, H., Lewis, M. J., Orba, Y., Sheehy, N., Yamamoto, Y., et al. (2006). Thymus-derived leukemia-lymphoma in mice transgenic for the tax gene of human T-lymphotropic virus type I. *Nat. Med.* 12, 466–472. doi: 10.1038/nm1389

Hermine, O., Wattel, E., Gessain, A., and Bazarbachi, A. (1998). Adult T cell leukaemia. *BioDrugs* 10, 447–462. doi: 10.2165/00063030-199810060-00003

Hinrichs, S. H., Nerenberg, M., Reynolds, R. K., Khoury, G., and Jay, G. (1987). A transgenic mouse model for human neurofibromatosis. *Science* 237, 1340–1343. doi: 10.1126/science.2888191

Hinuma, Y., Nagata, K., Hanaoka, M., Nakai, M., Matsumoto, T., Kinoshita, K. I., et al. (1981). Adult T-cell leukemia: antigen in an ATL cell line and detection of antibodies to the antigen in human sera. *Proc. Natl. Acad. Sci. U.S.A.* 78, 6476–6480. doi: 10.1073/pnas.78.10.6476

Hiramatsu, H., Nishikomori, R., Heike, T., Ito, M., Kobayashi, K., Katamura, K., et al. (2003). Complete reconstitution of human lymphocytes from cord blood CD34+ cells using the NOD/SCID/gammacnull mice model. *Blood* 102, 873–880. doi: 10.1182/blood-2002-09-2755

Hironaka, N., Mochida, K., Mori, N., Maeda, M., Yamamoto, N., and Yamaoka, S. (2004). Tax-independent constitutive IkappaB kinase activation in adult T-cell leukemia cells. *Neoplasia* 6, 266–278. doi: 10.1593/neo.3388

Hiyoshi, M., Okuma, K., Tateyama, S., Takizawa, K., Saito, M., Kuramitsu, M., et al. (2015). Furin-dependent CCL17-fused recombinant toxin controls HTLV-I infection by targeting and eliminating infected CCR4-expressing cells *in vitro* and in vivo. *Retrovirology* 12:73. doi: 10.1186/s12977-015-0199-8

Ikebe, E., Kawaguchi, A., Tezuka, K., Taguchi, S., Hirose, S., Matsumoto, T., et al. (2013). Oral administration of an HSP90 inhibitor, 17-DMAG, intervenes tumor-cell infiltration into multiple organs and improves survival period for ATL model mice. *Blood Cancer J.* 3:e132. doi: 10.1038/bcj.2013.30

Ishida, T., Utsunomiya, A., Iida, S., Inagaki, H., Takatsuka, Y., Kusumoto, S., et al. (2003). Clinical significance of CCR4 expression in adult T-cell leukemia/lymphoma: its close association with skin involvement and unfavorable outcome. *Clin. Cancer Res.* 9, 3625–3634.

Ishihara, S., Tachibana, N., Okayama, A., Murai, K., Tsuda, K., and Mueller, N. (1992). Successful graft of HTLV-I-transformed human T-cells (MT-2) in severe combined immunodeficiency mice treated with anti-asialo GM-1 antibody. *Jpn. J. Cancer Res.* 83, 320–323. doi: 10.1111/j.1349-7006.1992.tb00108.x

Ishii, T., Ishida, T., Utsunomiya, A., Inagaki, A., Yano, H., Komatsu, H., et al. (2010). Defucosylated humanized anti-CCR4 monoclonal antibody KW-0761 as a novel immunotherapeutic agent for adult T-cell leukemia/lymphoma. *Clin. Cancer Res.* 16, 1520–1531. doi: 10.1158/1078-0432.CCR-09-2697

Ishikawa, F., Yasukawa, M., Lyons, B., Yoshida, S., Miyamoto, T., Yoshimoto, G., et al. (2005). Development of functional human blood and immune systems in NOD/SCID/IL2 receptor γ chain (null) mice. *Blood* 106, 1565–1573. doi: 10.1182/blood-2005-02-0516

Ishitsuka, K., Kunami, N., Katsuya, H., Nogami, R., Ishikawa, C., Yotsumoto, F., et al. (2012). Targeting bcl-2 family proteins in adult T-cell leukemia/lymphoma: *In vitro* and in vivo effects of the novel bcl-2 family inhibitor ABT-737. *Cancer Lett.* 317, 218–225. doi: 10.1016/j.canlet.2011.11.030

Ishitsuka, K., and Tamura, K. (2014). Human T-cell leukaemia virus type I and adult T-cell leukaemia-lymphoma. *Lancet Oncol.* 15, e517–e526. doi: 10.1016/S1470-2045(14)70202-5

Ito, A., Ishida, T., Utsunomiya, A., Sato, F., Mori, F., Yano, H., et al. (2009). Defucosylated anti-CCR4 monoclonal antibody exerts potent ADCC against primary ATLL cells mediated by autologous human immune cells in NOD/Shi-scid, IL-2Rγnull mice in vivo. *J. Immunol.* 183, 4782–4791. doi: 10.4049/jimmunol.0900699

Ito, M., Hiramatsu, H., Kobayashi, K., Suzue, K., Kawahata, M., Hioki, K., et al. (2002). NOD/SCID/gamma(c) (null) mouse: An excellent recipient mouse model for engraftment of human cells. *Blood* 100, 3175–3182. doi: 10.1182/blood-2001-12-0207

Ito, M., Kobayashi, K., and Nakahata, T. (2008). NOD/shi-scid IL2r gamma (null) (NOG) mice more appropriate for humanized mouse models. *Curr. Top. Microbiol. Immunol.* 324, 53–76. doi: 10.1007/978-3-540-75647-7_3

Iwai, K., Mori, N., Oie, M., Yamamoto, N., and Fujii, M. (2001). Human T-cell leukemia virus type 1 tax protein activates transcription through AP-1 site by inducing DNA binding activity in T cells. *Virology* 279, 38–46. doi: 10.1006/viro.2000.0669

Ju, W., Zhang, M., Petrus, M., Maeda, M., Pise-Masison, C. A., and Waldmann, T. A. (2014). Combination of 9-aminoacridine with campath-1H provides effective therapy for a murine model of adult T-cell leukemia. *Retrovirology* 11:43. doi: 10.1186/1742-4690-11-43

Kannian, P., and Green, P. L. (2010). Human T lymphotropic virus type 1 (HTLV-I): Molecular biology and oncogenesis. *Viruses* 2, 2037–2077. doi: 10.3390/v2092037

Karube, K., Ohshima, K., Tsuchiya, T., Yamaguchi, T., Kawano, R., Suzumiya, J., et al. (2004). Expression of FoxP3, a key molecule in CD4+ CD25+ regulatory T cells, in adult T-cell leukaemia/lymphoma cells. *Br. J. Haematol.* 126, 81–84. doi: 10.1111/j.1365-2141.2004.04999.x

Kasugai, Y., Yoshida, N., Ohshima, K., Matsuo, K., Seto, M., and Tsuzuki, S. (2016). New mouse model of acute adult T-cell leukemia generated by transplantation of AKT, BclxL, and HBZ-transduced T cells. *Cancer Sci.* 107, 1072–1078. doi: 10.1111/cas.12974

Kataoka, K., Nagata, Y., Kitanaka, A., Shiraishi, Y., Shimamura, T., Yasunaga, J., et al. (2015). Integrated molecular analysis of adult T cell leukemia/lymphoma. *Nat. Genet.* 47, 1304–1315. doi: 10.1038/ng.3415

Kchour, G., Tarhini, M., Kooshyar, M. M., El Hajj, H., Wattel, E., Mahmoudi, M., et al. (2009). Phase 2 study of the efficacy and safety of the combination of arsenic trioxide, interferon alpha, and zidovudine in newly diagnosed chronic adult T-cell leukemia/lymphoma (ATL). *Blood* 113, 6528–6532. doi: 10.1182/blood-2009-03-211821

Kfoury, Y., Nasr, R., Favre-Bonvin, A., El-Sabban, M., Renault, N., Giron, M. L., et al. (2008). Ubiquitylated tax targets and binds the IKK signalosome at the centrosome. *Oncogene* 27, 1665–1676. doi: 10.1038/sj.onc.1210804

Kfoury, Y., Nasr, R., Hermine, O., de The, H., and Bazarbachi, A. (2005). Proapoptotic regimes for HTLV-I-transformed cells: Targeting tax and the NF-kappaB pathway. *Cell Death Differ.* 12(Suppl. 1), 871–877. doi: 10.1038/sj.cdd.4401624

Kfoury, Y., Nasr, R., Journo, C., Mahieux, R., Pique, C., and Bazarbachi, A. (2012). The multifaceted oncoprotein tax: Subcellular localization, posttranslational modifications, and NF-kappaB activation. *Adv. Cancer Res.* 113, 85–120. doi: 10.1016/B978-0-12-394280-7.00003-8

Kinosada, H., Yasunaga, J. I., Shimura, K., Miyazato, P., Onishi, C., Iyoda, T., et al. (2017). HTLV-I bZIP factor enhances T-cell proliferation by impeding the suppressive signaling of co-inhibitory receptors. *PLoS Pathog.* 13:e1006120. doi: 10.1371/journal.ppat.1006120

Koller, B. H., and Smithies, O. (1989). Inactivating the beta 2-microglobulin locus in mouse embryonic stem cells by homologous recombination. *Proc. Natl. Acad. Sci. U.S.A.* 86, 8932–8935. doi: 10.1073/pnas.86.22.8932

Kondo, A., Imada, K., Hattori, T., Yamabe, H., Tanaka, T., Miyasaka, M., et al. (1993). A model of in vivo cell proliferation of adult T-cell leukemia. *Blood* 82, 2501–2509.

Kress, A. K., Grassmann, R., and Fleckenstein, B. (2011). Cell surface markers in HTLV-I pathogenesis. *Viruses* 3, 1439–1459. doi: 10.3390/v3081439

Kuhlmann, A. S., Villaudy, J., Gazzolo, L., Castellazzi, M., Mesnard, J. M., and Duc Dodon, M. (2007). HTLV-I HBZ cooperates with JunD to enhance transcription of the human telomerase reverse transcriptase gene (hTERT). *Retrovirology* 4:92. doi: 10.1186/1742-4690-4-92

Kuribayashi, W., Takizawa, K., Sugata, K., Kuramitsu, M., Momose, H., Sasaki, E., et al. (2016). Impact of the SCF signaling pathway on leukemia stem cell-mediated ATL initiation and progression in an HBZ transgenic mouse model. *Oncotarget* 7, 51027–51043. doi: 10.18632/oncotarget.10210

Kwon, H., Ogle, L., Benitez, B., Bohuslav, J., Montano, M., Felsher, D. W., et al. (2005). Lethal cutaneous disease in transgenic mice conditionally expressing type I human T cell leukemia virus tax. *J. Biol. Chem.* 280, 35713–35722. doi: 10.1074/jbc.M504848200

Larocca, D., Chao, L. A., Seto, M. H., and Brunck, T. K. (1989). Human T-cell leukemia virus minus strand transcription in infected T-cells. *Biochem. Biophys. Res. Commun.* 163, 1006–1013. doi: 10.1016/0006-291X(89)92322-X

Lee, T. H., Coligan, J. E., Sodroski, J. G., Haseltine, W. A., Salahuddin, S. Z., Wong-Staal, F., et al. (1984). Antigens encoded by the 3'-terminal region of human T-cell leukemia virus: evidence for functional gene. *Science* 226, 57–62. doi: 10.1126/science.6089350

Lemasson, I., Lewis, M. R., Polakowski, N., Hivin, P., Cavanagh, M. H., Thebault, S., et al. (2007). Human T-cell leukemia virus type 1 (HTLV-I) bZIP protein interacts with the cellular transcription factor CREB to inhibit HTLV-I transcription. *J. Virol.* 81, 1543–1553. doi: 10.1128/JVI.00480-06

Liu, Y., Dole, K., Stanley, J. R., Richard, V., Rosol, T. J., Ratner, L., et al. (2002). Engraftment and tumorigenesis of HTLV-I transformed T cell lines in SCID/bg and NOD/SCID mice. *Leuk. Res.* 26, 561–567. doi: 10.1016/S0145-2126(01)00169-2

Ma, G., Yasunaga, J., and Matsuoka, M. (2016). Multifaceted functions and roles of HBZ in HTLV-I pathogenesis. *Retrovirology* 13:16. doi: 10.1186/s12977-016-0249-x

Macaire, H., Riquet, A., Moncollin, V., Biemont-Trescol, M. C., Duc Dodon, M., Hermine, O., et al. (2012). Tax protein-induced expression of antiapoptotic bfl-1 protein contributes to survival of human T-cell leukemia virus type 1 (HTLV-I)-infected T-cells. *J. Biol. Chem.* 287, 21357–21370. doi: 10.1074/jbc.M112.340992

Maeda, N., Muta, H., Oflazoglu, E., and Yoshikai, Y. (2010). Susceptibility of human T-cell leukemia virus type I-infected cells to humanized anti-CD30 monoclonal antibodies *in vitro* and in vivo. *Cancer Sci.* 101, 224–230. doi: 10.1111/j.1349-7006.2009.01354

Maeda, N., Ohashi, T., Chagan-Yasutan, H., Hattori, T., Takahashi, Y., Harigae, H., et al. (2015). Osteopontin-integrin interaction as a novel molecular target for antibody-mediated immunotherapy in adult T-cell leukemia. *Retrovirology* 12:99. doi: 10.1186/s12977-015-0225-x

Mahgoub, M., Yasunaga, J. I., Iwami, S., Nakaoka, S., Koizumi, Y., Shimura, K., et al. (2018). Sporadic on/off switching of HTLV-1 Tax expression is crucial to maintain the whole population of virus-induced leukemic cells. *Proc. Natl. Acad. Sci. U.S.A.* 115, E1269–E1278. doi: 10.1073/pnas.1715724115

Mahieux, R. (2015). A vaccine against HTLV-I HBZ makes sense. *Blood* 126, 1052–1053. doi: 10.1182/blood-2015-06-652040

Marcais, A., Suarez, F., Sibon, D., Frenzel, L., Hermine, O., and Bazarbachi, A. (2013). Therapeutic options for adult T-cell leukemia/lymphoma. *Curr. Oncol. Rep.* 15, 457–464. doi: 10.1007/s11912-013-0332-6

Marriott, S. J., and Semmes, O. J. (2005). Impact of HTLV-I tax on cell cycle progression and the cellular DNA damage repair response. *Oncogene* 24, 5986–5995. doi: 10.1038/sj.onc.1208976

Masaki, A., Ishida, T., Suzuki, S., Ito, A., Mori, F., Sato, F., et al. (2013). Autologous tax-specific CTL therapy in a primary adult T cell leukemia/lymphoma cell-bearing NOD/shi-scid, IL-2Rgammanull mouse model. *J. Immunol.* 191, 135–144. doi: 10.4049/jimmunol.1202692

Matsumoto, J., Ohshima, T., Isono, O., and Shimotohno, K. (2005). HTLV-I HBZ suppresses AP-1 activity by impairing both the DNA-binding ability and the stability of c-jun protein. *Oncogene* 24, 1001–1010. doi: 10.1038/sj.onc.1208297

Matsuoka, M., and Jeang, K. T. (2007). Human T-cell leukaemia virus type 1 (HTLV-I) infectivity and cellular transformation. *Nat. Rev. Cancer* 7, 270–280. doi: 10.1038/nrc2111

Matsuoka, M., and Jeang, K. T. (2011). Human T-cell leukemia virus type 1 (HTLV-1) and leukemic transformation: viral infectivity, Tax, HBZ and therapy. *Oncogene* 30, 1379–1389. doi: 10.1038/onc.2010.537

Mitagami, Y., Yasunaga, J., Kinosada, H., Ohshima, K., and Matsuoka, M. (2015). Interferon-γ promotes inflammation and development of T-cell lymphoma in HTLV-I bZIP factor transgenic mice. *PLoS Pathog.* 11:e1005120. doi: 10.1371/journal.ppat.1005120

Mitra-Kaushik, S., Harding, J., Hess, J., Schreiber, R., and Ratner, L. (2004). Enhanced tumorigenesis in HTLV-I tax-transgenic mice deficient in interferon-gamma. *Blood* 104, 3305–3311. doi: 10.1182/blood-2004-01-0266

Miyazato, P., Yasunaga, J., Taniguchi, Y., Koyanagi, Y., Mitsuya, H., and Matsuoka, M. (2006). De novo human T-cell leukemia virus type 1 infection of human lymphocytes in NOD-SCID, common gamma-chain knockout mice. *J. Virol.* 80, 10683–10691. doi: 10.1128/JVI.01009-06

Mori, N., Fujii, M., Ikeda, S., Yamada, Y., Tomonaga, M., Ballard, D. W., et al. (1999). Constitutive activation of NF-kappaB in primary adult T-cell leukemia cells. *Blood* 93, 2360–2368.

Nakamura, D., Yoshimitsu, M., Kuroki, A., Hachiman, M., Kamada, Y., Ezinne, C. C., et al. (2015). A new ATL xenograft model and evaluation of pyrrolidine dithiocarbamate as a potential ATL therapeutic agent. *Exp. Hematol.* 43, 944–950. doi: 10.1016/j.exphem.2015.06.303

Nasr, R., Marcais, A., Hermine, O., and Bazarbachi, A. (2017). Overview of targeted therapies for adult T-cell leukemia/lymphoma. *Methods Mol. Biol.* 1582, 197–216. doi: 10.1007/978-1-4939-6872-5_15

Nasr, R., Rosenwald, A., El-Sabban, M. E., Arnulf, B., Zalloua, P., Lepelletier, Y., et al. (2003). Arsenic/interferon specifically reverses 2 distinct gene networks critical for the survival of HTLV-I-infected leukemic cells. *Blood* 101, 4576–4582. doi: 10.1182/blood-2002-09-2986

Nerenberg, M., Hinrichs, S. H., Reynolds, R. K., Khoury, G., and Jay, G. (1987). The tat gene of human T-lymphotropic virus type 1 induces mesenchymal tumors in transgenic mice. *Science* 237, 1324–1329. doi: 10.1126/science.2888190

Nerenberg, M. I., and Wiley, C. A. (1989). Degeneration of oxidative muscle fibers in HTLV-I tax transgenic mice. *Am. J. Pathol.* 135, 1025–1033.

Nicot, C., Mahieux, R., Takemoto, S., and Franchini, G. (2000). Bcl-X (L) is up-regulated by HTLV-I and HTLV-II *in vitro* and in ex vivo ATLL samples. *Blood* 96, 275–281.

Niewiesk, S. (2016). Animals models of human T cell leukemia virus type I leukemogenesis. *ILAR J.* 57, 3–11. doi: 10.1093/ilar/ilv045

Nitta, T., Tanaka, M., Sun, B., Sugihara, E., Kimura, M., Kamada, Y., et al. (2008). Reduction of human T-cell leukemia virus type-1 infection in mice lacking nuclear factor-kappaB-inducing kinase. *Cancer Sci.* 99, 872–878. doi: 10.1111/j.1349-7006.2008.00766.x

Nosaka, K., Iwanaga, M., Imaizumi, Y., Ishitsuka, K., Ishizawa, K., Ishida, Y., et al. (2017). Epidemiological and clinical features of adult T-cell leukemia-lymphoma in japan, 2010-2011: a nationwide survey. *Cancer Sci.* 108, 2478–2486. doi: 10.1111/cas.13398

Ohsugi, T. (2013). A transgenic mouse model of human T cell leukemia virus type 1-associated diseases. *Front. Microbiol.* 4:49. doi: 10.3389/fmicb.2013.00049

Ohsugi, T., Ishibashi, K., Shingu, M., and Nomura, T. (1994). Engraftment of HTLV-I-transformed human T-cell line into SCID mice with NK cell function. *J. Vet. Med. Sci.* 56, 601–603. doi: 10.1292/jvms.56.601

Ohsugi, T., Kumasaka, T., Ishida, A., Ishida, T., Horie, R., Watanabe, T., et al. (2006). *In vitro* and in vivo antitumor activity of the NF-kappaB inhibitor DHMEQ in the human T-cell leukemia virus type I-infected cell line, HuT-102. *Leuk. Res.* 30, 90–97. doi: 10.1016/j.leukres.2005.06.001

Ohsugi, T., Kumasaka, T., Okada, S., Ishida, T., Yamaguchi, K., Horie, R., et al. (2007a). Dehydroxymethylepoxyquinomicin (DHMEQ) therapy reduces tumor formation in mice inoculated with tax-deficient adult T-cell leukemia-derived cell lines. *Cancer Lett.* 257, 206–215. doi: 10.1016/j.canlet.2007.07.017

Ohsugi, T., Kumasaka, T., Okada, S., and Urano, T. (2007b). The tax protein of HTLV-I promotes oncogenesis in not only immature T cells but also mature T cells. *Nat. Med.* 13, 527–528. doi: 10.1038/nm0507-527

Okayama, A., Tachibana, N., Ishihara, S., Nagatomo, Y., Murai, K., Okamoto, M., et al. (1997). Increased expression of interleukin-2 receptor alpha on peripheral blood mononuclear cells in HTLV-I tax/rex mRNA-positive asymptomatic carriers. *J. Acquir. Immune Defic. Syndr. Hum. Retrovirol.* 15, 70–75. doi: 10.1097/00042560-199705010-00011

Panfil, A. R., Al-Saleem, J. J., and Green, P. L. (2013). Animal models utilized in HTLV-I research. *Virology* 4, 49–59. doi: 10.4137/VRT.S12140

Panfil, A. R., Dissinger, N. J., Howard, C. M., Murphy, B. M., Landes, K., Fernandez, S. A., et al. (2016). Functional comparison of HBZ and the related APH-2 protein provides insight into human T-cell leukemia virus type 1 pathogenesis. *J. Virol.* 90, 3760–3772. doi: 10.1128/JVI.03113-15

Parrula, C., Fernandez, S. A., Zimmerman, B., Lairmore, M., and Niewiesk, S. (2011). Measles virotherapy in a mouse model of adult T-cell leukaemia/lymphoma. *J. Gen. Virol.* 92(Pt 6), 1458–1466. doi: 10.1099/vir.0.028910-0

Pearson, T., Greiner, D. L., and Shultz, L. D. (2008). Humanized SCID mouse models for biomedical research. *Curr. Top. Microbiol. Immunol.* 324, 25–51. doi: 10.1007/978-3-540-75647-7_2

Phillips, K. E., Herring, B., Wilson, L. A., Rickford, M. S., Zhang, M., Goldman, C. K., et al. (2000). IL-2Ralpha-directed monoclonal antibodies provide effective therapy in a murine model of adult T-cell leukemia by a mechanism other than blockade of IL-2/IL-2Ralpha interaction. *Cancer Res.* 60, 6977–6984.

Pise-Masison, C. A., Choi, K. S., Radonovich, M., Dittmer, J., Kim, S. J., and Brady, J. N. (1998). Inhibition of p53 transactivation function by the human T-cell lymphotropic virus type 1 tax protein. *J. Virol.* 72, 1165–1170.

Poiesz, B. J., Ruscetti, F. W., Gazdar, A. F., Bunn, P. A., Minna, J. D., and Gallo, R. C. (1980). Detection and isolation of type C retrovirus particles from fresh and cultured lymphocytes of a patient with cutaneous T-cell lymphoma. *Proc. Natl. Acad. Sci. U.S.A.* 77, 7415–7419. doi: 10.1073/pnas.77.12.7415

Portis, T., Grossman, W. J., Harding, J. C., Hess, J. L., and Ratner, L. (2001). Analysis of p53 inactivation in a human T-cell leukemia virus type 1 tax transgenic mouse model. *J. Virol.* 75, 2185–2193. doi: 10.1128/JVI.75.5.2185-2193.2001

Rauch, D., Gross, S., Harding, J., Bokhari, S., Niewiesk, S., Lairmore, M., et al. (2009a). T-cell activation promotes tumorigenesis in inflammation-associated cancer. *Retrovirology* 6:116. doi: 10.1186/1742-4690-6-116

Rauch, D., Gross, S., Harding, J., Niewiesk, S., Lairmore, M., Piwnica-Worms, D., et al. (2009b). Imaging spontaneous tumorigenesis: Inflammation precedes development of peripheral NK tumors. *Blood* 113, 1493–1500. doi: 10.1182/blood-2008-07-166462

Rauch, D. A., Harding, J. C., and Ratner, L. (2014). IL-15 deficient tax mice reveal a role for IL-1alpha in tumor immunity. *PLoS One* 9:e85028. doi: 10.1371/journal.pone.0085028

Ruddle, N. H., Li, C. B., Horne, W. C., Santiago, P., Troiano, N., Jay, G., et al. (1993). Mice transgenic for HTLV-I LTR-tax exhibit tax expression in bone, skeletal alterations, and high bone turnover. *Virology* 197, 196–204. doi: 10.1006/viro.1993.1580

Russell, S. J., Peng, K. W., and Bell, J. C. (2012). Oncolytic virotherapy. *Nat. Biotechnol.* 30, 658–670. doi: 10.1038/nbt.2287

Saito, M., Matsuzaki, T., Satou, Y., Yasunaga, J., Saito, K., Arimura, K., et al. (2009). In vivo expression of the HBZ gene of HTLV-I correlates with proviral load, inflammatory markers and disease severity in HTLV-I associated myelopathy/tropical spastic paraparesis (HAM/TSP). *Retrovirology* 6:19. doi: 10.1186/1742-4690-6-19

Saitoh, Y., Hamano, A., Mochida, K., Kakeya, A., Uno, M., Tsuruyama, E., et al. (2016). A20 targets caspase-8 and FADD to protect HTLV-I-infected cells. *Leukemia* 30, 716–727. doi: 10.1038/leu.2015.267

Satou, Y., Nosaka, K., Koya, Y., Yasunaga, J. I., Toyokuni, S., and Matsuoka, M. (2004). Proteasome inhibitor, bortezomib, potently inhibits the growth of adult T-cell leukemia cells both in vivo and *in vitro*. *Leukemia* 18, 1357–1363. doi: 10.1038/sj.leu.2403400

Satou, Y., Yasunaga, J., Yoshida, M., and Matsuoka, M. (2006). HTLV-I basic leucine zipper factor gene mRNA supports proliferation of adult T cell leukemia cells. *Proc. Natl. Acad. Sci. U.S.A.* 103, 720–725. doi: 10.1073/pnas.0507631103

Satou, Y., Yasunaga, J., Zhao, T., Yoshida, M., Miyazato, P., Takai, K., et al. (2011). HTLV-I bZIP factor induces T-cell lymphoma and systemic inflammation in vivo. *PLoS Pathog.* 7:e1001274. doi: 10.1371/journal.ppat.1001274

Shimoyama, M. (1991). Diagnostic criteria and classification of clinical subtypes of adult T-cell leukaemia-lymphoma. A report from the lymphoma study group (1984-87). *Br. J. Haematol.* 79, 428–437. doi: 10.1111/j.1365-2141.1991.tb08051.x

Shimoyama, M., Minato, K., Tobinai, K., Nagai, M., Setoya, T., Takenaka, T., et al. (1983). Atypical adult T-cell leukemia-lymphoma: Diverse clinical manifestations of adult T-cell leukemia-lymphoma. *Jpn. J. Clin. Oncol.* 13(Suppl. 2), 165–187.

Shirinian, M., Kambris, Z., Hamadeh, L., Grabbe, C., Journo, C., Mahieux, R., et al. (2015). A transgenic Drosophila melanogaster model to study human T-lymphotropic virus oncoprotein tax-1-driven transformation in vivo. *J. Virol.* 89, 8092–8095. doi: 10.1128/JVI.00918-15

Shultz, L. D., Schweitzer, P. A., Christianson, S. W., Gott, B., Schweitzer, I. B., Tennent, B., et al. (1995). Multiple defects in innate and adaptive immunologic function in NOD/LtSz-scid mice. *J. Immunol.* 154, 180–191.

Stewart, S. A., Feuer, G., Jewett, A., Lee, F. V., Bonavida, B., and Chen, I. S. (1996). HTLV-1 gene expression in adult T-cell leukemia cells elicits an NK cell response *in vitro* and correlates with cell rejection in SCID Mice. *Virology* 226, 167–175. doi: 10.1006/viro.1996.0643

Sugata, K., Satou, Y., Yasunaga, J., Hara, H., Ohshima, K., Utsunomiya, A., et al. (2012). HTLV-I bZIP factor impairs cell-mediated immunity by suppressing production of Th1 cytokines. *Blood* 119, 434–444. doi: 10.1182/blood-2011-05-357459

Sugata, K., Yasunaga, J., Mitobe, Y., Miura, M., Miyazato, P., Kohara, M., et al. (2015). Protective effect of cytotoxic T lymphocytes targeting HTLV-I bZIP factor. *Blood* 126, 1095–1105. doi: 10.1182/blood-2015-04-641118

Sun, S. C., Elwood, J., Beraud, C., and Greene, W. C. (1994). Human T-cell leukemia virus type I tax activation of NF-kappa B/rel involves phosphorylation and degradation of I kappa B alpha and RelA (p65)-mediated induction of the c-rel gene. *Mol. Cell. Biol.* 14, 7377–7384. doi: 10.1128/MCB.14.11.7377

Suzuki, T., Kitao, S., Matsushime, H., and Yoshida, M. (1996). HTLV-I tax protein interacts with cyclin-dependent kinase inhibitor p16INK4A and counteracts its inhibitory activity towards CDK4. *EMBO J.* 15, 1607–1614.

Swaims, A. Y., Khani, F., Zhang, Y., Roberts, A. I., Devadas, S., Shi, Y., et al. (2010). Immune activation induces immortalization of HTLV-I LTR-tax transgenic CD4+ T cells. *Blood* 116, 2994–3003. doi: 10.1182/blood-2009-07-231050

Takajo, I., Umeki, K., Morishita, K., Yamamoto, I., Kubuki, Y., Hatakeyama, K., et al. (2007). Engraftment of peripheral blood mononuclear cells from human T-lymphotropic virus type 1 carriers in NOD/SCID/gamma c (null) (NOG) mice. *Int. J. Cancer* 121, 2205–2211. doi: 10.1002/ijc.22972

Takatsuki, K., Uchiayama, T., Sagawa, K., and Yodoi, J. (1977). "Adult T-cell leukemia in Japan," in *Topics in Hematology*, eds S. Seno, F. Takaku, and S. Irino (Amsterdam: Exccrpta Medica), 73–77.

Takeda, S., Maeda, M., Morikawa, S., Taniguchi, Y., Yasunaga, J., Nosaka, K., et al. (2004). Genetic and epigenetic inactivation of tax gene in adult T-cell leukemia cells. *Int. J. Cancer* 109, 559–567. doi: 10.1002/ijc.20007

Tanaka, A., Takahashi, C., Yamaoka, S., Nosaka, T., Maki, M., and Hatanaka, M. (1990). Oncogenic transformation by the tax gene of human T-cell leukemia virus type I *in vitro*. *Proc. Natl. Acad. Sci. U.S.A.* 87, 1071–1075. doi: 10.1073/pnas.87.3.1071

Tezuka, K., Xun, R., Tei, M., Ueno, T., Tanaka, M., Takenouchi, N., et al. (2014). An animal model of adult T-cell leukemia: Humanized mice with HTLV-I-specific immunity. *Blood* 123, 346–355. doi: 10.1182/blood-2013-06-508861

Thebault, S., Basbous, J., Hivin, P., Devaux, C., and Mesnard, J. M. (2004). HBZ interacts with JunD and stimulates its transcriptional activity. *FEBS Lett.* 562, 165–170. doi: 10.1016/S0014-5793(04)00225-X

Traggiai, E., Chicha, L., Mazzucchelli, L., Bronz, L., Piffaretti, J. C., Lanzavecchia, A., et al. (2004). Development of a human adaptive immune system in cord blood cell-transplanted mice. *Science* 304, 104–107. doi: 10.1126/science.1093933

Uchiyama, T. (1996). ATL and HTLV-I: *In vivo* cell growth of ATL cells. *J. Clin. Immunol.* 16, 305–314.

Villaudy, J., Wencker, M., Gadot, N., Gillet, N. A., Scoazec, J. Y., Gazzolo, L., et al. (2011). HTLV-I propels thymic human T cell development in "human immune system", Rag2 (-)/(-) gamma c(-)/(-) mice. *PLoS Pathog.* 7, e1002231. doi: 10.1371/journal.ppat.1002231

Waldmann, T. A., Leonard, W. J., Depper, J. M., Kronke, M., Goldman, C. K., Oh, T., et al. (1984). Interleukin-2 receptor expression in retrovirus associated adult T-cell leukemia. *Princess Takamatsu Symp.* 15, 259–268.

Wang, C., Long, W., Peng, C., Hu, L., Zhang, Q., Wu, A., et al. (2016). HTLV-I tax functions as a ubiquitin E3 ligase for direct IKK activation via synthesis of mixed-linkage polyubiquitin chains. *PLoS Pathog.* 12:e1005584. doi: 10.1371/journal.ppat.1005584

Watanabe, T. (2017). Adult T-cell leukemia: Molecular basis for clonal expansion and transformation of HTLV-I-infected T cells. *Blood* 129, 1071–1081. doi: 10.1182/blood-2016-09-692574

Yamagishi, M., Nakano, K., Miyake, A., Yamochi, T., Kagami, Y., Tsutsumi, A., et al. (2012). Polycomb-mediated loss of miR-31 activates NIK-dependent NF-kappaB pathway in adult T cell leukemia and other cancers. *Cancer Cell* 21, 121–135. doi: 10.1016/j.ccr.2011.12.015

Yamamoto-Taguchi, N., Satou, Y., Miyazato, P., Ohshima, K., Nakagawa, M., Katagiri, K., et al. (2013). HTLV-I bZIP factor induces inflammation through labile Foxp3 expression. *PLoS Pathog.* 9:e1003630. doi: 10.1371/journal.ppat.1003630

Yamazaki, J., Mizukami, T., Takizawa, K., Kuramitsu, M., Momose, H., Masumi, A., et al. (2009). Identification of cancer stem cells in a tax-transgenic (tax-tg) mouse model of adult T-cell leukemia/lymphoma. *Blood* 114, 2709–2720. doi: 10.1182/blood-2008-08-174425

Yano, H., Ishida, T., Imada, K., Sakai, T., Ishii, T., Inagaki, A., et al. (2008). Augmentation of antitumour activity of defucosylated chimeric anti-CCR4 monoclonal antibody in SCID mouse model of adult T-cell leukaemia/lymphoma using G-CSF. *Br. J. Haematol.* 140, 586–589. doi: 10.1111/j.1365-2141.2007.06947.x

Yoshida, M., Miyoshi, I., and Hinuma, Y. (1982). Isolation and characterization of retrovirus from cell lines of human adult T-cell leukemia and its implication in the disease. *Proc. Natl. Acad. Sci. U.S.A.* 79, 2031–2035. doi: 10.1073/pnas.79.6.2031

Yu, P., Petrus, M. N., Ju, W., Zhang, M., Conlon, K. C., Nakagawa, M., et al. (2015). Augmented efficacy with the combination of blockade of the NOTCH-1 pathway, bortezomib and romidepsin in a murine MT-1 adult T cell leukemia model. *Leukemia* 29, 556–566. doi: 10.1038/leu.2014.241

Zhang, L. L., Wei, J. Y., Wang, L., Huang, S. L., and Chen, J. L. (2017). Human T-cell lymphotropic virus type 1 and its oncogenesis. *Acta Pharmacol. Sin.* 38, 1093–1103. doi: 10.1038/aps.2017.17

Zhang, M., Zhang, Z., Garmestani, K., Goldman, C. K., Ravetch, J. V., Brechbiel, M. W., et al. (2004). Activating fc receptors are required for antitumor efficacy of the antibodies directed toward CD25 in a murine model of adult t-cell leukemia. *Cancer Res.* 64, 5825–5829. doi: 10.1158/0008-5472.CAN-04-1088

Zhang, M., Zhang, Z., Goldman, C. K., Janik, J., and Waldmann, T. A. (2005). Combination therapy for adult T-cell leukemia-xenografted mice: Flavopiridol and anti-CD25 monoclonal antibody. *Blood* 105, 1231–1236. doi: 10.1182/blood-2004-05-1709

Zhang, Z., Zhang, M., Garmestani, K., Talanov, V. S., Plascjak, P. S., Beck, B., et al. (2006). Effective treatment of a murine model of adult T-cell leukemia using 211At-7G7/B6 and its combination with unmodified anti-tac (daclizumab) directed toward CD25. *Blood* 108, 1007–1012. doi: 10.1182/blood-2005-11-4757

Zhang, Z., Zhang, M., Ravetch, J. V., Goldman, C., and Waldmann, T. A. (2003). Effective therapy for a murine model of adult T-cell leukemia with the humanized anti-CD2 monoclonal antibody, MEDI-507. *Blood* 102, 284–288. doi: 10.1182/blood-2002-11-3601

Zhao, L. J., and Giam, C. Z. (1992). Human T-cell lymphotropic virus type I (HTLV-I) transcriptional activator, tax, enhances CREB binding to HTLV-I 21-base-pair repeats by protein-protein interaction. *Proc. Natl. Acad. Sci. U.S.A.* 89, 7070–7074. doi: 10.1073/pnas.89.15.7070

Zhao, T., Satou, Y., and Matsuoka, M. (2014). Development of T cell lymphoma in HTLV-I bZIP factor and tax double transgenic mice. *Arch. Virol.* 159, 1849–1856. doi: 10.1007/s00705-014-2099-y

Zhao, T., Yasunaga, J., Satou, Y., Nakao, M., Takahashi, M., Fujii, M., et al. (2009). Human T-cell leukemia virus type 1 bZIP factor selectively suppresses the classical pathway of NF-kappaB. *Blood* 113, 2755–2764. doi: 10.1182/blood-2008-06-161729

Zimmerman, B., Sargeant, A., Landes, K., Fernandez, S. A., Chen, C. S., and Lairmore, M. D. (2011). Efficacy of novel histone deacetylase inhibitor, AR42, in a mouse model of, human T-lymphotropic virus type 1 adult T cell lymphoma. *Leuk. Res.* 35, 1491–1497. doi: 10.1016/j.leukres.2011.07.015

16

Expression of miR-34a in T-Cells Infected by Human T-Lymphotropic Virus 1

Varun K. Sharma[1†‡], Vittoria Raimondi[2‡], Katia Ruggero[1†], Cynthia A. Pise-Masison[3], Ilaria Cavallari[2], Micol Silic-Benussi[2], Vincenzo Ciminale[1,2] and Donna M. D'Agostino[4]*

[1] Department of Surgery, Oncology and Gastroenterology, University of Padova, Padova, Italy, [2] Veneto Institute of Oncology IOV – IRCCS, Padova, Italy, [3] Animal Models and Retroviral Vaccines Section, National Cancer Institute, National Institutes of Health, Bethesda, MD, United States, [4] Department of Biomedical Sciences, University of Padova, Padova, Italy

***Correspondence:**
Vincenzo Ciminale
v.ciminale@unipd.it

†Present address:
Varun K. Sharma,
Noida International University,
Uttar Pradesh, India
Katia Ruggero,
Catalan Institute of Oncology,
Bellvitge Institute for Biomedical
Research, Barcelona, Spain

‡These authors have contributed equally to this work.

Human T-lymphotropic virus 1 (HTLV-1) immortalizes T-cells and is the causative agent of adult T-cell leukemia/lymphoma (ATLL). HTLV-1 replication and transformation are governed by multiple interactions between viral regulatory proteins and host cell factors that remain to be fully elucidated. The present study investigated the impact of HTLV-1 infection on the expression of miR-34a, a microRNA whose expression is downregulated in many types of cancer. Results of RT-PCR assays showed that five out of six HTLV-1-positive cell lines expressed higher levels of miR-34a compared to normal PBMC or purified CD4+ T-cells. ATLL cell line ED, which did not express miR-34a, showed methylation of the miR-34a promoter. Newly infected PBMC and samples from 10 ATLL patients also showed a prominent increase in miR-34a expression compared to PBMC controls. The primary miR-34a transcript expressed in infected cell line C91PL contained binding motifs for NF-κB and p53. Pharmacological inhibition of NF-κB with Bay 11-7082 indicated that this pathway contributes to sustain miR-34a levels in infected cells. Treatment of infected cell lines with the p53 activator nutlin-3a resulted in a further increase in miR-34a levels, thus confirming it as a transcriptional target of p53. Nutlin-3a-treated cells showed downregulation of known miR-34a targets including the deacetylase SIRT1, which was accompanied by increased acetylation of p53, a substrate of SIRT1. Transfection of C91PL cells with a miR-34a mimic also led to downregulation of mRNA targets including SIRT1 as well as the pro-apoptotic factor BAX. Unlike nutlin-3a, the miR-34a mimic did not cause cell cycle arrest or reduce cell viability. On the other hand, sequestration of miR-34a with a sponge construct resulted in an increase in death of C91PL cells. These findings provide evidence for a functional role for miR-34a in fine-tuning the expression of target genes that influence the turnover of HTLV-1-infected cells.

Keywords: HTLV-1, miR-34a, p53, nutlin-3a, adult T-cell leukemia/lymphoma

INTRODUCTION

Human T-lymphotropic virus 1 (HTLV-1) infects approximately 5–10 million people worldwide (Gessain and Cassar, 2012). About 5% of infected individuals develop an aggressive malignancy of mature CD4+ T-cells termed adult T-cell leukemia/lymphoma (ATLL) or a progressive demyelinating neurological disease termed tropical spastic paraparesis/HTLV-associated myelopathy (TSP/HAM) (Poiesz et al., 1980; Yoshida et al., 1982; Gessain et al., 1985; Osame et al., 1986). The transforming potential of HTLV-1 is attributable primarily to the viral proteins Tax (reviewed by Currer et al., 2012) and HBZ (reviewed by Ma et al., 2016), which are each capable of inducing hematological malignancies when expressed as transgenes in mice (Hasegawa et al., 2006; Satou et al., 2011). However, the long clinical latency and low penetrance of ATLL suggest that other viral and cellular factors contribute to determine the fate of HTLV-1-infected cells (for a recent review on HTLV-1, see Bangham, 2017).

The emerging importance of microRNAs (miRNAs) as fine-tuners of gene expression has prompted studies of the interplay between HTLV-1 and the miRNA network. The first such study showed that HTLV-1-infected cell lines express high levels of miR-21, miR-24, miR-146a, and miR-155 and reduced levels of miR-223 compared to normal CD4+ T-cells and uninfected T-cell lines (Pichler et al., 2008). ATLL cells also exhibit important differences in miRNA expression compared to normal peripheral blood mononuclear cells (PBMC) or CD4+ T-cells (Yeung et al., 2008; Bellon et al., 2009; Yamagishi et al., 2012). The first miRNA profiling study of ATLL samples revealed elevated expression of miR-93 and miR-130b, which target the mRNA coding for the pro-apoptotic protein TP53INP1 (Yeung et al., 2008). An analysis of a large panel of ATLL samples revealed an overall downregulation of miRNAs, including miR-31, an important target of which is NIK, an activator of the NF-κB pathway (Yamagishi et al., 2012). miR-145 is downregulated in HTLV-1-positive cell lines and ATLL samples, a property that correlates with poor ATLL patient survival; furthermore, miR-145 has growth suppressive effects when expressed in an ATLL cell line (Xia et al., 2014). Downregulation of miR-150 and miR-223 in HTLV-1-transformed cells results in loss of control of their target STAT1 (Moles et al., 2015). The viral proteins Tax, Rex, and HBZ interfere with the production of mature miRNAs by promoting degradation of Drosha (Tax; Van Duyne et al., 2012) and by inhibiting Dicer expression (HBZ; Gazon et al., 2016) and Dicer activity (Rex; Abe et al., 2010). On the other hand, Tax can upregulate expression of miR-130b (Yeung et al., 2008), miR-146a (Pichler et al., 2008; Tomita et al., 2009), and miR-155 (Tomita, 2012); TRAF6, an adaptor protein involved in a variety of signal transduction pathways, was identified as a target of miR-146a in HTLV-1-infected cell lines (Tomita et al., 2009). HBZ upregulates miR-17 and miR-21, which target proteins that control chromatin remodeling (Vernin et al., 2014). Binding of the HTLV-1 genomic RNA by miR-28-3p interferes with reverse transcription, thereby blocking productive infection (Bai and Nicot, 2015).

In a previous analysis of small RNA libraries, we observed that miR-34a, a highly conserved miRNA that is a component of the p53 pathway and is downregulated in many types of cancer (reviewed by Hermeking, 2010; Slabakova et al., 2017), was more abundant in HTLV-1-infected cell lines C91PL and MT-2 compared to control CD4+ cells (Ruggero et al., 2014). As described below, further investigation of the expression of miR-34a confirmed its upregulation in the context of HTLV-1 infection and identified targets with key roles in cell survival and death.

MATERIALS AND METHODS

Cell Culture

HTLV-1-positive T-cell lines HUT-102 (Poiesz et al., 1980), C91PL (Popovic et al., 1983), MT-2 (Popovic et al., 1983), C8166 (Bhat et al., 1993), ATL-2 (Maeda et al., 1985), and ED-40515(–) [an IL-2-independent subclone (Arima et al., 1992) of ED-40515 (Maeda et al., 1985), referred to as ED in the present study] and the T-ALL cell line Jurkat were maintained in RPMI-1640 (Sigma-Aldrich or Euroclone) supplemented with 10% fetal bovine serum (FBS, Invitrogen), 100 units/mL penicillin and 20 units/mL streptomycin (Euroclone, complete RPMI). HeLa cells were maintained in Dulbecco's Modified Eagles Medium (DMEM, Sigma-Aldrich or Euroclone) supplemented with 10% FBS and penicillin/streptomycin. Genetic profiling was carried out on the cell lines with the PowerPlex 16 HS kit (Promega, Raimondi et al., 2017). CD4+ cells were isolated from normal peripheral blood mononuclear cells (PBMC) using the MACS CD4+ T cell Isolation Kit II or MACS CD4 Microbeads (Miltenyi Biotec). The resulting preparations contained ≥90% CD4+ cells measured by flow cytometry using a PE-labeled anti-CD4 antibody (Becton Dickinson). PBMC samples from ATLL patients (described in Pise-Masison et al., 2009) were collected in the context of National Cancer Institute Institutional Review Board-approved studies (IRB nos. 00-C-0030 and 03-C-0194) with informed consent obtained from all subjects in accordance with the Declaration of Helsinki.

Infection of PBMC

C91PL cells were lethally irradiated (69.5 Gy). PMBC from two blood donors were cultured for 3 h and monocytes/macrophages were depleted by adhesion. 2×10^6 irradiated C91PL cells were cultured together with 2.5×10^6 PBMC in a total volume of 4.5 mL complete RPMI supplemented with 500 μg/mL phytohemagglutinin (PHA, Sigma-Aldrich). Control cultures containing only irradiated C91PL cells or PBMC stimulated with PHA were set up in parallel. After 2 days, cultures were supplemented interleukin-2 (IL-2, 20 U/mL, Roche). The control irradiated C91PL cells were dead after 2 weeks. RNA was isolated from aliquots of the stimulated PBMC after 8 days' culture and from aliquots of the co-cultures at days 30 and 72. Genomic DNA was isolated at the 72-day time point using a salting-out method (Qiagen Flexigene kit) and analyzed by PCR with primers specific for the HTLV-1 U5-gag region (Ruggero et al., 2014); PCR with primers specific for the pri/pre-miR-34a region (see below) served as a control. PCR products were separated in non-denaturing polyacrylamide gels, stained with GelRed (Biotium) and photographed using a Cambridge UVTEC system.

RNA Isolation

Total RNA was isolated using TRIZol (Life Technologies) or TriReagent (Sigma-Aldrich) according to the manufacturer's protocol and quantified using a Nanodrop or Implen spectrophotometer.

Sequencing of the TP53 Open Reading Frame

Total RNA was reverse-transcribed using the Superscript VILO cDNA Synthesis Kit (Invitrogen) according to the manufacturer's protocol. The resulting cDNA was PCR-amplified using TP53-specific primers described elsewhere (Muscolini et al., 2008). Amplicons were subjected to Sanger sequencing (Bigdye kit, Applied Biosystems) and analyzed on a 3730xl DNA Analyzer (Applied Biosystems). The resulting sequences were compared to NCBI p53 transcript reference sequence NM_000546.5 using Finch TV Version 1.4.0 (Geospiza).

miR-34a Promoter Methylation Analysis

Genomic DNA (500 ng) was subjected to bisulfite conversion, desulfonation and column-purification (EZ DNA Methylation-Gold kit, Zymo Research). One-tenth of the eluted DNA was PCR-amplified using primers specific for methylated and unmethylated CpG sites in a 170-bp segment spanning chr1 nt 9182497-9182328 (complementary strand; Chim et al., 2010). PCR products were separated by non-denaturing PAGE in 2% agarose gels. The efficiency of bisulfite conversion was evaluated by subjecting the methylated-MSP product obtained for cell line ED to Sanger sequencing as described above.

RACE

Total RNA isolated from C91PL cells was subjected to RACE (rapid identification of cDNA ends) using the FirstChoice RLM RACE kit (Ambion) according to the manufacturer's instructions. The 5′RACE was carried out using outer primer 5′RACE P1 (AGAGCTTCCGAAGTCCTGG) and inner primer 5′RACE P2 (TTGCTCACAACAACCAGCTAAGA) described by Navarro et al. (Navarro et al., 2009). 3′RACE was performed using outer primer 3′RACE P1 (GGACTTCGGAAGCTCTTCTG) and inner primer 3′RACE P2 (TGGGAAAGTGAGCTCCAGG). Second-round PCR products were separated on non-denaturing acrylamide gels. The most abundant product was eluted and sequenced using primer 5′RACE P2 or 3′RACE P2.

Quantitative RT-PCR

Total RNA was subjected to reverse transcription and qPCR using Taqman microRNA assays (Applied Biosystems) and a 7900HT Fast (Applied Biosystems) or LightCycler 480 (Roche) thermal cycler. RNU44 served as an endogenous control. Results were analyzed by relative quantification using the $2^{-\Delta\Delta CT}$ method and a calibrator specified in the figure legends. A threshold cycle (C_T) of 35 was considered "undetermined." For mRNA expression analysis, RNA was treated with DNase I (Invitrogen) and then reverse-transcribed using SuperScript II reverse transcriptase (Invitrogen) and random hexamers (Applied Biosystems) (**Figures 5**, **7** and Supplementary Figures 1, 4) or with the Superscript VILO cDNA Synthesis Kit (without DNAse treatment) (Supplementary Figure 3). cDNA was PCR-amplified by using a SYBR Green master mix (Roche or Thermo Scientific) and the following primer pairs: CDKN1A-s (AGACTCTCAGGGTCGAAAAC) and CDKN1A-as (TTCCAGGACTGCAGGCTTC); pri/pre-miR-34a-s (TGGCAGTGTCTTAGCTGGTTG) and pri/pre-miR-34a-as (GGCAGTATACTTGCTGATTGCTT) (Jiang et al., 2005); TP53INP1-s (CTTCCTCCAACCAAGAACCAG) and TP53INP1-as (CAAGCACTCAAGAGATGCCG); BAX-s (GTCTTTTTCCGAGTGGCAGC) and BAX-as (AGGAAGTCCAATGTCCAGCC); BIRC5-s (TTCTCAAGGACCACCGCATC) and BIRC5-as (TGAAGCAGAAGAAACACTGGG); SIRT1-s (ACATAGACACGCTGGAACAGG) and SIRT1-as (GATAGCAAGCGGTTCATCAGC); TP53-s (TGGAAGGAAATTTGCGTGTGG) and TP53-as (CCAGTGTGATGATGGTGAGG); ACTB-s (AGCACAGAGCCTCGCCTTTG) and ACTB-as (GGAATCCTTCTGACCCATGC); B2M-s (TGACTTTGTCACAGCCCAAG) and B2M-as (TTCAAACCTCCATGATGCTG); GAPDH-s (GAAGGTGAAGGTCGGAGTC) and GAPDH-as (GAAGATGGTGATGGGATTTC). PCR reactions were set up in duplicate or triplicate and amplified in a LightCycler 480 thermal cycler; ACTB (β-actin), B2M (β-2 microglobulin) or GAPDH was used as an endogenous control. RT-PCR to detect viral transcripts was performed as described elsewhere (Rende et al., 2011). Primers were synthesized by Sigma-Genosys.

Plasmids and Transfections

Plasmid pGFP-miR-34a-sponge was constructed by inserting the GFP coding sequence followed by four miR-34a target sequences into pcDNA3.1 (Invitrogen); a control plasmid lacking the miR-34a target sequences (GFP-control) was also cloned. The inserts were obtained from previously described retroviral vectors (Rao et al., 2010). In **Figure 4C**, HeLa cells were transfected using PolyJet transfection reagent (SignaGen Laboratories). In **Figure 7**, C91PL cells were electroporated as described (Silic-Benussi et al., 2010). DNA transfection mixtures and incubation times are indicated in the figure legends.

Drug Treatments

Bay 11-7082 and nutlin-3a (Sigma-Aldrich) were dissolved in dimethyl sulfoxide (DMSO, Hybrimax; Sigma-Aldrich). Cells were seeded in tissue culture plates at 300,000 cells/mL and treated with the drugs or with the same volume of DMSO (final dilution, 0.1%) for 48 h. Nutlin-3a was substituted with nutlin-3 (Tocris Bioscience) in some replicates; no substantial difference was noted in the effects of the two preparations.

Immunoblotting

Cells were lysed in Cell Disruption Buffer (Ambion) containing inhibitors of proteases and phosphatases (Complete and PhosSTOP, Roche); samples to be analyzed for p53 acetylation were supplemented with 10 mM nicotinamide (Sigma-Aldrich) to inhibit deacetylases. Samples were analyzed with a Bradford protein assay (Bradford, 1976), balanced for total protein and subjected to SDS-PAGE followed by electrotransfer to nitrocellulose membrane (GE Healthcare). Blots were saturated in non-fat milk (Euroclone) and incubated with rabbit anti-GAPDH antibody (Genetex), rabbit anti-β-actin polyclonal antibody (Sigma-Aldrich), rabbit anti-acetylated p53 antibody (Lysine-382, Cell Signaling), goat anti-p53 polyclonal antibody, and rabbit anti-SIRT1 polyclonal antibody (both from Santa Cruz Biotechnology) followed by horseradish peroxidase-conjugated secondary antibodies (Pierce or GE Healthcare). Immunoreactive bands were detected using Femto (Pierce) or LiteAblot Turbo (Euroclone) chemiluminescence reagent and a digital imager (BioRad ChemiDoc XRS or Cambridge UVTEC).

Analyses of Cell Turnover

For cell cycle analysis, cells were fixed in ethanol (Sigma-Aldrich), stained with propidium iodide (Sigma-Aldrich) in the presence of RNase A (Qiagen) and analyzed by flow cytometry using a FACSCalibur (BD, FL2-A setting) and ModFit software. Uptake of Live-Dead Far Red (Molecular Probes) was measured according to the manufacturer's instructions. Conversion of 3-(4, 5-dimethylthiazol-3-yl)-2,5-diphenyl tetrazolium bromide (MTT, Sigma-Aldrich) to blue formazan (MTT test; Mosmann, 1983) was measured using a standard protocol.

RESULTS

miR-34a Expression Is Increased in HTLV-1-Positive Cell Lines and Samples From ATLL Patients

Figure 1A shows the results of quantitative RT-PCR (qRT-PCR) to detect miR-34a in normal PBMC samples and in the HTLV-1-positive cell lines C91PL, MT-2, HUT-102, C8166, ATL-2, and ED, and in the uninfected cell lines HeLa and Jurkat; results were scaled against values measured in normal PBMC. These assays revealed that, with the exception of ED cells, all of the HTLV-1-positive cell lines expressed much higher levels of miR-34a compared to PBMC, HeLa, and Jurkat cells. Additional assays carried out on 11 samples of purified CD4+ cells from healthy donors indicated variable levels of miR-34a, which were lower than those measured in unfractionated PBMC (**Figure 1B**).

Results of qRT-PCR on PBMC isolated from 10 ATLL patients (described in **Figure 1C** and Pise-Masison et al., 2009) revealed increased levels of miR-34a in all of the patients' samples compared to the PBMC calibrator (mean 32-fold increase; **Figure 1C**).

Analysis of miR-34a Promoter Methylation Status in HTLV-1-Positive Cell Lines

The miR-34a gene is coded on the complementary strand of chromosome 1p36.22, a region that is frequently deleted in cancer (Calin et al., 2004). The miR-34a primary RNA (pri-miRNA) contains two or more exons. The region upstream of exon 1 contains many CpG dinucleotides that can be methylated, an event that contributes to the silencing of miR-34a expression in various tumor-derived cell lines and solid cancers, and in some hematological malignancies (Lodygin et al., 2008; Chim et al., 2010; Craig et al., 2011). The substantial difference in miR-34a levels detected in the cell lines shown in **Figure 1A** led us to investigate the methylation status of the miR-34a promoter by performing methylation-specific PCR (MSP) on a 170-bp segment of the miR-34a promoter region as described (Ng et al., 2014).

Results of MSP showed that the miR-34a promoter was substantially methylated in Jurkat and ED cells, but not in cell lines C91PL, MT-2, HUT-102, C8166, or ATL-2 (**Figure 2**). HeLa cells and normal PBMC also did not yield methylated products. Sequencing analysis of the methylated PCR product obtained for ED cells confirmed efficient bisulfite conversion and revealed methylation of 13 CpGs lying internal to the MSP PCR primers (data not shown).

miR-34a Is Upregulated in Newly Infected PBMC

To further examine the link between HTLV-1 infection and miR-34a expression, PBMC from two healthy donors were infected with HTLV-1 through co-cultivation with lethally irradiated C91PL cells and then analyzed by qRT-PCR. Results showed a progressive increase in miR-34a levels after 30 and 72 days of culture compared to the uninfected PBMC (left-hand graph, **Figure 3A**). The infected cultures also showed an increase in the levels of miR-146a, a miRNA that is known to be upregulated by Tax-mediated NF-κB stimulation (Pichler et al., 2008; Tomita et al., 2009; right-hand graph, **Figure 3A**). Results of end-point PCR on genomic DNA isolated at day 72 of the experiment using primers specific for the HTLV-1 *gag* gene confirmed that both co-cultures were HTLV-1-infected (**Figure 3B**). Short tandem repeat (STR) analysis yielded distinct profiles for the two co-cultures and the C91PL cell line, thus verifying that the co-cultures no longer contained input C91PL cells (table in **Figure 3B**).

Identification of a Spliced pri-miR-34a in C91PL Cells

Alternatively spliced miR-34a pri-miRNAs have been identified in different cell contexts. 5′RACE and 3′RACE on total RNA isolated from C91PL cells followed by sequencing analysis yielded a 2-exon pri-miR-34a of 894 nt (**Figure 4A**). This pri-miRNA is similar to a miR-34a precursor designated EF592573.1 that was previously identified in HeLa cells (**Figure 4A**; Chang et al., 2007). It is noteworthy that exon 1 of this pri-miRNA contains a binding site for NF-κB and a binding site for p53 that were shown to be engaged by these transcription factors in other cell systems (Tarasov et al., 2007; Li et al., 2012). Another miR-34a pri-miRNA designated EF570048.1, which was identified in a lung cancer cell line induced to express p53 (Tarasov et al., 2007), does not contain the p53- and NF-κB binding sites in exon 1 and contains a longer exon 2 sequence (**Figure 4A**).

The presence of an NF-κB binding site in the miR-34a pri-miRNA identified in C91PL cells was of interest, as constitutive activation of the NF-κB pathway is a hallmark of HTLV-1 infection/transformation (reviewed by Sun and Yamaoka, 2005). This led us to test the effects of the NF-κB inhibitor Bay 11-7082 (Pierce et al., 1997) on miR-34a expression in C91PL and MT-2 cells. As shown in **Figure 4B**, treatment with Bay 11-7082 led to a dose-dependent reduction in miR-34a levels in both cell

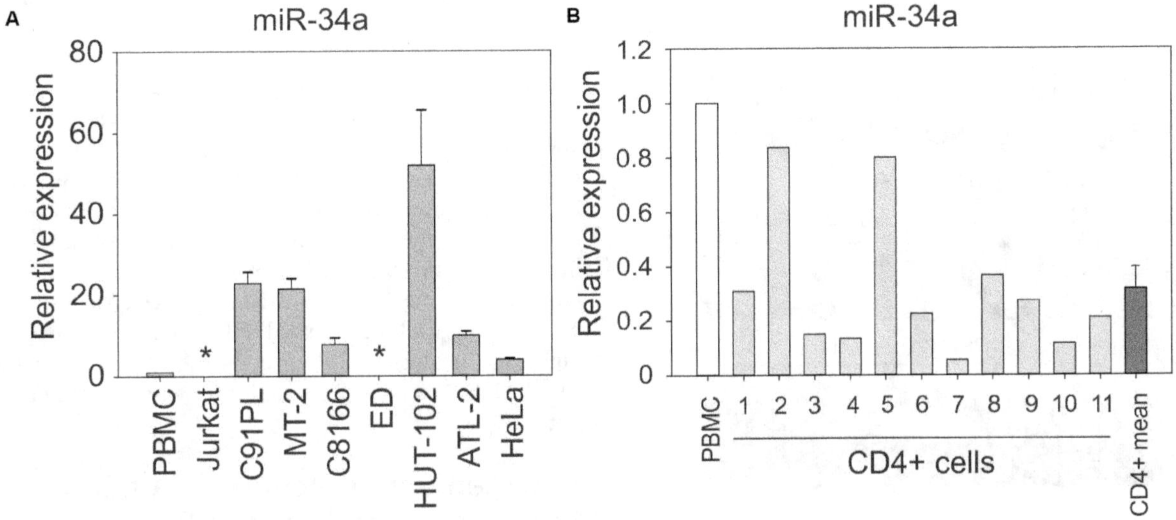

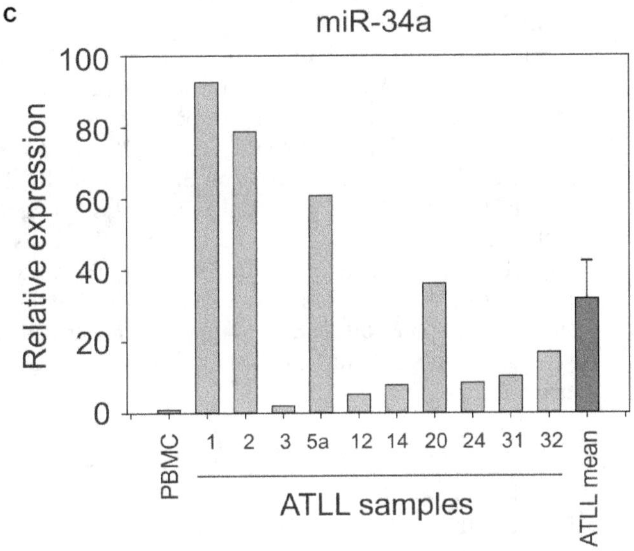

ATLL Patient	Gender/ Age	%CD4+ CD25+	Therapy
1	M/48	78	RIT
2	M/41	80	Zenapax
3	M/53	76	Zenapax
5a	F/50	82	CHOP
12	M/41	47	CVP
14	M/34	94	None
20	F/62	78	None
24	F/36	22	CHOP
31	M/45	92	None
32	M/54	77	None

FIGURE 1 | Expression of miR-34a in normal CD4+ cells, HTLV-1-infected cell lines, and ATLL samples. miR-34a was detected by qRT-PCR as described in the Section "Materials and Methods," with values measured in PBMC samples from healthy donors used as a calibrator (set at 1 in graphs). Panel **(A)** shows miR-34a levels in the indicated cell lines. Bars represent mean values from three to seven samples of cells with standard error. *miR-34a was below the limit of detection in cell lines ED and Jurkat. Panel **(B)** shows the expression of miR-34a in 11 samples of normal unstimulated CD4+ cells. The mean relative expression value for miR-34a in the CD4+ cell samples was 0.316 (bar labeled CD4+ mean). Panel **(C)** shows expression of miR-34a in 10 ATLL samples. The mean relative expression value for miR-34a in the ATLL samples was 31.91 (bar labeled ATLL mean). The table reports characteristics of the ATLL patients. Patient number 24 had ATLL manifested as a lymphoma. M, male; F, female; RIT, radioimmunotherapy; Zenapax, humanized monoclonal antibody against IL-2Rα; CHOP, cyclophosphamide, doxorubicin, vincristine, and prednisone; CVP, cyclophosphamide, vincristine, and prednisone.

lines, thus suggesting that NF-κB contributes to sustain miR-34a expression. A previous study showed that treatment of HTLV-1-infected cell lines and primary ATLL cells with Bay 11-7082 caused a reduction in the expression of NF-κB-responsive genes, accompanied by reduced cell viability and increased apoptosis (Mori et al., 2002). In line with these findings, we observed a dose-dependent loss of viability in the Bay 11-7082-treated C91PL and MT-2 cultures (**Figure 4B**).

miR-34a was previously observed to be upregulated in the Epstein–Barr virus (EBV)-infected B-cells through LMP1-mediated stimulation of the NF-κB pathway (Forte et al., 2012). The ability of Tax to activate NF-κB, CREB, and other transcription factors led us to test its effects on the expression of miR-34a in HeLa cells transfected with wildtype Tax and Tax mutants defective for activation of NF-κB or CREB. Results showed that miR-34a expression was increased by about twofold by wildtype Tax, and by about 1.5-fold by NF-κB-pathway-defective TaxM22, while TaxM47, defective for CREB activation, did not substantially affect miR-34a levels (**Figure 4C**). In contrast, miR-146a, known to be regulated by NF-κB, was strongly induced by wildtype Tax and TaxM47, but not by TaxM22.

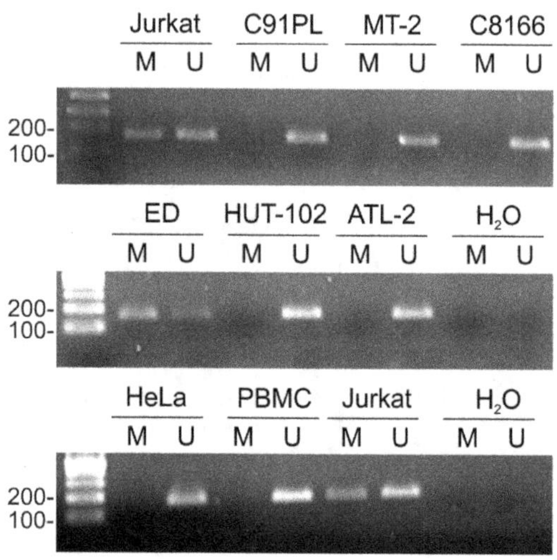

FIGURE 2 | miR-34a promoter methylation in cell lines. Genomic DNA from normal PBMC and the indicated cell lines was subjected to MSP as described in the Section "Materials and Methods." Shown are resulting PCR products after separation in 2% agarose gels. The first lane in each image contains a 100-bp DNA marker (Sharpmass 100, Euroclone). Identical results were obtained for a second DNA sample of each cell type; Jurkat cells are shown as an example.

p53 Activation Enhances miR-34a Expression in HTLV-1-Positive Cell Lines

We next investigated the effects of activation of p53 on miR-34a expression in C91PL and MT-2 cells, which were reported to produce wildtype p53 that is however functionally defective (Cereseto et al., 1996; Kamihira et al., 2001; Hasegawa et al., 2009). To activate p53 we treated the cell lines with nutlin-3a, which stabilizes p53 through inhibition of MDM2 (Vassilev et al., 2004). Results of qRT-PCR showed that nutlin-3a treatment resulted in increased expression of the p53 target genes CDKN1A (coding for p21Waf1/Cip1), TP53INP1, pri/pre-miR-34a, and mature miR-34a (**Figure 5A**).

qRT-PCR analysis to compare the expression of 12 other miRNAs in C91PL cells (Supplementary Figure 1) revealed that most of the tested miRNAs were downregulated by nutlin-3a, including miR-93 and miR-130b, whose levels are elevated in HTLV-1-positive cell lines and ATLL samples (Yeung et al., 2008). Of note, miR-125b, which is known to repress p53 expression in neuroblastoma cells and fibroblasts (Le et al., 2011), was upregulated with nutlin-3a treatment (Supplementary Figure 1).

We next investigated the expression of miR-34a target mRNAs coding for the NAD^+-dependent protein deacetylase SIRT1 (Yamakuchi et al., 2008), the inhibitor of apoptosis (IAP) family member BIRC5 (Survivin; Chen et al., 2010; Kaller et al., 2011; Shen et al., 2012) and the pro-apoptotic protein BAX (Fan and Wang, 2016). qRT-PCR results demonstrated a ~50% reduction in the SIRT1 mRNA, a more substantial reduction in BIRC5, and a marginal increase in the BAX mRNA (**Figure 5B**). The strong reduction in BIRC5 was likely due to the combined effect of miR-34a and transcriptional repression p53 (Hoffman et al., 2002), while the modest increase in BAX could be explained by the fact that BAX is both a target for repression by miR-34a and for transcriptional upregulation by p53 (Miyashita et al., 1994). Consistent with previous studies (Hasegawa et al., 2009) and with its effects on CDKN1A expression, nutlin-3a caused a block of C91PL and MT-2 cells in G0/G1 (**Figure 5C**), a reduction in viability (measured using the MTT assay **Figure 5D**), and increased staining with annexin V, a marker of apoptosis (data not shown). Increased levels of CDKN1A and miR-34a and a reduction in the levels of SIRT1 and cell viability were also observed in nutlin-3a-treated HUT-102 cells (Supplementary Figure 2). Results of cDNA sequencing confirmed that cell lines C91PL, MT-2, and HUT-102 coded for wildtype p53 protein (data not shown).

Engagement of the miR-34a-p53 Feedback Loop in Nutlin-3a-Treated Cells

Substrates of SIRT1 include lysine 382 of p53 (Nakamura et al., 2000; reviewed by Martinez-Redondo and Vaquero, 2013), whose deacetylation interferes with the tumor suppressor's functional activity (reviewed by Reed and Quelle, 2014). We therefore measured the effects of nutlin-3a on the levels of SIRT1 protein, total p53, and lysine 382-acetylated p53. Results showed that nutlin-3a treatment produced a reduction in the levels of SIRT1 (**Figure 6A**), and an increase in total p53 and acetylated p53 (**Figure 6B**), with a relative increase in acetylation on lysine 382 (graph, **Figure 6B**). This observation indicated that nutlin-3a engages the p53-miR-34a-SIRT1 positive feedback loop in infected cells.

Effects of an miR-34a Mimic and Sponge in C91PL Cells

To verify that SIRT1, BIRC5, and BAX represent direct targets of miR-34a in our cell system, we electroporated C91PL cells with a synthetic miR-34a mimic or control RNA. Results of RT-PCR revealed that the mimic-transfected cells had increased levels of miR-34a and reduced levels of the SIRT1, BIRC5, and BAX mRNAs (**Figure 7A**), thus providing direct evidence for targeting of these mRNAs by miR-34a. Immunoblot analysis confirmed the downregulation of SIRT1 protein in the miR-34 mimic-transfected cells (**Figure 7A**, right panel). However, the miR-34a mimic did not have a substantial effect on the levels of mRNAs coding for the p53 targets CDKN1A and TP53INP1, nor did it affect the cell cycle profile or cell viability (**Figure 7B**).

Studies of miR-34a in EBV-infected cells showed that its functional knockdown results in increased cell death (Forte et al., 2012). In an analogous experiment, we electroporated C91PL cells with a GFP expression plasmid containing four binding sites for miR-34a in its 3′UTR ("GFP-sponge") or with a control GFP plasmid lacking the miR-34a binding sites; all transfections included a plasmid expressing LNGFR as a transfection standard. Results of flow cytometry analyses after 3 days of culture showed that while the control- and

A

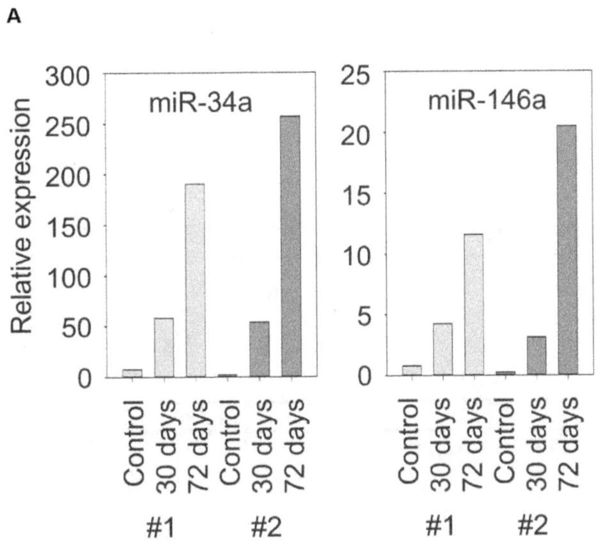

B

End-point PCR on genomic DNA

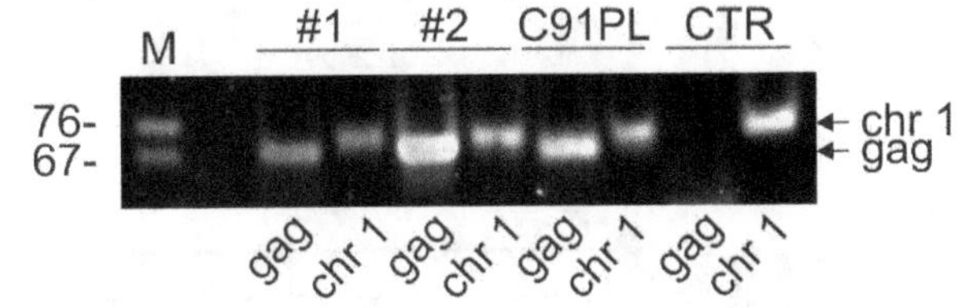

STR analysis of genomic DNA

Marker	#1	#2	C91PL
D3S1358	15,16	17,18	15,19
TH01	8	7,9.3	8,9
D21S11	31.2	28,29	29,31.2
D18S51	15	15,17	12,13
Penta E	7,10	7,17	16,20

FIGURE 3 | Changes in miR-34a and miR-146a levels upon infection of PBMC. Panel **(A)** shows miR-34a and miR-146a levels in PBMC from two healthy donors (distinguished by light- and dark gray bars) harvested 7 days after stimulation with PHA and IL-2 (Controls) and 30 or 72 days after PHA/IL-2 stimulation and co-cultivation with lethally irradiated C91PL cells. Relative expression values were calculated using the control PBMC as calibrators (set at 1 in the graphs). Panel **(B)** shows results of PCR to detect HTLV-1 sequences (lanes labeled HTLV-1) in the genomic DNA of infected PBMC cultures #1 and #2 at day 72. PCR to detect the pre-miR-34a genomic region (lanes labeled pre-miR-34a) served as a control for amplification of genomic DNA, and genomic DNA from C91PL cells and from PBMC of an uninfected donor (CTR) served as positive and negative controls for amplification of HTLV-1 DNA. The table reports values for selected short tandem repeat markers that distinguished the infected PBMC cultures from input C91PL cells.

sponge-transfected cells expressed similar levels of LNGFR, the expression of the GFP-miR-34a sponge was considerably lower than that of the GFP control (**Figure 7C**, left panel), an indication that endogenous miR-34a was silencing the sponge construct. An analysis of Live/Dead Far Red staining revealed an increase in death in the cultures transfected with the miR-34a-sponge compared to cells transfected with the control plasmid (**Figure 7C**, right panel). These observations provide evidence that, in analogy to observations made in EBV-infected B-cells, miR-34a favors the survival of C91PL cells.

DISCUSSION

Accumulated data indicate that HTLV-1 infection has an important impact on the pattern of microRNA expression in the host cell (reviewed by Moles and Nicot, 2015). In the present study we provide evidence indicating that HTLV-1 infection results in a substantial upregulation of miR-34a expression (**Figure 3**) that is sustained in most HTLV-1-positive cell lines (**Figure 1A**). Interestingly, miR-34a is also elevated in primary PBMC samples from ATLL patients (**Figure 1C**), suggesting that high miR-34a levels provide a selective advantage to HTLV-1-infected cells *in vivo* that persists during the complex process of transformation.

The hypothesis that, rather than representing a functionally irrelevant side effect of infection, upregulation of miR-34a contributes a pro-survival advantage to HTLV-1-infected cells is supported by the observation that its functional knockdown resulted in an increase in death of C91PL cells (**Figure 7C**) and is in line with observations made in EBV-infected cells (Forte et al., 2012).

The presence of an NF-κB binding site in the spliced pri-miR-34a identified in C91PL cells (**Figure 4A**) along with the observation that pharmacological inhibition of NF-κB with Bay 11-7082 resulted in a reduction of miR-34a expression in C91PL and MT-2 cells (**Figure 4B**) suggested that this pathway contributes to sustain miR-34a levels in HTLV-1-infected cells. However, we cannot exclude the possibility that the effect of Bay 11-7082 on miR-34a levels in part reflected a general inhibition of gene expression that accompanied the substantial reduction in cell viability (**Figure 4B**).

We considered Tax to be a likely candidate for activating miR-34a, given its ability to activate the NF-κB pathway. It was therefore surprising that both wildtype Tax and Tax defective for NF-κB activation (TaxM22) produced a modest increase in miR-34a in HeLa cells, while the CREB pathway-defective mutant (TaxM47) did not induce miR-34a (**Figure 4C**). While this result may reflect limitations of this cell line as an experimental system for studying miR-34a regulation, we must also consider the possibility that Tax affects miR-34a through its interactions with CREB (reviewed by Nyborg et al., 2010) or with other transcription factor complexes such as AP-1 (reviewed by Gazon et al., 2017). Furthermore, other viral and cellular factors besides Tax are likely to play a role in miR-34a expression. This latter possibility is supported by the fact that ATLL cells express little or no Tax (Takeda et al., 2004; Yamagishi et al., 2012). In contrast, ATLL cells consistently express HBZ (Satou and Matsuoka, 2007). Results of qRT-PCR assays on five of the

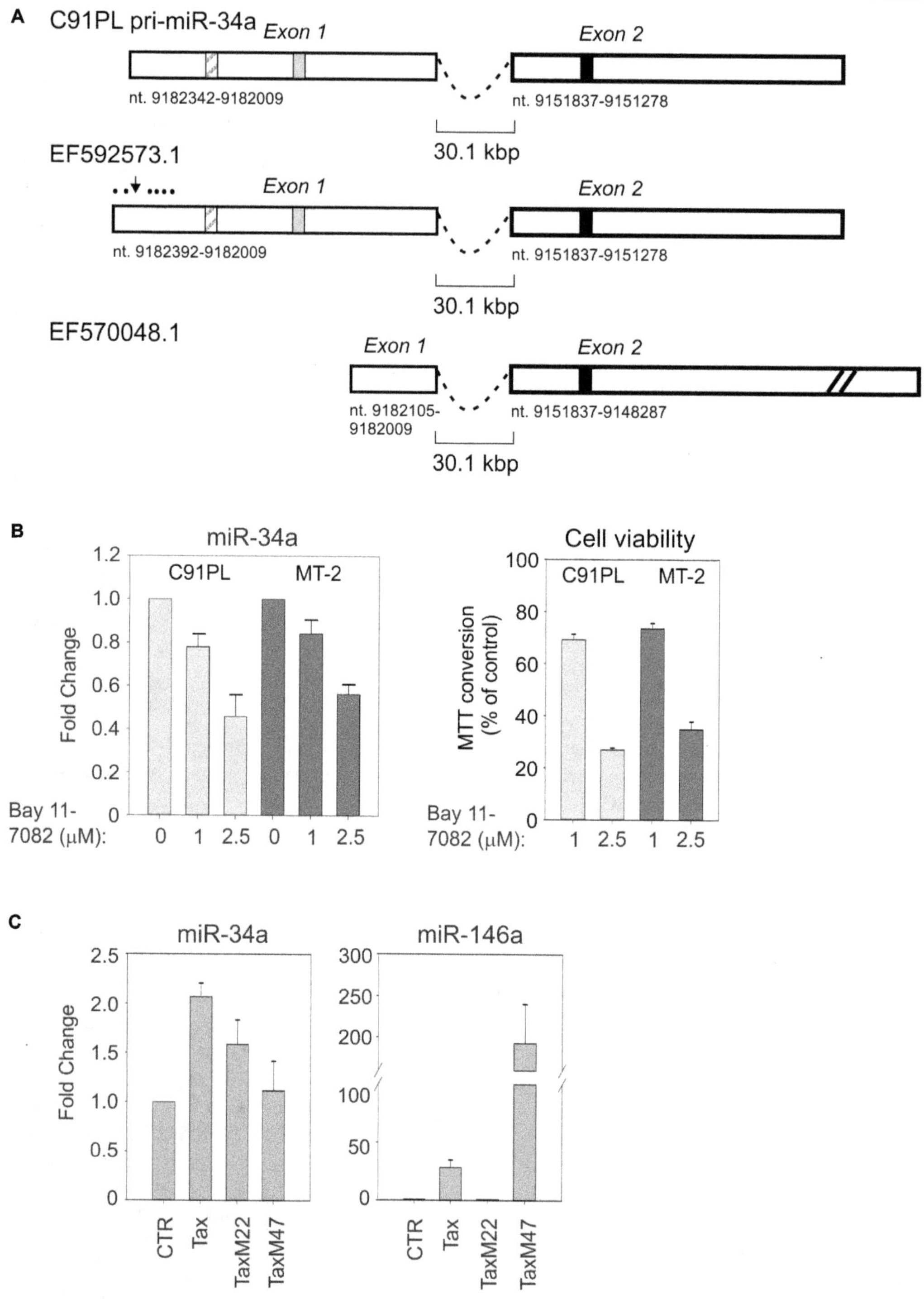

FIGURE 4 | C91PL cells produce a pri-miR-34a transcript that contains binding sites for NF-κB and p53. Panel **(A)** shows the spliced pri-miR-34a identified by RACE in the present study of C91PL cells, in a previous study of HeLa cells (EF592573.1; Chang et al., 2007) and in a lung cancer cell line engineered to produce p53 (EF570048.1; Tarasov et al., 2007). Two additional spliced pri-miR-34a transcripts identified in phorbol ester-treated K562 cells contain 2 exons located 5′ to exon 1 (not shown; Navarro et al., 2009). Numbering is according to the GenBank GRCh38.p12 primary assembly (minus strand). The box with diagonal lines indicates an NF-κB binding site (nt 9182264-9182255; Li et al., 2012), the gray box indicates a p53 binding site (nt 9182163-9182144; Raver-Shapira et al., 2007) and the black box indicates the position of mature miR-34a (nt 9151756-9151735). In EF570048.1, exon 2 is drawn at reduced scale (indicated by diagonal lines). Panel **(B)** shows expression of miR-34a and cell viability in C91PL and MT-2 cells after treatment for 48 h with the indicated concentrations of the NF-κB inhibitor Bay 11-7082; control cultures were treated with the same volume of DMSO (set at 1). Cell viability was measured as MTT conversion. In Panel **(C)**, HeLa cells were transfected with a plasmid coding for wildtype Tax, TaxM22, or TaxM47 (Smith and Greene, 1990) or with pBluescript KS+ (Stratagene; CTR, set at 1) and analyzed for expression of miR-34a and miR-146a. All graphs show mean values from three experimental replicates with standard error bars, scaled against controls.

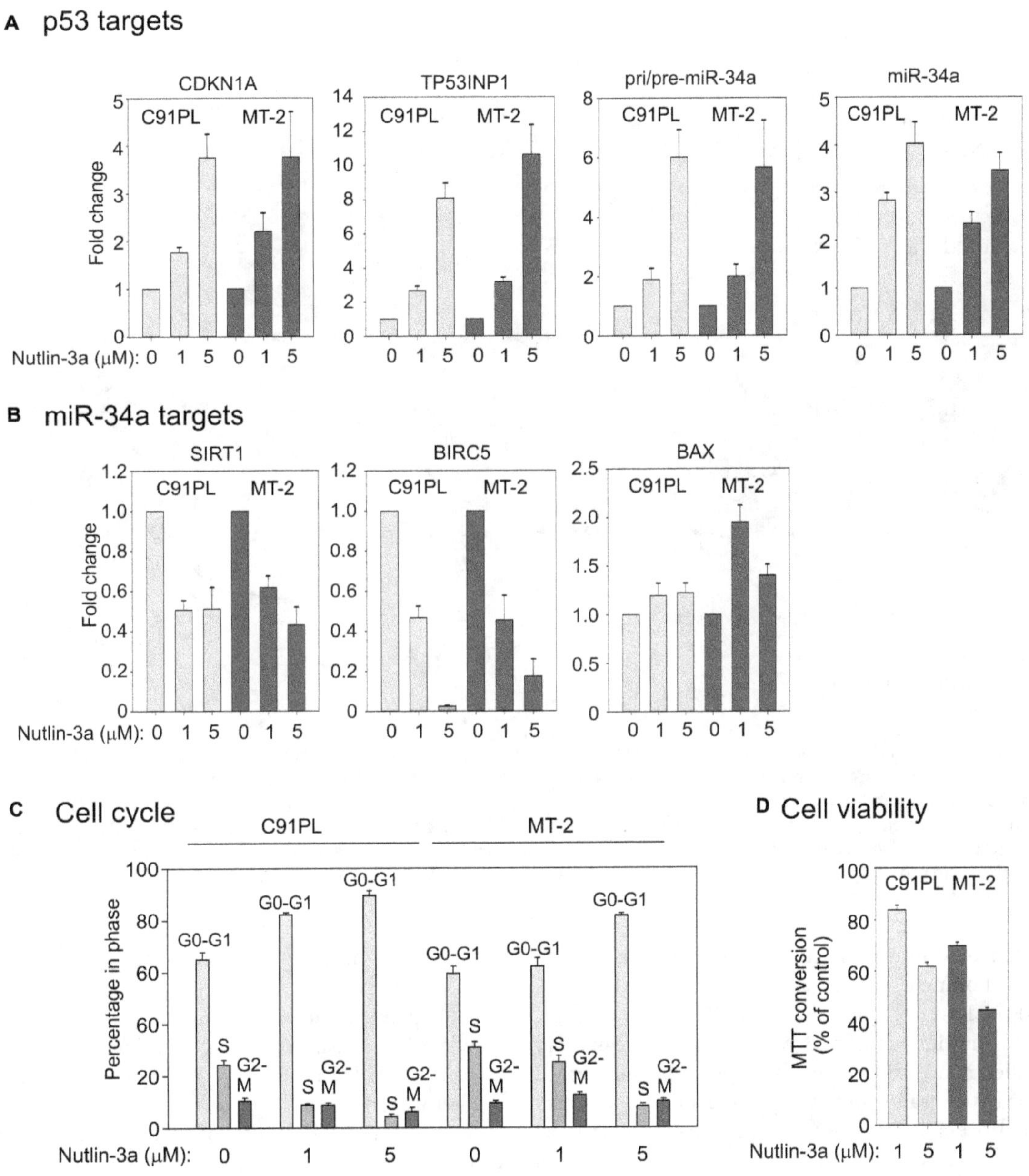

FIGURE 5 | Effects of nutlin-3a in C91PL and MT-2 cells. C91PL and MT-2 cells were treated with 1 or 5 μM nutlin-3a or with the same volume of DMSO for 48 h and then analyzed for expression of p53-responsive mRNAs, including pri-/pre-miR-34a and mature miR-34a [three experiments, Panel **(A)**] and miR-34a target transcripts [three experiments, Panel **(B)**]. Relative expression values were scaled against DMSO-treated controls (set at 1). Panel **(C)** shows cell cycle analysis measured in three experiments. Panel **(D)** shows cell viability measured by MTT conversion (six experiments). Graphs show mean values with standard error.

ATLL samples analyzed in the present study (nos. 1, 3, 14, 20, 31) confirmed the presence of HBZ mRNA in all of the samples, while Tax/Rex mRNA was undetectable (Supplementary Figures 3A,B). The levels of HBZ mRNA in these samples did not appear to correlate with their differences in miR-34a expression (**Figure 1C**).

Our experiments with nutlin-3a confirmed that miR-34a is a transcriptional target of p53 in C91PL, MT-2, and Hut-102 cells (**Figure 5A** and Supplementary Figure 2), and provided evidence that, by repressing SIRT1, miR-34a reinforces p53 activation (**Figures 5B, 6**). In contrast, treatment of ED cells with nutlin-3a did not result in an increase in p53 protein, and miR-34a remained undetectable (data not shown). This cell line, which was derived from leukemic cells of an ATLL patient, is defective for p53 expression (Ju et al., 2014) and contains a premature stop codon in the Tax open reading frame (Takeda et al., 2004). An analysis of ATLL sample nos. 1, 3, 14, 20, 31 for p53 mRNA revealed varying levels of expression (Supplementary Figure 3C) that did not appear to correlate with levels of miR-34a (**Figure 1C**).

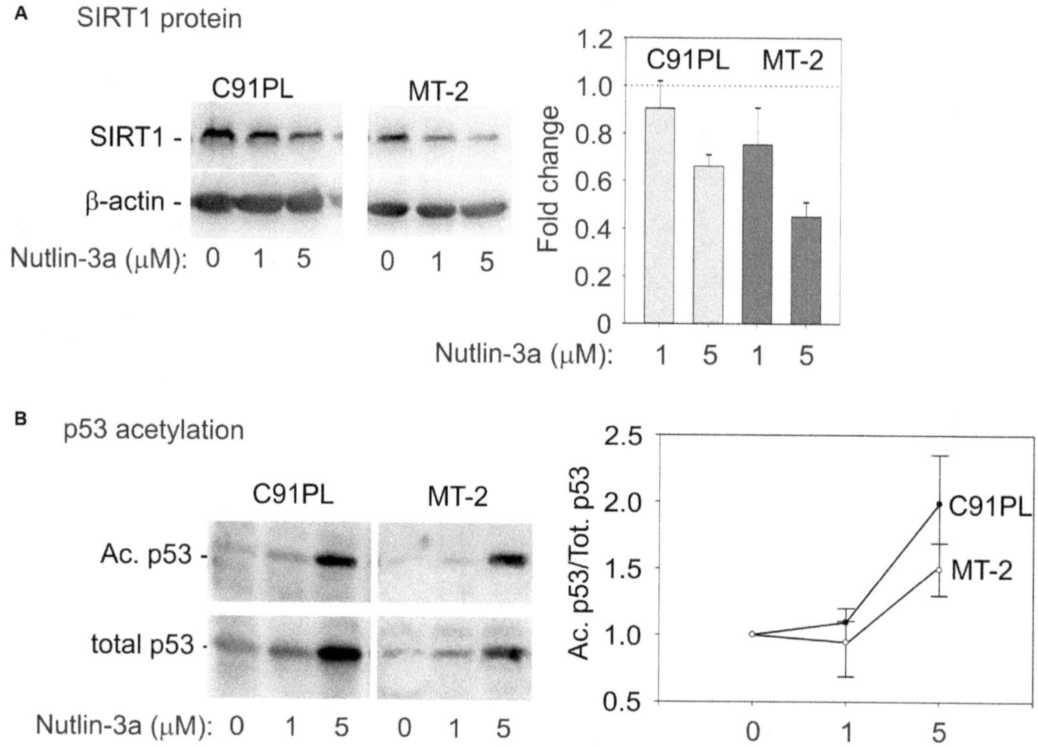

FIGURE 6 | Activation of the p53-miR-34a-SIRT1 feedback loop in C91PL and MT-2 cells. Panel **(A)** shows composite images of immunoblots to detect SIRT1 protein in lysates of control and nutlin-3a-treated cells, with β-actin serving as a loading control. The graph shows the mean fold change in SIRT1 protein expression normalized against β-actin measured in 3 experiments with standard error bars. In Panel **(B)**, lysates were immunoblotted with an antibody specific for p53 acetylated on lysine 382. The blots were then stripped and incubated with an anti-p53 antibody. Signal intensities were used to calculate ratios of acetylated p53 to total p53 in each sample. Values were scaled against the ratio calculated for the untreated cells to express a fold-change of acetylated/total p53 in the treated samples compared to untreated controls. The plot shows mean scaled ratios measured in three experiments with standard error bars.

Following the description of miR-34a as a transcriptional target of p53 (Bommer et al., 2007; Chang et al., 2007; He et al., 2007; Raver-Shapira et al., 2007; Tazawa et al., 2007), studies of the impact of p53 on the miRNA regulatory network identified many miRNAs whose expression is increased due to p53-mediated transcriptional upregulation or through p53-enhanced processing of miRNA precursors (reviewed by Hermeking, 2012). miR-145, a miRNA that is upregulated by p53 through its effects on Drosha-mediated pri-miRNA processing (Suzuki et al., 2009), is of interest, given its downregulation in the context of HTLV-1 and ATLL (Xia et al., 2014). Results of qRT-PCR experiments indicated that miR-145 is not expressed in untreated or nutlin-3a-treated C91PL and HUT-102 cells (data not shown). miR-107, which is upregulated by p53 and regulates hypoxic signaling in the colon cancer cell line HCT116 (Yamakuchi et al., 2010), showed a slight reduction in nutlin-3a-treated C91PL cells (Supplementary Figure 1). The lack of an increase in the levels of miR-145 and miR-107 upon nutlin-3a treatment suggests that p53 status might not be a major determinant controlling expression of these miRNAs in the context of HTLV-1-infected cells.

In addition to NF-κB and p53, other transcription factors including the p53 family member TAp73 (Agostini et al., 2011), ELK1 (Christoffersen et al., 2010), and transcription factors induced by phorbol esters (Navarro et al., 2009) are capable of upregulating miR-34a expression in different cell contexts (reviewed by Slabakova et al., 2017). The possible impact of these factors in HTLV-1-associated upregulation of miR-34a merits further investigation.

Results of early studies of miR-34a demonstrated that its ectopic expression induced cell cycle arrest or apoptosis in many cell lines (Chang et al., 2007; He et al., 2007; Raver-Shapira et al., 2007; Tarasov et al., 2007; Welch et al., 2007). These and other studies prompted the development of strategies to employ miR-34a to treat cancer (Beg et al., 2017; reviewed by Saito et al., 2015). Electroporation of a miR-34a mimic in C91PL cells led to a reduction in the expression of SIRT1, BIRC5 (Survivin), and BAX, thus providing evidence for a direct role for miR-34a in fine-tuning these targets in the context of HTLV-1 infection (**Figure 7A**). However, in our experiments the synthetic miR-34a mimic was not able to rescue p53 function or reduce cell viability (**Figure 7B**).

Studies of miR-34a have placed emphasis on its many targets that have an impact on cell proliferation and survival, such as MYC (Christoffersen et al., 2010) and MYCN (Cole et al., 2008), MET (He et al., 2007; Li et al., 2009), CCND1 and CDK6 (Sun et al., 2008), BCL2 (Cole et al., 2008), and

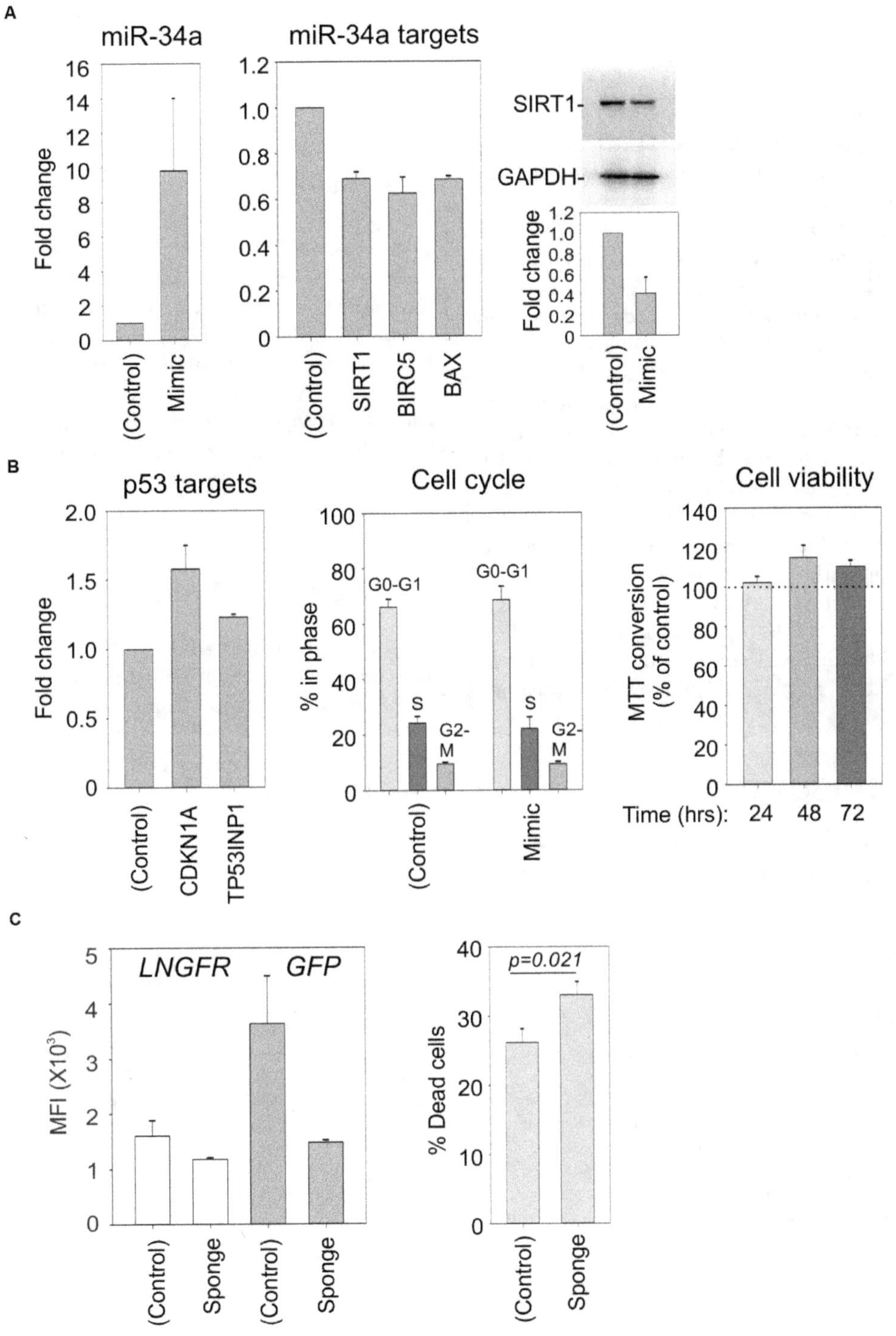

FIGURE 7 | Effects of a miR-34a mimic and sponge in C91PL cells. In Panels **(A)** and **(B)**, C91PL cells (4×10^6 cells, 4 μg DNA in 100 μL Buffer R) were electroporated with a miR-34a-mimic or a control synthetic RNA and analyzed 72 h later for changes in miR-34a and the indicated mRNAs (3–6 experimental replicates), SIRT1 protein (four replicates), and cell cycle (five replicates). Cell viability measured with the MTT test (four replicates) at the indicated time points after electroporation was scaled against values obtained for control-electroporated cultures at the corresponding time point. Graphs show mean values and standard error bars. The immunoblots show a representative experiment to quantify the change in SIRT1 normalized against GAPDH protein. In Panel **(C)**, C91PL cells were electroporated with pGFP-control or pGFP-miR-34a-sponge (see section "Materials and Methods") together with pMACS-LNGFR (Miltenyi Biotec, coding for truncated nerve growth factor receptor) and pBluescript KS+ (included as carrier). Three days later, aliquots of cells were labeled with APC-conjugated anti-LNGFR antibody (Miltenyi Biotec) or with Live/Dead Far Red (Invitrogen) and FITC-conjugated anti-LNGFR antibody (Miltenyi Biotec) and analyzed with a BD FACSalibur. The left-hand plot shows the mean fluorescence intensities (MFI) measured for the LNGFR and GFP signals in the GFP+LNGFR double-positive populations from three transfections with standard error bars. The right-hand plot shows mean % cell death values [=(% of Live/Dead Far Red+LNGFR-labeled cells/total % LNGFR-positive cells) × 100] from 11 transfections with standard error bars ($p = 0.021$, Mann–Whitney rank sum test).

NOTCH1 (Ji et al., 2008). The possibility that miR-34a favors survival of HTLV-1-infected cells by modulating expression of BAX, a known tumor suppressor (Yin et al., 1997), merits further investigation.

It is noteworthy that SIRT1 is upregulated in HTLV-1-transformed cells, and particularly in acute ATLL (Kozako et al., 2012). Experiments carried out in chronically infected cell lines and ATLL cells showed that siRNA-mediated knockdown of SIRT1 expression or treatment with sirtuin inhibitors results in apoptotic death, suggesting that SIRT1 is important for the survival of HTLV-1-transformed cells (Kozako et al., 2015; Tang et al., 2015). On the other hand, another study indicated that SIRT1 interferes with the ability of Tax to transactivate the LTR promoter (Tang et al., 2015). It is thus possible that, by modulating SIRT1 expression, miR-34a might enhance transcription of the viral genome. Along these lines, it is interesting to note that nutlin-3a substantially increased the levels of viral mRNAs in C91PL and MT-2 cells (Supplementary Figure 4A) and in HUT-102 cells (Supplementary Figure 2B). However, such an increase was not observed upon electroporation of the miR-34a mimic into C91PL cells (Supplementary Figure 4B). A thorough understanding of the impact of miR-34a on HTLV-1-infected T-cells is worthy of further study, given the current interest in the use of miR-34a mimics, nutlin-3a analogs, and SIRT1 inhibitors to treat various cancers.

AUTHOR CONTRIBUTIONS

VC and DMD designed the study. CP-M designed the experiments and provided the ATL samples. VKS, VR, KR, IC, and MS-B performed the experiments. All authors interpreted the data and prepared the manuscript.

FUNDING

This work was supported by the Associazione Italiana per la Ricerca sul Cancro (AIRC, Grant No. 13378 to VC), AIRC-Cariverona (regional grant to VC), and the University of Padova (Grant No. CPDA124913/12 to VC, Young Investigator Grant to MS-B, institutional funding to DMD). VKS was supported by a doctoral fellowship from the Fondazione Cassa di Risparmio di Padova e Rovigo (CARIPARO) and MS-B was supported by a fellowship from the Pezcoller Foundation.

ACKNOWLEDGMENTS

We thank Luigi Chieco-Bianchi, Katia Basso, Stefania Bortoluzzi, and Martina Pigazzi for valuable discussions, Paola Dalla Pria for technical assistance, Sonia Minuzzo for genetic profiling, Masao Matsuoka for providing information about cell line ED, and Dinesh Rao, Charles Bangham, and Roberto Accolla for providing reagents.

REFERENCES

Abe, M., Suzuki, H., Nishitsuji, H., Shida, H., and Takaku, H. (2010). Interaction of human T-cell lymphotropic virus type I Rex protein with Dicer suppresses RNAi silencing. *FEBS Lett.* 584, 4313–4318. doi: 10.1016/j.febslet.2010.09.031

Agostini, M., Tucci, P., Killick, R., Candi, E., Sayan, B. S., Rivetti di Val Cervo, P., et al. (2011). Neuronal differentiation by TAp73 is mediated by microRNA-34a regulation of synaptic protein targets. *Proc. Natl. Acad. Sci. U.S.A.* 108, 21093–21098. doi: 10.1073/pnas.1112061109

Arima, N., Daitoku, Y., Hidaka, S., Yamamoto, Y., Fujimoto, K., Matsushita, K., et al. (1992). Interleukin-2 production by primary adult T cell leukemia tumor cells is macrophage dependent. *Am. J. Hematol.* 41, 258–263. doi: 10.1002/ajh.2830410407

Bai, X. T., and Nicot, C. (2015). miR-28-3p is a cellular restriction factor that inhibits human T cell leukemia virus, type 1 (HTLV-1) replication and virus infection. *J. Biol. Chem.* 290, 5381–5390. doi: 10.1074/jbc.M114.626325

Bangham, C. R. M. (2017). Human T cell leukemia virus type 1: persistence and pathogenesis. *Annu. Rev. Immunol.* doi: 10.1146/annurev-immunol-042617-053222 [Epub ahead of print].

Beg, M. S., Brenner, A. J., Sachdev, J., Borad, M., Kang, Y. K., Stoudemire, J., et al. (2017). Phase I study of MRX34, a liposomal miR-34a mimic, administered twice weekly in patients with advanced solid tumors. *Invest. New Drugs* 35, 180–188. doi: 10.1007/s10637-016-0407-y

Bellon, M., Lepelletier, Y., Hermine, O., and Nicot, C. (2009). Deregulation of microRNA involved in hematopoiesis and the immune response in HTLV-I adult T-cell leukemia. *Blood* 113, 4914–4917. doi: 10.1182/blood-2008-11-189845

Bhat, N. K., Adachi, Y., Samuel, K. P., and Derse, D. (1993). HTLV-1 gene expression by defective proviruses in an infected T-cell line. *Virology* 196, 15–24. doi: 10.1006/viro.1993.1450

Bommer, G. T., Gerin, I., Feng, Y., Kaczorowski, A. J., Kuick, R., Love, R. E., et al. (2007). p53-mediated activation of miRNA34 candidate tumor-suppressor genes. *Curr. Biol.* 17, 1298–1307. doi: 10.1016/j.cub.2007.06.068

Bradford, M. M. (1976). A rapid and sensitive method for the quantitation of microgram quantities of protein utilizing the principle of protein-dye binding. *Anal. Biochem.* 72, 248–254. doi: 10.1016/0003-2697(76)90527-3

Calin, G. A., Sevignani, C., Dumitru, C. D., Hyslop, T., Noch, E., Yendamuri, S., et al. (2004). Human microRNA genes are frequently located at fragile sites and genomic regions involved in cancers. *Proc. Natl. Acad. Sci. U.S.A.* 101, 2999–3004. doi: 10.1073/pnas.0307323101

Cereseto, A., Diella, F., Mulloy, J. C., Cara, A., Michieli, P., Grassmann, R., et al. (1996). p53 functional impairment and high p21waf1/cip1 expression in human T-cell lymphotropic/leukemia virus type I-transformed T cells. *Blood* 88, 1551–1560.

Chang, T. C., Wentzel, E. A., Kent, O. A., Ramachandran, K., Mullendore, M., Lee, K. H., et al. (2007). Transactivation of miR-34a by p53 broadly influences gene expression and promotes apoptosis. *Mol. Cell.* 26, 745–752. doi: 10.1016/j.molcel.2007.05.010

Chen, Y., Bathula, S. R., Yang, Q., and Huang, L. (2010). Targeted nanoparticles deliver siRNA to melanoma. *J. Invest. Dermatol.* 130, 2790–2798. doi: 10.1038/jid.2010.222

Chim, C. S., Wong, K. Y., Qi, Y., Loong, F., Lam, W. L., Wong, L. G., et al. (2010). Epigenetic inactivation of the miR-34a in hematological malignancies. *Carcinogenesis* 31, 745–750. doi: 10.1093/carcin/bgq033

Christoffersen, N. R., Shalgi, R., Frankel, L. B., Leucci, E., Lees, M., Klausen, M., et al. (2010). p53-independent upregulation of miR-34a during oncogene-induced senescence represses MYC. *Cell Death. Differ.* 17, 236–245. doi: 10.1038/cdd.2009.109

Cole, K. A., Attiyeh, E. F., Mosse, Y. P., Laquaglia, M. J., Diskin, S. J., Brodeur, G. M., et al. (2008). A functional screen identifies miR-34a as a candidate neuroblastoma tumor suppressor gene. *Mol. Cancer Res.* 6, 735–742. doi: 10.1158/1541-7786.MCR-07-2102

Craig, V. J., Cogliatti, S. B., Imig, J., Renner, C., Neuenschwander, S., Rehrauer, H., et al. (2011). Myc-mediated repression of microRNA-34a promotes high-grade transformation of B-cell lymphoma by dysregulation of FoxP1. *Blood* 117, 6227–6236. doi: 10.1182/blood-2010-10-312231

Currer, R., Van Duyne, R., Jaworski, E., Guendel, I., Sampey, G., Das, R., et al. (2012). HTLV tax: a fascinating multifunctional co-regulator of viral and cellular pathways. *Front. Microbiol.* 3:406. doi: 10.3389/fmicb.2012.00406

Fan, N., and Wang, J. (2016). MicroRNA 34a contributes to virus-mediated apoptosis through binding to its target gene Bax in influenza A virus infection. *Biomed. Pharmacother.* 83, 1464–1470. doi: 10.1016/j.biopha.2016.08.049

Forte, E., Salinas, R. E., Chang, C., Zhou, T., Linnstaedt, S. D., Gottwein, E., et al. (2012). The Epstein-Barr virus (EBV)-induced tumor suppressor microRNA MiR-34a is growth promoting in EBV-infected B cells. *J. Virol.* 86, 6889–6898. doi: 10.1128/JVI.07056-11

Gazon, H., Barbeau, B., Mesnard, J. M., and Peloponese, J. M. Jr. (2017). Hijacking of the AP-1 signaling pathway during development of ATL. *Front. Microbiol.* 8:2686. doi: 10.3389/fmicb.2017.02686

Gazon, H., Belrose, G., Terol, M., Meniane, J. C., Mesnard, J. M., Cesaire, R., et al. (2016). Impaired expression of DICER and some microRNAs in HBZ expressing cells from acute adult T-cell leukemia patients. *Oncotarget* 7, 30258–30275. doi: 10.18632/oncotarget.7162

Gessain, A., Barin, F., Vernant, J. C., Gout, O., Maurs, L., Calender, A., et al. (1985). Antibodies to human T-lymphotropic virus type-I in patients with tropical spastic paraparesis. *Lancet* 2, 407–410. doi: 10.1016/S0140-6736(85)92734-5

Gessain, A., and Cassar, O. (2012). Epidemiological aspects and world distribution of HTLV-1 infection. *Front. Microbiol.* 3:388. doi: 10.3389/fmicb.2012.00388

Hasegawa, H., Sawa, H., Lewis, M. J., Orba, Y., Sheehy, N., Yamamoto, Y., et al. (2006). Thymus-derived leukemia-lymphoma in mice transgenic for the Tax gene of human T-lymphotropic virus type I. *Nat. Med.* 12, 466–472. doi: 10.1038/nm1389

Hasegawa, H., Yamada, Y., Iha, H., Tsukasaki, K., Nagai, K., Atogami, S., et al. (2009). Activation of p53 by Nutlin-3a, an antagonist of MDM2, induces apoptosis and cellular senescence in adult T-cell leukemia cells. *Leukemia* 23, 2090–2101. doi: 10.1038/leu.2009.171

He, L., He, X., Lim, L. P., de Stanchina, E., Xuan, Z., Liang, Y., et al. (2007). A microRNA component of the p53 tumour suppressor network. *Nature* 447, 1130–1134. doi: 10.1038/nature05939

Hermeking, H. (2010). The miR-34 family in cancer and apoptosis. *Cell Death Differ.* 17, 193–199. doi: 10.1038/cdd.2009.56

Hermeking, H. (2012). MicroRNAs in the p53 network: micromanagement of tumour suppression. *Nat. Rev. Cancer* 12, 613–626. doi: 10.1038/nrc3318

Hoffman, W. H., Biade, S., Zilfou, J. T., Chen, J., and Murphy, M. (2002). Transcriptional repression of the anti-apoptotic survivin gene by wild type p53. *J. Biol. Chem.* 277, 3247–3257. doi: 10.1074/jbc.M106643200

Ji, Q., Hao, X., Meng, Y., Zhang, M., Desano, J., Fan, D., et al. (2008). Restoration of tumor suppressor miR-34 inhibits human p53-mutant gastric cancer tumorspheres. *BMC Cancer* 8:266. doi: 10.1186/1471-2407-8-266

Jiang, J., Lee, E. J., Gusev, Y., and Schmittgen, T. D. (2005). Real-time expression profiling of microRNA precursors in human cancer cell lines. *Nucleic Acids Res.* 33, 5394–5403. doi: 10.1093/nar/gki863

Ju, W., Zhang, M., Petrus, M., Maeda, M., Pise-Masison, C. A., and Waldmann, T. A. (2014). Combination of 9-aminoacridine with Campath-1H provides effective therapy for a murine model of adult T-cell leukemia. *Retrovirology* 11:43. doi: 10.1186/1742-4690-11-43

Kaller, M., Liffers, S. T., Oeljeklaus, S., Kuhlmann, K., Roh, S., Hoffmann, R., et al. (2011). Genome-wide characterization of miR-34a induced changes in protein and mRNA expression by a combined pulsed SILAC and microarray analysis. *Mol. Cell. Proteomics* 10:M111010462. doi: 10.1074/mcp.M111.010462

Kamihira, S., Yamada, Y., Hirakata, Y., Tomonaga, M., Sugahara, K., Hayashi, T., et al. (2001). Aberrant expression of caspase cascade regulatory genes in adult T-cell leukaemia: survivin is an important determinant for prognosis. *Br. J. Haematol.* 114, 63–69. doi: 10.1046/j.1365-2141.2001.02902.x

Kozako, T., Aikawa, A., Shoji, T., Fujimoto, T., Yoshimitsu, M., Shirasawa, S., et al. (2012). High expression of the longevity gene product SIRT1 and apoptosis induction by sirtinol in adult T-cell leukemia cells. *Int. J. Cancer* 131, 2044–2055. doi: 10.1002/ijc.27481

Kozako, T., Suzuki, T., Yoshimitsu, M., Uchida, Y., Kuroki, A., Aikawa, A., et al. (2015). Novel small-molecule SIRT1 inhibitors induce cell death in adult T-cell leukaemia cells. *Sci. Rep.* 5:11345. doi: 10.1038/srep11345

Le, M. T., Shyh-Chang, N., Khaw, S. L., Chin, L., Teh, C., Tay, J., et al. (2011). Conserved regulation of p53 network dosage by microRNA-125b occurs through evolving miRNA-target gene pairs. *PLoS Genet* 7:e1002242. doi: 10.1371/journal.pgen.1002242

Li, J., Wang, K., Chen, X., Meng, H., Song, M., Wang, Y., et al. (2012). Transcriptional activation of microRNA-34a by NF-kappa B in human esophageal cancer cells. *BMC Mol. Biol.* 13:4. doi: 10.1186/1471-2199-13-4

Li, Y., Guessous, F., Zhang, Y., Dipierro, C., Kefas, B., Johnson, E., et al. (2009). MicroRNA-34a inhibits glioblastoma growth by targeting multiple oncogenes. *Cancer Res.* 69, 7569–7576. doi: 10.1158/0008-5472.CAN-09-0529

Lodygin, D., Tarasov, V., Epanchintsev, A., Berking, C., Knyazeva, T., Korner, H., et al. (2008). Inactivation of miR-34a by aberrant CpG methylation in multiple types of cancer. *Cell Cycle* 7, 2591–2600. doi: 10.4161/cc.7.16.6533

Ma, G., Yasunaga, J., and Matsuoka, M. (2016). Multifaceted functions and roles of HBZ in HTLV-1 pathogenesis. *Retrovirology* 13:16. doi: 10.1186/s12977-016-0249-x

Maeda, M., Shimizu, A., Ikuta, K., Okamoto, H., Kashihara, M., Uchiyama, T., et al. (1985). Origin of human T-lymphotrophic virus I-positive T cell lines in adult T cell leukemia. Analysis of T cell receptor gene rearrangement. *J. Exp. Med.* 162, 2169–2174. doi: 10.1084/jem.162.6.2169

Martinez-Redondo, P., and Vaquero, A. (2013). The diversity of histone versus nonhistone sirtuin substrates. *Genes Cancer* 4, 148–163. doi: 10.1177/1947601913483767

Miyashita, T., Krajewski, S., Krajewska, M., Wang, H. G., Lin, H. K., Liebermann, D. A., et al. (1994). Tumor suppressor p53 is a regulator of bcl-2 and Bax gene expression in vitro and in vivo. *Oncogene* 9, 1799–1805.

Moles, R., Bellon, M., and Nicot, C. (2015). STAT1: a novel target of miR-150 and miR-223 is involved in the proliferation of HTLV-I-transformed and ATL cells. *Neoplasia* 17, 449–462. doi: 10.1016/j.neo.2015.04.005

Moles, R., and Nicot, C. (2015). The emerging role of miRNAs in HTLV-1 infection and ATLL pathogenesis. *Viruses* 7, 4047–4074. doi: 10.3390/v7072805

Mori, N., Yamada, Y., Ikeda, S., Yamasaki, Y., Tsukasaki, K., Tanaka, Y., et al. (2002). Bay 11-7082 inhibits transcription factor NF-kappaB and induces apoptosis of HTLV-I-infected T-cell lines and primary adult T-cell leukemia cells. *Blood* 100, 1828–1834. doi: 10.1182/blood-2002-01-0151

Mosmann, T. (1983). Rapid colorimetric assay for cellular growth and survival: application to proliferation and cytotoxicity assays. *J. Immunol. Methods* 65, 55–63. doi: 10.1016/0022-1759(83)90303-4

Muscolini, M., Cianfrocca, R., Sajeva, A., Mozzetti, S., Ferrandina, G., Costanzo, A., et al. (2008). Trichostatin A up-regulates p73 and induces Bax-dependent apoptosis in cisplatin-resistant ovarian cancer cells. *Mol. Cancer Ther.* 7, 1410–1419. doi: 10.1158/1535-7163.MCT-08-0299

Nakamura, S., Roth, J. A., and Mukhopadhyay, T. (2000). Multiple lysine mutations in the C-terminal domain of p53 interfere with MDM2-dependent protein degradation and ubiquitination. *Mol. Cell. Biol.* 20, 9391–9398. doi: 10.1128/MCB.20.24.9391-9398.2000

Navarro, F., Gutman, D., Meire, E., Caceres, M., Rigoutsos, I., Bentwich, Z., et al. (2009). miR-34a contributes to megakaryocytic differentiation of K562 cells independently of p53. *Blood* 114, 2181–2192. doi: 10.1182/blood-2009-02-205062

Ng, H. Y., Wan, T. S., So, C. C., and Chim, C. S. (2014). Epigenetic inactivation of DAPK1, p14ARF, mir-34a and -34b/c in acute promyelocytic leukaemia. *J. Clin. Pathol.* 67, 626–631. doi: 10.1136/jclinpath-2014-202276

Nyborg, J. K., Egan, D., and Sharma, N. (2010). The HTLV-1 Tax protein: revealing mechanisms of transcriptional activation through histone acetylation and nucleosome disassembly. *Biochim. Biophys. Acta* 1799, 266–274. doi: 10.1016/j.bbagrm.2009.09.002

Osame, M., Usuku, K., Izumo, S., Ijichi, N., Amitani, H., Igata, A., et al. (1986). HTLV-I associated myelopathy, a new clinical entity. *Lancet* 1, 1031–1032. doi: 10.1016/S0140-6736(86)91298-5

Pichler, K., Schneider, G., and Grassmann, R. (2008). MicroRNA miR-146a and further oncogenesis-related cellular microRNAs are dysregulated in HTLV-1-transformed T lymphocytes. *Retrovirology* 5:100. doi: 10.1186/1742-4690-5-100

Pierce, J. W., Schoenleber, R., Jesmok, G., Best, J., Moore, S. A., Collins, T., et al. (1997). Novel inhibitors of cytokine-induced IkappaBalpha phosphorylation and endothelial cell adhesion molecule expression show anti-inflammatory effects in vivo. *J. Biol. Chem.* 272, 21096–21103. doi: 10.1074/jbc.272.34.21096

Pise-Masison, C. A., Radonovich, M., Dohoney, K., Morris, J. C., O'Mahony, D., Lee, M. J., et al. (2009). Gene expression profiling of ATL patients: compilation of disease-related genes and evidence for TCF4 involvement in BIRC5 gene expression and cell viability. *Blood* 113, 4016–4026. doi: 10.1182/blood-2008-08-175901

Poiesz, B. J., Ruscetti, F. W., Gazdar, A. F., Bunn, P. A., Minna, J. D., and Gallo, R. C. (1980). Detection and isolation of type C retrovirus particles from fresh and cultured lymphocytes of a patient with cutaneous T-cell lymphoma. *Proc. Natl. Acad. Sci. U.S.A.* 77, 7415–7419. doi: 10.1073/pnas.77.12.7415

Popovic, M., Lange-Wantzin, G., Sarin, P. S., Mann, D., and Gallo, R. C. (1983). Transformation of human umbilical cord blood T cells by human T-cell leukemia/lymphoma virus. *Proc. Natl. Acad. Sci. U.S.A.* 80, 5402–5406. doi: 10.1073/pnas.80.17.5402

Raimondi, V., Minuzzo, S., Ciminale, V., and D'Agostino, D. M. (2017). STR profiling of HTLV-1-infected cell lines. *Methods Mol. Biol.* 1582, 143–154. doi: 10.1007/978-1-4939-6872-5_11

Rao, D. S., O'Connell, R. M., Chaudhuri, A. A., Garcia-Flores, Y., Geiger, T. L., and Baltimore, D. (2010). MicroRNA-34a perturbs B lymphocyte development by repressing the forkhead box transcription factor Foxp1. *Immunity* 33, 48–59. doi: 10.1016/j.immuni.2010.06.013

Raver-Shapira, N., Marciano, E., Meiri, E., Spector, Y., Rosenfeld, N., Moskovits, N., et al. (2007). Transcriptional activation of miR-34a contributes to p53-mediated apoptosis. *Mol. Cell.* 26, 731–743. doi: 10.1016/j.molcel.2007.05.017

Reed, S. M., and Quelle, D. E. (2014). p53 acetylation: regulation and consequences. *Cancers* 7, 30–69. doi: 10.3390/cancers7010030

Rende, F., Cavallari, I., Corradin, A., Silic-Benussi, M., Toulza, F., Toffolo, G. M., et al. (2011). Kinetics and intracellular compartmentalization of HTLV-1 gene expression: nuclear retention of HBZ mRNA. *Blood* 117, 4855–4859. doi: 10.1182/blood-2010-11-316463

Ruggero, K., Guffanti, A., Corradin, A., Sharma, V. K., De Bellis, G., Corti, G., et al. (2014). Small noncoding RNAs in cells transformed by human T-cell leukemia virus type 1: a role for a tRNA fragment as a primer for reverse transcriptase. *J. Virol.* 88, 3612–3622. doi: 10.1128/JVI.02823-13

Saito, Y., Nakaoka, T., and Saito, H. (2015). microRNA-34a as a therapeutic agent against human cancer. *J. Clin. Med.* 4, 1951–1959. doi: 10.3390/jcm4111951

Satou, Y., and Matsuoka, M. (2007). Implication of the HTLV-I bZIP factor gene in the leukemogenesis of adult T-cell leukemia. *Int. J. Hematol.* 86, 107–112. doi: 10.1532/IJH97.07103

Satou, Y., Yasunaga, J., Zhao, T., Yoshida, M., Miyazato, P., Takai, K., et al. (2011). HTLV-1 bZIP factor induces T-cell lymphoma and systemic inflammation in vivo. *PLoS Pathog.* 7:e1001274. doi: 10.1371/journal.ppat.1001274

Shen, Z., Zhan, G., Ye, D., Ren, Y., Cheng, L., Wu, Z., et al. (2012). MicroRNA-34a affects the occurrence of laryngeal squamous cell carcinoma by targeting the antiapoptotic gene survivin. *Med. Oncol.* 29, 2473–2480. doi: 10.1007/s12032-011-0156-x

Silic-Benussi, M., Cavallari, I., Vajente, N., Vidali, S., Chieco-Bianchi, L., Di Lisa, F., et al. (2010). Redox regulation of T-cell turnover by the p13 protein of human T-cell leukemia virus type 1: distinct effects in primary versus transformed cells. *Blood* 116, 54–62. doi: 10.1182/blood-2009-07-235861

Slabakova, E., Culig, Z., Remsik, J., and Soucek, K. (2017). Alternative mechanisms of miR-34a regulation in cancer. *Cell Death Dis.* 8:e3100. doi: 10.1038/cddis.2017.495

Smith, M. R., and Greene, W. C. (1990). Identification of HTLV-I tax trans-activator mutants exhibiting novel transcriptional phenotypes. *Genes Dev.* 4, 1875–1885. doi: 10.1101/gad.4.11.1875

Sun, F., Fu, H., Liu, Q., Tie, Y., Zhu, J., Xing, R., et al. (2008). Downregulation of CCND1 and CDK6 by miR-34a induces cell cycle arrest. *FEBS Lett.* 582, 1564–1568. doi: 10.1016/j.febslet.2008.03.057

Sun, S. C., and Yamaoka, S. (2005). Activation of NF-kappaB by HTLV-I and implications for cell transformation. *Oncogene* 24, 5952–5964. doi: 10.1038/sj.onc.1208969

Suzuki, H. I., Yamagata, K., Sugimoto, K., Iwamoto, T., Kato, S., and Miyazono, K. (2009). Modulation of microRNA processing by p53. *Nature* 460, 529–533. doi: 10.1038/nature08199

Takeda, S., Maeda, M., Morikawa, S., Taniguchi, Y., Yasunaga, J., Nosaka, K., et al. (2004). Genetic and epigenetic inactivation of tax gene in adult T-cell leukemia cells. *Int. J. Cancer* 109, 559–567. doi: 10.1002/ijc.20007

Tang, H. M., Gao, W. W., Chan, C. P., Cheng, Y., Deng, J. J., Yuen, K. S., et al. (2015). SIRT1 suppresses human T-cell leukemia virus type 1 transcription. *J. Virol.* 89, 8623–8631. doi: 10.1128/JVI.01229-15

Tarasov, V., Jung, P., Verdoodt, B., Lodygin, D., Epanchintsev, A., Menssen, A., et al. (2007). Differential regulation of microRNAs by p53 revealed by massively parallel sequencing: miR-34a is a p53 target that induces apoptosis and G1-arrest. *Cell Cycle* 6, 1586–1593. doi: 10.4161/cc.6.13.4436

Tazawa, H., Tsuchiya, N., Izumiya, M., and Nakagama, H. (2007). Tumor-suppressive miR-34a induces senescence-like growth arrest through modulation of the E2F pathway in human colon cancer cells. *Proc. Natl. Acad. Sci. U.S.A.* 104, 15472–15477. doi: 10.1073/pnas.0707351104

Tomita, M. (2012). Important roles of cellular microRNA miR-155 in leukemogenesis by human T-cell leukemia virus type 1 infection. *ISRN Microbiol.* 2012:978607. doi: 10.5402/2012/978607

Tomita, M., Tanaka, Y., and Mori, N. (2009). MicroRNA miR-146a is induced by HTLV-1 Tax and increases the growth of HTLV-1-infected T-cells. *Int. J. Cancer* 130, 2300–2309. doi: 10.1002/ijc.25115

Van Duyne, R., Guendel, I., Klase, Z., Narayanan, A., Coley, W., Jaworski, E., et al. (2012). Localization and sub-cellular shuttling of HTLV-1 tax with the miRNA machinery. *PLoS One* 7:e40662. doi: 10.1371/journal.pone.0040662

Vassilev, L. T., Vu, B. T., Graves, B., Carvajal, D., Podlaski, F., Filipovic, Z., et al. (2004). In vivo activation of the p53 pathway by small-molecule antagonists of MDM2. *Science* 303, 844–848. doi: 10.1126/science.1092472

Vernin, C., Thenoz, M., Pinatel, C., Gessain, A., Gout, O., Delfau-Larue, M. H., et al. (2014). HTLV-1 bZIP factor HBZ promotes cell proliferation and genetic instability by activating OncomiRs. *Cancer Res.* 74, 6082–6093. doi: 10.1158/0008-5472.CAN-13-3564

Welch, C., Chen, Y., and Stallings, R. L. (2007). MicroRNA-34a functions as a potential tumor suppressor by inducing apoptosis in neuroblastoma cells. *Oncogene* 26, 5017–5022. doi: 10.1038/sj.onc.1210293

Xia, H., Yamada, S., Aoyama, M., Sato, F., Masaki, A., Ge, Y., et al. (2014). Prognostic impact of microRNA-145 down-regulation in adult T-cell leukemia/lymphoma. *Hum. Pathol.* 45, 1192–1198. doi: 10.1016/j.humpath.2014.01.017

Yamagishi, M., Nakano, K., Miyake, A., Yamochi, T., Kagami, Y., Tsutsumi, A., et al. (2012). Polycomb-mediated loss of miR-31 activates NIK-dependent NF-κB pathway in adult T cell leukemia and other cancers. *Cancer Cell* 21, 121–135. doi: 10.1016/j.ccr.2011.12.015

Yamakuchi, M., Ferlito, M., and Lowenstein, C. J. (2008). miR-34a repression of SIRT1 regulates apoptosis. *Proc. Natl. Acad. Sci. U.S.A.* 105, 13421–13426. doi: 10.1073/pnas.0801613105

Yamakuchi, M., Lotterman, C. D., Bao, C., Hruban, R. H., Karim, B., Mendell, J. T., et al. (2010). P53-induced microRNA-107 inhibits HIF-1 and tumor angiogenesis. *Proc. Natl. Acad. Sci. U.S.A.* 107, 6334–6339. doi: 10.1073/pnas.0911082107

Yeung, M. L., Yasunaga, J., Bennasser, Y., Dusetti, N., Harris, D., Ahmad, N., et al. (2008). Roles for microRNAs, miR-93 and miR-130b, and tumor protein 53-induced nuclear protein 1 tumor suppressor in cell growth dysregulation by human T-cell lymphotrophic virus 1. *Cancer Res.* 68, 8976–8985. doi: 10.1158/0008-5472.CAN-08-0769

Yin, C., Knudson, C. M., Korsmeyer, S. J., and Van Dyke, T. (1997). Bax suppresses tumorigenesis and stimulates apoptosis in vivo. *Nature* 385, 637–640. doi: 10.1038/385637a0

Yoshida, M., Miyoshi, I., and Hinuma, Y. (1982). Isolation and characterization of retrovirus from cell lines of human adult T-cell leukemia and its implication in the disease. *Proc. Natl. Acad. Sci. U.S.A.* 79, 2031–2035. doi: 10.1073/pnas.79.6.2031

17

How to Control HTLV-1-Associated Diseases: Preventing *de Novo* Cellular Infection Using Antiviral Therapy

Amandine Pasquier[1,2,3], Sandrine Alais[1,2], Loic Roux[4†], Maria-Isabel Thoulouze[5], Karine Alvarez[4], Chloé Journo[1,2], Hélène Dutartre[1,2‡] and Renaud Mahieux[1,2‡]*

[1] International Center for Research in Infectiology, Retroviral Oncogenesis Laboratory, INSERM U1111 – Université Claude Bernard Lyon 1, CNRS, UMR5308, Ecole Normale Supérieure de Lyon, Université Lyon, Lyon, France, [2] Equipe Labellisée Ligue Nationale Contre le Cancer, Paris, France, [3] Ecole Pratique des Hautes Etudes, Paris, France, [4] CNRS UMR 7257, Architecture et Fonction des Macromolecules Biologiques, Aix-Marseille Université, Marseille, France, [5] "Biofilm and Viral Transmission" Team, Structural Virology Unit, Department of Virology, CNRS UMR 3569, Institut Pasteur, Paris, France

***Correspondence:**
Renaud Mahieux
renaud.mahieux@ens-lyon.fr

†Present address:
Loic Roux,
RNA Therapeutics Institute,
Worcester, MA, United States

‡These authors have contributed equally to this work.

Five to ten million individuals are infected by Human T-cell Leukemia Virus type 1 (HTLV-1). HTLV-1 is transmitted through prolonged breast-feeding, by sexual contacts and by transmission of infected T lymphocytes through blood transfusion. One to ten percent of infected carriers will develop a severe HTLV-1-associated disease: Adult-T-cell leukemia/lymphoma (ATLL), or a neurological disorder named Tropical Spastic Paraparesis/HTLV-1 Associated Myelopathy (TSP/HAM). *In vivo*, HTLV-1 is mostly detected in $CD4^+$ T-cells, and to a lesser extent in $CD8^+$ T cells and dendritic cells. There is a strong correlation between HTLV-1 proviral load (PVL) and clinical status of infected individuals. Thus, reducing PVL could be part of a strategy to prevent or treat HTLV-1-associated diseases among carriers. Treatment of ATLL patients using conventional chemotherapy has very limited benefit. Some chronic and acute ATLL patients are, however, efficiently treated with a combination of interferon α and zidovudine (IFN-α/AZT), to which arsenic trioxide is added in some cases. On the other hand, no efficient treatment for TSP/HAM patients has been described yet. It is therefore crucial to develop therapies that could either prevent the occurrence of HTLV-1-associated diseases or at least block the evolution of the disease in the early stages. *In vivo*, reverse transcriptase (RT) activity is low in infected cells, which is correlated with a clonal mode of viral replication. This renders infected cells resistant to nucleoside RT inhibitors such as AZT. However, histone deacetylase inhibitors (HDACi) associated to AZT efficiently induces viral expression and prevent *de novo* cellular infection. In asymptomatic STLV-1 infected non-human primates, HDACi/AZT combination allows a strong decrease in the PVL. Unfortunately, rebound in the PVL occurs when the treatment is stopped, highlighting the need for better antiviral compounds. Here, we review previously used strategies targeting HTLV-1 replication. We also tested a series of HIV-1 RT inhibitors in an *in vitro* anti-HTLV-1 screen, and report that bis-POM-PMEA (adefovir dipivoxil) and bis-POC-PMPA (tenofovir disoproxil) are much more efficient compared to AZT to decrease HTLV-1 cell-to-cell transmission *in vitro*. Our results suggest that revisiting already established antiviral drugs is an interesting approach to discover new anti-HTLV-1 drugs.

Keywords: HTLV-1, antiviral therapy, proviral load, acyclic nucleoside phosphonates and thiophosphonates, AZT, prodrugs, cell–cell transmission

INTRODUCTION

Five to ten million individuals are infected worldwide by the oncogenic Human T-cell Leukemia Virus type 1 (HTLV-1) (Gessain and Cassar, 2012). This retrovirus is mainly present in Japan, Sub-Saharan Africa, the Caribbean region and Brazil. HTLV-1 is transmitted from mother-to-child through prolonged breast-feeding, by sexual contacts mostly from man to woman and by transmission of infected T lymphocytes through blood transfusion (Willems et al., 2017). HTLV-1 is mostly detected in $CD4^+$ T-lymphocytes, and to a lesser extent $CD8^+$ T-cells and dendritic cells *in vivo*. Interestingly, the differences in the number of HTLV-1-infected cells among distinct cell types are not due to the ability of HTLV-1 to infect these cells, but rather to its ability to persist and eventually transform the infected cells. Indeed, *in vitro*, human monocyte-derived dendritic cells are more susceptible to HTLV-1 infection than autologous $CD4^+$ T-cells (Alais et al., 2015). In animal models, although it infects $CD8^+$ T-cells and dendritic cells in the early phase of infection, HTLV-1 does not transform $CD8^+$ T-cells (Rahman et al., 2010; Valeri et al., 2010; Kannian et al., 2013). Nevertheless, there is a strong correlation between HTLV-1 proviral load (PVL, i.e., the number of single integrated copies of the viral genome in cells), and the number and abundance of HTLV-1 infected $CD4^+$ T-cells clones (Melamed et al., 2015), and the clinical status of the individuals (Gillet et al., 2013; Cook et al., 2014). A biomarker that would predict which carrier will develop an HTLV-1-associated pathology has not been described so far.

Difficulties to Treat HTLV-1 Patients: Two Associated Diseases and Treatments

One to ten percent of infected carriers develop a severe HTLV-1-associated disease during their life: Adult-T-cell leukemia/lymphoma (ATLL), a $CD4^+$ T lymphoproliferation of very poor prognosis with a mean survival time of 6 months in the acute form, or a progressive neurological disorder named Tropical Spastic Paraparesis/HTLV-1-Associated Myelopathy (TSP/HAM) (Yamano and Sato, 2012; Bangham et al., 2015). About 1000 cases of ATLL are diagnosed each year in Japan (Iwanaga et al., 2012). In the Martinique Island (French West Indies), a 14-year follow-up study allowed the detection of 123 TSP/HAM cases for only 400,000 inhabitants (Olindo et al., 2006). Because HTLV-1-infected cells are resistant to apoptosis-inducing agents, treatment of ATLL patients using conventional chemotherapy has very limited benefit (Utsunomiya et al., 2015). In Japan, when possible, ATLL patients undergo allogeneic hematopoietic stem cell transplantation (Utsunomiya et al., 2001). In Europe and United States, chronic and acute ATLL patients are treated with an anti-viral combination of recombinant interferon alpha (IFN-α) and AZT (zidovudine), combined in some cases with arsenic trioxide (El-Sabban et al., 2000; Nasr et al., 2011). It is worth noting that these results were convincingly confirmed by a world-wide meta analysis (Bazarbachi et al., 2010). However, this treatment is efficient only in a subset of patients, despite the efficacy of arsenic trioxide for inducing the death of HTLV-1-infected cells *in vitro* (El-Sabban et al., 2000; Mahieux and Hermine, 2005). While one report described the use of cyclosporin A for treating TSP/HAM patients with some benefits (Martin et al., 2012), treatment of these patients is in fact still an issue. Other drugs have been tried in clinic, although they generally have a very limited effect. As an example, a trial involving IFN-α had a modest but significant effect (Izumo et al., 1996). Different open trials [summarized by (Nakagawa et al., 1996)] have shown clinical benefit for glucocorticoids, followed by IFN-α, azathioprine and high-dose vitamin C. On the other hand, antiviral effects and/or a decrease in PVL, as well as immunomodulatory effects have been demonstrated for IFN-α (Saito et al., 2004; Rafatpanah et al., 2012), IFN-β (Oh et al., 2005), vitamin C (Moens et al., 2012), cyclosporine (Martin et al., 2012), danazol (Harrington et al., 1991), HDAC inhibitors (Lezin et al., 2007) in HAM/TSP *ex vivo* or *in vivo*.

Viral Amplification Through Clonal Expansion in Addition to *de Novo* Infection: A Difference With HIV-1 Propagation

Contrary to other viruses, HTLV-1 cannot be transmitted efficiently through cell-free viral particles. Using a cell-free experimental system, it was shown that compared to HIV-1, HTLV-1 had a low infectivity (at least 1000-fold lower luciferase activity) and that this was linked to some properties of the viral core and to post-entry processes that are still unclear (Derse et al., 2001). In contrast, HTLV-1 is efficiently transmitted following contacts between an infected donor cell and an uninfected target cell through the establishment of viral synapses and the transfer of viral biofilm (**Figure 1**, left part) (Igakura et al., 2003; Pais-Correia et al., 2010; Thoulouze and Alcover, 2011; Alais et al., 2015). Of note, cell-associated viral transmission of HIV-1 through nanotubes, filopodes or viral synapses is also more efficient than the cell-free infection protocol that is commonly used in *in vitro* experiments (Jolly and Sattentau, 2005; Sherer et al., 2007; Sowinski et al., 2008; Rudnicka et al., 2009; Zhong et al., 2013).

HTLV-1 infection then leads to the clonal expansion of infected cells (**Figure 1**, right part) (Wattel et al., 1995; Bangham

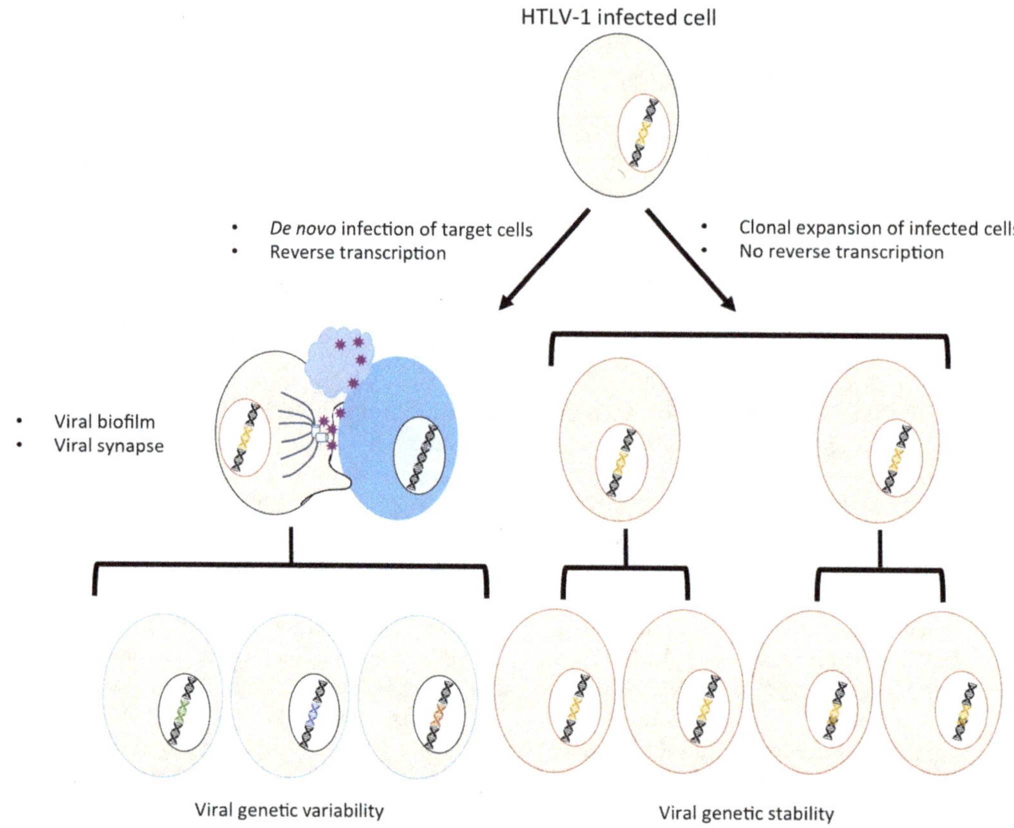

FIGURE 1 | Schematic representation of the two modes of HTLV-1 amplification. Left: HTLV-1 transmission occurs through *de novo* cellular infection, which requires production of viral particles that are transmitted via viral biofilm and viral synapses and involves a reverse transcription step. The use of RT might lead to sequence variability. Right: HTLV-1 infection promotes clonal expansion of infected cells, associated with a stability in the proviral sequence.

et al., 2014; Turpin et al., 2017; Watanabe, 2017). Because the reverse transcriptase (RT) is not involved in replication by clonal expansion, this phenomenon may explain the very low genetic variability of the virus despite the low fidelity of its RT (Mansky, 2000). Interestingly, HIV-1 clonal expansion also occurs (Maldarelli et al., 2014; Boritz et al., 2016) and has been suggested to allow the virus to overcome antibody neutralization and surface retention by the tetherin restriction factor (Zhong et al., 2013), although the consequences of such a phenomenon remain to be fully understood.

Targeting Viral Replication With the Use of Antivirals: The HIV Example

Nowadays, 25 antiretroviral agents classified in six classes have been approved to treat HIV infections (Cihlar and Fordyce, 2016). The antiretroviral therapy (cART) involves combinations of drugs to achieve maximal response and is generally composed of two nucleoside/nucleotide reverse transcriptase inhibitors (NRTIs) and a third active antiviral from a different class (NNRTI: non-nucleosidic RT inhibitor, INSTI: integrase inhibitor, PI: protease inhibitor, EI: entry inhibitor).

Nucleotide reverse transcriptase inhibitors were the first class of compounds to be used in HIV therapy, by the approval of zidovudine (AZT) in 1987 (Yarchoan et al., 1986), initially discovered as potent anti-cancer agent (Furmanski et al., 1980). NRTIs are compounds that become active after being phosphorylated into their triphosphate forms, in a process that involves three distinct phosphorylation steps catalyzed by cellular kinases. While they are generally poor substrates for cellular polymerases, triphosphorylated NRTIs compete with natural triphosphate nucleotides for incorporation into growing viral DNA by HIV RT (Furman et al., 1986), resulting in DNA chain termination by blocking further DNA extension (Mitsuya and Broder, 1987).

Historically, AZT and ddI (didanosine) were first shown to block HIV replication in T cells (Mitsuya et al., 1985) and subsequently were shown to suppress HIV replication in monocytes and macrophages *in vitro* (Perno et al., 1988). Other nucleoside analogs without oxacyclopentane sugar moiety, as well as 2′,3′-dideoxynucleosides such as d4T (Balzarini et al., 1987), or carbocyclic nucleosides such as Abacavir (ABC) (Daluge et al., 1997), were then identified to be active against HIV, resulting in the emergence of structure activity relationship (SAR) knowledge (Balzarini et al., 1987). However, nucleoside analogs were not equivalent in either activity or toxicity profiles when used as therapeutic drugs *in vitro* and *in vivo* (Mitsuya et al., 1990), and in particular in their capacity to be activated through phosphorylation in cells. Indeed, during this indispensable process, the first phosphorylation step is often rate-limiting and thus an important bottleneck that limits NRTI efficacy in cells.

To circumvent this drawback, acyclic nucleoside phosphonates (ANPs) have revolutionized the antiviral drug field (Holy, 2003; De Clercq and Holy, 2005). ANPs are nucleotide analogs with a phosphonate group attached to the nucleoside moiety, which require only two phosphorylation steps to be converted into their fully phosphorylated active form (Sharma et al., 2004). The anti-HIV properties of Tenofovir (PMPA, 9-[(R)-2-(Phosphonomethoxy)Propyl]Adenine) were first described in Balzarini et al. (1993) and 8 years later, the oral prodrug, namely Tenofovir disoproxil (TDF) was licensed for clinical use for the treatment of HIV infection. As such, TDF has remained a cornerstone for 14 years and has been replaced recently by a new oral prodrug of the same compound, namely Tenofovir Alafenamide (TAF), which allows a better pharmacokinetic profile in patient fluids (Ray et al., 2016).

Undesirable emergence of viral resistance, toxicity, drug-related adverse effects and drug-drug interactions associated with treatment using NRTIs have challenged the antiviral therapy. Over the last 30 years, it has led to the discovery and approval of several new compounds (Cihlar and Fordyce, 2016) with new ones still in ongoing development (Sax et al., 2015). This intense antiviral research provided the clinic with different anti-HIV active drugs that are now used in combination and have markedly decreased mortality and morbidity from HIV-1 infections in the developed world (Palella et al., 1998). This illustrates that the effort to provide the clinic with new drug regiments results in successful therapies that benefit to patients.

In 2017, the NRTIs approved by the FDA for the treatment of HIV-1 include Retrovir® (Zidovudine, AZT, 1987), Videx® (didanosine, ddI, 1991), Zerit® (stavudine, d4T, 1994), Epivir® (lamivudine, 3TC, 1995), Combivir® (AZT + 3TC, 1997), Ziagen® (abacavir sulfate, ABC, 1998), Trizivir® (AZT + 3TC + ABC, 2000), Viread® (tenofovir disoproxil fumarate, TDF, 2001), Emtriva® (emtricitabine, FTC, 2003), Epzicom® (3TC + ABC, 2004), Truvada® (TDF + FTC, 2004), Atripla® (TDF + FTC + Efavirenz (NNRTI), 2006), Complera® [TDF + FTC + Elvitegravir (INSTI) + booster, 2011], Triumeq® [3TC + ABC + dolutegravir (INSTI), 2014], Genvoya® [FTC + Tenofovir Alafenamide (TAF) + Elvitegravir (INSTI) + booster, 2015], Odefsey® [FTC + TAF + rilpivirine (NNRTI), 2016], Descovy® (FTC + TAF, 2016). Hivid® (zalcitabine, ddC, 1992) was discontinued due to genotoxicity. This antiviral success prompted clinician and researchers to test whether some of the anti-HIV drugs, and especially the acyclic phosphonates ones, could also be active against infections in which viral replication use a viral DNA polymerase enzyme, such as HBV, HSV, EBV, or CMV (De Clercq and Holy, 2005). Interestingly, although the acyclic phosphonate-including class has a broad-spectrum efficacy against several divergent DNA virus families, each acyclic phosphonate within the class has its own specificity. For example, HPMPC is able to inhibit the replication of *Papovaviridae*, *Adenoviridae* and some viruses from the *Herpesviridae* family, while PMEA and PMPA are able to inhibit *Hepadnaviridae* and *Retroviridae*, and PMEA and HPMPC are able to inhibit the replication of several, but not all *Herpesviridae* (De Clercq and Holy, 2005). Thus, some inhibitors first described as active against HIV may have a broader antiviral spectrum, even on viruses belonging to different classes, strongly supporting the concept of re-positioning of existing drugs.

HIV-NRTI Repositioning on HTLV-1

Because replication of both HIV-1 and HTLV-1 retroviruses (whether it is a frequent step or not, as discussed above) requires RNA reverse transcription performed by phylogenetically related RTs, it is tempting to hypothesize that some anti-HIV-1 drugs targeting HIV-1 RT might also be efficient against HTLV-1 RT. Indeed, using cellular assays of viral transmission, it was shown that AZT, 3TC, phosphonated carbocyclic 2′-Oxa-3′-Aza-nucleoside and Tenofovir each reduced HTLV-1 transmission through cell-cell contact (see **Table 1**) (Macchi et al., 1997; Zhang et al., 2001; Balestrieri et al., 2002, 2005, 2008; Chiacchio et al., 2005, 2007; Pais-Correia et al., 2010). Furthermore, *in vitro* RT assays using HTLV-1 virions confirmed that these NRTIs target the enzymatic RT activity of HTLV-1 (Garcia-Lerma et al., 2001; Balestrieri et al., 2005, 2008; Macchi et al., 2011). The first inhibitory values were obtained using a test that is not quantitative but allows only to determine the drug efficiency (Yamamoto et al., 1996). However, a clinical trial involving AZT as a putative HTLV-1 RT inhibitor, showed only a very limited efficacy in reducing HTLV-1 PVL and no immunological or clinical responses could be demonstrated (Taylor et al., 2006). Furthermore, other data suggested that AZT could in fact inhibit telomerase and lead to cell death, irrespective of the infected status of treated cells (Datta et al., 2006).

A number of studies were designed to decipher why nucleoside analogs, contrary to HIV-1, are so inefficient against HTLV-1. One obvious reason, already mentioned, is the pseudo-latency of the virus (Taylor and Matsuoka, 2005; Taylor et al., 2006). Indeed, in chronically infected asymptomatic carriers, HTLV-1 does not frequently replicate by reverse transcription but rather replicates through clonal expansion of the infected cells (Wattel et al., 1995; Gillet et al., 2011, 2017). Another explanation lies in the mode of viral cell-to-cell transmission. As discussed above, HTLV-1 transmission through cell-free viral particles does not seem efficient. In contrast, cell-to-cell transmission mostly occurs through cell–cell contacts between an infected cell and an uninfected target (Derse et al., 2001; Ceccaldi et al., 2006; Mazurov et al., 2010; Alais et al., 2015). HTLV-1 transmission after cell-to-cell contacts is thought to lead to a high multiplicity of infection and to drug resistance, as was previously described in the case of HIV (Sigal et al., 2011; Zhong et al., 2013). Interestingly, while AZT is indisputably recognized as an efficient HIV-1 replication inhibitor, reported IC_{50} are consistently higher when inhibition assays are performed using cell-to-cell infection settings instead of cell-free virus (Sigal et al., 2011; Agosto et al., 2014), suggesting that cell-to-cell transmitted viruses have an inherent resistance to reverse transcription inhibitors, which may not be linked to the molecular enzymatic resistance of the RT. Thus, efficacy of

TABLE 1 | Summary of reported inhibition tests using NRTI in cell-to-cell transmission **(A)** or in *in vitro* RT assays **(B)**.

(A) Cell-to-cell transmission.

	Hill et al., 2003	**Zhang et al., 2001**	**Macchi et al., 1997**	**Balestrieri et al., 2008**	**Balestrieri et al., 2008**
	Single-cycle infection (VLP and cell-to-cell transmission)	Cell-to-cell transmission with MT2 (donors) and cord blood PBMCs (targets)	Cell-to-cell transmission with MT2 (donors) and blood PBMCs (targets)		
	24 h p.i. single cycle infection (IC_{50})	**4 days p.i. viral RNA (IC_{50})**	**4–7 days viral RNA (IC_{50})**	**4 weeks (p19 EC_{50})**	**4–7 days viral RNA (EC_{100})**
Tenofovir	5 nM	*Not tested*	*Not tested*	0.73 μM	2.3 μM
AZT	0.11 μM	0.3 ± 0.15 μM DNA PCR and 0.01 ± 0.00 μM RT-PCR	0.01 ± 0.0 μM RNA dot blot and 0.03 ± 0.01 μM DNA dot blot	ND	0.7 μM
ddC	0.27 μM	*Not tested*	*Not tested*	*Not tested*	*Not tested*
Abacavir	4.6 μM	*Not tested*	*Not tested*	*Not tested*	*Not tested*
d4T	14.5 μM	*Not tested*	*Not tested*	*Not tested*	*Not tested*
3TC	22 μM	*Not tested*	*Not tested*	*Not tested*	*Not tested*

(B) *In vitro* RT activity.

	Garcia-Lerma et al., 2001	**Balestrieri et al., 2008**			**Balestrieri et al., 2005**	**Macchi et al., 2011**	
	Indirect RT activity assay on virions from infected cells. The amount of RT in the viral lysates was indirectly assessed through the antigen capture assay by measuring the amount of p19 in the supernatant from infected cells. The RT added to the assay was equivalent to a standardized amount of 32 pg of p19. RNA template = RNA extracted from transfected cells ectopically expressing the glycoprotein D of herpes simplex virus type 1 (HSV-1gD, 1 μg). The assayed compounds phosphorylated by crude extract were added at concentration ranging from 0.1 to 1000 nM; 1 mM dNTP, and DNA-PCR was carried out for 35 cycles (Yamamoto et al., 1996).	Indirect RT activity assay on virions from infected cells (RT added was equivalent to a standardized amount of 32 pg of p19), using RNA template corresponding to total RNA extracted from HSV-D glycoprotein transfected cells (HSV-1gD 1 μg). Inhibitors were phosphorylated by crude extract (0.1–1000 nM) and assays were carried out with 1 mM dNTP, and 35 cycles DNA-PCR.					
	RT from MT-2; Hut102 and patients' isolates (EC_{100})	**MT2 (EC_{100})**	**C91 (EC_{100})**	**Hut102 (EC_{100})**	**MT2 (EC_{100})**	**MT2 (EC_{100})**	**Patients' isolates (EC_{100})**
Tenofovir	*Not tested*	0.4 nM	1 nM	10 nM	1 nM	0.4 nM	10–1000 nM (TDF)
AZT	0.12–0.38 μM	0.4 nM	*Not tested*	*Not tested*	10 nM	0.4 nM	4–8 nM
3TC	>10 μM, i.e., no inhibition	*Not tested*	*Not tested*	*Not tested*	*Not tested*	*Not tested*	*Not tested*

IC_{50} is inhibitory concentration 50%, EC_{50} is effective concentration 50%. EC_{100} is effective concentration 100%.

AZT or other NRTIs against HTLV-1 might be further limited by the almost exclusive cell-to-cell transmission mode of the virus.

Several parameters might further account for the discrepancies between results obtained *in vitro* and *in vivo*. First, the efficacy of AZT against HTLV-1 as measured by the IC_{50} might be dependent on the experimental settings. This is illustrated by a study that reported a 1000-fold lower efficacy of compounds when used in cell-to-cell transmission assays (inhibition range of 1 μM) compared to *in vitro* RT inhibition assays performed on virions from supernatant of infected cells (inhibition range of around 1 nM) [**Table 1**, (Balestrieri et al., 2008)]. Thus, IC_{50} determined *in vitro* or in cellular assays may not be relevant to what occurs *in vivo*. Furthermore, sequence variation within the RT gene, in addition to its influence on replication efficiency (Mitchell et al., 2007), might also be responsible for different sensitivity of RT from virions produced by different HTLV-1 infected cells lines (Balestrieri et al., 2008) or RT derived from patient isolates (Macchi et al., 2011) to NRTIs (**Table 1**). Even if inhibition is measured in cell–cell transmission assays using the same infected donor cells, the use of different target cells might explain the variability in NRTI efficacy. For example, AZT efficacy was measured on HTLV-1 transmission after co-culture of chronically infected MT2 cells with primary cord PBMCs (Zhang et al., 2001) or adult blood PBMCs (Macchi et al., 1997; Balestrieri et al., 2008) (**Table 1**). In these assays, AZT was estimated to inhibit HTLV-1 transmission after 4–7 days of culture, as measured by viral DNA or viral RNA detected in primary PBMCs. AZT IC_{50} values were 0.03 ± 0.01 and 0.01 ± 0.00 μM for DNA and RNA, respectively, in PBMCs and 0.30 ± 0.15 and 0.01 ± 0.00 μM for DNA and RNA in cord blood, respectively (**Table 1**). The lower susceptibility to AZT inhibition of PBMCs from adult blood highlights that besides the direct effect of AZT on RT activity in virions released from MT2 infected cells (**Table 1**), other mechanisms exist in the target cells that modulate the efficacy of inhibition. This could include drug uptake and intracellular phosphorylation that have been shown to be cell-type dependent (Deville-Bonne et al., 2010; Roux et al., 2013), or RT activity in the target cell that could differ depending on the cellular environment. Very little is known regarding RT activity in cells during the early steps of the viral cycle. Yet, because cell-to-cell infection is much more efficient than cell-free infection, it is possible that either a higher number of virions is delivered to the target cell during cell-to-cell contact or that contact-dependent signals are required to enhance virus replication. Such signals could also modulate the sensitivity of target cells to RT inhibition, in a target cell-specific manner.

Of note, although co-culture assays probably mimic HTLV-1 transmission *in vivo*, inhibition is measured after several days of culture. This long-term culture of infected cells could select for T-cells susceptible to infection and able to survive after HTLV-1 infection, due to the immortalization potency of the virus. Furthermore, several rounds of infection could occur that may not be suitable for quantitative analysis of infection and replication. Alternative experimental systems have been developed to measure HTLV-1 spread after a single-cycle replication (Derse et al., 2001) and to measure viral inhibition using NRTI (Hill et al., 2003) (**Table 1**). This elegant system [reviewed in (Aloia et al., 2013)] is based on virus-like particles (VLPs) that contain HTLV-1 RT and that deliver a surrogate retroviral genome designed to express only a reporter gene product in infected cells. These VLPs are generated by co-transfecting cells (usually adherent cell lines such as HEK293T) with a packaging plasmid (allowing the expression of HTLV-1 structural, enzymatic and regulatory proteins under the control of a CMV promoter), a transfer reporter vector and an envelope-encoding plasmid [reviewed in (Derse et al., 2001)]. In these settings, the inhibition properties of several nucleoside analogs were determined (**Table 1**) and results confirmed AZT IC_{50} to be around 0.1 μM (Hill et al., 2003). It is worth noting that efficiencies of all compounds tested against HTLV-1 were at least 100-fold lower than that tested against HIV-1, further suggesting that although both RTs have similar sequence and enzymatic organization, the catalytic properties of the two RTs are different and/or differently regulated by the cellular environment of the infected cells. Moreover, although useful to monitor single-cycle infections, this system uses VLPs that may be different from WT virions and more importantly, it uses transfected HEK293T donor cells that are very different from the natural infected T-cells expected to transfer the virus *in vivo*. Knowing that cell–cell contacts may regulate HTLV-1 RT activity, the nature of the infected cell used to transfer the virus is expected to modulate the infectivity of the transferred virions as well as the efficacy of NRTIs.

Strategies to Control HTLV-1 PVL *in Vivo*: Replication Reactivation and Use of NRTIs

As discussed above, HTLV-1 pseudo-latency renders this retrovirus usually insensitive to RT inhibitors. Interestingly, viral expression and RT-dependent *de novo* cellular infection can be induced using histone deacetylase inhibitors (HDACi) such as valproic acid (VPA) (Lezin et al., 2007; Olindo et al., 2011). Importantly, long-term treatment of HTLV-1-infected patients with VPA is proven to be safe (Lezin et al., 2007; Olindo et al., 2011). In an earlier work conducted in non-human primates naturally infected with Simian T-Lymphotropic Virus type 1 (STLV-1), the simian equivalent of HTLV-1, we tested a combination of VPA, used to increase viral replication, and AZT, used to inhibit *de novo* cellular infection by targeting the RT activity (Afonso et al., 2010). Because the PVL is higher in TSP/HAM patients than in asymptomatic carriers and can therefore be considered as a prognostic marker (Yakova et al., 2005), we selected asymptomatic STLV-1-infected animals with a high PVL (Afonso et al., 2010). Our results showed for the first time that the AZT/VPA combination leads to a very strong and significant PVL decrease in all tested animals. However, STLV-1 PVL rebounded after the treatment was stopped, thus demonstrating that the virus had not been cleared during the treatment, a disappointing result reminiscent of what is described for HIV-1. As for HIV-1, this rebound of HTLV-1 viral load suggests (i) the existence of a putative viral reservoir inaccessible to the drugs, or (ii) that AZT is not efficient enough to block *de novo* HTLV-1 infection that occurs at a level below the detection threshold, or (iii) that viral escape mutants are selected during treatment, leading to resistance to AZT, although this is unlikely given that resistance mutation were not reported in human patients undergoing AZT. Nevertheless, these results demonstrate that preventing *de novo* infection is an option for controlling the virus and foster the use of more efficient drugs that would better control HTLV-1 dissemination within infected individuals.

Here we show a series of results obtained with already tested drugs as well as with newly compounds. We report for the first time that bis-POM-PMEA (adefovir dipivoxil) and bis-POC-PMPA (tenofovir disoproxil) are much more efficient compared to AZT to decrease HTLV-1 cell-to-cell transmission *in vitro*.

MATERIALS AND METHODS

Cells

Jurkat, Jurkat-LTR-Luc (Pais-Correia et al., 2010), and HTLV-1-infected T-cell line (C91PL) were maintained in RPMI 1640 medium supplemented with 10% fetal calf serum and penicillin-streptomycin (100 μg/ml, Gibco Life Technology). All cells were grown at 37°C in 5% CO_2. Jurkat-LTR-Luc cells, stably transfected with a plasmid encoding luciferase, under the control of the HTLV-1 long terminal repeat (LTR) promoter are maintained under hygromycin (450 μg/ml, Sigma) selection.

Drugs

AZT, d4T, 3TC, tenofovir alafenamide, tenofovir disoproxyl and tenofovir disoproxyl fumarate were purchased from Selleckchem. PMPA, PMEA were provided by Dr. Alvarez (AFMB, Marseille). S-PMPA, S-PMEA, bis-POC-S-PMPA, bis-POM-PMEA, and bis-POM-S-PMEA were synthesized and provided by Dr Alvarez (AFMB, Marseille). All drugs were resuspended in DMSO or water according to the furnisher, aliquoted and stored at −80°C until used. Aliquots were used only once to avoid drug degradation under thawing-freezing cycles.

Inhibition of Viral Transmission Test

Jurkat reporter cells (JK-LTR-Luc, 50.000 cells) were seeded in 96-well plates and treated for 18 h in presence or absence of the different drugs (1–50 μM, diluted in medium). Then, cells were washed once and HTLV-1-infected cells (C91PL, 10.000 cells) or control cells (Jurkat, 10.000 cells) were added together with or without an additional ½ dose of drugs, and the co-culture was resumed for 24 h. Cell viability was measured by trypan blue exclusion counting. The mortality index was determined as the percentage of dead cells in each condition normalized to the percentage of dead cells in the control condition set to 1. A mortality index higher than 1.5 signs drug toxicity. Reporter activities were assayed using the luciferase reporter assay system (Promega). Luciferase activity was normalized according to the amount of total proteins in the co-culture determined by Bradford (Bio-Rad). Results were then expressed as a percentage of control in the absence of drugs.

Statistical Analysis

One-way analysis of variance (ANOVA) with Bonferroni's *post hoc* multiple comparison test was used to determine statistically significant differences. Differences were considered significant if the p-value was < 0.05.

RESULTS

New Strategies for the Repositioning of Anti-HIV Inhibitors as Efficient Anti-HTLV-1 Drugs

High-throughput screening efforts against HIV-1 followed by structure-based drug design have allowed the discovery of several drugs that are more potent than AZT and related nucleoside analogs, and are still ongoing (Taylor et al., 2016). Highly active non-nucleoside inhibitors eliciting hardly any resistant mutants have been designed when purified target enzymes (RT, protease, integrase) were made available, providing essential structural and functional models. In strong contrast, discovery of anti-HTLV-1 specific drugs is currently significantly impeded by the unavailability of purified viral enzymes. In addition, and as stated above, because the virus is almost exclusively transmitted through cell–cell contacts, cell infection cannot be easily induced *in vitro* using cell-free particles as a source of infectious viruses (Alais et al., 2015). This further challenges the development of drug screening protocols. Since HIV-1 and HTLV-1 use very similar mechanisms of reverse transcription, we hypothesized that anti-HIV-1 NRTIs might be efficient against HTLV-1 RT. NNRTIs, which are highly specific to HIV-1 RT, were thus excluded from our screen.

We thus set up an *in vitro* cell co-culture assay that allows for the monitoring of HTLV-1 *de novo* infection in LTR-luciferase reporter Jurkat T-cells using HTLV-1 chronically infected C91PL T-cells as a source of virus (Alais et al., 2017) (**Figure 2A**). This system was used to test the efficacy of several NRTIs known to inhibit HIV-1 RT (**Figure 2B**). In addition, we selected pro-drugs of these NRTIs and evaluated their efficacy and toxicity. AZT (1) was used as a positive control. 2′,3′-dideoxynucleosides such as 3TC (2) and d4T (3) were also used as controls (**Figure 2B**). Because highly variable data are reported in the literature on the effectiveness of these compounds (**Table 1**), we also included them to calibrate the stringency of our assay. While the ANP Tenofovir (PMPA, 5) has been tested against HTLV-1 previously, its congener Adefovir (PMEA, 4) has never been evaluated on HTLV-1 so far, although its efficiency to inhibit HIV and HBV infections is described (De Clercq and Holy, 2005). Both compounds were included in our screen. We also included new ANP analogs, named thiophosphonates, S-PMEA (6) and S-PMPA (7), which have shown potent activity against HIV and HBV infections (Barral et al., 2011; Roux et al., 2013; Priet et al., 2015).

As discussed above, ANPs are NRTIs that contain a stable phosphonate group. After penetration into the target cells, ANPs are phosphorylated into their diphosphate forms which serve as substrates for RT and thus exert the antiviral effect. Because ANPs are negatively charged at physiological pH, they are less efficient in passing through cell membranes than neutral, lipophilic species. To address this issue, these compounds can be prepared as prodrugs. The activity of ANPs is dramatically increased by the use of their biolabile prodrugs: Adefovir dipivoxil (Hepsera®) was licensed for the treatment of chronic HBV infections during several years (Merle and Trepo, 2001) and Tenofovir is commercialized as two prodrugs, which are much more potent than their parent counterparts.

We therefore also evaluated Adefovir dipivoxil (bis-POM-PMEA, 8), its thiophosphonate equivalent bis-POM-S-PMEA (9), Tenofovir disoproxil (bis-POC-PMPA, 10), Tenofovir disoproxil fumarate (bis-POC-PMPA, 11), Tenofovir alafenamide (12) and the thiophosphonate congener bis-POC-S-PMPA (13).

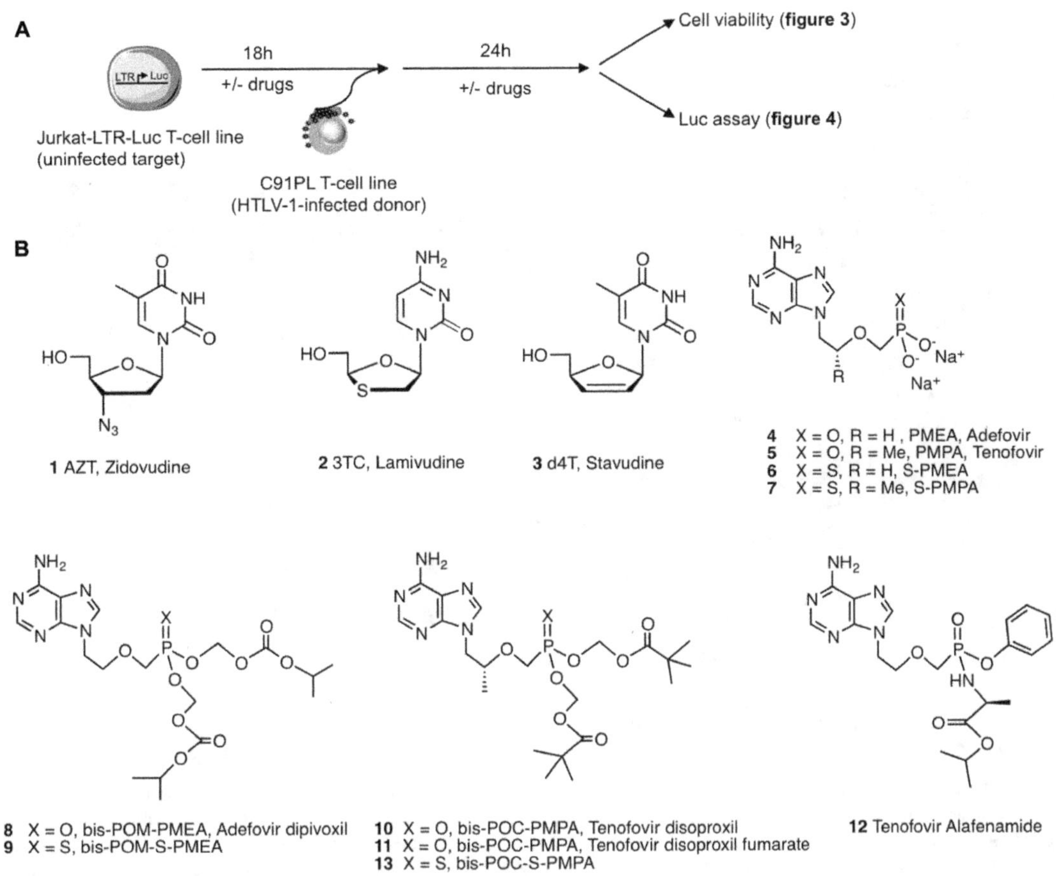

FIGURE 2 | Experimental assay and chemical structures of compounds used in this study. **(A)** Schematic representation of the experimental procedure. Jurkat-LTR-Luc [reporter cells for HTLV-1 productive infection in which expression of the luciferase gene is under the control of the viral Tax-dependent promoter (LTR)] are pretreated or not with drugs for 18 h before their co-culture with C91PL (HTLV-1 infected T-cell line) for 24 h. Cell viability and luciferase activity are then measured. **(B)** AZT (1) is used as the reference drug during all experiments, while 3TC (2) and d4T (3) are used as controls; the acyclic nucleoside phosphonate PMEA (4), its thio-derivative S-PMEA (6) and their prodrugs (8) and (9), respectively; the acyclic nucleoside phosphonate PMPA (5), its thio derivative S-PMPA (7) and their pro-drugs (10), (11), (12), and (13), respectively. Structures were drawn using ChemDraw. The molecule names (when available) are indicated under the structure.

First, we tested whether these compounds had any impact on target cell viability (**Figure 3**). The concentration of these drugs varied between 1 and 50 μM. The viability of treated cells was represented relative to that of the untreated cells, set to 1. An index > 1 means that the mortality of the treated cells is higher than that of the untreated cells. We considered that cell viability was impaired if the index was >1.5 and thus excluded all the drug concentrations leading to a mortality index > 1,5. None of the drugs, i.e., AZT (1), 3TC (2), d4T (3) (**Figure 3A**) as well as the PMEA (4–6) and PMPA (5–7) families had any significant effect on cell viability, i.e., their mortality index was lower than 1.5 (**Figure 3B**). As already reported (Roux et al., 2013), Bis-POM-PMEA (8) and Bis-POM-S-PMEA (9) were toxic when used at 10 μM (**Figure 3C**), as well as BIS-POC-PMPA (10) when used at 50 μM (**Figure 3C**). Thus, these concentrations were excluded for the subsequent evaluation of their ability to prevent HTLV-1 transmission to reporter T-cells after co-culture with infected donor cells (**Figure 4**). LTR-luciferase Jurkat reporter T-cells were used as target cells (Alais et al., 2017). Given that Tax is absent from the viral particle, transcription from the viral Tax-dependent LTR requires productive infection and Tax expression in the target cells, which can occur after contact with the infected donor cells through the establishment of viral synapses or transfer of viral biofilm (Pais-Correia et al., 2010). Interestingly, this experimental system does not rely on cell-free infection and is therefore closer to *in vivo* viral transmission.

Consistent with previous reports, AZT (1) had a significant but limited effect on HTLV-1 transmission, even at the high concentration of 50 μM (inhibition of 10–20% **Figures 4A,B**), while 3TC (2) had no effect, independently of the concentration used (**Figure 4A**). Surprisingly, AZT (1), had the same significant inhibitory activity for all the doses tested, without any dose-dependent effect (**Figure 4B**). d4T (3) also had no effect (**Figure 4A**), suggesting that previously published cell-free experiments using virion-associated enzymatic RT assays might not be relevant for determining that a drug has an anti-HTLV-1 effect (Garcia-Lerma et al., 2001). We then tested phosphonate PMEA (4) and PMPA (5) (**Figure 4B**) and compared their effect

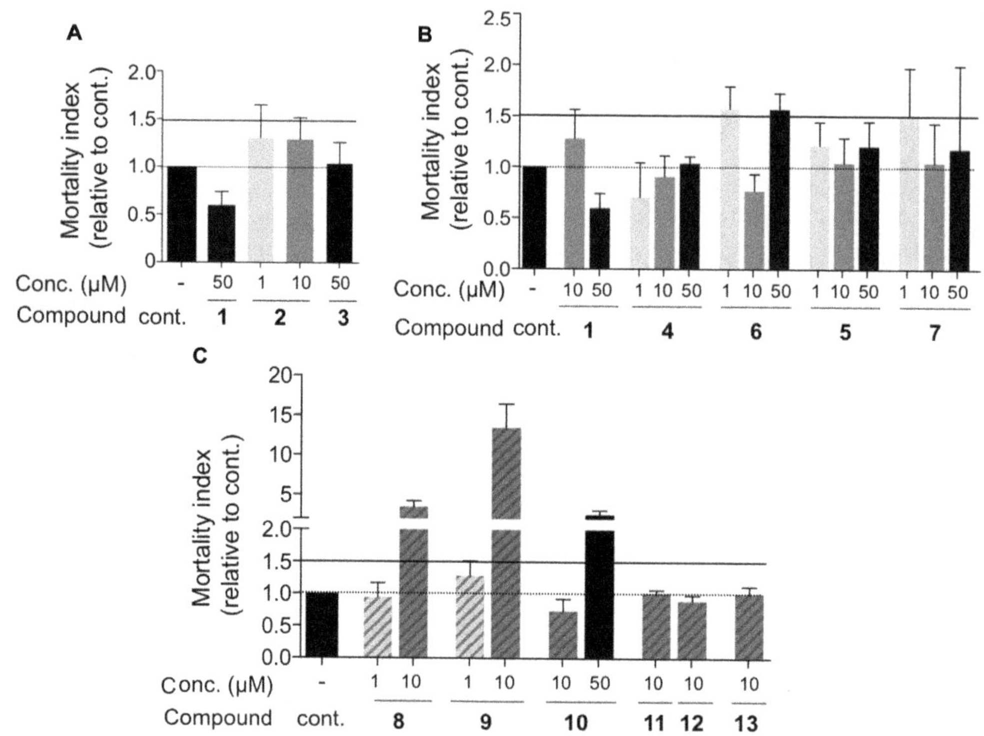

FIGURE 3 | Control of cell viability during drug treatment. Jurkat-LTR-Luc reporter cells were treated for 42 h and the proportion of dead cells was determined by trypan blue dye exclusion test. Data were normalized to the proportion of dead cells in untreated control conditions, set to 1, to obtain a mortality index. A mortality index higher than 1 indicate a drug-induced mortality. Drugs were used in the following assays if the mortality index was below 1.5. Drug concentration and compound number are indicated. Results were grouped based on the drug family: **(A)** nucleoside analogs, **(B)** acyclic nucleoside phosphonates, **(C)** prodrugs of acyclic nucleoside phosphonates. No significant differences were observed (one-way ANOVA), $n = 2–4$ independent experiments with triplicate samples.

to that of AZT. Contrary to AZT, PMEA (4) had a statistically significant dose-dependent inhibitory activity (**Figure 4B**). When used at 50 μM, PMEA was statistically more efficient than AZT (70% vs. 30% of viral transmission decrease, respectively, $p < 0.0001$). Of note, PMEA's active concentration on HTLV-1 was much higher than the dose reported to inhibit HIV-1 cell-free transmission, which is around 5 μM (Barral et al., 2011). Importantly, it is worth noting that when HIV-1 inhibition assays were performed using cell-to-cell infection instead of infection with cell-free virus, reported IC_{50} were also consistently higher (Sigal et al., 2011; Agosto et al., 2014). PMPA (5) did not show any inhibitory effect compared to control, AZT (1) or PMEA (4), independently of the concentration used (**Figure 4B**). This is somehow different from a previous report which showed that overnight Tenofovir treatment (1 μM) before co-culture had some effect on viral transmission from chronically-infected MT2 cells to PBMCs, while the same product at 0.01 μM had no effect (Balestrieri et al., 2005). We also tested S-PMEA (6) and S-PMPA (7) (**Figure 4B**), which showed a limited effect for (6) at the highest concentrations, yet not significantly better than (1) and significantly worse than (4). Thus, modifying PMEA into the thio-derivative S-PMEA (6) results in a strong decrease in HTLV-1 transmission inhibition [compare (4) vs. (6)], consistent with the results on HIV replication for which IC_{50} is threefold higher than that of the parental derivative (Barral et al., 2011). Interestingly, modifying PMPA into its thio-derivative (7) resulted in a limited but significant better inhibition than 5, at 1 and 10 μM (25% of inhibition $p < 0.01$), an effect that is lost when the concentration is increased to 50 μM (**Figure 4B**), probably due to the higher toxicity of 7 at 50 μM (**Figure 3B**).

We then tested ANP pro-drugs since they were reported to display an active concentration that was better than that of their parent compound due to the higher ability in passing through cell membranes (**Figure 4C**). bis-POM-PMEA (8) showed a statistically better, although modest, inhibition of LTR activation at 1 μM than the parent compound PMEA (4) tested at the same concentration (compare **Figure 4C**). Although S-PMEA (6) had a significant yet modest inhibitory effect at 1 μM (20% of inhibition $p < 0.001$, **Figure 4C**), its bis-POM modification (9) results in a loss of this effect when used at the same concentration. This suggests that although the cell-membrane penetration is increased, it does not translate into a better efficacy of the compound. This could be explained by the higher toxicity of the pro-drug (6) at 1 μM (**Figure 3C**), which within the 42 h of treatment could impair the metabolism of the pro-drug after its penetration into the cells. Higher concentration could not be used because of compound 9 toxicity (see **Figure 3C**).

For the PMPA compound family, no effect was observed when pro-drugs (10, 11, or 12) were used at 1 μM. This is consistent with the absence of inhibition observed with the parent PMPA (5)

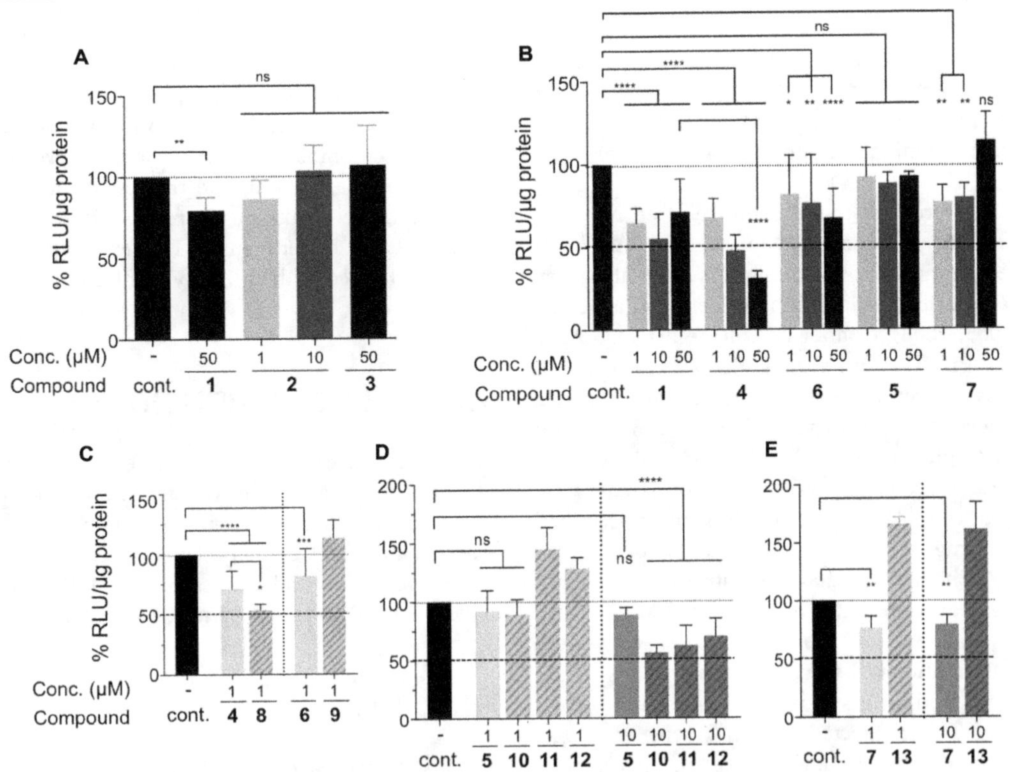

FIGURE 4 | Inhibition efficacy of drugs on HTLV-1 cell-to-cell transmission. **(A–E)** Luciferase activity was normalized to protein concentration (in µg, quantified by Bradford assay) and expressed as percent of untreated (control) cells set as 100%. Drug concentrations and drug names are indicated. Results were grouped based on the drug family: **(A)** AZT (1), 3TC (2) and d4T (3) nucleoside analogs, **(B)** AZT (1) and acyclic nucleoside phosphonates, **(C)** PMEA and S-PMEA and their prodrugs, **(D)** PMPA and a series of PMPA pro-drugs, **(E)** S-PMPA and its pro-drug. Statistical differences were determined using one-way ANOVA (ns, not significant, $^{*}p < 0.05$, $^{**}p < 0.01$, $^{***}p < 0.001$, $^{****}p < 0.0001$, n = 2–4 independent experiments with triplicate samples).

compound (**Figure 4D**). However, when used at 10 µM, prodrugs (10–12) had a higher effect than PMPA (5) itself and induced a significant reduction (50%, $p < 0.0001$) of HTLV-1 replication.

Finally, bis-POC-S-PMPA (13) did not show any improvement over S-PMPA (7) and even had a statistically significant positive effect on viral transmission (**Figure 4E**). This could be explained by the lower de-protection rate of (13) in cell, thus leading to a lower concentration of active S-PMPA within cells (Roux et al., 2013). Surprisingly, the pro-drug effect was lower than expected from the results observed in HIV tests in which IC_{50} are decreased by at least 120-fold (Roux et al., 2013) by bis-POC or bis-POM modifications. This suggests that in our settings, the rate of pro-drugs metabolism and their subsequent use as substrates by HTLV-1 RT might be very low, although they could increase over time, leading to a potential better inhibition.

Altogether, our results show that PMPA (5) is not active when HTLV-1 infection occurs through cell–cell contacts, although (5) inhibits HTLV-1 RT in *in vitro* enzymatic assays (see **Table 1**). However, we demonstrate that several pro-drugs of PMPA are efficient (10, 11, 12), confirming earlier studies (Macchi et al., 2011) and suggesting that they could be used in clinical trials. More importantly, our results extend the inhibitory potential of phosphonate derivatives by showing that PMEA (4) is more efficient than PMPA (5) to block HTLV-1 viral transmission, without any associated-toxicity. In addition, PMEA pro-drug (8) is also efficient, and at a 10-fold lower concentration than all PMPA pro-drugs. However, its associated-toxicity might compromise its use for clinical trial. Finally, the thio-modification of PMEA (6) does not increase PMEA (4) inhibitory potential, while PMPA thio-modification (7) increases PMPA (5) efficiency, although this effect is lost for the pro-drug (13).

DISCUSSION

HTLV-1-induced diseases, i.e., ATLL and TSP/HAM, are of bad prognosis. Despite the early discovery of HTLV-1 more than 30 years ago, and despite the outstanding genetic stability of the virus (which could be a topic of interest for the vaccine industry), specific anti-HTLV-1 drugs have not been developed yet. In addition, and contrary to the HIV-1 field, asymptomatic HTLV-1-infected individuals are not treated, although a high PVL is established to be a valuable risk factor that could be used as a biomarker to identify individuals at risk for developing HTLV-1-associated diseases. Rather, patients are treated once they develop the disease, which is probably too late. Clonal populations of infected T-cells are detected *in vivo*, suggesting that the virus mostly replicates through clonal expansion (Bangham et al., 2014), although viral expression, potentially leading to viral production, may occur upon metabolic changes occurring in protected body areas (Kulkarni et al., 2017). In addition, HTLV-1 does not replicate only through clonal expansion given that AZT as an *in vivo* effect in asymptomatic individuals (Afonso et al., 2010).

Anti-HIV-1 drugs were first tested using cell-free infection systems in which it is easy to quantify viral production. This is the reason why reverse-transcriptase inhibitors have also been tested using HTLV-1 purified virions (Garcia-Lerma et al., 2001; Balestrieri et al., 2005, 2008), or after co-culture between MT2 infected cell line and PBMCs from normal blood donor (Macchi et al., 2011). This is, however, an issue for translating results into the clinic. Indeed, when it occurs, HTLV-1 transmission goes through cell–cell contacts rather than through cell-free viral particles, a transmission mode that was proven to increase viral resistance to drugs in the HIV-1 situation. Previously described co-culture assays for drug screening used MT2 cells as donor cells (Macchi et al., 1997; Zhang et al., 2001; Balestrieri et al., 2008). Of note, this cell line was recently demonstrated to produce poorly infectious viral particles, even when these cells were used for viral transfer after cell-to-cell contacts (Alais et al., 2015). We have developed an easy-to-use assay relying on the co-culture between HTLV-1-infected C91PL T-cells and LTR-luciferase reporter Jurkat T-cells (Pais-Correia et al., 2010). Here, we demonstrate that this assay is useful to screen a series of drugs, in experimental settings that most closely model the *in vivo* viral transmission mode. Our results using stavudine d4T and lamivudine 3TC are consistent with the literature and show that, for some unexplained reasons, these drugs have no effect on HTLV-1 RT. More interestingly, we show here for the first time that Adefovir (PMEA) as well as its prodrug (Adefovir dipivoxil) work much better than AZT. In the case of Tenofovir (PMPA), the drug has no effect, however, the prodrug (Tenofovir disoproxil) has a significant inhibitory effect when used at 10 μM. This encourages us to validate its inhibitory potential *in vivo*, in naturally infected and currently asymptomatic STLV-1 infected non-human primates. The ongoing pre-clinical test combines Tenofovir disoproxil fumarate (Viread) to valproate, in a protocol similar to that used in the Afonso study (Afonso et al., 2010), in which the combination of AZT with valproate led to a strong PVL decrease as long as the treatment did go on. From these results and the results of the present study, it can therefore be expected to get better results with Tenofovir disoproxil fumarate (Viread). This could help HTLV-1 infected patients' treatment in the future.

AUTHOR CONTRIBUTIONS

AP and SA performed the experiments. LR and KA synthesized all phosphonates derivatives except the bis-POC-PMPA fumarate and PMPA Alafenamide which were purchased from Selleckchem. M-IT contributed with technical developments for some experiments and discussion. RM, HD, and KA designed the experiments. RM, CJ, HD, and KA wrote the article.

FUNDING

AP was supported by INSERM. RM and CJ are supported by Ecole Normale Supérieure de Lyon. RM is part of the French Laboratory of Excellence project ECOFECT (ANR-11-LABX-0048). SA and HD are supported by INSERM. KA is supported by CNRS and Aix-Marseille Université. LR was supported by Sidaction (Ph.D. grant). M-IT research was supported by AXA Research fund as well as by institutional grants from Institut Pasteur and the Centre National de la Recherche Scientifique (CNRS). The authors acknowledge the support of ARC, La Ligue Contre le Cancer (Equipe Labellisée EL2013-3 Mahieux) and ANRS.

REFERENCES

Afonso, P. V., Mekaouche, M., Mortreux, F., Toulza, F., Moriceau, A., Wattel, E., et al. (2010). Highly active antiretroviral treatment against STLV-1 infection combining reverse transcriptase and HDAC inhibitors. *Blood* 116, 3802–3808. doi: 10.1182/blood-2010-02-270751

Agosto, L. M., Zhong, P., Munro, J., and Mothes, W. (2014). Highly active antiretroviral therapies are effective against HIV-1 cell-to-cell transmission. *PLoS Pathog.* 10:e1003982. doi: 10.1371/journal.ppat.1003982

Alais, S., Dutartre, H., and Mahieux, R. (2017). Quantitative analysis of human T-lymphotropic virus type 1 (HTLV-1) infection using co-culture with Jurkat LTR-luciferase or Jurkat LTR-GFP reporter cells. *Methods Mol. Biol.* 1582, 47–55. doi: 10.1007/978-1-4939-6872-5_4

Alais, S., Mahieux, R., and Dutartre, H. (2015). Viral source-independent high susceptibility of dendritic cells to human T-cell leukemia virus type 1 infection compared to that of T lymphocytes. *J. Virol.* 89, 10580–10590. doi: 10.1128/JVI.01799-15

Aloia, A. L., Duffy, L., Pak, V., Lee, K. E., Sanchez-Martinez, S., Derse, D., et al. (2013). A reporter system for replication-competent gammaretroviruses: the inGluc-MLV-DERSE assay. *Gene Ther.* 20, 169–176. doi: 10.1038/gt.2012.18

Balestrieri, E., Forte, G., Matteucci, C., Mastino, A., and Macchi, B. (2002). Effect of lamivudine on transmission of human T-cell lymphotropic virus type 1 to adult peripheral blood mononuclear cells in vitro. *Antimicrob. Agents Chemother.* 46, 3080–3083. doi: 10.1128/AAC.46.9.3080-3083.2002

Balestrieri, E., Matteucci, C., Ascolani, A., Piperno, A., Romeo, R., Romeo, G., et al. (2008). Effect of phosphonated carbocyclic 2′-oxa-3′-aza-nucleoside on human T-cell leukemia virus type 1 infection in vitro. *Antimicrob. Agents Chemother.* 52, 54–64. doi: 10.1128/AAC.00470-07

Balestrieri, E., Sciortino, M. T., Mastino, A., and Macchi, B. (2005). Protective effect of the acyclic nucleoside phosphonate tenofovir toward human T-cell leukemia/lymphotropic virus type 1 infection of human peripheral blood mononuclear cells in vitro. *Antiviral Res.* 68, 154–162. doi: 10.1016/j.antiviral.2005.09.001

Balzarini, J., Holy, A., Jindrich, J., Naesens, L., Snoeck, R., Schols, D., et al. (1993). Differential antiherpesvirus and antiretrovirus effects of the (S) and (R) enantiomers of acyclic nucleoside phosphonates: potent and selective in vitro and in vivo antiretrovirus activities of (R)-9-(2-phosphonomethoxypropyl)-2,6-diaminopurine. *Antimicrob. Agents Chemother.* 37, 332–338. doi: 10.1128/AAC.37.2.332

Balzarini, J., Kang, G. J., Dalal, M., Herdewijn, P., De Clercq, E., Broder, S., et al. (1987). The anti-HTLV-III (anti-HIV) and cytotoxic activity of 2′,3′-didehydro-2′,3′-dideoxyribonucleosides: a comparison with their parental 2′,3′-dideoxyribonucleosides. *Mol. Pharmacol.* 32, 162–167.

Bangham, C. R., Araujo, A., Yamano, Y., and Taylor, G. P. (2015). HTLV-1-associated myelopathy/tropical spastic paraparesis. *Nat. Rev. Dis. Primers* 1:15012. doi: 10.1038/nrdp.2015.12

Bangham, C. R., Cook, L. B., and Melamed, A. (2014). HTLV-1 clonality in adult T-cell leukaemia and non-malignant HTLV-1 infection. *Semin. Cancer Biol.* 26, 89–98. doi: 10.1016/j.semcancer.2013.11.003

Barral, K., Weck, C., Payrot, N., Roux, L., Durafour, C., Zoulim, F., et al. (2011). Acyclic nucleoside thiophosphonates as potent inhibitors of HIV and HBV replication. *Eur. J. Med. Chem.* 46, 4281–4288. doi: 10.1016/j.ejmech.2011.06.034

Bazarbachi, A., Plumelle, Y., Carlos Ramos, J., Tortevoye, P., Otrock, Z., Taylor, G., et al. (2010). Meta-analysis on the use of zidovudine and interferon-alfa in adult T-cell leukemia/lymphoma showing improved survival in the leukemic subtypes. *J. Clin. Oncol.* 28, 4177–4183. doi: 10.1200/JCO.2010.28.0669

Boritz, E. A., Darko, S., Swaszek, L., Wolf, G., Wells, D., Wu, X., et al. (2016). Multiple origins of virus persistence during natural control of HIV infection. *Cell* 166, 1004–1015. doi: 10.1016/j.cell.2016.06.039

Ceccaldi, P. E., Delebecque, F., Prevost, M. C., Moris, A., Abastado, J. P., Gessain, A., et al. (2006). DC-SIGN facilitates fusion of dendritic cells with human T-cell leukemia virus type 1-infected cells. *J. Virol.* 80, 4771–4780. doi: 10.1128/JVI.80.10.4771-4780.2006

Chiacchio, U., Balestrieri, E., Macchi, B., Iannazzo, D., Piperno, A., Rescifina, A., et al. (2005). Synthesis of phosphonated carbocyclic 2′-oxa-3′-aza-nucleosides: novel inhibitors of reverse transcriptase. *J. Med. Chem.* 48, 1389–1394. doi: 10.1021/jm049399i

Chiacchio, U., Rescifina, A., Iannazzo, D., Piperno, A., Romeo, R., Borrello, L., et al. (2007). Phosphonated carbocyclic 2′-oxa-3′-azanucleosides as new antiretroviral agents. *J. Med. Chem.* 50, 3747–3750. doi: 10.1021/jm070285r

Cihlar, T., and Fordyce, M. (2016). Current status and prospects of HIV treatment. *Curr. Opin. Virol.* 18, 50–56. doi: 10.1016/j.coviro.2016.03.004

Cook, L. B., Melamed, A., Niederer, H., Valganon, M., Laydon, D., Foroni, L., et al. (2014). The role of HTLV-1 clonality, proviral structure, and genomic integration site in adult T-cell leukemia/lymphoma. *Blood* 123, 3925–3931. doi: 10.1182/blood-2014-02-553602

Daluge, S. M., Good, S. S., Faletto, M. B., Miller, W. H., St. Clair, M. H., Boone, L. R., et al. (1997). 1592U89, a novel carbocyclic nucleoside analog with potent, selective anti-human immunodeficiency virus activity. *Antimicrob. Agents Chemother.* 41, 1082–1093.

Datta, A., Bellon, M., Sinha-Datta, U., Bazarbachi, A., Lepelletier, Y., Canioni, D., et al. (2006). Persistent inhibition of telomerase reprograms adult T-cell leukemia to p53-dependent senescence. *Blood* 108, 1021–1029. doi: 10.1182/blood-2006-01-0067

De Clercq, E., and Holy, A. (2005). Acyclic nucleoside phosphonates: a key class of antiviral drugs. *Nat. Rev. Drug Discov.* 4, 928–940. doi: 10.1038/nrd1877

Derse, D., Hill, S. A., Lloyd, P. A., Chung, H., and Morse, B. A. (2001). Examining human T-lymphotropic virus type 1 infection and replication by cell-free infection with recombinant virus vectors. *J. Virol.* 75, 8461–8468. doi: 10.1128/JVI.75.18.8461-8468.2001

Deville-Bonne, D., El Amri, C., Meyer, P., Chen, Y., Agrofoglio, L. A., and Janin, J. (2010). Human and viral nucleoside/nucleotide kinases involved in antiviral drug activation: structural and catalytic properties. *Antiviral Res.* 86, 101–120. doi: 10.1016/j.antiviral.2010.02.001

El-Sabban, M. E., Nasr, R., Dbaibo, G., Hermine, O., Abboushi, N., Quignon, F., et al. (2000). Arsenic-interferon-alpha-triggered apoptosis in HTLV-I transformed cells is associated with tax down-regulation and reversal of NF-kappa B activation. *Blood* 96, 2849–2855.

Furman, P. A., Fyfe, J. A., St Clair, M. H., Weinhold, K. Rideout, J. L., Freeman, G. A., et al. (1986). Phosphorylation of 3′-azido-3′-deoxythymidine and selective interaction of the 5'-triphosphate with human immunodeficiency virus reverse transcriptase. *Proc. Natl. Acad. Sci. U.S.A.* 83, 8333–8337. doi: 10.1073/pnas.83.21.8333

Furmanski, P., Bourguignon, G. J., Bolles, C. S., Corombos, J. D., and Das, M. R. (1980). Inhibition by 2′,3′-dideoxythymidine of retroviral infection of mouse and human cells. *Cancer Lett.* 8, 307–315. doi: 10.1016/0304-3835(80)90146-9

Garcia-Lerma, J. G., Nidtha, S., and Heneine, W. (2001). Susceptibility of human T cell leukemia virus type 1 to reverse-transcriptase inhibitors: evidence for resistance to lamivudine. *J. Infect. Dis.* 184, 507–510. doi: 10.1086/322785

Gessain, A., and Cassar, O. (2012). Epidemiological aspects and world distribution of HTLV-1 infection. *Front. Microbiol.* 3:388. doi: 10.3389/fmicb.2012.00388

Gillet, N. A., Cook, L., Laydon, D. J., Hlela, C., Verdonck, K., Alvarez, C., et al. (2013). Strongyloidiasis and infective dermatitis alter human T lymphotropic virus-1 clonality in vivo. *PLoS Pathog.* 9:e1003263. doi: 10.1371/journal.ppat.1003263

Gillet, N. A., Malani, N., Melamed, A., Gormley, N., Carter, R., Bentley, D., et al. (2011). The host genomic environment of the provirus determines the abundance of HTLV-1-infected T-cell clones. *Blood* 117, 3113–3122. doi: 10.1182/blood-2010-10-312926

Gillet, N. A., Melamed, A., and Bangham, C. R. (2017). High-throughput mapping and clonal quantification of retroviral integration sites. *Methods Mol. Biol.* 1582, 127–141. doi: 10.1007/978-1-4939-6872-5_10

Harrington, W. J. Jr., Sheremata, W. A., Snodgrass, S. R., Emerson, S., Phillips, S., and Berger, J. R. (1991). Tropical spastic paraparesis/HTLV-1-associated myelopathy (TSP/HAM): treatment with an anabolic steroid danazol. *AIDS Res. Hum. Retroviruses* 7, 1031–1034. doi: 10.1089/aid.1991.7.1031

Hill, S. A., Lloyd, P. A., McDonald, S., Wykoff, J., and Derse, D. (2003). Susceptibility of human T cell leukemia virus type I to nucleoside reverse transcriptase inhibitors. *J. Infect. Dis.* 188, 424–427. doi: 10.1086/376531

Holy, A. (2003). Phosphonomethoxyalkyl analogs of nucleotides. *Curr. Pharm. Des.* 9, 2567–2592. doi: 10.2174/1381612033453668

Igakura, T., Stinchcombe, J. C., Goon, P. K., Taylor, G. P., Weber, J. N., Griffiths, G. M., et al. (2003). Spread of HTLV-I between lymphocytes by virus-induced polarization of the cytoskeleton. *Science* 299, 1713–1716. doi: 10.1126/science.1080115

Iwanaga, M., Watanabe, T., and Yamaguchi, K. (2012). Adult T-cell leukemia: a review of epidemiological evidence. *Front. Microbiol.* 3:322. doi: 10.3389/fmicb.2012.00322

Izumo, S., Goto, I., Itoyama, Y., Okajima, T., Watanabe, S., Kuroda, Y., et al. (1996). Interferon-alpha is effective in HTLV-I-associated myelopathy: a multicenter, randomized, double-blind, controlled trial. *Neurology* 46, 1016–1021. doi: 10.1212/WNL.46.4.1016

Jolly, C., and Sattentau, Q. J. (2005). Human immunodeficiency virus type 1 virological synapse formation in T cells requires lipid raft integrity. *J. Virol.* 79, 12088–12094. doi: 10.1128/JVI.79.18.12088-12094.2005

Kannian, P., Fernandez, S., Jones, K. S., and Green, P. L. (2013). Human T lymphotropic virus type 1 SU residue 195 plays a role in determining the preferential $CD4^+$ T cell immortalization/transformation tropism. *J. Virol.* 87, 9344–9352. doi: 10.1128/JVI.01079-13

Kulkarni, A., Mateus, M., Thinnes, C. C., McCullagh, J. S., Schofield, C. J., Taylor, G. P., et al. (2017). Glucose metabolism and oxygen availability govern reactivation of the latent human retrovirus HTLV-1. *Cell Chem. Biol.* 24, 1377.e3–1387.e3. doi: 10.1016/j.chembiol.2017.08.016

Lezin, A., Gillet, N., Olindo, S., Signate, A., Grandvaux, N., Verlaeten, O., et al. (2007). Histone deacetylase mediated transcriptional activation reduces proviral loads in HTLV-1 associated myelopathy/tropical spastic paraparesis patients. *Blood* 110, 3722–3728. doi: 10.1182/blood-2007-04-085076

Macchi, B., Balestrieri, E., Ascolani, A., Hilburn, S., Martin, F., Mastino, A., et al. (2011). Susceptibility of primary HTLV-1 isolates from patients with HTLV-1-associated myelopathy to reverse transcriptase inhibitors. *Viruses* 3, 469–483. doi: 10.3390/v3050469

Macchi, B., Faraoni, I., Zhang, J., Grelli, S., Favalli, C., Mastino, A., et al. (1997). AZT inhibits the transmission of human T cell leukaemia/lymphoma virus type I to adult peripheral blood mononuclear cells in vitro. *J. Gen. Virol.* 78(Pt 5), 1007–1016. doi: 10.1099/0022-1317-78-5-1007

Mahieux, R., and Hermine, O. (2005). In vivo and *in vitro* treatment of HTLV-1 and HTLV-2 infected cells with arsenic trioxide and interferon-alpha. *Leuk. Lymphoma* 46, 347–355. doi: 10.1080/10428190400019966

Maldarelli, F., Wu, X., Su, L., Simonetti, F. R., Shao, W., Hill, S., et al. (2014). HIV latency. Specific HIV integration sites are linked to clonal expansion and persistence of infected cells. *Science* 345, 179–183. doi: 10.1126/science.1254194

Mansky, L. M. (2000). In vivo analysis of human T-cell leukemia virus type 1 reverse transcription accuracy. *J. Virol.* 74, 9525–9531. doi: 10.1128/JVI.74.20.9525-9531.2000

Martin, F., Castro, H., Gabriel, C., Adonis, A., Fedina, A., Harrison, L., et al. (2012). Ciclosporin A proof of concept study in patients with active, progressive

HTLV-1 associated myelopathy/tropical spastic paraparesis. *PLoS Negl. Trop. Dis.* 6:e1675. doi: 10.1371/journal.pntd.0001675

Mazurov, D., Ilinskaya, A., Heidecker, G., Lloyd, P., and Derse, D. (2010). Quantitative comparison of HTLV-1 and HIV-1 cell-to-cell infection with new replication dependent vectors. *PLoS Pathog.* 6:e1000788. doi: 10.1371/journal.ppat.1000788

Melamed, A., Laydon, D. J., Al Khatib, H., Rowan, A. G., Taylor, G. P., and Bangham, C. R. (2015). HTLV-1 drives vigorous clonal expansion of infected $CD8^{+}$ T cells in natural infection. *Retrovirology* 12:91. doi: 10.1186/s12977-015-0221-1

Merle, P., and Trepo, C. (2001). Therapeutic management of hepatitis B-related cirrhosis. *J. Viral Hepat.* 8, 391–399. doi: 10.1046/j.1365-2893.2001.00299.x

Mitchell, M. S., Bodine, E. T., Hill, S., Princler, G., Lloyd, P., Mitsuya, H., et al. (2007). Phenotypic and genotypic comparisons of human T-cell leukemia virus type 1 reverse transcriptases from infected T-cell lines and patient samples. *J. Virol.* 81, 4422–4428. doi: 10.1128/JVI.02660-06

Mitsuya, H., and Broder, S. (1987). Strategies for antiviral therapy in AIDS. *Nature* 325, 773–778. doi: 10.1038/325773a0

Mitsuya, H., Weinhold, K. J., Furman, P. A., St Clair, M. H., Lehrman, S. N., Gallo, R. C., et al. (1985). 3′-Azido-3′-deoxythymidine (BW A509U): an antiviral agent that inhibits the infectivity and cytopathic effect of human T-lymphotropic virus type III/lymphadenopathy-associated virus in vitro. *Proc. Natl. Acad. Sci. U.S.A.* 82, 7096–7100. doi: 10.1073/pnas.82.20.7096

Mitsuya, H., Yarchoan, R., and Broder, S. (1990). Molecular targets for AIDS therapy. *Science* 249, 1533–1544. doi: 10.1126/science.1699273

Moens, B., Decanine, D., Menezes, S. M., Khouri, R., Silva-Santos, G., Lopez, G., et al. (2012). Ascorbic acid has superior ex vivo antiproliferative, cell death-inducing and immunomodulatory effects over IFN-alpha in HTLV-1-associated myelopathy. *PLoS Negl. Trop. Dis.* 6:e1729. doi: 10.1371/journal.pntd.0001729

Nakagawa, M., Nakahara, K., Maruyama, Y., Kawabata, M., Higuchi, I., Kubota, H., et al. (1996). Therapeutic trials in 200 patients with HTLV-I-associated myelopathy/ tropical spastic paraparesis. *J. Neurovirol.* 2, 345–355. doi: 10.3109/13550289609146899

Nasr, R., El Hajj, H., Kfoury, Y., de The, H., Hermine, O., and Bazarbachi, A. (2011). Controversies in targeted therapy of adult T cell leukemia/lymphoma: ON target or OFF target effects? *Viruses* 3, 750–769. doi: 10.3390/v3060750

Oh, U., Yamano, Y., Mora, C. A., Ohayon, J., Bagnato, F., Butman, J. A., et al. (2005). Interferon-beta1a therapy in human T-lymphotropic virus type I-associated neurologic disease. *Ann. Neurol.* 57, 526–534. doi: 10.1002/ana.20429

Olindo, S., Belrose, G., Gillet, N., Rodriguez, S., Boxus, M., Verlaeten, O., et al. (2011). Safety of long-term treatment of HAM/TSP patients with valproic acid. *Blood* 118, 6306–6309. doi: 10.1182/blood-2011-04-349910

Olindo, S., Cabre, P., Lezin, A., Merle, H., Saint-Vil, M., Signate, A., et al. (2006). Natural history of human T-lymphotropic virus 1-associated myelopathy: a 14-year follow-up study. *Arch. Neurol.* 63, 1560–1566. doi: 10.1001/archneur.63.11.1560

Pais-Correia, A. M., Sachse, M., Guadagnini, S., Robbiati, V., Lasserre, R., Gessain, A., et al. (2010). Biofilm-like extracellular viral assemblies mediate HTLV-1 cell-to-cell transmission at virological synapses. *Nat. Med.* 16, 83–89. doi: 10.1038/nm.2065

Palella, F. J. Jr., Delaney, K. M., Moorman, A. C., Loveless, M. O., Fuhrer, J., Satten, G. A., et al. (1998). Declining morbidity and mortality among patients with advanced human immunodeficiency virus infection. HIV outpatient study investigators. *N. Engl. J. Med.* 338, 853–860. doi: 10.1056/NEJM199803263381301

Perno, C. F., Yarchoan, R., Cooney, D. A., Hartman, N. R., Gartner, S., Popovic, M., et al. (1988). Inhibition of human immunodeficiency virus (HIV-1/HTLV-IIIBa-L) replication in fresh and cultured human peripheral blood monocytes/macrophages by azidothymidine and related 2′,3′-dideoxynucleosides. *J. Exp. Med.* 168, 1111–1125. doi: 10.1084/jem.168.3.1111

Priet, S., Roux, L., Saez-Ayala, M., Ferron, F., Canard, B., and Alvarez, K. (2015). Enzymatic synthesis of acyclic nucleoside thiophosphonate diphosphates: effect of the alpha-phosphorus configuration on HIV-1 RT activity. *Antiviral Res.* 117, 122–131. doi: 10.1016/j.antiviral.2015.03.003

Rafatpanah, H., Rezaee, A., Etemadi, M. M., Hosseini, R. F., Khorram, B., Afsahr, L., et al. (2012). The impact of interferon-alpha treatment on clinical and immunovirological aspects of HTLV-1-associated myelopathy in northeast of Iran. *J. Neuroimmunol.* 250, 87–93. doi: 10.1016/j.jneuroim.2012.05.004

Rahman, S., Manuel, S. L., Khan, Z. K., Wigdahl, B., Acheampong, E., Tangy, F., et al. (2010). Depletion of dendritic cells enhances susceptibility to cell-free infection of human T cell leukemia virus type 1 in CD11c-diphtheria toxin receptor transgenic mice. *J. Immunol.* 184, 5553–5561. doi: 10.4049/jimmunol.0903226

Ray, A. S., Fordyce, M. W., and Hitchcock, M. J. (2016). Tenofovir alafenamide: a novel prodrug of tenofovir for the treatment of human immunodeficiency virus. *Antiviral Res.* 125, 63–70. doi: 10.1016/j.antiviral.2015.11.009

Roux, L., Priet, S., Payrot, N., Weck, C., Fournier, M., Zoulim, F., et al. (2013). Ester prodrugs of acyclic nucleoside thiophosphonates compared to phosphonates: synthesis, antiviral activity and decomposition study. *Eur. J. Med. Chem.* 63, 869–881. doi: 10.1016/j.ejmech.2013.02.039

Rudnicka, D., Feldmann, J., Porrot, F., Wietgrefe, S., Guadagnini, S., Prevost, M. C., et al. (2009). Simultaneous cell-to-cell transmission of human immunodeficiency virus to multiple targets through polysynapses. *J. Virol.* 83, 6234–6246. doi: 10.1128/JVI.00282-09

Saito, M., Nakagawa, M., Kaseda, S., Matsuzaki, T., Jonosono, M., Eiraku, N., et al. (2004). Decreased human T lymphotropic virus type I (HTLV-I) provirus load and alteration in T cell phenotype after interferon-alpha therapy for HTLV-I-associated myelopathy/tropical spastic paraparesis. *J. Infect. Dis.* 189, 29–40. doi: 10.1086/380101

Sax, P. E., Wohl, D., Yin, M. T., Post, F., DeJesus, E., Saag, M., et al. (2015). Tenofovir alafenamide versus tenofovir disoproxil fumarate, coformulated with elvitegravir, cobicistat, and emtricitabine, for initial treatment of HIV-1 infection: two randomised, double-blind, phase 3, non-inferiority trials. *Lancet* 385, 2606–2615. doi: 10.1016/S0140-6736(15)60616-X

Sharma, P. L., Nurpeisov, V., Hernandez-Santiago, B., Beltran, T., and Schinazi, R. F. (2004). Nucleoside inhibitors of human immunodeficiency virus type 1 reverse transcriptase. *Curr. Top. Med. Chem.* 4, 895–919. doi: 10.2174/1568026043388484

Sherer, N. M., Lehmann, M. J., Jimenez-Soto, L. F., Horensavitz, C., Pypaert, M., and Mothes, W. (2007). Retroviruses can establish filopodial bridges for efficient cell-to-cell transmission. *Nat. Cell Biol.* 9, 310–315. doi: 10.1038/ncb1544

Sigal, A., Kim, J. T., Balazs, A. B., Dekel, E., Mayo, A., Milo, R., et al. (2011). Cell-to-cell spread of HIV permits ongoing replication despite antiretroviral therapy. *Nature* 477, 95–98. doi: 10.1038/nature10347

Sowinski, S., Jolly, C., Berninghausen, O., Purbhoo, M. A., Chauveau, A., Kohler, K., et al. (2008). Membrane nanotubes physically connect T cells over long distances presenting a novel route for HIV-1 transmission. *Nat. Cell Biol.* 10, 211–219. doi: 10.1038/ncb1682

Taylor, B. S., Olender, S. A., Tieu, H. V., and Wilkin, T. J. (2016). CROI 2016: advances in antiretroviral therapy. *Top. Antivir. Med.* 24, 59–81.

Taylor, G. P., Goon, P., Furukawa, Y., Green, H., Barfield, A., Mosley, A., et al. (2006). Zidovudine plus lamivudine in human T-lymphotropic virus type-I-associated myelopathy: a randomised trial. *Retrovirology* 3:63. doi: 10.1186/1742-4690-3-63

Taylor, G. P., and Matsuoka, M. (2005). Natural history of adult T-cell leukemia/lymphoma and approaches to therapy. *Oncogene* 24, 6047–6057. doi: 10.1038/sj.onc.1208979

Thoulouze, M. I., and Alcover, A. (2011). Can viruses form biofilms? *Trends Microbiol.* 19, 257–262. doi: 10.1016/j.tim.2011.03.002

Turpin, J., Alais, S., Marcais, A., Bruneau, J., Melamed, A., Gadot, N., et al. (2017). Whole body clonality analysis in an aggressive STLV-1 associated leukemia (ATLL) reveals an unexpected clonal complexity. *Cancer Lett.* 389, 78–85. doi: 10.1016/j.canlet.2016.12.022

Utsunomiya, A., Choi, I., Chihara, D., and Seto, M. (2015). Recent advances in the treatment of adult T-cell leukemia-lymphomas. *Cancer Sci.* 106, 344–351. doi: 10.1111/cas.12617

Utsunomiya, A., Miyazaki, Y., Takatsuka, Y., Hanada, S., Uozumi, K., Yashiki, S., et al. (2001). Improved outcome of adult T cell leukemia/lymphoma with allogeneic hematopoietic stem cell transplantation. *Bone Marrow Transplant.* 27, 15–20. doi: 10.1038/sj.bmt.1702731

Valeri, V. W., Hryniewicz, A., Andresen, V., Jones, K., Fenizia, C., Bialuk, I., et al. (2010). Requirement of the human T-cell leukemia virus p12 and p30 products for infectivity of human dendritic cells and macaques but not rabbits. *Blood* 116, 3809–3817. doi: 10.1182/blood-2010-05-284141

Watanabe, T. (2017). Adult T-cell leukemia: molecular basis for clonal expansion and transformation of HTLV-1-infected T cells. *Blood* 129, 1071–1081. doi: 10.1182/blood-2016-09-692574

Wattel, E., Vartanian, J. P., Pannetier, C., and Wain-Hobson, S. (1995). Clonal expansion of human T-cell leukemia virus type I-infected cells in asymptomatic and symptomatic carriers without malignancy. *J. Virol.* 69, 2863–2868.

Willems, L., Hasegawa, H., Accolla, R., Bangham, C., Bazarbachi, A., Bertazzoni, U., et al. (2017). Reducing the global burden of HTLV-1 infection: an agenda for research and action. *Antiviral Res.* 137, 41–48. doi: 10.1016/j.antiviral.2016.10.015

Yakova, M., Lezin, A., Dantin, F., Lagathu, G., Olindo, S., Jean-Baptiste, G., et al. (2005). Increased proviral load in HTLV-1-infected patients with rheumatoid arthritis or connective tissue disease. *Retrovirology* 2:4. doi: 10.1186/1742-4690-2-4

Yamamoto, S., Folks, T. M., and Heneine, W. (1996). Highly sensitive qualitative and quantitative detection of reverse transcriptase activity: optimization, validation, and comparative analysis with other detection systems. *J. Virol. Methods* 61, 135–143. doi: 10.1016/0166-0934(96)02078-2

Yamano, Y., and Sato, T. (2012). Clinical pathophysiology of human T-lymphotropic virus-type 1-associated myelopathy/tropical spastic paraparesis. *Front. Microbiol.* 3:389. doi: 10.3389/fmicb.2012.00389

Yarchoan, R., Klecker, R. W., Weinhold, K. J., Markham, P. D., Lyerly, H. K., Durack, D. T., et al. (1986). Administration of 3′-azido-3′-deoxythymidine, an inhibitor of HTLV-III/LAV replication, to patients with AIDS or AIDS-related complex. *Lancet* 1, 575–580. doi: 10.1016/S0140-6736(86)92808-4

Zhang, J., Balestrieri, E., Grelli, S., Matteucci, C., Pagnini, V., D'Agostini, C., et al. (2001). Efficacy of 3′-azido 3′deoxythymidine (AZT) in preventing HTLV-1 transmission to human cord blood mononuclear cells. *Virus Res.* 78, 67–78. doi: 10.1016/S0168-1702(01)00285-4

Zhong, P., Agosto, L. M., Ilinskaya, A., Dorjbal, B., Truong, R., Derse, D., et al. (2013). Cell-to-cell transmission can overcome multiple donor and target cell barriers imposed on cell-free HIV. *PLoS One* 8:e53138. doi: 10.1371/journal.pone.0053138

18

HTLV Deregulation of the NF-κB Pathway: An Update on Tax and Antisense Proteins Role

*Stefania Fochi, Simona Mutascio, Umberto Bertazzoni, Donato Zipeto and Maria G. Romanelli**

Department of Neurosciences, Biomedicine and Movement Sciences, University of Verona, Verona, Italy

Correspondence:
Maria G. Romanelli
mariagrazia.romanelli@univr.it

Human T-cell lymphotropic virus type 1 (HTLV-1) is the causative agent of adult T-cell leukemia (ATL), an aggressive CD4$^+$/CD25$^+$ T-cell malignancy and of a severe neurodegenerative disease, HTLV-1 associated myelopathy/tropical spastic paraparesis (HAM/TSP). The chronic activation or deregulation of the canonical and non-canonical nuclear factor kappa B (NF-κB) pathways play a crucial role in tumorigenesis. The HTLV-1 Tax-1 oncoprotein is a potent activator of the NF-κB transcription factors and the NF-κB response is required for promoting the development of HTLV-1 transformed cell lines. The homologous retrovirus HTLV-2, which also expresses a Tax-2 transforming protein, is not associated with ATL. In this review, we provide an updated synopsis of the role of Tax-1 in the deregulation of the NF-κB pathway, highlighting the differences with the homologous Tax-2. Special emphasis is directed toward the understanding of the molecular mechanisms involved in NF-κB activation resulting from Tax interaction with host factors affecting several cellular processes, such as cell cycle, apoptosis, senescence, cell proliferation, autophagy, and post-translational modifications. We also discuss the current knowledge on the role of the antisense viral protein HBZ in down-regulating the NF-κB activation induced by Tax, and its implication in cellular senescence. In addition, we review the recent studies on the mechanism of HBZ-mediated inhibition of NF-κB activity as compared to that exerted by the HTLV-2 antisense protein, APH-2. Finally, we discuss recent advances aimed at understanding the role exerted in the development of ATL by the perturbation of NF-κB pathway by viral regulatory proteins.

Keywords: HTLV, NF-κB, Tax, HBZ, APH-2, adult T-cell leukemia, cell proliferation, apoptosis

INTRODUCTION

Human T-cell lymphotropic/leukemia virus type 1 (HTLV-1) is the etiological agent of adult T-cell leukemia (ATL), a malignancy of CD4$^+$/CD25$^+$ T cells and of a chronic inflammatory disease called HTLV-1 associated myelopathy/tropical spastic paraparesis (HAM/TSP) (Poiesz et al., 1980; Hinuma et al., 1981; Gessain et al., 1985; Gallo et al., 2017). It is estimated that at least 20 million people worldwide are infected with HTLV-1 (Gessain and Cassar, 2012; Willems et al., 2017) and approximately 5% of HTLV-1 carriers develop ATL after a latency of 20–50 years from infection (Zhang et al., 2017). HTLV-1 provirus encodes, among others, a regulatory protein, Tax and an accessory antisense strand product HTLV-1 bZip protein (HBZ), which are pivotal factors in

HTLV-1 pathogenesis (Yasuma et al., 2016). Tax is a transcriptional activator of the viral long terminal repeat (LTR) with the capability to unsettle several cellular signal transduction pathways. HBZ is an inhibitor of 5′ LTR Tax-1 transactivation and is required for viral persistence (Barbeau et al., 2013). HBZ is a potent viral oncoprotein which plays an important role in deregulating several cellular processes in concerted action with Tax, affecting cell proliferation, apoptosis, autophagy, and immune escape (Zhao, 2016). Both these viral regulatory proteins promote T-cell proliferation. However, the exact mechanism underlying their role in inducing cell proliferation is still not clearly understood. The genetically related HTLV type 2 virus, although its association with ATL has not been established, encodes a homolog Tax-2 regulatory protein that induces T-cell proliferation *in vitro* and an antisense protein, named antisense protein HTLV-2 (APH-2) that, unlike HBZ, is dispensable for HTLV-2 infection and persistence (Yin et al., 2012). Their structural properties are shown in **Figures 1A,B.** Comparative studies between HTLV-1 and HTLV-2 have contributed to highlight differences in the virus-host interaction that may have key roles in tumorigenesis (Higuchi and Fujii, 2009; Bertazzoni et al., 2011; Romanelli et al., 2013).

Persistent activation of NF-κB by Tax is a key event for the T-cell transformation and development of ATL (Qu and Xiao, 2011; Zhang et al., 2017). Accumulating evidence indicates that the HTLVs have evolved specific strategies mediated by Tax and antisense proteins to deregulate NF-κB signaling pathways. While HBZ is consistently expressed in all ATL cells, Tax is not expressed in approximately 60% of them, even though the HTLV-1 proviral genome is integrated and NF-κB is constitutively activated (Zhao, 2016). This suggests that additional factors contribute to sustain the persistent activation of NF-κB, in the absence of Tax, in ATL cells (Matsuoka and Jeang, 2007). The alteration of the NF-κB signaling pathway could also be involved in the inflammatory state observed in HAM/TSP (Peloponese et al., 2006). An interesting aspect of Tax and HBZ functions is their opposite effect on the regulation of cellular signaling pathways (Zhao and Matsuoka, 2012; Ma et al., 2016) as further discussed here.

In this review, we summarize the recent advances in understanding the molecular mechanisms involved in NF-κB deregulation, mediated by Tax and antisense proteins, through the interaction with host factors and their roles in cell survival and proliferation.

TAX-MEDIATED NF-κB ACTIVATION

Two distinct pathways lead to NF-κB activation, known as the canonical and the non-canonical pathways that involve different upstream, intermediate, and effector factors. A common step of both pathways is the activation of a complex that contains a serine-specific IκB kinase (IKK) composed by two catalytic kinase subunits, IKKα and IKKβ, and the regulatory non-enzymatic scaffold protein NEMO (known as IKKγ). In the canonical pathway, adaptor proteins (TRAFs) are recruited to the cytoplasmic domain of the cell membrane tumor necrosis factor receptor (TNF-R) and activate the IKK complex thus inducing the phosphorylation of IκB inhibitor and the seclusion of NF-κB precursors within the cytoplasm (Sun, 2017). This phenomenon leads to IκB degradation and nuclear translocation of the p50/RelA transcriptional effectors. At variance with the canonical pathway, the non-canonical one involves an IKK complex that does not contain NEMO, but two IKKα subunits. The NF-κB-inducing kinase (NIK) activates the IKK complex, leading to p100 processing and the final release in the nucleus of p52/RelB active heterodimer (Durand and Baldwin, 2017).

Based on the study of the molecular mechanisms of NF-κB activation driven by Tax-1, two relevant aspects emerged: the recruitment of Tax in cellular protein complexes (Bertazzoni et al., 2011; Qu and Xiao, 2011) and their post-translational modifications (Lavorgna and Harhaj, 2014). Studies comparing Tax-1 and Tax-2 have highlighted relevant differences in their activation of the NF-κB pathway as a result of protein interaction: both proteins activate the classical pathway, but only Tax-1 activates the non-canonical one; Tax-1, unlike Tax-2, triggers the activation of the non-canonical pathway recruiting NEMO and IKKα to p100, promoting the processing of p100 to p52 (Xiao et al., 2001; Higuchi et al., 2007; Shoji et al., 2009); both Tax proteins interact with TAB2 and NEMO/IKKγ stimulating the translocation of the p50/RelA heterodimers into the nucleus, but only Tax-1 interacts with TRAF6, an E3 ligase that triggers the ubiquitination and activation of the downstream NF-κB signaling cascade (Avesani et al., 2010; Journo et al., 2013). Furthermore, only Tax-1 interacts with the p52/p100 and RelB factors of the non-canonical pathway, inducing the expression of OX40L, a T-cell co-stimulatory molecule of the tumor necrosis factor family implicated in the adaptive immunity (Motai et al., 2016).

We have recently shown that Tax-1 and Tax-2 form complexes with two homologous non-canonical IκB kinases, IKKε and TBK1, which are not component of IKK complexes, but are implicated in the activation of NF-κB, STAT3 and induction of IFNα (Shen and Hahn, 2011; Diani et al., 2015). An additional study demonstrating the presence of Tax and TBK1 in lipid raft microdomains along with canonical IκB supports the role of Tax-1 as a promoter of the molecular crosstalk between the canonical IKKs and additional signaling pathways involved in cell survival and proliferation (Zhang et al., 2016). Interestingly, it has also been reported that Tax-1 forms complexes with the ubiquitin-conjugating enzyme Ubc13, NEMO, Tax1 binding protein1 (TAX1BP) and NRP/Optineurin in the membrane lipid rafts microdomain. In these complexes, the cell adhesion molecule 1 (CADM1) acts as a molecular scaffold recruiting Tax-1 (Pujari et al., 2015). This interaction contributes to the activation of the IKK complex and the inactivation of the NF-κB negative regulator A20 enzyme, thus maintaining a persistent NF-κB activation. An additional consequence of the Tax reorganization of the component of the lipid raft is the deregulation of autophagy. Tax-1, in fact, participates to the connection of the IKK complex to the autophagy molecular complexes by interacting directly with Beclin1 and PI3KC3 and contributing to the assembly of autophagosomes (Ren et al., 2012, 2015; Chen et al., 2015). Tax-1 induction of NF-κB also increases the expression of inhibitors of apoptosis, such as the anti-apoptotic

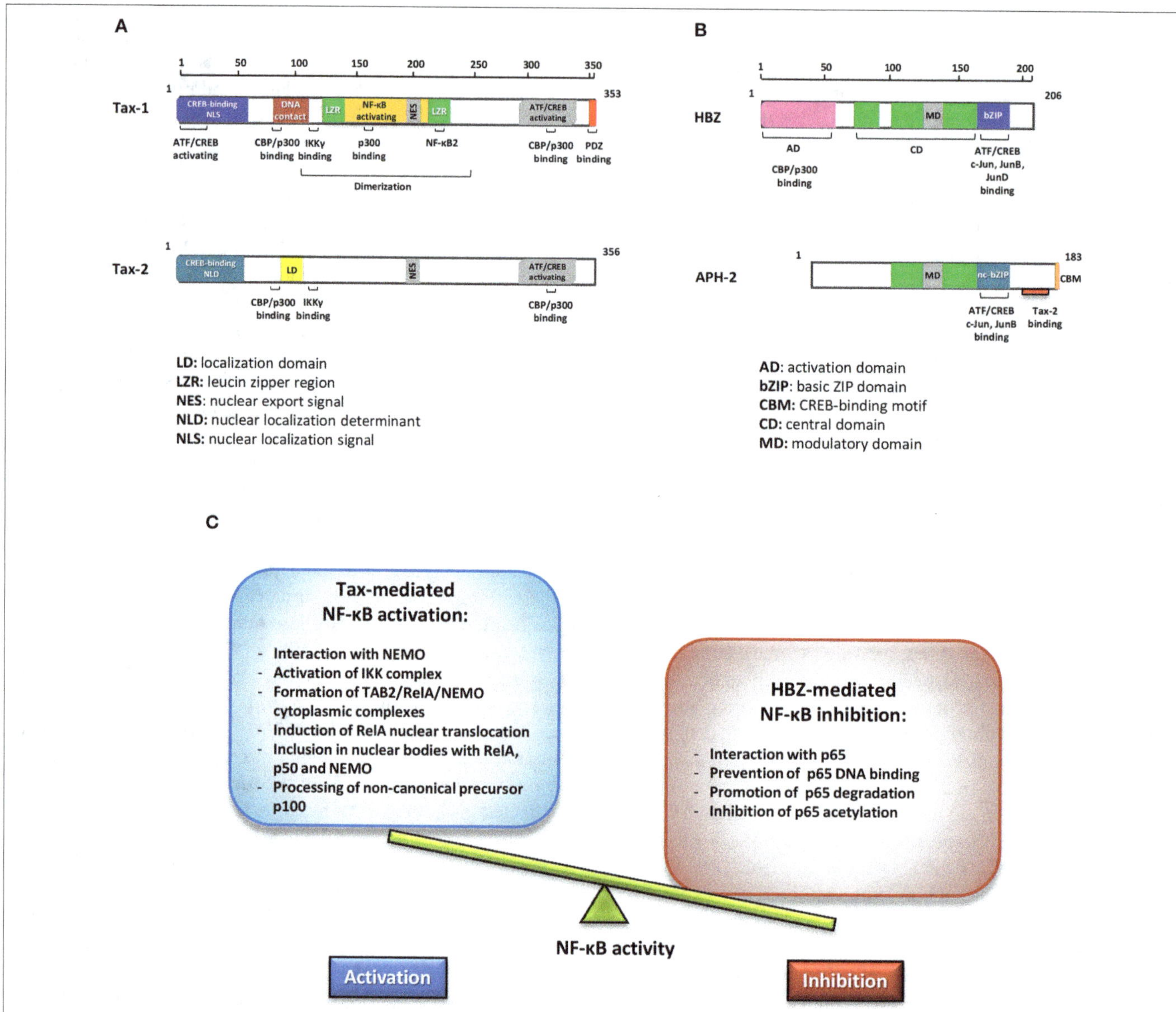

FIGURE 1 | Schematic representations of Tax-1 and Tax-2 **(A)** and of HBZ and APH-2 functional domains **(B)**. Different colors highlight structural differences. **(C)** Effects of Tax and HBZ on NF-κB. Schematic representation of the opposite effects in regulating NF-κB activity exerted by Tax-1 and HBZ.

c-Flip gene, and of genes involved in cell cycle progression, including cyclin D2, cyclin E, E2F1, CDK2, CDK4, and CDK6 (Wang et al., 2014; Bangham and Matsuoka, 2017; Karimi et al., 2017).

It has been recently reported that Tax-activation of NF-κB can be suppressed by host factors. Among them, the transcriptional regulator of the major histocompatibility complex class II (CIITA) impairs the nuclear translocation of RelA and directly interacts with Tax-1/RelA in nuclear bodies, preventing Tax-1 mediated activation of NF-κB-responsive promoters (Forlani et al., 2013, 2016). In addition, the apoptotic regulator Bcl-3 has been demonstrated to inhibit RelA nuclear translocation and its DNA binding activity, resulting in a downregulation of Tax-induced NF-κB activation (Wang et al., 2013). The decrease in Tax-NF-κB activation could also be due to Tax proteasomal degradation induced by host factor interaction (Lavorgna and Harhaj, 2014). Tax-1 interaction with the molecular chaperone HSP90 was shown to protect Tax from proteasomal degradation (Gao and Harhaj, 2013), whereas the interaction with PDLIM2 (PDZ-LIM domain-containing protein) within the nuclear matrix induces its polyubiquitination-mediated proteasomal degradation (Yan et al., 2009; Fu et al., 2010). Furthermore, two tumor suppressor genes, MDFIC and MDF, have been recently identified as Tax-1 interactors that alter its subcellular distribution and stability, reducing Tax-dependent activation of NF-κB (Kusano et al., 2015).

The second major mechanism required for Tax-1 and Tax-2 NF-κB activation is the process of post-translational modification, which includes ubiquitination, SUMOylation and phosphorylation. It is well established that Tax phosphorylation is required for its nuclear translocation and stabilization in the nuclear bodies containing RelA (Bex et al., 1999; Turci et al., 2006). The requirements of ubiquitination and SUMOylation are more complex to define. Both the E2 enzyme Ubc13 and the E3 Ring Finger Protein 8 (RNF8) promote Tax K63-linked polyubiquitination and are essential for the activation of the IKK complex (Shembade et al., 2007; Ho et al., 2015). Other proteins, including E3 ubiquitin ligases, TRAF2, 5, or 6, can potentiate Tax polyubiquitination (Yu et al., 2008). SUMOylated Tax has been demonstrated to bind p300, RelA and NEMO in nuclear bodies (Nasr et al., 2006). In addition, SUMOylation of Tax may be involved in the regulation of Tax stability and NF-κB pathway activation (Kfoury et al., 2011). We have described that SUMOylation and ubiquitination influence Tax proteins intracellular localization, as well as the interaction with NF-κB factors and their transactivating activity (Turci et al., 2012). However, the role of Tax SUMOylation in NF-κB activation remains controversial, given that Tax-induced IKK activation has been shown to correlate with the level of Tax ubiquitination, but not with Tax SUMOylation (Bonnet et al., 2012; Pène et al., 2014). A recent study suggests that Tax itself may function as an ubiquitin E3 ligase that, in association with the ubiquitin-conjugating enzyme E2, catalyzes the assembly of mixed polyUb chains (Wang et al., 2016). However, a more recent study does not attribute to Tax an E3 ligase activity, while suggesting that multivalent interactions between NEMO proteins and Ub-chains can lead to the formation of a macromolecular Taxisome and consequently to the activation of the IKK complex (Shibata et al., 2017).

An additional mechanism that operates within the cells to maintain the NF-κB activation induced by Tax-1 is the positive feedback loop derived by NF-κB target genes. A recent report describes that the over-expression of the early growth response protein 1 (EGR1) induced by Tax-1 activation of NF-κB, results in the stabilization of EGR1 by direct interaction with Tax and nuclear translocation of p65, enhancing NF-κB activation (Huang et al., 2017). A similar positive loop is fostered by the overexpression of the interleukin receptor IL-17RB. Tax-1 promotes the expression of IL-17RB by NF-κB activation and establishes an IL-17RB-NF-κB feed-forward autocrine loop that maintains persistent NF-κB activation (Lavorgna et al., 2014).

TAX AND HBZ INTERPLAY ON NF-κB DEREGULATION

HBZ can promote viral latency by antagonizing many of the activities mediated by Tax. HBZ inhibits the activation of the HTLV-1 5′ LTR preventing the formation of the Tax transactivation complex (Gaudray et al., 2002; Clerc et al., 2008). The activation of the classical NF-κB pathway by Tax is inhibited selectively by HBZ expression (Zhao et al., 2009; Wurm et al., 2012). This inhibition is connected to the following properties of HBZ as shown in **Figure 1C**: (a) the interaction with p65; (b) the inhibition of p65 DNA binding; (c) the enhanced degradation of p65 through PDLIM2 E3 ubiquitin ligase; (d) the reduction of p65 acetylation. All these processes result in the reduction of the expression of several NF-κB target genes. A typical example is the *cyclin D1* promoter gene, an essential regulator of the G1/S phase transition of the cell cycle that is overexpressed by Tax-mediated NF-κB activation, while it is downregulated by HBZ interaction with p65 (Ma et al., 2017).

The HBZ inhibition of NF-κB has been proposed to be a critical step in the oligoclonal expansion of HTLV-1-infected cells by downregulating the senescence process (Giam and Semmes, 2016). NF-κB hyper-activation induced by Tax leads to the over-expression of the cyclin-dependent kinase inhibitors, p21 and p27, thus promoting an arrest of cell proliferation that triggers senescence. The proposed model envisages that in HTLV-1 infected cells, in which the p21/p27 functions is impaired, the HBZ downregulation of NF-κB may contrast the senescence induced by Tax hence promoting the expansion of the infected cells (Kuo and Giam, 2006; Zhang et al., 2009; Zhi et al., 2011).

In contrast to HBZ, the HTLV-2 homolog protein APH-2 is dispensable for HTLV infection and persistence and does not promote T-cell proliferation *in vitro* (Yin et al., 2012; Barbeau et al., 2013). In addition, APH-2 expression correlates with the proviral load in HTLV-2 infected subjects and, contrary to HBZ, does not promote lymphocytosis (Saito et al., 2009; Douceron et al., 2012). Of note, HBZ and APH-2 also diverge in the interaction with Tax, since HBZ does not bind Tax-1, whereas Tax-2 interacts with APH-2 (Marban et al., 2012). A recent study has shown that despite HBZ and APH-2 interact with p65/RelA and repress its transactivation activity in transfected cells, they diverge in the induction of p65 degradation since this is not detected in the presence of APH-2 (Panfil et al., 2016). This different effect suggests that the two proteins may adopt different mechanisms to interfere with NF-κB activation. The differences between regulatory proteins of HTLV-1 and HTLV-2 in deregulating NF-κB are outlined in **Table 1**.

ROLE OF TAX AND HBZ IN ATL DEVELOPMENT

The opposite functions of Tax and HBZ in the regulation of signaling pathways and their effects in survival and proliferation appear as relevant steps during HTLV-1 cellular transformation and tumorigenesis (Giam and Semmes, 2016; Bangham and Matsuoka, 2017; Zhang et al., 2017). The absence of Tax expression in the late stages of the infection is linked to *tax* gene mutations and DNA methylation of the 5′ LTR provirus (Furukawa et al., 2001; Koiwa et al., 2002). On the opposite, the 3′ LTR negative strand remains intact and non-methylated, allowing HBZ to be systematically expressed in ATL cells (Taniguchi et al., 2005; Miyazaki et al., 2007). Unlike HBZ, Tax-1 is highly immunogenic and its inactivation may represent a fundamental strategy to evade the host immune system, a critical step in ATL development (Kogure and Kataoka, 2017). HBZ, like Tax-1, deregulates cell proliferation by targeting key factors implicated in cell survival. HBZ, in fact, binds to ATF3/p53 complexes and inhibits the p53 expression induced by ATF3, thus

TABLE 1 | Comparative effects of HTLV regulatory proteins on NF-κB pathways.

	Tax-1	Tax-2	References
Canonical NF-κB transactivation	+	+	Sun et al., 1994
Non-canonical NF-κB transactivation	+	−	Higuchi et al., 2007
NF-κB transactivation (lipid raft translocation of IKK)	+	−	Huang et al., 2009
Interaction with p100/p52	+	−	Shoji et al., 2009
Interaction with p65	+	+	Zhao et al., 2009; Panfil et al., 2016
	HBZ	**APH-2**	**References**
Canonical NF-κB inhibition	+	+	Zhao et al., 2009; Panfil et al., 2016
Non-canonical NF-κB inhibition	−	nd	Zhao et al., 2009
Interaction with p65	+	+	Zhao et al., 2009; Panfil et al., 2016
Inhibition of Tax-mediated transactivation of NF-κB	+	nd	Zhao et al., 2009
Binding to Tax	−	+	Marban et al., 2012
Inhibition of p65 DNA binding capacity	+	nd	Zhao et al., 2009
p65 degradation	+	−	Panfil et al., 2016
Inhibition of p65 acetylation	+	nd	Wurm et al., 2012

nd, Not determined.

promoting ATL cells proliferation (Hagiya et al., 2011). HBZ also induces the expression of the anti-apoptotic genes *BCL2* and *Flip*, interacting with C/EBPα and deregulating the C/EBP signaling (Zhao et al., 2013). Both Tax-1 and HBZ are involved in the inhibition of the tumor suppressor p53. In particular, Tax inhibits p53 activity through the p65 subunit of NF-κB or by sequestering p300/CBP from p53 (Ariumi et al., 2000; Karimi et al., 2017). Recent studies revealed that HBZ, by binding p300/CBP, inhibits p53 acetylation and decreases the p53 activity (Wright et al., 2016).

The selectivity of HBZ in inhibiting the classical NF-κB pathway opens an interesting area of investigation on the role of the non-canonical NF-κB pathway in tumorigenesis. During ATL development, HBZ might downmodulate the classical NF-κB pathway more efficiently when Tax expression is silenced, leading to predominant activation of the alternative pathway (Zhao et al., 2009). It has also been demonstrated that freshly isolated ATL cells display high expression levels of NIK, persistent phosphorylation of IκBα, aberrant processing of p52, and nuclear translocation of p50, p52, and RelB, despite the absence of Tax-1 expression (Chan and Greene, 2012).

Genetic and epigenetic alterations, including miRNAs expression profile, have been intensively investigated in the genome of ATL patients (Yeung et al., 2008; Bellon et al., 2009; Yamagishi and Watanabe, 2012; Watanabe, 2017). It has been proposed that the genomic instability may derive from Tax inhibition of DNA double-strand break repair and induction of micronuclei formation. ATL cells are characterized by frequent gain-of-function alterations of genes involved in the T-cell receptor/NF-κB signaling pathway, such as PLCG1, PKCB, and CARD11 or loss-of-function mutations in upstream factors, such as TRAF3 (Cook et al., 2017; Kogure and Kataoka, 2017). Mutations or intragenic deletions of these genes result in NF-κB induction in the absence of Tax-1. A progressive epigenetic downregulation of *miR-31* has been demonstrated in ATL (Fujikawa et al., 2016). Of note, miR-31 negatively regulates the expression of NIK and miR-31 loss in ATL triggers the persistent activation of NF-κB, inducing apoptosis resistance and contributing to the abnormal proliferation of cancer cells (Yamagishi et al., 2012). In addition, Fujikawa et al. (2016) showed that Tax-mediated NF-κB activation induces the over-expression of the histone-lysine methyltransferase, EZH2, leading to host epigenetic machinery deregulation. It has been proposed that EZH2 may contribute to NF-κB activation through miR-31 silencing and consequently NIK induction, in a positive feedback loop (Sasaki et al., 2011; Fujikawa et al., 2016). Genetic mutations have been also suggested to cause IL-17RB overexpression which triggers classical NF-κB activation by an autocrine-loop in a subset of Tax-negative ATL cell lines (Lavorgna et al., 2014).

CONCLUSIONS AND PERSPECTIVES

HTLV-1 appears to benefit from the antagonistic functions of Tax and HBZ in the deregulation of cellular signaling pathways, resulting in the loss of control of many biological processes such as proliferation and survival of HTLV-1-infected cells. The interplay between Tax and HBZ on NF-κB regulation has a prominent role in viral persistence in ATL cells, thus contributing to leukemic transformation. The intensive studies conducted in recent years aimed at understanding the effect of Tax constitutive activation and HBZ inhibition of NF-κB have contributed to further elucidate the molecular mechanism of NF-κB activation. However, several open questions about its functional role in ATL development still need to be addressed: the exact role of the persistent NF-κB activation in ATL cells; the contribution to tumorigenesis of the alternative pathway activation; the role of the different mechanisms that are adopted by HBZ and APH-2 to interfere with NF-κB activation; the dynamic organization of lipid raft complexes in HTLV-1 infected cell. It is hoped that the application of the CRISPR/Cas9 genome editing new technique will offer a useful tool to investigate the requirement of specific interactions of Tax and HBZ with cell factors that activate the mechanisms driving to tumorigenesis.

AUTHOR CONTRIBUTIONS

SF, SM, and MR wrote the review. SF, SM, UB, DZ, and MR participated in the conception and design of the review. All authors read and approved the final manuscript.

FUNDING

This study was supported by grants from University of Verona (FFO 2016 Romanelli) and University of Verona-Veneto Institute of Oncology. Joint Project 2016 grant (Romanelli).

REFERENCES

Ariumi, Y., Kaida, A., Lin, J. Y., Hirota, M., Masui, O., Yamaoka, S., et al. (2000). HTLV-1 Tax oncoprotein represses the p53-mediated trans-activation function through coactivator CBP sequestration. *Oncogene* 19, 1491–1499. doi: 10.1038/sj.onc.1203450

Avesani, F., Romanelli, M. G., Turci, M., Di Gennaro, G., Sampaio, C., Bidoia, C., et al. (2010). Association of HTLV Tax proteins with TAK1-binding protein 2 and RelA in calreticulin-containing cytoplasmic structures participates in Tax-mediated NF-κB activation. *Virology* 408, 39–48. doi: 10.1016/j.virol.2010.08.023

Bangham, C. R. M., and Matsuoka, M. (2017). Human T-cell leukaemia virus type 1: parasitism and pathogenesis. *Philos. Trans. R. Soc. Lond. B Biol. Sci.* 372:20160272. doi: 10.1098/rstb.2016.0272

Barbeau, B., Peloponese, J. M., and Mesnard, J. M. (2013). Functional comparison of antisense proteins of HTLV-1 and HTLV-2 in viral pathogenesis. *Front. Microbiol.* 4:226. doi: 10.3389/fmicb.2013.00226

Bellon, M., Lepelletier, Y., Hermine, O., and Nicot, C. (2009). Deregulation of microRNA involved in hematopoiesis and the immune response in HTLV-I adult T-cell leukemia. *Blood* 113, 4914–4917. doi: 10.1182/blood-2008-11-189845

Bertazzoni, U., Turci, M., Avesani, F., Di Gennaro, G., Bidoia, C., and Romanelli, M. G. (2011). Intracellular localization and cellular factors interaction of HTLV-1 and HTLV-2 tax proteins: similarities and functional differences. *Viruses* 3, 541–560. doi: 10.3390/v3050541

Bex, F., Murphy, K., Wattiez, R., Burny, A., and Gaynor, R. B. (1999). Phosphorylation of the human T-cell leukemia virus type 1 transactivator tax on adjacent serine residues is critical for tax activation. *J. Virol.* 73, 738–745.

Bonnet, A., Randrianarison-huetz, V., Nzounza, P., Nedelec, M., Chazal, M., Waast, L., et al. (2012). Low nuclear body formation and tax SUMOylation do not prevent NF-kappaB promoter activation, *Retrovirology* 9:77. doi: 10.1186/1742-4690-9-77

Chan, J. K., and Greene, W. C. (2012). Dynamic roles for NF-κB in HTLV-I and HIV-1 retroviral pathogenesis. *Immunol. Rev.* 246, 286–310. doi: 10.1111/j.1600-065X.2012.01094.x

Chen, L., Liu, D., Zhang, Y., Zhang, H., and Cheng, H. (2015). The autophagy molecule Beclin 1 maintains persistent activity of NF-κB and Stat3 in HTLV-1-transformed T lymphocytes. *Biochem. Biophys. Res. Commun.* 465, 739–745. doi: 10.1016/j.bbrc.2015.08.070

Clerc, I., Polakowski, N., André-Arpin, C., Cook, P., Barbeau, B., Mesnard, J. M., et al. (2008). An interaction between the human T cell leukemia virus type 1 basic leucine zipper factor (HBZ) and the KIX domain of p300/CBP contributes to the down-regulation of Tax-dependent viral transcription by HBZ. *J. Biol. Chem.* 283, 23903–23913. doi: 10.1074/jbc.M803116200

Cook, L., Melamed, A., Yaguchi, H., and Bangham, C. R. (2017). The impact of HTLV-1 on the cellular genome. *Curr. Opin. Virol.* 26, 125–131. doi: 10.1016/j.coviro.2017.07.013

Diani, E., Avesani, F., Bergamo, E., Cremonese, G., Bertazzoni, U., and Romanelli, M. G. (2015). HTLV-1 Tax protein recruitment into IKKε and TBK1 kinase complexes enhances IFN-I expression. *Virology* 476, 92–99. doi: 10.1016/j.virol.2014.12.005

Douceron, E., Kaidarova, Z., Miyazato, P., Matsuoka, M., Murphy, E. L., and Mahieux, R. (2012). HTLV-2 APH-2 expression is correlated with proviral load but APH-2 does not promote lymphocytosis. *J. Infect. Dis.* 205, 82–86. doi: 10.1093/infdis/jir708

Durand, J. K., and Baldwin, A. S. (2017). Targeting IKK and NF-κB for therapy. *Adv. Protein Chem. Struct. Biol.* 107, 77–115. doi: 10.1016/bs.apcsb.2016.11.006

Forlani, G., Abdallah, R., Accolla, R. S., and Tosi, G. (2013). The MHC-II transactivator CIITA, a restriction factor against oncogenic HTLV-1 and HTLV-2 retroviruses: similarities and differences in the inhibition of Tax-1 and Tax-2 viral transactivators. *Front. Microbiol.* 4:234. doi: 10.3389/fmicb.2013.00234

Forlani, G., Abdallah, R., Accolla, R. S., and Tosi, G. (2016). The major histocompatibility complex class II transactivator CIITA inhibits the persistent activation of NF-κB by the human T cell lymphotropic virus type 1 Tax-1 oncoprotein. *J. Virol.* 90, 3708–3721. doi: 10.1128/JVI.03000-15

Fu, J., Yan, P., Li, S., Qu, Z., and Xiao, G. (2010). Molecular determinants of PDLIM2 in suppressing HTLV-I Tax-mediated tumorigenesis. *Oncogene* 29, 6499–6507. doi: 10.1038/onc.2010.374

Fujikawa, D., Nakagawa, S., Hori, M., Kurokawa, N., Soejima, A., Nakano, K., et al. (2016). Polycomb-dependent epigenetic landscape in adult T-cell leukemia. *Blood* 127, 1790–1802. doi: 10.1182/blood-2015-08-662593

Furukawa, Y., Kubota, R., Tara, M., Izumo, S., and Osame, M. (2001). Existence of escape mutant in HTLV-I tax during the development of adult T-cell leukemia, *Blood* 97, 987–993. doi: 10.1182/blood.V97.4.987

Gallo, R. G., Willems, L., and Tagaya, Y. (2017). Time to go back to the original name. *Front. Microbiol.* 2017, 8:1800. doi: 10.3389/fmicb.2017.01800

Gao, L., and Harhaj, E. W. (2013). HSP90 protects the human T-cell leukemia virus type 1 (HTLV-1) tax oncoprotein from proteasomal degradation to support NF-κB activation and HTLV-1 replication. *J. Virol.* 87, 13640–13654. doi: 10.1128/JVI.02006-13

Gaudray, G., Gachon, F., Basbous, J., Biard-Piechaczyk, M., Devaux, C., and Mesnard, J. M. (2002). The complementary strand of the human T-cell leukemia virus type 1 RNA genome encodes a bZIP transcription factor that down-regulates viral transcription. *J. Virol.* 76, 12813–12822. doi: 10.1128/JVI.76.24.12813-12822.2002

Gessain, A., Barin, F., Vernant, J. C., Gout, O., Maurs, L., Calender, A., et al. (1985). Antibodies to human T-lymphotropic virus type-I in patients with tropical spastic paraparesis. *Lancet* 2, 407–410. doi: 10.1016/S0140-6736(85)92734-5

Gessain, A., and Cassar, O. (2012). Epidemiological aspects and world distribution of HTLV-1 infection. *Front. Microbiol.* 3:388. doi: 10.3389/fmicb.2012.00388

Giam, C. Z., and Semmes, O. J. (2016). HTLV-1 infection and adult T-cell leukemia/lymphoma-a tale of two proteins: Tax and HBZ. *Viruses* 8:E161. doi: 10.3390/v8060161

Hagiya, K., Yasunaga, J. I., Satou, Y., Ohshima, K., and Matsuoka, M. (2011). ATF3, an HTLV-1 bZip factor binding protein, promotes proli*ferati*on of adult T-cell leukemia cells. *Retrovirology* 8:19. doi: 10.1186/1742-4690-8-19

Higuchi, M., and Fujii, M. (2009). Distinct functions of HTLV-1 Tax1 from HTLV-2 Tax2 contribute key roles to viral pathogenesis. *Retrovirology* 6:117. doi: 10.1186/1742-4690-6-117

Higuchi, M., Tsubata, C., Kondo, R., Yoshida, S., Takahashi, M., Oie, M., et al. (2007). Cooperation of NF-kappaB2/p100 activation and the PDZ domain binding motif signal in human T-cell leukemia virus type 1 (HTLV-1) Tax1 but not HTLV-2 Tax2 is crucial for interleukin-2-independent growth transformation of a T-cell line. *J. Virol.* 81, 11900–11907. doi: 10.1128/JVI.00532-07

Hinuma, Y., Nagata, K., Hanaoka, M., Nakai, M., Matsumoto, T., Kinoshita, K. I., et al. (1981). Adult T-cell leukemia: antigen in an ATL cell line and detection of antibodies to the antigen in human sera. *Proc. Natl. Acad. Sci. U.S.A.* 78, 6476–6480. doi: 10.1073/pnas.78.10.6476

Ho, Y. K., Zhi, H., Bowlin, T., Dorjbal, B., Philip, S., Zahoor, M. A., et al. (2015). HTLV-1 Tax stimulates ubiquitin E3 ligase, ring finger protein 8, to assemble lysine 63-linked polyubiquitin chains for TAK1 and IKK activation. *PLoS Pathog.* 11, 1–19. doi: 10.1371/journal.ppat.1005102

Huang, J., Ren, T., Guan, H., Jiang, Y., and Cheng, H. (2009). HTLV-1 Tax is a critical lipid raft modulator that hijacks IkappaB kinases to the microdomains for persistent activation of NF-kappaB. *J. Biol. Chem.* 284, 6208–6217. doi: 10.1074/jbc.M806390200

Huang, Q., Niu, Z., Han, J., Liu, X., Lv, Z., Yuan, L., et al. (2017). HTLV-1 Tax upregulates early growth response protein 1 through nuclear factor-κB signaling. *Oncotarget* 8, 51123–51133. doi: 10.18632/oncotarget.17699

Journo, C., Bonnet, A., Favre-Bonvin, A., Turpin, J., Vinera, J., Côté, E., et al. (2013). Human T cell leukemia virus type 2 tax-mediated NF-κB activation involves a mechanism independent of Tax conjugation to ubiquitin and SUMO. *J. Virol.* 87, 1123–1136. doi: 10.1128/JVI.01792-12

Karimi, M., Mohammadi, H., Hemmatzadeh, M., Mohammadi, A., Rafatpanah, H., and Baradaran, B. (2017). Role of the HTLV-1 viral factors in the induction of apoptosis. *Biomed. Pharmacother.* 85, 334–347. doi: 10.1016/j.biopha.2016.11.034

Kfoury, Y., Setterblad, N., El-Sabban, M., Zamborlini, A., Dassouki, Z., El Hajj, H., et al. (2011). Tax ubiquitylation and SUMOylation control the dynamic shuttling of Tax and NEMO between Ubc9 nuclear bodies and the centrosome. *Blood* 117, 190–199. doi: 10.1182/blood-2010-05-285742

Kogure, Y., and Kataoka, K. (2017). Genetic alterations in adult T-cell leukemia/lymphoma. *Cancer Sci.* 108, 1719–1725. doi: 10.1111/cas.13303

Koiwa, T., Hamano-Usami, A., Ishida, T., Okayama, A., Yamaguchi, K., Kamihira, S., et al. (2002). 5′-long terminal repeat-selective CpG methylation of latent human T-cell leukemia virus type 1 provirus *in vitro* and *in vivo*. *J. Virol.* 76, 9389–9397. doi: 10.1128/JVI.76.18.9389-9397.2002

Kuo, Y. L., and Giam, C. Z. (2006). Activation of the anaphase promoting complex by HTLV-1 tax leads to senescence. *EMBO J.* 25, 1741–1752. doi: 10.1038/sj.emboj.7601054

Kusano, S., Yoshimitsu, M., Hachiman, M., and Ikeda, M. (2015). I-mfa domain proteins specifically interact with HTLV-1 Tax and repress its transactivating functions. *Virology* 486, 219–227. doi: 10.1016/j.virol.2015.09.020

Lavorgna, A., and Harhaj, E. W. (2014). Regulation of HTLV-1 tax stability, cellular trafficking and NF-κB activation by the ubiquitin-proteasome pathway. *Viruses* 6, 3925–3943. doi: 10.3390/v6103925

Lavorgna, A., Matsuoka, M., and Harhaj, E. W. (2014). A critical role for IL-17RB signaling in HTLV-1 tax-induced NF-κB activation and T-cell transformation. *PLoS Pathog.* 10:e1004418. doi: 10.1371/journal.ppat.1004418

Ma, G., Yasunaga, J., and Matsuoka, M. (2016). Multifaceted functions and roles of HBZ in HTLV-1 pathogenesis. *Retrovirology* 13:16. doi: 10.1186/s12977-016-0249-x

Ma, Y., Zhang, B., Wang, D., Qian, L., Song, X., Wang, X., et al. (2017). HTLV-1 basic leucine zipper factor downregulates cyclin D1 expression via interactions with NF-κB. *Int. J. Mol. Med.* 764–770. doi: 10.3892/ijmm.2017.2868

Marban, C., McCabe, A., Bukong, T. N., Hall, W. W., and Sheehy, N. (2012). Interplay between the HTLV-2 Tax and APH-2 proteins in the regulation of the AP-1 pathway. *Retrovirology* 9:98. doi: 10.1186/1742-4690-9-98

Matsuoka, M., and Jeang, K. T. (2007). Human T-cell leukaemia virus type 1 (HTLV-1) infectivity and cellular transformation. *Nat. Rev. Cancer* 7, 270–280. doi: 10.1038/nrc2111

Miyazaki, M., Yasunaga, J. I., Taniguchi, Y., Tamiya, S., Nakahata, T., and Matsuoka, M. (2007). Preferential selection of human T-cell leukemia virus type 1 provirus lacking the 5′ long terminal repeat during oncogenesis. *J. Virol.* 81, 5714–5723. doi: 10.1128/JVI.02511-06

Motai, Y., Takahashi, M., Takachi, T., Higuchi, M., Hara, T., Mizuguchi, M., et al. (2016). Human T-cell leukemia virus type 1 (HTLV-1) Tax1 oncoprotein but not HTLV-2 Tax2 induces the expression of OX40 ligand by interacting with p52/p100 and RelB. *Virus Genes* 52, 4–13. doi: 10.1007/s11262-015-1277-7

Nasr, R., Chiari, E., El-sabban, M., Mahieux, R., Kfoury, Y., Abdulhay, M., et al. (2006). Tax ubiquitylation and sumoylation control critical cytoplasmic and nuclear steps of NF-κB activation Tax ubiquitylation and sumoylation control critical cytoplasmic and nuclear steps of NF-κB activation. *Blood* 107, 4021–4029. doi: 10.1182/blood-2005-09-3572

Panfil, A. R., Dissinger, N. J., Howard, C. M., Murphy, B. M., Landes, K., Fernandez, S. A., et al. (2016). Functional comparison of HBZ and the related APH-2 protein provides insight into human T-cell leukemia virus type 1 pathogenesis. *J. Virol.* 90, 3760–3772. doi: 10.1128/JVI.03113-15

Peloponese, J. M., Yeung, M. L., and Jeang, K. T. (2006). Modulation of nuclear factor-kappaB by human T cell leukemia virus type 1 Tax protein: implications for oncogenesis and inflammation. *Immunol. Res.* 34, 1–12. doi: 10.1385/IR:34:1:1

Pène, S., Waast, L., Bonnet, A., Bénit, L., and Pique, C. (2014). A non-SUMOylated tax protein is still functional for NF-κB pathway activation. *J. Virol.* 88, 10655–10661. doi: 10.1128/JVI.01827-14

Poiesz, B. J., Ruscetti, F. W., Gazdar, A. F., Bunn, P. A., Minna, J. D., and Gallo, R. C. (1980). Detection and isolation of type C retrovirus particles from fresh and cultured lymphocytes of a patient with cutaneous T-cell lymphoma. *Proc. Nat. Acad. Sci. U.S.A.* 77, 7415–7419. doi: 10.1073/pnas.77.12.7415

Pujari, R., Hunte, R., Thomas, R., van der Weyden, L., Rauch, D., Ratner, L., et al. (2015). Human T-cell leukemia virus type 1 (HTLV-1) tax requires CADM1/TSLC1 for inactivation of the NF-κB inhibitor A20 and constitutive NF-κB signaling. *PLoS Pathog.* 11:e1004721. doi: 10.1371/journal.ppat.1004721

Qu, Z., and Xiao, G. (2011). Human T-cell lymphotropic virus: a model of NF-κB-associated tumorigenesis. *Viruses* 3, 714–749. doi: 10.3390/v3060714

Ren, T., Dong, W., Takahashi, Y., Xiang, D., Yuan, Y., Liu, X., et al. (2012). HTLV-2 tax immortalizes human CD4+ memory T lymphocytes by oncogenic activation and dysregulation of autophagy. *J. Biol. Chem.* 287, 34683–34693. doi: 10.1074/jbc.M112.377143

Ren, T., Takahashi, Y., Liu, X., Loughran, T. P., Sun, S. C., Wang, H. G., et al. (2015). HTLV-1 Tax deregulates autophagy by recruiting autophagic molecules into lipid raft microdomains. *Oncogene* 34, 334–345. doi: 10.1038/onc.2013.552

Romanelli, M. G., Diani, E., Bergamo, E., Casoli, C., Ciminale, V., Bex, F., et al. (2013). Highlights on distinctive structural and functional properties of HTLV Tax proteins. *Front. Microbiol.* 4:271. doi: 10.3389/fmicb.2013.00271

Saito, M., Matsuzaki, T., Satou, Y., Yasunaga, J. I., Saito, K., Arimura, K., et al. (2009). *In vivo* expression of the HBZ gene of HTLV-1 correlates with proviral load, inflammatory markers and disease severity in HTLV-1 associated myelopathy/tropical spastic paraparesis (HAM/TSP). *Retrovirology* 6:19. doi: 10.1186/1742-4690-6-19

Sasaki, D., Imaizumi, Y., Hasegawa, H., Osaka, A., Tsukasaki, K., Choi, Y. L., et al. (2011). Overexpression of enhancer of zeste homolog 2 with trimethylation of lysine 27 on histone H3 in adult T-cell leukemia/lymphoma as a target for epigenetic therapy. *Haematologica* 96, 712–719. doi: 10.3324/haematol.2010.028605

Shembade, N., Harhaj, N. S., Yamamoto, M., Akira, S., and Harhaj, E. W. (2007). The human T-cell leukemia virus type 1 Tax oncoprotein requires the ubiquitin-conjugating enzyme Ubc13 for NF-kappaB activation. *J. Virol.* 81, 13735–13742. doi: 10.1128/JVI.01790-07

Shen, R. R., and Hahn, W. C. (2011). Emerging roles for the non-canonical IKKs in cancer *Oncogene* 30, 631–641. doi: 10.1038/onc.2010.493

Shibata, Y., Tokunaga, F., Goto, E., Komatsu, G., Gohda, J., Saeki, Y., et al. (2017). HTLV-1 Tax induces formation of the active macromolecular IKK complex by generating Lys63- and Met1-linked hybrid polyubiquitin Chains. *PLoS Pathog.* 13, 1–25. doi: 10.1371/journal.ppat.1006162

Shoji, T., Higuchi, M., Kondo, R., Takahashi, M., Oie, M., Tanaka, Y., et al. (2009). Identification of a novel motif responsible for the distinctive transforming activity of human T-cell leukemia virus (HTLV) type 1 Tax1 protein from HTLV-2 Tax2. *Retrovirology* 6:83. doi: 10.1186/1742-4690-6-83

Sun, S. C. (2017). The non-canonical NF-κB pathway in immunity and inflammation. *Nat. Rev. Immunol.* 17, 545–558. doi: 10.1038/nri.2017.52

Sun, S. C., Elwood, J., Béraud, C., and Greene, W. C. (1994). Human T-cell leukemia virus type I Tax activation of NF-kappa B/Rel involves phosphorylation and degradation of I kappa B alpha and RelA (p65)-mediated induction of the c-rel gene. *Mol. Cell. Biol.* 14, 7377–7384. doi: 10.1128/MCB.14.11.7377

Taniguchi, Y., Nosaka, K., Yasunaga, J., Maeda, M., Mueller, N., Okayama, A., et al. (2005). Silencing of human T-cell leukemia virus type I gene transcription by epigenetic mechanisms. *Retrovirology* 2:64. doi: 10.1186/1742-4690-2-64

Turci, M., Lodewick, J., Di Gennaro, G., Rinaldi, A. S., Marin, O., Diani, E., et al. (2012). Ubiquitination and sumoylation of the HTLV-2 Tax-2B protein regulate its NF-κB activity: a comparative study with the HTLV-1 Tax-1 protein. *Retrovirology* 9:102. doi: 10.1186/1742-4690-9-102

Turci, M., Romanelli, M. G., Lorenzi, P., Righi, P., and Bertazzoni, U. (2006). Localization of human T-cell lymphotropic virus type II Tax protein is dependent upon a nuclear localization determinant in the N-terminal region. *Gene* 365, 119–124. doi: 10.1016/j.gene.2005.09.043

Wang, C., Long, W., Peng, C., Hu, L., Zhang, Q., Wu, A., et al. (2016). HTLV-1 Tax functions as a ubiquitin E3 ligase for direct IKK activation via synthesis of mixed-linkage polyubiquitin Chains. *PLoS Pathog.* 12, 1–25. doi: 10.1371/journal.ppat.1005584

Wang, J., Li, J., Huang, Y., Song, X., Niu, Z., Gao, Z., et al. (2013). Bcl-3 suppresses Tax-induced NF-κB activation through p65 nuclear translocation blockage in HTLV-1-infected cells. *Int. J. Oncol.* 42, 269–276. doi: 10.3892/ijo.2012.1685

Wang, W., Zhou, J., Shi, J., Zhang, Y., Liu, S., Liu, Y., et al. (2014). Human T-cell leukemia virus type 1 Tax-deregulated autophagy pathway and c-FLIP expression contribute to resistance against death receptor-mediated apoptosis. *J. Virol.* 88, 2786–2798. doi: 10.1128/JVI.03025-13

Watanabe, T. (2017). Adult T-cell leukemia: molecular basis for clonal expansion and transformation of HTLV-1-infected T cells. *Blood* 129, 1071–1081. doi: 10.1182/blood-2016-09-692574

Willems, L., Hasegawa, H., Accolla, R., Bangham, C., Bazarbachi, A., Bertazzoni, U., et al. (2017). Reducing the global burden of HTLV-1 infection: an agenda for research and action. *Antiviral Res.* 137, 41–48. doi: 10.1016/j.antiviral.2016.10.015

Wright, D. G., Marchal, C., Hoang, K., Ankney, J. A., Nguyen, S. T., Rushing, A. W., et al. (2016). Human T-cell leukemia virus type-1-encoded protein HBZ represses p53 function by inhibiting the acetyltransferase activity of p300/CBP and HBO1. *Oncotarget* 7, 1687–1706. doi: 10.18632/oncotarget.6424

Wurm, T., Wright, D. G., Polakowski, N., Mesnard, J. M., and Lemasson, I. (2012). The HTLV-1-encoded protein HBZ directly inhibits the acetyl transferase activity of p300/CBP. *Nucleic Acids Res.* 40, 5910–5925. doi: 10.1093/nar/gks244

Xiao, G., Cvijic, M. E., Fong, A., Harhaj, E. W., Uhlik, M. T., Waterfield, M., et al. (2001). Retroviral oncoprotein Tax induces processing of NF-kappaB2/p100 in T cells: evidence for the involvement of IKKalpha. *EMBO J.* 20, 6805–6815. doi: 10.1093/emboj/20.23.6805

Yamagishi, M., Nakano, K., Miyake, A., Yamochi, T., Kagami, Y., Tsutsumi, A., et al. (2012). Polycomb-mediated loss of miR-31 activates NIK-dependent NF-κB pathway in adult T cell leukemia and other cancers. *Cancer Cell* 21, 121–135. doi: 10.1016/j.ccr.2011.12.015

Yamagishi, M., and Watanabe, T. (2012). Molecular hallmarks of adult T cell leukemia. *Front. Microbiol.* 3:334. doi: 10.3389/fmicb.2012.00334

Yan, P., Fu, J., Qu, Z., Li, S., Tanaka, T., Grusby, M. J., et al. (2009). PDLIM2 suppresses human T-cell leukemia virus type I Tax-mediated tumorigenesis by targeting Tax into the nuclear matrix for proteasomal degradation. *Blood* 113, 4370–4380. doi: 10.1182/blood-2008-10-185660

Yasuma, K., Yasunaga, J. I., Takemoto, K., Sugata, K., Mitobe, Y., Takenouchi, N., et al. (2016). HTLV-1 bZIP factor impairs anti-viral immunity by inducing co-inhibitory molecule, T cell immunoglobulin and ITIM domain (TIGIT). *PLoS Pathog.* 12, 1–22. doi: 10.1371/journal.ppat.1005372

Yeung, M. L., Yasunaga, J., Bennasser, Y., Dusetti, N., Harris, D., Ahmad, N., et al. (2008). Roles for microRNAs, miR-93 and miR-130b, and tumor protein 53-induced nuclear protein 1 tumor suppressor in cell growth dysregulation by human T-cell lymphotrophic virus 1. *Cancer Res.* 68, 8976–8985. doi: 10.1158/0008-5472.CAN-08-0769

Yin, H., Kannian, P., Dissinger, N., Haines, R., Niewiesk, S., and Green, P. L. (2012). Human T-cell leukemia virus type 2 antisense viral protein 2 is dispensable for *in vitro* immortalization but functions to repress early virus replication *in vivo*. *J. Virol.* 86, 8412–8421. doi: 10.1128/JVI.00717-12

Yu, Q., Minoda, Y., Yoshida, R., Yoshida, H., Iha, H., Kobayashi, T., et al. (2008). HTLV-1 Tax-mediated TAK1 activation involves TAB2 adapter protein. *Biochem. Biophys. Res. Commun.* 365, 189–194. doi: 10.1016/j.bbrc.2007.10.172

Zhang, H., Chen, L., Cai, S. H., and Cheng, H. (2016). Identification of TBK1 and IKKε, the non-canonical IκB kinases, as crucial pro-survival factors in HTLV-1-transformed T lymphocytes. *Leuk. Res.* 46, 37–44. doi: 10.1016/j.leukres.2016.04.012

Zhang, L. L., Wei, J., Wang, L., Huang, S., and Chen, J. L. (2017). Human T-cell lymphotropic virus type 1 and its oncogenesis. *Acta Pharmacol. Sin.* 38, 1093–1103. doi: 10.1038/aps.2017.17

Zhang, L., Zhi, H., Liu, M., Kuo, Y. L., and Giam, C. Z. (2009). Induction of p21(CIP1/WAF1) expression by human T-lymphotropic virus type 1 Tax requires transcriptional activation and mRNA stabilization. *Retrovirology* 6:35. doi: 10.1186/1742-4690-6-35

Zhao, T. (2016). The role of HBZ in HTLV-1-induced oncogenesis. *Viruses* 8, 1–12. doi: 10.3390/v8020034

Zhao, T., Coutts, A., Xu, L., Yu, J., Ohshima, K., and Matsuoka, M. (2013). HTLV-1 bZIP factor supports proliferation of adult T cell leukemia cells through suppression of C/EBPα signaling. *Retrovirology* 10:159. doi: 10.1186/1742-4690-10-159

Zhao, T., and Matsuoka, M. (2012). HBZ and its roles in HTLV-1 oncogenesis. *Front. Microbiol.* 3:247. doi: 10.3389/fmicb.2012.00247

Zhao, T., Yasunaga, J., Satou, Y., Nakao, M., Takahashi, M., Fujii, M., et al. (2009). Human T-cell leukemia virus type 1 bZIP factor selectively suppresses the classical pathway of NF-kappaB. *Blood* 113, 2755–2764. doi: 10.1182/blood-2008-06-161729

Zhi, H., Yang, L., Kuo, Y. L., Ho, Y. K., Shih, H. M., and Giam, C. Z. (2011). NF-κB hyper-activation by HTLV-1 tax induces cellular senescence, but can be alleviated by the viral anti-sense protein HBZ. *PLoS Pathog.* 7, 1–12. doi: 10.1371/journal.ppat.1002025

19

A Preclinical Model for the ATLL Lymphoma Subtype with Insights into the Role of Microenvironment in HTLV-1-Mediated Lymphomagenesis

Mattia Vicario[1†‡], Adriana Mattiolo[1‡], Barbara Montini[1], Maria Assunta Piano[1], Ilaria Cavallari[1], Alberto Amadori[1,2], Luigi Chieco-Bianchi[2§] and Maria Luisa Calabrò[1*§]*

[1] Immunology and Molecular Oncology, Veneto Institute of Oncology, IOV – IRCCS, Padua, Italy, [2] Department of Surgery, Oncology and Gastroenterology, University of Padova, Padua, Italy

***Correspondence:**
Maria Luisa Calabrò
luisella.calabro@iov.veneto.it;
lcalabro@unipd.it
Luigi Chieco-Bianchi
luigi.chiecobianchi@unipd.it

†Present address:
Mattia Vicario,
Department of Biomedical Sciences,
University of Padova, Padua, Italy

‡These authors have contributed equally to this work as first authors.

§These authors have contributed equally to this work as last authors.

Adult T cell Leukemia/Lymphoma (ATLL) is a mature T cell malignancy associated with Human T cell Leukemia Virus type 1 (HTLV-1) infection. Among its four main clinical subtypes, the prognosis of acute and lymphoma variants remains poor. The long latency (3–6 decades) and low incidence (3–5%) of ATLL imply the involvement of viral and host factors in full-blown malignancy. Despite multiple preclinical and clinical studies, the contribution of the stromal microenvironment in ATLL development is not yet completely unraveled. The aims of this study were to investigate the role of the host microenvironment, and specifically fibroblasts, in ATLL pathogenesis and to propose a murine model for the lymphoma subtype. Here we present evidence that the oncogenic capacity of HTLV-1-immortalized C91/PL cells is enhanced when they are xenotransplanted together with human foreskin fibroblasts (HFF) in immunocompromised BALB/c Rag2$^{-/-}\gamma_C{}^{-/-}$ mice. Moreover, cell lines derived from a developed lymphoma and their subsequent *in vivo* passages acquired the stable property to induce aggressive T cell lymphomas. In particular, one of these cell lines, C91/III cells, consistently induced aggressive lymphomas also in NOD/SCID/IL2Rγ_C KO (NSG) mice. To dissect the mechanisms linked to this enhanced tumorigenic ability, we quantified 45 soluble factors released by these cell lines and found that 21 of them, mainly pro-inflammatory cytokines and chemokines, were significantly increased in C91/III cells compared to the parental C91/PL cells. Moreover, many of the increased factors were also released by human fibroblasts and belonged to the known secretory pattern of ATLL cells. C91/PL cells co-cultured with HFF showed features reminiscent of those observed in C91/III cells, including a similar secretory pattern and a more aggressive behavior *in vivo*. On the whole, our data provide evidence that fibroblasts, one of the major stromal components, might enhance tumorigenesis of HTLV-1-infected and immortalized T cells, thus throwing light on the role of microenvironment contribution in ATLL pathogenesis. We also propose that the lymphoma induced in NSG mice by injection with C91/III cells represents a new murine preclinical ATLL model that could be adopted to test novel therapeutic interventions for the aggressive lymphoma subtype.

Keywords: HTLV-1, ATLL, lymphoma, microenvironment, fibroblasts, preclinical model

INTRODUCTION

According to the criteria of the World Health Organization [WHO] (2008) for classification of tumors originating from hematopoietic and lymphoid tissues, Adult T cell Leukemia/Lymphoma (ATLL) is defined as a peripheral T cell neoplasia linked to infection with Human T cell Leukemia Virus type 1 (HTLV-1), a virus identified first by Gallo and co-workers in 1980 (Poiesz et al., 1980). ATLL incidence is therefore linked to the prevalence of HTLV-1 infection, which is not uniformly distributed around the world; clusters of high HTLV-1 prevalence are mainly found in Southwestern Japan, the Caribbean Basin and some parts of Africa and South America (Gessain and Cassar, 2012). ATLL develops almost exclusively in adults, although a few cases in children have been reported (de Oliveira et al., 1990; Watanabe, 2017). The diagnosis is mainly based on clinical features, presence of HTLV-1 infection and, when detected, peculiar tumor cell morphology, represented by atypical circulating lymphocytes with petal-shaped nuclei (Matutes, 2007). From the clinical point of view, ATLL is characterized by a high heterogeneity in its presentation and clinical course. Four main subtypes have been described (Shimoyama, 1991), two aggressive and fast-growing (acute leukemia, lymphoma) and two indolent (smoldering, chronic) variants. A new classification has recently been proposed that combines clinical features with specific genetic aberrations shown to have a prognostic value (Kataoka et al., 2018). Treatment strategies are based on ATLL subtype, and antiviral therapy is usually effective in slowly progressing subtypes. Prognosis of fast-growing variants remains poor in spite of various attempts of pharmacologic treatments (Hermine, 2015; Katsuya et al., 2015).

The very prolonged latent period (decades) between primary infection and development of full-blown disease and the relative rare occurrence of ATLL among asymptomatic HTLV-1 carriers (3–5%) suggest that additional genetic and/or epigenetic changes in infected T cells as well as host factors are necessary for ATLL induction. Besides ATLL, HTLV-1-infected subjects may also feature opportunistic infections (Nakada et al., 1987) and develop chronic inflammatory disease such as myelopathy (Gessain et al., 1985; Osame and Igata, 1989), arthropathy, myositis, uveitis, and dermatitis (Yamano and Sato, 2012). Thus, circumstantial evidence indicates that the infection with HTLV-1 induces deficiency and dysregulation of host immunity.

HTLV-1 belongs to the deltaretrovirus genus, together with the simian T lymphotropic virus and bovine leukemia virus. HTLV-1 is a complex retrovirus whose genome comprises structural, regulatory and accessory genes. Accumulating evidence indicates that the regulatory gene *tax* is crucially involved in ATLL pathogenesis. In fact, Tax protein exhibits pleiotropic functions (Romanelli et al., 2013); besides transcriptionally activating its long terminal repeats (Felber et al., 1985; Seiki et al., 1986), it interacts with cellular transcription factors (NF-kB, CREB, and AP-1) and upregulates the expression of multiple cellular genes involved in cell proliferation and genomic instability (Armstrong et al., 1993; Baranger et al., 1995; Munoz and Israel, 1995; Fujii et al., 2000; Grassmann et al., 2005; Fochi et al., 2018). However, in the majority of cases, ATLL cells show a Tax-low or Tax-negative phenotype, suggesting that Tax, while critical for T cell immortalization and transformation, may be not crucial in late stages of ATLL (Takeda et al., 2004). In contrast, another viral gene, the HTLV-1 basic leucine zipper factor (HBZ) encoded in the minus strand of the viral genome, appears to be transcribed in all cases of ATLL (Gaudray et al., 2002). Furthermore, it has been reported that HBZ mRNA, but not HBZ protein, could induce T cell proliferation and promote cell survival (Satou et al., 2006). Thus, a current hypothesis is that transactivation by Tax is needed to initiate the processes that lead to ATLL, while HBZ is responsible for maintaining the neoplastic phenotype of ATLL cells (Matsuoka and Jeang, 2007; Giam and Semmes, 2016).

Bidirectional communication of potentially oncogenic cells with surrounding stroma creates a tissue microenvironment permissive to disease initiation and progression. Among the stromal components, fibroblasts play a prominent role; the tumor-promoting activity of cancer-associated fibroblasts has been extensively studied in epithelial cancers (Olumi et al., 1999; Erez et al., 2010). Within the tumor stroma, fibroblasts, likely through a transforming growth factor beta signaling, acquire an activated phenotype which is mainly characterized by their expression of alpha-smooth muscle actin similar to that observed in the wound healing process (Vaughan et al., 2000). These cells are provided with an armamentarium of released factors that can alter tissue homeostasis, promote angiogenesis, cancer cell proliferation and invasiveness, and recruit immunosuppressive and tumor-promoting cells (Kalluri and Zeisberg, 2006; Mueller et al., 2007; Erez et al., 2010; Bissell and Hines, 2011). Specific bone marrow stromal niches have been identified for leukemia development (Jin et al., 2008; Zhang et al., 2012). Concerning ATLL, *in vitro* interaction with epithelial and fibroblastic cell lines was shown to induce apoptosis resistance in primary ATLL cells and ATLL cell lines (Miyatake et al., 2013, 2015) as well as viral latency (Kinpara et al., 2009), highlighting the role of stromal components in ATLL pathogenesis.

Different preclinical models have been developed to better understand the pathogenesis of ATLL and HTLV-linked degenerative/inflammatory diseases (Panfil et al., 2013). Good models for HTLV-1 infection have been established in non-human primates, rabbits and rats (Dodon et al., 2012; Hajj et al., 2012). However, mice are preferred as they are more manageable and cost-effective models to study the virus/host factors critical for ATLL induction and for evaluation of its specific treatments. Xenografts of ATLL cells or some ATLL-derived cell lines in immunocompromised mice successfully replicated features of ATLL (Parrula et al., 2009). On the other hand, T cells immortalized *in vitro* by HTLV-1 showed no or poor growth, depending mainly on the host constitutive immunodeficiency degree, the innate immunity by natural killer (NK) cell antitumor activity being particularly critical for restraining the engraftment (Ishihara et al., 1992; Feuer et al., 1993, 1995; Imada et al., 1995; Liu et al., 2002).

The aim of the present study was to analyze the evolution of HTLV-1-infected T lymphocytes from the immortalized status, commonly observed in asymptomatic carriers, to the quite rare neoplastic transformation leading to clinically overt ATLL. Among the multiple factors involved in this oncogenic switch, including genetic and epigenetic cell alterations, our attention was focused on the influence of one of the major microenvironment components, specifically fibroblasts. Moreover, we established of a highly lymphomagenic C91/PL-derived cell line that, when xenotransplanted into immunodeficient NSG mice, may constitute a new preclinical mouse model for the lymphoma variant of ATLL.

MATERIALS AND METHODS

Cell Lines

The HTLV-1-immortalized C91/PL cell line was established by co-cultivation of umbilical cord blood T cells with leukemic T cells from an ATLL patient (PL) (Popovic et al., 1983), and was originally obtained from Prof. Robin Weiss (Chester Beatty Laboratories, London). This cell line and its *ex vivo*-derived cell lines (designated as C91/I, C91/II and C91/III) were grown in RPMI 1640 (Sigma-Aldrich, Munich, Germany) supplemented with 10% fetal calf serum (FCS, Gibco, Foster City, CA, United States), 2 mM L-glutamine (Gibco) and 50 μg/mL gentamycin (Sigma-Aldrich) (complete medium). Human foreskin fibroblasts (HFF) were a kind gift from Dr. Abatangelo and Dr. Zavan (University of Padova). HFF were propagated in Dulbecco's modified Eagle Culture Medium (DMEM, Sigma-Aldrich) supplemented with 10% FCS, 2 mM L-glutamine and 50 μg/mL gentamycin.

Analysis of the clonal T cell receptor (TCR) beta chain and gamma chain gene rearrangements in C91/PL cells and in C91/PL-derived cells was carried out using the IdentiClone™ TCRB + TCRG T cell Clonality Assay (Invivoscribe Technologies, San Diego, CA, United States), according to the manufacturer's instructions. Short tandem repeat (STR) profile of C91/PL and C91/PL-derived cell lines was carried out with the PowerPlex 18D System (Promega, Madison, WI, United States) using an Applied Biosystems 3130XL genetic analyzer and Genemapper ID Ver. 3.2.1 software (BMR Genomics S.r.l., Padua, Italy). As an STR profile of C91/PL cells is not available in DMSZ, ATCC, and COG databases, C91/PL cells and their more tumorigenic counterparts (C91/II and C91/III) were authenticated by comparing the STR profile obtained with 18 genetic markers to that determined using the C91/PL cells received by the National Institute for Biological Standards and Controls (NIBSC), United Kingdom. The presence of the HTLV-1 provirus was confirmed in all *ex vivo* cell lines by qualitative single-round PCR with primer pairs specific for the *tax* region as previously reported (Calabro et al., 1993). All cell lines were mycoplasma-free, as confirmed by periodical PCR check.

Mice

BALB/c Rag2/Common γ chain (γ_c)-double KO ($Rag2^{-/-}\gamma_c^{-/-}$) mice were originally received from the Freiburg University Medical Center, Germany. Null mutation of the RAG2 gene prevents B and T lymphocyte development in these mice, while absence of the cytokine receptor common gamma chain, required for signal transduction of multiple cytokines [including interleukin (IL)-2, IL-4, IL-7, IL-9, IL-15, and IL-21] prevents NK cell maturation. Mice were inbreed and maintained in our animal facility, and were used in the first set of experiments and for the *in vivo* passages of the C91/PL-derived cells. Because of a temporary closure of our animal SPF facility for enlargement and renovation, all mouse colonies were eliminated and subsequently, after a 7 month-gap, restored with new stocks. Therefore, the second set of experiments was conducted with BALB/c $Rag2^{-/-}\gamma_c^{-/-}$ (BRG) mice ("excluded flora," i.e., free of rodent pathogens and seven selected opportunistic pathogens as well as Segmented Filamentous Bacteria) obtained from Taconic (Germantown, NY). Other experiments were performed using NOD/SCID/IL2Rγ_c KO (NSG) mice, obtained from Charles River (Charles River Laboratories, Calco, Italy), that combine the NOD/SCID background to the lack of a functional common gamma chain.

Ethics Statement

This study was carried out in accordance with the institutional guidelines that comply with the Italian Animal Welfare Law (D.L. No. 116/1992; and subsequent documents). The project was evaluated and approved by the local ethics committee of the University of Padova (Comitato Etico di Ateneo per la Sperimentazione Animale, CEASA) and communicated to the relevant Italian authority (Italian Ministry of Health, VI Office) (Project No. 32/2009; Permit Protocol No. 51740, 15/09/2010). The project had to be renewed, and it was evaluated and approved by the Italian Ministry of Health (Project and Permit protocol No. 932/2016, 10/10/2016).

Assessment of the Contribution of Fibroblasts to Lymphoma Growth *in Vivo*

Exponentially growing HFF and C91/PL cells mixed in a 1:2 ratio were suspended in a final density of 6×10^7 total cells/mL, and 0.1 mL of this cell suspension (i.e., 2×10^6 HFF and 4×10^6 C91/PL) was injected intraperitoneally (i.p.) into each mouse. Control mice were i.p. injected with 4×10^6 C91/PL cells. Mice were checked biweekly for cachexia and presence of abdominal masses. For ethical reasons, tumor-bearing animals were killed when presenting signs of suffering, and each mouse was considered to have died from tumor progression on this date. At necroscopy, the finding of organ involvement, abdominal and pelvic tumor masses were considered lymphoma growth, which was subsequently confirmed by histological examination.

Assessment of Engraftment and Tumorigenesis of C91/PL-Derived Cell Lines

Fragments of masses from $Rag2^{-/-}\gamma_c^{-/-}$ mice affected by lymphomatous growth were processed under sterile conditions and the obtained cell suspension was used to set up *ex vivo* cultures. After *in vitro* growth for 3–4 weeks, these cells were used as inoculum to assess their ability to induce a lymphomatous growth in mice. Different cell lines were thus obtained from the subsequent *in vivo* passages: C91/I, C91/II and C91/III, corresponding to the first, second and third *in vivo* passage, respectively. Each cell line was then tested for lymphoma induction in different doses and ages of injected mice.

Engraftment efficiency and tumorigenesis of C91/III cells were also assessed in NSG mice. Six-day-old mice were i.p. injected with 4×10^6 C91/III cells/mouse. Four-week-old mice were injected with 1×10^6 and 4×10^6 C91/III cells/mice (five mice per group). Moreover, we further analyzed the contribution of HFF to lymphoma growth by injecting C91/PL cells after *in vitro* co-culture with HFF. To this end, 5-day-old NSG mice were injected i.p. with 4×10^6 co-cultured C91/PL cells (nine mice). Control mice were injected i.p. with 4×10^6 C91/PL cells (seven mice). Mice were checked, culled and analyzed as reported above.

Histology and Immunohistochemistry

Fragments of infiltrated murine organs and tissues were fixed in 10% formalin and embedded in paraffin. Sections were cut (4-mm thick) from *ex vivo* samples and prepared for appropriate staining. Hematoxylin–eosin staining was used for histological diagnosis. Sections were immunostained for CD25 (Leica Biosystems, Nussloch, Germany) and for Ki67 (Dako, Glostrup, Denmark), as previously reported (Di Stefano et al., 2008).

Immunophenotypic and Viral Characterization of C91/PL, C91/III and HFF-Co-cultured C91/PL Cell Lines

C91/PL, C91/III and HFF-co-cultured C91/PL were analyzed for the expression of surface antigens by flow cytometry. The following anti-human monoclonal antibodies were used: fluorescein isothiocyanate (FITC)-conjugated anti-CD1a (BD Pharmingen, Franklin Lakes, NJ, United States), phycoerythrin (PE)-conjugated anti-CD2 (Life Technologies, Carlsbad, CA, United States), Alexa488-conjugated anti-CD3, anti-CD4, and anti-CD25 (Bio-Rad Laboratories, Inc., Hercules, CA, United States), phycoerythrin-Cy5 (PC5)-conjugated anti-CD5 and anti CD-7 (Coulter, Fullerton, CA, United States), FITC-conjugated anti-CD34 (BD Pharmingen), PE-conjugated anti-CD117 (Miltenyi Biotec, Bergisch Gladbach, Germany), Allophycocyanin (APC)-conjugated anti-CD133 (Miltenyi Biotec), A488-conjugated anti-FOXP3 (AbD Serotec, Oxford, United Kingdom), PE-conjugated anti-CD54 (eBioscience, Inc., San Diego, CA, United States), and FITC-conjugated anti-cell adhesion molecule 1 (CADM1, Medical and Biological Laboratories, Nagoya, Japan). Samples were analyzed on a BD LSR II flow cytometer (BD Biosciences, Milano, Italy). All cytofluorimetric data were analyzed using Kaluza Analysis software Ver. 1.3 (Beckman Coulter, Brea, CA, United States).

Quantitative analysis of HTLV-1 transcripts in C91/PL, C91/III and HFF-co-cultured C91/PL cells was performed as previously described (Rende et al., 2011; Cavallari et al., 2013).

Quantitative Analyses of Soluble Factors

To measure soluble factors released by C91/PL, C91/III and HFF-co-cultured C91/PL cells, cells were seeded at a concentration of 1×10^6/mL. Supernatants were collected after 72 h and centrifuged for 5 min at 200 g. Cell pellets were discarded, and supernatants were centrifuged for 30 min at 2,800 *g* to eliminate cell debris. Supernatants were also collected from the three cell lines kept in culture to evaluate concentration fluctuations and consistency of detectable factors during standard passage. The secretory pattern of HFF was measured by seeding 3×10^5 cells in a six-well plate; supernatants were collected after 72 h and processed as described above. To compare the profile and amount of factors released by all cell lines, a Luminex xMAP approach was used (ProcartaPlex Human Cytokine/Chemokine/Growth Factor Panel 1 96 tests, Affymetrix eBioscience Ltd., Hatfield, United Kingdom) for the multianalyte detection of 45 secreted proteins. This assay detects the following proteins: IL-1 receptor antagonist (IL-1RA), IL-1alpha (IL-1α), IL-1beta (IL-1β), IL-2, IL-4, IL-5, IL-6, IL-7, IL-8/ C-X-C motif chemokine ligand 8 (CXCL8), IL-9, IL-10, IL-12p70, IL-13, IL-15, IL-17A, IL-18, IL-21, IL-22, IL-23, IL-27, IL-31, leukemia inhibitory factor (LIF), tumor necrosis factor alpha (TNFα), TNF beta/lymphotoxin-alpha (TNFβ/LTA), interferon alpha (IFNα), IFN gamma (IFNγ), growth-regulated oncogene-alpha GROα/CXCL1, eotaxin/C-C motif chemokine ligand 11 (Eotaxin/CCL11), IFNγ-induced protein 10 (IP10)/CXCL10, monocyte chemotactic protein-1 (MCP-1)/CCL2, macrophage inflammatory protein-1 alpha (MIP-1α)/CCL3, MIP-1 beta, MIP-1β/CCL4, regulated on activation normal T cell expressed and secreted (RANTES)/CCL5, stromal cell-derived factor 1 alpha (SDF-1α/CXCL12), brain-derived neurotrophic factor (BDNF), granulocyte-macrophage colony-stimulating factor (GM-CSF), hepatocyte growth factor (HGF), placental growth factor (PLGF), epidermal growth factor (EGF), fibroblast growth factor 2 (FGF-2), platelet-derived growth factor-BB (PDGF-BB), stem cell factor (SCF), nerve growth factor beta (βNGF), vascular endothelial growth factor A (VEGF-A), VEGF-D.

IL-8/CXCL8 and TNFα were also quantified using cytokine-specific assays (Human IL-8-ELISA and Human TNFα-ELISA Ready-SET-Go Kits, Affymetrix eBioscience Ltd.), according to the manufacturer's instructions.

Analysis of the Crosstalk Between C91/PL Cells and HFF

To assess whether HFF were able to modulate C91/PL cell turnover, 2×10^5 C91/PL cells were co-cultured in the presence or absence of semiconfluent HFF seeded onto a six-well plate. Induced apoptosis was measured after 24, 48, and 72 h of culture by flow cytometry after Annexin V/Propidium Iodide (PI) staining (Annexin-V-FLUOS Staining Kit, Roche Diagnostics, Mannheim, Germany). Apoptosis was induced by serum reduction (5% FCS). Proliferation was analyzed by flow cytometry after staining with carboxyfluorescein succinimidyl ester (Cell Trace™ CFSE, Life Technologies).

Long-term HFF-co-culture of C91/PL cells was carried out in T75 flasks to periodically assess morphological and phenotypic changes. Two-month co-culture of C91/PL cells was carried out, and then cells were removed from co-culture and maintained as long as 6 weeks in culture to ensure the absence of residual fibroblasts. The secretory pattern, the profile of HTLV-1 transcripts and the ability to induce *in vivo* lymphoma of these long-term co-cultured cells were analyzed at this time point.

To assess the contribution of soluble factors in the crosstalk between C91/PL cells and HFF, short-term co-cultures in six-well plates were set up, with C91/PL cells added directly to the HFF or placed into transwell chambers (BD Falcon Cell Culture Inserts, Durham, NC, United States; pore size 0.4 μm). The amount of secreted IL-8/CXCL8 and TNFα was measured in the supernatants collected after 3 and 10 days of co-culture.

Statistical Analyses

Curves reporting the percentage of survivors over time were estimated by the Kaplan–Meier method and compared with the log-rank test. Data analysis was performed using the MedCalc statistical software (Mariakerke, Belgium). Statistical data are presented as mean ± standard deviation. Two-sided Student's *t*-test was used to estimate statistical significance of differences between the two cell lines. *P*-values < 0.05 were considered significant.

RESULTS

Co-inoculation of Fibroblasts Promotes the Tumorigenesis of C91/PL Cells

Preliminary experiments conducted to xenograft, either i.p. or subcutaneously, the HTLV-1-immortalized C91/PL cell line at different doses (from 2 to 10×10^6 cells per mouse) in immunodeficient adult Rag2$^{-/-}\gamma_c^{-/-}$ mice did not lead to tumor development (data not shown). The evidence that fibroblasts are effective in promoting tumor initiation and progression in different mouse models of carcinogenesis (Bissell and Hines, 2011) prompted us to analyze whether co-injection of fibroblasts with C91/PL cells in newborn Rag2$^{-/-}\gamma_c^{-/-}$ mice could facilitate the engraftments of this cell line. Thus, we performed a first set of experiments using six 4–6-day-old Rag2$^{-/-}\gamma_c^{-/-}$ mice per group: control mice were i.p. injected with 4×10^6 C91/PL cells, whereas the experimental group received the same dose of C91/PL cells mixed with HFF in a 2:1 ratio. This experiment was repeated twice and each experiment was stopped after 200 days. As shown in **Table 1**, all control mice (12 of 12) were disease-free throughout the experiment and autopsy did not disclose any pathological sign. Conversely, among the 12 animals receiving C91/PL cells mixed with HFF, four developed lymphoma with a median survival time of 105 days.

TABLE 1 | Lymphomagenic efficiency of C91/PL cells in Rag2$^{-/-}\gamma_c^{-/-}$ mice.

	No. of positive mice/total No. of mice[a]
C91/PL	0/12
C91/PL + HFF	4/12[b]

[a]4-6-day-old mice were intraperitoneally (i.p.) inoculated with 4×10^6 C91/PL alone or with 2×10^6 human foreskin fibroblasts (HFF) cells. [b]Mice with lymphomatous abdominal infiltration were culled at 70, 90, 120, and 195 days after i.p. cell inoculation.

At necroscopy, tumor masses were found in the abdomen, often extending into the pelvic cavity. The tumor infiltrated the mesentery, the small and large intestine, the liver hilum and the pancreas. The spleen was only slightly enlarged and free from lymphoma invasion. Histologically, the neoplastic tissue was composed of large pleomorphic cells with abundant eosinophilic cytoplasm and vesicular basophilic, irregularly lobulated nucleus with prominent nucleoli. Numerous multinucleated syncytial giant cells were also observed and mitotic figures were common. Necrotic and hemorrhagic areas were sometimes found within the lymphomatous masses and organ infiltrates.

Thus, in this set of experiments, using mice deriving from an in-house long-lasting Rag2$^{-/-}\gamma_c^{-/-}$ colony, co-inoculation of fibroblasts was found to trigger the oncogenic potential of the HTLV-1-immortalized T cell line.

Ex Vivo C91/PL-Derived Cell Lines Induce Aggressive Lymphoma

The tumor mass from one of the four diseased mice of the first set of experiments, culled 90 days after cell injection, was removed and processed under sterile conditions and a continuous cell culture was obtained (C91/I), as shown in **Figure 1A**. These cells were then i.p inoculated into 4–5-day-old Rag2$^{-/-}\gamma_c^{-/-}$ mice at different doses and all (7 of 7) animals belonging to the group injected with the higher dose developed a very aggressive and invasive lymphoma with a short latency period (median survival, 15 days, **Table 2**). Two subsequent *in vivo* passages, followed by establishment of tumorigenic cell lines (C91/II and C91/III), were performed. C91/II cells engrafted into mice with high efficiency and also induced aggressive and diffused tumor masses in 2–4-week-old mice (**Table 2**). All mice (10 out of 10) inoculated with 4×10^6 C91/III cells at 4–8 days of age developed a lymphoma with a median latency of 32 days. C91/II and C91/III cell lines showed an increased lung tropism, resulting in frequent lung metastases as well as in lymphomatous lung involvement as primary presentation. Mice inoculated with these cell lines showed a pathological picture similar to that described above and shown in **Figure 1B**. Moreover, involvement by lymphomatous whitish tissue was also observed in the renal pericapsular fatty tissue, with invasion of cortical kidney parenchyma. In some animals, tumor infiltration was also found in ovary, uterine wall and testis. The abdominal and diaphragmatic peritoneal surface was frequently scattered with small neoplastic white nodules and, in two mice, injected with 4×10^6 C91/III cells at 4 days of age, slightly hemorrhagic ascitic fluid was also found in the peritoneal cavity. Furthermore, the muscles surrounding the lumbar spinal column often appeared infiltrated by whitish tumor tissue. Microscopically, mice showed histological and cytological features indicative of lymphomatous growth (**Figure 1C**); kidney, pancreas, liver and intestinal wall were constantly infiltrated (**Figures 1D–G**). Neoplastic white nodules or small metastatic cell embolic aggregates and single neoplastic cells (**Figure 1H**) within capillaries were found in the lung of 10 mice. Immunochemical staining of lung

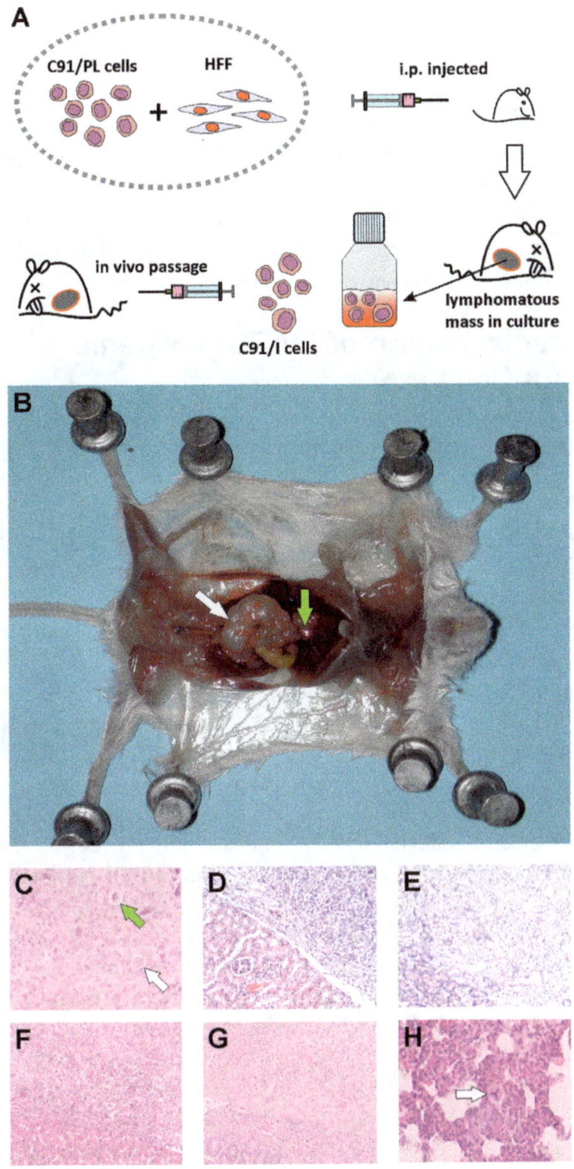

FIGURE 1 | Xenotransplantation of C91/PL-derived cell lines in Rag2$^{-/-}\gamma_c^{-/-}$ mice. **(A)** Schematic representation of the *in vivo* experiments. C91/PL cells intraperitoneally co-injected with human fibroblasts into a 5-day-old Rag2$^{-/-}\gamma_c^{-/-}$ mouse lead to lymphoma development. A fragment of the induced abdominal lymphomatous mass was processed under sterile conditions and cultured *in vitro* to establish a C91/PL-derived lymphomatous cell culture (C91/I). Injection of these cells into a Rag2$^{-/-}\gamma_c^{-/-}$ mouse led to an aggressive Adult T cell Leukemia/Lymphoma (ATLL)-like lymphoma. Additional *in vitro/in vivo* passages were done with the establishment of two other lymphomatous cell lines (C91/II and C91/III). **(B)** Macroscopic view of lymphomatous masses developed in a mouse injected with C91/III cells involving mesenteric nodes (white arrow) and the liver hilum (green arrow). **(C)** Hematoxylin and eosin staining of formalin-fixed, paraffin-embedded tissues shows tumor mass composed of pleomorphic cells mixed with syncytial cells (white arrow) with giant nuclei; the green arrow shows a pleomorphic cell with a giant nucleus. Original magnifications 400×. **(D–G)** Lymphomatous infiltration of kidney pericapsular area, pancreas, liver, and intestinal wall, respectively. Original magnification 200×. **(H)** Lung parenchyma with small embolic metastasis; arrow indicates abnormal tripolar mitosis. Original magnification 400×.

TABLE 2 | Lymphoma induction by C91/PL-derived cells in Rag2$^{-/-}\gamma_c^{-/-}$ mice.

C91/PL-derived cells	Age of mice (days)	No. of cells injected	No. of positive mice/total No. of mice	Median latency (Range, days)
C91/I	4–5	1×10^6	0/4	–
	4–5	4×10^6	7/7	15 (12–34)
C91/II	4–8	4×10^6	15/15	17 (9–41)
	16	4×10^5	3/3	112 (61–112)
	30	4×10^5	3/7	133 (53–202)
	30	4×10^6	3/6	34 (26–34)
C91/III	5	1×10^6	2/3	52 (41–62)
	4–8	4×10^6	10/10	32 (17–62)

metastases to detect human CD25 (i.e., IL-2 receptor alpha chain) showed the presence of a consistent number of neoplastic cells (**Supplementary Figure S1**). The spleen was free from lymphomatous involvement, being characterized by erythroid and myeloid hyperplasia of the red pulp.

These findings indicated that the highly tumorigenic capacity acquired by C91/PL cells following co-transplantation with HFF, and through *in vitro/in vivo* passages, is a stable and reproducible property.

In Vivo Validation of the Supporting Role of Fibroblasts

To further confirm the supporting activity exerted by fibroblasts, a second set of experiments was performed. As the previously tested Rag2$^{-/-}\gamma_c^{-/-}$ mice were no longer available in our animal house (see section "Materials and Methods"), we used brand new mice with identical knock-out genes (BALB/c Rag2$^{-/-}\gamma_c^{-/-}$, BRG), but obtained from a different source and with a well-defined, restricted microbiota, that were endowed with high susceptibility to tumor xenoengraftment (Yamamoto and Schiestl, 2014; Rosshart et al., 2017). The experimental setting was identical to the previous one and two groups of six, 5-day-old, BRG mice were similarly i.p. injected with C91/PL cells alone (control group) or mixed with HFF in a 2:1 ratio. Mice injected with HFF and C91/PL cells developed lymphoma as did control mice injected with C91/PL cells alone (**Figure 2A**). However, HFF co-inoculation significantly triggered lymphoma development compared to control mice, with a twofold decrease in survival time (log-rank test, $p = 0.0168$). The pathological findings of mice developing lymphoma were substantially similar to those observed in the first set of experiments.

Therefore, these data indicate that, in appropriate hosts, C91/PL cells may *per se* induce lymphoma in immunocompromised mice. However, HFF co-inoculation exerts a significant enhancement of the oncogenic potential of C91/PL cells *in vivo*, confirming the relevance of stromal contribution in HTLV-1-mediated lymphomagenesis.

C91/III Cells Efficiently Engraft in NSG Mice

To further confirm the constitutive capability of C91/III cells to induce lymphoma with high reproducible efficiency, a different

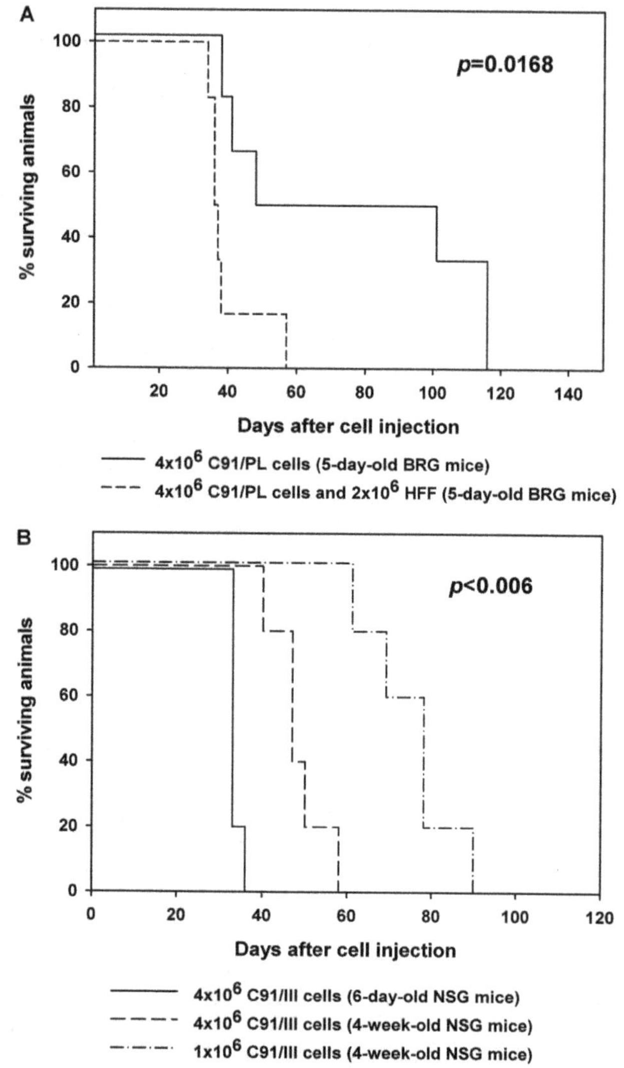

FIGURE 2 | Kaplan–Meier survival curves for mice injected with C91/PL cells, C91/PL cells with human foreskin fibroblasts (HFF), or C91/III cells. **(A)** Five-day-old BRG mice intraperitoneally injected with C91/PL cells and HFF showed a statistically significant decrease in the survival time compared to control mice injected with C91/PL cells (six mice per group). **(B)** Engraftment of C91/III cells in NSG mice. Survival curves for 6-day-old or 4-week-old mice intraperitoneally inoculated with different doses of C91/III cells (five mice per group). *P*-values reported in the figure were calculated by the log-rank test; *p*-value in **B** refers to all pairwise comparisons and was adjusted for multiple comparisons with Bonferroni correction.

mouse strain was employed. Therefore, mice of NSG strain were i.p. injected with C91/III cells. **Figure 2B** shows the Kaplan–Meier survival curves for mice injected with C91/III cells at 6 days or 4 weeks of age (five mice each group). Six-day-old mice injected with 4×10^6 cells had a median survival time (33 days) similar to that of baby $Rag2^{-/-}\gamma_c^{-/-}$, and showed at necroscopy diffused mesenteric lymphomatous masses; two of them showed macroscopic lung metastatic nodules. Interestingly, the same dose of C91/III cells induced lymphoma in all young adult mice with a median survival time of 47 days. Moreover, even at a dose of 1×10^6 cells, all young adult mice died from a diffused tumor, with a median survival time of 78 days. At necroscopy, diseased mice showed abdominal tumor masses involving the peritoneal and pelvic organs and frequent lung involvement.

These data show that C91/III cells, even in low dose, engraft with high efficiency in NSG mice and induce an aggressive lymphoma with a relative short latency in young adult mice. Therefore, C91/III cells xenotransplanted in the adult NSG mouse setting might usefully be employed for preclinical evaluation of drug candidates for ATLL lymphoma variant.

Characterization of the Tumorigenic C91/III Cell Line

Based on the results obtained in NSG mice, C91/III cells were chosen as a representative tumorigenic cell line for further characterization. Analysis of clonality showed an identical monoclonal pattern in the rearrangement of the β and γ chains of the TCR gene in C91/PL and C91/III cells (not shown). STR profile of C91/III cells further confirmed derivation from the parental cell line (**Supplementary Table S1**). The presence of the HTLV-1 provirus was confirmed by a qualitative standard PCR using primer pairs specific for the *tax* region (not shown).

C91/III cells morphologically resembled C91/PL cells, although the percentage of giant cells was higher (**Figure 3A**). Phenotypic analyses did not disclose substantial differential expression of surface markers (**Table 3**). C91/III cells were all CD4+/CD25+ as the parental C91/PL cell line, they expressed similar levels of CD5 and, at high intensity, the two adhesion molecules CADM1 and CD54. C91/PL and C91/III cells did not stain for stem cell markers (CD34, CD117, and CD133). Intracellular staining for forkhead box P3 (FOXP3) transcriptional factor was similar in the two cell lines.

Comparative qRT-PCR analysis of the viral transcript profile in C91/PL and C91/III cells showed no significant variation in the expression of the gag, env, and tax/rex mRNAs. In contrast, all the transcripts coding for the accessory proteins showed a marked increase in C91/III cells, ranging from 4-fold (p21rex) to 18-fold (p13). Statistical analysis showed that these differences were significant ($p \leq 0.05$) except for HBZ Sp1 ($p = 0.056$) (**Figure 3B**).

Characterization of the Secretory Profile of C91/PL, C91/III Cells, and HFF

To assess whether C91/III cells were characterized by a differential profile of secreted soluble factors, we measured 45 cytokines, chemokines and growth factors in the supernatants of both cell lines (**Supplementary Table S2**). The secretory profile of C91/III cells did not qualitatively differ from that of the parental cell line, at least in the 45 soluble factors tested using the Luminex technology. However, the release of 22 factors was quantitatively altered in C91/III cells, as shown in **Figure 4A**. The amount of 21 soluble factors was significantly increased in C91/III cells, while only one (IP-10/CXCL10) was reduced. Many of the upregulated factors were direct pro-inflammatory cytokines (IL-1α, IL-6, TNFα) and α and β chemokines (**Supplementary Table S2**), able to amplify the inflammatory response. Interestingly, the highest

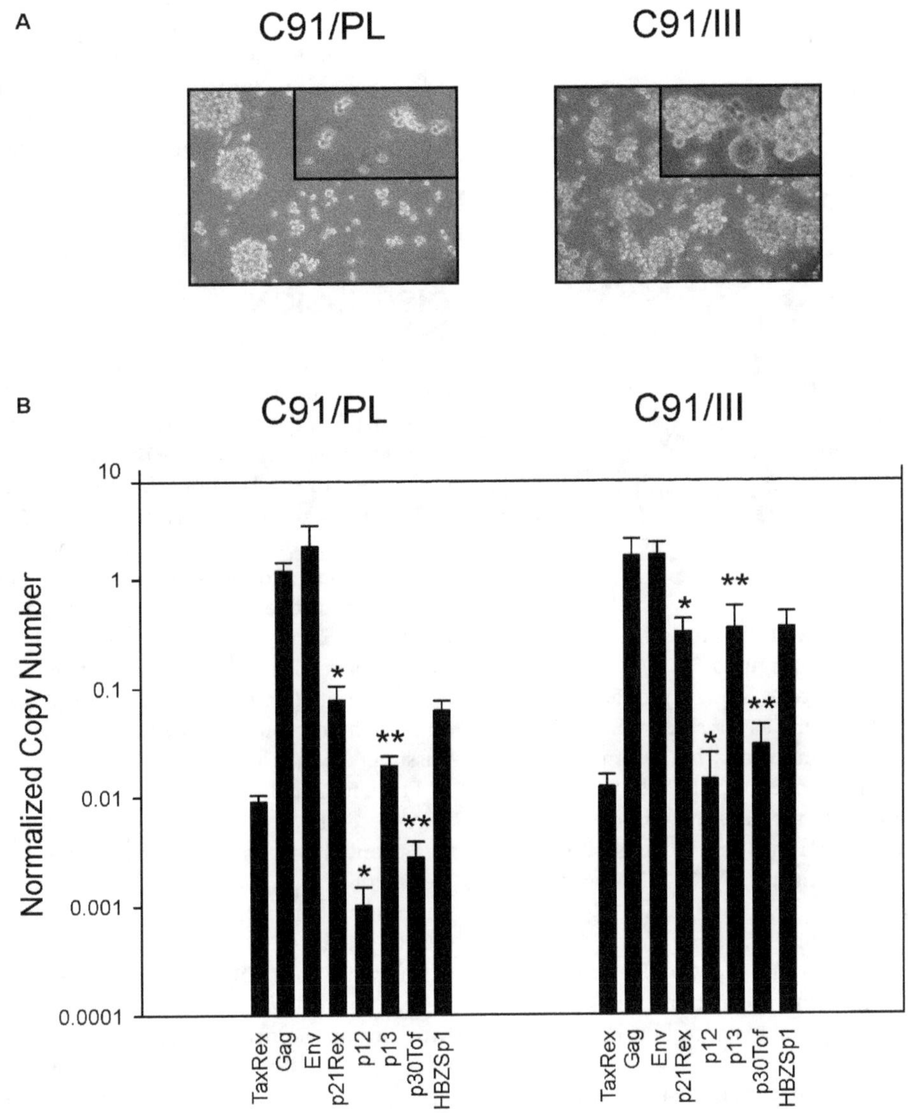

FIGURE 3 | Morphologic and viral characterization of C91/PL and C91/III cell lines. **(A)** Phase contrast images of C91/PL and C91/III cells in standard culture conditions. The tumorigenic C91/III cell line showed a higher percentage of multinucleated giant cells. Original magnification 100×; Upper right-end side insert 200×. **(B)** Profile of Human T cell Leukemia Virus type 1 (HTLV-1) transcripts. C91/III cells showed a statistically significant increase in the normalized copy number (NCN) of all transcripts encoding the viral accessory proteins compared to C91/PL cells. NCN values were calculated by normalizing the absolute copy number of each transcript for the copy number of the glyceraldehyde 3-phosphate dehydrogenase (GAPDH) transcripts. Values are the mean of three independent measurements performed in triplicate; standard error bars are shown. *$p < 0.05$, **$p < 0.01$.

increase was shown by IL-8/CXCL8, a chemokine involved in tumorigenesis through autocrine and paracrine signaling responsible for the triggering of invasiveness, angiogenesis and metastasis (Balkwill and Mantovani, 2001).

We also characterized the secretory pattern of HFF used in our *in vivo* experiments (**Supplementary Table S2**). The high release of many chemokines and IL-6 indicated that HFF partially overlapped the secretory profile of "activated" fibroblasts (Kalluri and Zeisberg, 2006). Among the growth factors, HGF, βNGF, and VEGF-A were found in measurable amounts, while EGF, FGF-2, and PDGF-BB, which are other soluble mediators secreted by activated fibroblasts, were not detected. Interestingly, some of the factors released by C91/III cells showing the most pronounced increase were characteristic of the HFF secretory profile (**Figure 4B**) suggesting that C91/III cells might have reinforced autonomous stimulatory activity, thus becoming independent from the microenvironment support.

Analysis of the Crosstalk Between C91/PL Cells and HFF

As HFF were found to trigger tumorigenesis of C91/PL cells, we studied the impact of HFF on C91/PL cells in a co-culture system. Results showed that co-cultivation with HFF

TABLE 3 | Immunophenotypic characterization of C91/PL and C91/III cells.

Marker	C91/PL	C91/III
CD1a	–[a]	–
CD2	–	–
CD3	–	–
CD4	++++[b]	+++/++++
CD5	++/+++	++/+++
CD7	–	–
CD25	+++	++/+++
CD34	–	–
CD117	–	–
CD133	–	–
CD54	++++	++++
CADM1	++++	++++
FOXP3	+++	+++

[a]Negative. [b]+ indicates that <25% of cells expressed the analyzed surface marker by cytofluorimetric analyses; ++ indicates 25–50%; +++ indicates 50–75%; ++++ indicates >75%.

significantly decreased the induced apoptosis in C91/PL cells after 48 and 72 h of co-culture (**Figures 5A,B**). On the other hand, HFF-co-cultured C91/PL cells showed proliferation rates comparable to those of C91/PL cells, as assessed by CFSE labeling (**Figure 5C**). Of interest, short-term and long-term co-culture with HFF led to morphologic changes in C91/PL cells resembling those of C91/III cells (**Figure 5D**, see **Figure 3A** for comparison), whereas immunophenotypic variations were not observed for the periodically checked surface antigens (CD4, CD25, and CD54, data not shown).

In vivo lymphomagenic activity was analyzed in 5-day-old NSG mice injected i.p. with co-cultured C91/PL cells, which showed a statistically significant reduction (log-rank test, $p = 0.0005$) in the survival compared to control mice i.p. injected with C91/PL cells (**Figure 5E**).

On the other hand, no significant variation in the expression of any of the viral mRNAs was detected (**Figure 5F**), suggesting that the co-culture microenvironment does not have an impact on either HTLV-1 rate of transcription or on the alternative splicing pattern of viral mRNAs.

Interestingly, the secretory pattern of co-cultured C91/PL cells was similar, albeit not identical, to that found in C91/III cells (**Supplementary Figure S2**), with a statistically significant increase in IL-8/CXCL8 and many other chemokines. Among pro-inflammatory factors, a significant increase was evidenced for TNFα, whereas IL-1α and IL-6 were augmented but not at a statistically significant level.

To assess the relevance of the paracrine crosstalk between C91/PL cells and HFF, two soluble factors were measured after 3 and 10 days of co-culture in direct contact or in transwell inserts. Transwell co-culture induced an increase, even though less pronounced, in secreted IL-8/CXCL8 and TNFα (**Supplementary Figure S3**), suggesting that soluble crosstalk is also implicated in this heterotypic cell interaction.

On the whole, these data indicated that HFF may increase the survival of C91/PL cells *in vitro* and may contribute to the induction of morphological and secretory changes similar to those of C91/III cells. Most importantly, co-cultured C91/PL cells were much more tumorigenic compared to the parental cells when injected into baby NSG mice. On the other hand, the remarkably overlapping viral expression patterns of C91/PL cells, with or without HFF, argue against a role of viral genes in the *in vivo* more aggressive behavior of co-cultured C91/PL cells and, likely, of C91/III cells.

DISCUSSION

To recapitulate ATLL pathogenesis *in vivo*, transgenic and humanized mouse models have been developed. Tax-transgenic mice mainly developed arthropathy and other inflammatory disorders (Iwakura et al., 1991; Yamamoto et al., 1993; Saggioro et al., 1997) or solid tumor (Hinrichs et al., 1987; Nerenberg et al., 1987) and leukemia (Grossman et al., 1995; Hasegawa et al., 2006; Ohsugi, 2013), depending on the promoter used. HBZ-transgenic mice develop not only T-cell lymphoma but also systemic chronic inflammation after a long latency period and with low/variable incidence (Satou et al., 2011). Thus, transgenic mice are useful tools to investigate the activities of the Tax and HBZ proteins but are limited in that the resulting phenotypes are dependent on a given promoter and the functions of other viral products are missing.

Further ATLL models exploiting the humanized mice were recently developed, based on immunodeficient animals transplanted with human hematopoietic stem cells from cord blood or fetal liver, thus able to reconstitute a human immune system permissive to HTLV-1 infection. Villaudy et al. (2011) generated humanized mice by intrahepatic engraftment of human cord blood $CD34^+$ cells into $Rag2^{-/-}\gamma_c{}^{-/-}$ mice. During thymocyte maturation, these mice were infected with intraperitoneal injection of irradiated MT-2 cells as virus donors. Some of the animals, after a long latency, showed pathological signs resembling the lymphoma variant of ATLL. A different clinical presentation was achieved in HTLV-1-infected humanized NOG mice (Tezuka et al., 2014), showing ATLL-like leukemogenesis, characterized by leukocytosis, hepatosplenomegaly, high plasmatic levels of cytokines and sporadic appearance of flower cells in peripheral blood. The humanized mouse model, although promising for investigating the natural history of HTLV-1 infection and the host-specific immune response, critical steps for eventual development of ATLL, is however not easily achievable because of the rather complex procedures including the ethics committee evaluation for the use of human primary hematopoietic stem cells and fetal tissues.

Engraftment of *in vitro* HTLV-1-infected and immortalized T cells gave variable results in SCID mice, depending on which cell line was tested (Feuer et al., 1993; Imada et al., 1995), and was shown to be facilitated by abrogation of the host NK function obtained through irradiation or treatment with an anti-asialo GM1 antibody (Ishihara et al., 1992; Feuer et al., 1995).

Subsequent experiments on xenotransplantation of *in vitro* HTLV-1-transformed T cell lines or peripheral blood

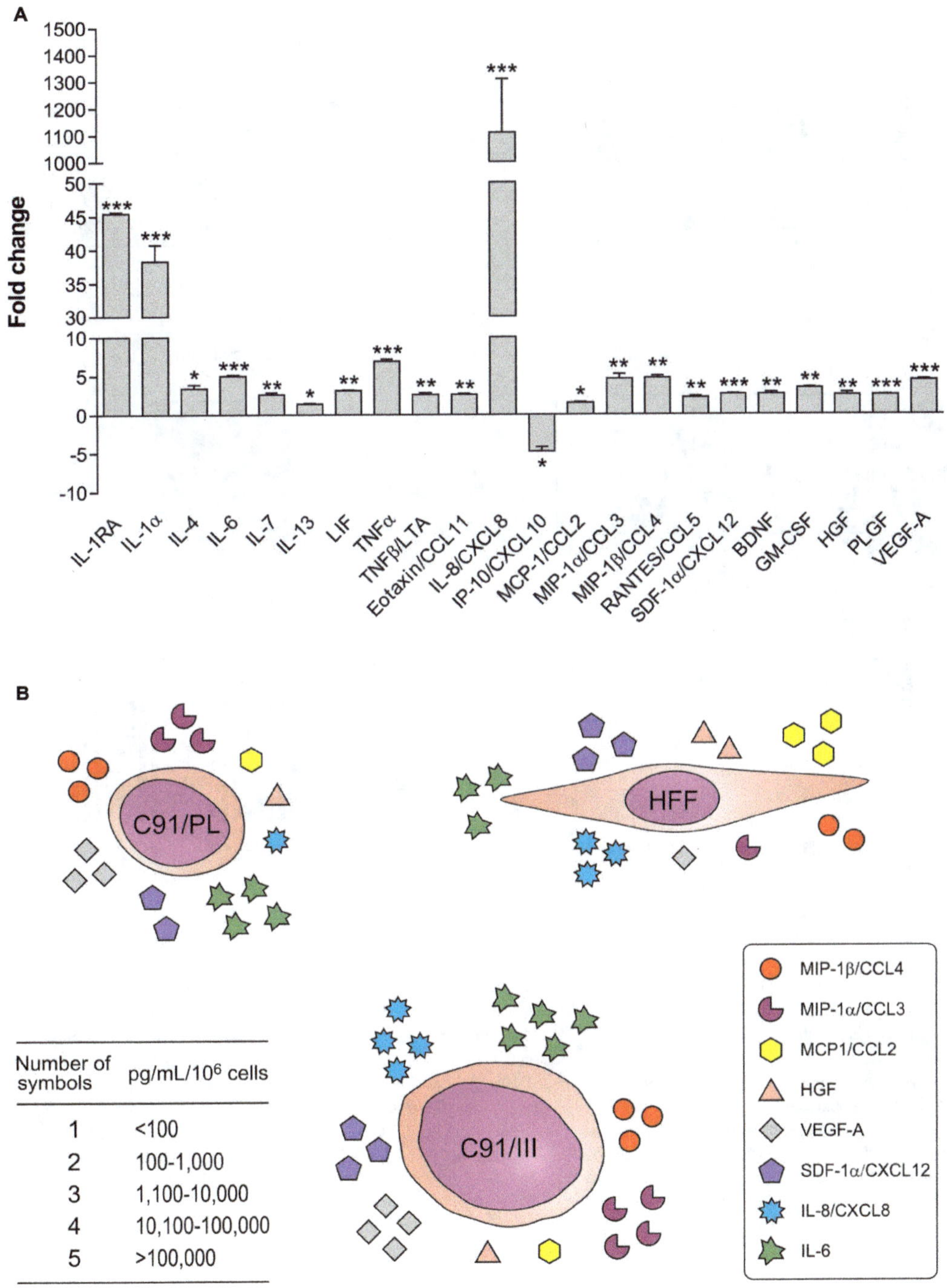

Number of symbols	pg/mL/10[6] cells
1	<100
2	100-1,000
3	1,100-10,000
4	10,100-100,000
5	>100,000

FIGURE 4 | Soluble factors differentially released by C91/III cells. **(A)** Data are reported as ratio between the mean of the values, expressed in pg/mL/10^6 cells, measured in the supernatants of C91/III cells and in the supernatants of C91/PL cells; standard deviations of the ratio are calculated according to the theory of error propagation (Calabro et al., 2009). Only significantly different soluble factors are shown, and statistical significance is referred to the comparison of the concentration values between the two cell lines, calculated by two-tailed Student's *t*-test. *$p < 0.05$, **$p < 0.01$, ***$p < 0.001$. **(B)** The figure schematically shows that some of the soluble factors increased in the lymphomagenic cells belong to the HFF secretory profile, suggesting that C91/III cells acquired an autocrine stimulatory loop, becoming independent from HFF support.

mononuclear cells from HTLV-1-infected asymptomatic carriers in mice with various degrees of immunodeficiency confirmed that lack of innate immunity is crucial for a favorable neoplastic infiltration and/or outgrowth (Liu et al., 2002; Takajo et al., 2007). This finding is consistent with the beneficial effect exerted by alloreactive donor NK cells, generated following hematopoietic stem cell transplantation, on the clinical outcome of leukemia patients (Gismondi et al., 2015).

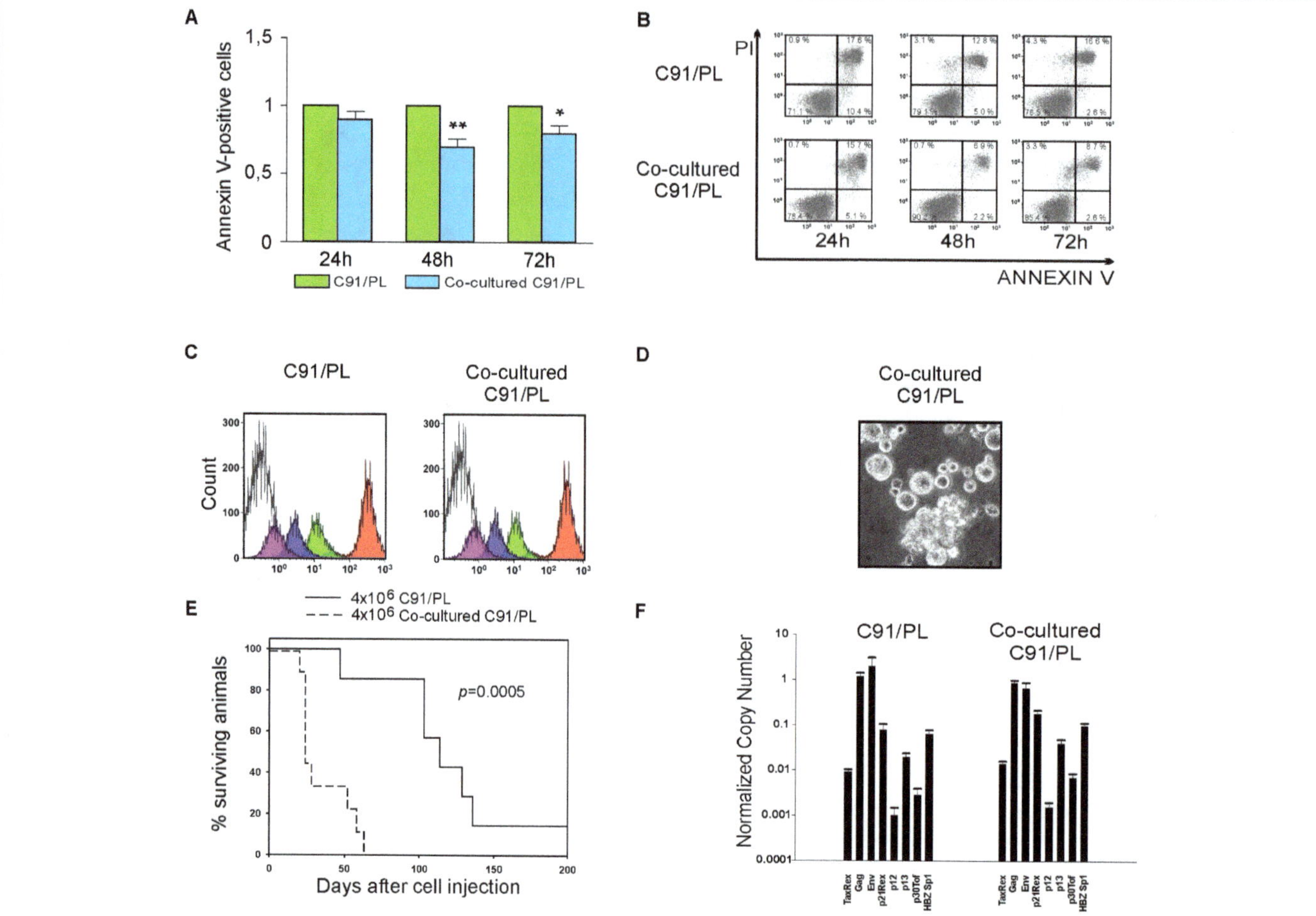

FIGURE 5 | Analysis of the *in vitro* crosstalk of C91/PL cells and HFF. **(A)** Co-culture with HFF reduced apoptosis in C91/PL cells. Apoptosis was analyzed in C91/PL cultured in standard conditions and in C91/PL cells co-cultured with HFF cells at 24, 48, and 72 h. Data are reported as ratio between the mean of the percentage of Annexin V–positive cells in two different co-cultures and the standard deviations (SD) of the ratio are calculated among four different experimental groups. Data are presented as mean ± SD. SD of the ratio was calculated according to the theory of error propagation. Statistical significance was determined by two-tailed Student's *t*-test. $^{*}p < 0.05$, $^{**}p < 0.01$. **(B)** Flow cytometry analysis of Annexin V/Propidium Iodide (PI) stained C91/PL cells cultured in standard conditions **(Upper)** and in co-culture with HFF **(Lower)** at three different time points. A representative experiment is shown and the percentage of cells in each quadrant of the flow plots is provided. **(C)** Measurement of the *in vitro* proliferation of C91/PL cells cultured in the absence **(Left)** or presence **(Right)** of HFF. Cells were labeled with carboxyfluorescein succinimidyl ester (CFSE) and analyzed with XL Epic cytofluorimeter after 0 h (red), 24 h (green), 48 h (blue), and 72 h (violet). The shaded histogram shows the unlabeled cells. The plots show that co-culture with HFF does not affect the proliferation rate of C91/PL cells. Histograms represent the data obtained in two independent experiments. **(D)** Phase contrast images of C91/PL cells after co-culture with HFF. Eight-week-co-culture with HFF induced an increase in giant cells, thus resembling the C91/III cell line. Original magnification 100×. **(E)** Kaplan–Meier survival curves for 5-day-old NSG mice i.p. injected with 4×10^6 C91/PL cells (seven mice) and with 4×10^6 co-cultured C91/PL cells (nine mice). A statistically significant reduction (log-rank test; $p = 0.0005$) in the overall survival of mice injected with co-cultured cells was observed. **(F)** Analysis of HTLV-1 transcripts. Quantitative analyses of viral transcripts showed no significant variation in any of the HTLV-1 mRNAs in co-cultured C91/PL cells compared to the C91/PL cell line, suggesting that the increased survival and *in vivo* lymphoma induction acquired by co-cultured cells were not likely due to virus-encoded factors. Data are reported as described in **Figure 3B**.

Therefore, to focus on the influence of host microenvironment other than immune cell components, BALB/c Rag2$^{-/-}\gamma_c^{-/-}$ and BRG ("excluded flora") mice, characterized by immunological dysfunction of T, B, and NK cells, were employed in the present study. By injecting intraperitoneally newborn mice with C91/PL cells together with HFF a significant triggering of lymphoma development was observed in two sets of experiments (**Table 1** and **Figure 2A**). The different percentage of mice developing lymphoma in the two series of experiments may likely be explained by mouse intra-strain genetic background variation (Eisener-Dorman et al., 2009) and/or different gut microbiota (Yamamoto and Schiestl, 2014; Rosshart et al., 2017). Unfortunately, experiments could not be continued in the Rag2$^{-/-}\gamma_c^{-/-}$ mice first employed in our experiments, as these animals are no longer available. C91/PL-derived cell lines were further established, which consistently induced neoplastic infiltration when injected alone in newborn and young adult Rag2$^{-/-}\gamma_c^{-/-}$ and NSG mice (**Table 2** and **Figure 2B**). This indicated that a stable and highly tumorigenic capacity was acquired by C91/PL cells following co-transplantation with HFF and subsequent *in vitro/in vivo* passages. Immunodeficient mice xenotransplanted with C91/I, C91/II, and C91/III cell lines exhibited lymphomatous masses diffused to lungs, abdominal and pelvic organs, reminiscent of the lymphoma variant of ATLL (**Figure 1**). Their engraftment efficiency was more reproducible and shorter in latency than that exhibited by the parental C91/PL cells. Interestingly, the frequent lymphomatous involvement of lungs in mice injected with C91/II and C91/III cells indicates an increased lung tropism, a peculiar feature of ATLL cells. In fact, pulmonary chronic inflammatory lesions were described in about 30% of HTLV-1 carriers and lung lymphomatous infiltration was observed in 45% of ATLL patients (Yoshioka et al., 1985; Okada et al., 2006).

The role of stromal microenvironment, and particularly of fibroblasts, its major cellular component, on tumor induction has been extensively documented in epithelial cancers, leading to the notion that fibroblasts exhibit a bimodal effect (Bhowmick et al., 2004; Kalluri and Zeisberg, 2006). In the initial phase of tumorigenesis, they constrain the oncogenic conversion by remodeling the stromal architecture, while in later stages they promote tumor progression and invasiveness, engaging an active crosstalk with the cancerous cells (Marsh et al., 2013; Heneberg, 2016). Complex crosstalk between bone marrow microenvironment and leukemia cells has also been reported, and therapeutic interventions targeting leukemia niches have been proposed (Jin et al., 2008; Zhang et al., 2012). Similarly, lymphoma cells may interact with stromal components to acquire a more aggressive and invasive behavior. In the setting of ATLL, co-culture of clinical samples of ATLL cells, *in vitro* HTLV-1-transformed and ATLL-derived cell lines with murine bone marrow stromal cells supported the growth of primary ATLL cells as well as established cell lines (Nagai et al., 2008). The proliferative enhancement was paralleled by downregulation of Tax, while the expression level of HBZ gene remained unchanged (Nagai et al., 2008). Kinpara et al. (2009) also found that co-culture of ATLL-derived or HTLV-1-transformed T cells with human epithelial-like cells (HEK293T cells) or with murine NIH3T3 fibroblasts did suppress viral p19 and *gag* mRNA, and this effect, albeit reversible, was mediated by IFN-α and IFN-β. ATLL cell lines and primary ATLL cells, after short-term co-culture with epithelial-like cells were found to be protected from apoptosis, became quiescent and acquired a cancer stem cell-like phenotype (Miyatake et al., 2013, 2015). All these findings highlight the relevant impact of microenvironment in ATLL pathogenesis, indicating that stromal signaling may contribute to the establishment of viral latency as well as may induce *in vitro* a more aggressive neoplastic phenotype. In the present study we show the role of human primary fibroblasts in T-cell lymphomagenesis using an *in vivo* preclinical setting.

Preliminary characterization of the lymphomagenic C91/III cell line, derived from the parental C91/PL cells, did not show relevant phenotypic changes (**Table 3**), but the analysis of its secretory repertoire indicated that mechanisms involved in epithelial cancers may also apply to ATLL. Indeed, a relevant role in lymphomagenesis is played by cytokines and chemokines which provide a milieu favorable to tumor growth and a signaling path for cell migration and tissue invasion (Hsu et al., 1993). The disseminated nature of the neoplasia developed in C91/III-injected mice was paralleled by a substantial alteration in the secretory profile of these cells, mainly characterized by the significant increase of several soluble factors, most of which equipped with a direct (IL-1α, IL-6, TNFα) or indirect (mainly chemokines) pro-inflammatory activity (**Figure 4**). The major fold increment was found in IL-8/CXCL8, an important mediator for tissue infiltration (Balkwill and Mantovani, 2001). Remarkably, some of the factors increased in C91/III cells also belonged to the secretory profile of ATLL cells (Yamada et al., 1996) as well as of HFF, and are known to facilitate tumor progression *in vivo* (Davalos et al., 2010). These data suggest that C91/III cells may autonomously induce a pro-inflammatory-like status that, through autocrine and paracrine loops, is able to promote tumorigenesis and organ infiltration. Interestingly, the *in vitro* crosstalk of C91/PL cells with HFF led to the acquisition of a pro-inflammatory secretory phenotype similar to that observed in the tumorigenic C91/III cells (**Supplementary Figure S2**). HFF-co-cultured C91/PL cells not only changed their turn over, acquiring resistance to induced apoptosis (**Figures 5A,B**) but, most importantly, acquired an enhanced lymphomagenic capability upon xenotransplantation (**Figure 5E**).

Transwell experiments also demonstrated a moderate increase in IL-8/CXCL8 and TNFα secretion, thus suggesting that, in addition to cell-to-cell contact, soluble factors play a role in the heterotypic cell interaction (**Supplementary Figure S3**). However, it should be mentioned that *in vivo* preliminary experiments by co-injection of conditioned medium from HFF with C91/PL cells failed to increase cell tumorigenesis (data not shown). Additional studies are needed to further confirm that HFF-secreted factors are not sufficient *per se* to modify the *in vivo* behavior of C91/PL cells. The overall increase in the expression of all viral accessory mRNAs observed in C91/III cells (**Figure 3B**) might suggest the acquisition of defective proviruses during the process of clonal evolution of these cells, and denotes another feature reminiscent of ATLL cells *in vivo* (Ohshima et al., 1991). However, HFF-co-cultured C91/PL cells showed no significant differences in the viral transcriptional profile (**Figure 5F**), suggesting that the increased tumorigenicity of co-cultured cells, and likely of C91/III cells, is not driven by viral factors.

Evolutionary models of cancer development imply the existence of silent precursor cells from which, through a stepwise process of accumulation of somatic mutations, aggressive malignancies eventually rise (Greaves, 2015). Actually, persistence of the pre-leukemic precursors, carrying a founder mutation, albeit not malignant *per se*, has recently been demonstrated in acute leukemia (Shlush et al., 2014). In the ATLL setting, it may be envisaged that the insertion of the HTLV-1 DNA provirus in the genome of T cells constitutes the "founder" mutation conferring them the properties of persistent, pre-malignant ancestors, which, upon additional favorable forces, may ultimately reach full oncogenic capacity. A well-known similar condition is represented by the long life persistence of B cells immortalized by the Epstein-Barr virus from which lymphomas may develop in immunosuppressed subjects.

CONCLUSION

The finding that fibroblasts possessing an activated-like, pro-inflammatory secretory phenotype enhance the tumorigenesis of C91/PL cells throws light on the triggering activity of the host microenvironment on HTLV-1-linked lymphomagenesis. More work is needed to clarify the basis of this triggering effect as well as to reveal whether genetic and epigenetic changes, produced within the HTLV-1-immortalized T cells, share the responsibility for full-blown malignancy.

Furthermore, our results show that the C91/III cell line, originating from a long-term *in vitro* HTLV-1-immortalized T cell line, is endowed with a highly reproducible lymphomagenic capacity in NSG mice that may be exploited as a simple murine preclinical model to investigate *in vivo* the efficacy of therapeutic interventions against the lymphoma variant of ATLL.

AUTHOR CONTRIBUTIONS

MV and AM designed the experiments, performed the research and data analysis, and contributed to paper writing. BM and MAP performed the research, analyzed the results, and contributed to paper writing. AA contributed to the interpretation of data and critically reviewed the paper. IC performed the research and data analysis for HTLV-1 transcriptional profiles, and critically revised the paper. LC-B conceived and designed the study, established the preclinical model, analyzed the results, and wrote the paper. MLC conceived and designed the study, established the new cell lines, analyzed the results, and wrote the paper.

FUNDING

This work was supported by grants from Associazione Italiana contro le Leucemie-Linfomi e Mieloma (Grant CUP No. C92I15002530007, to LC-B and AA) and from 5X1000-IOV2011 (Grant CUP No. J94G14000180001 to MLC). MV, AM, BM, MAP, and IC were recipients of a Ricerca Corrente fellowship, Italian Ministry of Health (IMH). IMH funding was also used for payment of the publication fee.

ACKNOWLEDGMENTS

The authors thank Giuseppe Opocher for constant support, Laura Lignitto, Gianni Esposito, Vito Barbieri, Francesca Rende, Chiara Menin, and Ambrogio Fassina for their help in various aspects of this work, and Christina Drace for help in preparing the manuscript. They also thank Robert C. Gallo for initial helpful comments and suggestions and Vincenzo Ciminale for critical revision.

SUPPLEMENTARY MATERIAL

FIGURE S1 | Representative photomicrographs of immunoreactions for CD25 showing lymphomatous infiltration in the lung parenchyma **(A)** and a nearby pulmonary bronchiole **(B)** observed in a C91/III cell-injected mouse. Original magnification 400×.

FIGURE S2 | Soluble factors released by C91/PL cells after co-culture with HFF. Cytokines, chemokines and growth factors secreted in culture supernatants by C91/PL cells removed after 2 months from the co-culture with HFF and maintained as long as 6 weeks in culture. Values are normalized to those of C91/PL cells. Data are reported as ratio between the mean of the values, expressed in pg/mL/10^6 cells, measured in co-cultured C91/PL cells and the mean of the values measured in the supernatants of C91/PL cells, and the standard deviations of the ratio are calculated according to the theory of error propagation (Calabro et al., 2009). Only significantly different soluble factors are shown, and statistical significance was determined by two-tailed Student's *t*-test. $^*p < 0.05$, $^{**}p < 0.01$, $^{***}p < 0.001$.

FIGURE S3 | IL-8/CXCL8 and TNFα released in culture supernatants after short-term co-culture of C91/PL cells with HFF. A significant increase in the secretion of IL-8/CXCL8 **(A)** and TNFα **(B)** was observed after 3 days of co-culture of C91/PL cells, either in direct contact or placed in transwell inserts, with HFF; IL-8/CXCL8 increment persisted after 10 days of co-culture. Control wells with C91/PL cells or HFF were set up and analyzed in parallel. As previously observed (**Supplementary Table S2**), HFF did not contribute to TNFα increase. The increase in IL-8/CXCL8 and TNFα in transwell co-cultures, albeit lower than that measured in direct co-cultures, indicated that the heterotypic crosstalk is also mediated by soluble factors. Data are expressed in pg/mL/10^6 cells. Statistical significance was calculated by two-tailed Student's *t*-test. $^*p < 0.05$; $^{**}p < 0.01$; $^{***}p < 0.001$; $^{****}p < 0.0001$.

TABLE S1 | Short Tandem Repeat (STR) profiles of cell lines.

TABLE S2 | Soluble factors released by C91/PL and C91/III cells and human foreskin fibroblasts (HFF).

REFERENCES

Armstrong, A. P., Franklin, A. A., Uittenbogaard, M. N., Giebler, H. A., and Nyborg, J. K. (1993). Pleiotropic effect of the human T-cell leukemia virus Tax protein on the DNA binding activity of eukaryotic transcription factors. *Proc. Natl. Acad. Sci. U.S.A.* 90, 7303–7307. doi: 10.1073/pnas.90.15.7303

Balkwill, F., and Mantovani, A. (2001). Inflammation and cancer: back to Virchow? *Lancet* 357, 539–545. doi: 10.1016/S0140-6736(00)04046-0

Baranger, A. M., Palmer, C. R., Hamm, M. K., Giebler, H. A., Brauweiler, A., Nyborg, J. K., et al. (1995). Mechanism of DNA-binding enhancement by the human T-cell leukaemia virus transactivator Tax. *Nature* 376, 606–608. doi: 10.1038/376606a0

Bhowmick, N. A., Chytil, A., Plieth, D., Gorska, A. E., Dumont, N., Shappell, S., et al. (2004). TGF-beta signaling in fibroblasts modulates the oncogenic potential of adjacent epithelia. *Science* 303, 848–851. doi: 10.1126/science.1090922

Bissell, M. J., and Hines, W. C. (2011). Why don't we get more cancer? A proposed role of the microenvironment in restraining cancer progression. *Nat. Med.* 17, 320–329. doi: 10.1038/nm.2328

Calabro, M. L., Gasperini, P., Di Gangi, I. M., Indraccolo, S., Barbierato, M., Amadori, A., et al. (2009). Antineoplastic activity of lentiviral vectors expressing interferon-alpha in a preclinical model of primary effusion lymphoma. *Blood* 113, 4525–4533. doi: 10.1182/blood-2008-09-180307

Calabro, M. L., Luparello, M., Grottola, A., Del Mistro, A., Fiore, J. R., Angarano, G., et al. (1993). Detection of human T lymphotropic virus type II/b in human immunodeficiency virus type 1-coinfected persons in southeastern Italy. *J. Infect. Dis.* 168, 1273–1277. doi: 10.1093/infdis/168.5.1273

Cavallari, I., Rende, F., Bender, C., Romanelli, M. G., D'Agostino, D. M., and Ciminale, V. (2013). Fine tuning of the temporal expression of HTLV-1 and HTLV-2. *Front. Microbiol.* 4:235. doi: 10.3389/fmicb.2013.00235

Davalos, A. R., Coppe, J. P., Campisi, J., and Desprez, P. Y. (2010). Senescent cells as a source of inflammatory factors for tumor progression. *Cancer Metastasis Rev.* 29, 273–283. doi: 10.1007/s10555-010-9220-9

de Oliveira, M. S., Matutes, E., Famadas, L. C., Schulz, T. F., Calabro, M. L., Nucci, M., et al. (1990). Adult T-cell leukaemia/lymphoma in Brazil and its relation to HTLV-I. *Lancet* 336, 987–990. doi: 10.1016/0140-6736(90)92432-H

Di Stefano, M., Calabro, M. L., Di Gangi, I. M., Cantatore, S., Barbierato, M., Bergamo, E., et al. (2008). In vitro and in vivo human herpesvirus 8 infection of placenta. *PLoS One.* 3:e4073. doi: 10.1371/journal.pone.0004073

Dodon, M. D., Villaudy, J., Gazzolo, L., Haines, R., and Lairmore, M. (2012). What we are learning on HTLV-1 pathogenesis from animal models. *Front. Microbiol.* 3:320. doi: 10.3389/fmicb.2012.00320

Eisener-Dorman, A. F., Lawrence, D. A., and Bolivar, V. J. (2009). Cautionary insights on knockout mouse studies: the gene or not the gene? *Brain Behav. Immun.* 23, 318–324. doi: 10.1016/j.bbi.2008.09.001

Erez, N., Truitt, M., Olson, P., Arron, S. T., and Hanahan, D. (2010). Cancer-associated fibroblasts are activated in incipient neoplasia to orchestrate tumor-promoting inflammation in an NF-κB-dependent manner. *Cancer Cell* 17, 135–147. doi: 10.1016/j.ccr.2009.12.041

Felber, B. K., Paskalis, H., Kleinman-Ewing, C., Wong-Staal, F., and Pavlakis, G. N. (1985). The pX protein of HTLV-I is a transcriptional activator of its long terminal repeats. *Science* 229, 675–679. doi: 10.1126/science.2992082

Feuer, G., Stewart, S. A., Baird, S. M., Lee, F., Feuer, R., and Chen, I. S. (1995). Potential role of natural killer cells in controlling tumorigenesis by human T-cell leukemia viruses. *J. Virol.* 69, 1328–1333.

Feuer, G., Zack, J. A., Harrington, W. J. Jr., Valderama, R., Rosenblatt, J. D., Wachsman, W., et al. (1993). Establishment of human T-cell leukemia virus type I T-cell lymphomas in severe combined immunodeficient mice. *Blood* 82, 722–731.

Fochi, S., Mutascio, S., Bertazzoni, U., Zipeto, D., and Romanelli, M. G. (2018). HTLV deregulation of the NF-kappaB pathway: an update on tax and antisense proteins role. *Front. Microbiol.* 9:285. doi: 10.3389/fmicb.2018.00285

Fujii, M., Iwai, K., Oie, M., Fukushi, M., Yamamoto, N., Kannagi, M., et al. (2000). Activation of oncogenic transcription factor AP-1 in T cells infected with human T cell leukemia virus type 1. *AIDS Res. Hum. Retroviruses* 16, 1603–1606. doi: 10.1089/08892220050193029

Gaudray, G., Gachon, F., Basbous, J., Biard-Piechaczyk, M., Devaux, C., and Mesnard, J. M. (2002). The complementary strand of the human T-cell leukemia virus type 1 RNA genome encodes a bZIP transcription factor that down-regulates viral transcription. *J. Virol.* 76, 12813–12822. doi: 10.1128/JVI.76.24.12813-12822.2002

Gessain, A., Barin, F., Vernant, J. C., Gout, O., Maurs, L., Calender, A., et al. (1985). Antibodies to human T-lymphotropic virus type-I in patients with tropical spastic paraparesis. *Lancet* 2, 407–410. doi: 10.1016/S0140-6736(85)92734-5

Gessain, A., and Cassar, O. (2012). Epidemiological aspects and world distribution of HTLV-1 infection. *Front. Microbiol.* 3:388. doi: 10.3389/fmicb.2012.00388

Giam, C. Z., and Semmes, O. J. (2016). HTLV-1 infection and adult T-Cell leukemia/lymphoma-A tale of two proteins: tax and HBZ. *Viruses* 8:E161. doi: 10.3390/v8060161

Gismondi, A., Stabile, H., Nisti, P., and Santoni, A. (2015). Effector functions of natural killer cell subsets in the control of hematological malignancies. *Front. Immunol.* 6:567. doi: 10.3389/fimmu.2015.00567

Grassmann, R., Aboud, M., and Jeang, K. T. (2005). Molecular mechanisms of cellular transformation by HTLV-1 Tax. *Oncogene* 24, 5976–5985. doi: 10.1038/sj.onc.1208978

Greaves, M. (2015). Evolutionary determinants of cancer. *Cancer Discov.* 5, 806–820. doi: 10.1158/2159-8290.CD-15-0439

Grossman, W. J., Kimata, J. T., Wong, F. H., Zutter, M., Ley, T. J., and Ratner, L. (1995). Development of leukemia in mice transgenic for the tax gene of human T-cell leukemia virus type I. *Proc. Natl. Acad. Sci. U.S.A.* 92, 1057–1061. doi: 10.1073/pnas.92.4.1057

Hajj, H. E., Nasr, R., Kfoury, Y., Dassouki, Z., Nasser, R., Kchour, G., et al. (2012). Animal models on HTLV-1 and related viruses: what did we learn? *Front. Microbiol.* 3:333. doi: 10.3389/fmicb.2012.00333

Hasegawa, H., Sawa, H., Lewis, M. J., Orba, Y., Sheehy, N., Yamamoto, Y., et al. (2006). Thymus-derived leukemia-lymphoma in mice transgenic for the Tax gene of human T-lymphotropic virus type I. *Nat. Med.* 12, 466–472. doi: 10.1038/nm1389

Heneberg, P. (2016). Paracrine tumor signaling induces transdifferentiation of surrounding fibroblasts. *Crit. Rev. Oncol. Hematol.* 97, 303–311. doi: 10.1016/j.critrevonc.2015.09.008

Hermine, O. (2015). ATL treatment: is it time to change? *Blood.* 126, 2533–2534. doi: 10.1182/blood-2015-10-670489

Hinrichs, S. H., Nerenberg, M., Reynolds, R. K., Khoury, G., and Jay, G. (1987). A transgenic mouse model for human neurofibromatosis. *Science* 237, 1340–1343. doi: 10.1126/science.2888191

Hsu, S. M., Waldron, J. W. Jr., Hsu, P. L., et al. (1993). Cytokines in malignant lymphomas: review and prospective evaluation. *Hum. Pathol.* 24, 1040–1057. doi: 10.1016/0046-8177(93)90183-H

Imada, K., Takaori-Kondo, A., Akagi, T., Shimotohno, K., Sugamura, K., Hattori, T., et al. (1995). Tumorigenicity of human T-cell leukemia virus type I-infected cell lines in severe combined immunodeficient mice and characterization of the cells proliferating in vivo. *Blood* 86, 2350–2357.

Ishihara, S., Tachibana, N., Okayama, A., Murai, K., Tsuda, K., and Mueller, N. (1992). Successful graft of HTLV-I-transformed human T-cells (MT-2) in severe combined immunodeficiency mice treated with anti-asialo GM-1 antibody. *Jpn. J. Cancer Res.* 83, 320–323. doi: 10.1111/j.1349-7006.1992.tb00108.x

Iwakura, Y., Tosu, M., Yoshida, E., Takiguchi, M., Sato, K., Kitajima, I., et al. (1991). Induction of inflammatory arthropathy resembling rheumatoid arthritis in mice transgenic for HTLV-I. *Science* 253, 1026–1028. doi: 10.1126/science.1887217

Jin, L., Tabe, Y., Konoplev, S., Xu, Y., Leysath, C. E., Lu, H., et al. (2008). CXCR4 up-regulation by imatinib induces chronic myelogenous leukemia (CML) cell migration to bone marrow stroma and promotes survival of quiescent CML cells. *Mol. Cancer Ther.* 7, 48–58. doi: 10.1158/1535-7163.MCT-07-0042

Kalluri, R., and Zeisberg, M. (2006). Fibroblasts in cancer. *Nat. Rev. Cancer* 6, 392–401. doi: 10.1038/nrc1877

Kataoka, K., Iwanaga, M., Yasunaga, J. I., Nagata, Y., Kitanaka, A., Kameda, T., et al. (2018). Prognostic relevance of integrated genetic profiling in adult T-cell leukemia/lymphoma. *Blood* 131, 215–225. doi: 10.1182/blood-2017-01-761874

Katsuya, H., Ishitsuka, K., Utsunomiya, A., Hanada, S., Eto, T., Moriuchi, Y., et al. (2015). Treatment and survival among 1594 patients with ATL. *Blood* 126, 2570–2577. doi: 10.1182/blood-2015-03-632489

Kinpara, S., Hasegawa, A., Utsunomiya, A., Nishitsuji, H., Furukawa, H., Masuda, T., et al. (2009). Stromal cell-mediated suppression of human T-cell leukemia virus type 1 expression in vitro and in vivo by type I interferon. *J. Virol.* 83, 5101–5108. doi: 10.1128/JVI.02564-08

Liu, Y., Dole, K., Stanley, J. R., Richard, V., Rosol, T. J., Ratner, L., et al. (2002). Engraftment and tumorigenesis of HTLV-1 transformed T cell lines in SCID/bg and NOD/SCID mice. *Leuk. Res.* 26, 561–567. doi: 10.1016/S0145-2126(01)00169-2

Marsh, T., Pietras, K., and McAllister, S. S. (2013). Fibroblasts as architects of cancer pathogenesis. *Biochim. Biophys. Acta* 1832, 1070–1078. doi: 10.1016/j.bbadis.2012.10.013

Matsuoka, M., and Jeang, K. T. (2007). Human T-cell leukaemia virus type 1 (HTLV-1) infectivity and cellular transformation. *Nat. Rev. Cancer* 7, 270–280. doi: 10.1038/nrc2111

Matutes, E. (2007). Adult T-cell leukaemia/lymphoma. *J. Clin. Pathol.* 60, 1373–1377. doi: 10.1136/jcp.2007.052456

Miyatake, Y., Oliveira, A. L., Jarboui, M. A., Ota, S., Tomaru, U., Teshima, T., et al. (2013). Protective roles of epithelial cells in the survival of adult T-cell leukemia/lymphoma cells. *Am. J. Pathol.* 182, 1832–1842. doi: 10.1016/j.ajpath.2013.01.015

Miyatake, Y., Sheehy, N., Ikeshita, S., Hall, W. W., and Kasahara, M. (2015). Anchorage-dependent multicellular aggregate formation induces CD44 high cancer stem cell-like ATL cells in an NF-kappaB- and vimentin-dependent manner. *Cancer Lett.* 357, 355–363. doi: 10.1016/j.canlet.2014.11.055

Mueller, L., Goumas, F. A., Affeldt, M., Sandtner, S., Gehling, U. M., Brilloff, S., et al. (2007). Stromal fibroblasts in colorectal liver metastases originate from resident fibroblasts and generate an inflammatory microenvironment. *Am. J. Pathol.* 171, 1608–1618. doi: 10.2353/ajpath.2007.060661

Munoz, E., and Israel, A. (1995). Activation of NF-kappa B by the Tax protein of HTLV-1. *Immunobiology* 193, 128–136. doi: 10.1016/S0171-2985(11)80535-8

Nagai, K., Jinnai, I., Hata, T., Usui, T., Sasaki, D., Tsukasaki, K., et al. (2008). Adhesion-dependent growth of primary adult T cell leukemia cells with down-regulation of HTLV-I p40Tax protein: a novel in vitro model of the growth of acute ATL cells. *Int. J. Hematol.* 88, 551–564. doi: 10.1007/s12185-008-0207-z

Nakada, K., Yamaguchi, K., Furugen, S., Nakasone, T., Nakasone, K., Oshiro, Y., et al. (1987). Monoclonal integration of HTLV-I proviral DNA in patients with strongyloidiasis. *Int. J. Cancer* 40, 145–148. doi: 10.1002/ijc.2910400203

Nerenberg, M., Hinrichs, S. H., Reynolds, R. K., Khoury, G., and Jay, G. (1987). The tat gene of human T-lymphotropic virus type 1 induces mesenchymal tumors in transgenic mice. *Science* 237, 1324–1329. doi: 10.1126/science.2888190

Ohshima, K., Kikuchi, M., Masuda, Y., Kobari, S., Sumiyoshi, Y., Eguchi, F., et al. (1991). Defective provirus form of human T-cell leukemia virus type I in adult T-cell leukemia/lymphoma: clinicopathological features. *Cancer Res.* 51, 4639–4642.

Ohsugi, T. (2013). A transgenic mouse model of human T cell leukemia virus type 1-associated diseases. *Front. Microbiol.* 4:49. doi: 10.3389/fmicb.2013.00049

Okada, F., Ando, Y., Yoshitake, S., Yotsumoto, S., Matsumoto, S., Wakisaka, M., et al. (2006). Pulmonary CT findings in 320 carriers of human T-lymphotropic virus type 1. *Radiology* 240, 559–564. doi: 10.1148/radiol.2402050886

Olumi, A. F., Grossfeld, G. D., Hayward, S. W., Carroll, P. R., Tlsty, T. D., and Cunha, G. R. (1999). Carcinoma-associated fibroblasts direct tumor progression of initiated human prostatic epithelium. *Cancer Res.* 59, 5002–5011.

Osame, M., and Igata, A. (1989). The history of discovery and clinico-epidemiology of HTLV-I-associated myelopathy(HAM). *Jpn. J. Med.* 28, 412–414. doi: 10.2169/internalmedicine1962.28.412

Panfil, A. R., Al-Saleem, J. J., and Green, P. L. (2013). Animal models utilized in HTLV-1 research. *Virology* 4, 49–59. doi: 10.4137/VRT.S12140

Parrula, C., Zimmerman, B., Nadella, P., Shu, S., Rosol, T., Fernandez, S., et al. (2009). Expression of tumor invasion factors determines systemic engraftment and induction of humoral hypercalcemia in a mouse model of adult T-cell leukemia. *Vet. Pathol.* 46, 1003–1014. doi: 10.1354/vp.08-VP-0254-N-FL

Poiesz, B. J., Ruscetti, F. W., Gazdar, A. F., Bunn, P. A., Minna, J. D., and Gallo, R. C. (1980). Detection and isolation of type C retrovirus particles from fresh and cultured lymphocytes of a patient with cutaneous T-cell lymphoma. *Proc. Natl. Acad. Sci. U.S.A.* 77, 7415–7419. doi: 10.1073/pnas.77.12.7415

Popovic, M., Lange-Wantzin, G., Sarin, P. S., Mann, D., and Gallo, R. C. (1983). Transformation of human umbilical cord blood T cells by human T-cell leukemia/lymphoma virus. *Proc. Natl. Acad. Sci. U.S.A.* 80, 5402–5406. doi: 10.1073/pnas.80.17.5402

Rende, F., Cavallari, I., Corradin, A., Silic-Benussi, M., Toulza, F., Toffolo, G. M., et al. (2011). Kinetics and intracellular compartmentalization of HTLV-1 gene expression: nuclear retention of HBZ mRNAs. *Blood* 117, 4855–4859. doi: 10.1182/blood-2010-11-316463

Romanelli, M. G., Diani, E., Bergamo, E., Casoli, C., Ciminale, V., Bex, F., et al. (2013). Highlights on distinctive structural and functional properties of HTLV Tax proteins. *Front. Microbiol.* 4:271. doi: 10.3389/fmicb.2013.00271

Rosshart, S. P., Vassallo, B. G., Angeletti, D., Hutchinson, D. S., Morgan, A. P., Takeda, K., et al. (2017). Wild mouse gut microbiota promotes host fitness and improves disease resistance. *Cell* 171, 1015–1028 e13. doi: 10.1016/j.cell.2017.09.016

Saggioro, D., Rosato, A., Esposito, G., Rosenberg, M. P., Harrison, J., Felber, B. K., et al. (1997). Inflammatory polyarthropathy and bone remodeling in HTLV-I Tax-transgenic mice. *J. Acquir. Immune Defic. Syndr. Hum. Retrovirol.* 14, 272–280. doi: 10.1097/00042560-199703010-00012

Satou, Y., Yasunaga, J., Yoshida, M., and Matsuoka, M. (2006). HTLV-I basic leucine zipper factor gene mRNA supports proliferation of adult T cell leukemia cells. *Proc. Natl. Acad. Sci. U.S.A.* 103, 720–725. doi: 10.1073/pnas.0507631103

Satou, Y., Yasunaga, J., Zhao, T., Yoshida, M., Miyazato, P., Takai, K., et al. (2011). HTLV-1 bZIP factor induces T-cell lymphoma and systemic inflammation in vivo. *PLoS Pathog.* 7:e1001274. doi: 10.1371/journal.ppat.1001274

Seiki, M., Inoue, J., Takeda, T., and Yoshida, M. (1986). Direct evidence that p40x of human T-cell leukemia virus type I is a trans-acting transcriptional activator. *EMBO J.* 5, 561–565.

Shimoyama, M. (1991). Diagnostic criteria and classification of clinical subtypes of adult T-cell leukaemia-lymphoma. A report from the Lymphoma Study Group (1984-87). *Br. J. Haematol.* 79, 428–437. doi: 10.1111/j.1365-2141.1991.tb08051.x

Shlush, L. I., Zandi, S., Mitchell, A., Chen, W. C., Brandwein, J. M., Gupta, V., et al. (2014). Identification of pre-leukaemic haematopoietic stem cells in acute leukaemia. *Nature* 506, 328–333. doi: 10.1038/nature13038

Takajo, I., Umeki, K., Morishita, K., Yamamoto, I., Kubuki, Y., Hatakeyama, K., et al. (2007). Engraftment of peripheral blood mononuclear cells from human T-lymphotropic virus type 1 carriers in NOD/SCID/gammac(null) (NOG) mice. *Int. J. Cancer* 121, 2205–2211. doi: 10.1002/ijc.22972

Takeda, S., Maeda, M., Morikawa, S., Taniguchi, Y., Yasunaga, J., Nosaka, K., et al. (2004). Genetic and epigenetic inactivation of tax gene in adult T-cell leukemia cells. *Int. J. Cancer* 109, 559–567. doi: 10.1002/ijc.20007

Tezuka, K., Xun, R., Tei, M., Ueno, T., Tanaka, M., Takenouchi, N., et al. (2014). An animal model of adult T-cell leukemia: humanized mice with HTLV-1-specific immunity. *Blood* 123, 346–355. doi: 10.1182/blood-2013-06-508861

Vaughan, M. B., Howard, E. W., and Tomasek, J. J. (2000). Transforming growth factor-beta1 promotes the morphological and functional differentiation of the myofibroblast. *Exp. Cell Res.* 257, 180–189. doi: 10.1006/excr.2000.4869

Villaudy, J., Wencker, M., Gadot, N., Gillet, N. A., Scoazec, J. Y., Gazzolo, L., et al. (2011). HTLV-1 propels thymic human T cell development in "human immune system" Rag2$^{-/-}$ gamma c$^{-/-}$ mice. *PLoS Pathog.* 7:e1002231. doi: 10.1371/journal.ppat.1002231

Watanabe, T. (2017). Adult T-cell leukemia: molecular basis for clonal expansion and transformation of HTLV-1-infected T cells. *Blood* 129, 1071–1081. doi: 10.1182/blood-2016-09-692574

World Health Organization [WHO] (2008). "WHO classification of tumours of haematopoietic and lymphoid tissues," in *WHO Classification of Tumours*, 4th Edn, eds S. H. Swerdlow, E. Campo, N. L. Harris, E. S. Jaffe, S. A. Pileri, H. Stein, et al. (Lyon: IARC).

Yamada, Y., Ohmoto, Y., Hata, T., Yamamura, M., Murata, K., Tsukasaki, K., et al. (1996). Features of the cytokines secreted by adult T cell leukemia (ATL) cells. *Leuk. Lymphoma.* 21, 443–447. doi: 10.3109/10428199609093442

Yamamoto, H., Sekiguchi, T., Itagaki, K., Saijo, S., and Iwakura, Y. (1993). Inflammatory polyarthritis in mice transgenic for human T cell leukemia virus type I. *Arthritis Rheum.* 36, 1612–1620. doi: 10.1002/art.1780361117

Yamamoto, M. L., and Schiestl, R. H. (2014). Intestinal microbiome and lymphoma development. *Cancer J.* 20, 190–194. doi: 10.1097/PPO.0000000000000047

Yamano, Y., and Sato, T. (2012). Clinical pathophysiology of human T-lymphotropic virus-type 1-associated myelopathy/tropical spastic paraparesis. *Front. Microbiol.* 3:389. doi: 10.3389/fmicb.2012.00389

Yoshioka, R., Yamaguchi, K., Yoshinaga, T., and Takatsuki, K. (1985). Pulmonary complications in patients with adult T-cell leukemia. *Cancer* 55, 2491–2494. doi: 10.1002/1097-0142(19850515)55:10<2491::AID-CNCR2820551030>3.0.CO;2-I

Zhang, B., Ho, Y. W., Huang, Q., Maeda, T., Lin, A., Lee, S. U., et al. (2012). Altered microenvironmental regulation of leukemic and normal stem cells in chronic myelogenous leukemia. *Cancer Cell.* 21, 577–592. doi: 10.1016/j.ccr.2012.02.018

20

Future Perspectives on Drug Targeting in Adult T Cell Leukemia-Lymphoma

Francesca Marino-Merlo[1], Antonio Mastino[2,3], Sandro Grelli[4], Olivier Hermine[5], Ali Bazarbachi[6,7]† and Beatrice Macchi[1]†*

[1] IRCCS Centro Neurolesi "Bonino-Pulejo", Messina, Italy, [2] Department of Chemical, Biological, Pharmaceutical and Environmental Sciences, University of Messina, Messina, Italy, [3] Institute of Translational Pharmacology, The National Research Council, Rome, Italy, [4] Department of Experimental Medicine and Surgery, University of Rome "Tor Vergata", Rome, Italy, [5] INSERM U1163, CNRS ERL 8654, Department of Hematology, Imagine Institute, Hôpital Necker-Enfants Malades, Paris, France, [6] Department of Internal Medicine, American University of Beirut, Beirut, Lebanon, [7] Department of Anatomy, Cell Biology and Physiological Sciences, American University of Beirut, Beirut, Lebanon, [8] Department of System Medicine, University of Rome "Tor Vergata", Rome, Italy

***Correspondence:**
Antonio Mastino
antonio.mastino@unime.it

†These authors have contributed equally to this work as co-last authors.

Human T cell leukemia virus type 1 (HTLV-1) is the etiological agent of adult T cell leukemia/lymphoma (ATL), HTLV-1 associated myelopathy (HAM/TSP), and of a number of inflammatory diseases with an estimated 10–20 million infected individuals worldwide. Despite a number of therapeutic approaches, a cure for ATL is still in its infancy. Conventional chemotherapy has short-term efficacy, particularly in the acute subtype. Allogeneic stem cell transplantation offers long-term disease control to around one third of transplanted patients, but few can reach to transplant. This prompted, over the past recent years, the conduction of a number of clinical trials using novel treatments. Meanwhile, new data have been accumulated on biological and molecular bases of HTLV-1 transforming and infecting activity. These data offer new rational for targeted therapies of ATL. Taking into account the double-face of ATL as an hematologic malignancy as well as a viral infectious disease, this Mini-Review seeks to provide an up-to-date overview of recent efforts in the understanding of the mechanisms involved in already used therapeutic regimens showing promising results, and in selecting novel drug targets for ATL.

Keywords: HTLV-1, ATL, antiviral agents, biological therapy, targeted therapy

INTRODUCTION

Human T cell leukemia virus type 1 (HTLV-1) is the first identified human retrovirus endemic in southwestern Japan, the Caribbean islands, inter-tropical Africa, South America, Romania and the Middle East, with an estimated 10–20 million infected individuals worldwide. HTLV-1 is known to cause adult T cell leukemia/lymphoma (ATL), HTLV-1 associated myelopathy (HAM/TSP), and a number of inflammatory diseases. Firstly detected and isolated from a cutaneous T cell lymphoma almost 40 years ago, HTLV-1 still represents a significant challenge for the scientific community engaged to disclose its oncogenic potential and to identify a focused therapy (Gallo, 2005; Tagaya and Gallo, 2017). HTLV-1 transforms CD4+ lymphocytes *in vitro* and *in vivo*, and complex mechanisms control virus spreading, expression of viral proteins and host immune response in infected individuals (**Figure 1**). As a consequence, ATL patients are often refractory to intensive, conventional chemotherapy regimens. Classically used regimen are CHOP, CHOEP dose-adjusted EPOCH, and hyper-CVAD, alternating with high-dose methotrexate and cytarabine

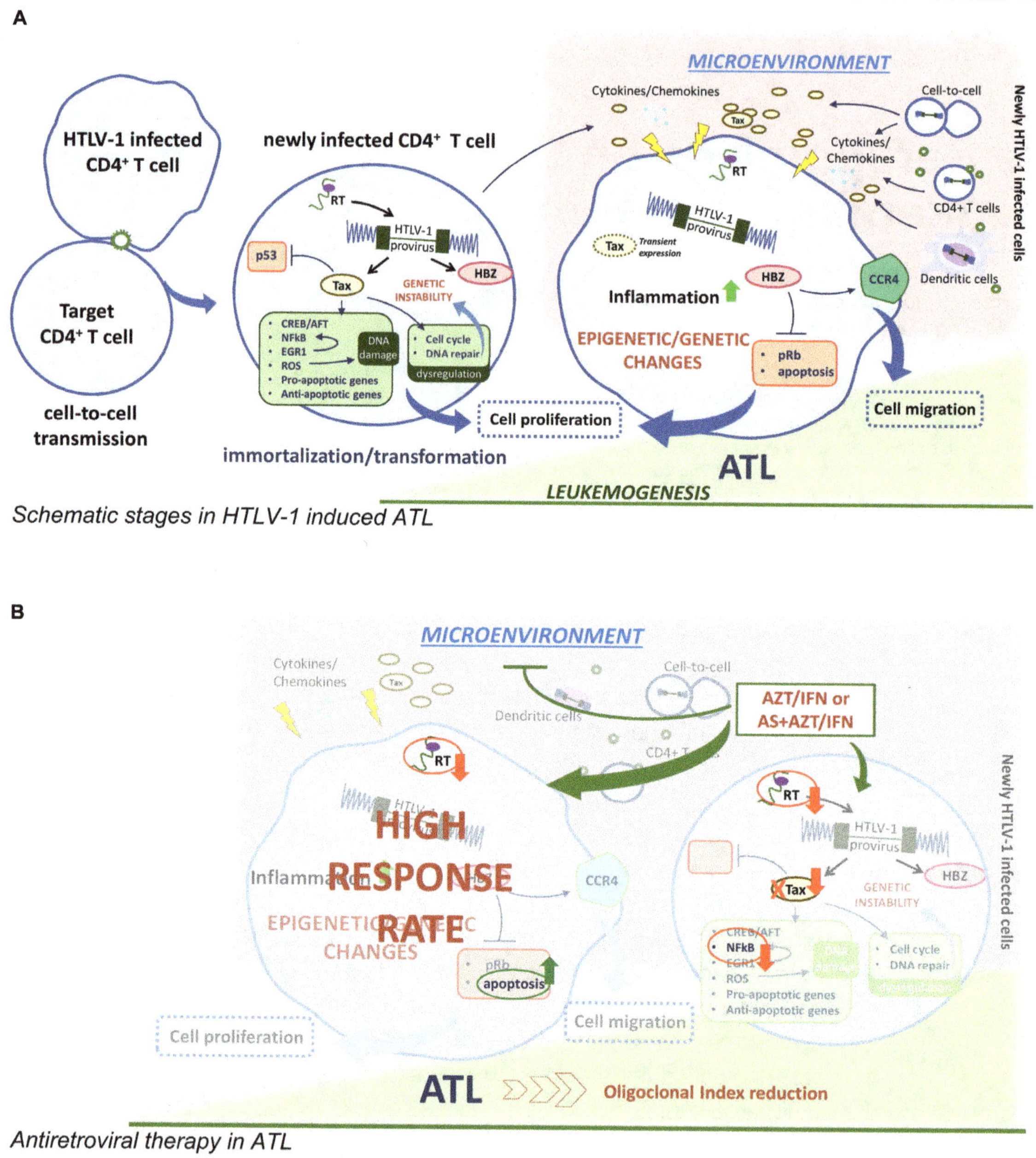

FIGURE 1 | HTLV-1 driven leukomogenesis and possible mechanism of antiviral therapy. **(A)** Main immortalization/transformation process activated by HTLV-1 regulatory protein Tax and HBZ in newly infected CD4+ cells transmit virus to uninfected CD4+ cells and leading to epigenetic and genetic changes in ATL transformed cells. Cell migration and cell-to-cell virus transmission, presumably involving also dendritic cells, favor the release of inflammatory cytokines and chemokines milieu in the microenvironment and contribute to the maintaining of the infected clones. **(B)** AZT/IFN with or without arsenic trioxide (AS) could interrupt the maintaining route of ATL.

(Dittus and Sloan, 2017). An intensive treatment consisting of VCAP-AMP-VECP with the prophylactic use of G-CSF, has been introduced in Japan (Watanabe et al., 2011). In addition, results of preclinical studies, such as the high expression of CCR4 in ATL cells (Ishida et al., 2003), led to hypothesize new targets for biological therapy in ATL. Indeed, clinical trials using humanized monoclonal antibodies such as mogamulizumab (anti-CCR4) (Ishida et al., 2004; Yamamoto et al., 2010), alemtuzumab (anti-CD52) (Sharma et al., 2017), or daclizumab (anti-CD25) (Berkowitz et al., 2014) have been conducted on ATL patients (**Figure 2**). However, an important hurdle emerged in practically all the completed studies, is the limited duration of the response. Allogeneic stem transplantation could represent an alternative, potentially curative approach (Zell et al., 2016). Unfortunately, its use is limited to a small percentage of ATL patients. Recently published review articles provide detailed information on the

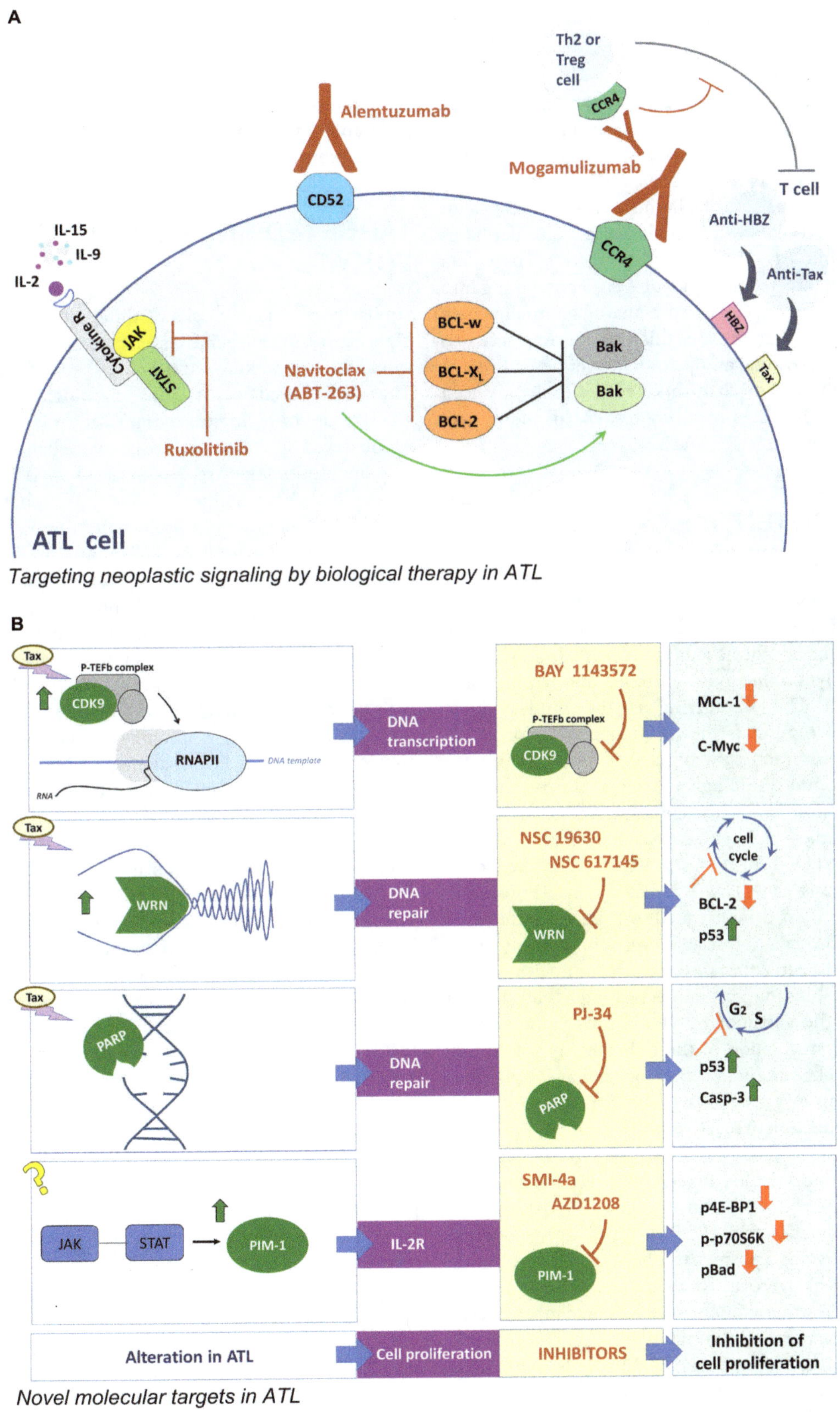

FIGURE 2 | Novel approaches for ATL therapy. **(A)** Biological therapy with monoclonal antibodies and corresponding targets involved in neoplastic signaling in ATL cells. **(B)** Compounds and corresponding targets potentially useful for innovative targeted therapies in ATL.

state-of-the-art of treatment strategies until today adopted in clinical trials for ATL patients and on related results (Kato and Akashi, 2015; Hermine et al., 2018), and this mini-review will not address these aspects.

What makes the design of therapeutic approaches for ATL problematic is the double routes of HTLV-1 transmission *in vivo*, i.e., the mitotic and the RT-dependent ones, that might require a combined approach targeting both chronically infected host cells and direct viral replication. Actually, in the past recent years a lot of data have been accumulated on biological and molecular bases of HTLV-1 transforming as well as infecting activity. These data offer new rational for targeted therapies of ATL. Thus, taking into account the double-face of ATL as an hematologic malignancy as well as a viral infectious disease, this mini-review seeks to specifically focus on providing an up-to-date overview of recent efforts in: (i) understanding mechanisms involved in already used therapeutic regimens showing promising results, (ii) identifying and selecting novel drug targets for ATL.

ONCOGENIC POTENTIAL OF HTLV-1

Adult T cell leukemia/lymphoma is a severe, aggressive leukemia, unequivocally associated to HTLV-1 infection, which develop in 2–4% of the infected individuals after a long time of latency. Four clinical forms/subtypes of ATL have been recognized: acute, lymphomatous, chronic, and smoldering. (Shimoyama, 1991). The smoldering is the mildest form while the acute type represents the most aggressive form, but the life expectancy for each subtype is very poor reaching a maximum of 24 months. Although the molecular mechanism of virus transformation is highly complex and not entirely clarified, several studies highlighted that the viral regulatory proteins Tax and HTLV-1 bZIP factor (HBZ) play key roles in driving oncogenesis in HTLV-1 infected cells (**Figure 1**). The Tax protein has been demonstrated to trigger a number of immortalization/transformation related events in the early phase after infection, such as activation of the interaction with cAMP-responsive element-binding protein/activating transcription factor (CREB/ATF), the activation of NF-κB transcription factor, inhibition of p53, dysregulation of the cell cycle by interfering with the cellular checkpoint, impairment of cellular DNA repair mechanisms resulting in genetic instability, induction of DNA damage through production of reactive oxygen species, induction of both pro-apoptotic and anti-apoptotic activities (Chlichlia and Khazaie, 2010). In particular, Tax has been recently shown to intervene at an early phase of cell transformation by upregulating a family of early growth factor1 (EGR1), which upregulate the NF-κB system, establishing a positive feedback loop (Huang et al., 2017). Thus, Tax is likely required to initiate leukemogenesis while its expression is not routinely detected later in fresh cells from at least 50% of patients with established ATL. Recent data, however, demonstrated that survival of ATL cells depend on transient *tax* expression (Dassouki et al., 2015; Mahgoub et al., 2018) Conversely, HBZ is persistently expressed, even if at low level, *in vivo* in ATL cells, and interacts with elongation factors, Rb/E2F-1 complex, for cell cycle progression (Kawatsuki et al., 2016), inhibits apoptosis and upregulates expression of CCR4, thus promoting proliferation and migration of T cells (Sugata et al., 2016) and finally inducing global epigenetic changes in infected cells. In addition epigenetic dysregulation plays a role in ATL transformation consisting in GpC methylation of cell cycle, p53, apoptotic genes and histone modification of epigenetic reprogramming genes (Watanabe, 2017).

TARGETED BIOLOGICAL THERAPY FOR ATL

Similarly to leukemic cells of different types, ATL cells exhibit high expression of genes associated to cell proliferation/death, cytokines, chemokines and/or markers of cell transformation. Therefore, shared potential pharmacological targets can justify in ATL the use of biological therapy set up for other malignancies. As observed in other neoplasia, the balance between pro and anti apoptotic response is subverted in ATL cells. Preclinical studies have shown that HTLV-1 infection *in vitro* gives rise, in a first phase, to high proliferation and a concomitant high apoptosis rate in infected cells until, in a successive phase, the selection of immortalized clones lead to outgrowth of cells preferentially exhibiting anti-apoptotic gene expression (Matteucci et al., 2004). Coherently, transformed clones from ATL patients over-express in culture the anti-apoptotic Bcl-2, Bcl-xL, and Bcl-w proteins and exhibit *ex vivo* a 10- to 20-fold higher sensitivity to navitoclax (ABT-263), an orally bio-available mimetic of the Bcl-2 homology domain 3 small molecule, as compared to non-HTLV-1-associated leukemic cells (Tse et al., 2008). Interestingly, molecular studies showed that the efficacy of navitoclax in ATL cells *in vitro* was increased by Tax induced upregulation of the pro-apoptotic Bax gene. However, the side effects of navitoclax limit its therapeutic use *in vivo*. Another crucial target in ATL can be detected in the complex network of autocrine (IL-2-IL-2Rα/IL-15-IL-15Rα) and paracrine (IL-9) loops, able to drive *ex vivo* spontaneous proliferation of ATL cells at an early stage (Chen et al., 2010). The three involved cytokines share in common a γc receptor whose expression is regulated by a family of kinase (JAK/STAT). Interestingly, JAK/STAT selective inhibitors suppressed the proliferation of smoldering/chronic ATL cells *ex vivo* (Ju et al., 2011). Given that combining inhibitors of the same signaling pathway can increase the chance to block cancer cell growth, a combination of navitoclax and of the JAK/STAT inhibitor ruxolitinib, was tested on ATL cells *ex vivo* and in animals. The combination provided additive/synergistic activity in inhibiting proliferation of ATL cells, delayed tumor growth and prolonged survival in tumor –bearing mice. This was associated to increasing inhibition of Bcl-xL which favored the upregulation of the pro apoptotic gene expression (Zhang et al., 2015). *In vivo*, the immune response has a remarkable impact on the turnover of HTLV-1 infected clones. Actually, cytotoxic T lymphocytes recognizing the immunodominant viral protein Tax and HBZ are both critical to determine the proviral load (Bangham and Matsuoka, 2017). Ruxolitinib is currently under evaluation in phase 2 clinical trials in ATL patients.

ANTIRETROVIRAL THERAPY FOR ATL

Resistance to conventional chemotherapy prompted a reconsideration of therapy in ATL. Taking into account the retroviral etiology of the disease and that antiretroviral drugs proved their effectiveness in counteracting HIV infection, the use of AZT was tested in ATL. The rationale for this therapeutic intervention was to keep a low level of virus spreading and, possibly, also to limit the onset of inflammatory processes. In fact, successful results were initially reported in two preliminary phase II studies, using the combination of AZT and alpha interferon (IFN) showing an unexpected high response rate, particularly in previously untreated acute ATL patients (Gill et al., 1995; Hermine et al., 1995, 1998). A prospective phase II study on 13 patients who received AZT/IFN treatment as initial therapy, showed nine complete response (CR) and four partial remission with mild toxicity. The CR patients survived more than three years, after which most of the patients relapse underlying the need for additional therapy with AZT/IFN (Hermine et al., 2002). The impact of first-line AZT/IFN therapy on long-term survival was reported in a worldwide meta-analysis showing a significant improvement of survival of the leukemic subtypes with an unprecedented 100% 5-year overall-survival in chronic and smoldering ATL, at a median follow-up time of 5 years (Bazarbachi et al., 2010). Although AZT/IFN therapy provided reasonable management of ATL, most patients relapse. To counteract this aspect, based on previous experience in acute promyelocytic leukemia, arsenic trioxide (AS) was tested in ATL. The arsenic/IFN combination induced proteasomal degradation of Tax through stepwise polysumoylation and SUMO-dependent ubiquitination (El-Sabban et al., 2000; Dassouki et al., 2015). This combination cured murine ATL derived from tax-transgenic through selective targeting of ATL leukemia initiating cells (El Hajj et al., 2010). The triple combination of arsenic, IFN and AZT resulted in a high rate of response in chronic ATL patients (Kchour et al., 2009). However, the mechanisms involved in the therapeutic effectiveness of AZT/IFN are not clear, although *in vitro* studies demonstrated that this combination differently affects HTLV-1 mRNA and viral protein expression and activates the p53 pathway and apoptosis in HTLV-1 infected cells (Kinpara et al., 2013). Nevertheless, no clear evidence was provided concerning the inhibition of viremia by AZT/IFN in ATL patients. However, viremia in ATL patients is low and viral load, reverse transcriptase (RT) activity and/or other virus related assays, carried out in lymphocytes from patients, could be more reliable parameters for assessing HTLV-1 replicative potential in ATL patients. Interestingly, we have recently reported that long-term *in vivo* therapy with AZT and IFN actually caused complete inhibition of RT activity, reduction of p19 release and viral mRNA, and a dramatic decrease of the oligoclonal index, in short-term cultures of PBMCs from ATL patients who responded to therapy, but not in those who did not respond (Macchi et al., 2017). Thus, the above reported data sustain that the therapeutic efficacy of AZT/IFN combination in ATL is actually mediated, at least in part, by the inhibition of RT-dependent viral replication. Consequently, we can hypothesize that the AZT/IFN combination in ATL patients targets viral replication presumably outside leukemic cells. This could occur in cells such as dendritic cells or in newly infected T lymphocytes, immediately after their first contact with the virus, or other cell types in which a dynamic, continuous viral replication occurs. In this case, AZT/IFN treatment can impede HTLV-1 viral replication that could provide a microenvironment that is mandatory for survival and/or renewal of ATL cells (**Figure 1**). This could occur through direct cell-to-cell communication or paracrine stimulation through secreted Tax or various cytokines/chemokines as reported for chronic lymphocytic leukemia (Bazarbachi et al., 2011). Moreover, these findings could explain the impossibility to set-up long-term culture of ATL cells *in vitro* and why the AZT/IFN combination exerts a beneficial effect *in vivo* but not *ex vivo* on ATL cells. A recent trial using EPOCH chemotherapy in combination with bortezomib, to block NF-κB activation, and raltegravir, as antiviral drug, in acute ATL and in ATL lymphoma showed that this regiment was well tolerated, leading to 67% response rate. Changes in RNA viral load and HBZ viral expression *ex vivo* were found as reliable parameters of response as well as inhibition of viral replication and repression of NF-κB activation through proteasome inhibition (Ratner et al., 2016). Thus, accumulating evidence sustains that controlling virus spread is a crucial aspect in ATL therapy.

Regarding *in vitro* studies, AZT and tenofovir inhibit virus transmission to PBMC in short-term cultures and interfere with immortalization in the long run at a drug concentration which was poorly toxic toward uninfected cells (Macchi et al., 1997; Balestrieri et al., 2005). Conversely, lamivudine was not effective in inhibiting HTLV-1 infection *in vitro*, presumably owed to the presence of an isoleucine at position 118 in HTLV-1 RT, conferring natural resistance to 3TC (Balestrieri et al., 2002; Toro et al., 2003). Different not licensed compounds, such as the PCOAN phosphonates, were able to inhibit cell-to-cell HTLV-1 transmission directly inhibiting the HTLV-1 RT activity, as demonstrated by a cell-free assay (Balestrieri et al., 2008b). Inhibition of HTLV-1 infection could also rely on other still unclear mechanisms, as shown in case of compounds of natural origin such as carbohydrate-binding agents (Balestrieri et al., 2008a) and an extract from the seeds of bergamot, which remarkably blocked virus horizontal transmission *in vitro* (Balestrieri et al., 2011). In addition, raltegravir and diketo acid, MK-2048, were active inhibitors of viral transmission as well as viral immortalization of HTLV-1 in lymphoid and non lymphoid cells, *in vitro* (Seegulam and Ratner, 2011).

NOVEL MOLECULAR TARGETS IN ATL

Further targets are being investigated to find new therapeutic approaches in ATL (**Figure 2**). Tax is known to remarkably affect the host cell proliferation by directly intervening in the processes regulating DNA transcription, replication and repair. On the basis of this regulatory role of Tax in HTLV-1 transformation, a few druggable targets have been demonstrated in preclinical studies. CDK9 is a component of a transcription factor, P-TEFb, essential for transcription of most MHC class II genes and for the transcriptional elongation by phosphorylating the C-terminal domain of RNA polymerase II. The importance of CDK9 for a targeted therapy was demonstrated in advanced stage of chronic lymphocytic leukemia and multiple myeloma (Tong et al., 2010). An additional reason to investigate CDK9 as a possible molecular target in ATL is that P-TEFb is present within HTLV-1 transformed cells in Tax-regulated complexes (Cho et al., 2010). In fact, the P-TEFb/CDK-9 inhibitor, BAY 1143572, was able to block the growth of ATL cells *ex vivo*, and to decrease MCL-1 and c-Myc expression levels. In addition, BAY 1143572 decreased ATL cell migration in the liver and bone marrow in a model of ATL *in vivo* xenograft, in immunocompromised NOG mice (Narita et al., 2017). Tax was reported to affect the replicative fork during DNA replication by blocking the progression of the process (Chaib-Mezrag et al., 2014). Helicases are deeply involved in DNA double-strand break repair through the homologous repair as well as the non-homologous end-joining pathway. In particular, the WRN helicases are mutated in cancer and are generally highly expressed in human leukemia (Sallmyr et al., 2008). This finding prompted to assay WRN helicases inhibitors in ATL cells. The results showed that the WRN inhibitors NSC 19630 and NSC 617145 efficiently killed HTLV-1-transformed and patient-derived cells, by inducing cell cycle arrest, downregulation of BCl-2, caspase 3 activation, and apoptosis in a dose-dependent manner, without affecting HTLV-1 expression (Moles et al., 2016). Tax was also recognized to induce genomic double strand breaks during DNA replication and alteration in the subsequent use of non-homologous end joining pathway for repair during the S phase (Baydoun et al., 2012). Thus, the genomic instability afforded by Tax represents a possible target within the repair enzymes family. PJ-34, a small molecule inhibitor of poly (ADP-ribose) polymerase (PARP), arrested cell cycle at S/G2M phase, inducing reactivation of p53 and caspase 3 activation in HTLV-1 infected cells. However, MT-2 chronically infected cells were resistant to PJ-34, showing reduced caspase 3 cleavage and increased RelA/p65 expression. These results suggest that the NF-κB system might be involved in resistance to PJ-34 (Bai et al., 2015). The inhibitors of JAK/STAT pathway regulating expression of IL-2R common γ chain reported as therapeutic option in ATL, were found to be highly immunosuppressive. Thus more recent data proposed a different JAK/STAT-pathway associated target, relying on the Pim 1 downstream target of JAK, whose expression is regulated by miRNA124a. The Pim 1 gene was found to be constitutively expressed in 71% of freshly isolated ATL cells and in chronically HTLV-1 infected cell lines, while PBMC from healthy donors were negative. Treatment with the Pim 1 inhibitors, SMI-4a or AZD1208, decreased ATL cells proliferation and decreased Pim 1 activity as demonstrated by downregulation of p4E-BP1, p-p70S6K, and p-Bad. The AZD 1208 was found more efficacious than SMI-4a. Moreover, AZD 1208 significantly inhibited ATL tumors in the pre-clinical NOG mice model (Bellon et al., 2016), showing that the JAK/STAT-Pim1 pathway could be a novel therapeutic target for the treatment of ATL. In addition the ATL cells exhibited a downregulation of miRNA and Dicer expression. The suitability of these target was demonstrated by the fact that the *in vitro* effect of deacetylase inhibitor, valproate, on ATL cells was owed to the rescue of the pre-miRNA maturation pathway (Gazon et al., 2016).

CONCLUSION

Theoretically, ATL cells exhibit a number of potential, different viral and cellular pharmacological targets. A number of studies suggests to take under consideration the suitability of numerous, known drugs to counteract ATL. However, it is hard to explain why, despite broad chances of potentially druggable targets, success is still limited in ATL clinical studies. Possible reasons could reside in the long latency of HTLV-1 infection, allowing the virus to escape host response, as well as the lack of suitable markers of disease progression. Hopefully, more deep knowledge of how the virus affects the regulation of host immune response and the metabolic requirements of transformed cells could represent new issues for future challenges in ATL therapy.

AUTHOR CONTRIBUTIONS

FM-M searched for literature, wrote the paper, and performed the elaboration and graphical representation of the figures. AM, AB, and BM searched for literature and wrote the paper. SG searched for literature. OH revised the paper.

FUNDING

This work was supported by the Italian Ministry of University and Research, Projects of National Interest (PRIN; AM and BM); Istituto Superiore di Sanità, AIDS Project; the University of Rome "Tor Vergata" (BM); and the University of Messina and the Institute of Translational Pharmacology (AM). FM-M is a recipient of a fellowship from Centro Neurolesi Bonino Pulejo (IRCCS).

REFERENCES

Bai, X. T., Moles, R., Chaib-Mezrag, H., and Nicot, C. (2015). Small PARP inhibitor PJ-34 induces cell cycle arrest and apoptosis of adult T-cell leukemia cells. *J. Hematol. Oncol.* 8:117. doi: 10.1186/s13045-015-0217-2

Balestrieri, E., Ascolani, A., Igarashi, Y., Oki, T., Mastino, A., Balzarini, J., et al. (2008a). Inhibition of cell-to-cell transmission of human T-cell lymphotropic virus type 1 in vitro by carbohydrate-binding agents. *Antimicrob. Agents Chemother.* 52, 2771–2779. doi: 10.1128/AAC.01671-07

Balestrieri, E., Forte, G., Matteucci, C., Mastino, A., and Macchi, B. (2002). Effect of lamivudine on transmission of human T-cell lymphotropic virus type 1 to adult peripheral blood mononuclear cells in vitro. *Antimicrob. Agents Chemother.* 46, 3080–3083. doi: 10.1128/AAC.46.9.3080-3083.2002

Balestrieri, E., Matteucci, C., Ascolani, A., Piperno, A., Romeo, R., Romeo, G., et al. (2008b). Effect of phosphonated carbocyclic 2′-oxa-3′-aza-nucleoside on human T-cell leukemia virus type 1 infection in vitro. *Antimicrob. Agents Chemother.* 52, 54–64.

Balestrieri, E., Pizzimenti, F., Ferlazzo, A., Giofre, S. V., Iannazzo, D., Piperno, A., et al. (2011). Antiviral activity of seed extract from *Citrus bergamia* towards human retroviruses. *Bioorg. Med. Chem.* 19, 2084–2089. doi: 10.1016/j.bmc.2011.01.024

Balestrieri, E., Sciortino, M. T., Mastino, A., and Macchi, B. (2005). Protective effect of the acyclic nucleoside phosphonate tenofovir toward human T-cell leukemia/lymphotropic virus type 1 infection of human peripheral blood mononuclear cells in vitro. *Antiviral Res.* 68, 154–162. doi: 10.1016/j.antiviral.2005.09.001

Bangham, C. R. M., and Matsuoka, M. (2017). Human T-cell leukaemia virus type 1: parasitism and pathogenesis. *Philos. Trans. R. Soc. Lond. B Biol. Sci.* 372:20160272. doi: 10.1098/rstb.2016.0272

Baydoun, H. H., Bai, X. T., Shelton, S., and Nicot, C. (2012). HTLV-I tax increases genetic instability by inducing DNA double strand breaks during DNA replication and switching repair to NHEJ. *PLoS One* 7:e42226. doi: 10.1371/journal.pone.0042226

Bazarbachi, A., Plumelle, Y., Carlos Ramos, J., Tortevoye, P., Otrock, Z., Taylor, G., et al. (2010). Meta-analysis on the use of zidovudine and interferon-alfa in adult T-cell leukemia/lymphoma showing improved survival in the leukemic subtypes. *J. Clin. Oncol.* 28, 4177–4183. doi: 10.1200/JCO.2010.28.0669

Bazarbachi, A., Suarez, F., Fields, P., and Hermine, O. (2011). How I treat adult T-cell leukemia/lymphoma. *Blood* 118, 1736–1745. doi: 10.1182/blood-2011-03-345702

Bellon, M., Lu, L., and Nicot, C. (2016). Constitutive activation of Pim1 kinase is a therapeutic target for adult T-cell leukemia. *Blood* 127, 2439–2450. doi: 10.1182/blood-2015-11-685032

Berkowitz, J. L., Janik, J. E., Stewart, D. M., Jaffe, E. S., Stetler-Stevenson, M., Shih, J. H., et al. (2014). Safety, efficacy, and pharmacokinetics/pharmacodynamics of daclizumab (anti-CD25) in patients with adult T-cell leukemia/lymphoma. *Clin. Immunol.* 155, 176–187. doi: 10.1016/j.clim.2014.09.012

Chaib-Mezrag, H., Lemacon, D., Fontaine, H., Bellon, M., Bai, X. T., Drac, M., et al. (2014). Tax impairs DNA replication forks and increases DNA breaks in specific oncogenic genome regions. *Mol. Cancer* 13:205. doi: 10.1186/1476-4598-13-205

Chen, J., Petrus, M., Bryant, B. R., Nguyen, V. P., Goldman, C. K., Bamford, R., et al. (2010). Autocrine/paracrine cytokine stimulation of leukemic cell proliferation in smoldering and chronic adult T-cell leukemia. *Blood* 116, 5948–5956. doi: 10.1182/blood-2010-04-277418

Chlichlia, K., and Khazaie, K. (2010). HTLV-1 Tax: linking transformation, DNA damage and apoptotic T-cell death. *Chem. Biol. Interact.* 188, 359–365. doi: 10.1016/j.cbi.2010.06.005

Cho, W. K., Jang, M. K., Huang, K., Pise-Masison, C. A., and Brady, J. N. (2010). Human T-lymphotropic virus type 1 Tax protein complexes with P-TEFb and competes for Brd4 and 7SK snRNP/HEXIM1 binding. *J. Virol.* 84, 12801–12809. doi: 10.1128/JVI.00943-10

Dassouki, Z., Sahin, U., El Hajj, H., Jollivet, F., Kfoury, Y., Lallemand-Breitenbach, V., et al. (2015). ATL response to arsenic/interferon therapy is triggered by SUMO/PML/RNF4-dependent Tax degradation. *Blood* 125, 474–482. doi: 10.1182/blood-2014-04-572750

Dittus, C., and Sloan, J. M. (2017). Adult T-cell leukemia/lymphoma: a problem abroad and at home. *Hematol. Oncol. Clin. North Am.* 31, 255–272. doi: 10.1016/j.hoc.2016.11.005

El Hajj, H., El-Sabban, M., Hasegawa, H., Zaatari, G., Ablain, J., Saab, S. T., et al. (2010). Therapy-induced selective loss of leukemia-initiating activity in murine adult T cell leukemia. *J. Exp. Med.* 207, 2785–2792. doi: 10.1084/jem.20101095

El-Sabban, M. E., Nasr, R., Dbaibo, G., Hermine, O., Abboushi, N., Quignon, F., et al. (2000). Arsenic-interferon-alpha-triggered apoptosis in HTLV-I transformed cells is associated with tax down-regulation and reversal of NF-kappa B activation. *Blood* 96, 2849–2855.

Gallo, R. C. (2005). The discovery of the first human retrovirus: HTLV-1 and HTLV-2. *Retrovirology* 2:17. doi: 10.1186/1742-4690-2-17

Gazon, H., Belrose, G., Terol, M., Meniane, J. C., Mesnard, J. M., Cesaire, R., et al. (2016). Impaired expression of DICER and some microRNAs in HBZ expressing cells from acute adult T-cell leukemia patients. *Oncotarget* 7, 30258–30275. doi: 10.18632/oncotarget.7162

Gill, P. S., Harrington, W. Jr., Kaplan, M. H., Ribeiro, R. C., Bennett, J. M., Liebman, H. A., et al. (1995). Treatment of adult T-cell leukemia-lymphoma with a combination of interferon alfa and zidovudine. *N. Engl. J. Med.* 332, 1744–1748. doi: 10.1056/NEJM199506293322603

Hermine, O., Allard, I., Levy, V., Arnulf, B., Gessain, A., Bazarbachi, A., et al. (2002). A prospective phase II clinical trial with the use of zidovudine and interferon-alpha in the acute and lymphoma forms of adult T-cell leukemia/lymphoma. *Hematol. J.* 3, 276–282. doi: 10.1038/sj.thj.6200195

Hermine, O., Bouscary, D., Gessain, A., Turlure, P., Leblond, V., Franck, N., et al. (1995). Brief report: treatment of adult T-cell leukemia-lymphoma with zidovudine and interferon alfa. *N. Engl. J. Med.* 332, 1749–1751. doi: 10.1056/NEJM199506293322604

Hermine, O., Ramos, J. C., and Tobinai, K. (2018). A review of new findings in adult t-cell leukemia-lymphoma: a focus on current and emerging treatment strategies. *Adv. Ther.* 35, 135–152. doi: 10.1007/s12325-018-0658-4

Hermine, O., Wattel, E., Gessain, A., and Bazarbachi, A. (1998). Adult T cell leukaemia: a review of established and new treatments. *BioDrugs* 10, 447–462. doi: 10.2165/00063030-199810060-00003

Huang, Q., Niu, Z., Han, J., Liu, X., Lv, Z., Li, H., et al. (2017). HTLV-1 Tax upregulates early growth response protein 1 through nuclear factor-kappaB signaling. *Oncotarget* 8, 51123–51133. doi: 10.18632/oncotarget.17699

Ishida, T., Iida, S., Akatsuka, Y., Ishii, T., Miyazaki, M., Komatsu, H., et al. (2004). The CC chemokine receptor 4 as a novel specific molecular target for immunotherapy in adult T-Cell leukemia/lymphoma. *Clin. Cancer Res.* 10, 7529–7539. doi: 10.1158/1078-0432.CCR-04-0983

Ishida, T., Utsunomiya, A., Iida, S., Inagaki, H., Takatsuka, Y., Kusumoto, S., et al. (2003). Clinical significance of CCR4 expression in adult T-cell leukemia/lymphoma: its close association with skin involvement and unfavorable outcome. *Clin. Cancer Res.* 9, 3625–3634.

Ju, W., Zhang, M., Jiang, J. K., Thomas, C. J., Oh, U., Bryant, B. R., et al. (2011). CP-690,550, a therapeutic agent, inhibits cytokine-mediated Jak3 activation and proliferation of T cells from patients with ATL and HAM/TSP. *Blood* 117, 1938–1946. doi: 10.1182/blood-2010-09-305425

Kato, K., and Akashi, K. (2015). Recent advances in therapeutic approaches for adult t-cell leukemia/lymphoma. *Viruses* 7, 6604–6612. doi: 10.3390/v7122960

Kawatsuki, A., Yasunaga, J. I., Mitobe, Y., Green, P. L., and Matsuoka, M. (2016). HTLV-1 bZIP factor protein targets the Rb/E2F-1 pathway to promote proliferation and apoptosis of primary CD4(+) T cells. *Oncogene* 35, 4509–4517. doi: 10.1038/onc.2015.510

Kchour, G., Tarhini, M., Kooshyar, M. M., El Hajj, H., Wattel, E., Mahmoudi, M., et al. (2009). Phase 2 study of the efficacy and safety of the combination of arsenic trioxide, interferon alpha, and zidovudine in newly diagnosed chronic adult T-cell leukemia/lymphoma (ATL). *Blood* 113, 6528–6532. doi: 10.1182/blood-2009-03-211821

Kinpara, S., Kijiyama, M., Takamori, A., Hasegawa, A., Sasada, A., Masuda, T., et al. (2013). Interferon-alpha (IFN-alpha) suppresses HTLV-1 gene expression and cell cycling, while IFN-alpha combined with zidovudine induces p53 signaling and apoptosis in HTLV-1-infected cells. *Retrovirology* 10:52. doi: 10.1186/1742-4690-10-52

Macchi, B., Balestrieri, E., Frezza, C., Grelli, S., Valletta, E., Marcais, A., et al. (2017). Quantification of HTLV-1 reverse transcriptase activity in ATL patients treated with zidovudine and interferon-alpha. *Blood Adv.* 1, 748–752. doi: 10.1182/bloodadvances.2016001370

Macchi, B., Faraoni, I., Zhang, J., Grelli, S., Favalli, C., Mastino, A., et al. (1997). AZT inhibits the transmission of human T cell leukaemia/lymphoma virus type I to adult peripheral blood mononuclear cells in vitro. *J. Gen. Virol.* 78, 1007–1016. doi: 10.1099/0022-1317-78-5-1007

Mahgoub, M., Yasunaga, J. I., Iwami, S., Nakaoka, S., Koizumi, Y., Shimura, K., et al. (2018). Sporadic on/off switching of HTLV-1 Tax expression is crucial to maintain the whole population of virus-induced leukemic cells. *Proc. Natl. Acad. Sci. U.S.A.* 115, E1269–E1278. doi: 10.1073/pnas.1715724115

Matteucci, C., Balestrieri, E., Macchi, B., and Mastino, A. (2004). Modulation of apoptosis during HTLV-1-mediated immortalization process in vitro. *J. Med. Virol.* 74, 473–483. doi: 10.1002/jmv.20201

Moles, R., Bai, X. T., Chaib-Mezrag, H., and Nicot, C. (2016). WRN-targeted therapy using inhibitors NSC 19630 and NSC 617145 induce apoptosis in HTLV-1-transformed adult T-cell leukemia cells. *J. Hematol. Oncol.* 9:121. doi: 10.1186/s13045-016-0352-4

Narita, T., Ishida, T., Ito, A., Masaki, A., Kinoshita, S., Suzuki, S., et al. (2017). Cyclin-dependent kinase 9 is a novel specific molecular target in adult T-cell leukemia/lymphoma. *Blood* 130, 1114–1124. doi: 10.1182/blood-2016-09-741983

Ratner, L., Rauch, D., Abel, H., Caruso, B., Noy, A., Barta, S. K., et al. (2016). Dose-adjusted EPOCH chemotherapy with bortezomib and raltegravir for human T-cell leukemia virus-associated adult T-cell leukemia lymphoma. *Blood Cancer J.* 6:e408. doi: 10.1038/bcj.2016.21

Sallmyr, A., Tomkinson, A. E., and Rassool, F. V. (2008). Up-regulation of WRN and DNA ligase IIIalpha in chronic myeloid leukemia: consequences for the repair of DNA double-strand breaks. *Blood* 112, 1413–1423. doi: 10.1182/blood-2007-07-104257

Seegulam, M. E., and Ratner, L. (2011). Integrase inhibitors effective against human T-cell leukemia virus type 1. *Antimicrob. Agents Chemother.* 55, 2011–2017. doi: 10.1128/AAC.01413-10

Sharma, K., Janik, J. E., O'Mahony, D., Stewart, D., Pittaluga, S., Stetler-Stevenson, M., et al. (2017). Phase II study of alemtuzumab (CAMPATH-1) in patients with HTLV-1-associated Adult T-cell leukemia/lymphoma. *Clin. Cancer Res.* 23, 35–42. doi: 10.1158/1078-0432.CCR-16-1022

Shimoyama, M. (1991). Diagnostic-criteria and classification of clinical subtypes of adult t-cell leukemia-lymphoma - a report from the lymphoma-study-group (1984-87). *Br. J. Haematol.* 79, 428–437. doi: 10.1111/j.1365-2141.1991.tb08051.x

Sugata, K., Yasunaga, J., Kinosada, H., Mitobe, Y., Furuta, R., Mahgoub, M., et al. (2016). HTLV-1 viral factor HBZ induces CCR4 to promote T-cell migration and proliferation. *Cancer Res.* 76, 5068–5079. doi: 10.1158/0008-5472.CAN-16-0361

Tagaya, Y., and Gallo, R. C. (2017). The exceptional oncogenicity of HTLV-1. *Front. Microbiol.* 8:1425. doi: 10.3389/fmicb.2017.01425

Tong, W. G., Chen, R., Plunkett, W., Siegel, D., Sinha, R., Harvey, R. D., et al. (2010). Phase I and pharmacologic study of SNS-032, a potent and selective Cdk2, 7, and 9 inhibitor, in patients with advanced chronic lymphocytic leukemia and multiple myeloma. *J. Clin. Oncol.* 28, 3015–3022. doi: 10.1200/JCO.2009.26.1347

Toro, C., Rodes, B., Mendoza, C., and Soriano, V. (2003). Lamivudine resistance in human T-cell leukemia virus type 1 may be due to a polymorphism at codon 118 (V->I) of the reverse transcriptase. *Antimicrob. Agents Chemother.* 47, 1774;author rely 1774–1775. doi: 10.1128/AAC.47.5.1774-1775.2003

Tse, C., Shoemaker, A. R., Adickes, J., Anderson, M. G., Chen, J., Jin, S., et al. (2008). ABT-263: a potent and orally bioavailable Bcl-2 family inhibitor. *Cancer Res.* 68, 3421–3428. doi: 10.1158/0008-5472.CAN-07-5836

Watanabe, T. (2017). Adult T-cell leukemia: molecular basis for clonal expansion and transformation of HTLV-1-infected T cells. *Blood* 129, 1071–1081. doi: 10.1182/blood-2016-09-692574

Watanabe, T., Tobinai, K., Shibata, T., Tsukasaki, K., Morishima, Y., Maseki, N., et al. (2011). Phase II/III study of R-CHOP-21 versus R-CHOP-14 for untreated indolent B-cell non-Hodgkin's lymphoma: JCOG 0203 trial. *J. Clin. Oncol.* 29, 3990–3998. doi: 10.1200/JCO.2011.34.8508

Yamamoto, K., Utsunomiya, A., Tobinai, K., Tsukasaki, K., Uike, N., Uozumi, K., et al. (2010). Phase I study of KW-0761, a defucosylated humanized anti-CCR4 antibody, in relapsed patients with adult T-cell leukemia-lymphoma and peripheral T-cell lymphoma. *J. Clin. Oncol.* 28, 1591–1598. doi: 10.1200/JCO.2009.25.3575

Zell, M., Assal, A., Derman, O., Kornblum, N., Battini, R., Wang, Y., et al. (2016). Adult T-cell leukemia/lymphoma in the Caribbean cohort is a distinct clinical entity with dismal response to conventional chemotherapy. *Oncotarget* 7, 51981–51990. doi: 10.18632/oncotarget.10223

Zhang, M., Mathews Griner, L. A., Ju, W., Duveau, D. Y., Guha, R., Petrus, M. N., et al. (2015). Selective targeting of JAK/STAT signaling is potentiated by Bcl-xL blockade in IL-2-dependent adult T-cell leukemia. *Proc. Natl. Acad. Sci. U.S.A.* 112, 12480–12485. doi: 10.1073/pnas.1516208112

21

HTLV-1 HBZ Viral Protein: A Key Player in HTLV-1 Mediated Diseases

*Marco Baratella, Greta Forlani and Roberto S. Accolla**

Laboratories of General Pathology and Immunology "Giovanna Tosi", Department of Medicine and Surgery, School of Medicine, University of Insubria, Varese, Italy

***Correspondence:**
Roberto S. Accolla
roberto.accolla@uninsubria.it

Human T cell leukemia virus type 1 (HTLV-1) is an oncogenic human retrovirus that has infected 10–15 million people worldwide. After a long latency, 3–5% of infected individuals will develop either a severe malignancy of CD4+ T cells, known as Adult T-cell Leukemia (ATL) or a chronic and progressive inflammatory disease of the nervous system designated Tropical Spastic Paraparesis/HTLV-1-Associated Myelopathy (HAM/TSP). The precise mechanism behind HTLV-1 pathogenesis still remains elusive. Two viral regulatory proteins, Tax-1 and HTLV-1 bZIP factor (HBZ) are thought to play a critical role in HTLV-1-associated diseases. Tax-1 is mainly involved in the onset of neoplastic transformation and in elicitation of the host's inflammatory responses; its expression may be lost during cell clonal proliferation and oncogenesis. Conversely, HBZ remains constantly expressed in all patients with ATL, playing a role in the proliferation and maintenance of leukemic cells. Recent studies have shown that the subcellular distribution of HBZ protein differs in the two pathologies: it is nuclear with a speckled-like pattern in leukemic cells and is cytoplasmic in cells from HAM/TSP patients. Thus, HBZ expression and distribution could be critical in the progression of HTLV-1 infection versus the leukemic state or the inflammatory disease. Here, we reviewed recent findings on the role of HBZ in HTLV-1 related diseases, highlighting the new perspectives open by the possibility of studying the physiologic expression of endogenous protein in primary infected cells.

Keywords: HTLV-1, ATL, HAM/TSP, HBZ, Tax-1

TAX-1 AND HBZ: THE TWO KEY PLAYERS IN HTLV-1-ASSOCIATED PATHOLOGIES

Human T-cell leukemia virus type 1 (HTLV-1) was the first oncogenic retrovirus identified in humans (Poiesz et al., 1980; Yoshida et al., 1982). It is currently estimated that HTLV-1 infects at least 10–15 million people worldwide. Large HTLV-1 endemic areas exist in Southern Japan, the Caribbean, Central and South America, the Middle East, Melanesia, and equatorial regions of Africa (Proietti et al., 2005; Gessain and Cassar, 2012). Although most HTLV-1 infected individuals remain asymptomatic carriers (AC) lifelong, about 3–5% of them develop, after many years of clinical latency, a severe malignancy of CD4+ T cells, known as Adult T-cell Leukemia (ATL).

Four clinical subtypes of HTLV-1 associated ATL are described: smoldering, chronic, lymphoma and acute (Shimoyama, 1991). Depending on the subtype and symptoms, the treatment is different and could include: 'watchful waiting,' chemotherapy, antiviral therapy, immunotherapy, allogeneic hematopoietic stem cell transplantation (alloHSCT) and targeted therapies

(Utsunomiya et al., 2015). However, despite the available treatments, the prognosis is still extremely poor (Nasr et al., 2017). HTLV-1 is also the causative agent of a neurological disease called tropical spastic paraparesis/HTLV-1-associated myelopathy (HAM/TSP) (Gessain et al., 1985), characterized by different neurological features such as spasticity, muscle weakness and sensory deficits. In general, the central function and cranial nerves are strongly compromised and the clinical course is progressive and without remission (Martin et al., 2014). Due to the wide spectrum of symptoms and the inefficiency of the antiviral therapies, the symptomatic treatment represents the current standard of therapy of this disorder.

The pathogenetic mechanisms at the basis of HTLV-1 infection progressing toward ATL or HAM/TSP are not clearly understood although there are sedimented indications that interactions between the viral encoded proteins and a variety of cellular targets are crucial. Two viral proteins, Tax-1 and HTLV-1 bZIP factor (HBZ), encoded by the sense and antisense viral transcripts, respectively, are largely responsible for the malignant transformation and immortalization of the infected T cells. Tax-1 is a potent activator of viral transcription and is involved in neoplastic transformation through modulation of the expression of cellular genes and deregulation of diverse cell signaling pathways involved in cell proliferation, DNA damage repair, and apoptosis (Hall and Fujii, 2005; Matsuoka and Jeang, 2007; Chlichlia and Khazaie, 2010; Tosi et al., 2011; Forlani et al., 2013) The oncogenic properties of Tax-1 are strongly associated to its ability to constitutively activate the nuclear factor kappa B (NF-kB) pathway (Petropoulos et al., 1996; Forlani et al., 2016). It is of note that the alteration of NF-kB signaling pathway could also be involved in the inflammatory state observed in HAM/TSP (Peloponese et al., 2006). Indeed, it has been suggested that pro-inflammatory CD8+ T lymphocytes (CTL), specific for Tax-1-expressing CD4+ T cells, could infiltrate the central nervous system (CNS), kill the target cells and produce inflammatory cytokines, in particular NF-kB-inducible cytokines, that may participate to tissue damage (Asquith and Bangham, 2000). The high immunogenicity of Tax-1 renders it the major target of effectors CTL, and this event may be responsible for the loss of Tax-1 expression during the long clinically latent period leading to overt ATL (Asquith et al., 2005). Genetic and epigenetic modification in the proviral genome are also responsible of Tax-1 silencing (Takeda et al., 2004). Indeed only 40% of ATL patients can express Tax-1 mRNA suggesting that this viral factor is dispensable for the maintenance of leukemia (Matsuoka and Jeang, 2007). Similar data were reported for the inflammatory diseases, in which it was found that 50% of HAM/TSP patiens can express Tax-1 mRNA. (Usui et al., 2008; Saito et al., 2009; Andrade et al., 2013). Interestingly, our recent findings in a restricted sample of patients demonstrated that, at least at protein level, Tax-1 is found in 75% of HAM/TSP cases and in 100% of HTLV-1 AC, but not in ATL cases (Baratella et al., 2017).

At variance with Tax-1, HBZ mRNA (Matsuoka and Green, 2009) and HBZ protein (Raval et al., 2015) are constantly expressed all ATL cases and in HTLV-1 infected individuals, indicating that HBZ is essential not only for cellular transformation but also for the maintenance of leukemic state. HBZ expression increases the proliferation of HTLV-1 infected T cells in culture, and more importantly induces both T-cell lymphomas and systemic inflammation in mice (Satou et al., 2006, 2011). Several studies have suggested a crucial function of HBZ also in HTLV-1-associated inflammatory disorders as it is always found at both mRNA and protein levels in TSP/HAM patients (Andrade et al., 2013). Interestingly, and at variance with Tax-1, most AC do express HBZ mRNA (Saito et al., 2009; Andrade et al., 2013) but not the protein (Baratella et al., 2017). The fact that HBZ protein is scarcely produced in infected cells may explain why is not, or only marginally, discriminated by HBZ specific CTLs in HTLV-1-infected individuals and in HAM/TSP patients (Hilburn et al., 2011; Rowan et al., 2014). The lower immunogenicity of HBZ compared to Tax-1, and the ability of HBZ to inhibit most of Tax-1 activity, could favor the virus immune escape, thus promoting the spreading of infection and the persistence of viral latency. Several studies have indicated a key function of HBZ in supporting and/or maintaining the proliferation of HTLV-1 infected cells and by this, the initiation and persistence of ATL. It has been reported that HBZ promotes ATL cell proliferation by inhibiting apoptosis through different mechanisms: it impairs the binding of AFT3 to p53, thus affecting the activation of p53-mediated apoptosis signaling (Hagiya et al., 2011); it inhibits the transcriptional activation of pro-apoptotic genes as Bim and Fas Ligand, by interfering with FoxO3 (Tanaka-Nakanishi et al., 2014). It was also reported that HBZ impairs anti-viral immunity responses: it binds to NFAT and inhibits the production of Th1 cytokines (particularly IFN-γ); it induces the expression of TIGIT on the cell surface in ATL (Yasuma et al., 2016) and promotes cells migration and proliferation by enhancing CCR4 expression on T-cell surface (Sugata et al., 2016).

Interestingly, several studies have demonstrated that HBZ exerts opposite effects with respect to Tax-1 on signaling pathways. Tax-1 activates while HBZ selectively inhibits the classical NF-κB pathway by affecting the binding of p65-RelA to its consensus DNA sequence and promoting its degradation. This inhibition leads to the activation of the alternative NF-κB pathway (Zhao et al., 2009). It suppresses Tax-1-mediated viral transactivation by interacting with the KIX domain of p300/CBP and impairing the binding of these cellular factors to Tax-1 (Clerc et al., 2008). HBZ suppresses, while Tax-1 activates, Wnt pathway by interacting with the disheveled-associating protein with a high frequency of Leucine residues (DAPLE) (Ma et al., 2013).

Taken together, all these data emphasize the critical role of HBZ in promoting cellular proliferation and the persistence of the viral infection.

HBZ: BIOCHEMICAL ASPECTS AND DISTINCT SUBCELLULAR DISTRIBUTION OF ENDOGENOUS PROTEIN IN ATL AND HAM/TSP

Since its discovery in 2002, HBZ has become a crucial hotspot in HTLV-1 research. HBZ gene is encoded by the minus strand of the HTLV-1 RNA genome (Gaudray et al., 2002) and transcribed by a functional promoter, contained in the U5 sequence of the 3′ Long Terminal Repeat (LTR) (Yoshida et al., 2008).

Three major HBZ transcriptional isoforms have been described: the unspliced (usHBZ) form and two alternative spliced form (SP1 and SP2) (Cavanagh et al., 2006; Murata et al., 2006). Regarding the two spliced HBZ isoforms, although both of them have been found in HTLV-1 infected cells, the SP2 variant occurred less frequently than SP1 (Cavanagh et al., 2006). The SP1 spliced and unspliced *HBZ* transcripts are translated into polypeptides of 206 and 209 amino acids, respectively, and they have almost identical sequences except for a stretch of seven amino acids at the N-terminus of the protein (MAAS for SP1 HBZ and MVNFVSA for usHBZ). However, the half-life of usHBZ protein is much shorter than that of SP1 HBZ (Yoshida et al., 2008) and the expression level of SP1 HBZ is four times higher than that of usHBZ in ATL cells (Usui et al., 2008). Nevertheless, it was reported that the two HBZ protein variants exhibit similar functions (Ma et al., 2016), as they were characterized by conserved functional domains: an N-terminal activation domain (AD), a central domain (CD) and a C-terminal basic ZIP domain (bZIP). HBZ contains three nuclear localization signals (NLS) responsible for its nuclear localization (Hivin et al., 2005; Zhao and Matsuoka, 2012) and two functional nuclear export signals (NES) within its N-terminal region (Mukai and Ohshima, 2014). Most of the reported sub-cellular localizations, biochemical interactions and functional aspects related to HBZ have been assessed in cells overexpressing tagged HBZ. Through its bZIP domain, HBZ was reported to interact with CREB/CREB-2, and this association was instrumental to suppress Tax-mediated HTLV-1 viral transcription (Gaudray et al., 2002). Similar experiments have shown that HBZ binds to different proteins of the JUN family via its bZIP domain. Upon binding to HBZ, JunB, and cJun were recruited in nuclear bodies and degraded, thus HBZ reduces the cJun/JunB-mediated transcriptional activation of a series of genes (Basbous et al., 2003; Thébault et al., 2004). Conversely, the association of HBZ to JunD did not inhibit the JunD-mediated transcriptional activation of target genes; indeed HBZ-JunD complex was reported to increase HBZ gene expression (Thébault et al., 2004; Gazon et al., 2012).

Most of the reported HBZ interactions, however, were assessed by artificial overexpressing systems (transfected cells) and thus it has been hard to extrapolate these results to the real situation encountered in infected cells or in leukemic cells from patients. Only recently, the availability of an anti-HBZ monoclonal antibody (mAb), 4D4-F3, isolated in our laboratory, has made it possible to assess endogenous HBZ expression, localization and interaction *in vivo* in HTLV-1 infected and in ATL patient (Raval et al., 2015). Indeed, endogenous HBZ interacts and co-localizes with p300 and JunD. Partial colocalization was observed also for CBP and CREB2 (Raval et al., 2015). By using the 4D4-F3 mAb we were able to quantify the HBZ protein demonstrating that the amount of HBZ in ATL patients is around 0.45×10^{-2} pg/cell corresponding to 17.461 molecules/cell, that is 20- to 50-fold lower than the amount expressed in HBZ transfected cells. Similar results were recently obtained by another group (Shiohama et al., 2016). Interestingly, our recent data generated by immunofluorescence with the 4D4-F3 mAb and careful confocal microscopy studies have demonstrated that HBZ protein is expressed in 80 to 100% of ATL cells, in 0.4 to 11% of PBMC of HAM/TSP cells, and very rarely, if any, in PBMC of asymptomatic HTLV-1 carriers (Raval et al., 2015; Baratella et al., 2017).

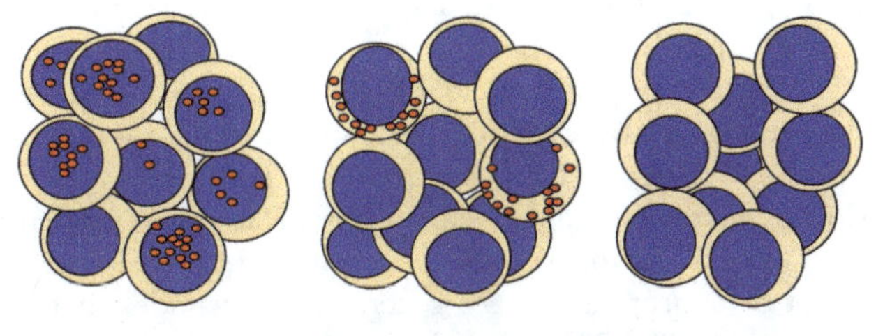

FIGURE 1 | HTLV-1 bZIP factor (HBZ) distribution in PBMC from Adult T-cell Leukemia (ATL), HAM/TSP and asymptomatic carriers (AC). HBZ protein localizes in distinct subcellular compartments and with different percentage in leukemic cells of ATL patients and PBMC of HAM/TSP patients. HBZ is expressed in the nucleus of 90% and in the cytoplasm of 20% of cells isolated from ATL or HAM/TSP patients, respectively. AC do not express detectable levels of HBZ.

These studies have also demonstrated that endogenous HBZ protein is localized in the nucleus of ATL cells (Raval et al., 2015; Shiohama et al., 2016) with a similar speckle-like distribution as the one observed in cells transfected with tagged HBZ protein (Gaudray et al., 2002). However, the HBZ nuclear aggregates found in cells overexpressing tagged-HBZ, in particular GFP-HBZ, were shown to be artifacts of chimeric proteins. Although the composition of these nuclear structures containing HBZ is still unknown it has been demonstrated that HBZ-specific nuclear bodies did not overlap with Promyelocytic Leukemia (PML) nuclear bodies (Raval et al., 2015), indicating that HBZ did not co-localize with PML protein. Of note, PML bodies were shown to act as co-activators of Tax-1, without binding to or co-localizing with the viral transactivator (Ariumi et al., 2003)

One of the most important finding of our studies was the demonstration of the distinct subcellular localization of HBZ protein in HAM/TSP as compared to ATL cells. Until now, HBZ has been always assumed to have an exclusive nuclear localization. Thus, it was relatively unexpected to find that HBZ is exclusively localized in the cytoplasm of PBMC of HAM/TSP patients (**Figure 1**). This cytoplasmic localization was not affected by the presence of leptomycin B, a nuclear export inhibitor, indicating that the viral factor did not shuttle in and out of the nucleus (Baratella et al., 2017). Furthermore, cells expressing cytoplasmic HBZ were almost exclusively found in the CD4+ T cell compartment, very rarely (less than 1%) in CD8+ T cells, and never in B cells or NK cells. Interestingly, the CD4+ HBZ cells, did not expressed the CD25 T cell activation marker

(Baratella et al., 2017), thus suggesting that they were either not in rapid proliferation or, if resting, not included in the classical regulatory T cell compartment. This goes in line with recent findings showing that the HBZ-specific humoral immune response correlated with reduced CD4+ T cell activation in HAM/TSP patients (Enose-Akahata et al., 2013). Within this context, the HBZ cytoplasmic localization in HAM/TSP patients may not be appropriate to the generation of peptides that can efficiently bind MHC class I molecules for presentation to, and scrutiny by CTLs. Of further interest was the finding that, at least in the patients' sample analyzed in our studies, the expression of HBZ and Tax-1 was mutually exclusive (Baratella et al., 2017).

Taken together, these observations let us to propose the cytoplasmic localization of HBZ protein in PBMC of HAM/TSP patients as the first molecular marker of disease, since until now the only associative parameters with the neurological disease were either clinical correlates or the rather non-specific high proviral load (Nagai et al., 1998). Recently, it has been reported that HBZ-transfected cell lines, including the T cell line Jurkat, may partially segregate HBZ in the cytoplasm as result of interaction with THEMIS (Kinosada et al., 2017). With the limitations expressed above on the physiological correlates of HBZ in overexpressing systems, THEMIS may represent a potential HBZ interactor contributing to the cytoplasmic segregation of HBZ also in HAM/TSP patients' cells. Future studies in patient settings should clarify this point.

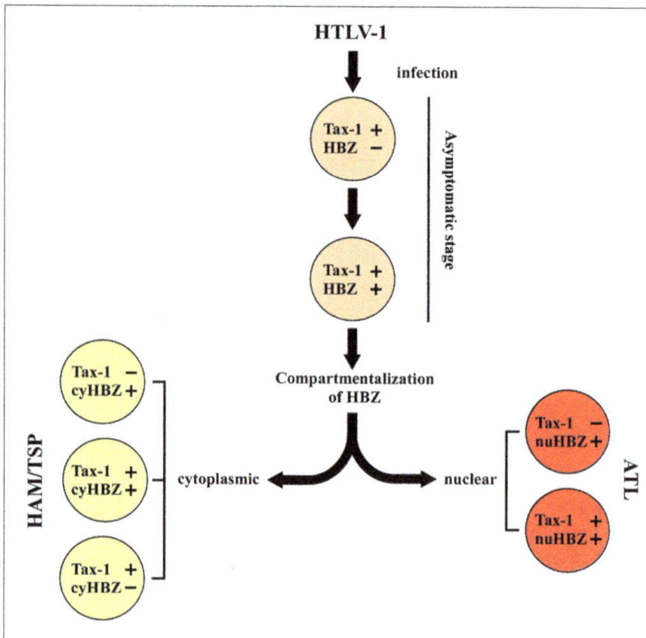

FIGURE 2 | A hypothetical model of disease progression in HTLV-1 infected people. The teory is based on the expression and localization of Tax-1 and HBZ proteins at the single cell level. Primary HTLV-1 infection is associated to the expression of Tax-1. The progression of the disease leads to a different stage characterized by the concominant expression of Tax-1 and HBZ. The subsequent localization of HBZ in the nucleus (nuHBZ) or cytoplasm (cyHBZ) points the disease versus the leukemic or the inflammatory state, as more detailed in the text.

CONCLUDING REMARKS AND FUTURE PERSPECTIVES: THE GROWING RELEVANCE OF HBZ IN THE ONSET OF HTLV-1 ASSOCIATED DISEASES

Although several studies have been focussed at clarifying the role of HBZ in HTLV-1-associated diseases, much research is still required to clearly define the molecular and cellular basis of the distinct outcome of viral infection, whether evolving versus the leukemia/lymphoma or versus the neuroinflammatory disorders. In this review, we have summarized the basic molecular aspects of HBZ expression and, importantly, the new findings related to the distinct subcellular localization of HBZ in HTLV-1 infected individuals, AC, affected by ATL or by HAM/TSP neurological disease.

A new paradigm is emerging that suggests a possible roadmap in the evolution of infection based on these new findings (**Figure 2**). We may think that primary HTLV-1 infection is characterized by the expression of Tax-1 (this stage may be observed in PBMC of still asymptomatic patients), followed by the possible transient coexpression of HBZ (yet to be demonstrated). As we have no specific demonstration of this intermediate step, we do not know whether HBZ protein can localize in the nucleus, cytoplasm or both cellular compartments at these early stages of infection. Immediately after, a compartmentalization of HBZ in the cytoplasm can take place in case of evolution of the infection toward HAM/TSP. At this stage, the evolving pathology can either present a concomitant Tax-1 expression in the same cells (rare), or a mutually exclusive expression of Tax-1 and HBZ in different cells, as we observed in our patients. Conversely, in case of evolution toward the leukemic stage, HBZ will be expressed always and only in the nucleus with or without the coexpression of Tax-1. This hypothetical scheme is useful to test several additional hypotheses and investigate previously unforeseen mechanisms. For example, can we pinpoint a stage in which HBZ is compartmentalized in the cytoplasm still without clinical signs of neurological disease and use this parameter also as a prognostic marker of future development to HAM/TSP? Moreover, what are the molecular mechanisms that drive the distinct nuclear or cytoplasmic localization of the HBZ protein? Are there specific interactors that retain HBZ in the two different cellular compartments? Are these mechanism related to the onset and/or the persistence of the distinct pathologies? These and other fascinating questions' are now, we believe, experimentally amenable to scrutiny.

AUTHOR CONTRIBUTIONS

MB, GF, and RA participated in the conception and design of the review and revised the manuscript. All the authors read, critiqued, and approved the final manuscript.

FUNDING

This work was supported by University of Insubria intramural grant FAR 2014-2016 to RA and GF; by The Institutional Grant 2017, University of Insubria, to RA; by European Community FP7 project HepaVAC Grant Number 602893 to RA (www.hepavac.eu).

ACKNOWLEDGMENTS

The authors wish to thank Dr. Goutham Raval, Dr. Alessandra Tedeschi, and Dr. Farah Bou Nasser Eddine for their participation to several experiments on HBZ localization that were the ground for the results discussed in this review.

REFERENCES

Andrade, R. G., Goncalves Pde, C., Ribeiro, M. A., Romanelli, L. C., Ribas, J. G., Torres, E. B., et al. (2013). Strong correlation between tax and HBZ mRNA expression in HAM/TSP patients: distinct markers for the neurologic disease. *J. Clin. Virol.* 56, 135–140. doi: 10.1016/j.jcv.2012.10.003

Ariumi, Y., Ego, T., Kaida, A., Matsumoto, M., Pandolfi, P. P., and Shimotohno, K. (2003). Distinct nuclear body components, PML and SMRT, regulate the trans-acting function of HTLV-1 Tax oncoprotein. *Oncogene* 22, 1611–1619. doi: 10.1038/sj.onc.1206244

Asquith, B., and Bangham, C. R. M. (2000). The role of cytotoxic T lymphocytes in human T-cell lymphotropic virus Type 1 infection. *J. Theor. Biol.* 207, 65–79. doi: 10.1006/jtbi.2000.2156

Asquith, B., Mosley, A. J., Heaps, A., Tanaka, Y., Taylor, G. P., McLean, A. R., et al. (2005). Quantification of the virus-host interaction in human T lymphotropic virus I infection. *Retrovirology* 2:75. doi: 10.1186/1742-4690-2-75

Baratella, M., Forlani, G., Raval, G. U., Tedeschi, A., Gout, O., Gessain, A., et al. (2017). Cytoplasmic localization of HTLV-1 HBZ protein: a biomarker of HTLV-1-associated myelopathy/tropical spastic paraparesis (HAM/TSP). *PLOS Negl. Trop. Dis.* 11:e0005285. doi: 10.1371/journal.pntd.0005285

Basbous, J., Arpin, C., Gaudray, G., Piechaczyk, M., Devaux, C., and Mesnard, J. M. (2003). The HBZ factor of human T-cell leukemia virus type I dimerizes with transcription factors JunB and c-Jun and modulates their transcriptional activity. *J. Biol. Chem.* 278, 43620–43627. doi: 10.1074/jbc.m307275200

Cavanagh, M. H., Landry, S., Audet, B., Arpin-André, C., Hivin, P., Paré, M. E., et al. (2006). HTLV-I antisense transcripts initiating in the 3'LTR are alternatively spliced and polyadenylated. *Retrovirology* 3:15. doi: 10.1186/1742-4690-3-15

Chlichlia, K., and Khazaie, K. (2010). HTLV-1 Tax: linking transformation, DNA damage and apoptotic T-cell death. *Chem. Biol. Interact.* 188, 359–365. doi: 10.1016/j.cbi.2010.06.005

Clerc, I., Polakowski, N., Andre-Arpin, C., Cook, P., Barbeau, B., Mesnard, J. M., et al. (2008). An interaction between the human T cell leukemia virus type 1 basic leucine zipper factor (HBZ) and the KIX domain of p300/CBP contributes to the down-regulation of tax-dependent viral transcription by HBZ. *J. Biol. Chem.* 283, 23903–23913. doi: 10.1074/jbc.M803116200

Enose-Akahata, Y., Abrams, A., Massoud, R., Bialuk, I., Johnson, K. R., Green, P. L., et al. (2013). Humoral immune response to HTLV-1 basic leucine zipper factor (HBZ) in HTLV-1-infected individuals. *Retrovirology* 10:19. doi: 10.1186/1742-4690-10-19

Forlani, G., Abdallah, R., Accolla, R. S., and Tosi, G. (2013). The MHC-II transactivator CIITA, a restriction factor against oncogenic HTLV-1 and HTLV-2 retroviruses: similarities and differences in the inhibition of Tax-1 and Tax-2 viral transactivators. *Front. Microbiol.* 4:234. doi: 10.3389/fmicb.2013.00234

Forlani, G., Abdallah, R., Accolla, R. S., and Tosi, G. (2016). The major histocompatibility complex class II transactivator CIITA inhibits the persistent activation of NF-kB by the human T cell lymphotropic virus Type 1 Tax-1 oncoprotein. *J. Virol.* 90, 3708–3721. doi: 10.1128/JVI.03000-15

Gaudray, G., Gachon, F., Basbous, J., Biard-Piechaczyk, M., Devaux, C., and Mesnard, J. M. (2002). The complementary strand of the human T-cell leukemia virus Type 1 RNA genome encodes a bZIP transcription factor that down-regulates viral transcription. *J. Virol.* 76, 12813–12822. doi: 10.1128/JVI.76.24.12813-12822.2002

Gazon, H., Lemasson, I., Polakowski, N., Cesaire, R., Matsuoka, M., Barbeau, B., et al. (2012). Human T-cell leukemia virus type 1 (HTLV-1) bZIP factor requires cellular transcription factor JunD to upregulate HTLV-1 antisense transcription from the 3' long terminal repeat. *J. Virol.* 86, 9070–9078. doi: 10.1128/JVI.00661-12

Gessain, A., and Cassar, O. (2012). Epidemiological aspects and world distribution of HTLV-1 infection. *Front. Microbiol.* 3:388. doi: 10.3389/fmicb.2012.00388

Gessain, A., Vernant, J. C., Maurs, L., Barin, F., Gout, O., Calender, A., et al. (1985). Antibodies to human t-lymphotropic virus type-i in patients with tropical spastic paraparesis. *Lancet* 326, 407–410. doi: 10.1016/S0140-6736(85)92734-5

Hagiya, K., Yasunaga, J., Satou, Y., Ohshima, K., and Matsuoka, M. (2011). ATF3, an HTLV-1 bZip factor binding protein, promotes proliferation of adult T-cell leukemia cells. *Retrovirology* 8:19. doi: 10.1186/1742-4690-8-19

Hall, W. W., and Fujii, M. (2005). Deregulation of cell-signaling pathways in HTLV-1 infection. *Oncogene* 24, 5965–5975. doi: 10.1038/sj.onc.1208975

Hilburn, S., Rowan, A., Demontis, M. A., MacNamara, A., Asquith, B., Bangham, C. R. M., et al. (2011). In vivo expression of human T-lymphotropic virus type 1 basic leucine-Zipper protein generates specific CD8+ and CD4+ T lymphocyte responses that correlate with clinical outcome. *J. Infect. Dis.* 203, 529–536. doi: 10.1093/infdis/jiq078

Hivin, P., Frédéric, M., Arpin-André, C., Basbous, J., Gay, B., Thébault, S., et al. (2005). Nuclear localization of HTLV-I bZIP factor (HBZ) is mediated by three distinct motifs. *J. Cell Sci.* 118(Pt 7), 1355–1362. doi: 10.1242/jcs.01727

Kinosada, H., Yasunaga, J.-I., Shimura, K., Miyazato, P., Onishi, C., Iyoda, T., et al. (2017). HTLV- enhances T-cell proliferation by impeding the suppressive signaling of co-inhibitory receptors. *PLOS Pathog.* 13:e1006120. doi: 10.1371/journal.ppat.1006120

Ma, G., Yasunaga, J., Fan, J., Yanagawa, S., and Matsuoka, M. (2013). HTLV-1 bZIP factor dysregulates the Wnt pathways to support proliferation and migration of adult T-cell leukemia cells. *Oncogene* 32, 4222–4230. doi: 10.1038/onc.2012

Ma, G., Yasunaga, J., and Matsuoka, M. (2016). Multifaceted functions and roles of HBZ in HTLV-1 pathogenesis. *Retrovirology* 13:16. doi: 10.1186/s12977-016-0249-x

Martin, F., Taylor, G. P., and Jacobson, S. (2014). Inflammatory manifestations of HTLV-1 and their therapeutic options. *Expert Rev. Clin. Immunol.* 10, 1531–1546. doi: 10.1586/1744666X.2014.966690

Matsuoka, M., and Green, P. L. (2009). The HBZ gene, a key player in HTLV-1 pathogenesis. *Retrovirology* 6:71. doi: 10.1186/1742-4690-6-71

Matsuoka, M., and Jeang, K. T. (2007). Human T-cell leukaemia virus type 1 (HTLV-1) infectivity and cellular transformation. *Nat. Rev. Cancer* 7, 270–280. doi: 10.1038/nrc2111

Mukai, R., and Ohshima, T. (2014). HTLV-1 HBZ positively regulates the mTOR signaling pathway via inhibition of GADD34 activity in the cytoplasm. *Oncogene* 33, 2317–2328. doi: 10.1038/onc.2013.181

Murata, K., Hayashibara, T., Sugahara, K., Uemura, A., Yamaguchi, T., Harasawa, H., et al. (2006). A novel alternative splicing isoform of human T-cell leukemia virus type 1 bZIP factor (HBZ-SI) targets distinct subnuclear localization. *J. Virol.* 80, 2495–2505. doi: 10.1128/jvi.80.5.2495-2505.2006

Nagai, M., Usuku, K., Matsumoto, W., Kodama, D., Takenouchi, N., Moritoyo, T., et al. (1998). Analysis of HTLV-I proviral load in 202 HAM/TSP patients and 243 asymptomatic HTLV-I carriers: high proviral load strongly predisposes to HAM/TSP. *J. Neurovirol.* 4, 586–593. doi: 10.3109/13550289809114225

Nasr, R., Marçais, A., Hermine, O., and Bazarbachi, A. (2017). Overview of targeted therapies for adult T-Cell leukemia/lymphoma. *Methods Mol. Biol.* 1582, 197–216. doi: 10.1007/978-1-4939-6872-5_15

Peloponese, J. M., Yeung, M. L., and Jeang, K. T. (2006). Modulation of nuclear factor-kappaB by human T cell leukemia virus type 1 tax protein: implications for oncogenesis and inflammation. *Immunol. Res.* 34, 1–12. doi: 10.1385/IR:34:1:1

Petropoulos, L., Lin, R., and Hiscott, J. (1996). Human T cell leukemia virus Type 1 tax protein increases NF-κB dimer formation and antagonizes the inhibitory activity of the IκBα regulatory protein. *Virology* 225, 52–64. doi: 10.1006/viro.1996.0574

Poiesz, B. J., Ruscetti, F. W., Gazdar, A. F., Bunn, P. A., Minna, J. D., and Gallo, R. C. (1980). Detection and isolation of type C retrovirus particles from fresh and cultured lymphocytes of a patient with cutaneous T-cell lymphoma. *Proc. Natl. Acad. Sci. U.S.A.* 77, 7415–7419. doi: 10.1073/pnas.77.12.7415

Proietti, F. A., Carneiro-Proietti, A. B., Catalan-Soares, B. C., and Murphy, E. L. (2005). Global epidemiology of HTLV-I infection and associated diseases. *Oncogene* 24, 6058–6068. doi: 10.1038/sj.onc.1208968

Raval, G. U., Bidoia, C., Forlani, G., Tosi, G., Gessain, A., and Accolla, R. S. (2015). Localization, quantification and interaction with host factors of endogenous HTLV-1 HBZ protein in infected cells and ATL. *Retrovirology* 12:59. doi: 10.1186/s12977-015-0186-0

Rowan, A. G., Suemori, K., Fujiwara, H., Yasukawa, M., Tanaka, Y., Taylor, G. P., et al. (2014). Cytotoxic T lymphocyte lysis of HTLV-1 infected cells is limited by weak HBZ protein expression, but non-specifically enhanced on induction of Tax expression. *Retrovirology* 11:116. doi: 10.1186/s12977-014-0116-6

Saito, M., Matsuzaki, T., Satou, Y., Yasunaga, J., Saito, K., Arimura, K., et al. (2009). In vivo expression of the HBZ gene of HTLV-1 correlates with proviral load, inflammatory markers and disease severity in HTLV-1 associated myelopathy/tropical spastic paraparesis (HAM/TSP). *Retrovirology* 6:19. doi: 10.1186/1742-4690-6-19

Satou, Y., Yasunaga, J., Zhao, T., Yoshida, M., Miyazato, P., Takai, K., et al. (2011). HTLV-1 bZIP factor induces T-cell lymphoma and systemic inflammation in vivo. *PLOS Pathog.* 7:e1001274. doi: 10.1371/journal.ppat.1001274

Satou, Y., Yasunaga, J.-I., Yoshida, M., and Matsuoka, M. (2006). HTLV-I basic leucine zipper factor gene mRNA supports proliferation of adult T cell leukemia cells. *Proc. Natl. Acad. Sci. U.S.A.* 103, 720–725. doi: 10.1073/pnas.0507631103

Shimoyama, M. (1991). Diagnostic criteria and classification of clinical subtypes of adult T-cell leukaemia-lymphoma. A report from the lymphoma study group (1984–87). *Br. J. Haematol.* 79, 428–437. doi: 10.1111/j.1365-2141.1991.tb08051.x

Shiohama, Y., Naito, T., Matsuzaki, T., Tanaka, R., Tomoyose, T., Takashima, H., et al. (2016). Absolute quantification of HTLV-1 basic leucine zipper factor (HBZ) protein and its plasma antibody in HTLV-1 infected individuals with different clinical status. *Retrovirology* 13:29. doi: 10.1186/s12977-016-0263-z

Sugata, K., Yasunaga, J., Kinosada, H., Mitobe, Y., Furuta, R., Mahgoub, M., et al. (2016). HTLV-1 viral factor HBZ induces CCR4 to promote T-cell migration and proliferation. *Cancer Res.* 76, 5068–5079. doi: 10.1158/0008-5472.CAN-16-0361

Takeda, S., Maeda, M., Morikawa, S., Taniguchi, Y., Yasunaga, J., Nosaka, K., et al. (2004). Genetic and epigenetic inactivation of tax gene in adult T-cell leukemia cells. *Int. J. Cancer* 109, 559–567. doi: 10.1002/ijc.20007

Tanaka-Nakanishi, A., Yasunaga, J., Takai, K., and Matsuoka, M. (2014). HTLV-1 bZIP factor suppresses apoptosis by attenuating the function of FoxO3a and altering its localization. *Cancer Res.* 74, 188–200. doi: 10.1158/0008-5472.CAN-13-0436

Thébault, S., Basbous, J., Hivin, P., Devaux, C., and Mesnard, J. M. (2004). HBZ interacts with JunD and stimulates its transcriptional activity. *FEBS Lett.* 562, 165–170. doi: 10.1016/s0014-5793(04)00225-x

Tosi, G., Forlani, G., Andresen, V., Turci, M., Bertazzoni, U., Franchini, G., et al. (2011). Major histocompatibility complex class II transactivator CIITA is a viral restriction factor that targets human T-cell lymphotropic virus type 1 Tax-1 function and inhibits viral replication. *J. Virol.* 85, 10719–10729. doi: 10.1128/JVI.00813-11

Usui, T., Yanagihara, K., Tsukasaki, K., Murata, K., Hasegawa, H., Yamada, Y., et al. (2008). Characteristic expression of HTLV-1 basic zipper factor (HBZ) transcripts in HTLV-1 provirus-positive cells. *Retrovirology* 5:34. doi: 10.1186/1742-4690-5-34

Utsunomiya, A., Choi, I., Chihara, D., and Seto, M. (2015). Recent advances in the treatment of adult T-cell leukemia-lymphomas. *Cancer Sci.* 106, 344–351. doi: 10.1111/cas.12617

Yasuma, K., Yasunaga, J., Takemoto, K., Sugata, K., Mitobe, Y., Takenouchi, N., et al. (2016). HTLV-1 bZIP factor impairs anti-viral immunity by inducing co-inhibitory molecule, T cell immunoglobulin and ITIM domain (TIGIT). *PLOS Pathog.* 12:e1005372. doi: 10.1371/journal.ppat.1005372

Yoshida, M., Miyoshi, I., and Hinuma, Y. (1982). Isolation and characterization of retrovirus from cell lines of human adult T-cell leukemia and its implication in the disease. *Proc. Natl. Acad. Sci. U.S.A.* 79, 2031–2035. doi: 10.1073/pnas.79.6.2031

Yoshida, M., Satou, Y., Yasunaga, J., Fujisawa, J., and Matsuoka, M. (2008). Transcriptional control of spliced and unspliced human T-cell leukemia virus type 1 bZIP factor (HBZ) gene. *J. Virol.* 82, 9359–9368. doi: 10.1128/JVI.00242-08

Zhao, T., and Matsuoka, M. (2012). HBZ and its roles in HTLV-1 oncogenesis. *Front. Microbiol.* 3:247. doi: 10.3389/fmicb.2012.00247

Zhao, T., Yasunaga, J., Satou, Y., Nakao, M., Takahashi, M., Fujii, M., et al. (2009). Human T-cell leukemia virus type 1 bZIP factor selectively suppresses the classical pathway of NF-κB. *Blood* 113, 2755–2764. doi: 10.1182/blood-2008-06-161729

22

The Role of Chemokines in the Pathogenesis of HTLV-1

Razieh Zargari[1†], Maryam Mahdifar[1†], Asadollah Mohammadi[2], Zohreh Vahidi[1], Gholamhossein Hassanshahi[3] and Houshang Rafatpanah[1*]

[1] Immunology Research Center, Inflammation and Inflammatory Diseases Division, Mashhad University of Medical Sciences, Mashhad, Iran, [2] Cellular and Molecular Research Center, Research Institute for Health Development, Kurdistan University of Medical Sciences, Sanandaj, Iran, [3] Molecular Medicine Research Center, Rafsanjan University of Medical Sciences, Rafsanjan, Iran

***Correspondence:**
Houshang Rafatpanah
RafatpanahH@mums.ac.ir

† These authors have contributed equally to this work and share first authorship

Human T cell leukemia virus type 1 (HTLV-1) is a human retrovirus that is associated with two main diseases: HTLV-associated myelopathy/tropical spastic paraparesis (HAM/TSP) and adult T cell leukemia/lymphoma (ATL). Chemokines are highly specialized groups of cytokines that play important roles in organizing, trafficking, homing, and in the migration of immune cells to the bone marrow, lymphoid organs and sites of infection and inflammation. Aberrant expression or function of chemokines, or their receptors, has been linked to the protection against or susceptibility to specific infectious diseases, as well as increased the risk of autoimmune diseases and malignancy. Chemokines and their receptors participate in pathogenesis of HTLV-1 associated diseases from inflammation in the central nervous system (CNS) which occurs in cases of HAM/TSP to T cell immortalization and tissue infiltration observed in ATL patients. Chemokines represent viable effective prognostic biomarkers for HTLV-1-associated diseases which provide the early identification of high-risk, treatment possibilities and high-yielding clinical trials. This review focuses on the emerging roles of these molecules in the outcome of HTLV-1-associated diseases.

Keywords: chemokine, HTLV-1, ATL, HAM/TSP, inflammatory response

INTRODUCTION

Human T cell leukemia virus type 1 (HTLV-1) belongs to the Retroviridae family, the orthoretrovirinae subfamily and to the deltaretrovirus genus (Poiesz et al., 1980). HTLV-1 has been recognized as a causative agent of two important diseases, HTLV-1 associated myelopathy/tropical spastic paraparesis (HAM/TSP) and adult T cell leukemia/lymphoma (ATL) (Yoshida et al., 1982; Gessain et al., 1985; Osame et al., 1986). The virus not only induces HAM/TSP and ATL in a small proportion of HTLV-1 carriers, but it is also associated

Abbreviations: ACs, asymptomatic carriers; BALF, bronchoalveolar lavage fluid; CFA, cryptogenic fibrosing alveolitis; CLA^{+}, cutaneous lymphocyte antigen positive; CTCL, cutaneous T cell lymphoma; CTLs, cytotoxic T lymphocytes; ELC, EBI1 ligand chemokine; FOXP3, Forkhead Box P3; GIT, gastrointestinal tract; GPCRs, G Protein–Coupled Receptors; HBZ, HTLV-1 bZIP factor; HCs, healthy controls; IL, interleukin; IP-10, IFN-γ inducible Protein-10; LFA-1, lymphocyte function-associated antigen 1; MCP-1, monocyte chemotactic protein-1; MDC, macrophage-derived chemokine; MIP, macrophage inflammatory protein; pML, primary murine lymphoblastoid; RANTES, regulated on activation normal T expressed and secreted; SCID, severe combined immunodeficiency; SDF-1α, Stromal cell-Derived Factor-1; SLC, secondary lymphoid-tissue chemokine; TARC, thymus activation-regulated chemokine.

with several diseases such as HTLV-1 associated arthropathy (HAAP), exocrinopathy, HTLV-1 uveitis (HU), cutaneous T cell lymphoma (CTCL), dermatitis, Graves' disease, lymphadenitis, polymyositis, and chronic respiratory diseases (Kimura et al., 1986; Green et al., 1989; Morgan et al., 1989; Nishioka et al., 1989; LaGrenade et al., 1990; Hall et al., 1991; Mochizuki et al., 1992; Ohshima et al., 1992; Yamaguchi et al., 1994).

The exact mechanism of the development and pathogenesis of HAM/TSP and ATL among less than 3–5% of HTLV-1 carriers has not fully understood. However, it has been reported that genetic background such as human leukocyte antigen (HLA) and HTLV-1 genomic variation are associated with the risk of HAM/TSP among HTLV-1 carriers (Mahieux et al., 1995; Jeffery et al., 1999; Furukawa et al., 2000; Sabouri et al., 2005; Rafatpanah et al., 2007). The major criteria that differ HAM/TSP patients from HTLV-1-infected ACs are high titers of anti-HTLV-1 antibody, increased proviral loads, higher frequency of HTLV-1 specific cytotoxic T lymphocytes (CTLs), and elevation in activated lymphocytes (Kira et al., 1992; Elovaara et al., 1993; Nagai et al., 1998). Evidence also suggests viral and environmental factors, the efficiency of immune response and the genetic background of the HTLV-1 infected individual affect the outcome of HTLV-1 infection.

In HAM/TSP patients, high CTL responses may lead to the production of inflammatory cytokines and bystander damage. Based on this criterion, HTLV-1 specific $CD8^+$ clones in HAM/TSP patients produce a wide variety of members of cyto/chemokine family, such as Interferon-gamma (IFN-γ), Tumor necrosis factor-alpha (TNF-α), Interleukin (IL)-2, Macrophage inflammatory protein 1-alpha (MIP-1α)/CCL3, beta (MIP-1β)/CCL4, and Matrix metalloproteinase-9 (MMP-9), which promote the inflammatory response (Elovaara et al., 1993). HTLV-1 infected cells have the potential to produce IFN-γ and migration toward tissues, including central nervous system (CNS), which may induce a harmful non-specific immune response within the CNS (Elovaara et al., 1993). Furthermore, Tax-inducible production of TNF-α, IL-1β, and IL-6 by macrophages and microglia may enhance the proinflammatory response in the CNS (Giraudon et al., 1995; Ahuja et al., 2007).

The immune response and the number of infected cells in the cerebrospinal fluid (CSF) are important factors for the induction of chronic inflammation observed in the CNS of HAM/TSP patients (Bangham et al., 2015; Kubota, 2017). Plasma level of CCL24 that is related to type 2 immune response has been reported to be reduced in HAM/TSP. The low CCL24 plasma levels support the theory of enhanced type 1 immune response in HAM/TSP patients (Guerreiro et al., 2006). It seems that the number of chemokines and their receptors in the peripheral blood and in the site of inflammation act as a risk factor for the development of HAM/TSP. Previous studies have revealed that chemokines and their receptors play a critical role in HTLV-1 infection and migration of HTLV-1 infected cells into the CNS through the blood-brain barrier (BBB); playing an essential role in the onset of HAM/TSP (Ando et al., 2013). It was also shown that cerebral endothelial cells are susceptible to HTLV-1 infection, leading to HTLV-1 passage through the BBB and subsequent BBB breakdown (Afonso et al., 2007, 2008).

ATL is a peripheral T-cell malignancy with poor prognosis (Katsuya et al., 2015). In contrast to HAM/TSP, ATL patients exhibit immunosuppressive conditions (Uchiyama et al., 1977; Bunn et al., 1983; Shimoyama and Group, 1991). The immune system in ATL patients is directly suppressed by cytokines such as IL-10 and transforming growth factor-beta (TGF-B), which might be attributed to IL-10 production by Tax and HTLV-1 basic leucine zipper factor (HBZ) promotion (Mori et al., 1996; Inagaki et al., 2006; Matsuoka and Yasunaga, 2013).

Studies have shown that ATL cells obtained from peripheral blood and skin tumors of ATL patients are $CD4^+CD25^+$ $FOxp3^{low}$ and express cytotoxic T lymphocyte-associated antigen (CTLA)-4 with no regulatory activity (Shimauchi et al., 2008; Satou et al., 2012). Therefore, the pattern of cytokine production provides evidence for the pathogenesis of HTLV-1 associated diseases. The other factor, which is involved in the migration and tissue microenvironmental localization of lymphocytes, is adhesion molecules that are expressed on the surface of certain cells (El-Sabban et al., 2002; Banerjee et al., 2007). Furthermore, chemokines and their associated receptors are also involved in the pathogenesis of ATL as it infiltrates ATL cells into the lymph nodes, spleen, liver, skin and GIT (Hasegawa et al., 2000; Bangham et al., 2015; Kubota, 2017). Here, we discuss the role of chemokines and their receptors in the development and pathogenesis of HAM/TSP, ATL and other HTLV-1 related diseases.

CHEMOKINES AND THEIR RECEPTORS

Chemokines or chemotactic cytokines are a group of small molecules (weighing ~8–14 kDa) which regulate migration and trafficking of various types of immune cells through interactions with a subset of cell surface seven-transmembrane G protein-coupled receptors (GPCRs) (Zlotnik and Yoshie, 2000). To date, about 50 chemokines and 20 chemokine receptors have been identified in humans (Blanchet et al., 2012). Based on a new classification, chemokine ligands and receptors are named as 'L' and 'R,' respectively (Zlotnik and Yoshie, 2000). Chemokines play pivotal roles in the development, homeostasis, and functions of the immune system (Zlotnik and Yoshie, 2000). They mainly act to traffic monocytes, lymphocytes, neutrophils, and eosinophils to the site of injury, infection, and inflammation, and thus play an important role in host defense mechanisms (Zlotnik and Yoshie, 2000). Chemokines are involved in the pathobiology of chronic inflammations, autoimmune diseases, tumorigenesis, and tumor metastasis. In fact, a disease often occurs when the physiological role of chemokines is disrupted (Raman et al., 2011).

In addition to their role in the immune system, chemokines also have a wide range of effects on many different cell types including various types of the CNS or endothelial cells, where they result in either angiogenesis or angiostasis (Strieter et al., 1995; Ma et al., 1998). They are involved in cell migration during embryogenesis and also play important roles in wound healing, angiogenesis/angiostasis and tumor development and metastasis (McGrath et al., 1999; Mashino et al., 2002; Li et al., 2004; Yang and Richmond, 2004; Burns et al., 2006; Ishida et al., 2008). The role played by chemokine/receptor axes in tumor pathophysiology is complex as some chemokines enhance tumor development and metastasis, while others associate with an effective anti-tumor immune response (Singh et al., 2007).

On the basis of the arrangement of the two N-terminal cysteine residues in their biochemical structure, chemokines are further divided into four major subfamilies, CXC, CX3C, CC, and C, depending on whether the first two cysteine residues have either one or more amino acids between them (CXC, CX3C) or are adjacent (CC) (Miller and Mayo, 2017). CXC chemokines have the ability to activate and attract neutrophils and T lymphocytes, while CC chemokines are effective on multiple leukocyte subtypes including monocytes, eosinophils, T lymphocytes, basophils, dendritic cells (DCs), natural killer (NK) cells, and neutrophils. CX3C (Fraktaline) induces the migration of T cells, NK cells, and monocytes (Luster, 1998; Lee et al., 2018). Since the receptors for CC and CXC chemokines determine the viral tropism and mediate the efficient entry of human immunodeficiency virus (HIV) into host cells, CC and CXC chemokines could suppress HIV infection (Owais and Arya, 1999). CC chemokines mostly inhibit the entry and replication of macrophage-tropic strains; nonetheless, CXC chemokines mostly inhibit the entry of T-tropic strains (Owais and Arya, 1999).

According to their function and pattern of expression, chemokines have also been classified into homeostatic and inflammatory groups or both. The homeostatic chemokines are constitutively expressed and important in many physiological processes and are involved in normal circulation and trafficking of leukocytes in lymph nodes and thymus, while the production of inflammatory group is induced by infection or other proinflammatory stimuli resulting in recruitment of leukocytes to the insulted regions, in particular, due to infection (Cyster, 1999; Sharma, 2010).

Chemokine receptors are subdivided into CCR, CXCR, CX3CR, and CR based on the ligand to which they bind. The interesting fact about the chemokine receptors is the great level of redundancy observed in their effects on targets in which most of them can bind to several chemokines and most chemokines utilize more than one receptor (Zlotnik and Yoshie, 2000). During viral infection, the expression of CC chemokines is more prevalent than CXC chemokines (Melchjorsen et al., 2003; Dahm et al., 2016). CCL3 and CCL5 appear to be mostly related to viral infections (Rösler et al., 1998; Bonville et al., 1999; Haeberle et al., 2001; Domachowske et al., 2002). Among CXCL chemokines which are produced during viral infection, CXCL10 expression is observed in many viral infections (Dawson et al., 2000; Haeberle et al., 2001; Liu et al., 2001; Peterson et al., 2001). CCL2 and CXCL10 are expressed at large measures during human viral meningitis and influenza infections, while in the case of hepatitis C virus (HCV), CCL5 and CXCL9 are more expressed (Lahrtz et al., 1997; Dawson et al., 2000; Haeberle et al., 2001; Apolinario et al., 2002).

CHEMOKINES AND HTLV-1 VIRAL TRANSCRIPTIONAL REGULATORS

HTLV-1 Tax and HBZ viral products are associated with oncogenic transformation and ATL pathogenesis (Pozzatti et al., 1990; Tanaka et al., 1990; Satou et al., 2006). Tax is a key regulatory protein encoded by the pX region of the HTLV-1 genome, while HBZ is encoded by the antisense strand of the virus genome (Nagashima et al., 1986; Larocca et al., 1989; Cavanagh et al., 2006). Tax and HBZ regulate transcription of viral proteins, host factors and cellular signaling pathways, resulting in development of HTLV-1 associated diseases (Currer et al., 2012; Zhao, 2016). The HTLV-1 Tax protein has been known as a potent activator of a variety of transcription factors including NF-kB, CREB, SRF, and AP-1 (Fujii et al., 1992, 1995, 2000; Bex et al., 1998; Jeang, 2001). It was shown that the transcription factors NF-kB and AP-1 are involved in Tax-mediated transactivation of IL-8 (Mori et al., 1998). Extracellular Tax has the ability to enhance maturation of human monocyte-derived DCs and exerts changes in their function, which is accompanied by constant antigen presentation to T cells, leading to uncontrolled T cell proliferation and disease progression (Jain et al., 2007). Furthermore, Tax induces the production of proinflammatory cytokines and chemokines involved in the activation of DCs (Jain et al., 2007). The production of CXCL10, CCL1, CCL3, and CCL4 is induced by Tax (Baba et al., 1996). CCL3 and CCL4 are among CC chemokines that attract DCs into the lymph nodes and play an important role in the effector phase of lymphocyte response and are capable of recruiting T cells with regulatory properties to the inflammation site (Sallusto et al., 1998b; Moser and Loetscher, 2001; Jain et al., 2007). CCL20/MIP-3α is a CC chemokine with a crucial role in the initiation of inflammation (Imaizumi et al., 2002).

It has been reported that Tax elevates the levels of CCL20 mRNA in HTLV-1 infected cell lines and human T cell lines. The specific receptor for this chemokine is CCR6, which is also increased in HTLV-1 infected cell lines suggesting an autocrine/paracrine mechanism for the development of HTLV-1-associated diseases (Baba et al., 1997; Imaizumi et al., 2002). Among various leukocyte subsets, CCR6 was detected and strongly up-regulated by IL-2 in lymphocytes ($CD4^+$ and $CD8^+$ T cells and B cells) but not in NK cells, monocytes, or granulocytes (Baba et al., 1997). CCL20 is responsible for chemo-attraction of effector/memory T cells and B cells through its receptor, CCR6 (Baba et al., 1997). Furthermore, CCL20 recruits immature DCs to the inflammation site and is crucial in the regulation of inflammation toward the skin and mucosal surfaces under homeostatic and inflammatory conditions, as well as in pathological conditions, including cancer and autoimmune disorders such as rheumatoid arthritis (Mantovani, 1999; Schutyser et al., 2003).

Tax induces CC chemokines including CCL5, CCL4 and, CCL3 in cultured peripheral blood mononuclear cells (PBMCs) from uninfected subjects (Barrios et al., 2011, 2014). These chemokines are known as HIV inhibitory factors and expression of CCL5, CCL4, CCL3, CCR5 ligands, is associated with the shift of HIV infection from macrophage to the T cell tropic (Moriuchi et al., 1998; Llano and Esté, 2005; Wang et al., 2017).

HBZ also plays a crucial role in the pathogenesis of ATL and HAM/TSP (Satou et al., 2006; Enose-Akahata et al., 2017). It has been shown that HBZ interacts with cAMP response element-binding protein (CREB) transcription factor, suppresses the Tax-mediated transactivation, and inhibits the canonical NF-kB pathway (Gaudray et al., 2002; Lemasson et al., 2007; Zhao et al., 2009). It has been reported that HBZ induces the production of CCL21 and CCR4 in HTLV-1 infected cells, therefore, supporting the proliferation and migration of infected T cells (Sugata et al., 2016; Naito et al., 2018).

HTLV-1 also incorporates viral protein Tax, and viral mRNAs including Tax, HBZ, and env, into exosomes derived from infected cells, which contribute to enhanced cellular FLICE-like inhibitory protein (cFLIP)-dependent survival of the recipient cell with induction of NF-kB, and resistance to apoptosis, by activation of AKT. Therefore, the exosomes derived from HTLV-1 infected cells might associate with HTLV-1 pathogenesis (Jaworski et al., 2014).

CHEMOKINES AND HAM/TSP PATHOGENESIS

In HAM/TSP patients, $CD4^{+}$ $CD25^{+}$ $CCR4^{+}$ $Foxp3^{-}$ T cells which serve as reservoirs for HTLV-1 (Yamano et al., 2009), express low levels of TGF-β and IL-10, while they express high levels of IFN-γ, thus, this pattern of cytokine production may contribute to disease pathogenesis (Yamano et al., 2009; Quaresma et al., 2016). High levels of the proinflammatory cytokines, chemokines, and MMP in HAM/TSP patients, is likely to be associated with HAM/TSP pathogenesis (Biddison et al., 1997; Lepoutre et al., 2009). The frequency of HTLV-1 specific CTLs is higher in HAM/TSP patients compared to ACs (Elovaara et al., 1993; Kubota et al., 1998). These CTLs express chemokine receptors with high migratory capacity (Bieganowska et al., 1999) involved in the migration of leukocytes to the site of inflammation and might play a crucial role in the pathogenesis of HAM/TSP. The neurological damages in CNS of HAM/TSP patients is caused by recognition of $CD4^{+}$HTLV-1 infected cells in the CNS cells by specific $CD8^{+}$ CTL, resulting in the production of inflammatory cytokines such as IFN-γ, TNF-α and MMP (Jacobson et al., 1990; Wucherpfennig et al., 1992; Kubota et al., 1994; Umehara et al., 1994, 1998). Furthermore, the interaction between HTLV-1-specific CTL and HTLV1-infected CD4+ T cells leads to apoptosis in surrounding neural cells such as oligodendrocytes and finally demyelination of the CNS (Ijichi et al., 1993; Hanon et al., 2000). Further studies have demonstrated that chemokines might contribute to the HAM/TSP pathogens and we summarize their results. According to the potential ability of chemokines to direct leukocyte trafficking, homing, and migration, in this study chemokines and their receptors are further classified based on their function.

Trafficking

CXCL9 and CXCL10

CXCL9 and CXCL10 induce the recruitment of CXCR3 expressing cells including the $CXCR3^{+}$ IFN-γ producing HTLV-1 infected $CD4^{+}$ T cells that promote the production of CXCL10 by the astrocytes, creating a positive feedback loop that results in the CNS chronic inflammation (Futsch et al., 2018). In Multiple Sclerosis (MS), which is also a demyelinating neuroinflammatory disease, it is thought that CXCL9 and CXCL10 in the CSF are involved in the recruitment of $CXCR3^{+}$ T cells into CNS and contribute to MS pathogenesis (Müller et al., 2010; Cheng and Chen, 2014; Vazirinejad et al., 2014; Koper et al., 2018).

CXCL9 and CXCL10 are presented in high levels in serum and CSF of patients with HAM/TSP and contribute to the pathogenesis of HAM/TSP (Sato et al., 2013; Guerra et al., 2018). Chaves et al. showed that the plasma levels of CXCL8 and CXCL9 were greater in HAM/TSP patients compared to carriers (Chaves et al., 2016). Moreover, a correlation was observed between plasma levels of CXCL8, CXCL9, and CXCL10 and proviral load, in which the HTLV-1 infected individuals with higher proviral loads had increased plasma levels of these chemokines (Chaves et al., 2016). CXCL10 in the CSF, as indicators of CNS inflammation, correlate with disease progression and has been suggested as a candidate for the prognosis of HAM/TSP development (Guerreiro et al., 2006; Sato et al., 2013; Lima et al., 2017). These markers might even be more potent than the proviral load for the prediction of disease progression in HTLV-1 infected individuals (Sato et al., 2013). Elevated CSF and serum levels of CXCL9 and CXCL10, compared to carriers, alongside low levels of CCL2 are considered as pivotal markers of HAM/TSP (Guerreiro et al., 2006; Sato et al., 2013; Starling et al., 2015; Lima et al., 2017). Amorim et al. (2014) showed that the production of both CXCL9 and CCL5 is elevated in HTLV-1 infected subjects compared with HCs. This high chemokine production might lead to increased inflammation in HTLV-1 infected individuals.

Sato et al. (2018b) showed that according to the concentrations of CSF CXCL10 and neopterin levels, HAM/TSP patients are classified into three groups based on disease activity and clinical progression rate. Further studies showed that low-dose oral prednisolone maintenance therapy decreased the CSF CXCL10 levels and is associated with good functional prognosis, suggesting CSF CXCL10 is a promising surrogate marker for HAM/TSP treatment (Tamaki et al., 2019).

CXCL8

CXCL8 might be associated with BBB disruption through the attraction of neutrophils within the CNS (Chaves et al., 2016). CXCL8 and its receptors, CXCR1 and CXCR2, are involved in the infiltration of leukocytes into the CNS (Almasi et al., 2013). CXCR1 and CXCR2 are expressed on neutrophils, monocytes, NK cells, $CD4^{+}$ and $CD8^{+}$T cells (Chuntharapai et al., 1994; Traves et al., 2004). CXCR2 has a role in inflammation and myelin disorders (Charo and Ransohoff, 2006). The levels of CXCR2 and its ligands are elevated in various CNS inflammatory conditions (Hosking et al., 2010). CXCR2 mRNA is highly expressed in HAM/TSP and CXCR2 $CD8^{+}$ T cells have high frequencies in HAM/TSP patients. The high frequency of CXCR2 $CD8^+$ T cells in HAM/TSP may contribute to demyelination and inflammation. However, CXCR1 $CD8^+$ T cells are lower in HAM/TSP patients when compared to ACs. Since there is a negative correlation between proviral load and the percentage of T cells expressing CXCR1, it has been suggested that the high frequency of CXCR1 $CD8^+$ T cells in ACs is associated with an efficient cell-mediated immune response that leads to limiting the spread of the virus (Rajaei et al., 2018). Thus, CXCL8, CXCL9, and CXCL10 are thought to be associated with HAM/TSP pathogenesis as well.

CCL2

CCL2 is thought to be associated with the migration of T cells and monocytes into the CNS of HAM/TSP patients. In HAM/TSP patients, CCL2, expressed in the CNS from vascular endothelial cells and mononuclear cells which have been infiltrated into the CNS, attracts T cells and monocytes into the CNS (Umehara et al., 1996). The lower levels of CCL2, which polarizes the immune response toward Th2 (Sallusto et al., 1998a), in HAM/TSP patients compared to HCs, also reflect a Th1 dominant immune response in HAM/TSP (Gu et al., 2000; Narikawa et al., 2005). It is likely that decreased levels of CCL2, which is associated with type two immune response, reflect the polarization of immune response toward a type 1 cytokine profile (Sallusto et al., 1998a; Gu et al., 2000). da Silva-Malta et al. reported that -46C/C polymorphism in the antigen receptor for chemokines (DARC) is associated with leukopenia, neutropenia and lower levels of CCL2, however, this polymorphism is not a risk factor for HAM/TSP (da Silva-Malta et al., 2017).

CCL20

The attraction of leukocytes to the site of inflammation may be conducted by CCL20/MIP-3α (Rossi et al., 1997). HTLV-1 infected individuals may develop inflammatory diseases according to CCL20 expression in HTLV-1 infected and uninfected T cells, which are induced by Tax through the NF-kB pathway. Since HTLV-1-infected T cells also express CCR6, the CCL20 receptor, an autocrine/paracrine mechanism may be involved in the pathogenesis of HTLV-1 associated diseases (Imaizumi et al., 2002; Rafatpanah et al., 2017). CCL20 is also a potent chemokine for the attraction of DCs into the CNS, suggesting infected DCs might play pivotal roles in HAM/TSP pathogenesis by induction of autoreactive T cells (Imaizumi et al., 2002).

CCL1

The role of CCL1 in the pathogenesis of experimental autoimmune encephalomyelitis (EAE) has been clearly investigated. It has been demonstrated that proteolipid protein-specific T-cell clones and production of the murine CCL1 homolog associates with EAE (Kuchroo et al., 1993; Teuscher et al., 1999). CCL1 is mainly produced by HTLV-1 infected cells of HAM/TSP patients in a Tax inducible and dependent manner and CCR8, the specific CCL1 receptor, is expressed on both HTLV-1 infected and uninfected T cells. The production of CCL1 is dependent on CREB and AP1 and the plasma levels of CCL1 are higher in HAM/TSP patients than the carriers and healthy subjects (Saito et al., 2017). The CCL1-CCR8 signaling axis may contribute to the pathogenesis of HAM/TSP, which might be suggested as a potential therapeutic target for immunotherapy of the disease (Saito et al., 2017).

CXCL11

CXCL11 or Interferon inducible T cell alpha chemoattractant (ITAC) enhances the chemotaxis of activated T cells and NK cells and its gene expression is induced by IFN-γ. CXCR3, the CXCL11 receptor, is also highly expressed on activated T cells, memory T cells, and NK cells and IFN-γ promotes its gene expression. It is also reported that CXCL11 is involved in the pathogenesis of cell mediated immunity (Müller et al., 2010). HAM/TSP patients have high levels of CXCL11 in plasma and CSF and the CSF CXCL11 can be used as a predictor of HAM/TSP development (Romanelli et al., 2018; Rosa et al., 2018).

Homing

CCR4

$CCR4^+$ T cells are selectively infected by HTLV-1 and converted into abnormal cells leading to the chronic inflammation in HAM/TSP patients (Araya et al., 2014; Yamano and Coler-Reilly, 2017). Elevated frequency of IFN-γ producing HTLV-1 infected $CD4^+CD25^+CCR4^+$ T cells with proinflammatory features is found in HAM/TSP patients and this is associated with the disease pathogenesis (Araya et al., 2014; Yamano and Coler-Reilly, 2017). HTLV-1 induces transcriptional changes via Tax in infected T cells and downregulates the expression of FOXP3 in $CD4^+CD25^+CCR4^+$ Tregs to lose expression of the transcription factor FOXP3 and increases the production of IFN-γ in these cells, which in turn leads to development of inflammation (Araya et al., 2014, 2015). The HTLV-1 Tax-expressing infected $CD4^+CD25^+CCR4^+$ T cells exert changes in infected cells that may phenotypically become IFN-γ producing Th1-like cells, which probably enhance the inflammatory state (Araya et al., 2014, 2015). HBZ play a pivotal role in HAM/TSP by increasing the number of inducible Treg cells (iTreg) and induction of IFN-γ production (Yamamoto-Taguchi et al., 2013). Administration of Mogamulizumab/KW-0761, as a humanized anti-CCR4 monoclonal antibody (mAb), in HAM/TSP, has reduced the number of HTLV-1 infected cells and inflammatory markers. However, it was accompanied by adverse effects including rash, lymphopenia and leukopenia (Sato et al., 2018a).

Migration

CCL3/MIP-1α and CCL4/MIP-1β

MIP can be selectively expressed and secreted in a *tax* transfected Jurkat cell line (Sharma and May, 1999). Elevated levels of MIP has been also demonstrated in the CSF of HAM/TSP patients (Miyagishi et al., 1995). HTLV-1-transformed T cells released CCL3/MIP-1α as a major monocyte chemoattractant, suggesting this molecule might play a pivotal role in the outcome of HTLV-1-related-diseases (Bertini et al., 1995).

In addition to Monocyte chemoattraction (Schall et al., 1990), CCL3 and CCL4 may attract $CD8^+$ and $CD4^+$ T cells (Taub et al., 1993). HTLV-1 specific CTLs produce CCL3 and CCL4, which may be related to inflammation, observed in HAM/TSP patients (Biddison et al., 1997).

CHEMOKINES AND ATL PATHOGENESIS

ATL is an aggressive peripheral T cell neoplasm associated with HTLV-1 (Poiesz et al., 1981; Yoshida et al., 1982). ATL generally has a very poor prognosis and shorter overall survival (OS) compared to other peripheral T cell lymphoma (PTCL) (Vose et al., 2008). Clonal proliferation of HTLV-1 infected $CD4^+$ T cells mediated by HTLV-1 viral factors, specifically Tax and HBZ promotes cellular transformation and leads to the development of ATL (Tanaka et al., 1990; Smith and Greene, 1991; Satou et al., 2006). Tax is able to stimulate cell proliferation in ATL and inhibits cell apoptosis (Yoshida, 2001; Matsuoka and Yasunaga, 2013; Mühleisen et al., 2014). In fact, Tax induces expression of anti-apoptotic proteins and genes which are involved in cell proliferation and consistently inactivates tumor suppressor proteins (Yoshida, 2001; Matsuoka and Yasunaga, 2013; Mühleisen et al., 2014). The increased cellular proliferation along with inhibited apoptosis results in prolonged cell survival and transformation of HTLV-1 infected cells (Yoshida, 2001). Since Tax is HTLV-1 specific major antigen that is recognized by CTLs, expression of Tax is lost in most of the ATL cases in order to escape host immune response (Kannagi et al., 1993). Despite Tax that is inactivated in most of the cases, HBZ is always expressed in all cases and plays an important role in leukemogenesis of HTLV-1 infected cells (Satou et al., 2006). In fact, Tax associates with the initiation of transformation, while HBZ is needed to maintain the transformation when Tax is silenced (Ma et al., 2016). HBZ also contributes to ATL oncogenesis by inhibition of apoptosis and supporting the proliferation and migration of ATL cells (Ma et al., 2013). The process of tissue infiltration of ATL cells and HTLV-1 infected T cells is likely to be regulated by chemokines, chemokine receptors, and adhesion molecules (Mori et al., 2000; Sugata et al., 2016). Here, we focus on chemokines and chemokine receptors involved in tissue infiltration of HTLV-1 infected and enhancement of proliferation, survival, and immortalization of ATL cells.

Trafficking

CCL1/I-309 and CCR8

CCL1/I-309 is a chemokine with the ability of monocyte attraction (Miller and Krangel, 1992a,b). CCL1 observed in the supernatant of cultured ATL cells exerts anti-apoptotic effects against ATL cells (Ruckes et al., 2001). In addition to CCL1 production, ATL cells also express CCR8, the CCL1 chemokine receptor (Ruckes et al., 2001). The resistance of ATL cells to apoptosis might be attributed to the consequence of an autocrine loop between CCL1 and CCR8 (Ruckes et al., 2001).

CCL2/MCP-1

The elevated levels of monocyte chemoattractant protein 1 (MCP-1)/CCL2 mRNA in HTLV-1-infected T cell lines compared with uninfected cells have been reported (Mori et al., 2000). It is also shown that Tax induces the endogenous CCL2 through activation of the 5′ transcriptional regulatory region of the CCL2 gene in the human Jurkat T -cell line (Mori et al., 2000). Tax induces NF-κB binding to both CCL2 κB sites in order to transactivate the CCL2 gene via induction of NF-κB. Thus, the CCL2 gene regulation is disrupted by Tax and CCL2 is constitutively expressed in HTLV-1-infected cells (Mori et al., 2000). CCL2 also modulates the expression of leukocyte adhesion molecules and thus associates with tissue infiltration of leukocytes. Therefore, high levels of CCL2 expression in ATL cells may affect cell adhesion and tissue infiltration of ATL cells (Jiang et al., 1992). These findings might have important implications for our understanding of HTLV-1-associated diseases.

Homing

CXCL12/SDF-1 α and CXCR4/CXCR7

The interaction between stromal cell-derived factor 1 (SDF1)/CXCL12 and CXCR4 has been identified to play a paramount role in the migration of ATL cells (Arai et al., 1998). Expression of CXCL12 in HTLV-1 infected T cell lines have been proven to be induced by Tax (Arai et al., 1998). It was also demonstrated that Tax enhances the CXCL12/CXCR4 axis in ATL cells (Twizere et al., 2007; Kawaguchi et al., 2009). Additionally, ADM3100 which serves as a CXCR4 antagonist prevents migration of human ATL cells and primary murine lymphoblastoid (pML) cells toward CXCL12 and also inhibits organ infiltration by lymphomatous cells in severe combined immunodeficient (SCID) mice (Kawaguchi et al., 2009). However, a study by Hasegawa et al. (2000) showed that the intensity of CXCR4 expression is regardless of ATL tissue infiltration.

Another chemokine receptor which is overexpressed in HTLV-1 infection is CXCR7 (Jin et al., 2009). CXCR7 serves as the CXCL12 receptor and is frequently expressed by various tumor cell lines and is also supposed to support cell survival and growth (Burns et al., 2006; Infantino et al., 2006; Wang et al., 2008). CXCR7 has been shown to be expressed on some human PBMCs including B cells, monocytes and DCs, while T cells mostly lack expression of CXCR7 (Burns et al., 2006; Infantino et al., 2006; Wang et al., 2008). Although CXCR7 is not expressed on T cells (Burns et al., 2006), a study by Jin et al. (2009) showed that CXCR7 expression on T cells is induced by Tax, suggesting contributes in enhancement of cell survival, growth, and immortalization of HTLV-1 infected cells. It has been shown that CCX754, the synthetic CXCR7 antagonist, inhibits the growth of HTLV-1 infected cells (Jin et al., 2009).

CXCL8

The CXCL8 mRNA expression has been detected in peripheral leukemic cells of ATL patients and in cultured HTLV-1-infected T cell lines, including HUT-102, MT-1, SALT-3, and SKT-1B (Mori et al., 1995). CXCL8 protein has also been measured in the culture medium of these cells and in the extracellular fluids of ATL patients (Mori et al., 1995). Based on these findings HTLV-1 tax gene may transactivate the CXCL8 gene through the NF-κB pathway in HTLV-1-infected cells. CXCL8 is rarely expressed in

T cells, but its expression is increased in ATL cells and HTLV-1-infected cells, therefore, the potential role of CXCL8 in ATL pathogenesis remains to be determined (Mori et al., 1995).

CCL18

CCL18 is significantly expressed by DCs and mainly attracts naive T cells (Adema et al., 1997; Pivarcsi et al., 2004). Hence, it may critically be involved in the initiation of an immune response (Schutyser et al., 2005). A study conducted by Shimizu et al. reported that cases with HTLV-1 associated lymphadenitis type and Hodgkin's like-type significantly express high levels of CCL18 compared to the non-specific lymphadenitis subjects (Shimizu et al., 2007). CCL18 can be associated with an efficient immune response in ATL patients through activation and attraction of leukocytes and thus plays a pivotal role in the initiation of the immune response against ATL cells (Shimizu et al., 2007).

Migration

CCL25 and CCR9

The characteristic feature of ATL is organ infiltration by leukemic cells (Matsuoka, 2003). ATL cells from patients with GIT involvement have been found to express high levels of CCR9 which enables ATL cells to infiltrate into intestinal mucosa where CCL25, the only ligand for CCR9, is abundantly produced (Utsunomiya et al., 1988; Nagakubo et al., 2007). Nagakubo et al. (2007) reported that the expression of CCR9 on ATL cells is induced by Tax. Immunohistochemical studies have provided pieces of evidence that CCR9 is expressed on ATL cells that were infiltrated into GIT (Nagakubo et al., 2007). Additionally, according to the anti-apoptotic properties of CCL25 against CCR9 expressing cells, the CCL25/CCR9 interaction axis may be involved in ATL cells immortalization process (Youn et al., 2001; Nagakubo et al., 2007). However, the freshly isolated ATL cells have been reported to scarcely express CCR9 (Nagakubo et al., 2007).

CCL17/TARC, CCL22/MDC, and CCR4

CCR4 is highly expressed by most of ATL cells and is involved in the invasion of ATL cells into the skin tissues (Yoshie et al., 2002; Ishida et al., 2003; Hieshima et al., 2008). CCR4 is well established to be expressed on Th2 cells, Treg cells, and a fraction of memory T cells in which most of them are cutaneous lymphocyte antigen positive (CLA$^+$) skin-homing memory T cells (Yoshie et al., 2001; Hieshima et al., 2008). Skin lesions are frequently observed in ATL and infiltration of ATL cells into skin can be directed by the interaction of CCR4 and its ligands, namely thymus and activation-regulated chemokine (TARC)/CCL17 as well as macrophage-derived chemokine (MDC)/CCL22 (Yoshie et al., 2002; Ishida et al., 2003; Shimauchi et al., 2005; Hiyoshi et al., 2015). Interestingly, a study by Shimauchi et al. (2005) demonstrated that CCR4$^+$ ATL cells also produce CCR4 ligands, CCL17 and CCL22.

Tax induces CCL22 expression by infected T cells and facilitates the interaction between infected cells and CCR4$^+$ uninfected cells. The efficient attraction of uninfected CCR4$^+$ T cells into CCL22 expressing infected cells, alongside with formation of viral synapse results in HTLV-1 transmission, indicating that ATL cells and HTLV-1-infected T cells may originate preferentially from CCR4$^+$ T cells (Hieshima et al., 2008). Additionally, the expression of CCR4 on HTLV-1 infected T cells is not Tax inducible. Therefore, the consistent CCR4 expression on ATL cells and HTLV-1 transformed T cells may be attributable to the perception that HTLV-1 preferentially infects CCR4$^+$ T cells through the recruitment of CCR4$^+$ T cells via CCL22 and/or due to some growth advantages of HTLV-1–infected CCR4$^+$ T cells *in vivo* as well as *in vitro* (Yoshie et al., 2002). However, Sugata et al. (2016) showed that HBZ upregulates CCR4 expression in ATL and HTLV-1 infected cells and thus, migration and proliferation of those cells are increased, thus CCR4 may not be initially expressed on the infected cells (Percher et al., 2017). CD4$^+$ FOXP3$^+$ cells are observed at a high frequency in ATL patients (Toulza et al., 2009). The plasma levels of CCL22 in HTLV-1 infected individuals is elevated and correlates with the frequency of FOXP3$^+$ cells that express CCR4. FOXP3$^+$ cells play a dual role in ATL and they enhance the suppression of immune response and reduce disease progression (Toulza et al., 2010). CCL22, which is produced by HTLV-1 infected cells, may attract CD4$^+$ CCR4$^+$Foxp3$^+$ T cells with regulatory functions (Toulza et al., 2010). These CD4$^+$ CCR4$^+$ Foxp3$^+$ T cells may retard the progression of HTLV-1 associated diseases and also contribute to the immunosuppressive state in HTLV-1 infection (Toulza et al., 2008; Toulza et al., 2009, 2010).

Somatic mutations in the CCR4 gene occur in about 26% of ATL cases and might enhance cell migration toward CCL17 and CCL22 gradient which may contribute to ATL progression. Thus, CCR4 signaling plays an important role in ATL pathogenesis and it is a potential target for ATL therapy (Nakagawa et al., 2014). Several reports demonstrated that Mogamulizumab/KW-0761, as a humanized anti-CCR4 monoclonal antibody, has a potential efficacy against relapsed CCR4$^+$ ATL cases and induces a long-term survival in cases who received Mogamulizumab monotherapy (Yamamoto et al., 2010; Ishida et al., 2017). According to previous studies, Mogamulizumab has increased antibody-dependent cell cytotoxicity (ADCC) of CCR4$^+$ATL cells. This cytotoxicity is thought to be mainly mediated by NK cells (Ishida et al., 2012). Furthermore, a shortened form of Pseudomonas exotoxin, PE38, fused to CCL17, has been found to be able to kill HTLV-1-infected cells. This effect depends on the expression of both CCR4 and furin, a proprotein convertase, which are both elevated in HTLV-1-infected cells (Hiyoshi et al., 2015).

CCL19/ELC, CCL21/SLC and CCR7

Both EBI1-ligand chemokine (ELC)/CCL19 and secondary lymphoid-tissue chemokine (SLC)/CCL21 are well defined to play a role in the migration and homing of lymphocytes to lymphoid tissues via their common receptor, the CCR7 (Kozai et al., 2017). It has been shown that CCR7 is highly expressed on ATL cells from patients with lymphoid organ infiltration, thus it seems that CCR7 may be involved in the attraction of ATL cells into lymphoid tissues (Hasegawa et al., 2000). ATL can be further characterized by CCR7 expression within CD4$^+$ CCR4$^+$T cells, by which the CCR7$^+$ phenotype refers to aggressive ATL, while CCR7$^-$ phenotype refers to progressive indolent ATL (Kagdi et al., 2017). Gain-of-function mutations in CCR7 gene occurring in about 11% of ATL cases induce surface CCR7 expression and chemotaxis toward CCL19 and CCL21, and also enhances PI3K/AKT signaling, implicating the association of these activating mutations in the pathogenesis of ATL (Kataoka et al., 2015).

CCL5/RANTES

Regulated on activation, normal T cell expressed and secreted (RANTES)/CCL5 serves as a chemotactic factor for memory T cells that are produced by HTLV-1-infected cells and ATL cells in a Tax inducible manner (Mori et al., 2004). Although CCL5 is related to Th1 cells (Schrum et al., 1996), it is thought to be involved in ATL pathogenesis and tissue infiltration of infected cells, through CCR5 as its receptor (Mori et al., 2004). ATL and HTLV-1-infected cell lines produce high levels of CCL5 (Mori et al., 2004). Furthermore, HTLV-1 infected T cells have been found to particularly express the CCL5 gene as well as generate CCL5 (Mori et al., 2004). A recombinant vector containing the Tax-1 gene, *pCDNA3.1-TAX*, is also thought to be able to induce CCR5 expression in the K562 cell line (Haghnazari Sadaghiani et al., 2019). Furthermore, CCL5 is expressed by leukemic cells in peripheral blood and lymph nodes of ATL patients. It was shown that Tax induces the expression of CCL5 mRNA in a human Jurkat T cell line through NF-κB activation pathway (Mori et al., 2004). The increased expression of CCL5 by HTLV-1-infected T cells may contribute to pathophysiology of HTLV-1-associated diseases. However, its potential role remains to be determined.

CCL3/MIP-1α and CCL4/MIP-1β

CCL3/MIP-1α and CCL4/MIP-1β are highly produced in the culture supernatant and cytoplasmic fractions of tumor-infiltrating lymphocytes (Tanaka et al., 1998). It has been demonstrated that proinflammatory cytokines, including TNF-α, IL-1β, and IL-6 show increased levels in ATL cells. Tax enhances CCL3 and CCL4 production in ATL cells (Bertini et al., 1995; Tanaka et al., 1998), and it is thought that the elevated production of CCL3 and CCL4 chemokines in ATL cells might induce integrin-dependent adhesion of ATL cells to the endothelium and therefore may contribute to tissue infiltration of ATL cells. Pretreatment of ATL cells with antibodies against CCL3 and CCL4 led to a decrease of tissue infiltration of ATL cells through the reduction in integrin mediated adhesion (Tanaka et al., 1998).

CX3CL1/Fractalkine

Fractalkine or CX3CL1 is the only member of the CX3C chemokine subfamily (Zlotnik and Yoshie, 2000; Jones et al., 2010). CX3CL1 is found both in membrane-bound and soluble forms. The membrane-bound form acts as a chemoattractive factor and its expression is induced on endothelial cells by inflammatory cytokines (Bazan et al., 1997). The CX3CL1 receptor, CX3CR1, is expressed on a number of cells including cytotoxic effector lymphocytes like NK cells and cytotoxic effector T cells, mature monocytes/macrophages, and mucosal dendritic cells, which all are involved in elimination of cancer cells (Bazan et al., 1997; Imai et al., 2005). The gene expression of CX3CR1 is downregulated in ATL cells (Shimizu et al., 2007). In the Shimizu et al. study it was shown that the CX3CR1 expression in HTLV-1 associated subjects was lower than the non-specific lymphadenitis cases, suggesting HTLV-1 infection associates with host immunity through a reduction in the number of cytotoxic cells (Shimizu et al., 2007). These results proposed that downregulation of CX3CR1 plays an important part in immune response against the ATL cells.

CHEMOKINES AND OTHER HTLV-1-ASSOCIATED DISEASES

It is well documented that ATL is associated with severe immunosuppression and patients with ATL are prone to opportunistic infections (Uchiyama et al., 1977; Bunn et al., 1983; Shimoyama and Group, 1991). However, the immune response controlling infection in HAM/TSP patients may become more harmful and participate in HAM/TSP development (Asquith and Bangham, 2008). Changes in immune response may also be related to other HTLV-1 associated diseases. Pieces of evidence indicate that different types of skin lesions observed in HTLV-1 carriers may be associated with a distinct imbalance of inflammatory response. The serum levels of CCL5, CXCL8, CXCL9, and CXCL10 have shown alterations in HTLV-1 infected individuals with or without skin lesions. It has been reported that HTLV-1 infected cases with autoimmune lesions have elevated levels of CCL5 and CXCL8 compared to those with infectious lesions. CXCL10 level has also been found to be elevated in HTLV-1 infected subjects with autoimmune lesions compared to the cases without lesions (Coelho-dos-Reis et al., 2013). Infection of lung epithelial cells by HTLV-1 was tested by co-culturing A549 alveolar and NCIH292 tracheal epithelial cell lines with MT-2 cell line which is an HTLV-1 infected T cell line. The HTLV-1 infected lung epithelial cells produce inflammatory chemokines such as CCL2, CCL5, and CCL20 (Teruya et al., 2008), which are all able to contribute to the pathogenesis of HTLV-1 associated pulmonary disorders (Teruya et al., 2008).

CCL5 is highly chemotactic for memory T cells and monocytes (Schall et al., 1990). This chemokine is expressed in HTLV-1-infected cell lines and primary ATL cells and may have a role in tissue infiltration (Mori et al., 2004). Monocytes have been suggested to be involved in immune regulation and progression of HTLV-1 associated diseases (de Castro-Amarante et al., 2016). HTLV-1 provirus has been detected in all three subsets of monocyte (classical, intermediate, non-classical) and found to have a profound impact on surface receptors, migratory function, and the frequency of all monocyte subsets. In HTLV-1 infected subjects, classical monocytes express high levels of CCR5, CXCR3 and CX3CR1 and a correlation has been observed between the proviral load and the migratory capacity of classical monocytes toward CCL2, CCL5, and CX3CL1, suggesting that migration of monocyte into inflammatory sites is related with persistence of HTLV-1 infection and progression of HTLV-1 associated inflammatory diseases, such as Myositis, Uveitis, and HAM/TSP (de Castro-Amarante et al., 2016).

The levels of some chemokines including CXCL9 and CXCL10 are undetectable in neutrophils from HTLV-1 infected and uninfected subjects (Bezerra et al., 2011). Furthermore, although neutrophils are of the main sources of CXCL8 and CCL4, these two chemokines have similar production in neutrophils from both infected and uninfected subjects, indicating that neutrophils effector functions are not impaired in these two groups (Bezerra et al., 2011). Pulmonary disorders can develop in ACs as well as HTLV-1 infected patients (Seki et al., 1999; Yamazato et al., 2003; Yamamoto et al., 2004; Nakayama et al., 2013). High levels of CCL3 and CCL5 in bronchoalveolar lavage fluid (BALF) have been detected in HTLV-1 carriers (Seki et al., 1999; Yamazato et al., 2003). Since CCL3 and CCL5 are involved in the attraction of T cells and memory T cells, high levels of these β-chemokines in HTLV-1 carriers may play a role in developing pulmonary disorders (Seki et al., 1999). It has been reported that in HTLV-1 positive cryptogenic fibrosing alveolitis (CFA) patients, a chronic inflammatory interstitial lung disease, the BALF levels of CCL3 (Taub et al., 1996; Matsuyama et al., 2003) and CXCL10, T cell chemoattractants and activators (Loetscher et al., 1996; Matsuyama et al., 2003), are higher than those of without HTLV-1 infection. Furthermore, CCL3 and CXCL10 BALF levels were associated with an increased number of activated T cells. Moreover, HTLV-1 infection may also contribute in the development of CFA (Matsuyama et al., 2003).

HTLV-1 infected individuals with overactive bladder, have similar serum concentrations of CXCL9 and CXCL10 compared to the carriers (Santos et al., 2012). Adding exogenous IL-10 and TGF-β to the cell cultures from HTLV-1 overactive bladder patients decreased the production of IFN-γ (Santos et al., 2012). The proviral load and levels of some immunological factors such as TNF-α and IL-17 in overactive bladder patients are similar to HAM/TSP patients. According to these observations, the overactive bladder is a common condition between carriers and HAM/TSP patients, and it was proposed that patients with overactive bladder may further develop HAM/TSP (Santos et al., 2012). **Supplementary Table 1** represents the chemokines and chemokine receptors associated with HTLV-1 pathogenesis.

The following issues influence the limitation of the present review including: omission of relevant research, the authors' point of view, and misconception in the translation of data from the primary literature. Controversial results of the study may have been observed due to an analysis of chemokine levels in serum/plasma, CSF and supernatant of HTLV-1 cell lines and PBMCs culture, sample size, disease duration, clinical status and heterogeneity of patients and HTLV-1 carriers. To the best of our knowledge, this review has been one of the first attempts to thoroughly explain the role of chemokines and their receptors in HTLV-1 infection; as such, a wide range of data regarding the chemokines and receptors are a strength of the study.

In this study, the status of chemokines and chemokine receptors was discussed in HTLV-1 infection focusing on two major diseases associated with the virus: ATL and HAM/TSP. According to the available studies reviewed in this work, it is clear that chemokines play a critical role in inflammation and regulating the trafficking, migration, and tissue infiltration of leukocytes, and the dysregulation of various chemokines could play an important role in the pathophysiology of HTLV-1 associated disease. Clinical studies in HAM/TSP and ATL patients have been provided promising results to be hopeful to develop and recommend more effective inhibitors of chemokines and their receptors. Therefore, development of chemokine-directed therapeutic strategies for targeting leukocyte recruitment pathways could be a promising therapeutic approach for HTLV-1 infection.

CONCLUSION

In this review, we have discussed the potential role of chemokines and their receptors in the outcome of HTLV-1 associated diseases. A great number of chemokines and their respective receptors have been implicated in the migration of HTLV-1 infected cells into CNS and skin lesions are a vital aspect in pathogenesis of HTLV-1-associated disease. Elevation of chemokines levels in serum and CSF in HTLV-1 infection has been considered as prognostic biomarkers and a promising surrogate marker for HAM/TSP treatment. The complexity of the chemokine network is perplexing. Thus, further investigation is required to target chemokine network that could open up a new door to therapy and prophylaxis against HAM/TSP and ATL.

AUTHOR CONTRIBUTIONS

HR designed the study and wrote the manuscript. RZ designed and drafted the manuscript. MM drafted and wrote the manuscript. AM and GH revised the manuscript. ZV drafted the manuscript.

ACKNOWLEDGMENTS

We would like to thank of Chancellor for research of Mashhad University of Medical Sciences, Mashhad, Iran, for financial support.

REFERENCES

Adema, G. J., Hartgers, F., Verstraten, R., de Vries, E., Marland, G., Menon, S., et al. (1997). A dendritic-cell-derived C-C chemokine that preferentially attracts naive T cells. *Nature* 387, 713–717. doi: 10.1038/42716

Afonso, P. V., Ozden, S., Cumont, M. C., Seilhean, D., Cartier, L., Rezaie, P., et al. (2008). Alteration of blood-brain barrier integrity by retroviral infection. *PLoS Pathog.* 4:e1000205. doi: 10.1371/journal.ppat.1000205

Afonso, P. V., Ozden, S., Prevost, M.-C., Schmitt, C., Seilhean, D., Weksler, B., et al. (2007). Human blood-brain barrier disruption by retroviral-infected lymphocytes: role of myosin light chain kinase in endothelial tight-junction disorganization. *J. Immunol.* 179, 2576–2583. doi: 10.4049/jimmunol.179.4.2576

Ahuja, J., Lepoutre, V., Wigdahl, B., Khan, Z. K., and Jain, P. (2007). Induction of pro-inflammatory cytokines by human T-cell leukemia virus type-1 Tax protein as determined by multiplexed cytokine protein array analyses of human dendritic cells. *Biomed. Pharmacother.* 61, 201–208. doi: 10.1016/j.biopha.2007.02.006

Almasi, S., Aliparasti, M. R., Farhoudi, M., Babaloo, Z., Baradaran, B., Zamani, F., et al. (2013). Quantitative evaluation of CXCL8 and its receptors (CXCR1 and CXCR2) gene expression in Iranian patients with multiple sclerosis. *Immunol. Investig.* 42, 737–748. doi: 10.3109/08820139.2013.812652

Amorim, C. F., Souza, A. S., Diniz, A. G., Carvalho, N. B., Santos, S. B., and Carvalho, E. M. (2014). Functional activity of monocytes and macrophages in HTLV-1 infected subjects. *PLoS Negl. Trop. Dis.* 8:e3399. doi: 10.1371/journal.pntd.0003399

Ando, H., Sato, T., Tomaru, U., Yoshida, M., Utsunomiya, A., Yamauchi, J., et al. (2013). Positive feedback loop via astrocytes causes chronic inflammation in virus-associated myelopathy. *Brain* 136, 2876–2887. doi: 10.1093/brain/awt183

Apolinario, A., Majano, P. L., Alvarez-Pérez, E., Saez, A., Lozano, C., Vargas, J., et al. (2002). Increased expression of T cell chemokines and their receptors in chronic hepatitis C: relationship with the histological activity of liver disease. *Am. J. Gastroenterol.* 97, 2861–2870. doi: 10.1111/j.1572-0241.2002.07054.x

Arai, M., Ohashi, T., Tsukahara, T., Murakami, T., Hori, T., Uchiyama, T., et al. (1998). Human T-cell leukemia virus type 1 Tax protein induces the expression of lymphocyte chemoattractant SDF-1/PBSF. *Virology* 241, 298–303. doi: 10.1006/viro.1997.8968

Araya, N., Sato, T., Ando, H., Tomaru, U., Yoshida, M., Coler-Reilly, A., et al. (2014). HTLV-1 induces a Th1-like state in CD4+CCR4+ T cells. *J. Clin. Invest.* 124, 3431–3442. doi: 10.1172/jci75250

Araya, N., Sato, T., Tomaru, U., Coler-Reilly, A., Yagishita, N., Yamauchi, J., et al. (2015). HTLV-1 Tax induces Th1 master regulator T-bet and thus IFN-γ in CD4+CCR4+ T-cells of virus-associated myelopathy patients. *Retrovirology* 12:44. doi: 10.1186/1742-4690-12-s1-p44

Asquith, B., and Bangham, C. R. (2008). How does HTLV-I persist despite a strong cell-mediated immune response? *Trends Immunol.* 29, 4–11. doi: 10.1016/j.it.2007.09.006

Baba, M., Imai, T., Nishimura, M., Kakizaki, M., Takagi, S., Hieshima, K., et al. (1997). Identification of CCR6, the specific receptor for a novel lymphocyte-directed CC chemokine LARC. *J. Biol. Chem.* 272, 14893–14898. doi: 10.1074/jbc.272.23.14893

Baba, M., Imai, T., Yoshida, T., and Yoshie, O. (1996). Constitutive expression of various chemokine genes in human T-cell lines infected with human T-cell leukemia virus type 1: role of the viral transactivator Tax. *Int. J. Cancer* 66, 124–129. doi: 10.1002/(sici)1097-0215(19960328)66:1<124::aid-ijc21>3.0.co;2-c

Banerjee, P., Feuer, G., and Barker, E. (2007). Human T-cell leukemia virus type 1 (HTLV-1) p12I down-modulates ICAM-1 and -2 and reduces adherence of natural killer cells, thereby protecting HTLV-1-infected primary CD4+ T cells from autologous natural killer cell-mediated cytotoxicity despite the reduction of major histocompatibility complex class I molecules on infected cells. *J. Virol.* 81, 9707–9717. doi: 10.1128/jvi.00887-07

Bangham, C. R., Araujo, A., Yamano, Y., and Taylor, G. P. (2015). HTLV-1-associated myelopathy/tropical spastic paraparesis. *Nat. Rev. Dis. Primers.* 1:15012. doi: 10.1038/nrdp.2015.12

Barrios, C. S., Abuerreish, M., Lairmore, M. D., Castillo, L., Giam, C. Z., and Beilke, M. A. (2011). Recombinant human T-cell leukemia virus types 1 and 2 Tax proteins induce high levels of CC-chemokines and downregulate CCR5 in human peripheral blood mononuclear cells. *Viral. Immunol.* 24, 429–439. doi: 10.1089/vim.2011.0037

Barrios, C. S., Castillo, L., Zhi, H., Giam, C. Z., and Beilke, M. (2014). Human T cell leukaemia virus type 2 tax protein mediates CC-chemokine expression in peripheral blood mononuclear cells via the nuclear factor kappa B canonical pathway. *Clin. Exp. Immunol.* 175, 92–103. doi: 10.1111/cei.12213

Bazan, J. F., Bacon, K. B., Hardiman, G., Wang, W., Soo, K., Rossi, D., et al. (1997). A new class of membrane-bound chemokine with a CX3C motif. *Nature* 385, 640–644. doi: 10.1038/385640a0

Bertini, R., Luini, W., Sozzani, S., Bottazzi, B., Ruggiero, P., Boraschi, D., et al. (1995). Identification of MIP-1α/LD78 as a monocyte chemoattractant released by the HTLV-I-transformed cell line MT4. *AIDS Res. Hum. Retroviruses* 11, 155–160. doi: 10.1089/aid.1995.11.155

Bex, F., Yin, M.-J., Burny, A., and Gaynor, R. B. (1998). Differential transcriptional activation by human T-cell leukemia virus type 1 Tax mutants is mediated by distinct interactions with CREB binding protein and p300. *Mol. Cell. Bio.* 18, 2392–2405. doi: 10.1128/mcb.18.4.2392

Bezerra, C. A., Cardoso, T. M., Giudice, A., Porto, A. F., Santos, S. B., Carvalho, E. M., et al. (2011). Evaluation of the microbicidal activity and cytokines/chemokines profile released by neutrophils from HTLV-1-Infected Individuals. *Scand. J. Immunol.* 74, 310–317. doi: 10.1111/j.1365-3083.2011.02579.x

Biddison, W. E., Kubota, R., Kawanishi, T., Taub, D. D., Cruikshank, W. W., Center, D. M., et al. (1997). Human T cell leukemia virus type I (HTLV-I)-specific CD8+ CTL clones from patients with HTLV-I-associated neurologic disease secrete proinflammatory cytokines, chemokines, and matrix metalloproteinase. *J. Immunol.* 159, 2018–2025.

Bieganowska, K., Hollsberg, P., Buckle, G. J., Lim, D. G., Greten, T. F., Schneck, J., et al. (1999). Direct analysis of viral-specific CD8+ T cells with soluble HLA-A2/Tax11-19 tetramer complexes in patients with human T cell lymphotropic virus-associated myelopathy. *J. Immunol.* 162, 1765–1771.

Blanchet, X., Langer, M., Weber, C., Koenen, R. R., and von Hundelshausen, P. (2012). Touch of chemokines. *Front. Immunol.* 3:175. doi: 10.3389/fimmu.2012.00175

Bonville, C., Rosenberg, H., and Domachowske, J. (1999). Macrophage inflammatory protein-1α and RANTES are present in nasal secretions during ongoing upper respiratory tract infection. *Pediatr. Allergy Immunol.* 10, 39–44. doi: 10.1034/j.1399-3038.1999.101005.x

Bunn, P. A. Jr., Schechter, G. P., Jaffe, E., Blayney, D., Young, R. C., Matthews, M. J., et al. (1983). Clinical course of retrovirus-associated adult T-cell lymphoma in the United States. *N. Engl. J. Med* 309, 257–264. doi: 10.1056/nejm198308043090501

Burns, J. M., Summers, B. C., Wang, Y., Melikian, A., Berahovich, R., Miao, Z., et al. (2006). A novel chemokine receptor for SDF-1 and I-TAC involved in cell survival, cell adhesion, and tumor development. *J. Exp. Med.* 203, 2201–2213. doi: 10.1084/jem.20052144

Cavanagh, M.-H., Landry, S., Audet, B., Arpin-André, C., Hivin, P., Paré, M.-E., et al. (2006). HTLV-I antisense transcripts initiating in the 3′LTR are alternatively spliced and polyadenylated. *Retrovirology* 3, 15–15. doi: 10.1186/1742-4690-3-15

Charo, I. F., and Ransohoff, R. M. (2006). The many roles of chemokines and chemokine receptors in inflammation. *N. Engl. J. Med.* 354, 610–621. doi: 10.1056/NEJMra052723

Chaves, D. G., Sales, C. C., de Cassia Goncalves, P., da Silva-Malta, M. C., Romanelli, L. C., Ribas, J. G., et al. (2016). Plasmatic proinflammatory chemokines levels are tricky markers to monitoring HTLV-1 carriers. *J. Med. Virol.* 88, 1438–1447. doi: 10.1002/jmv.24481

Cheng, W., and Chen, G. (2014). Chemokines and chemokine receptors in multiple sclerosis. *Med. Inflamm.* 2014:659206. doi: 10.1155/2014/659206

Chuntharapai, A., Lee, J., Hebert, C. A., and Kim, K. J. (1994). Monoclonal antibodies detect different distribution patterns of IL-8 receptor A and IL-8 receptor B on human peripheral blood leukocytes. *J. Immunol.* 153, 5682–5688.

Coelho-dos-Reis, J. G. A., Passos, L., Duarte, M. C., Araújo, M. G., Campi-Azevedo, A. C., Teixeira-Carvalho, A., et al. (2013). Immunological profile of HTLV-1-infected patients associated with infectious or autoimmune dermatological disorders. *PLoS Negl. Trop. Dis.* 7:e2328. doi: 10.1371/journal.pntd.0002328

Currer, R., Van Duyne, R., Jaworski, E., Guendel, I., Sampey, G., Das, R., et al. (2012). HTLV tax: a fascinating multifunctional co-regulator of viral and cellular pathways. *Front. Microbiol.* 3:406. doi: 10.3389/fmicb.2012.00406

Cyster, J. G. (1999). Chemokines and cell migration in secondary lymphoid organs. *Science* 286, 2098–2102. doi: 10.1126/science.286.5447.2098

da Silva-Malta, M. C. F., Sales, C. C., Guimaraes, J. C., de Cassia Goncalves, P., Chaves, D. G., Santos, H. C., et al. (2017). The Duffy null genotype is associated with a lower level of CCL2, leukocytes and neutrophil count but not with the clinical outcome of HTLV-1 infection. *J. Med. Microbiol.* 66, 1207–1216. doi: 10.1099/jmm.0.000539

Dahm, T., Rudolph, H., Schwerk, C., Schroten, H., and Tenenbaum, T. (2016). Neuroinvasion and Inflammation in Viral Central Nervous System Infections. *Mediators Inflamm.* 2016:8562805. doi: 10.1155/2016/8562805

Dawson, T. C., Beck, M. A., Kuziel, W. A., Henderson, F., and Maeda, N. (2000). Contrasting effects of CCR5 and CCR2 deficiency in the pulmonary inflammatory response to influenza A virus. *Am. J. Pathol.* 156, 1951–1959. doi: 10.1016/s0002-9440(10)65068-7

de Castro-Amarante, M. F., Pise-Masison, C. A., McKinnon, K., Parks, R. W., Galli, V., Omsland, M., et al. (2016). Human T Cell leukemia virus type 1 infection of the three monocyte subsets contributes to viral burden in humans. *J. Virol.* 90, 2195–2207. doi: 10.1128/JVI.02735-15

Domachowske, J. B., Bonville, C. A., Easton, A. J., and Rosenberg, H. F. (2002). Differential expression of proinflammatory cytokine genes in vivo in response to pathogenic and nonpathogenic pneumovirus infections. *J. Infect. Dis.* 186, 8–14. doi: 10.1086/341082

Elovaara, I., Koenig, S., Brewah, A. Y., Woods, R., Lehky, T., and Jacobson, S. (1993). High human T cell lymphotropic virus type 1 (HTLV-1)-specific precursor cytotoxic T lymphocyte frequencies in patients with HTLV-1-associated neurological disease. *J. Exp. Med.* 177, 1567–1573. doi: 10.1084/jem.177.6.1567

El-Sabban, M. E., Merhi, R. A., Haidar, H. A., Arnulf, B., Khoury, H., Basbous, J., et al. (2002). Human T-cell lymphotropic virus type 1-transformed cells induce angiogenesis and establish functional gap junctions with endothelial cells. *Blood* 99, 3383–3389. doi: 10.1182/blood.v99.9.3383

Enose-Akahata, Y., Vellucci, A., and Jacobson, S. (2017). Role of HTLV-1 Tax and HBZ in the Pathogenesis of HAM/TSP. *Front. Microbiol.* 8:2563. doi: 10.3389/fmicb.2017.02563

Fujii, M., Chuhjo, T., Minamino, T., Masaaki, N., Miyamoto, K.-I., and Seiki, M. (1995). Identification of the Tax interaction region of serum response factor that mediates the aberrant induction of immediate early genes through CArG boxes by HTLV-I Tax. *Oncogene* 11, 7–14.

Fujii, M., Iwai, K., Oie, M., Fukushi, M., Yamamoto, N., Kannagi, M., et al. (2000). Activation of oncogenic transcription factor AP-1 in T cells infected with human T cell leukemia virus type 1. *AIDS Res. Hum. Retroviruses* 16, 1603–1606. doi: 10.1089/08892220050193029

Fujii, M., Tsuchiya, H., Chuhjo, T., Akizawa, T., and Seiki, M. (1992). Interaction of HTLV-1 Tax1 with p67SRF causes the aberrant induction of cellular immediate early genes through CArG boxes. *Genes Dev.* 6, 2066–2076. doi: 10.1101/gad.6.11.2066

Furukawa, Y., Yamashita, M., Usuku, K., Izumo, S., Nakagawa, M., and Osame, M. (2000). Phylogenetic subgroups of human t cell lymphotropic virus (HTLV) type i in the tax gene and their association with different risks for HTLV-I—associated myelopathy/tropical spastic paraparesis. *J. Infect. Dis.* 182, 1343–1349. doi: 10.1086/315897

Futsch, N., Prates, G., Mahieux, R., Casseb, J., and Dutartre, H. (2018). Cytokine networks dysregulation during HTLV-1 Infection and Associated Diseases. *Viruses* 10:691. doi: 10.3390/v10120691

Gaudray, G., Gachon, F., Basbous, J., Biard-Piechaczyk, M., Devaux, C., and Mesnard, J. M. (2002). The complementary strand of the human T-cell leukemia virus type 1 RNA genome encodes a bZIP transcription factor that down-regulates viral transcription. *J. Virol.* 76, 12813–12822. doi: 10.1128/jvi.76.24.12813-12822.2002

Gessain, A., Barin, F., Vernant, J. C., Gout, O., Maurs, L., Calender, A., et al. (1985). Antibodies to human T-lymphotropic virus type-I in patients with tropical spastic paraparesis. *Lancet* 2, 407–410. doi: 10.1016/s0140-6736(85)92734-5

Giraudon, P., Bernard, A., Malcus, C., Dufay, N., Desgranges, C., and Belin, M. F. (1995). Retroviral Infection (HTLV-I) induces cytokine-regulated immunomodulation and cytotoxicity of medulloblastoma cells. *J. Neuropathol. Exp. Neurol.* 54, 165–174. doi: 10.1097/00005072-199503000-00003

Green, J. E., Hinrichs, S. H., Vogel, J., and Jay, G. (1989). Exocrinopathy resembling Sjogren's syndrome in HTLV-1 tax transgenic mice. *Nature* 341, 72–74. doi: 10.1038/341072a0

Gu, L., Tseng, S., Horner, R. M., Tam, C., Loda, M., and Rollins, B. J. (2000). Control of TH2 polarization by the chemokine monocyte chemoattractant protein-1. *Nature* 404, 407–411. doi: 10.1038/35006097

Guerra, M., Luna, T., Souza, A., Amorim, C., Carvalho, N. B., Carvalho, L., et al. (2018). Local and systemic production of proinflammatory chemokines in the pathogenesis of HAM/TSP. *Cell Immunol.* 334, 70–77. doi: 10.1016/j.cellimm.2018.09.009

Guerreiro, J., Santos, S., Morgan, D., Porto, A., Muniz, A., Ho, J., et al. (2006). Levels of serum chemokines discriminate clinical myelopathy associated with human T lymphotropic virus type 1 (HTLV-1)/tropical spastic paraparesis (HAM/TSP) disease from HTLV-1 carrier state. *Clin. Exp. Immunol.* 145, 296–301. doi: 10.1111/j.1365-2249.2006.03150.x

Haeberle, H. A., Kuziel, W. A., Dieterich, H.-J., Casola, A., Gatalica, Z., and Garofalo, R. P. (2001). Inducible expression of inflammatory chemokines in respiratory syncytial virus-infected mice: role of MIP-1α in lung pathology. *J. Virol.* 75, 878–890. doi: 10.1128/jvi.75.2.878-890.2001

Haghnazari Sadaghiani, N., Pirayeshfard, L., Aghaie, A., and Sharifi, Z. (2019). The Effect of TAX-1 gene of human T-cell leukemia virus Type-1 on the Expression of CCR5 in K562 Cell Line. *Avicenna J. Med. Biotechnol.* 11, 67–71.

Hall, W. W., Liu, C. R., Schneewind, O., Takahashi, H., Kaplan, M. H., Roupe, G., et al. (1991). Deleted HTLV-I provirus in blood and cutaneous lesions of patients with mycosis fungoides. *Science* 253, 317–320. doi: 10.1126/science.1857968

Hanon, E., Hall, S., Taylor, G. P., Saito, M., Davis, R., Tanaka, Y., et al. (2000). Abundant tax protein expression in CD4+ T cells infected with human T-cell lymphotropic virus type I (HTLV-I) is prevented by cytotoxic T lymphocytes. *Blood* 95, 1386–1392. doi: 10.1182/blood.v95.4.1386.004k22_1386_1392

Hasegawa, H., Nomura, T., Kohno, M., Tateishi, N., Suzuki, Y., Maeda, N., et al. (2000). Increased chemokine receptor CCR7/EBI1 expression enhances the infiltration of lymphoid organs by adult T-cell leukaemia cells. *Blood* 95, 30–38. doi: 10.1182/blood.v95.1.30.001k09_30_38

Hieshima, K., Nagakubo, D., Nakayama, T., Shirakawa, A.-K., Jin, Z., and Yoshie, O. (2008). Tax-inducible production of CC chemokine ligand 22 by human T cell leukemia virus type 1 (HTLV-1)-infected T cells promotes preferential transmission of HTLV-1 to CCR4-expressing CD4+ T cells. *J. Immunol.* 180, 931–939. doi: 10.4049/jimmunol.180.2.931

Hiyoshi, M., Okuma, K., Tateyama, S., Takizawa, K., Saito, M., Kuramitsu, M., et al. (2015). Furin-dependent CCL17-fused recombinant toxin controls HTLV-1 infection by targeting and eliminating infected CCR4-expressing cells in vitro and in vivo. *Retrovirology* 12:73. doi: 10.1186/s12977-015-0199-8

Hosking, M. P., Tirotta, E., Ransohoff, R. M., and Lane, T. E. (2010). CXCR2 signaling protects oligodendrocytes and restricts demyelination in a mouse model of viral-induced demyelination. *PloS One* 5:e11340. doi: 10.1371/journal.pone.0011340

Ijichi, S., Izumo, S., Eiraku, N., Machigashira, K., Kubota, R., Nagai, M., et al. (1993). An autoaggressive process against bystander tissues in HTLV-I-infected individuals: a possible pathomechanism of HAM/TSP. *Med. Hypotheses* 41, 542–547. doi: 10.1016/0306-9877(93)90111-3

Imai, T., Nishimura, M., Nanki, T., and Umehara, H. (2005). [Fractalkine and inflammatory diseases]. *Nihon Rinsho Meneki Gakkai Kaishi* 28, 131–139. doi: 10.2177/jsci.28.131

Imaizumi, Y., Sugita, S., Yamamoto, K., Imanishi, D., Kohno, T., Tomonaga, M., et al. (2002). Human T cell leukemia virus type-I Tax activates human macrophage inflammatory protein-3α/CCL20 gene transcription via the NF-κB pathway. *Int. Immunol.* 14, 147–155. doi: 10.1093/intimm/14.2.147

Inagaki, A., Ishida, T., Ishii, T., Komatsu, H., Iida, S., Ding, J., et al. (2006). Clinical significance of serum Th1-, Th2- and regulatory T cells-associated cytokines in adult T-cell leukemia/lymphoma: high interleukin-5 and -10 levels are significant unfavorable prognostic factors. *Int. J Cancer* 118, 3054–3061. doi: 10.1002/ijc.21688

Infantino, S., Moepps, B., and Thelen, M. (2006). Expression and regulation of the orphan receptor RDC1 and its putative ligand in human dendritic and B cells. *J. Immuno.* 176, 2197–2207. doi: 10.4049/jimmunol.176.4.2197

Ishida, T., Joh, T., Uike, N., Yamamoto, K., Utsunomiya, A., Yoshida, S., et al. (2012). Defucosylated anti-CCR4 monoclonal antibody (KW-0761) for relapsed adult T-cell leukemia-lymphoma: a multicenter phase II study. *J. Clin. Oncol.* 30, 837–842. doi: 10.1200/jco.2011.37.3472

Ishida, T., Utsunomiya, A., Iida, S., Inagaki, H., Takatsuka, Y., Kusumoto, S., et al. (2003). Clinical significance of CCR4 expression in adult T-cell leukemia/lymphoma: its close association with skin involvement and unfavorable outcome. *Clin. Cancer Res.* 9(10 Pt 1), 3625–3634.

Ishida, T., Utsunomiya, A., Jo, T., Yamamoto, K., Kato, K., Yoshida, S., et al. (2017). Mogamulizumab for relapsed adult T-cell leukemia-lymphoma: updated follow-up analysis of phase I and II studies. *Cancer Sci.* 108, 2022–2029. doi: 10.1111/cas.13343

Ishida, Y., Gao, J.-L., and Murphy, P. M. (2008). Chemokine receptor CX3CR1 mediates skin wound healing by promoting macrophage and fibroblast accumulation and function. *J. Immunol.* 180, 569–579. doi: 10.4049/jimmunol.180.1.569

Jacobson, S., Shida, H., McFarlin, D. E., Fauci, A. S., and Koenig, S. (1990). Circulating CD8+ cytotoxic T lymphocytes specific for HTLV-I pX in patients with HTLV-I associated neurological disease. *Nature* 348, 245–248. doi: 10.1038/348245a0

Jain, P., Ahuja, J., Khan, Z. K., Shimizu, S., Meucci, O., Jennings, S. R., et al. (2007). Modulation of dendritic cell maturation and function by the Tax protein of human T cell leukemia virus type 1. *J. Leukocyte Biol.* 82, 44–56. doi: 10.1189/jlb.1006641

Jaworski, E., Narayanan, A., Van Duyne, R., Shabbeer-Meyering, S., Iordanskiy, S., Saifuddin, M., et al. (2014). Human T-lymphotropic virus type 1-infected cells secrete exosomes that contain Tax protein. *J. Biol. Chem.* 289, 22284–22305. doi: 10.1074/jbc.M114.549659

Jeang, K.-T. (2001). Functional activities of the human T-cell leukemia virus type I Tax oncoprotein: cellular signaling through NF-κB. *Cytokine Growth Factor Rev.* 12, 207–217. doi: 10.1016/s1359-6101(00)00028-9

Jeffery, K. J., Usuku, K., Hall, S. E., Matsumoto, W., Taylor, G. P., Procter, J., et al. (1999). HLA alleles determine human T-lymphotropic virus-I (HTLV-I) proviral load and the risk of HTLV-I-associated myelopathy. *Proc. Natl.Acad. Sci. U.S.A.*96, 3848–3853. doi: 10.1073/pnas.96.7.3848

Jiang, Y., Beller, D., Frendl, G., and Graves, D. (1992). Monocyte chemoattractant protein-1 regulates adhesion molecule expression and cytokine production in human monocytes. *J. Immunol.* 148, 2423–2428.

Jin, Z., Nagakubo, D., Shirakawa, A. K., Nakayama, T., Shigeta, A., Hieshima, K., et al. (2009). CXCR7 is inducible by HTLV-1 Tax and promotes growth and survival of HTLV-1-infected T cells. *Int. J. Cancer* 125, 2229–2235. doi: 10.1002/ijc.24612

Jones, B. A., Beamer, M., and Ahmed, S. (2010). Fractalkine/CX3CL1: a potential new target for inflammatory diseases. *Mol. Interv.* 10, 263–270. doi: 10.1124/mi.10.5.3

Kagdi, H. H., Demontis, M. A., Fields, P. A., Ramos, J. C., Bangham, C. R., and Taylor, G. P. (2017). Risk stratification of adult T-cell leukemia/lymphoma using immunophenotyping. *Cancer Med.* 6, 298–309. doi: 10.1002/cam4.928

Kannagi, M., Matsushita, S., and Harada, S. (1993). Expression of the target antigen for cytotoxic T lymphocytes on adult T-cell-leukemia cells. *Int. J. Cancer* 54, 582–588. doi: 10.1002/ijc.2910540411

Kataoka, K., Nagata, Y., Kitanaka, A., Shiraishi, Y., Shimamura, T., Yasunaga, J., et al. (2015). Integrated molecular analysis of adult T cell leukemia/lymphoma. *Nat. Genet.* 47, 1304–1315. doi: 10.1038/ng.3415

Katsuya, H., Ishitsuka, K., Utsunomiya, A., Hanada, S., Eto, T., Moriuchi, Y., et al. (2015). Treatment and survival among 1594 patients with ATL. *Blood* 126, 2570–2577. doi: 10.1182/blood-2015-03-632489

Kawaguchi, A., Orba, Y., Kimura, T., Iha, H., Ogata, M., Tsuji, T., et al. (2009). Inhibition of the SDF-1α-CXCR4 axis by the CXCR4 antagonist AMD3100 suppresses the migration of cultured cells from ATL patients and murine lymphoblastoid cells from HTLV-I Tax transgenic mice. *Blood* 114, 2961–2968. doi: 10.1182/blood-2008-11-189308

Kimura, I., Tsubota, T., Tada, S., and Sogawa, J. (1986). Presence of antibodies against adult T cell leukemia antigen in the patients with chronic respiratory diseases. *Acta Med. Okayama* 40, 281–284. doi: 10.18926/amo/31916

Kira, J., Nakamura, M., Sawada, T., Koyanagi, Y., Ohori, N., Itoyama, Y., et al. (1992). Antibody titers to HTLV-I-p40tax protein and gag-env hybrid protein in HTLV-I-associated myelopathy/tropical spastic paraparesis: correlation with increased HTLV-I proviral DNA load. *J. Neurol. Sci.* 107, 98–104. doi: 10.1016/0022-510x(92)90215-7

Koper, O. M., Kaminska, J., Sawicki, K., and Kemona, H. (2018). CXCL9, CXCL10, CXCL11, and their receptor (CXCR3) in neuroinflammation and neurodegeneration. *Adv. Clin. Exp. Med.* 27, 849–856. doi: 10.17219/acem/68846

Kozai, M., Kubo, Y., Katakai, T., Kondo, H., Kiyonari, H., Schaeuble, K., et al. (2017). Essential role of CCL21 in establishment of central self-tolerance in T cells. *J. Exp. Med.* 214, 1925–1935. doi: 10.1084/jem.20161864

Kubota, R. (2017). Pathogenesis of human T-lymphotropic virus type 1-associated myelopathy/tropical spastic paraparesis. *Clin. Exp. Neuroimmunol.* 8, 117–128. doi: 10.1111/cen3.12395

Kubota, R., Kawanishi, T., Matsubara, H., Manns, A., and Jacobson, S. (1998). Demonstration of human T lymphotropic virus type I (HTLV-I) tax-specific CD8+ lymphocytes directly in peripheral blood of HTLV-I-associated myelopathy/tropical spastic paraparesis patients by intracellular cytokine detection. *J. Immunol.* 161, 482–488.

Kubota, R., Umehara, F., Izumo, S., Ijichi, S., Matsumuro, K., Yashiki, S., et al. (1994). HTLV-I proviral DNA amount correlates with infiltrating CD4+ lymphocytes in the spinal cord from patients with HTLV-I-associated myelopathy. *J. Neuroimmunol.* 53, 23–29. doi: 10.1016/0165-5728(94)90060-4

Kuchroo, V. K., Martin, C. A., Greer, J. M., Ju, S. T., Sobel, R. A., and Dorf, M. E. (1993). Cytokines and adhesion molecules contribute to the ability of myelin proteolipid protein-specific T cell clones to mediate experimental allergic encephalomyelitis. *J. Immunol.* 151, 4371–4382.

LaGrenade, L., Hanchard, B., Fletcher, V., Cranston, B., and Blattner, W. (1990). Infective dermatitis of Jamaican children: a marker for HTLV-I infection. *Lancet* 336, 1345–1347. doi: 10.1016/0140-6736(90)92896-p

Lahrtz, F., Piali, L., Nadal, D., Pfister, H. W., Spanaus, K. S., Baggiolini, M., et al. (1997). Chemotactic activity on mononuclear cells in the cerebrospinal fluid of patients with viral meningitis is mediated by interferon-γ inducible protein-10 and monocyte chemotactic protein-1. *Eur. J. Immunol.* 27, 2484–2489. doi: 10.1002/eji.1830271004

Larocca, D., Chao, L. A., Seto, M. H., and Brunck, T. K. (1989). Human T-cell leukemia virus minus strand transcription in infected T-cells. *Biochem. Biophys. Res. Commun.* 163, 1006–1013. doi: 10.1016/0006-291x(89)92322-x

Lee, M., Lee, Y., Song, J., Lee, J., and Chang, S.-Y. (2018). Tissue-specific role of CX3CR1 expressing immune cells and their relationships with human disease. *Immune Network* 18:e5. doi: 10.4110/in.2018.18.e5

Lemasson, I., Lewis, M. R., Polakowski, N., Hivin, P., Cavanagh, M. H., Thebault, S., et al. (2007). Human T-cell leukemia virus type 1 (HTLV-1) bZIP protein interacts with the cellular transcription factor CREB to inhibit HTLV-1 transcription. *J. Virol.* 81, 1543–1553. doi: 10.1128/jvi.00480-06

Lepoutre, V., Jain, P., Quann, K., Wigdahl, B., and Khan, Z. K. (2009). Role of resident CNS cell populations in HTLV-1-associated neuroinflammatory disease. *Front. Biosci.* 14:1152–1168.

Li, Y. M., Pan, Y., Wei, Y., Cheng, X., Zhou, B. P., Tan, M., et al. (2004). Upregulation of CXCR4 is essential for HER2-mediated tumor metastasis. *Cancer Cell* 6, 459–469. doi: 10.1016/j.ccr.2004.09.027

Lima, L. M., Cardoso, L. S., Santos, S. B., Oliveira, R. R., Oliveira, S. C., Goes, A. M., et al. (2017). Schistosoma antigens downregulate CXCL9 production by PBMC of HTLV-1-infected individuals. *Acta Trop.* 167, 157–162. doi: 10.1016/j.actatropica.2016.12.030

Liu, M. T., Armstrong, D., Hamilton, T. A., and Lane, T. E. (2001). Expression of Mig (monokine induced by interferon-γ) is important in T lymphocyte recruitment and host defense following viral infection of the central nervous system. *J. Immunol.* 166, 1790–1795. doi: 10.4049/jimmunol.166.3.1790

Llano, A., and Esté, J. (2005). Chemokines and other cytokines in human immunodeficiency virus type 1 (HIV-1) infection. *Inmunología* 24, 246–260.

Loetscher, M., Gerber, B., Loetscher, P., Jones, S. A., Piali, L., Clark-Lewis, I., et al. (1996). Chemokine receptor specific for IP10 and mig: structure, function, and expression in activated T-lymphocytes. *J. Exp. Med.* 184, 963–969. doi: 10.1084/jem.184.3.963

Luster, A. D. (1998). Chemokines—chemotactic cytokines that mediate inflammation. *N. Engl. J. Med.* 338, 436–445. doi: 10.1056/nejm199802123380706

Ma, G., Yasunaga, J., Fan, J., Yanagawa, S., and Matsuoka, M. (2013). HTLV-1 bZIP factor dysregulates the Wnt pathways to support proliferation and migration of adult T-cell leukemia cells. *Oncogene* 32, 4222–4230. doi: 10.1038/onc.2012.450

Ma, G., Yasunaga, J., and Matsuoka, M. (2016). Multifaceted functions and roles of HBZ in HTLV-1 pathogenesis. *Retrovirology* 13:16. doi: 10.1186/s12977-016-0249-x

Ma, Q., Jones, D., Borghesani, P. R., Segal, R. A., Nagasawa, T., Kishimoto, T., et al. (1998). Impaired B-lymphopoiesis, myelopoiesis, and derailed cerebellar neuron migration in CXCR4-and SDF-1-deficient mice. *Proc. Natl. Acad. Sci. U.S.A.* 95, 9448–9453. doi: 10.1073/pnas.95.16.9448

Mahieux, R., de Thé, G., and Gessain, A. (1995). The tax mutation at nucleotide 7959 of human T-cell leukemia virus type 1 (HTLV-1) is not associated with tropical spastic paraparesis/HTLV-1-associated myelopathy but is linked to the cosmopolitan molecular genotype. *J. Virol.* 69, 5925–5927. doi: 10.1128/jvi.69.9.5925-5927.1995

Mantovani, A. (1999). The chemokine system: redundancy for robust outputs. *Immunol. Today* 20, 254–257. doi: 10.1016/s0167-5699(99)01469-3

Mashino, K., Sadanaga, N., Yamaguchi, H., Tanaka, F., Ohta, M., Shibuta, K., et al. (2002). Expression of chemokine receptor CCR7 is associated with lymph node metastasis of gastric carcinoma. *Cancer Res.* 62, 2937–2941.

Matsuoka, M. (2003). Human T-cell leukemia virus type I and adult T-cell leukemia. *Oncogene* 22:5131. doi: 10.1038/sj.onc.1206551

Matsuoka, M., and Yasunaga, J. (2013). Human T-cell leukemia virus type 1: replication, proliferation and propagation by Tax and HTLV-1 bZIP factor. *Curr. Opin. Virol.* 3, 684–691. doi: 10.1016/j.coviro.2013.08.010

Matsuyama, W., Kawabata, M., Mizoguchi, A., Iwami, F., Wakimoto, J., and Osame, M. (2003). Influence of human T lymphotrophic virus type I on cryptogenic fibrosing alveolitis — HTLV-I associated fibrosing alveolitis: proposal of a new clinical entity. *Clin. Expe. Immunol.* 133, 397–403. doi: 10.1046/j.1365-2249.2003.02240.x

McGrath, K. E., Koniski, A. D., Maltby, K. M., McGann, J. K., and Palis, J. (1999). Embryonic expression and function of the chemokine SDF-1 and its receptor. CXCR4. *Dev. Biol.* 213, 442–456. doi: 10.1006/dbio.1999.9405

Melchjorsen, J., Sørensen, L. N., and Paludan, S. R. (2003). Expression and function of chemokines during viral infections: from molecular mechanisms to in vivo function. *J. Leukocyte Biol.* 74, 331–343. doi: 10.1189/jlb.1102577

Miller, M. C., and Mayo, K. H. (2017). Chemokines from a Structural Perspective. *Int. J. Mol. Sci.* 18:E2088. doi: 10.3390/ijms18102088

Miller, M. D., and Krangel, M. S. (1992a). Biology and biochemistry of the chemokines: a family of chemotactic and inflammatory cytokines. *Crit. Rev. Immunol.* 12, 17–46.

Miller, M. D., and Krangel, M. S. (1992b). The human cytokine I-309 is a monocyte chemoattractant. *Proc. Natl. Acad. Sci. U.S.A.* 89, 2950–2954. doi: 10.1073/pnas.89.7.2950

Miyagishi, R., Kikuchi, S., Fukazawa, T., and Tashiro, K. (1995). Macrophage inflammatory protein-1α in the cerebrospinal fluid of patients with multiple sclerosis and other inflammatory neurological diseases. *J. Neurol. Sci.* 129, 223–227. doi: 10.1016/0022-510x(95)00004-l

Mochizuki, M., Watanabe, T., Yamaguchi, K., Takatsuki, K., Yoshimura, K., Shirao, M., et al. (1992). HTLV-I uveitis: a distinct clinical entity caused by HTLV-I. *Jpn .J .Cancer Res.* 83, 236–239. doi: 10.1111/j.1349-7006.1992.tb00092.x

Morgan, O. S., Rodgers-Johnson, P., Mora, C., and Char, G. (1989). HTLV-1 and polymyositis in Jamaica. *Lancet* 2, 1184–1187. doi: 10.1016/s0140-6736(89)91793-5

Mori, N., Gill, P. S., Mougdil, T., Murakami, S., Eto, S., and Prager, D. (1996). Interleukin-10 gene expression in adult T-cell leukemia. *Blood* 88, 1035–1045. doi: 10.1182/blood.v88.3.1035.bloodjournal8831035

Mori, N., Krensky, A. M., Ohshima, K., Tomita, M., Matsuda, T., Ohta, T., et al. (2004). Elevated expression of CCL5/RANTES in adult T-cell leukemia cells: possible transactivation of the CCL5 gene by human T-cell leukemia virus type I tax. *Int. J. Cancer* 111, 548–557. doi: 10.1002/ijc.20266

Mori, N., Mukaida, N., Ballard, D. W., Matsushima, K., and Yamamoto, N. (1998). Human T-cell leukemia virus type I Tax transactivates human interleukin 8 gene through acting concurrently on AP-1 and nuclear factor-kappaB-like sites. *Cancer Res.* 58, 3993–4000.

Mori, N., Murakami, S., Oda, S., Prager, D., and Eto, S. (1995). Production of interleukin 8 in adult T-cell leukemia cells: possible transactivation of the interleukin 8 gene by human T-cell leukemia virus type I tax. *Cancer Res.* 55, 3592–3597.

Mori, N., Ueda, A., Ikeda, S., Yamasaki, Y., Yamada, Y., Tomonaga, M., et al. (2000). Human T-cell leukemia virus type I tax activates transcription of the human monocyte chemoattractant protein-1 gene through two nuclear factor-kappaB sites. *Cancer Res.* 60, 4939–4945.

Moriuchi, H., Moriuchi, M., and Fauci, A. S. (1998). Factors secreted by human T lymphotropic virus type I (HTLV-I)–infected cells can enhance or inhibit replication of HIV-1 in HTLV-I–uninfected cells: implications for in vivo coinfection with HTLV-I and HIV-1. *J. Exp. Med.* 187, 1689–1697. doi: 10.1084/jem.187.10.1689

Moser, B., and Loetscher, P. (2001). Lymphocyte traffic control by chemokines. *Nat. Immunol.* 2, 123–128. doi: 10.1038/84219

Mühleisen, A., Giaisi, M., Köhler, R., Krammer, P. H., and Li-Weber, M. (2014). Tax contributes apoptosis resistance to HTLV-1-infected T cells via suppression of Bid and Bim expression. *Cell Death Dis.* 5:e1575. doi: 10.1038/cddis.2014.536

Müller, M., Carter, S., Hofer, M., and Campbell, I. (2010). Review: the chemokine receptor CXCR3 and its ligands CXCL9, CXCL10 and CXCL11 in neuroimmunity – a tale of conflict and conundrum. *Neuropathol. Appl. Neurobiol.* 36, 368–387. doi: 10.1111/j.1365-2990.2010.01089.x

Nagai, M., Usuku, K., Matsumoto, W., Kodama, D., Takenouchi, N., Moritoyo, T., et al. (1998). Analysis of HTLV-I proviral load in 202 HAM/TSP patients and 243 asymptomatic HTLV-I carriers: high proviral load strongly predisposes to HAM/TSP. *J. Neurovirol.* 4, 586–593. doi: 10.3109/13550289809114225

Nagakubo, D., Jin, Z., Hieshima, K., Nakayama, T., Shirakawa, A. K., Tanaka, Y., et al. (2007). Expression of CCR9 in HTLV-1+ T cells and ATL cells expressing Tax. *Int. J. Cancer* 120, 1591–1597. doi: 10.1002/ijc.22483

Nagashima, K., Yoshida, M., and Seiki, M. (1986). A single species of pX mRNA of human T-cell leukemia virus type I encodes trans-activator p40x and two other phosphoproteins. *J. Virol.* 60, 394–399. doi: 10.1128/jvi.60.2.394-399.1986

Naito, T., Yasunaga, J. I., Mitobe, Y., Shirai, K., Sejima, H., Ushirogawa, H., et al. (2018). Distinct gene expression signatures induced by viral transactivators of different HTLV-1 subgroups that confer a different risk of HAM/TSP. *Retrovirology* 15:72. doi: 10.1186/s12977-018-0454-x

Nakagawa, M., Schmitz, R., Xiao, W., Goldman, C. K., Xu, W., Yang, Y., et al. (2014). Gain-of-function CCR4 mutations in adult T cell leukemia/lymphoma. *J. Exp. Med.* 211, 2497–2505. doi: 10.1084/jem.20140987

Nakayama, Y., Yamazato, Y., Tamayose, M., Atsumi, E., Yara, S., Higa, F., et al. (2013). Increased expression of HBZ and Foxp3 mRNA in bronchoalveolar lavage cells taken from human T-lymphotropic virus type 1-associated lung disorder patients. *Intern. Med.* 52, 2599–2609. doi: 10.2169/internalmedicine.52.0845

Narikawa, K., Fujihara, K., Misu, T., Feng, J., Fujimori, J., Nakashima, I., et al. (2005). CSF-chemokines in HTLV-I-associated myelopathy: CXCL10 up-regulation and therapeutic effect of interferon-alpha. *J. Neuroimmunol.* 159, 177–182. doi: 10.1016/j.jneuroim.2004.10.011

Nishioka, K., Maruyama, I., Sato, K., Kitajima, I., Nakajima, Y., and Osame, M. (1989). Chronic inflammatory arthropathy associated with HTLV-I. *Lancet* 1:441. doi: 10.1016/s0140-6736(89)90038-x

Ohshima, K., Kikuchi, M., Masuda, Y., Sumiyoshi, Y., Eguchi, F., Mohtai, H., et al. (1992). Human T-cell leukemia virus type I associated lymphadenitis. *Cancer* 69, 239–248.

Osame, M., Usuku, K., Izumo, S., Ijichi, N., Amitani, H., Igata, A., et al. (1986). HTLV-I associated myelopathy, a new clinical entity. *LancetLondon* 1, 1031–1032. doi: 10.1016/s0140-6736(86)91298-5

Owais, M., and Arya, S. K. (1999). Antiviral chemokines: intracellular life of recombinant C-C chemokine RANTES. *J. Hum. Virol.* 2, 270–282.

Percher, F., Curis, C., Pérès, E., Artesi, M., Rosewick, N., Jeannin, P., et al. (2017). HTLV-1-induced leukotriene B4 secretion by T cells promotes T cell recruitment and virus propagation. *Nat. Commun.* 8:15890. doi: 10.1038/ncomms15890

Peterson, K. E., Robertson, S. J., Portis, J. L., and Chesebro, B. (2001). Differences in cytokine and chemokine responses during neurological disease induced by polytropic murine retroviruses map to separate regions of the viral envelope gene. *J. Virol.* 75, 2848–2856. doi: 10.1128/jvi.75.6.2848-2856.2001

Pivarcsi, A., Gombert, M., Dieu-Nosjean, M.-C., Lauerma, A., Kubitza, R., Meller, S., et al. (2004). CC chemokine ligand 18, an atopic dermatitis-associated and dendritic cell-derived chemokine, is regulated by staphylococcal products and allergen exposure. *J. Immunol.* 173, 5810–5817. doi: 10.4049/jimmunol.173.9.5810

Poiesz, B. J., Ruscetti, F. W., Gazdar, A. F., Bunn, P. A., Minna, J. D., and Gallo, R. C. (1980). Detection and isolation of type C retrovirus particles from fresh and cultured lymphocytes of a patient with cutaneous T-cell lymphoma. *Proc. Natl.Acad. Sci. U.S.A.* 77, 7415–7419. doi: 10.1073/pnas.77.12.7415

Poiesz, B. J., Ruscetti, F. W., Reitz, M. S., Kalyanaraman, V. S., and Gallo, R. C. (1981). Isolation of a new type C retrovirus (HTLV) in primary uncultured cells of a patient with Sezary T-cell leukaemia. *Nature* 294, 268–271. doi: 10.1038/294268a0

Pozzatti, R., Vogel, J., and Jay, G. (1990). The human T-lymphotropic virus type I tax gene can cooperate with the ras oncogene to induce neoplastic transformation of cells. *Mol. Cell Biol.* 10, 413–417. doi: 10.1128/mcb.10.1.413

Quaresma, J. A. S., Yoshikawa, G. T., Koyama, R. V. L., Dias, G. A. S., Fujihara, S., and Fuzii, H. T. (2016). HTLV-1, immune response and autoimmunity. *Viruses* 8:5. doi: 10.3390/v8010005

Rafatpanah, H., Felegari, M., Azarpazhooh, M. R., Vakili, R., Rajaei, T., Hampson, I., et al. (2017). Altered expression of CXCR3 and CCR6 and their ligands in HTLV-1 carriers and HAM/TSP patients. *J. Med. Virol.* 89, 1461–1468. doi: 10.1002/jmv.24779

Rafatpanah, H., Pravica, V., FaridHosseini, R., Tabatabaei, A., Ollier, W., Poulton, K., et al. (2007). Association between HLA-DRB1* 01 and HLA-Cw* 08 and outcome following HTLV-I infection. *Iran. J.Immunol.* 4, 94–100.

Rajaei, T., Farajifard, H., Rezaee, S. A., Azarpazhooh, M. R., Mahmoudi, M., Valizadeh, N., et al. (2018). Different roles of CXCR1 and CXCR2 in HTLV-1 carriers and HTLV-1-associated myelopathy/tropical spastic paraparesis (HAM/TSP) patients. *Med. Microbiol. Immunol.* 208, 641–650. doi: 10.1007/s00430-018-0568-8

Raman, D., Sobolik-Delmaire, T., and Richmond, A. (2011). Chemokines in health and disease. *Exp. Cell Res.* 317, 575–589. doi: 10.1016/j.yexcr.2011.01.005

Romanelli, L. C. F., Miranda, D. M., Carneiro-Proietti, A. B. F., Mamede, M., Vasconcelos, H. M. M., Martins, M. L., et al. (2018). Spinal cord hypometabolism associated with infection by human T-cell lymphotropic virus type 1(HTLV-1). *PLoS Negl. Trop. Dis.* 12:e0006720. doi: 10.1371/journal.pntd.0006720

Rosa, D. V., Magno, L. A., Pereira, N. C., Romanelli, L. C., Albuquerque, M. R., Martins, M. L., et al. (2018). Plasma and cerebrospinal fluid levels of cytokines as disease markers of neurologic manifestation in long-term HTLV-1 infected individuals. *Biomark Med.* 12, 447–454. doi: 10.2217/bmm-2017-0313

Rösler, A., Pohl, M., Braune, H.-J., Oertel, W., Gemsa, D., and Sprenger, H. (1998). Time course of chemokines in the cerebrospinal fluid and serum during herpes simplex type 1 encephalitis. *J Neurol. Sci.* 157, 82–89. doi: 10.1016/s0022-510x(98)00061-6

Rossi, D. L., Vicari, A. P., Franz-Bacon, K., McClanahan, T. K., and Zlotnik, A. (1997). Identification through bioinformatics of two new macrophage proinflammatory human chemokines: MIP-3alpha and MIP-3beta. *J. Immunol.* 158, 1033–1036.

Ruckes, T., Saul, D., Van Snick, J., Hermine, O., and Grassmann, R. (2001). Autocrine antiapoptotic stimulation of cultured adult T-cell leukemia cells by overexpression of the chemokine I-309. *Blood* 98, 1150–1159. doi: 10.1182/blood.v98.4.1150

Sabouri, A. H., Saito, M., Usuku, K., Bajestan, S. N., Mahmoudi, M., Forughipour, M., et al. (2005). Differences in viral and host genetic risk factors for development of human T-cell lymphotropic virus type 1 (HTLV-1)-associated myelopathy/tropical spastic paraparesis between Iranian and Japanese HTLV-1-infected individuals. *J. Gen. Virol.* 86, 773–781. doi: 10.1099/vir.0.80509-0

Saito, M., Sejima, H., Naito, T., Ushirogawa, H., Matsuzaki, T., Matsuura, E., et al. (2017). The CC chemokine ligand (CCL) 1, upregulated by the viral transactivator Tax, can be downregulated by minocycline: possible implications for long-term treatment of HTLV-1-associated myelopathy/tropical spastic paraparesis. *Virol. J.* 14:234. doi: 10.1186/s12985-017-0902-6

Sallusto, F., Lanzavecchia, A., and Mackay, C. R. (1998a). Chemokines and chemokine receptors in T-cell priming and Th1/Th2-mediated responses. *Immunol. Today* 19, 568–574. doi: 10.1016/s0167-5699(98)01346-2

Sallusto, F., Schaerli, P., Loetscher, P., Schaniel, C., Lenig, D., Mackay, C. R., et al. (1998b). Rapid and coordinated switch in chemokine receptor expression during dendritic cell maturation. *Eur. J. Immunol.* 28, 2760–2769. doi: 10.1002/(sici)1521-4141(199809)28:09<2760::aid-immu2760>3.0.co;2-n

Santos, S. B., Oliveira, P., Luna, T., Souza, A., Nascimento, M., Siqueira, I., et al. (2012). Immunological and viral features in patients with overactive bladder associated with Human T-Cell Lymphotropic Virus Type 1 Infection. *J. Med. Virol.* 84, 1809–1817. doi: 10.1002/jmv.23341

Sato, T., Coler-Reilly, A., Utsunomiya, A., Araya, N., Yagishita, N., Ando, H., et al. (2013). CSF CXCL10. *PLoS Negl. Trop. Dis.* 7:e2479. doi: 10.1371/journal.pntd.0002479

Sato, T., Coler-Reilly, A. L. G., Yagishita, N., Araya, N., Inoue, E., Furuta, R., et al. (2018a). Mogamulizumab (Anti-CCR4) in HTLV-1-Associated Myelopathy. *N. Engl. J. Med.* 378, 529–538. doi: 10.1056/NEJMoa1704827

Sato, T., Yagishita, N., Tamaki, K., Inoue, E., Hasegawa, D., Nagasaka, M., et al. (2018b). Proposal of classification criteria for HTLV-1-associated myelopathy/tropical spastic paraparesis disease activity. *Front. Microbiol.* 9:1651. doi: 10.3389/fmicb.2018.01651

Satou, Y., Utsunomiya, A., Tanabe, J., Nakagawa, M., Nosaka, K., and Matsuoka, M. (2012). HTLV-1 modulates the frequency and phenotype of FoxP3+CD4+ T cells in virus-infected individuals. *Retrovirology* 9:46. doi: 10.1186/1742-4690-9-46

Satou, Y., Yasunaga, J.-I., Yoshida, M., and Matsuoka, M. (2006). HTLV-I basic leucine zipper factor gene mRNA supports proliferation of adult T cell leukemia cells. *Proc. Natl.Acad. Sci. U.S.A.* 103, 720–725. doi: 10.1073/pnas.0507631103

Schall, T. J., Bacon, K., Toy, K. J., and Goeddel, D. V. (1990). Selective attraction of monocytes and T lymphocytes of the memory phenotype by cytokine RANTES. *Nature* 347:669. doi: 10.1038/347669a0

Schrum, S., Probst, P., Fleischer, B., and Zipfel, P. F. (1996).). Synthesis of the CC-chemokines MIP-1alpha, MIP-1beta, and RANTES is associated with a type 1 immune response. *J. Immunol.* 157, 3598–3604.

Schutyser, E., Richmond, A., and Van Damme, J. (2005). Involvement of CC chemokine ligand 18 (CCL18) in normal and pathological processes. *J. Leukoc Biol.* 78, 14–26. doi: 10.1189/jlb.1204712

Schutyser, E., Struyf, S., and Van Damme, J. (2003). The CC chemokine CCL20 and its receptor CCR6. *Cytokine Growth Factor Rev.* 14, 409–426. doi: 10.1016/s1359-6101(03)00049-2

Seki, M., Kadota, J. I., Higashiyama, Y., Iida, K., Iwashita, T., Sasaki, E., et al. (1999). Elevated levels of beta-chemokines in bronchoalveolar lavage fluid (BALF) of individuals infected with human T lymphotropic virus type-1 (HTLV-1). *Clin. Exp. Immunol.* 118, 417–422. doi: 10.1046/j.1365-2249.1999.01093.x

Sharma, M. (2010). Chemokines and their receptors: orchestrating a fine balance between health and disease. *Crit. Rev. Biotechnol.* 30, 1–22. doi: 10.1080/07388550903187418

Sharma, V., and May, C. C. (1999). Human T-cell lymphotrophic virus type-I tax gene induces secretion of human macrophage inflammatory protein-1α. *Biochem. Biophys. Res. Commun.* 262, 429–432. doi: 10.1006/bbrc.1999.1242

Shimauchi, T., Imai, S., Hino, R., and Tokura, Y. (2005). Production of thymus and activation-regulated chemokine and macrophage-derived chemokine by CCR4+ adult T-cell leukemia cells. *Clin. Cancer Res.* 11, 2427–2435. doi: 10.1158/1078-0432.ccr-04-0491

Shimauchi, T., Kabashima, K., and Tokura, Y. (2008). Adult T-cell leukemia/lymphoma cells from blood and skin tumors express cytotoxic T lymphocyte-associated antigen-4 and Foxp3 but lack suppressor activity toward autologous CD8+ T cells. *Cancer Sci.* 99, 98–106.

Shimizu, K., Karube, K., Arakawa, F., Nomura, Y., Komatani, H., Yamamoto, K., et al. (2007). Upregulation of CC chemokine ligand 18 and downregulation of CX3C chemokine receptor 1 expression in human T-cell leukemia virus type 1-associated lymph node lesions: Results of chemokine and chemokine receptor DNA chip analysis. *Cancer Sci.* 98, 1875–1880. doi: 10.1111/j.1349-7006.2007.00627.x

Shimoyama, M., and Group, L. S. (1991). Diagnostic criteria and classification of clinical subtypes of adult T-cell leukaemia-lymphoma: a report from the lymphoma study group (1984–87). *Br. J. Haematol.* 79, 428–437. doi: 10.1111/j.1365-2141.1991.tb08051.x

Singh, S., Sadanandam, A., and Singh, R. K. (2007). Chemokines in tumor angiogenesis and metastasis. *Cancer Metastasis Rev.* 26, 453–467. doi: 10.1007/s10555-007-9068-9

Smith, M. R., and Greene, W. C. (1991). Molecular biology of the type I human T-cell leukemia virus (HTLV-I) and adult T-cell leukemia. *J. Clin. Investig.* 87, 761–766. doi: 10.1172/jci115078

Starling, A. L. B., Coelho-dos-Reis, J. G. A., Peruhype-Magalhães, V., Pascoal-Xavier, M. A., Gonçalves, D. U., Béla, S. R., et al. (2015). Immunological signature of the different clinical stages of the HTLV-1 infection: establishing serum biomarkers for HTLV-1-associated disease morbidity. *Biomarkers* 20, 502–512. doi: 10.3109/1354750X.2015.1094141

Strieter, R. M., Polverini, P. J., Kunkel, S. L., Arenberg, D. A., Burdick, M. D., Kasper, J., et al. (1995). The functional role of the ELR motif in CXC chemokine-mediated angiogenesis. *J. Biol. Chem.* 270, 27348–27357. doi: 10.1074/jbc.270.45.27348

Sugata, K., Yasunaga, J.-I., Kinosada, H., Mitobe, Y., Furuta, R., Mahgoub, M., et al. (2016). HTLV-1 Viral Factor HBZ Induces CCR4 to Promote T-cell Migration and Proliferation. *Cancer Res.* 76, 5068–5079. doi: 10.1158/0008-5472.CAN-16-0361

Tamaki, K., Sato, T., Tsugawa, J., Fujioka, S., Yagishita, N., Araya, N., et al. (2019). Cerebrospinal fluid CXCL10 as a candidate surrogate marker for HTLV-1-associated myelopathy/tropical spastic paraparesis. *Front. Microbiol.* 10:2110. doi: 10.3389/fmicb.2019.02110

Tanaka, A., Takahashi, C., Yamaoka, S., Nosaka, T., Maki, M., and Hatanaka, M. (1990). Oncogenic transformation by the tax gene of human T-cell leukemia virus type I in vitro. *Proc. Natl.Acade. Sci. U.S.A.* 87, 1071–1075. doi: 10.1073/pnas.87.3.1071

Tanaka, Y., Mine, S., Figdor, C. G., Wake, A., Hirano, H., Tsukada, J., et al. (1998). Constitutive chemokine production results in activation of leukocyte function-associated antigen-1 on Adult T-Cell leukemia cells. *Blood* 91, 3909–3919. doi: 10.1182/blood.v91.10.3909.3909_3909_3919

Taub, D. D., Conlon, K., Lloyd, A. R., Oppenheim, J. J., and Kelvin, D. J. (1993). Preferential migration of activated CD4+ and CD8+ T cells in response to MIP-1 alpha and MIP-1 beta. *Science* 260, 355–358. doi: 10.1126/science.7682337

Taub, D. D., Turcovski-Corrales, S. M., Key, M. L., Longo, D. L., and Murphy, W. J. (1996). Chemokines and T lymphocyte activation: I. Beta chemokines costimulate human T lymphocyte activation in vitro. *J. Immunol.* 156, 2095–2103.

Teruya, H., Tomita, M., Senba, M., Ishikawa, C., Tamayose, M., Miyazato, A., et al. (2008). Human T-cell leukemia virus type I infects human lung epithelial cells and induces gene expression of cytokines, chemokines and cell adhesion molecules. *Retrovirology* 5:86. doi: 10.1186/1742-4690-5-86

Teuscher, C., Butterfield, R. J., Ma, R. Z., Zachary, J. F., Doerge, R. W., and Blankenhorn, E. P. (1999). Sequence polymorphisms in the chemokines Scya1 (TCA-3), Scya2 (monocyte chemoattractant protein (MCP)-1), and Scya12 (MCP-5) are candidates for eae7, a locus controlling susceptibility to monophasic remitting/nonrelapsing experimental allergic encephalomyelitis. *J. Immunol.* 163, 2262–2266.

Toulza, F., Heaps, A., Tanaka, Y., Taylor, G. P., and Bangham, C. R. (2008). High frequency of CD4+FoxP3+ cells in HTLV-1 infection: inverse correlation with HTLV-1-specific CTL response. *Blood* 111, 5047–5053. doi: 10.1182/blood-2007-10-118539

Toulza, F., Nosaka, K., Takiguchi, M., Pagliuca, T., Mitsuya, H., Tanaka, Y., et al. (2009). FoxP3+ regulatory T cells are distinct from leukemia cells in HTLV-1-associated adult T-cell leukemia. *Int. J. Cancer* 125, 2375–2382. doi: 10.1002/ijc.24664

Toulza, F., Nosaka, K., Tanaka, Y., Schioppa, T., Balkwill, F., Taylor, G. P., et al. (2010). Human T-lymphotropic virus type 1-induced CC chemokine ligand 22 maintains a high frequency of functional FoxP3+ regulatory T cells. *J. Immunol.* 185, 183–189. doi: 10.4049/jimmunol.0903846

Traves, S. L., Smith, S. J., Barnes, P. J., and Donnelly, L. E. (2004). Specific CXC but not CC chemokines cause elevated monocyte migration in COPD: a role for CXCR2. *J. Leukocyte Biol.* 76, 441–450. doi: 10.1189/jlb.1003495

Twizere, J.-C., Springael, J.-Y., Boxus, M., Burny, A., Dequiedt, F., Dewulf, J.-F., et al. (2007). Human T-cell leukemia virus type-1 Tax oncoprotein regulates G-protein signaling. *Blood* 109, 1051–1060. doi: 10.1182/blood-2006-06-026781

Uchiyama, T., Yodoi, J., Sagawa, K., Takatsuki, K., and Uchino, H. (1977). Adult T-cell leukemia: clinical and hematologic features of 16 cases. *Blood* 50, 481–492. doi: 10.1182/blood.v50.3.481.bloodjournal503481

Umehara, F., Izumo, S., Ronquillo, A. T., Matsumuro, K., Sato, E., and Osame, M. (1994). Cytokine expression in the spinal cord lesions in HTLV-I-associated myelopathy. *J. Neuropathol. Exp. Neurol.* 53, 72–77. doi: 10.1097/00005072-199401000-00009

Umehara, F., Izumo, S., Takeya, M., Sato, E., Osame, M., and Takahasi, K. (1996). Expression of adhesion molecules and monocyte chemoattractant protein-1 (MCP-1) in the spinal cord lesions in HTLV-I-associated myelopathy. *Acta Neuropathol.* 91, 343–350. doi: 10.1007/s004010050435

Umehara, F., Okada, Y., Fujimoto, N., Abe, M., Izumo, S., and Osame, M. (1998). Expression of matrix metalloproteinases and tissue inhibitors of metalloproteinas HTLV-I-associated myelopathy. *J. Neuropathol. Expe. Neurol.* 57, 839–849. doi: 10.1097/00005072-199809000-00005

Utsunomiya, A., Hanada, S., Terada, A., Kodama, M., Uematsu, T., Tsukasa, S., et al. (1988). Adult T-cell leukemia with leukemia cell infiltration into the gastrointestinal tract. *Cancer* 61, 824–828. doi: 10.1002/1097-0142(19880215)61:4<824::aid-cncr2820610430>3.0.co;2-6

Vazirinejad, R., Ahmadi, Z., Kazemi Arababadi, M., Hassanshahi, G., and Kennedy, D. (2014). The biological functions, structure and sources of CXCL10 and its outstanding part in the pathophysiology of multiple sclerosis. *Neuroimmunomodulation* 21, 322–330. doi: 10.1159/000357780

Vose, J., Armitage, J., and Weisenburger, D. (2008). International peripheral T-cell and natural killer/T-cell lymphoma study: pathology findings and clinical outcomes. *J. Clin. Oncol.* 26, 4124–4130. doi: 10.1200/jco.2008.16.4558

Wang, J., Shiozawa, Y., Wang, J., Wang, Y., Jung, Y., Pienta, K. J., et al. (2008). The role of CXCR7/RDC1 as a chemokine receptor for CXCL12/SDF-1 in prostate cancer. *J. Biol. Chem.* 283, 4283–4294. doi: 10.1074/jbc.m707465200

Wang, Z., Shang, H., and Jiang, Y. (2017). Chemokines and chemokine receptors: accomplices for human immunodeficiency virus infection and latency. *Front. Immunol.* 8:1274. doi: 10.3389/fimmu.2017.01274

Wucherpfennig, K. W., Hollsberg, P., Richardson, J. H., Benjamin, D., and Hafler, D. A. (1992). T-cell activation by autologous human T-cell leukemia virus type I-infected T-cell clones. *Proc. Natl. Acad. Sci. U.S.A.* 89, 2110–2114. doi: 10.1073/pnas.89.6.2110

Yamaguchi, K., Mochizuki, M., Watanabe, T., Yoshimura, K., Shirao, M., Araki, S., et al. (1994). Human T lymphotropic virus type 1 uveitis after Graves disease. *Br. J. Ophthalmol.* 78, 163–166. doi: 10.1136/bjo.78.3.163

Yamamoto, K., Utsunomiya, A., Tobinai, K., Tsukasaki, K., Uike, N., Uozumi, K., et al. (2010). Phase I study of KW-0761, a defucosylated humanized anti-CCR4 antibody, in relapsed patients with adult T-cell leukemia-lymphoma and peripheral T-cell lymphoma. *J. Clin. Oncol.* 28, 1591–1598. doi: 10.1200/JCO.2009.25.3575

Yamamoto, M., Matsuyama, W., Oonakahara, K., Watanabe, M., Higashimoto, I., Kawabata, M., et al. (2004). Influence of human T lymphotrophic virus type I on diffuse pan-bronchiolitis. *Clin. Exp. Immunol,* 136, 513–520. doi: 10.1111/j.1365-2249.2004.02485.x

Yamamoto-Taguchi, N., Satou, Y., Miyazato, P., Ohshima, K., Nakagawa, M., Katagiri, K., et al. (2013). HTLV-1 bZIP factor induces inflammation through labile Foxp3 expression. *PLoS pathog.* 9:e1003630. doi: 10.1371/journal.ppat.1003630

Yamano, Y., Araya, N., Sato, T., Utsunomiya, A., Azakami, K., Hasegawa, D., et al. (2009). Abnormally High Levels of Virus-Infected IFN-γ(+)CCR4(+)CD4(+)CD25(+) T cells in a retrovirus-associated neuroinflammatory disorder. *PLoS One* 4:e6517. doi: 10.1371/journal.pone.0006517

Yamano, Y., and Coler-Reilly, A. (2017). HTLV-1 induces a Th1-like state in CD4(+)CCR4(+) T cells that produces an inflammatory positive feedback loop via astrocytes in HAM/TSP. *J. Neuroimmunol.* 304, 51–55. doi: 10.1016/j.jneuroim.2016.08.012

Yamazato, Y., Miyazato, A., Kawakami, K., Yara, S., Kaneshima, H., and Saito, A. (2003). High expression of p40(tax) and pro-inflammatory cytokines and chemokines in the lungs of human T-lymphotropic virus type 1-related bronchopulmonary disorders. *Chest* 124, 2283–2292. doi: 10.1378/chest.124.6.2283

Yang, J., and Richmond, A. (2004). The angiostatic activity of interferon-inducible protein-10/CXCL10 in human melanoma depends on binding to CXCR3 but not to glycosaminoglycan. *Mol. Ther. J. Am. Soc. Gene Ther.* 9, 846–855. doi: 10.1016/j.ymthe.2004.01.010

Yoshida, M. (2001). Multiple viral strategies of HTLV-1 for dysregulation of cell growth control. *Annu. Rev. Immunol.* 19, 475–496. doi: 10.1146/annurev.immunol.19.1.475

Yoshida, M., Miyoshi, I., and Hinuma, Y. (1982). Isolation and characterization of retrovirus from cell lines of human adult T-cell leukemia and its implication in the disease. *Proc. Natl. Acad. Sci. U.S.A.* 79, 2031–2035. doi: 10.1073/pnas.79.6.2031

Yoshie, O., Fujisawa, R., Nakayama, T., Harasawa, H., Tago, H., Izawa, D., et al. (2002). Frequent expression of CCR4 in adult T-cell leukemia and human

T-cell leukemia virus type 1–transformed T cells. *Blood* 99, 1505–1511. doi: 10.1182/blood.v99.5.1505

Yoshie, O., Imai, T., and Nomiyama, H. (2001). Chemokines in immunity. *Adv. Immunol.* 78, 57–110.

Youn, B.-S., Kim, Y. J., Mantel, C., Yu, K.-Y., and Broxmeyer, H. E. (2001). Blocking of c-FLIPL–independent cycloheximide-induced apoptosis or Fas-mediated apoptosis by the CC chemokine receptor 9/TECK interaction. *Blood* 98, 925–933. doi: 10.1182/blood.v98.4.925

Zhao, T. (2016). The Role of HBZ in HTLV-1-Induced Oncogenesis. *Viruses* 8:E34. doi: 10.3390/v8020034

Zhao, T., Yasunaga, J., Satou, Y., Nakao, M., Takahashi, M., Fujii, M., et al. (2009). Human T-cell leukemia virus type 1 bZIP factor selectively suppresses the classical pathway of NF-kappaB. *Blood* 113, 2755–2764. doi: 10.1182/blood-2008-06-161729

Zlotnik, A., and Yoshie, O. (2000). Chemokines: a new classification system and their role in immunity. *Immunity* 12, 121–127. doi: 10.1016/s1074-7613(00)80165-x

Permissions

All chapters in this book were first published by Frontiers; hereby published with permission under the Creative Commons Attribution License or equivalent. Every chapter published in this book has been scrutinized by our experts. Their significance has been extensively debated. The topics covered herein carry significant findings which will fuel the growth of the discipline. They may even be implemented as practical applications or may be referred to as a beginning point for another development.

The contributors of this book come from diverse backgrounds, making this book a truly international effort. This book will bring forth new frontiers with its revolutionizing research information and detailed analysis of the nascent developments around the world.

We would like to thank all the contributing authors for lending their expertise to make the book truly unique. They have played a crucial role in the development of this book. Without their invaluable contributions this book wouldn't have been possible. They have made vital efforts to compile up to date information on the varied aspects of this subject to make this book a valuable addition to the collection of many professionals and students.

This book was conceptualized with the vision of imparting up-to-date information and advanced data in this field. To ensure the same, a matchless editorial board was set up. Every individual on the board went through rigorous rounds of assessment to prove their worth. After which they invested a large part of their time researching and compiling the most relevant data for our readers.

The editorial board has been involved in producing this book since its inception. They have spent rigorous hours researching and exploring the diverse topics which have resulted in the successful publishing of this book. They have passed on their knowledge of decades through this book. To expedite this challenging task, the publisher supported the team at every step. A small team of assistant editors was also appointed to further simplify the editing procedure and attain best results for the readers.

Apart from the editorial board, the designing team has also invested a significant amount of their time in understanding the subject and creating the most relevant covers. They scrutinized every image to scout for the most suitable representation of the subject and create an appropriate cover for the book.

The publishing team has been an ardent support to the editorial, designing and production team. Their endless efforts to recruit the best for this project, has resulted in the accomplishment of this book. They are a veteran in the field of academics and their pool of knowledge is as vast as their experience in printing. Their expertise and guidance has proved useful at every step. Their uncompromising quality standards have made this book an exceptional effort. Their encouragement from time to time has been an inspiration for everyone.

The publisher and the editorial board hope that this book will prove to be a valuable piece of knowledge for researchers, students, practitioners and scholars across the globe.

List of Contributors

Fabio E. Leal
Oncovirology Program, Instituto Nacional de Câncer (INCA), Rio de Janeiro, Brazil
Microbiology Immunology and Tropical Medicine, George Washington University, Washington, DC, United States

Soraya Maria Menezes, Tim Dierckx, Jurgen Vercauteren and Johan Van Weyenbergh
Department of Microbiology and Immunology, Rega Institute for Medical Research, KU Leuven, Leuven, Belgium

Emanuela A. S. Costa, Aluisio C. Segurado and Esper G. Kallas
Departamento de Moléstias Infecciosas e Parasitárias, Faculdade de Medicina, Universidade de São Paulo, São Paulo, Brazil

Phillip M. Brailey, Douglas F. Nixon and Rui Andre Saraiva Raposo
Oncovirology Program, Instituto Nacional de Câncer (INCA), Rio de Janeiro, Brazil

Lucio Gama
Department of Molecular and Comparative Pathobiology, Johns Hopkins University School of Medicine, Baltimore, MD, United States

Ricardo Khouri
Department of Microbiology and Immunology, Rega Institute for Medical Research, KU Leuven, Leuven, Belgium
Fundação Oswaldo Cruz, Instituto Gonçalo Moniz (IGM), Salvador-Bahia, Brazil

Bernardo Galvão-Castro
Escola Bahiana de Medicina e Saúde Pública, Salvador-Bahia, Brazil

Hélène Gazon, Pradeep Chauhan, Malik Hamaidia, Clotilde Hoyos, Lin Li, Roghaiyeh Safari and Luc Willems
National Fund for Scientific Research, Molecular and Cellular Epigenetics, Interdisciplinary Cluster for Applied Genoproteomics, Liège, Belgium
Molecular Biology, TERRA, Gemboux Agro-Bio Tech, Gembloux, Belgium

Eisuke Inoue
Medical Informatics, St. Marianna University School of Medicine, Kawasaki, Japan

Yoshimi Enose-Akahata, Ashley Vellucci and Steven Jacobson
Viral Immunology Section, Division of Neuroimmunology and Neurovirology, National Institute of Neurological Disorders and Stroke, National Institutes of Health, Bethesda, MD, United States

Tomoo Sato, Naoko Yagishita, Daisuke Hasegawa, Hiroko Suzuki, Natsumi Araya and Ariella Coler-Reilly
Department of Rare Diseases Research, Institute of Medical Science, St. Marianna University School of Medicine, Kawasaki, Japan

Keiko Tamaki and Yoshio Tsuboi
Department of Neurology, Fukuoka University, Fukuoka, Japan

Misako Nagasaka
Department of Oncology, Karmanos Cancer Institute, Wayne State University, Detroit, MI, United States
Department of Advanced Medical Innovation, St. Marianna University Graduate School of Medicine, Kawasaki, Japan

Yasuhiro Hasegawa
Department of Neurology, St. Marianna University School of Medicine, Kawasaki, Japan

Ayako Takata
Department of Preventive Medicine, St. Marianna University School of Medicine, Kawasaki, Japan

Yoshihisa Yamano
Department of Rare Diseases Research, Institute of Medical Science, St. Marianna University School of Medicine, Kawasaki, Japan
Department of Advanced Medical Innovation, St. Marianna University Graduate School of Medicine, Kawasaki, Japan

Norbert Donhauser, Stefanie Heym and Andrea K. Thoma-Kress
Institute of Clinical and Molecular Virology, Universitätsklinikum Erlangen, Friedrich-Alexander-Universität Erlangen-Nürnberg, Erlangen, Germany

Yutaka Tagaya and Robert C. Gallo
Division of Basic Science, Institute of Human Virology, University of Maryland School of Medicine, Baltimore, MD, United States

Amanda R. Panfil, Jacob Al-Saleem and Cory M. Howard
Department of Veterinary Biosciences, Center for Retrovirus Research, The Ohio State University, Columbus, OH, United States

Nikoloz Shkriabai and Mamuka Kvaratskhelia
Division of Infectious Diseases, School of Medicine, University of Colorado Denver, Aurora, CO, United States

Patrick L. Green
Department of Veterinary Biosciences, Center for Retrovirus Research, The Ohio State University, Columbus, OH, United States
Comprehensive Cancer Center and Solove Research Institute, The Ohio State University, Columbus, OH, United States

Anurag Kulkarni and Charles R. M. Bangham
Section of Virology, Division of Infectious Diseases, Department of Medicine, Imperial College London, London, United Kingdom

Maria A. F. Queiroz, Vânia N. Azevedo, Ednelza da S. G. Amoras, Tuane C. F. Moura, Marluísa de O. Guimarães Ishak, Ricardo Ishak, Antonio C. R. Vallinoto and Rosimar N. Martins Feitosa
Laboratory of Virology, Institute of Biological Sciences, Federal University of Pará, Belém, Brazil

Azusa Tanaka
Department of Drug Discovery Medicine, Graduate School of Medicine, Kyoto University, Kyoto, Japan

Masao Matsuoka
Department of Hematology, Rheumatology and Infectious Diseases, Faculty of Life Sciences, Kumamoto University, Kumamoto, Japan
Institute for Frontier Life and Medical Sciences, Kyoto University, Kyoto, Japan

Makoto Yamagishi, Dai Fujikawa and Kaoru Uchimaru
Department of Computational Biology and Medical Sciences, Graduate School of Frontier Sciences, The University of Tokyo, Tokyo, Japan

Toshiki Watanabe
The Institute of Medical Science, The University of Tokyo, Tokyo, Japan

Hélène Gazon
Belgium Molecular and Cellular Epigenetics, Interdisciplinary Cluster for Applied Genoproteomics, University of Liège, Liège, Belgium

Benoit Barbeau
Département des Sciences Biologiques and Centre de Recherche BioMed, Université du Québec à Montréal, Montréal, QC, Canada

Jean-Michel Mesnard and Jean-Marie Peloponese Jr.
Institut de Recherche en Infectiologie de Montpellier, Centre National de la Recherche Scientifique, Université de Montpellier, Montpellier, France

Robert C. Gallo and Yutaka Tagaya
Division of Basic Science, Institute of Human Virology, University of Maryland School of Medicine, Baltimore, MD, United States
HTLV-1 Task Force, Global Virus Network, Baltimore, MD, United States

Luc Willems
HTLV-1 Task Force, Global Virus Network, Baltimore, MD, United States
Research Director, National Fund for Scientific Research at University of Liège, Liège, Belgium

Ilaria Cavallari, Micol Silic-Benussi and Vittoria Raimondi
Veneto Institute of Oncology IOV-IRRCS, Padova, Italy

Gloria Scattolin
Department of Surgery, Oncology, and Gastroenterology, University of Padova, Padova, Italy

Donna M. D'Agostino
Department of Biomedical Sciences, University of Padova, Padova, Italy

Vincenzo Ciminale
Veneto Institute of Oncology IOV-IRRCS, Padova, Italy
Department of Surgery, Oncology, and Gastroenterology, University of Padova, Padova, Italy

Sara Moodad and Ali Bazarbachi
Department of Internal Medicine, Faculty of Medicine, American University of Beirut, Beirut, Lebanon
Department of Anatomy, Cell Biology and Physiological Sciences, Faculty of Medicine, American University of Beirut, Beirut, Lebanon

Abdou Akkouche and Rita Hleihel
Department of Internal Medicine, Faculty of Medicine, American University of Beirut, Beirut, Lebanon

Nadine Darwiche
Department of Biochemistry and Molecular Genetics, Faculty of Medicine, American University of Beirut, Beirut, Lebanon

Marwan El-Sabban
Department of Anatomy, Cell Biology and Physiological Sciences, Faculty of Medicine, American University of Beirut, Beirut, Lebanon

Hiba El Hajj
Department of Internal Medicine, Faculty of Medicine, American University of Beirut, Beirut, Lebanon
Department of Experimental Pathology, Immunology and Microbiology, Faculty of Medicine, American University of Beirut, Beirut, Lebanon

Varun K. Sharma and Katia Ruggero
Department of Surgery, Oncology and Gastroenterology, University of Padova, Padova, Italy

Vittoria Raimondi, Ilaria Cavallari and Micol Silic-Benussi
Veneto Institute of Oncology IOV - IRCCS, Padova, Italy

Cynthia A. Pise-Masison
Animal Models and Retroviral Vaccines Section, National Cancer Institute, National Institutes of Health, Bethesda, MD, United States

Vincenzo Ciminale
Department of Surgery, Oncology and Gastroenterology, University of Padova, Padova, Italy
Veneto Institute of Oncology IOV - IRCCS, Padova, Italy

Donna M. D'Agostino
Department of Biomedical Sciences, University of Padova, Padova, Italy

Amandine Pasquier
International Center for Research in Infectiology, Retroviral Oncogenesis Laboratory, INSERM U1111 - Université Claude Bernard Lyon 1, CNRS, UMR5308, Ecole Normale Supérieure de Lyon, Université Lyon, Lyon, France
Equipe Labellisée Ligue Nationale Contre le Cancer, Paris, France
Ecole Pratique des Hautes Etudes, Paris, France

Sandrine Alais, Chloé Journo, Hélène Dutartre and Renaud Mahieux
International Center for Research in Infectiology, Retroviral Oncogenesis Laboratory, INSERM U1111 - Université Claude Bernard Lyon 1, CNRS, UMR5308, Ecole Normale Supérieure de Lyon, Université Lyon, Lyon, France
Equipe Labellisée Ligue Nationale Contre le Cancer, Paris, France

Loic Roux and Karine Alvarez
CNRS UMR 7257, Architecture et Fonction des Macromolecules Biologiques, Aix-Marseille Université, Marseille, France

Maria-Isabel Thoulouze
"Biofilm and Viral Transmission" Team, Structural Virology Unit, Department of Virology, CNRS UMR 3569, Institut Pasteur, Paris, France

Stefania Fochi, Simona Mutascio, Umberto Bertazzoni, Donato Zipeto and Maria G. Romanelli
Department of Neurosciences, Biomedicine and Movement Sciences, University of Verona, Verona, Italy

Mattia Vicario, Adriana Mattiolo, Barbara Montini, Maria Assunta Piano, Ilaria Cavallari and Maria Luisa Calabrò
Immunology and Molecular Oncology, Veneto Institute of Oncology, IOV - IRCCS, Padua, Italy

Alberto Amadori
Immunology and Molecular Oncology, Veneto Institute of Oncology, IOV - IRCCS, Padua, Italy
Department of Surgery, Oncology and Gastroenterology, University of Padova, Padua, Italy

Luigi Chieco-Bianchi
Department of Surgery, Oncology and Gastroenterology, University of Padova, Padua, Italy

Francesca Marino-Merlo and Beatrice Macchi
IRCCS Centro Neurolesi "Bonino-Pulejo", Messina, Italy

Antonio Mastino
Department of Chemical, Biological, Pharmaceutical and Environmental Sciences, University of Messina, Messina, Italy
Institute of Translational Pharmacology, The National Research Council, Rome, Italy

Sandro Grelli
Department of Experimental Medicine and Surgery, University of Rome "Tor Vergata", Rome, Italy

Olivier Hermine
INSERM U1163, CNRS ERL 8654, Department of Hematology, Imagine Institute, Hôpital Necker-Enfants Malades, Paris, France

Ali Bazarbachi
Department of Internal Medicine, American University of Beirut, Beirut, Lebanon
Department of Anatomy, Cell Biology and Physiological Sciences, American University of Beirut, Beirut, Lebanon
Department of System Medicine, University of Rome "Tor Vergata", Rome, Italy

Marco Baratella, Greta Forlani and Roberto S. Accolla
Laboratories of General Pathology and Immunology "Giovanna Tosi", Department of Medicine and Surgery, School of Medicine, University of Insubria, Varese, Italy

Razieh Zargari, Maryam Mahdifar, Zohreh Vahidi and Houshang Rafatpanah
Immunology Research Center, Inflammation and Inflammatory Diseases Division, Mashhad University of Medical Sciences, Mashhad, Iran

Asadollah Mohammadi
Cellular and Molecular Research Center, Research Institute for Health Development, Kurdistan University of Medical Sciences, Sanandaj, Iran

Gholamhossein Hassanshahi
Molecular Medicine Research Center, Rafsanjan University of Medical Sciences, Rafsanjan, Iran

Index

www.ingramcontent.com/pod-product-compliance
Lightning Source LLC
LaVergne TN
LVHW081028210926
848485LV00004B/972